Structural Analysis

Transamerica Pyramid, San Francisco, California. (Courtesy of the Kaiser Steel Corporation.)

Structural Analysis

Fourth Edition

Jack C. McCormac

CLEMSON UNIVERSITY

1817

HARPER & ROW, PUBLISHERS, New York

Cambridge, Philadelphia, San Francisco,
London, Mexico City, São Paulo, Sydney

Other books by Jack C. McCormac available from Harper & Row

Design of Reinforced Concrete

Structural Steel Design, Third Edition

Sponsoring Editor: Cliff Robichaud
Project Editor: David Nickol
Designer: Robert Sugar
Production Manager: Marion Palen
Compositor: Science Press
Printer: and binder: The Maple Press Company
Art Studio: Fine Line Illustrations Inc.

Cover: Blue Cross and Blue Shield Service Center. Designed by Odell Associates Inc., Charlotte. Constructed by the Building Division of the Nello L. Teer Company, Durham. Photograph courtesy of Blue Cross and Blue Shield of North Carolina.

STRUCTURAL ANALYSIS, Fourth Edition

Library of Congress Cataloging in Publication Data

McCormac, Jack C.
Structural analysis.

Includes bibliographical references and index.
1. Structures, Theory of. I. Title.
TA645.M3 1984 624.1'71 83-12994
ISBN 0-06-044342-1

Contents

Preface

The purpose of this book is unchanged: to introduce the beginning student to the elementary fundamentals of structural analysis for beams, trusses, and rigid frames. Sufficient information is included to provide an understanding of statically determinate structures, and the book may be used for a first course in statically indeterminate structures as well.

Chapters 1 to 14 discuss statically determinate structures. The analysis of statically indeterminate structures, their geometry, and the elastic properties of their components are presented in Chapters 15 to 25. Chapter 26 deals with the plastic analysis of steel structures, and Chapter 27 introduces matrix analysis. Throughout, the stress is on basic fundamentals that apply to all cases, with rather little emphasis on learning formulas, fixed notation systems, or procedures for determining forces in particular types of structures. Although it would be possible in several instances to combine two or more of these generally quite short chapters, assignments are simpler if they are kept separate.

In this new edition the following changes have been made. In Chapter 1, there is an expanded discussion of the loads that structures have to support. A brief introduction to cables has been added in Chapter 3. A large percentage of the homework problems throughout the text have been revised. Many new problems have been added, including some with SI units. Chapter 27 on matrix analysis has been extensively revised.

The book includes 680 homework problems as well as many illustrative examples. By solving several problems in each chapter, the student should be able to firmly fix in his or her mind the theory involved. Answers to even-numbered problems are provided.

For many of us involved in structural engineering, structural analysis is a fascinating field, and it is my hope in presenting this book that its readers will find the field as attractive as we do.

Grateful acknowledgment is given for aid received from several sources. I am indebted to C. N. Antoni, W. A. Brown, R. C. Brinker, J. P. Cook, R. E. Elling, R. W. Fitzgerald, J. F. Fleming, W. Kilpatrick, S. P. Maggard, D. H. McLean, J. A. Mueller, J. C. Smith, I. A. Trively, H. A. Turner, J. T. Watkins, B. A. Whisler, and the late W. L. Lowry, who have by their suggestions and criticisms directly contributed to the preparation of this manuscript, and to my own professors, B. B. Williams, J. B. Wilbur, M. J. Holley, the late C. H. Norris, and the late W. M. Fife, who patiently instructed me in their structures classes. The books used in the courses taught by the above men naturally have influenced the enclosed material to some extent. These books were *Structural Theory* by Sutherland and Bowman, *Theory of Simple Structures* by Shedd and Vawter, and *Elementary Structural Analysis* by Norris and Wilbur. Finally thanks are due to Mrs. Mannetta Shusterman who helped type the manuscript.

Jack C. McCormac

1 Introduction

1.1 STRUCTURAL ANALYSIS AND DESIGN

The applications of loads to a structure cause forces and deformations in the structure. The determination of these forces and deformations is called *structural analysis*.

Structural design includes the arrangement and proportioning of structures and their parts so that they will satisfactorily support the loads to which they may be subjected. In detail structural design involves the following: the general layout of structures; studies of the possible structural forms or types that may provide feasible solutions; consideration of loading conditions; preliminary structural analyses and designs of the possible solutions; the selection of a solution; and the final structural analysis and design of the structure including the preparation of design drawings.

This book is devoted to structural analysis with only occasional remarks concerning the other phases of structural design. Structural analysis can be so interesting to many persons that they become completely attached to it and have the feeling that they want to become 100% involved in the subject. Although analyzing and predicting the behavior of structures and their parts is an extremely important part of structural design it is only one of several important and interrelated steps. As a result it is rather unusual for a person to be employed completely as a structural analyst. He or she will in almost all probability be involved in several or all phases of structural design.

It is said that Robert Louis Stevenson for a time studied structural engineering but he apparently found the "science of stresses and strains" too dull for his lively imagination, and as a result devoted his life to the writing of prose and poetry [1]. After reading some of his work (*Treasure Island*, *Kidnapped*, etc.) most of us would agree that the world is a better place

White Bird Canyon Bridge, White Bird, Idaho. (Courtesy of the American Institute of Steel Construction, Inc.)

because of his decision. Nevertheless there are a great number of us who feel that structural analysis and design are extremely interesting topics. It is hoped that this book will add to the number.

1.2 TYPES OF STRUCTURES

Structural engineering embraces an extensive variety of structures other than bridges and buildings. There are stadiums, power poles, radio and television towers, cables, arches, water tanks, concrete pavements, and many others. The sizes range from small frames consisting of a few beams and columns to the 1450-ft (foot) Sears Tower in Chicago and the Humber Estuary Bridge in England with its 4626-ft suspended span.

To face this wide range of sizes and types of structures, it seems unwise for those entering the structural field to learn to handle only one or two special cases. They should learn the basic fundamentals, which apply not only to all of the structures mentioned in the preceding paragraph but also to structures of types not necessarily considered to lie within the civil engineering field—ships and airplanes, for example.

The laws of statics, which are the fundamentals of all structural analysis, are stressed throughout the book. This emphasis should give readers a solid foundation for more advanced study and convince them that structural theory is not difficult and that it is unnecessary to memorize special cases. Some of the structures to be analyzed may seem to have rather weird shapes. These are included not to confuse, but rather to bring out the fact that the basic principles apply to all structures regardless of shape or size.

1.3 STRUCTURAL MEMBERS

The primary types of structural members to be considered follow.

Beams are those members that are subjected to bending or flexure. They are usually thought of as being in horizontal positions and loaded with gravity or vertical loads.

Ties are members that are subjected to axial tension only.

Struts (also referred to as columns or posts) are members that are subjected to axial compression only.

1.4 FRAMED STRUCTURES

The truss and rigid frame are the two basic types of structural frames formed from the structural members.

A ***truss*** consists of a group of ties and struts so designed and connected that they form a structure which acts as a large beam. The members usually form one or more triangles in a single plane and are so arranged that the external loads are applied at the joints and theoretically cause only axial tension or axial compression in the members. The members are assumed to be connected at their joints with frictionless hinges or pins, which allow the ends of the members freedom to rotate slightly. A common type of truss is shown in Fig. 1.1.

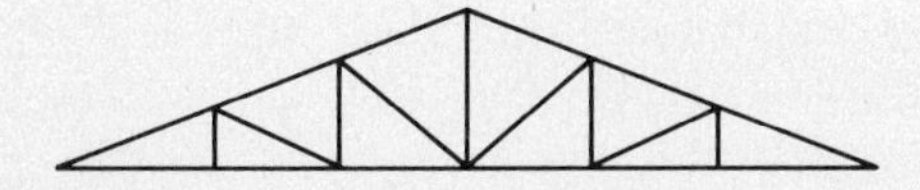

Figure 1.1

Trusses lying in one plane will be considered initially; however, framed structures having members not lying in a common plane are discussed in Chap. 14. These frames are called three-dimensional structures or space frames. (Transmission towers and framed domes are two types of space frames frequently seen.)

A ***rigid frame*** is a structure having moment-resisting joints. The members are rigidly connected at their ends so that no joint translation is possible (that is, the members at a joint may rotate as a group but may not move with respect to each other). The load-carrying ability of the structure is increased by the ability of the joints to resist moment. Several rigid frames are shown in Fig. 1.2.

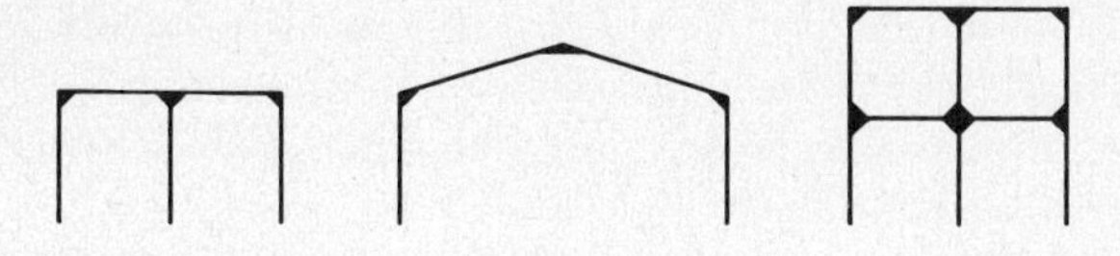

Figure 1.2

1.5 LOADS

Perhaps the most important and most difficult task faced by the structural designer is the accurate estimation of the loads which may be applied to a structure during its life. No loads that may reasonably be expected to occur may be overlooked. After loads are estimated the next problem is to decide the worst possible combinations of these loads which might occur at one time. For instance, would a highway bridge completely covered with ice and snow be simultaneously subjected to fast moving lines of heavily loaded trailer trucks in every lane and to a 90-mile lateral wind, or is some lesser combination of these loads more feasible?

The next two sections of this chapter provide a brief introduction to the types of loads with which the structural designer needs to be familiar. The purpose of these sections is not to discuss loads in great detail but rather to give the reader a "feel" for the subject. As will be seen, loads are classed as being dead loads or live loads.

1.6 DEAD LOADS

Dead loads are loads of constant magnitude that remain in one position. They consist of the structural frame's own weight and other loads that are permanently attached to the frame. For a steel-frame building some dead loads are the frame, walls, floors, roof, plumbing, and fixtures.

To design a structure it is necessary for the weights or dead loads of the various parts to be estimated for use in the analysis. The exact sizes and weights of the parts are not known until the structural analysis is made and the members of the structure selected. The weights, as determined from the actual design, must be compared with the estimated weights. If large discrepancies are present, it will be necessary to repeat the analysis and design using better estimated weights.

Reasonable estimates of structure weights may be obtained by referring to similar type structures or to various formulas and tables available in most civil engineering handbooks. An experienced designer can estimate very closely the weights of most structures and will spend little time repeating designs because of poor estimates.

1.7 LIVE LOADS

Live loads are loads that may change in position and magnitude. Simply stated, all loads that are not dead loads are live loads. Live loads that move under their own power are said to be moving loads, such as trucks, people, and cranes, whereas those loads that may be moved are movable loads, such as furniture, warehouse materials, and snow. Other live loads include those caused by construction operations, wind, rain, earthquakes, blasts, soils, and temperature changes. A brief description of some of these loads follows:

1. *Snow and ice*. In the colder states snow and ice loads are often quite important. One inch of snow is equivalent to approximately 0.5 psf (pounds per

Cold-storage warehouse, Grand Junction, Colorado. (Courtesy of the American Institute of Steel Construction, Inc.)

square foot) but it may be higher at lower elevations where snow is denser. For roof designs, snow loads of from 10 to 40 psf are used, the magnitude depending primarily on the slope of the roof and to a lesser degree on the character of the roof surface. The larger values are used for flat roofs, the smaller ones for sloped roofs. Snow tends to slide off sloped roofs, particularly those with metal or slate surfaces. A load of approximately 10 psf might be used for 45° (degree) slopes and a 40-psf load for flat roofs. Studies of snowfall records in areas with severe winters may indicate the occurrence of snow loads much greater than 40 psf with values as high as 80 psf in northern Maine.

Snow is a variable load which may cover an entire roof or only part of it. There may be drifts against walls or buildup in valleys or between parapets. Snow may slide off one roof onto a lower one. The wind may blow it off one side of a sloping roof or the snow may crust over and remain in position even during very heavy winds.

Bridges are generally not designed for snow loads, since the loads are usually not appreciable. In any case it is doubtful that a full load of snow and maximum traffic would be present at the same time. Bridges and towers are sometimes covered with layers of ice from 1 to 2 in (inches) thick. The weight of the ice runs up to about 10 psf. Another factor to be considered is the increased surface area of the ice-coated members as it pertains to wind loads.

2. *Rain*. Though snow loads are a more severe problem than rain loads for the usual roof the situation may be reversed for flat roofs, particularly those in warmer climates. If water on a flat roof accumulates faster than it runs off, the

result is called *ponding* because the increased load causes the roof to deflect into a dish shape that can hold more water, which causes greater deflections, and so on. This process continues until equilibrium is reached or until collapse occurs. Ponding is a serious matter as illustrated by the large number of flat-roof failures that occur due to ponding every year in the United States.

Ponding will occur on almost any flat roof to a certain degree even though roof drains are present. Roof drains may very well be used but they may be inadequate during severe storms or they may become partially or completely clogged. The best method of preventing ponding is to have an appreciable slope on the roof ($\frac{1}{4}$ in/ft or more) together with good drainage facilities. In addition to the usual ponding another problem may occur for very large flat roofs (with perhaps an acre or more of surface area). During heavy rainstorms strong winds frequently occur. If there is a great deal of water on the roof a strong wind may very well push a large quantity of it towards one end. The result can be a dangerous water depth as regards the load in psf on that end. For such situations *scuppers* are sometimes used. These are large holes or tubes in walls or parapets that enable water above a certain depth to quickly drain off the roof.

3. *Traffic loads for bridges*. Bridges are subjected to series of concentrated loads of varying magnitude caused by groups of truck or train wheels. A detailed discussion of these types of loadings is presented in Chap. 13.

4. *Impact loads*. Impact loads are caused by the vibration of moving or movable loads. It is obvious that a crate dropped on the floor of a warehouse or a truck bouncing on uneven pavement of a bridge causes greater stresses than would occur if the loads were applied gently and gradually. Impact loads are equal to the difference between the magnitude of the loads actually caused and the magnitude of the loads had they been dead loads. Several expressions commonly used for estimating impact are given in Chap. 13.

5. *Lateral loads*. Lateral loads are of two main types: wind and earthquake. A survey of engineering literature for the past 150 years reveals many references to structural failures caused by wind. Perhaps the most infamous of these have been bridge failures such as those of the Tay Bridge in Scotland in 1879 (which caused the deaths of 75 persons) and the Tacoma Narrows Bridge (Tacoma, Washington) in 1940. But there have also been many disastrous building failures due to wind during the same period such as that of the Union Carbide Building in Toronto in 1958. It is important to realize that a large percentage of building failures due to wind have occurred during their erection [2].

a. *Wind loads*. A geat deal of research has been conducted in recent years on the subject of wind loads. Nevertheless a great deal more work needs to be done as the estimation of these forces can by no means be classified as an exact science. The magnitudes of wind loads vary with geographical locations, heights above ground, types of terrain surrounding the buildings including other nearby structures, and other factors.

Wind pressures are usually assumed to be uniformly applied to the windward surfaces of buildings and are assumed to be capable of coming from

Access bridge, Renton, Washington. (Courtesy of the Bethlehem Steel Corporation.)

any direction. These assumptions are not very accurate because wind pressures are not very even over large areas, the pressures near the corners of buildings being probably greater than elsewhere due to wind rushing around the corners, and so on. From a practical standpoint, therefore, all of the possible variations cannot be considered in design although today's specifications are becoming more and more detailed in their requirements.

When the designer working with large stationary buildings makes poor wind estimates the results are probably not too serious but this is not the case when tall slender buildings (or long flexible bridges) are being considered. The usual practice for buildings is to ignore wind forces unless their heights are at least twice their least lateral dimensions. For such cases as these it is felt that the floors and walls give the frame sufficient lateral stiffness to eliminate the need for a definite wind bracing system. Should the buildings have their walls and floors constructed of very modern light materials or should they be located in areas of unusual wind conditions the 2 to 1 ratio will probably not be followed. Building codes do not usually provide for estimated forces during tornadoes. The forces created directly in the paths of these storms are so violent that it is not considered economically feasible to design buildings to resist them.

Wind forces act as pressures on vertical windward surfaces, pressures or suction on sloping windward surfaces (depending on the slope) and suction on flat surfaces and on leeward vertical and sloping surfaces (due to the creation of negative pressures or vacuums). The student may have noticed this definite suction effect where shingles or other roof coverings have been lifted from the leeward roof surfaces of buildings. Suction or uplift can easily be demonstrated by holding a piece of paper at two of its corners and blowing above it. For some common structures uplift may be as large as 20 to 30 psf or even more.

During the passing of a tornado or hurricane, a sharp reduction in atmospheric pressure occurs. This decrease in pressure is not reflected inside airtight buildings, and the inside pressures, being greater than the external pressures, cause outward forces against the roofs and walls. Nearly everyone has heard stories of the walls of a building "exploding" outward during a storm.

The average building code in the United States makes no reference to wind velocities in the area or to shapes of the building or to other factors. It just probably requires the use of some specified wind pressure in design such as 20 psf on projected area in elevation up to 300 ft with an increase of 2.5 psf for each additional 100-ft increase in height. The values given in these codes are thought to be rather inaccurate for modern structural design.

For a period of several years the ASCE Task Committee on Wind Forces made a detailed study of existing information concerning wind forces. A splendid report of this information entitled *Wind Forces on Structures* [3] was presented by the committee in 1961. As stated in the report, its purpose was to provide a compact source of information which could be practically used by the design profession.

In the report there is much information presented concerning wind-pressure coefficients for various types of structures, information concerning maximum wind velocities for particular geographical areas, and much additional data which can be of great value in realistically estimating wind forces.

The wind pressure on a building can be estimated with the expression to follow in which p is the pressure in pounds per square foot acting on vertical surfaces, C_s is a shape coefficient, and V is the basic wind velocity in miles per hour estimated from weather bureau records for that area of the country.

$$p = 0.002558 C_s V^2$$

The coefficient C_s depends on the shape of the structure, primarily the roof. For box type structures C_s is 1.3 of which 0.8 is for the pressure on the windward side and 0.5 is for suction on the leeward side. For such a building the total pressure on the two surfaces equals 20 psf for a wind velocity of 77.8 mph.

As described in the foregoing paragraphs, the accurate determination of the most critical wind loads on a building or bridge is an extremely involved problem; however, sufficient information is available today to permit satisfactory estimates on a reasonably simple basis.

b. *Earthquake loads*. Many areas of the world including the western part of the United States, fall in earthquake territory and in these areas it is necessary to consider seismic forces in design for tall or short buildings. During an earthquake there is an acceleration of the ground surface. This acceleration can be broken down into vertical and horizontal components. Usually the vertical component of the acceleration is assumed to be negligible but the horizontal component can be severe.

Most buildings can be designed with little extra expense to withstand the forces caused during an earthquake of fairly severe intensity. On the other hand, earthquakes during recent years have clearly shown that the average building that is not designed for earthquake forces can be destroyed by earthquakes that are not particularly severe. The usual practice is to design buildings for additional lateral loads (representing the estimate of earthquake forces) that are equal to some percentage (5 to 10%) of the weight of the building and its contents. An excellent reference on the subject of seismic forces is a publication by the Structural Engineers Association of California entitled *Recommended Lateral Force Requirements and Commentary* [4].

It should be noted that many persons look upon the seismic loads to be used in designs as being merely percentage increases of the wind loads. This thought is not really correct, however, as seismic loads are different in their action and as they are not proportional to the exposed area but to the building weight above the level in question.

The effect of the horizontal acceleration increases with the distance above the ground because of the "whipping effect" of the earthquake, and design loads should be increased accordingly. Obviously, towers, water tanks, and penthouses on building roofs occupy precarious positions during an earthquake.

6. *Longitudinal loads*. Longitudinal loads are another type of load that needs to be considered in designing some structures. Stopping a train on a railroad bridge or a truck on a highway bridge causes longitudinal forces to be applied. It is not difficult to imagine the tremendous longitudinal force

Hungry Horse Dam and Reservoir, Rocky Mountains, in northwestern Montana. (Courtesy of the Montana Travel Promotion Bureau.)

developed when the driver of a 40-ton trailer truck traveling 60 mph suddenly has to apply the brakes while crossing a highway bridge. There are other longitudinal load situations, such as ships running into docks and the movement of traveling cranes that are supported by building frames.

7. *Other live loads.* Among the other types of live loads with which the structural designer will have to contend are *soil pressures* (such as the exertion of lateral earth pressures on walls or upward pressures on foundations); *hydrostatic pressures* (as water pressure on dams, inertia forces of large bodies of water during earthquakes and uplift pressures on tanks and basement structures); *blast loads* (caused by explosions, sonic booms, and military weapons); *thermal forces* (due to changes in temperature resulting in structural deformations and resulting structural forces); and *centrifugal forces* (as those caused on curved bridges by trucks and trains or similar effects on roller coasters, etc.).

1.8 SELECTION OF DESIGN LOADS

To assist the designer in estimating the magnitudes of live loads with which he or she should proportion structures, various records have been assembled through the years in the form of building codes and specifications. These publications provide conservative estimates of live load magnitudes for various situations. One of the most widely used design load specifications for buildings is that published by the American National Standards Institute (abbreviated ANSI) [5]. Some other commonly used specifications are:

1. For railroad bridges, American Railway Engineering Association (AREA) [6].
2. For highway bridges, American Association of State Highway and Transportation Officials (AASHTO) [7].
3. For buildings, the Uniform Building Code [8].

These specifications will on many occasions clearly prescribe the loads for which structures are to be designed. Despite the availability of this information, the designer's ingenuity and knowledge of the situation are often needed to predict what loads a particular structure will have to support in years to come. Over the past several decades insufficient estimates of future traffic loads by bridge designers have resulted in a great amount of replacement with wider and stronger structures.

1.9 SI UNITS

The International Bureau of Weights and Measures has as its goal the establishment of a rational and coherent worldwide system of units. In 1960 they named this system the "International System of Units," with the abbreviation SI in all languages. SI units, which are currently being adopted by quite a few countries including most English-speaking nations (Britain, Australia, Canada, South Africa, and New Zealand), differ from the metric system now being used in most European countries and in many other parts of the world.

The SI system has the very important advantage that only one unit is given for each physical quantity, such as the meter (m) for length, the kilogram (kg) for mass, the second (s) for time, the newton (N) for force, and so on. From these basic units other units are derived as follows.

area	square meter (m^2)
acceleration	meter per second squared (m/s^2)
force	kilogram meter per second squared ($kg \cdot m/s^2$) = newton (N)
stress	newton per square meter (N/m^2) = pascal (Pa)

The multiples of these units are in the decimal system and are expressed in powers of 10 that are multiples of three, as shown in Table 1.1, where length units are illustrated. In other words, multiple and submultiple steps of 1000 are recommended. It will be noted that the centimeter is not shown because its value (which would be 10^{-2} in the table) is inconsistent with the theory of having prefixes that are ternary powers of 10.

Table 1.1 SI DECIMAL MULTIPLES

SI SYMBOL	NAME	MULTIPLIER	EXAMPLE (METERS)
G	giga	1 000 000 000	Gm = 1 000 000 000
M	mega	1 000 000	Mm = 1 000 000
k	kilo	1 000	km = 1 000
m	milli	0.001	mm = 0.001
μ	micro	0.000 001	μm = 0.000 001
n	nano	0.000 000 001	nm = 0.000 000 001

In many countries of the world the comma is used to indicate a decimal; thus to avoid confusion in the SI system, spaces rather than commas are used. For a number having four or more digits, the digits are separated into groups of threes, counting both right and left from the decimal. For example, 3,245,621 is written as 3 245 621 and 2,015.3216 is written as 2 015.321 6.

When units are to be multiplied together, a dot is used to separate them. Bending moment is a term frequently used in structural design. In SI units it is expressed in newton meters. It is to be written as N · m (not mN, which represents millinewton). There is a great premium in the SI system on symbology. As a result, the designer must be exceptionally careful to use the correct symbols. For example, it is correct to write 100 meters, but in abbreviation 100 m must be used and not 100 ms, which would be 100 milliseconds.

Table 1.2 gives conversion values for some customary measurements to the SI system. Perhaps if you were to memorize a few approximate values from this table, you would begin to get a feel for the relative values between the two

Steel girder approach spans to Milwaukee Harbor Bridge. (Courtesy of the Wisconsin Department of Transportation.)

systems. For instance, 25 mm is approximately equal to 1 in, and 300 mm is approximately equal to 1 ft.

The unit of stress in the SI system is the newton per square meter (N/m^2), which is also called a pascal (Pa). To have a feel for the magnitude of this number, a megapascal (MPa) is equal to 0.145 kip per square inch. Thus, 138 MPa is equal to 20 ksi.

In the same fashion, the terms in-k and ft-k are constantly used. In the SI system they are usually expressed in kN · m. One kN · m is equal to 0.738 ft-k. From Table 1.2 it can be seen that 1 lb is equal to 4.448 222 N. Thus, 1 N is equal to about one-fourth of a pound, which is roughly equal to the weight of one apple.

Table 1.2 COMMON VALUES FOR SOME COMMON UNITS

CUSTOMARY UNITS	SI UNITS
1 in	25.400 mm = 0.025 400 m
1 in^2	645.16 mm^2 = 6.451 600 $m^2 \times 10^{-4}$
1 ft	304.800 mm = 0.304 800 m
1 lb	4.448 222 N
1 kip	4 448.222 N = 4.448 222 kN
1 psi	6.894 757 kN/m^2 = 0.006 895 MN/m^2 = 0.006 895 N/mm^2
1 psf	47.880 N/m^2 = 0.047 880 kN/m^2
1 ksi	6.894 757 MN/m^2 = 6.894 757 MPa
1 in-lb	0.112 985 N · m
1 ft-lb	1.355 818 N · m
1 in-k	112.985 N · m
1 ft-k	1 355.82 N · m = 1.355 82 kN · m

In a few places throughout the book the author has shown the SI units in parentheses. He particularly wants the reader to understand that from the standpoint of the designer there is no problem in learning and using SI units. The few units (mm, m, MPa = N/mm^2, and kN · m) can be learned in a few minutes and can be applied with no difficulties. Companies involved in the manufacture of items used in construction such as beams, reinforcing bars, window frames, and so on, however, are in a different situation as they face a difficult and expensive problem in converting their sizes, shapes, tables, and so on, into the new units with possible changes in actual sizes to rounded or standard dimensions in SI units.

1.10 LINE DIAGRAMS

To calculate the forces in a structure with reasonable simplicity, it is necessary to replace the actual structure with a series of lines representing the center lines of the members. This convenient representation does cause discrepancies from actual conditions, but the errors are generally within reasonable limits.

Line diagrams represent actual structures, examples of which can probably be found in the reader's neighborhood, wherever he or she lives, and the reader is encouraged to attempt to visualize the makeup of the actual structures so represented. Figure 1.3(a) shows a series of wood joists supported by a wood beam which itself is supported by concrete-block walls. A line diagram for the beam, its supports, and the loads applied to it is drawn in Fig. 1.3(b). The load transferred from a joist to the beam is spread over the width of the joist, but the line diagram has only an arrow acting at the center of the joist. A similar condition is the support supplied by the walls to the beam.

A drawing of a portion of a bolted steel roof truss as it is frequently assembled can be seen in Fig. 1.3(c), whereas (d) represents the line diagram for the truss.

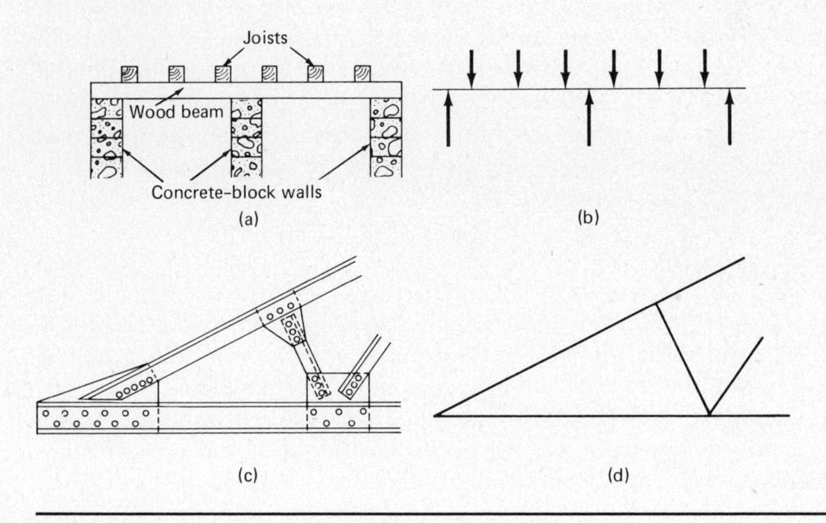

Figure 1.3

1.11 CALCULATION ACCURACY

A most important point that many students with their superb pocket calculators have difficulty in understanding is that structural analysis is not an exact science for which answers can confidently be calculated to eight or more places. Computations equivalent to only three places are probably far more accurate than the estimates of material strengths and magnitudes of loads used for structural analysis and design. The common materials dealt with in structures (wood, steel, concrete, and a few others) have ultimate strengths that can only be estimated. The loads applied to structures may be known within a few hundred pounds or no better than a few thousand pounds. It therefore seems inconsistent to require force computations to more than three significant digits.

Several partly true assumptions will be made about the construction of trusses: Truss members are connected with frictionless pins; the deformation of truss members under load is so slight as to cause no effect on member forces; and so forth. These deviations from actual conditions emphasize that it is of little advantage to carry structural analysis to many decimal places. Furthermore, retention of calculations to more than two or three significant places may be harmfully misleading, giving a fictitious sense of precision.

1.12 CHECKS ON PROBLEMS

A definite advantage of structural analysis is the possibility of making mathematical checks on the analysis by some method other than the one initially used, or by the same method from some other position on the structure. The reader should be able in nearly every situation to determine if his or her work has been done correctly.

All of us, unfortunately, have the weakness of making exasperating mistakes, and the best that can be done is to keep them to the absolute minimum. The application of the simple arithmetical checks suggested in the following chapters will eliminate many of these costly blunders. The best structural designer is not necessarily the one who makes the fewest mistakes initially but is probably the one who discovers the largest percentage of his or her mistakes and corrects them.

1.13 COMPUTERS

When a student completes formal studies today and goes to work for a structural firm he or she will be confronted with a wide assortment of computers, minicomputers, hand-held computers, programmable pocket calculators, and so on, which can be programmed to handle structural analysis and design problems. Furthermore, the available equipment and programs are expanding at a very rapid rate (seemingly almost on a daily basis).

As a result of this amazing list of available computer tools, the reader may very well ask the question "Why do I need to know the theories behind structural analysis when I can get the work done with an existing computer program in a few minutes?" If a person does not have a knowledge of structural analysis he or she will have no way of proving that the computer output is correct. The results will be just a group of numbers which may be reasonable or may be completely unreasonable. These wonderful devices may be used to take away a great deal of the tedious and detailed calculations involved in structures, but they cannot relieve the designer of his or her responsibility for their structures. He or she must make decisions concerning the applicability of a particular program and the reliability of the output.

The use of electronic computers for solving structural problems began during the mid-1950s. In the brief period since then, their application to everyday structural problems has become a reality and today most large firms use them routinely for many types of problems. Nevertheless the profession has probably witnessed only the very beginning of the computer revolution, and many designers still have little or no conception of the tremendous impact computers will have on them and their profession. So many analysis and design problems can be handled with such unbelievable efficiency that the consulting firm that does not take advantage of computers may already be noncompetitive.

The immediate future of computer design is so promising it is not necessary to gaze off into distant decades to see tremendous developments for the benefit of structural design. Almost all types and sizes of trusses, towers, buildings, bridges, and so on, are commonly analyzed and designed with computers. Who can say that before the end of the decade the designer will not be able to sit at his or her desk and send data to a central computer requesting the analysis and design of a multistory building or a large bridge and have supplied within a few minutes the complete design together with takeoff quantities and other data?

1.14 ACCESS TO COMPUTERS

Only the largest firms can afford to have "in-house" computers, as their cost may run as high as several thousand dollars per month. Many organizations have found that the expense of having mainframe computers in their offices is far too great for the amount of time the computers are actually needed for fast-running engineering programs. Designers do, however, need access to computers for specific programs, with the result that many firms have turned either to service bureaus or to time-sharing companies and lately to microcomputers and to programmable calculators.

When service bureaus are used, the designer does not need to be familiar with computers or computer programs: he only needs to furnish the data to the bureaus and they select an appropriate program and provide the solution. The data furnished include such things as dimensions, materials, specifications, loading conditions, and types of supports. Time requirements for this type of service usually vary from one or two days to one week, and the charge to the customer is generally on an hourly rate.

In between "in-house" computers and service bureaus lies time sharing, which many experts claim is so imporant it is going to affect drastically the lives of everyone in the United States in the near future. Time sharing is growing so rapidly that by the end of the century it may be one of the world's largest industries. A design firm may have a terminal installed in their office that will enable them to use a remotely located computer at a small fraction of the cost required to install their own computer. The time-sharing system actually consists of a large number of remote terminals that are connected by telephone to the computer.

Usually the terminal device is rented at a monthly rate and an hourly rate is charged for the actual time the computer is used with a certain minimum amount of use guaranteed per month. (It should be realized that many companies act both as service bureaus and as time-sharing companies.)

In the United States today there are several dozen general-purpose time-sharing companies, the one best known to the reader probably being the Airline Reservation System.

Regardless of where a program is obtained, almost any designer would want to take a sufficiently general problem or two which he or she has previously solved (from his or her knowledge of structural analysis) to check out the correctness of the program. There are few designers who would be willing to blindly stake their professional reputations on the claims made by the representative of a service bureau, time-sharing company, or other organization without making a preliminary check.

Throughout this book an attempt has been made to analyze structures with only a few unknowns so that the theories of structural analysis are illustrated without hiding them under a maze of calculations. The longer problems (as well as the shorter ones) can be handled with computers. In various places the author comments on calculations where computer usage is particularly helpful.

REFERENCES

[1] *Proceedings of the First United States Conference on Prestressed Concrete* (Cambridge, Mass.: Massachusetts Institute of Technology, 1951), p. 1.

[2] Wind Forces on Structures, Task Committee on Wind Forces, Committee on Loads and Stresses, Structural Division, ASCE, Final Report, *Transactions ASCE* 126, Part II (1961): 1124–1125.

[3] Ibid, pp. 1124–1198.

[4] *Recommended Lateral Force Requirements and Commentary* (San Francisco, Calif.: Structural Engineers Association of California (SEAOC), 1975).

[5] *American National Standard Building Code Requirements for Minimum Design Loads in Buildings and Other Structures* (New York: American Standards Institute, 1972), ANSI A58.1-1982.

[6] *Specifications for Steel Railway Bridges* (Chicago, Ill.: American Railway Engineering Association (AREA), 1965).

[7] *Standard Specifications for Highway Bridges* 12th ed., (Washington, D.C.: American Association of State Highway and Transportation Officials (AASHTO), 1977).

[8] *Uniform Building Code* (Whittier, Calif.: International Conference of Building Officials, 1979).

2 Reactions

2.1 EQUILIBRIUM

A body at rest is said to be in *equilibrium*. The resultant of the external loads on the body and the supporting forces or reactions is zero. Not only must the sum of all forces (or their components) acting in any possible direction be zero but also the sum of the moments of all forces about any axis must be zero.

If a structure or part thereof is to be in equilibrium under the action of a system of loads, it must satisfy the six statics equilibrium equations. Using the Cartesian x, y, and z system the equations can be written as

$$\Sigma F_x = 0 \qquad \Sigma F_y = 0 \qquad \Sigma F_z = 0$$

$$\Sigma M_x = 0 \qquad \Sigma M_y = 0 \qquad \Sigma M_z = 0$$

For purposes of analysis and design the large majority of structures can be considered as being plane structures without loss of accuracy. For these cases the sum of the forces in the x and y directions and the sum of the moments about an axis perpendicular to the plane must be zero

$$\Sigma F_x = 0 \qquad \Sigma F_y = 0 \qquad \Sigma M_z = 0$$

These equations are commonly written as

$$\Sigma H = 0 \qquad \Sigma V = 0 \qquad \Sigma M = 0$$

These equations cannot be proved algebraically; they are merely statements of Sir Isaac Newton's observation that for every action on a body at rest there is an equal and opposite reaction. Whether the structure under consideration is a beam, a truss, a rigid frame, or some other type of assembly supported by various reactions, the equations of statics must apply if the body is to remain in equilibrium.

The structures considered for the first 13 chapters in this text are considered to be coplanar, whereas three-dimensional structures or space frames are considered in Chap. 14.

2.2 MOVING BODIES

The statement was made in the preceding section that a body at rest is in equilibrium. It is possible, however, for an entire structure to move and yet be in equilibrium. An airplane or ship moves, but its individual parts do not move with respect to each other.

For a moving body, additional forces must be included for the equations of statics to be applicable. These are the inertia forces, and with them included the body may be considered to be acted upon by a set of forces in equilibrium for purposes of structural analysis.

2.3 CALCULATION OF UNKNOWNS

To identify a force completely, there are three unknowns that must be determined: They are the magnitude, direction, and line of action of the force. All of these values are known for external loads, but for a reaction only the point of application and perhaps the direction are known.

The total number of unknowns that can be determined by the equations of statics is controlled by the number of equations available. It does not make any difference how many reactions a structure has or how many unknowns each reaction has; there are three equations of statics and they can be used to determine but three unknowns for each structure. The determination of more

City Park Lake Bridge, Baton Rouge, Louisiana. (Courtesy of the American Institute of Steel Construction, Inc.)

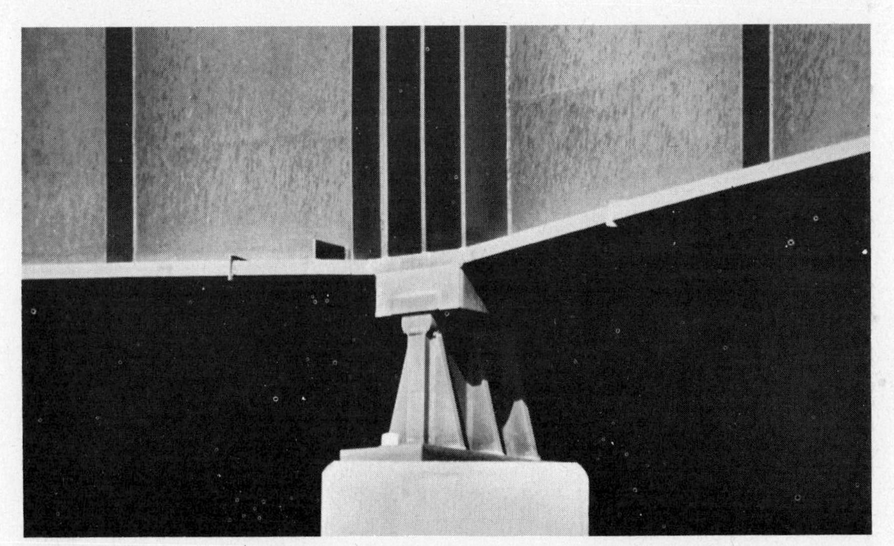

Hinged support for a bridge girder. (Courtesy of the Bethlehem Steel Corporation.)

than three unknowns requires additional equations or methods to use in conjunction with the statics equations. It will be seen that in some few instances, owing to special construction features, equations of condition are available in addition to the usual equations.

2.4 TYPES OF SUPPORT

Structural frames may be supported by hinges, rollers, fixed ends, or links. These supports are discussed in the following paragraphs.

A *hinge* or pin-type support (represented herein by the symbol ⩘) is assumed to be connected to the structure with a frictionless pin. This type of support prevents movement in a horizontal and a vertical direction but does not prevent slight rotation about the hinge. There are two unknown forces at a hinge: the magnitude of the force required to prevent horizontal movement and the magnitude of the force required to prevent vertical movement. (The support supplied at a hinge may also be referred to as an inclined force which is the resultant of the horizontal force and the vertical force at the support. Two unknowns remain: the magnitude and direction of the inclined resultant.)

A *roller* type of support (represented herein by the symbol ⚲) is assumed to offer resistance to movement only in a direction perpendicular to the supporting surface beneath the roller. There is no resistance to slight rotation about the roller or to movement parallel to the supporting surface. The magnitude of the force required to prevent movement perpendicular to the supporting surface is the one unknown. **Rollers may be installed in such a manner that they can resist movement either toward or away from the supporting surface.**

A *fixed-end* support (represented herein by the symbol ⊢) is assumed to offer resistance to rotation about the support and to movement vertically and horizontally. There are three unknowns: the magnitude of the force to prevent horizontal movement, the magnitude of the force to prevent vertical movement, and the magnitude of the force to prevent rotation.

A *link* type of support (represented herein by the symbol o—o) is quite similar to the roller in its action because the pins at each end are assumed to be frictionless. The line of action of the supporting force must be in the direction of the link and through the two pins. One unknown is present: the magnitude of the force in the direction of the link.

2.5 STATICALLY DETERMINATE STRUCTURES

The discussion of supports showed there are three unknown reaction components at a fixed end, two at a hinge, and one at a roller or link. If, for a particular structure, the total number of reaction components equals the number of equations available, the unknowns may be calculated, and the structure is then said to be statically determinate externally. Should the number of unknowns be greater than the number of equations available, the structure is statically indeterminate externally; *if less, it is unstable externally.*

The internal arrangement of some structures is such that one or more equations of condition is available. The arch of Fig. 3.5 has an internal pin (or

The Chesapeake Bay Bridge illustrating beam spans, plate girder spans, deck truss, cantilever construction, and suspension construction, all included in this structure. (Courtesy of the Bethlehem Steel Corporation.)

hinge) at C. The internal moment at this "frictionless" pin is zero because no rotation can be transferred between the adjacent parts of the structure. A special condition exists because the internal moment at the pin must be zero regardless of the loading. A similar statement cannot be made for any continuous section of the beam.

By definition, a hinge transmits no rotation, and the three equations of statics plus a $\Sigma M = 0$ equation at hinge C are available to find the four unknown reaction components at A and B. The omission of a member in the truss of Fig. 15.3 will be shown to give another condition equation. If the number of condition equations plus the three equations of statics equals the number of unknowns, the structure is statically determinate; if more, it is unstable; and if less, it is statically indeterminate.

Several structures are classified in Table 2.1 as to their statical condition. The beam labeled c in the table supported on its ends with rollers is stable under vertical loads but is unstable under inclined loads. A structure may be stable under one arrangement of loads, but if it is not stable under any other conceivable set of loads, it is unstable. This condition is sometimes referred to as ***unstable equilibrium.*** The structure labeled j has two internal hinges and thus two equations of condition. There are five equations available and five unknown reaction components; the structure is statically determinate. If one of the supporting hinges were changed to a roller, the structure would become unstable.

2.6 GEOMETRIC INSTABILITY

The ability of a structure to support adequately the loads applied to it is dependent not only on the number of reaction components but also on the arrangement of those components. It is possible for a structure to have as many or more reaction components than there are equations available and yet be unstable. This condition is referred to as *geometric instability*.

The frame of Fig. 2.1(a) has three reaction components, and three equations available for their solution; however, a study of the moment situation at B shows the structure to be unstable. The line of action of the reaction at A passes through B, and unless the line of action of the force P passes through the

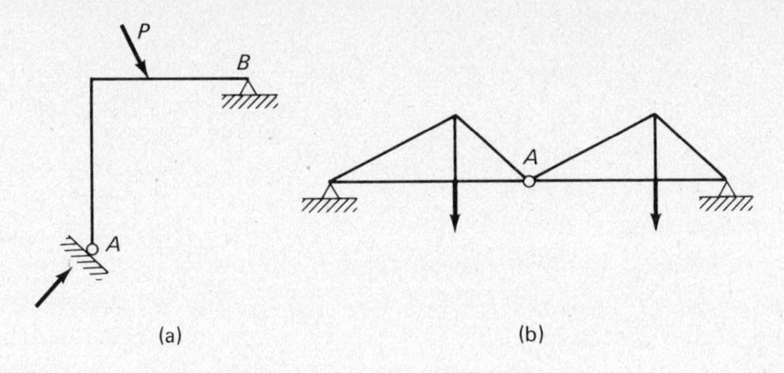

Figure 2.1

Table 2.1 STATICAL CLASSIFICATION OF STRUCTURES

STRUCTURE	NUMBER OF UNKNOWNS	NUMBER OF EQUATIONS	STATICAL CONDITION
a	 3	3	Statically determinate (referred to as a simple beam)
b	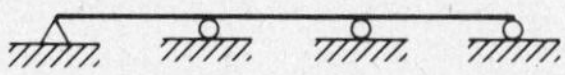 5	3	Statically indeterminate to second degree (a continuous beam)
c	2	3	Unstable
d	 3	3	Statically determinate (a cantilever beam)
e	 6	3	Statically indeterminate to third degree (a fixed-ended beam)
f	 4	3	Statically indeterminate to first degree (a propped beam)
g	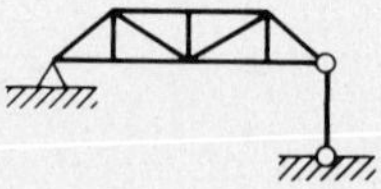 3	3	Statically determinate
h Hinge	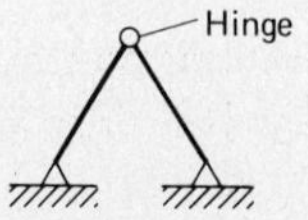 4	4	Statically determinate
i	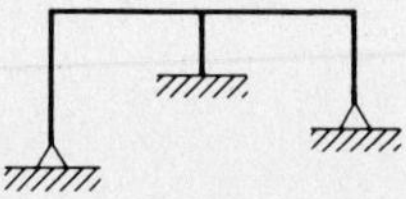 7	3	Statically indeterminate to fourth degree
j	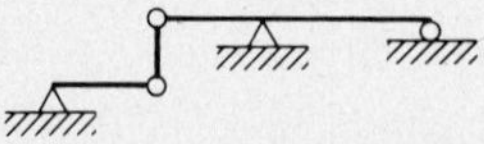 5	5	Statically determinate

same point, the sum of the moments about B cannot equal zero. There is no resistance to rotation about B, and the frame will immediately begin to rotate. It may not collapse, but it will rotate until a stable situation is developed when the line of action of the reaction at A passes some little distance from B. Of prime importance to the engineer is for a structure to hold its position under load. One that does not do so is unstable.

Mississippi River Bridge, St. Paul, Minnesota. (Courtesy of Kenneth M. Wright Studios, St. Paul, Minnesota.)

Another geometrically unstable structure is shown in Fig. 2.1(b). Four equations are available to compute the four unknown reaction components, but rotation will instantaneously occur about the hinge at A. After a slight deflection vertically of A, the structure will probably become stable.

2.7 SIGN CONVENTION

The particular sign convention that is used for tension, compression, and so forth is of little consequence as long as a consistent system is used. The author uses the following signs in his computations.

1. For *tension* a positive sign is used, the thought being that pieces in tension become longer or have plus lengths.
2. A negative sign is used for pieces in *compression* because they are compressed or shortened and therefore have minus lengths.
3. For clockwise *moments* a positive sign is usually used; for counterclockwise moments, a negative sign. This system is of even less importance than the tension and compression system, the important thing being to use the same sign convention in taking moments throughout each complete problem to avoid confusion.
4. On many occasions it is possible to determine the direction of a *reaction* by inspection, but where it is not possible, a direction is assumed and the appropriate statics equation written. If on solution of the equation the numerical value for the reaction is positive, the assumed direction was correct; if negative, the assumed direction was incorrect.

2.8 HORIZONTAL AND VERTICAL COMPONENTS

It is a good plan to compute the vertical and horizontal components of inclined forces for use in making calculations. If this practice is not followed, the perpendicular distances from the lines of action of inclined forces to the point where moment is being taken will have to be found. The calculation of these distances is often difficult, and the possibility of making mistakes in setting up the equations is greatly increased.

2.9 FREE-BODY DIAGRAMS

For a structure to be in equilibrium each and every part of the structure must be in equilibrium. If the statics equations are applicable to an entire structure, they must also be applicable to any part of the structure, no matter how large or how small.

It is therefore possible to draw a diagram of any part of a structure and apply the statics equations to that part. The result, called a *free-body diagram*, must include all of the forces applied to that portion of the structure. These forces are the external reactions and loads and the internal forces applied from the adjoining parts of the structure.

A simple beam is cut into two free bodies in Fig. 2.2. The internal forces—shear, moment, and axial—are assumed to be in the directions indicated by the arrows on the left free body. The corresponding forces on the right free body are by necessity in opposite directions from those on the left. Should the left-hand side of a beam tend to move up with respect to the right-hand side, the right side must pull down with an equal and opposite force if equilibrium is present.

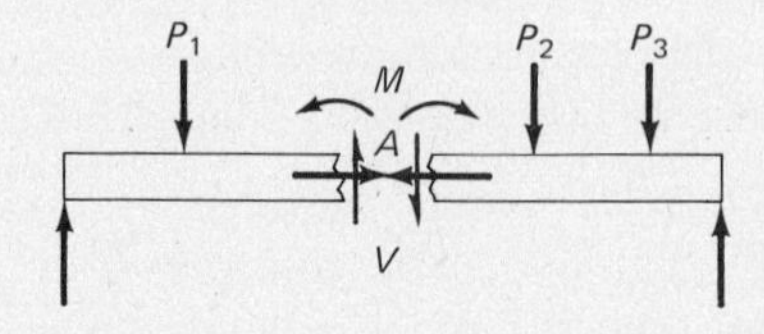

Figure 2.2

Isolating certain sections of structures and considering the forces applied to those sections is the basis of all structural analysis. It is doubtful that this procedure can be overemphasized to the reader, who will discover over and over that free-body diagrams open the way to the solution of structural problems.

2.10 REACTIONS BY PROPORTIONS

The calculation of reactions is fundamentally a matter of proportions. To illustrate this point, reference is made to Fig. 2.3(a). The load P is three-fourths of the distance from the left-hand support A to the right-hand support B. By

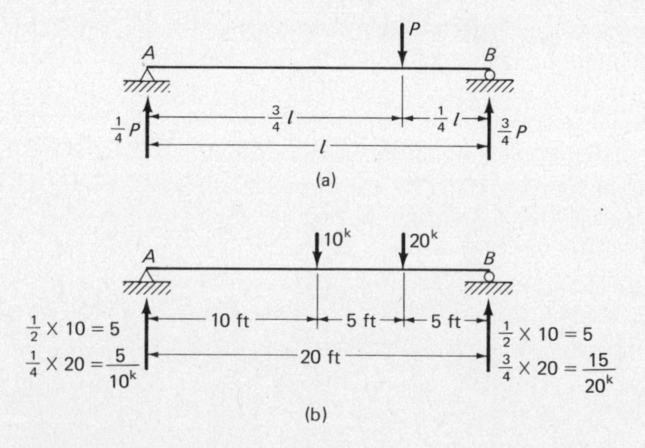

Figure 2.3

proportions, the right-hand support will carry three-fourths of the load and the left-hand support will carry the remaining one-fourth of the load.

Similarly, for the beam of Fig. 2.3(b), the 10-kip (k or kip is the abbreviation for kilopound) load is one-half of the distance from A to B and each support will carry half of it, or 5 kip. The 20-kip load is three-fourths of the distance from A to B. The B support will carry three-fourths of it, or 15 kip, and the A support will carry one-fourth or 5 kip. In this manner the total reaction at the A support was found to be 10 kip and the total reaction at the B support 20 kip.

2.11 REACTIONS CALCULATED BY EQUATIONS OF STATICS

Reaction calculations by the equations of statics are illustrated by Examples 2.1 to 2.3. In applying the $\Sigma M = 0$ equation, a point may usually be selected as the center of moments so that the lines of action of all but one of the unknowns pass through the point. The unknown is determined from the moment equation, and the other reaction components are found by applying the $\Sigma H = 0$ and $\Sigma V = 0$ equations.

The beam of Example 2.1 has three unknown reaction components, vertical and horizontal ones at A and a vertical one at B. Moments are taken about A to find the value of the vertical component at B. All of the vertical forces are equated to zero, and the vertical reaction component at A is found. A similar equation is written for the horizontal forces applied to the structure, and the horizontal reaction component at A is found to be zero.

The solutions of reaction problems may be checked by taking moments about the other support, as illustrated in Example 2.1. For future examples space is not taken to show the checking calculations. **A problem, however, should be considered incomplete until a mathematical check of this nature is made.**

The roller of the frame of Example 2.3 is supported by an inclined surface. The statics equations are still applicable, because the direction of the reaction at B is known (perpendicular to the supporting surface). If the direction of the reaction is known, the relationship between the vertical component, the horizontal component, and the reaction itself is known. Here the reaction has a slope of four vertically to three horizontally (4:3), which is the reverse of the slope of the supporting surface of three to four (3:4). Moments are taken about the left support, which gives an equation including the horizontal and vertical components of the reaction at the inclined roller. But both components are in terms of that reaction; therefore only one unknown, R_B, is present in the equation and its value is easily obtained.

Example 2.1

Compute the reaction components for the beam shown in Fig. 2.4.

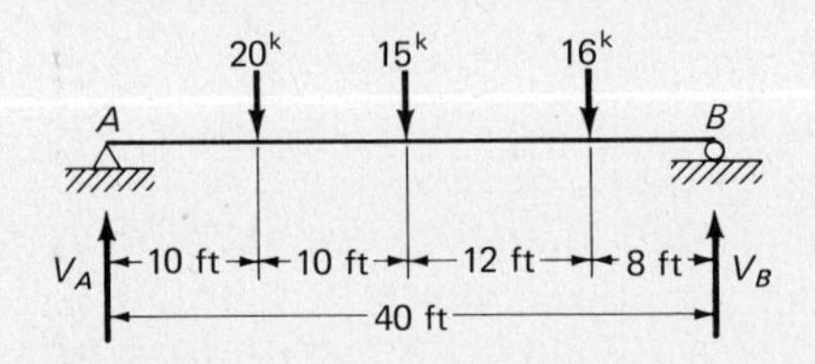

Figure 2.4

SOLUTION

$\Sigma M_A = 0$

$$(20)(10) + (15)(20) + (16)(32) - 40\ V_B = 0$$

$$V_B = 25.3^k \uparrow$$

$\Sigma V = 0$

$$20 + 15 + 16 - 25.3 - V_A = 0$$

$$V_A = 25.7^k \uparrow$$

Checking: $\Sigma M_B = 0$

$$(V_A)(40) - (20)(30) - (15)(20) - (16)(8) = 0$$

$$V_A = 25.7^k \uparrow$$

Example 2.2

Find all reaction components in the structure shown in Fig. 2.5.

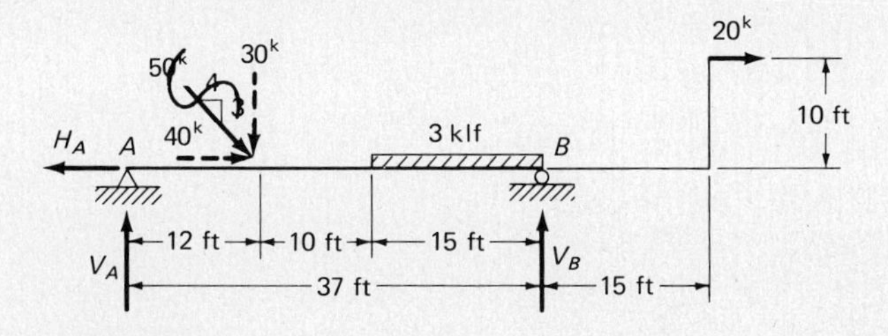

Figure 2.5

SOLUTION

$$\Sigma M_A = 0$$

$$(30)(12) + (3 \times 15)(29.5) + (20)(10) - (V_B)(37) = 0$$

$$V_B = 51^k \uparrow$$

$$\Sigma V = 0$$

$$30 + 45 - 51 - V_A = 0$$

$$V_A = 24^k \uparrow$$

$$\Sigma H = 0$$

$$40 + 20 - H_A = 0$$

$$H_A = 60^k \leftarrow$$

Example 2.3

Compute the reactions for the frame shown in Fig. 2.6.

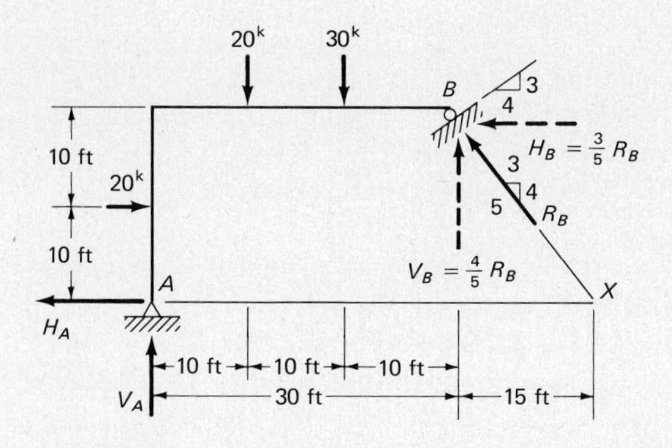

Figure 2.6

SOLUTION

$$\Sigma M_A = 0$$

$$(20)(10) + (20)(10) + (30)(20) - (H_B)(20) - (V_B)(30) = 0$$

$$200 + 200 + 600 - (\tfrac{3}{5}R_B)(20) - (\tfrac{4}{5})(R_B)(30) = 0$$

$$R_B = 27.8^k \nwarrow$$

$$V_B = \tfrac{4}{5}R_B = 22.2^k \uparrow$$

$$H_B = \tfrac{3}{5}R_B = 16.7^k \leftarrow$$

$$\Sigma V = 0$$

$$20 + 30 - 22.2 - V_A = 0$$

$$V_A = 27.8^k \uparrow$$

$$\Sigma H = 0$$

$$20 - 16.7 - H_A = 0$$

$$H_A = 3.3^k \leftarrow$$

ALTERNATE SOLUTION

Should the line of action of R_B be extended until it intersected a horizontal line through A at point X, another convenient location for taking moments would be available. Moments are taken at point X, and only one unknown (V_A) appears in the equation. This method may be simpler than the previous solution.

$$\Sigma M_x = 0$$

$$(V_A)(45) + (20)(10) - (20)(35) - (30)(25) = 0$$

$$V_A = 27.8^k \uparrow$$

2.12 PRINCIPLE OF SUPERPOSITION

As a student proceeds with his or her study of structural analysis he or she will encounter structures subject to large numbers of forces and to different kinds of forces (concentrated, uniform, triangular, dead, live, impactive, etc.). To assist in handling such situations there is available an extremely useful tool called the *principle of superposition.*

The principle follows: **if the structural behavior is linearly elastic the forces acting on a structure may be separated or divided in any convenient fashion and the structure analyzed for the separate cases. The final results can then be obtained by adding up the individual results.** The author previously made use of this principle in Fig. 2.3(b) when the reactions for the two loads were determined separately by proportions and then were added together to obtain the final values. The principle applies not only to reactions but also to shears, moments, stresses, strains, and displacements.

There are two important situations where the principle of superposition is not valid. The first occurs where the geometry of the structure is appreciably changed under the action of the loads. The second occurs where the structure consists of a material for which stresses are not directly proportional to strains. This latter situation can occur when the material is stressed beyond its elastic limit or when the material does not follow Hooke's law for any part of its stress strain curve.

PROBLEMS

For Prob. 2.1, determine which of the structures shown in the accompanying illustration are statically determinate, statically indeterminate (including the degree of indeterminancy), and unstable as regards outer forces.

Problem 2.1

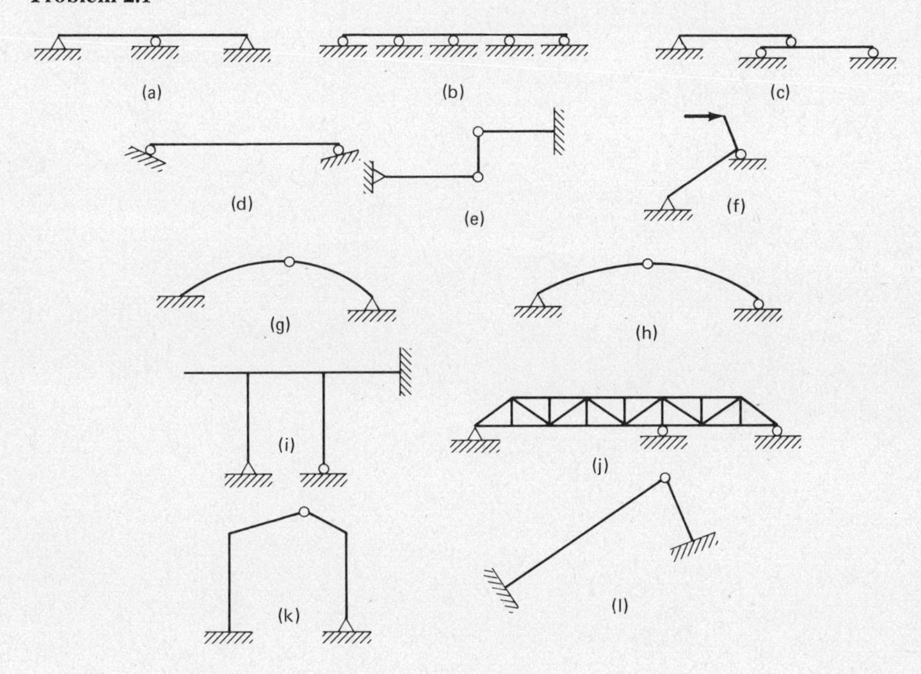

For Probs. 2.2 to 2.30, compute the reactions for the structures.

Problem 2.2 (*Ans.* $V_L = 42^k \uparrow$, $V_R = 48^k \uparrow$)

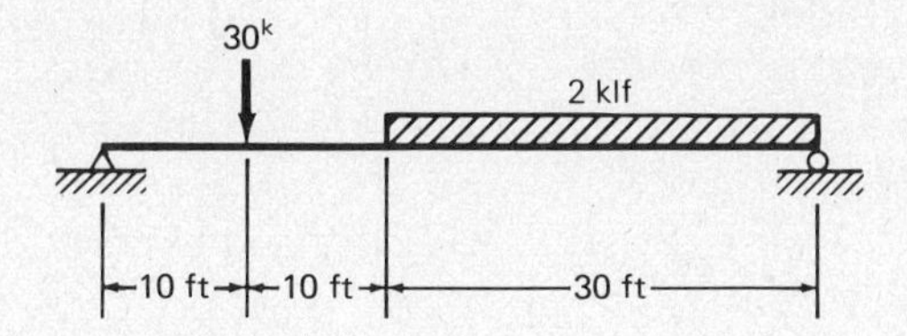

Problem 2.3

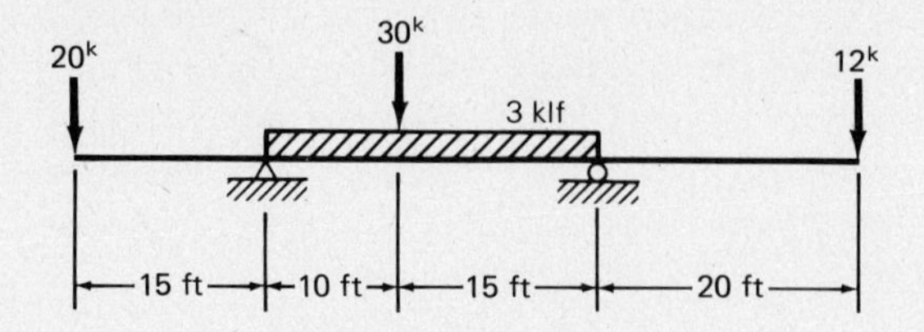

Problem 2.4 (*Ans.* $V_L = 26.33^k \uparrow$, $V_R = 75.67^k \uparrow$, $H_R = 20^k \rightarrow$)

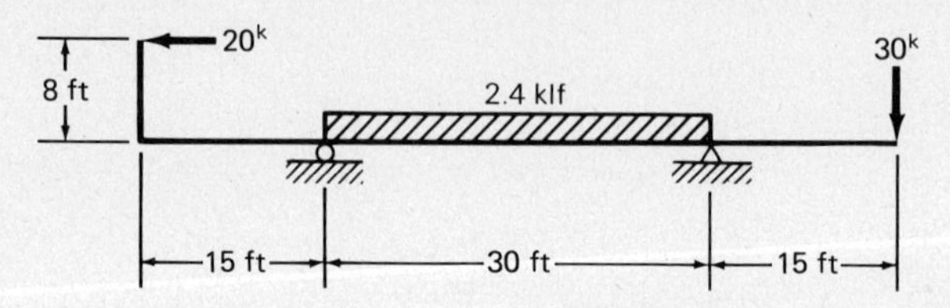

Problem 2.5

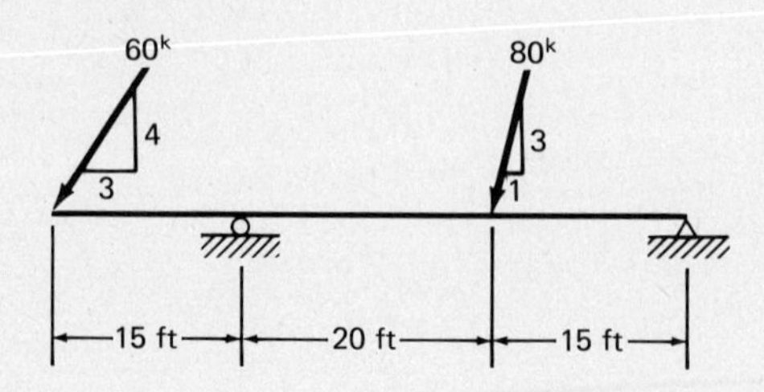

Problem 2.6 (*Ans.* $V_L = 10^k \downarrow$, $H_L = 50^k \leftarrow$, $V_R = 154^k \uparrow$)

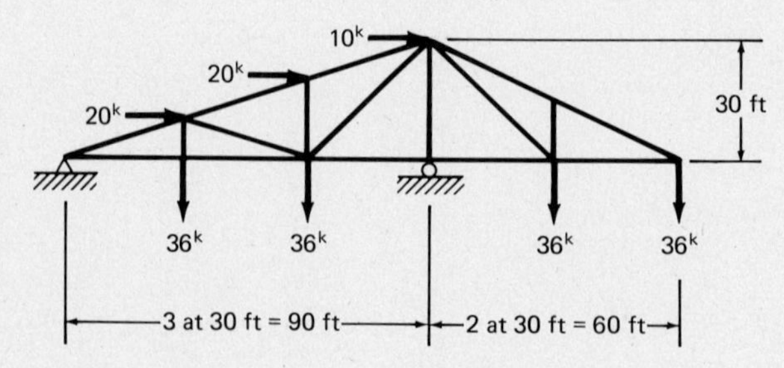

Problem 2.7

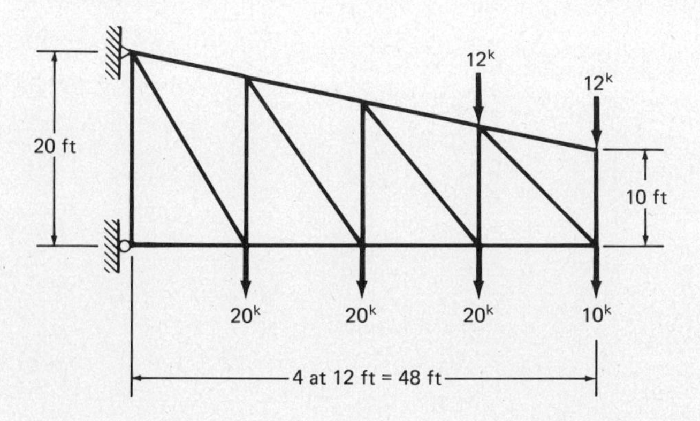

Problem 2.8 (*Ans.* $V_L = 100.95$ kN ↑, $H_L = 19.40$ kN ←, $V_R = 44.19$ kN ↑)

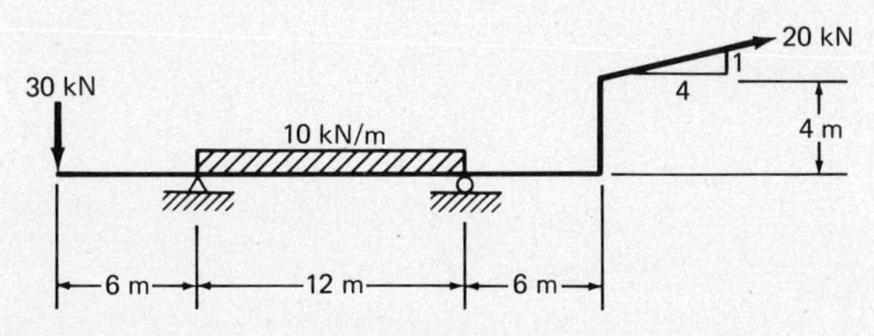

Problem 2.9

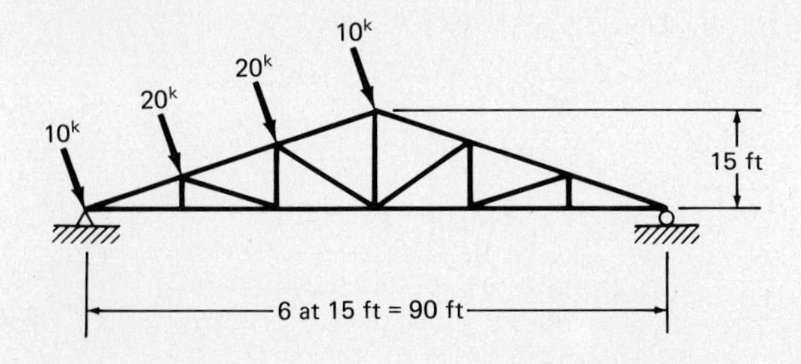

Problem 2.10 (*Ans.* $V_L = 11.4^k$ ↑, $H_L = 16.1^k$ ←, $V_R = 0.67^k$ ↓)

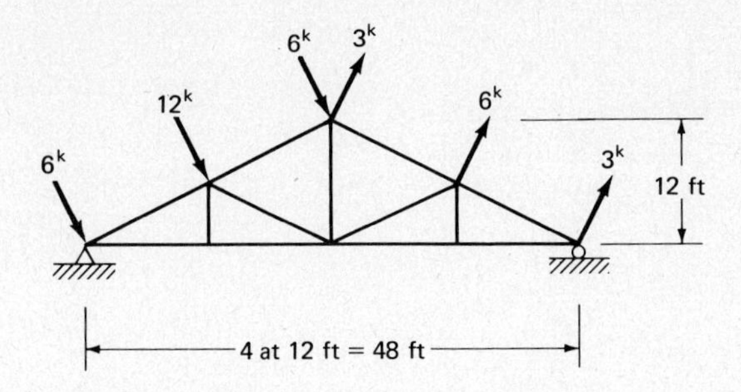

Problem 2.11

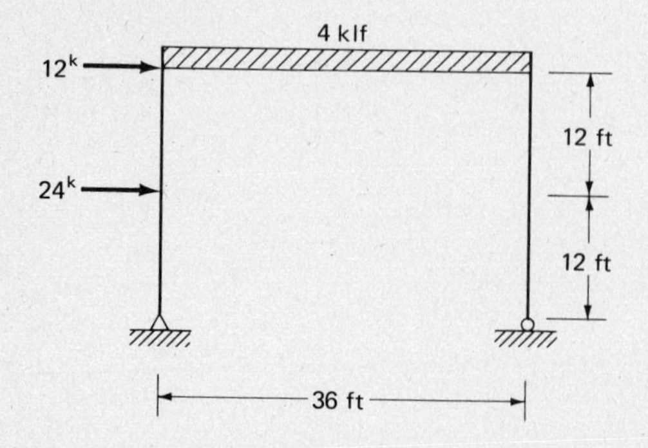

Problem 2.12 (*Ans.* $V_L = 69.1^k \uparrow$, $V_R = 53.1^k \uparrow$, $H_R = 16.1^k \leftarrow$)

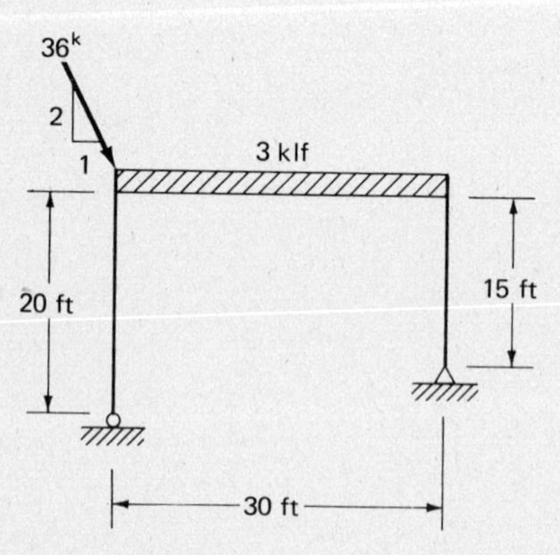

Problem 2.13

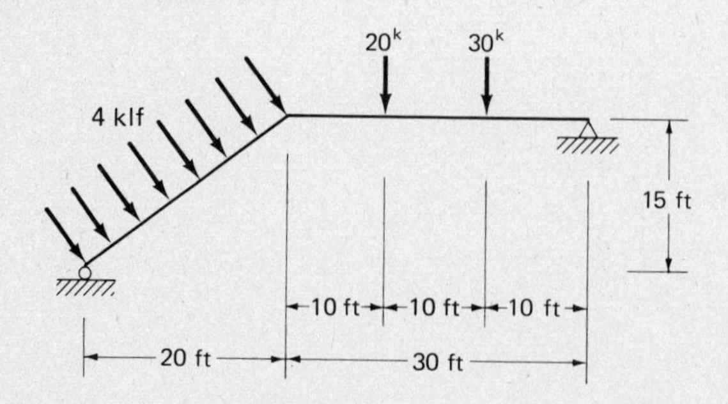

Problem 2.14 (*Ans.* $H_L = 131.9^k \rightarrow$, $H_R = 71.9^k \leftarrow$, $V_R = 125^k \uparrow$)

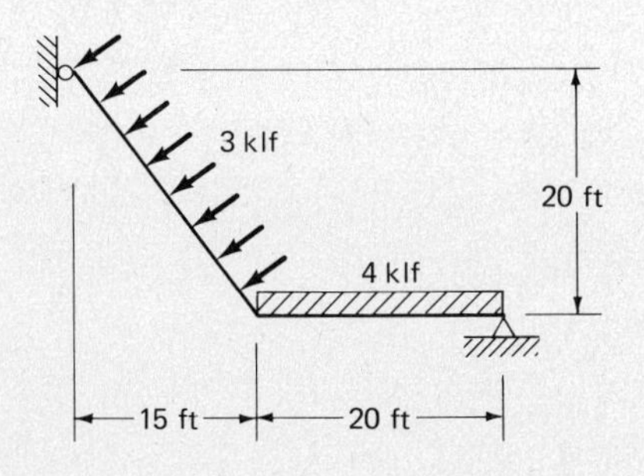

Problem 2.15

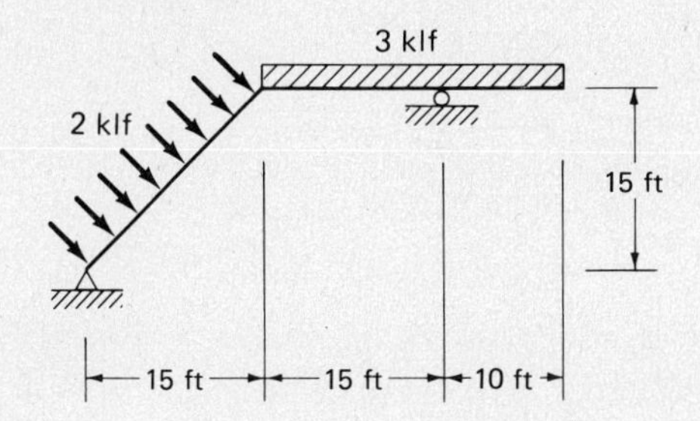

Problem 2.16 (*Ans.* $V_L = 31.3^k \uparrow$, $V_R = 35.7^k \uparrow$, $H_L = 10^k \leftarrow$)

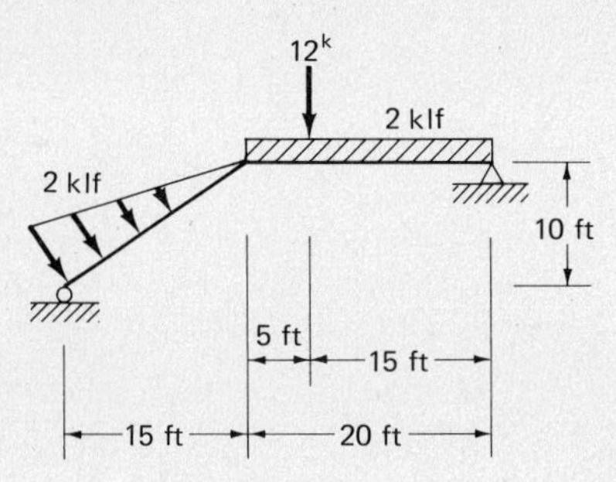

Problem 2.17

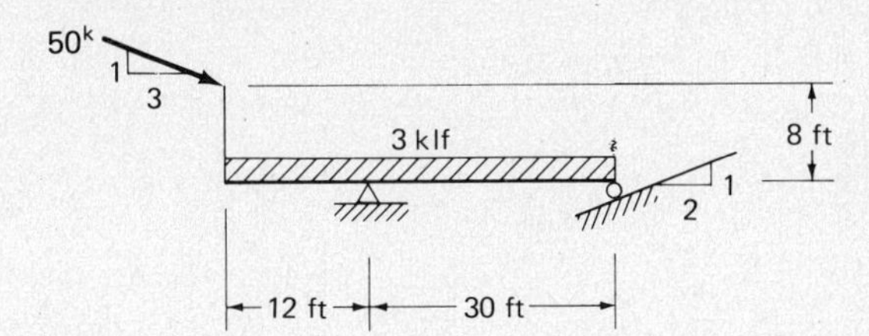

Problem 2.18 (*Ans.* $V_L = 25.2^k \uparrow$, $H_L = 12.6^k \rightarrow$, $V_R = 20.8^k \uparrow$, $H_R = 0.6^k \leftarrow$)

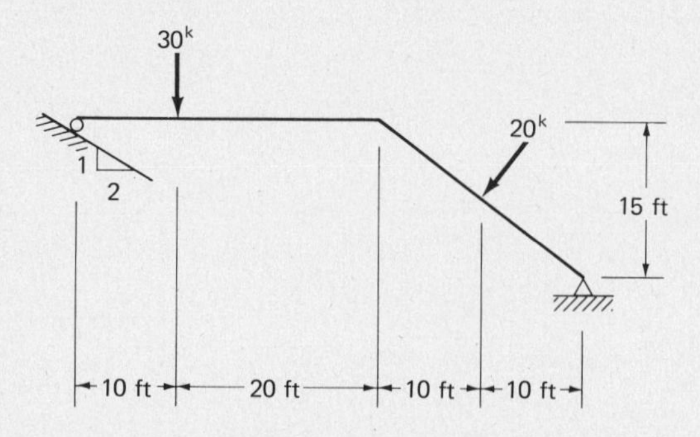

Problem 2.19

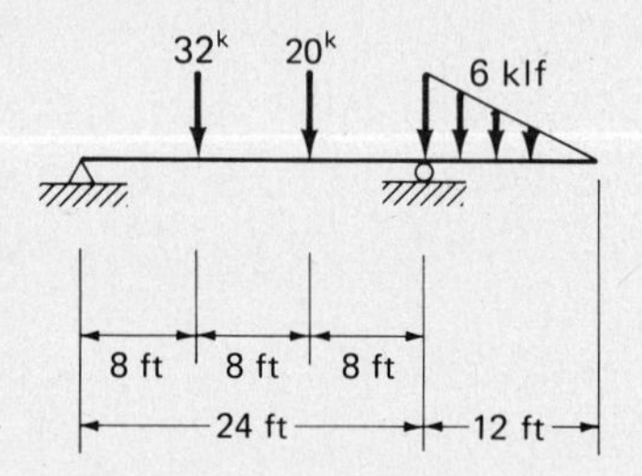

Problem 2.20 (*Ans.* $V_L = 33.3^k \uparrow$, $V_R = 6.7^k \uparrow$)

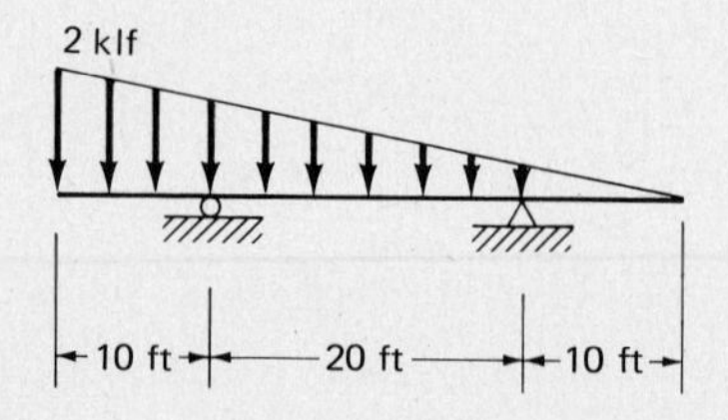

Problem 2.21

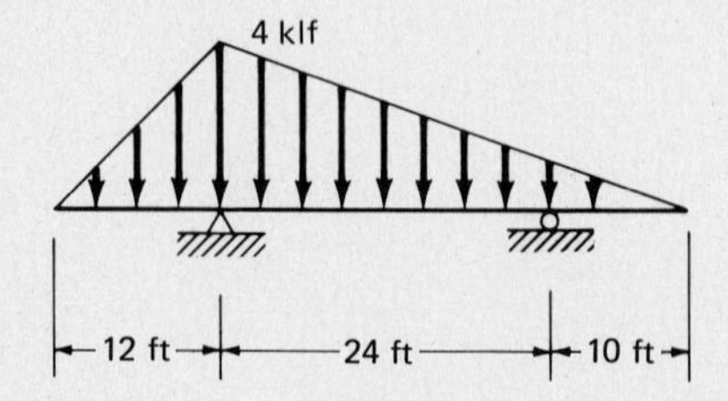

Problem 2.22 Consider only 1 ft length of frame. (*Ans.* $V_L = 65.0^k \uparrow$, $H_L = 21.7^k \rightarrow$, $V_R = 85.0^k \uparrow$, $H_R = 6.5^k \rightarrow$)

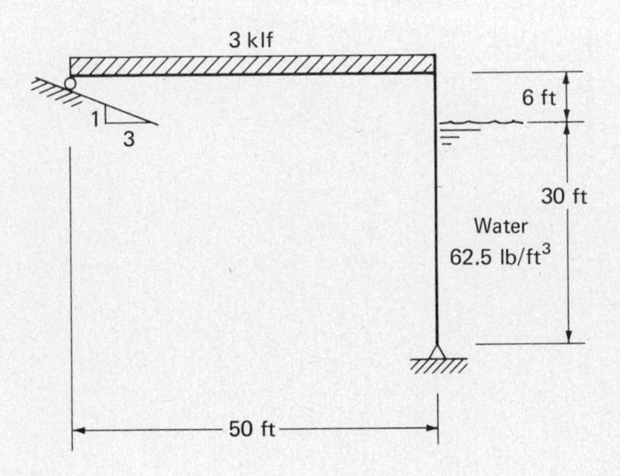

Problem 2.23

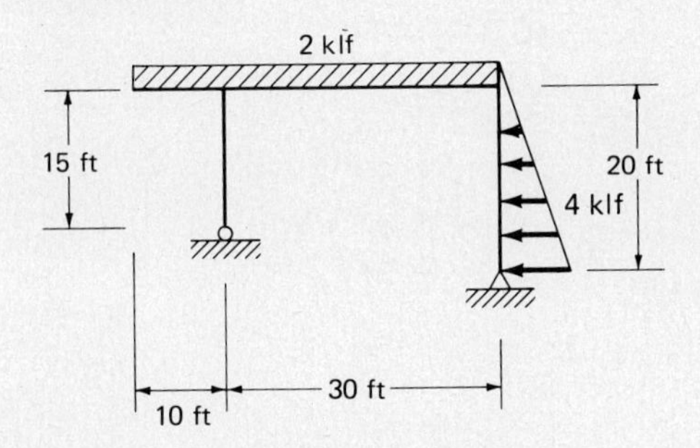

Problem 2.24 (*Ans.* $V_L = 22.5^k \uparrow$, $H_L = 30^k \rightarrow$, $V_R = 42.5^k \uparrow$)

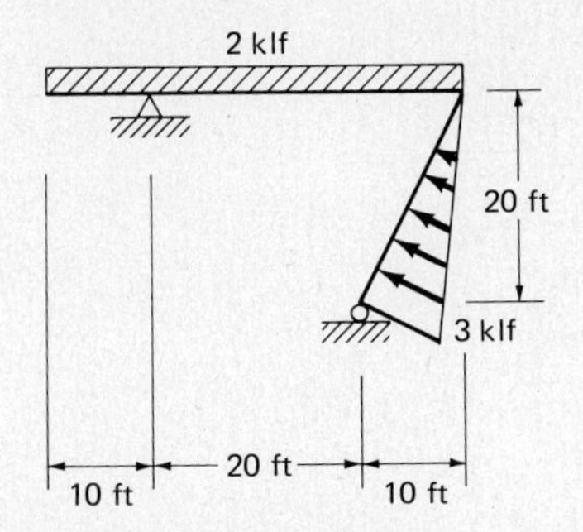

Problem 2.25

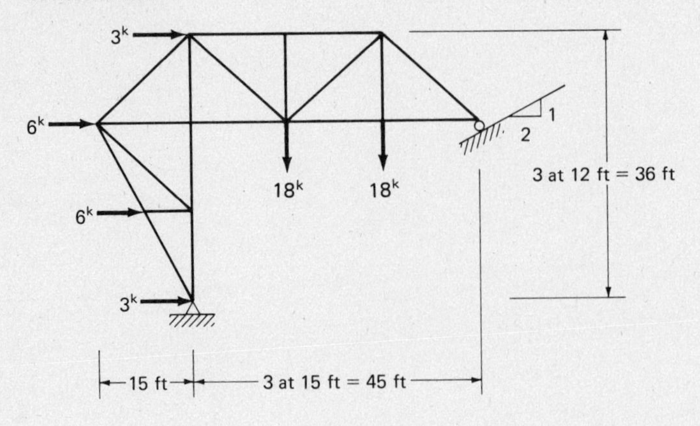

Problem 2.26 (*Ans.* $V_L = 120^k \uparrow$, $H_L = 26.67^k \leftarrow$, $H_R = 26.67^k \rightarrow$)

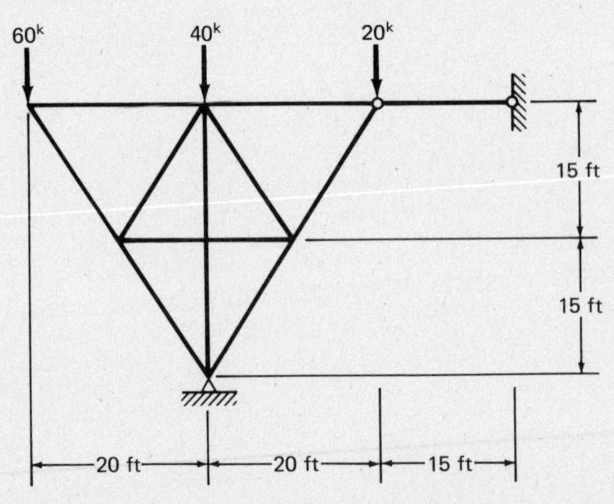

Problem 2.27

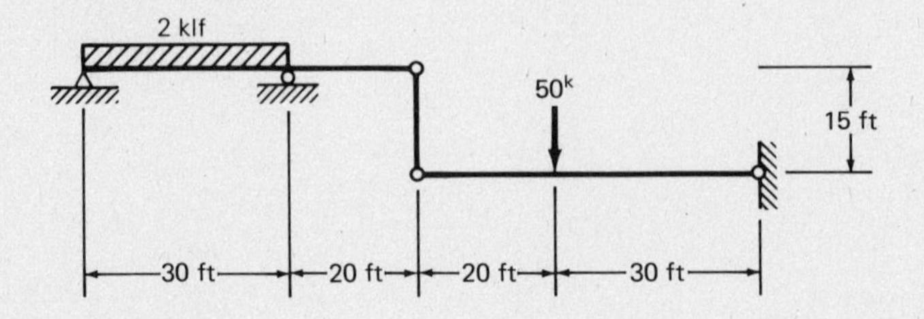

Problem 2.28 (*Ans.* $V_L = 141.6^k \uparrow$, $V_R = 159.9^k \uparrow$)

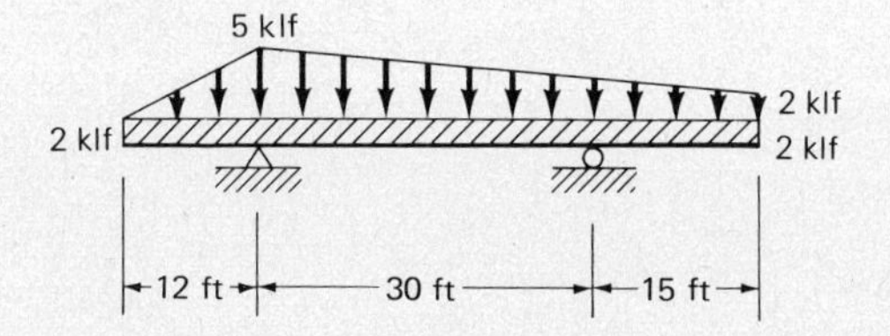

Problem 2.29

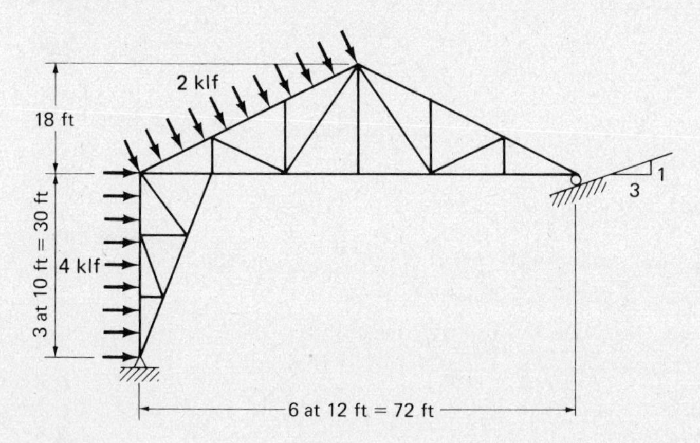

Problem 2.30 (*Ans.* $V_L = 69.5^k \uparrow$, $H_L = 34.7^k \rightarrow$, $V_R = 13.2^k \uparrow$, $H_R = 47.9^k \rightarrow$)

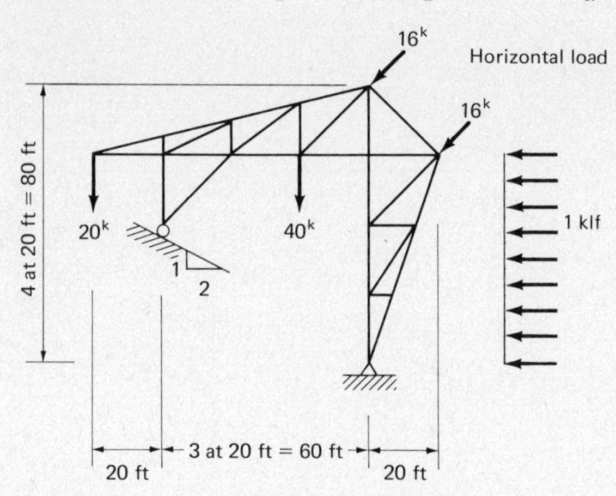

3 Reactions for Cantilever and Arch-Type Construction

3.1 THE SIMPLE CANTILEVER

The simple cantilever pictured in Fig. 3.1 has three unknown reaction components supporting it at the fixed end; they are the forces required to resist horizontal movement, vertical movement, and rotation. They may be determined with the equations of statics as illustrated in Example 3.1.

Example 3.1

Find all reaction components for the cantilever beam shown in Fig. 3.1.

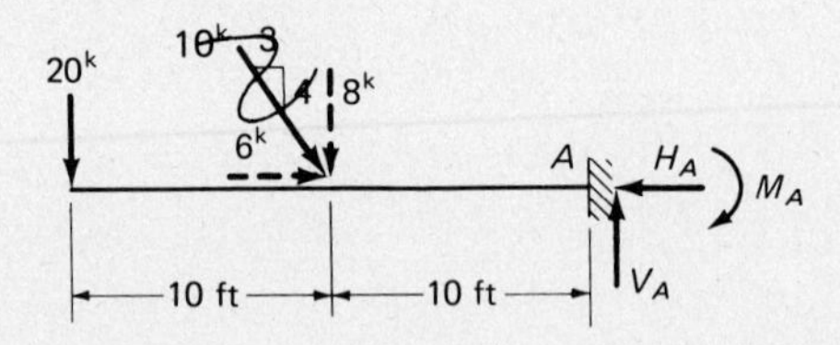

Figure 3.1

SOLUTION

$$\Sigma V = 0$$

$$20 + 8 - V_A = 0$$

$$V_A = 28^k \uparrow$$

United Airlines hangar, San Francisco, California. (Courtesy of the Lincoln Electric Company.)

$$\Sigma H = 0$$

$$6 - H_A = 0$$

$$H_A = 6^k \leftarrow$$

$$\Sigma M_A = 0$$

$$(20)(20) + (8)(10) - M_A = 0$$

$$M_A = 480 \text{ ft k} \circlearrowright$$

3.2 CANTILEVER STRUCTURES

Moments in structures that are simply supported increase rapidly as the spans become longer. It will be seen that bending increases approximately in proportion to the square of the span length. Stronger and more expensive structures are required to resist the greater moments. For very long spans, moments are so large that it becomes economical to introduce special types of structures that will reduce the moments. One of these types is cantilever construction as illustrated in Fig. 3.2

A cantilever type of structure is substituted for the three simple beams of Fig. 3.2(a) by making the beam continuous over the interior supports B and C and introducing hinges in the center span as indicated in (b). An equation of

North Dakota State Teachers College fieldhouse, Valley City, North Dakota. (Courtesy of the American Institute of Timber Construction.)

condition ($\Sigma M_{\text{hinge}} = 0$) is available at each of the hinges so introduced, giving a total of five equations and five unknowns. The structure is statically determinate.

The moment advantage of cantilever construction is illustrated in Fig. 3.3 The diagrams shown give the variation of moment in each of the structures of Fig. 3.2 due to a uniform load of 3 klf (kip or kilopounds per linear foot) for the entire spans. The maximum moment for the cantilever type is seen to be considerably less than that for the simple spans, and this permits lighter and cheaper construction. The plotting of moment diagrams is fully explained in the next chapter.

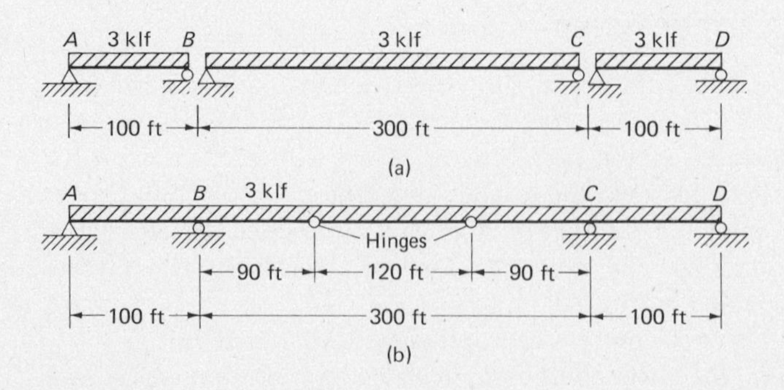

Figure 3.2

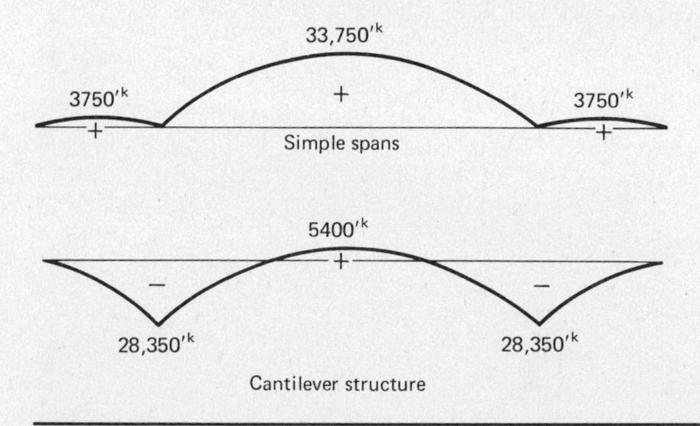

Figure 3.3

3.3 REACTION CALCULATIONS FOR CANTILEVER STRUCTURES

Cantilever construction consists essentially of two simple beams, each with an overhanging or cantilevered end as follows:

with another simple beam in between supported by the cantilevered ends:

The first step in determining the reactions for a structure of this type is to isolate the center simple beam and compute the forces necessary to support it at each end. Second, these forces are applied as downward loads on the respective cantilevers, and as a final step the end beam reactions are determined individually. Example 3.2 illustrates the entire process.

Example 3.2

Calculate all reactions for the cantilever structure of Fig. 3.4.

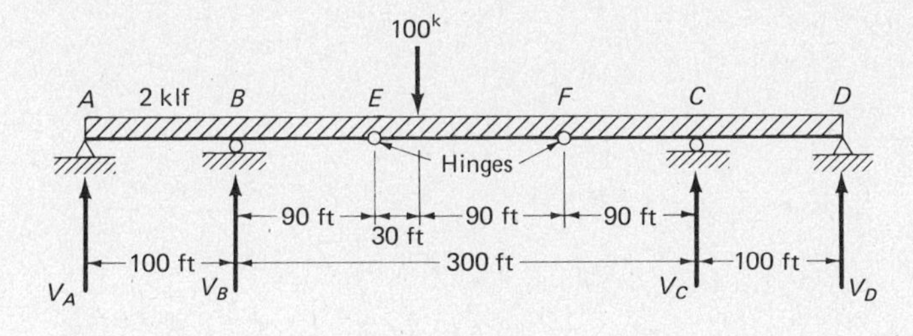

Figure 3.4

Structural arch construction, U.S.A.F. hangar, Edwards Air Force Base, California. (Courtesy of the Bethlehem Steel Corporation.)

SOLUTION

By isolating the center section from the two end sections,

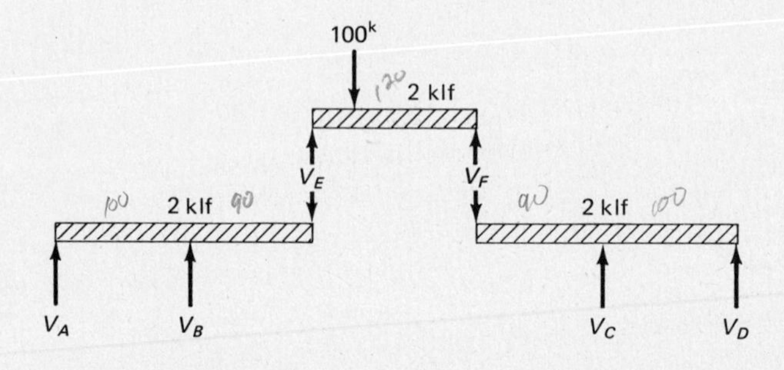

By computing reactions for center beam,

$$\Sigma M_E = 0$$

$$(100)(30) + (2 \times 120)(60) - 120V_F = 0$$

$$V_F = 145^k$$

$$\Sigma V = 0$$

$$100 + 240 - 145 - V_E = 0$$

$$V_E = 195^k$$

By computing reactions for left-end beam,

$$\Sigma M_A = 0$$

$$(2 \times 190)(95) + (195)(190) - 100V_B = 0$$

$$V_B = 731.5^k \uparrow$$

$$\Sigma V = 0$$

$$380 + 195 - 731.5 - V_A = 0$$

$$V_A = 156.5^k \downarrow$$

Similarly, the reactions for the right-end beam are found to equal

$$V_C = 636.5^k \uparrow$$

$$V_D = -111.5^k \downarrow$$

An examination of the reactions obtained for the left-end and right-end sections of this beam show why cantilever bridges are often called "seesaw" bridges. These structures are primarily supported by the first interior supports on each end where the reactions are quite large. The end supports may very well have to provide downward reaction components. Thus an end section of a cantilever structure seems to act as a seesaw over the first interior support with downward loads on both sides.

3.4 THREE-HINGED ARCHES

The reactions for structures discussed previously, which had horizontal supports, were vertical and parallel under vertical loading. Arches are structures that produce horizontal converging reactions under vertical load. They tend to flatten out under load and must be fixed against horizontal movement at their supports. (There is an old saying to the effect that "an arch never sleeps," the thought being that the horizontal reactions are always present due to the structure's own weight even when no live loads are applied.)

Arches may be constructed with three hinges, two hinges, one hinge (very rare), or no hinges (quite common in concrete construction). The three-hinged type is discussed here because it is the only statically determinate one.

Examination of the three-hinged arch pictured in Fig. 3.5 reveals there are two reaction components at each support, or a total of four. Three

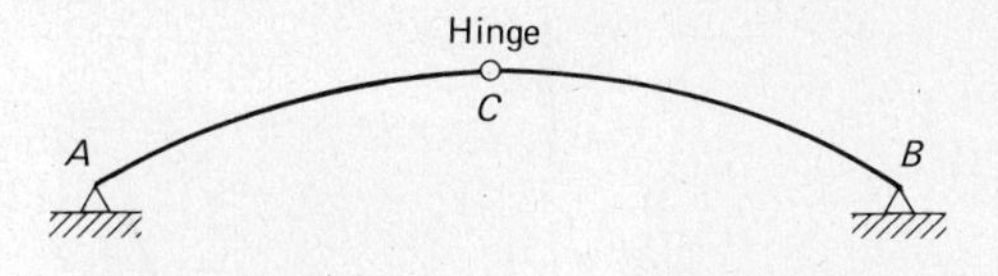

Figure 3.5

equations of statics and one condition equation ($\Sigma M_C = 0$, where the subscript c means crown hinge) are available to find the unknowns.

The arch of Example 3.3 is handled by taking moments at one of the supports to obtain the vertical reaction component at the second support. Because the supports are on the same level, the horizontal reaction component at the second support passes through the point where moments are being taken. When one vertical reaction component has been found, the other may be obtained with the $\Sigma V = 0$ equation. The horizontal reaction components are obtained by taking moments at the crown hinge of the forces either to the left or to the right. The only unknown appearing in either equation is the horizontal reaction component on that side, and the equation is solved for its value. The other horizontal component is found by writing the $\Sigma H = 0$ equation for the entire structure. Once the reactions are determined it is easily possible to compute the moment and axial force in the arch at any point by statics.

Laminated timber beams, Maumee, Ohio. (Courtesy of Koppers Company, Inc.)

The computation of reactions for the arch of Example 3.4 is slightly more complicated because the supports are not on the same level. Taking moments at one support results in an equation involving both the horizontal and vertical components of reaction at the other support. Moments may be taken about the crown hinge of the forces on the same side as those two unknowns. The resulting equation contains the same two unknowns. Solving the equations simultaneously gives the values of the components, and the application of the $\Sigma H = 0$ and $\Sigma V = 0$ equations results in the values of the remaining components.

Example 3.3

Find all reaction components for the three-hinged arch shown in Fig. 3.6.

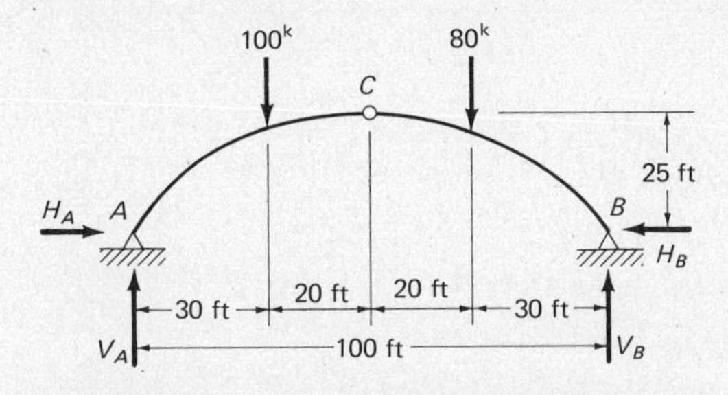

Figure 3.6

SOLUTION

$$\Sigma M_A = 0$$

$$(100)(30) + (80)(70) - 100V_B = 0$$

$$V_B = 86^k \uparrow$$

$$\Sigma V = 0$$

$$100 + 80 - 86 - V_A = 0$$

$$V_A = 94^k \uparrow$$

By using the free-body sketch shown,

$$\Sigma M_C \text{ to left} = 0$$

$$(94)(50) - (100)(20) - 25H_A = 0$$

$$H_A = 108^k \rightarrow$$

$$\Sigma H = 0$$

$$108 - H_B = 0$$

$$H_B = 108^k \leftarrow$$

Example 3.4

For the structure diagrammed in Fig. 3.7 determine reaction components at both supports.

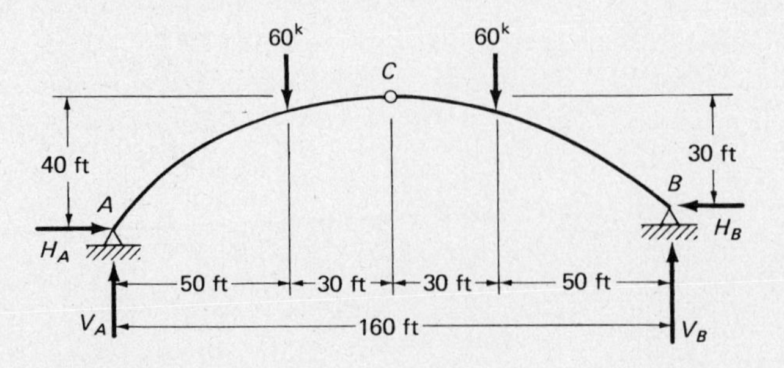

Figure 3.7

SOLUTION:

$$\Sigma M_A = 0$$

$$(60)(50) + (60)(110) - 10H_B - 160V_B = 0 \qquad (1)$$

$$10H_B + 160V_B = 9600$$

$$\Sigma M_C \text{ to right} = 0$$

$$(60)(30) + 30H_B - 80V_B = 0 \qquad (2)$$

$$30\,H_B - 80V_B = -1800$$

Solving Eqs. (1) and (2) simultaneously gives

$$H_B = 85.7^k \leftarrow$$

$$V_B = 54.6^k \uparrow$$

$$\Sigma H = 0$$

$$85.7 - H_A = 0$$

$$H_A = 85.7^k \rightarrow$$

$$\Sigma V = 0$$

$$60 + 60 - 54.6 - V_A = 0$$

$$V_A = 65.4^k \uparrow$$

The reactions for the arch of Example 3.4 may be computed without using simultaneous equations. The horizontal and vertical axes (on the basis of which the $\Sigma H = 0$ and $\Sigma V = 0$ equations are written) are so rotated that the horizontal axis passes through the two supporting hinges (Fig. 3.8). Components of forces can be computed parallel to the X' and Y' axes, and the $\Sigma X' = 0$ and $\Sigma Y' = 0$ equations applied, but the computations are rather inconvenient.

Glued laminated three-hinged arches, The Jai Alai Fronton, Riveria Beach, Florida. (Courtesy of the Southern Pine Association.)

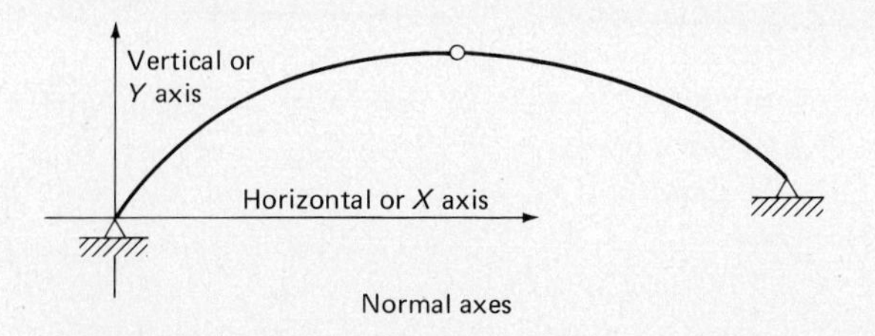

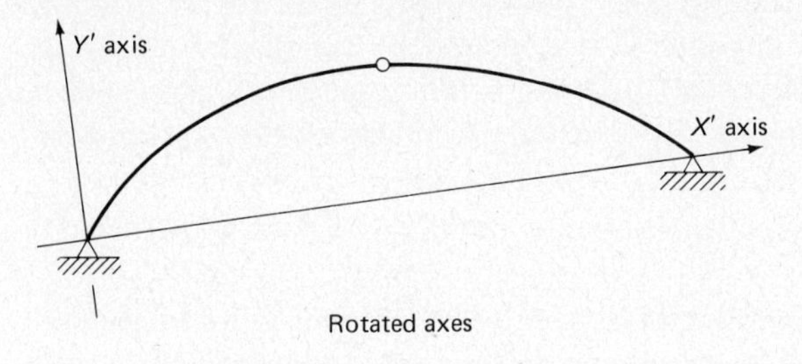

Figure 3.8

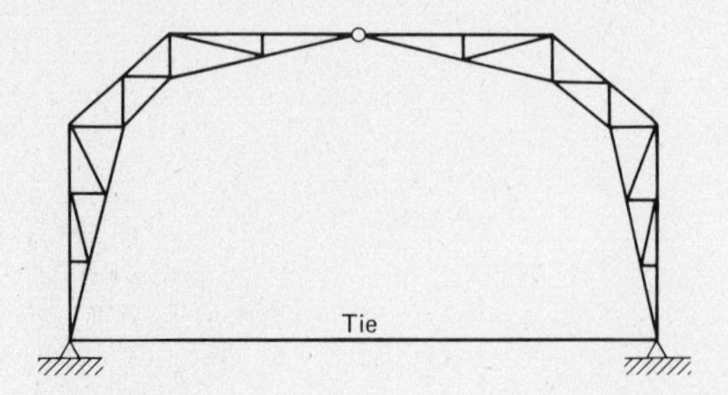

Figure 3.9

3.5 USES OF ARCHES AND CANTILEVER STRUCTURES

Three-hinged steel arches are used for short and medium-length bridge spans of up to approximately 600 ft. They are used for buildings where large clear spans are required underneath, as for hangars, field houses, and armories. Steel two-hinged arches are generally economical for bridges from 600 to 900 ft in length, with a few exceptional spans being over 1600 ft long. Concrete hingeless arches are used for bridges of from 100- to 400-ft spans. Cantilever-type bridges are used for spans of from approximately 500 ft up to very long spans such as the 1800-ft center span of the Quebec Bridge.

An arch is a structure that requires foundations capable of resisting the large thrusts at the supports. In arches for buildings it is possible to carry the thrusts by tying the supports together with steel rods as illustrated in Fig. 3.9, with steel sections, or even with specially designed floors. These are referred to as *tied arches*. For many locations the three-hinged arch is selected over the statically indeterminate arches because of poor foundation conditions with the possibility of settlement. It will become evident in later chapters that foundation settlement may cause severe stress changes in statically indeterminate structures. Ease of erection is another advantage of three-hinged arches. It is often convenient to assemble and ship the two halves of a precast concrete, structural steel or laminated timber arch separately and assemble them on the job.

The fact that cantilever-type construction reduces bending moments for long spans has been previously demonstrated. Arch-type construction also reduces moments, because the reactions at the supports tend to cause bending in the arch in a direction opposite to that caused by the downward loads. Because of this characteristic of having small bending moments, arches were admirably suited for masonry construction as practiced by the ancient builders.

3.6 CABLES

Steel cables are used for supporting loads over long spans such as for suspension bridges, large open roofs, cable car systems, and so on. In addition they are used as guys for derricks, radio towers, and similar structures. Steel cables are economically manufactured from high-strength steel wires providing perhaps the lowest cost/strength ratio of any common structural members. They are easily handled and positioned even for very long spans.

For the discussion to follow the cable weight is neglected. When a cable of a given length is suspended between two supports the shape it takes is determined by the applied loads. As a matter of fact, when vertical loads are supported a cable will take the shape of the bending–moment diagram (see Chap. 4) of a simple beam of the same span subjected to the same loads. For instance, if a cable is subjected to a set of concentrated loads such as those shown in Fig. 3.10, the segments of the cable will fall along straight lines. (This shape is actually inverted from the shape of the moment diagrams drawn in later chapters of this book.)

A cable supporting a horizontal uniform load will assume a parabolic shape. Cables supporting the roadway of suspension bridges are usually assumed to fall into this class (though the loads are applied to the cables as closely spaced concentrated loads from the hangers). A cable supporting a load that is uniform along its length (such as a transmission line) will take the form of a catenary. Unless the sag of such a cable is quite large in proportion to its length it may be assumed to be parabolic in shape with the result that its analysis is appreciably simplified.

Cables are assumed to be so flexible that they cannot resist bending or compression. As a result they are in direct tension and a condition equation ($\Sigma M = 0$ at any point in the cable) is available for analysis. Should the position or sag of a cable at a particular point be known the reactions at the cable ends and the sag at any other point in the cable can be determined with these equations. A numerical example of this type is presented in Example 3.5. The weight of the cable is assumed to be negligible in this case.

Example 3.5

Determine the reactions for the cable in Fig. 3.10 and the sag at the 40-kip load.

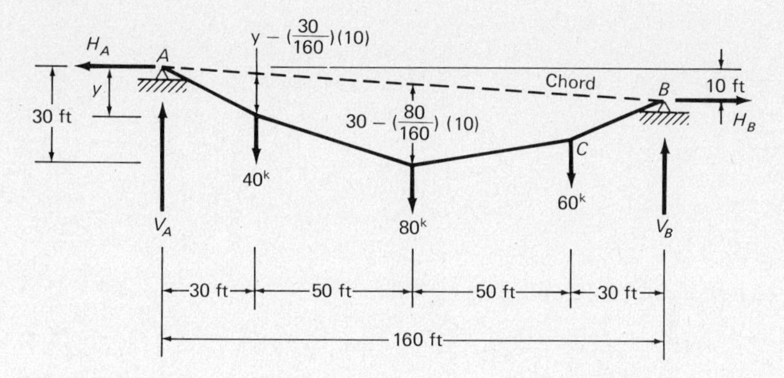

Figure 3.10

Cable-stayed Sitka Harbor Bridge, Sitka, Alaska. (Courtesy of the Alaska Department of Transportation.)

SOLUTION

$$\Sigma M_B = 0$$

$$160V_A - 10H_A - (60)(30) - (80)(80) - (40)(130) = 0 \quad (1)$$

$$160V_A - 10H_A = 13{,}400$$

$$\Sigma M_{80^k} \text{ load to left} = 0$$

$$80V_A - 30H_A - (40)(50) = 0 \quad (2)$$

$$80V_A - 30H_A = 2000$$

Solving Eqs. (1) and (2) simultaneously gives

$$H_A = 188^k \leftarrow$$

$$V_A = 95.5^k \uparrow$$

By $\Sigma V = 0$ and $\Sigma H = 0$,

$$H_B = 188^k \rightarrow$$

$$V_b = 84.5^k \uparrow$$

ΣM to left of 40^k load letting the sag there $= y$

$$30\,V_A - H_A y = 0$$

$$(30)(95.5) - (188)(y) = 0$$

$$y = 15.24 \text{ ft}$$

The resultant tension at any point can be obtained from the following equation in which H and V are the horizontal and vertical components of tensile force in the cable at that point.

$$T = \sqrt{(H)^2 + (V)^2}$$

From this equation it can be seen that the tension varies as one moves along the cable. If only vertical loads are present the value of H will be constant throughout and the maximum tensile force occurring can be determined by substituting the maximum value of the vertical force into the equation.

The distortion or deflection of most structures is assumed to be negligible in computing the forces produced in those structures. Such an assumption is not correct, however, for many cable structures, particularly the flat ones where a little sag can drastically affect cable tensions. This topic is not considered herein but is described very well in a book by Firmage [1]. Flat cables cause very large horizontal reaction components and thus have very high tensile forces.

PROBLEMS

For Probs. 3.1 to 3.24 determine the reactions for the structures.

Problem 3.1

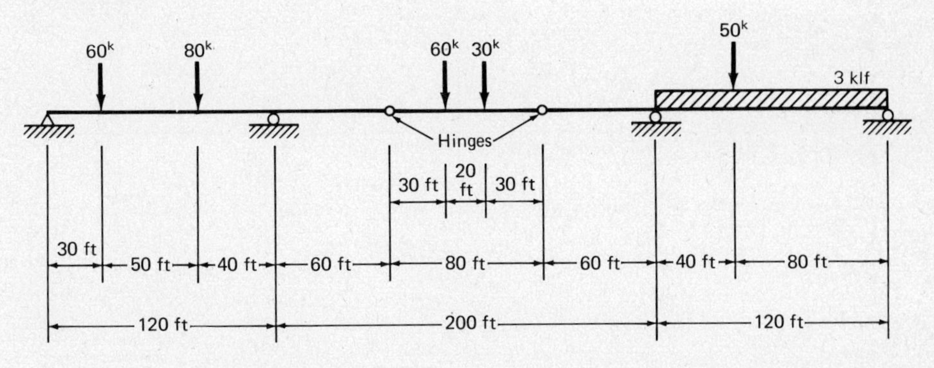

Problem 3.2 (*Ans.* $V_A = 45^k\downarrow$, $V_B = 250^k\uparrow$, $V_C = 220^k\uparrow$, $V_D = 55^k\downarrow$)

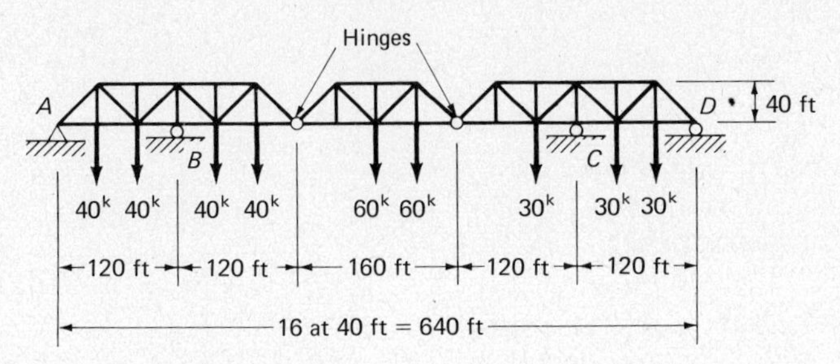

Problem 3.3

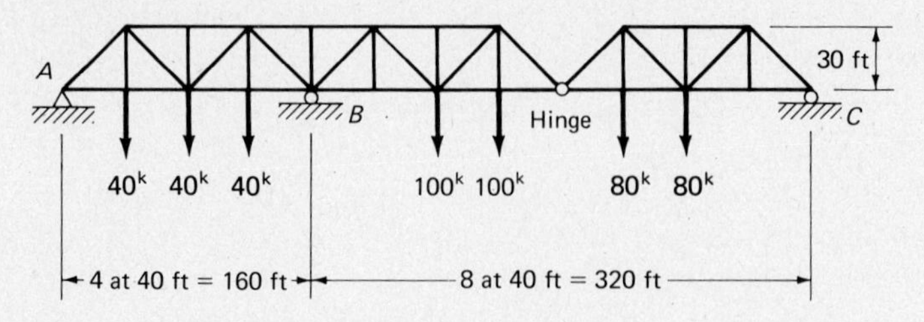

Problem 3.4 (*Ans.* $V_A = 210^k\uparrow$, $V_B = 790^k\uparrow$, $V_C = 20^k\uparrow$)

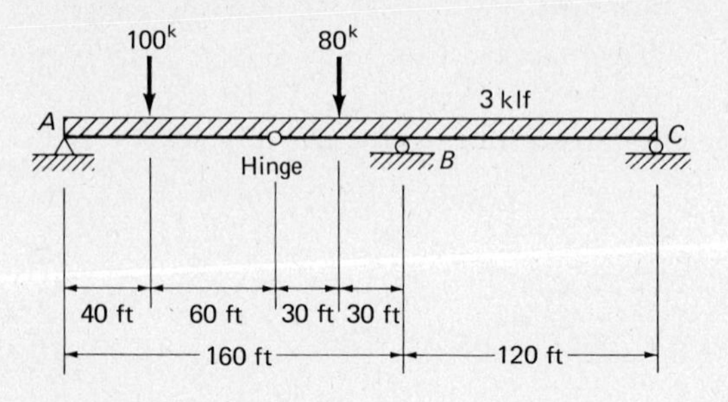

Problem 3.5

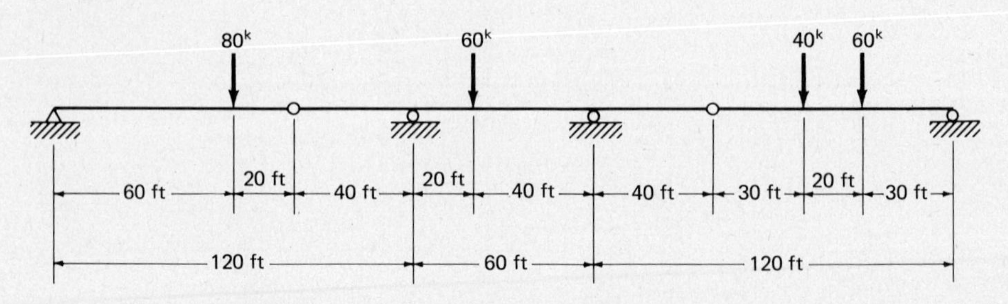

Problem 3.6 (*Ans.* $V_L = 55^k\uparrow$, $H_L = 50^k\rightarrow$, $V_R = 65^k\uparrow$, $H_R = 50^k\leftarrow$)

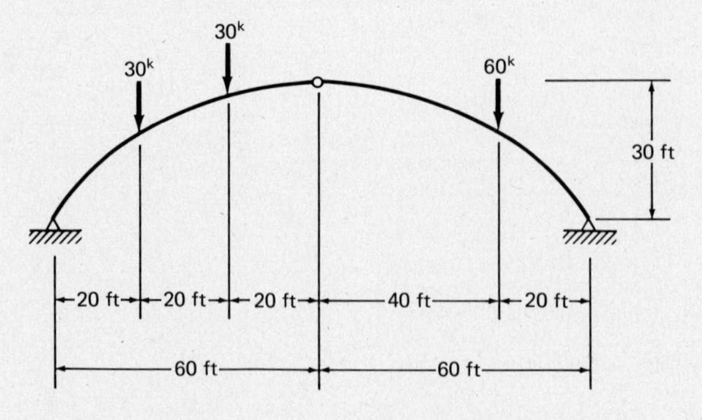

Problem 3.7

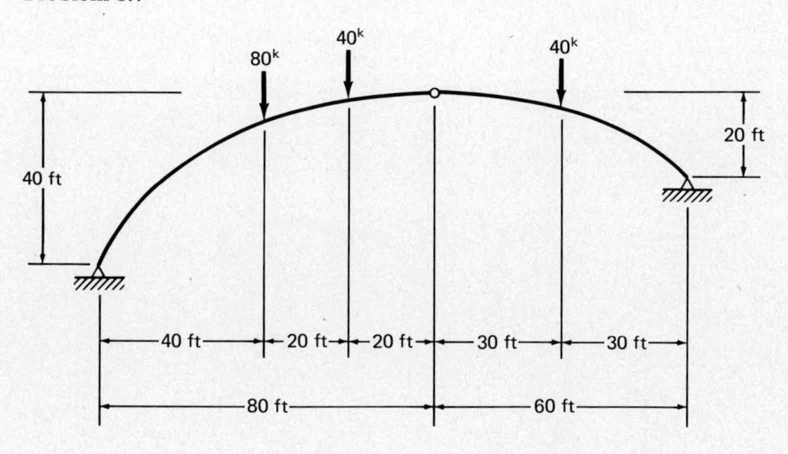

Problem 3.8 (*Ans.* $V_L = 57.9^k\uparrow$, $H_L = 75.7^k\rightarrow$, $V_R = 69.1^k\uparrow$, $H_R = 16.8^k\leftarrow$)

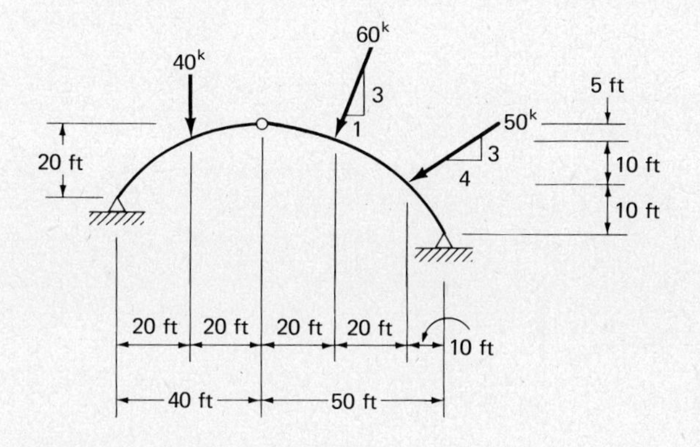

Problem 3.9

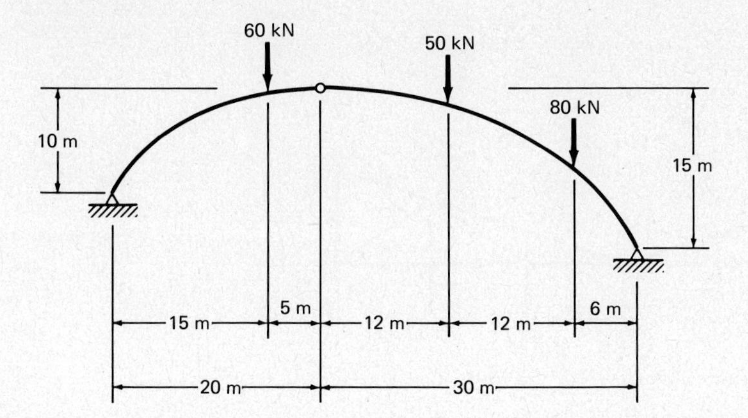

Problem 3.10 (*Ans.* $V_L = 68.8^k\uparrow$, $H_L = 155.4^k\rightarrow$, $V_R = 71.2^k\uparrow$, $H_R = 185.4^k\leftarrow$)

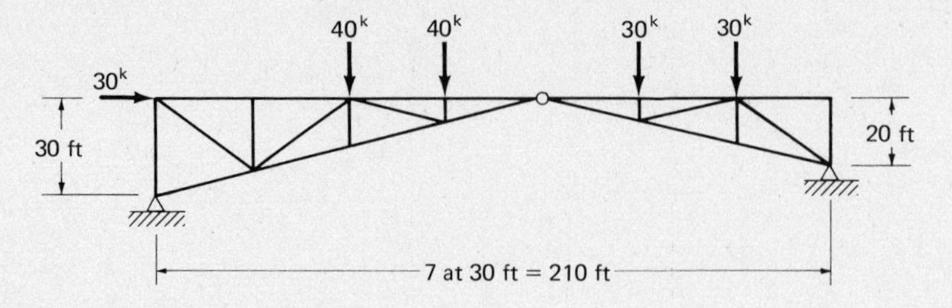

Problem 3.11

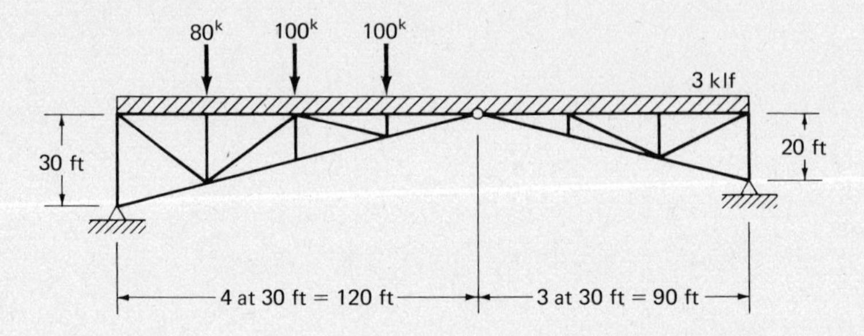

Problem 3.12 (*Ans.* $V_L = 36.7^k\uparrow$, $H_L = 100^k\rightarrow$, $V_R = 63.3^k\uparrow$, $H_R = 100^k\leftarrow$)

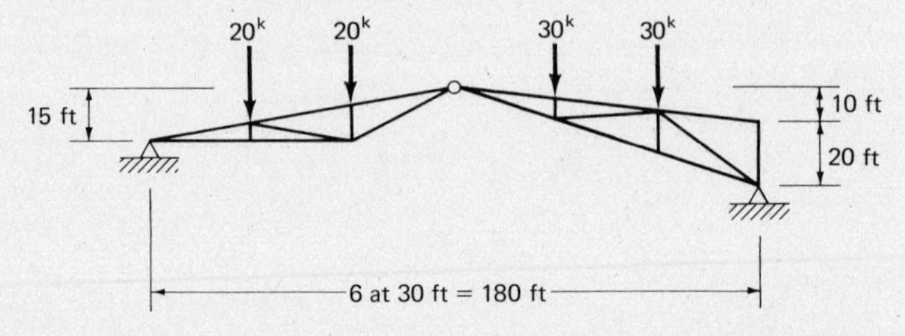

Problem 3.13

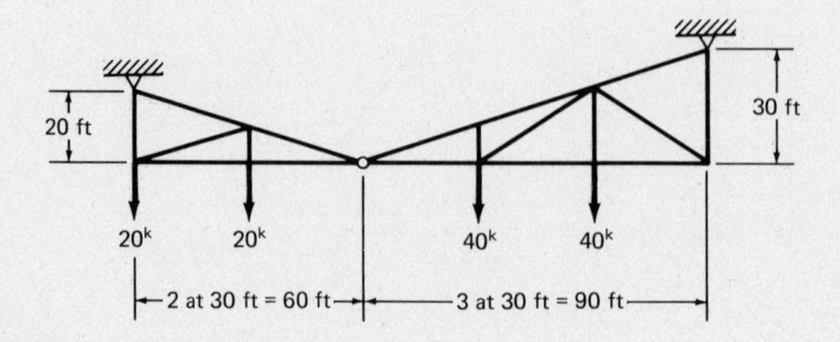

Problem 3.14 (*Ans.* $V_L = 293.2$ kN$\uparrow$, $H_L = 429.4$ kN$\rightarrow$, $V_R = 126.8$ kN$\uparrow$, $H_R = 429.4$ kN$\leftarrow$)

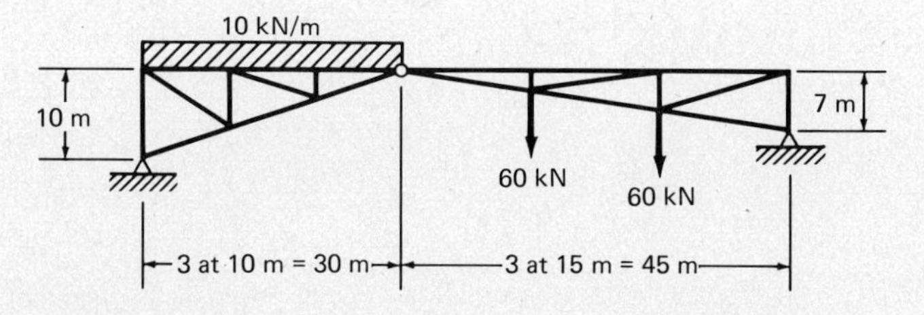

Problem 3.15

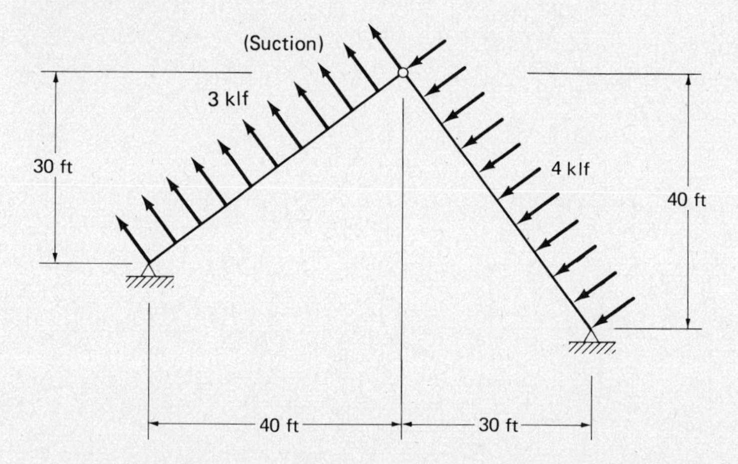

3.16. Repeat Prob. 3.15 if the 3-klf load is reduced to 2 klf and the 4-klf load is increased to 6 klf. (*Ans.* $V_L = 50^k\uparrow$, $H_L = 150^k\rightarrow$, $V_R = 50^k\uparrow$, $H_R = 150^k\rightarrow$)

3.17. Repeat Prob. 3.16 if the 2-klf load is removed.

Problem 3.18 (*Ans.* $V_L = 47.9^k\uparrow$, $H_L = 48.0^k\rightarrow$, $V_R = 32.1^k\uparrow$, $H_R = 72.0^k\rightarrow$)

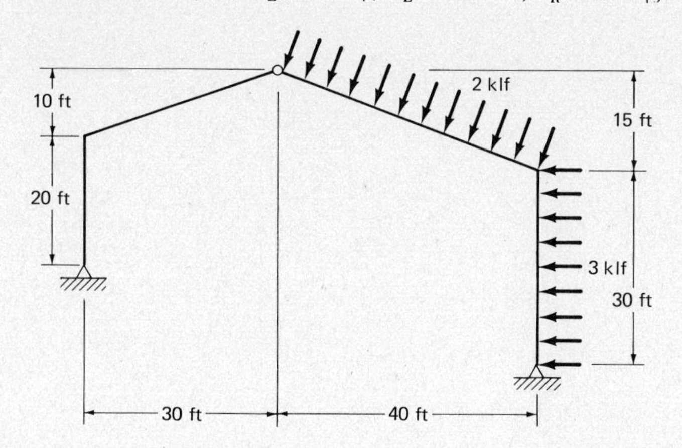

Problem 3.19

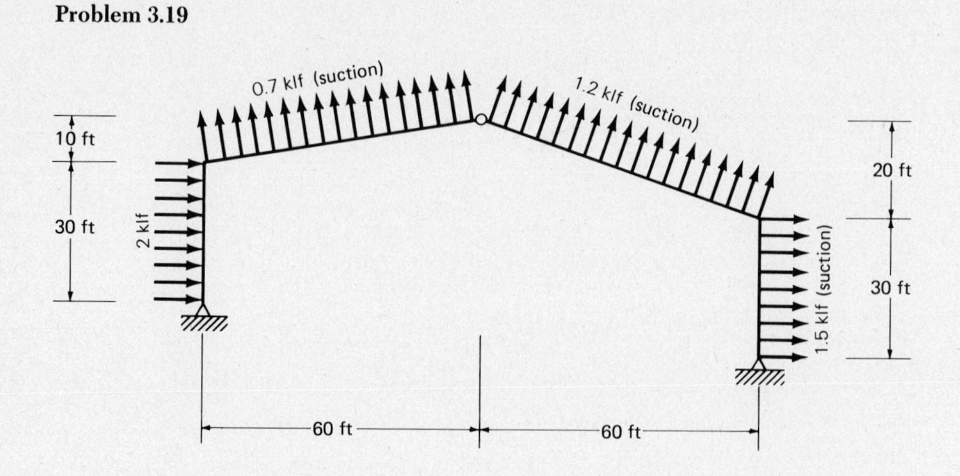

Problem 3.20 (*Ans.* $V_L = 87.5^k\uparrow$, $H_L = 112.5^k\rightarrow$, $V_R = 82.5^k\uparrow$, $H_R = 112.5^k\leftarrow$)

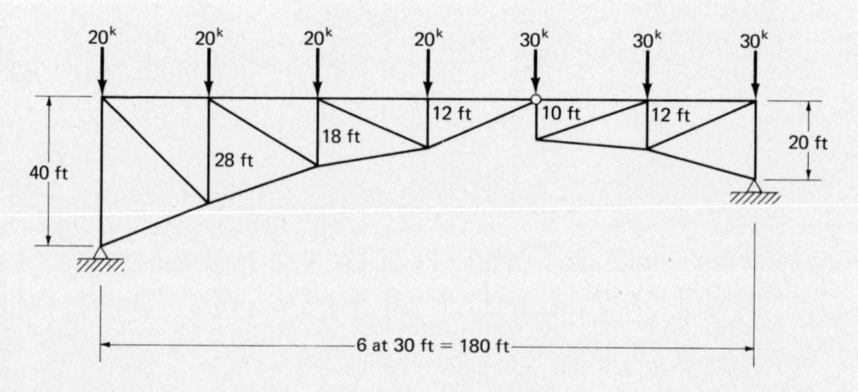

Problem 3.21

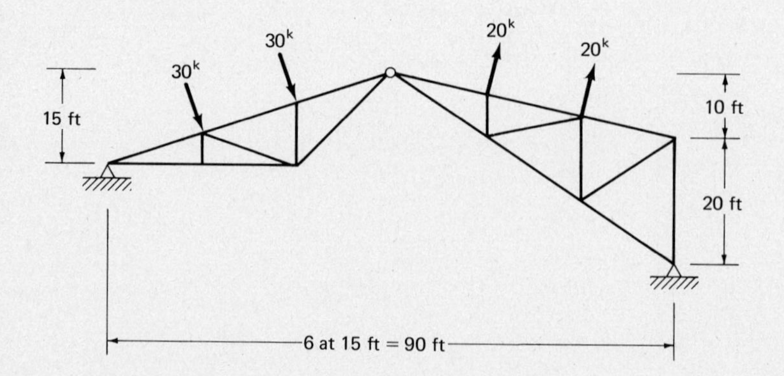

Problem 3.22 (*Ans.* $V_A = 76.7^k \downarrow$, $V_B = 346.7^k \uparrow$, $V_C = 30^k \uparrow$)

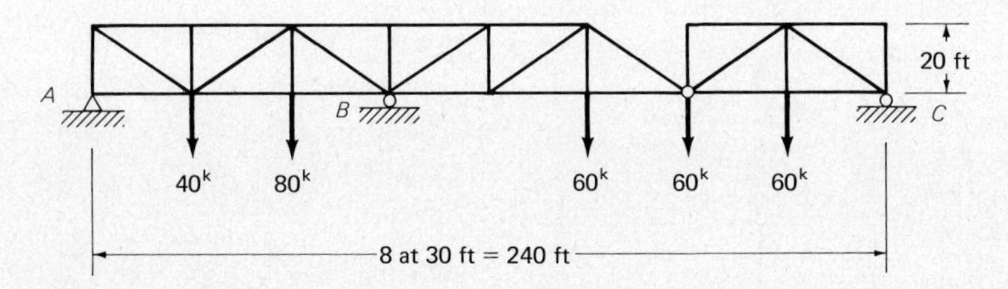

Problem 3.23

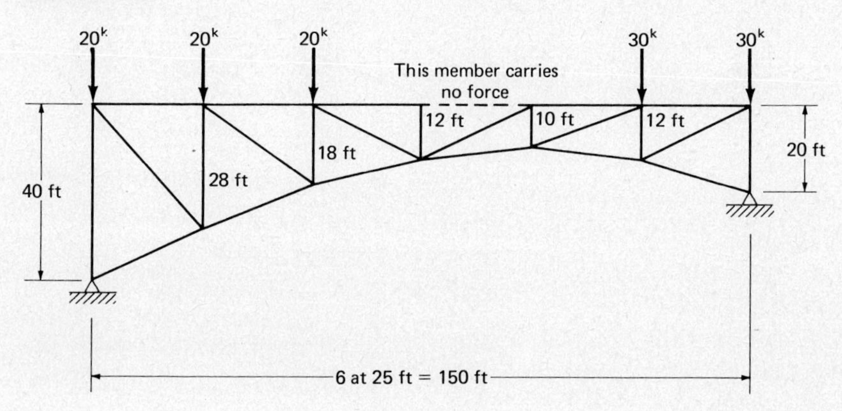

Problem 3.24 (*Ans.* $V_L = 34.3^k \uparrow$, $H_L = 11.5^k \rightarrow$, $V_R = 36^k \uparrow$, $H_R = 31.7^k \leftarrow$)

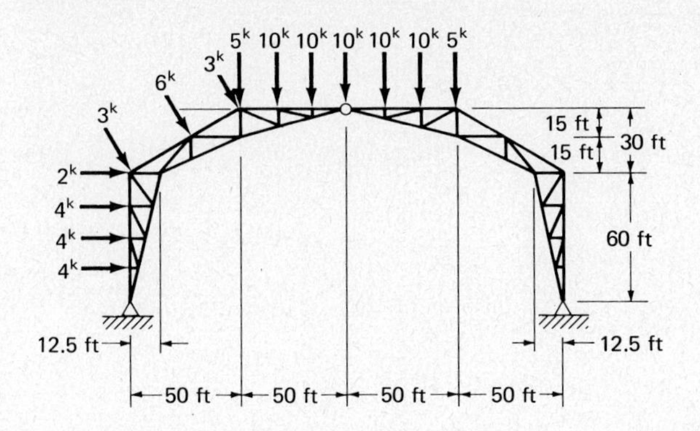

3.25. Determine the reaction components, cable sag at the 15- and 20-kip loads, and the maximum tensile force in the cable.

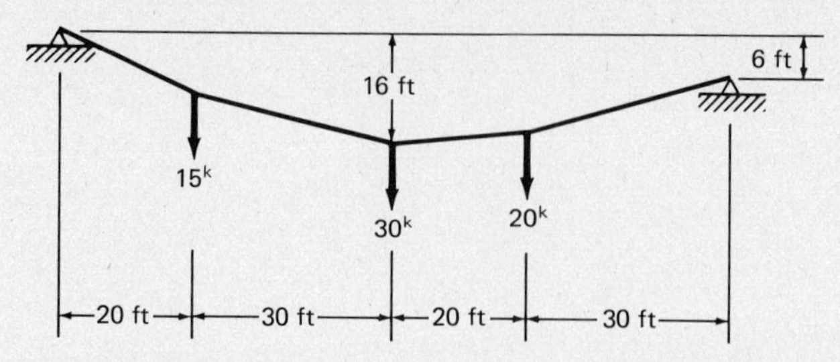

3.26. Determine the reaction components, cable sag at the 30-kN load, and the maximum tensile force in the cable. (*Ans.* $V_A = 35.7$ kN↑, $V_B = 44.3$ kN↑, $H_A = 110.8$ kN←, $y = 5.22$ m)

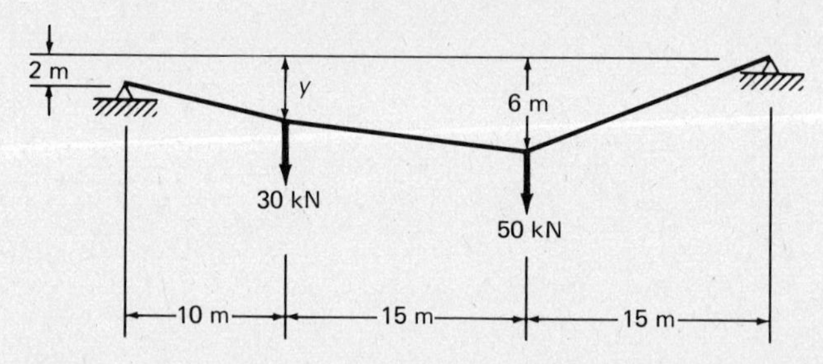

3.27. Determine the reaction components for the sag shown. Repeat the calculation if the center line sag is reduced to 4 ft.

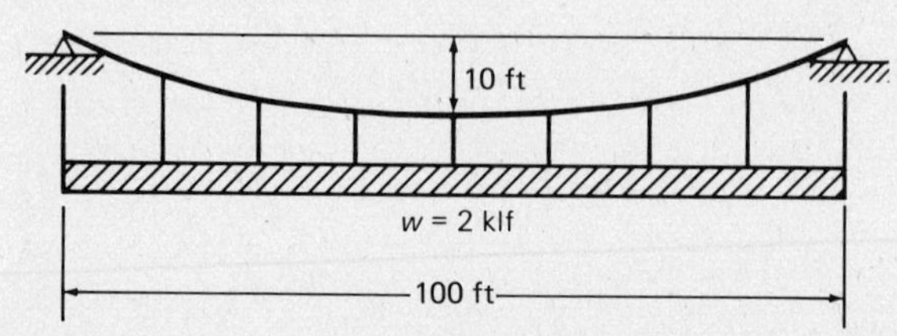

REFERENCE

[1] D.A. Firmage, *Fundamental Theory of Structures* (New York: Wiley, 1963), pp. 256–265.

4
Shear and Moment Diagrams

4.1 INTRODUCTION

An important phase of structural engineering is the understanding of shear and moment diagrams and their construction. It is doubtful that there is any other point at which careful study will give more reward in structural knowledge. These diagrams, from which values of shear and moment at any point in a beam are immediately available, are very convenient in design.

To examine the internal conditions of a structure, a free body must be taken out and studied to see what forces have to be present if the body is to be held in place, or in equilibrium. Shear and bending moments are two actions of the external loads on a structure that need to be understood to study properly the internal forces.

Shear is defined as the algebraic summation of the external forces to the left or to the right of a section that are perpendicular to the axis of the beam. In this book it is considered to be positive if the sum of the forces to the left is up or the sum of the forces to the right is down. The calculations for shear at two sections in a simple beam are given in Example 4.1. In each case the summations are made both to the left and to the right to prove that identical results are obtained. It is often convenient to find a shear by considering the side of the section that has the least number of forces.

Example 4.1

Find the shear at sections *a–a* and *b–b*, Fig. 4.1

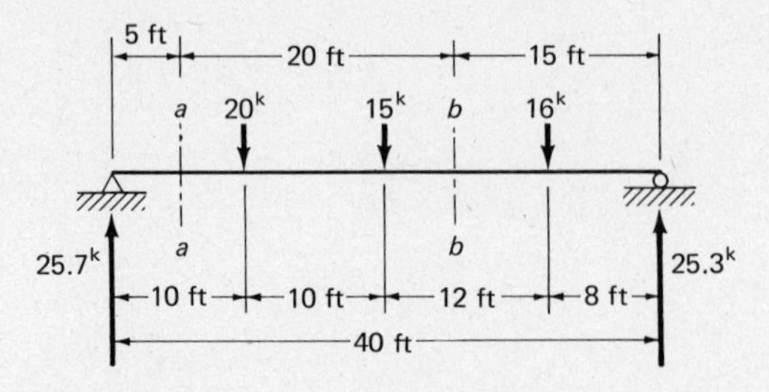

Figure 4.1

SOLUTION

Shear at section *a–a*:

$$V_{a\text{-}a} \text{ to left} = 25.7^{k} \uparrow, \text{ or } +25.7^{k}$$

$$V_{a\text{-}a} \text{ to right} = 20 + 15 + 16 - 25.3 = 25.7^{k} \downarrow, \text{ or } +25.7^{k}$$

Shear at section *b–b*

$$V_{b\text{-}b} \text{ to left} = 25.7 - 20 - 15 = 9.3^{k} \downarrow = -9.3^{k}$$

$$V_{b\text{-}b} \text{ to right} = 16 - 25.3 = 9.3^{k} \uparrow = -9.3^{k}$$

Bending moment is the algebraic sum of the moments of all of the external forces to the left or to the right of a particular section, the moments being taken about an axis through the centroid of the cross section. A positive sign herein indicates that the moment to the left is clockwise or the moment to the right is counterclockwise. A study of Fig. 4.7 shows that positive moment at a section causes tension in the bottom fibers and compression in the top fibers of the beam at the section. Should the member under consideration be a vertical member, a fairly standard sign convention is to consider the right-hand side of the member to be the bottom side.

The bending moments at sections *a–a* and *b–b* in the beam of Example 4.1 are computed as follows:

Moment at section *a–a*:

$$M_{a\text{-}a} \text{ to left} = (25.7)(5) = 128.5'^{k} \curvearrowright, \text{ or } +128.5'^{k}$$

$$M^{a\text{-}a} \text{ to right} = (25.3)(35) - (16)(27) - (15)(15) - (20)(5)$$
$$= 128.5'^{k} \curvearrowleft, \text{ or } +128.5'^{k}$$

Moment at section *b–b*:

$$M_{b\text{-}b} \text{ to left} = (25.7)(25) - (20)(15) - (15)(5) = 267.5'^{k} \curvearrowright, \text{ or } +267.5'^{k}$$

$$\mathrm{M}_{\mathrm{b\text{-}b}} \text{ to right} = (25.3)(15) - (16)(7) = 267.5'^{k} \curvearrowleft, \text{ or } +267.5'^{k}$$

The Tridge, a triple-span pedestrian bridge using glued laminated timber, Midland, Michigan. (Courtesy of Koppers Company, Inc.)

4.2 SHEAR DIAGRAMS

Shear diagrams are quite simple to draw in most cases. The standard method is to start with the left end of the structure and work to the right. As each concentrated load or reaction is encountered, a vertical line is drawn to represent the quantity and direction of the force involved. Between the forces a horizontal line is drawn to indicate no change in shear.

Where uniform loads are encountered, the shear is changing at a constant rate per foot and can be represented by a straight but inclined line on the diagram. When an ordinate on the shear diagram is above the line, a positive shear is indicated, because the sum of the forces to the left of that point is up. A shear diagram for a simple beam is drawn in Example 4.2.

Example 4.2

Draw a shear diagram for the beam shown in Fig. 4.2.

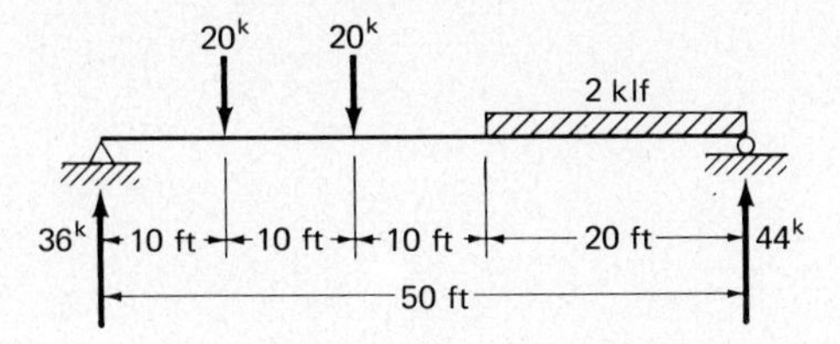

Figure 4.2

SOLUTION

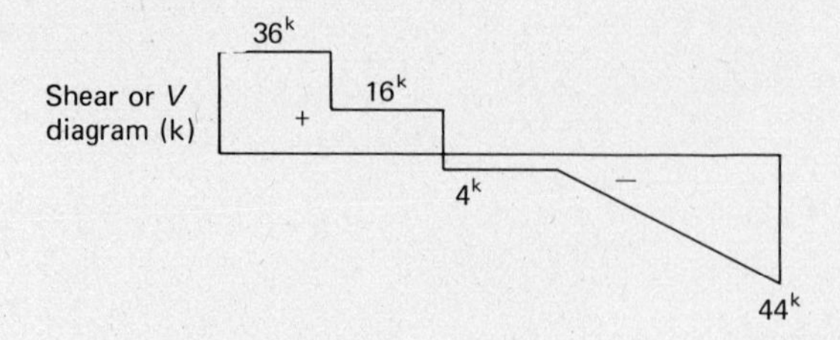

4.3 MOMENT DIAGRAMS

The moments at various points in a structure necessary for plotting a bending-moment diagram may be obtained algebraically by taking moments at those points, but the procedure is quite tedious if there are more than two or three loads applied to the structure. The method developed in the next section is much more practical.

4.4 RELATIONSHIPS OF LOADS, SHEARS, AND BENDING MOMENTS

There are significant mathematical relations between the loads, shears, and moments in a beam. These relations are discussed in the following paragraphs with reference to Fig. 4.3.

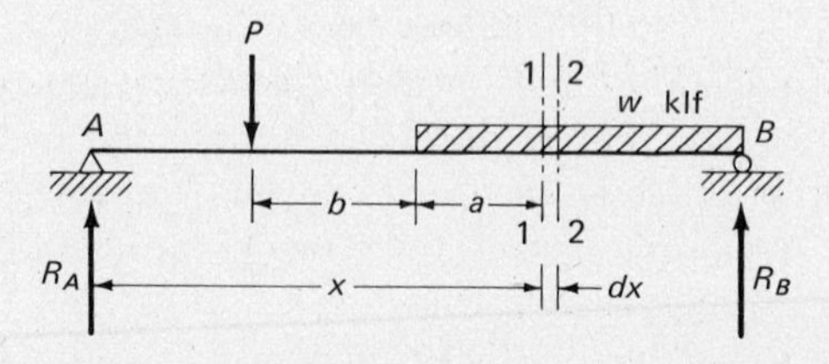

Figure 4.3

The shear and moment at section 1–1 may be written as follows:

$$V_{1\text{-}1} = R_A - P - wa$$

$$M_{1\text{-}1} = R_A x - P(a + b) - \frac{wa^2}{2}$$

The shear and moment at section 2–2, a distance dx to the right of section 1–1, are

$$V_{2\text{-}2} = V_{1\text{-}1} + dV = R_A - P - wa - w\,dx$$

$$M_{2\text{-}2} = M_{1\text{-}1} + dM = R_A x - P(a + b) - \frac{wa^2}{2} + V_{1\text{-}1}\,dx - \frac{w\,dx^2}{2}$$

From these equations the changes in shear and moment in a dx distance are seen to equal.

$$\frac{dV}{dx} = -w$$

$$\frac{dM}{dx} = V \qquad \text{omitting the infinitesimal } \frac{w\,dx^2}{2}$$

These two relationships are very useful to the structural designer. The first indicates that the rate of change of shear at any point equals the load per unit of distance at the point; meaning that the slope of the shear curve at any point is equal to the load at that point. The second equation indicates that the rate of change of moment at any point equals the shear. This relationship means that the slope of the bending-moment curve at any point equals the shear.

The procedure for drawing shear and moment diagrams, to be described in Sec. 4.5, is based on the above equations and is applicable to all structures regardless of loads or spans. Before the process is described, it may be well to examine the equations more carefully. A particular value of dV/dx or dM/dx is good only for the portion of the structure at which the function is continuous. For instance, in the beam of Example 4.4 the rate of change of shear from A to B equals the uniform load, 4 klf. At the 30-kip load, which is assumed to act at a point, the rate of change of shear and the slope of the shear diagram are infinite, and a vertical line is drawn on the shear diagram to represent a concentrated load. The rate of change of moment from A to B has been constant, but at B the shear changes decidedly, as does the rate of change of moment. In other words, an expression for shear or moment from A to B is not the same as the expression from B to C beyond the concentrated load. The equations of the diagrams are not continuous beyond a point where the function is discontinuous.

4.5 MOMENT DIAGRAMS DRAWN FROM SHEAR DIAGRAMS

The change in moment between two points on a structure has been shown to equal the shear between those points times the distance between them ($dM = V\,dx$); therefore the change in moment equals the area of the shear diagram between the points.

The relationship between shear and moment greatly simplifies the drawing of moment diagrams. To determine the moment at a particular section, it is only necessary to compute the total area beneath the shear curve, either to the left or to the right of the section, taking into account the algebraic signs of the various segments of the shear curve. Shear and moment diagrams are self-checking. If they are initiated at one end of a structure, usually the left, and check out to the proper value on the other end, the work is probably correct.

The author usually finds it convenient to compute the area of each part of a shear diagram and record that value on the part in question. This procedure is followed for the examples of this chapter where the values enclosed in the shear diagrams are areas. These recorded values appreciably simplify the construction of the moment diagrams.

The rate of change of moment at a point has been shown to equal the shear at that point ($dM/dx = V$). Whenever the shear passes through zero, the rate of change of moment must be zero ($dM/dx = 0$), and the moment is at a maximum or a minimum. If the moment diagram is being drawn from left to right and the shear diagram changes from positive to negative, the moment will reach a positive maximum at that point. Beyond the point it begins to diminish as the negative shear is added. If the shear diagram changes from negative to positive, the moment reaches a negative maximum and then begins to taper off as the positive shear area is added.

This theory, which indicates that maximum moment occurs where the shear is zero, is not always applicable. In some cases (at the end of a beam or at a point of discontinuity) the maximum moment can occur where the shear is not zero. Should a cantilever beam be loaded with gravity loads, the maximum shear and the maximum moment will both occur at the fixed end.

When shear and moment diagrams are drawn for inclined members, the components of loads and reactions perpendicular to the centroidal axes of the members are used and the diagrams are drawn parallel to the members. Examples 4.3 to 4.5 illustrate the procedure for drawing shear and moment diagrams for ordinary beams. In studying these diagrams particular emphasis should be given to their shapes under uniform loads, between concentrated loads, and so on.

Example 4.3

Draw shear and moment diagrams for the beam shown in Fig. 4.4.

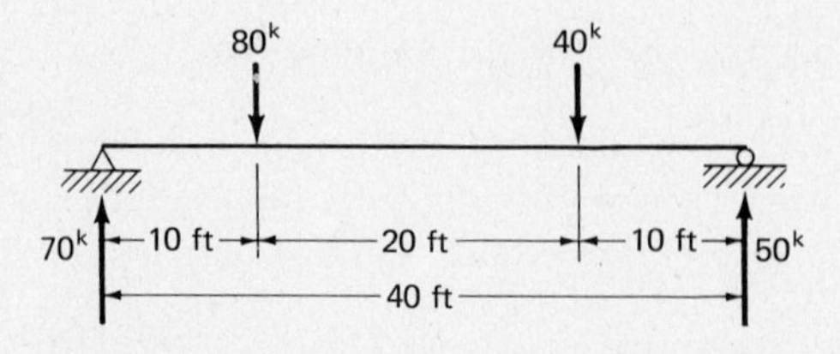

Figure 4.4

SOLUTION

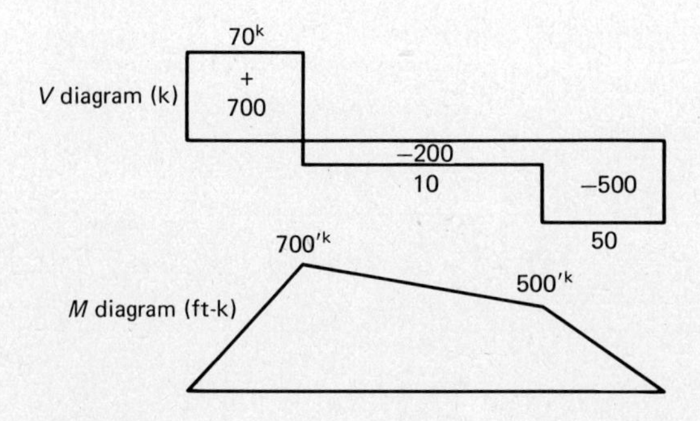

Example 4.4

Draw shear and moment diagrams for the structure shown in Fig. 4.5.

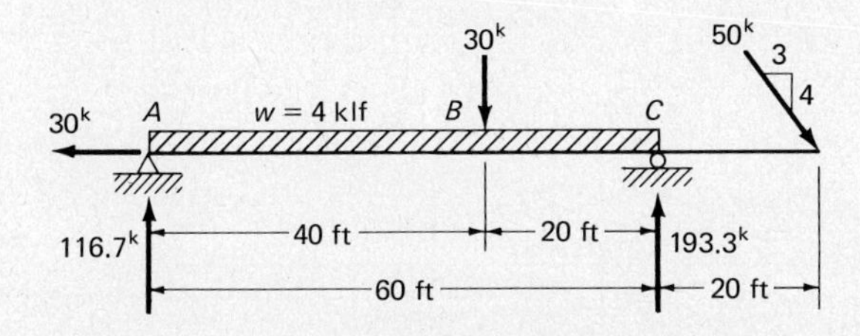

Figure 4.5

SOLUTION

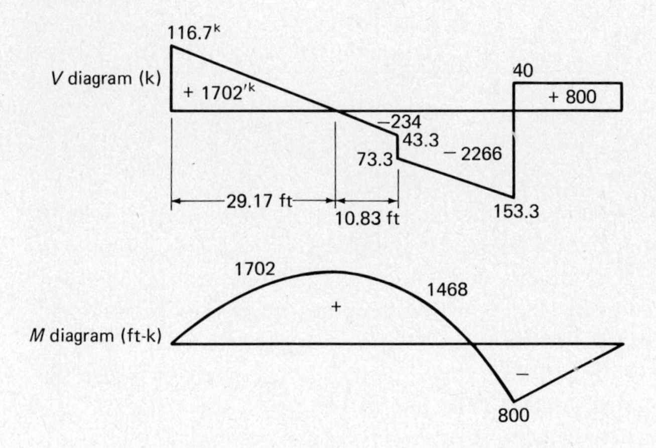

Example 4.5

Draw shear and moment diagrams for the cantilever-type structure shown in Fig. 4.6.

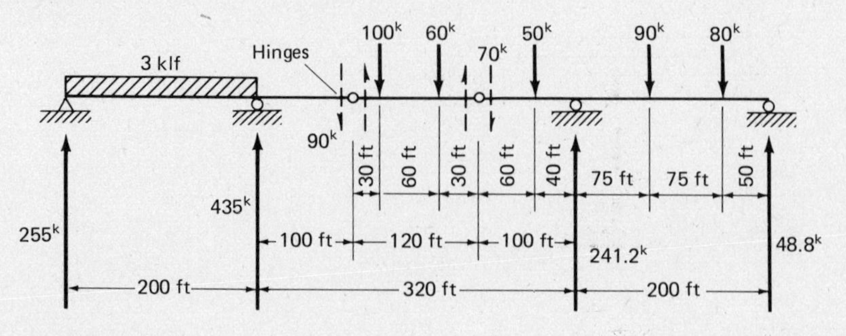

Figure 4.6

SOLUTION

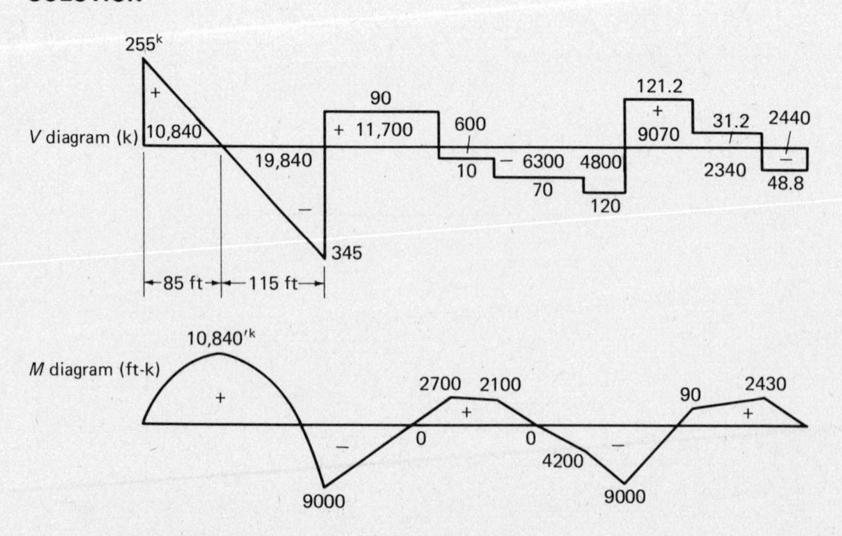

Some structures have rigid arms fastened to them. If horizontal or inclined loads are applied to these arms, a twist or moment will be suddenly induced in the structure at the point of attachment. The fact that moment is taken about an axis through the centroid of the section becomes important because the lever arms of the forces applied must be measured to that centroid. To draw the moment diagram at the point of attachment, it is necessary to figure the moment an infinitesimal distance to the left of the point and then add the moment applied by the arm. The moment exactly at the point of attachment is discontinuous and cannot be figured, but the moment immediately beyond that point is available.

The usual sign convention for positive and negative moments will apply in deciding whether to add or subtract the induced moment. It can be seen in Fig.

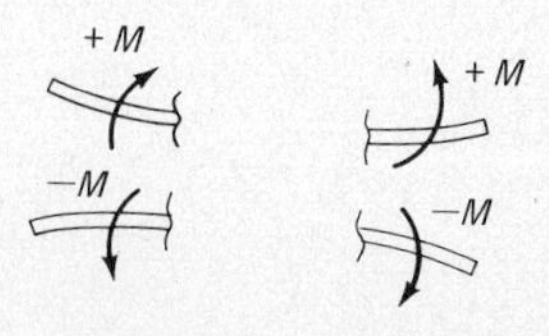

Figure 4.7

4.7 that forces to the left of a section that tend to cause clockwise moments produce tension in the bottom fibers (+ moment), whereas those forces to the left that tend to cause counterclockwise moments produce tension in the upper fibers (− moment). Similarly, a counterclockwise moment to the right of the section produces tension in the bottom fibers; a clockwise moment produces tension in the upper fibers.

Shear and moment diagrams are shown in Example 4.6 for a beam that has a moment induced at a point by a rigid arm to which a couple is applied. The moment diagram is drawn from left to right. Considering the moment of the forces to the left of a section through the beam immediately after the rigid arm is reached, it can be seen that the couple causes a clockwise or positive moment, and its value is added to the moment obtained by summation of the shear-diagram areas up to the attached arm.

Pedestrian bridge in Pullen Park, Raleigh, North Carolina. (Courtesy of the American Institute of Timber Construction.)

Example 4.6

Draw shear and moment diagrams for the beam shown in Fig. 4.8.

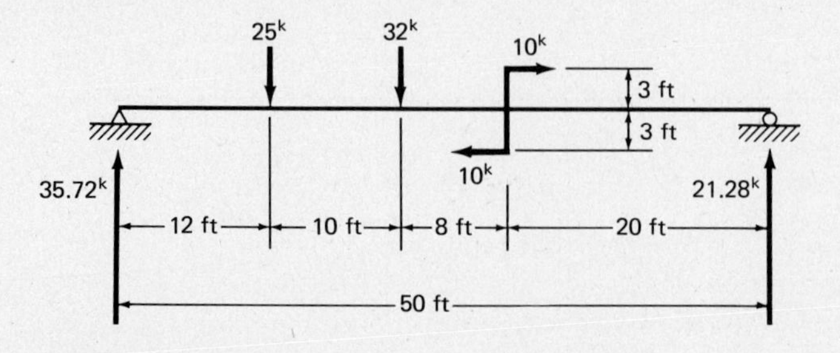

Figure 4.8

SOLUTION

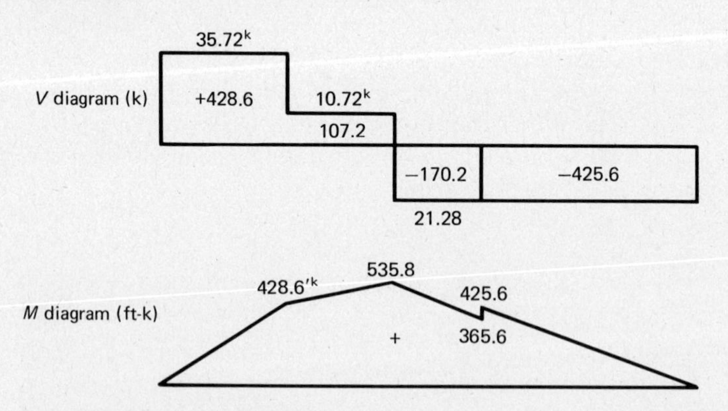

Students may at first have a little difficulty in drawing shear and moment diagrams for structures which are subjected to triangular loads. Example 4.7 is presented to demonstrate how to cope with them. The reactions are calculated for the beam shown in Fig. 4.9 and the shear diagram is sketched. It will, however, be noticed that the point of zero shear is unknown and is shown in the figure as being located a distance x from the left support. The ordinate on the load diagram at this point is labeled as y and can be expressed in terms of x by writing the following expression:

$$\frac{x}{30} = \frac{y}{2}$$

$$y = \frac{1}{15}x$$

At this point of zero shear the sum of the vertical forces to the left can be written as the upward reaction 10 kips minus the downward uniformly varying load to the left.

$$10 - \left(\frac{1}{2}\right)(x)\left(\frac{1}{15}x\right) = 0$$

$$x = 17.32 \text{ ft}$$

This value of x can also be determined by writing an expression for moment in the beam at a distance x from the support and then taking $dM/dx = 0$ of that expression and solving the result for x.

Finally the moment at a particular point can be determined by taking (to the left or right of the point) the sum of the forces times their respective lever arms. For this example the moment at 17.32 ft is

$$M = (10)(17.32) - (10)\left(\frac{17.32}{3}\right) = 115.5'^{k}$$

The student may wonder why the author (once x was determined) did not just sum up the area under the shear diagram from the left support to the point of zero shear. Such a procedure is correct but the person must be sure that he or she determines the area properly. When partial parabolas are involved it may be necessary to determine the areas by calculus instead of with the standard parabolic formulas. As a result it is often simpler just to take moments.

Example 4.7

Draw the shear and moment diagrams for the beam shown in Fig. 4.9.

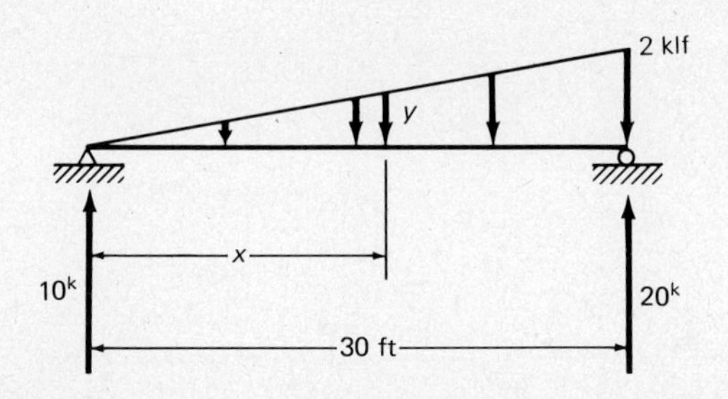

Figure 4.9

SOLUTION

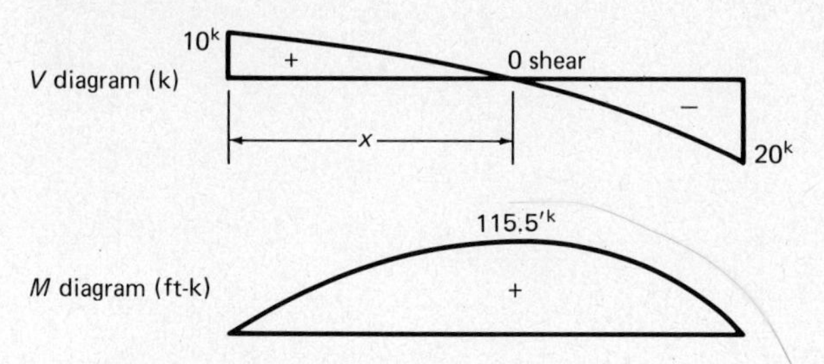

The beam of Example 4.8 is a continuous beam which cannot be analyzed by the equations of statics. The reactions have been computed by a method to be discussed in a later chapter, and the shear and moment diagrams have been drawn to show that the load, shear, and moment relationships are applicable to all structures.

Example 4.8

Draw the shear and moment diagrams for the continuous beam shown in Fig. 4.10 for which the reactions are given.

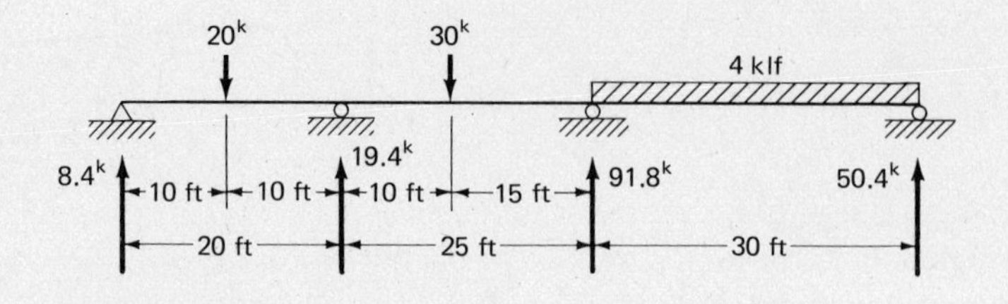

Figure 4.10

SOLUTION

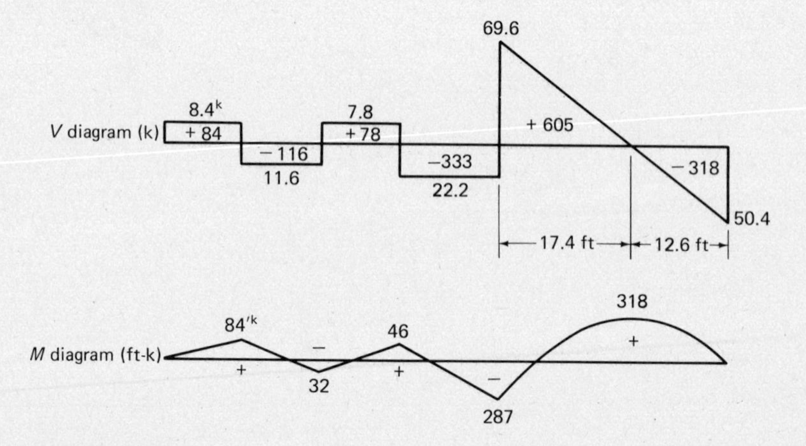

PROBLEMS

For Probs. 4.1 to 4.42 draw shear and moment diagrams for the structures.

Problem 4.1

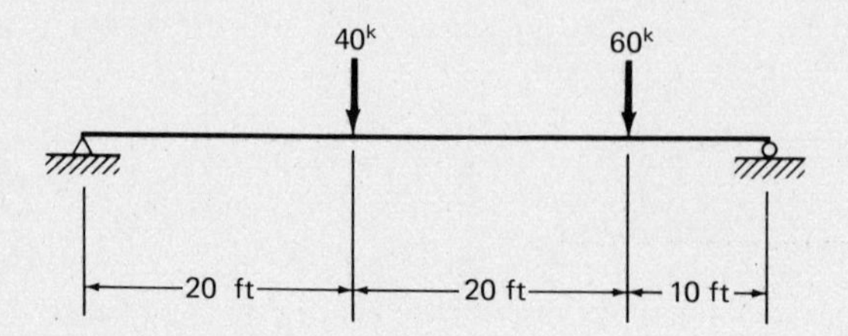

Problem 4.2 (*Ans.* Max $V = 40^k$, max $M = 300'^k$)

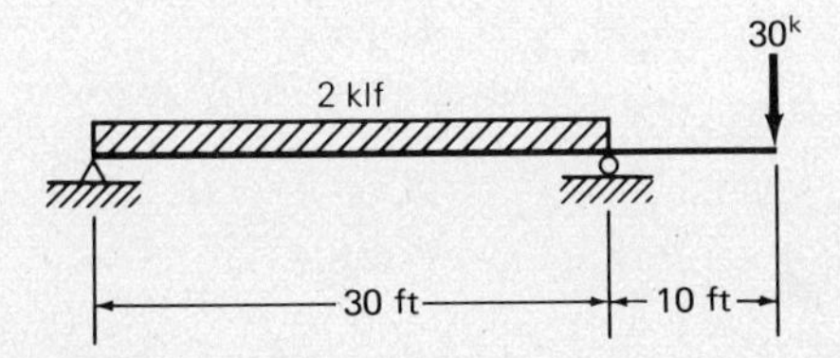

Problem 4.3

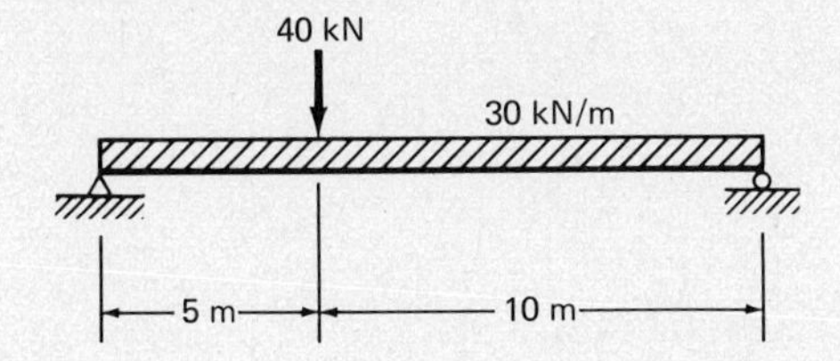

Problem 4.4 (*Ans.* max $V = 48^k$, max $M = 720'^k$)

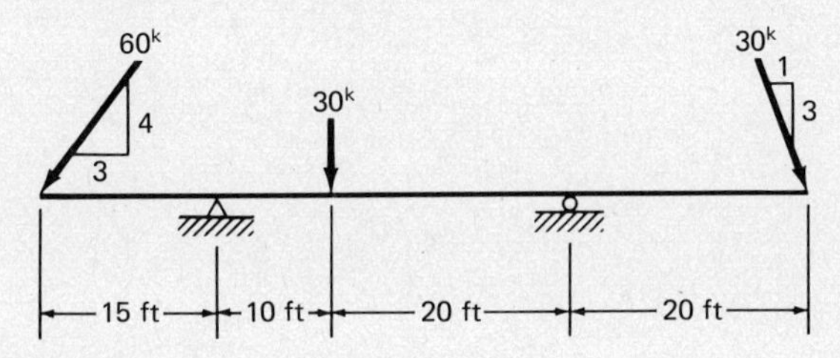

Problem 4.5

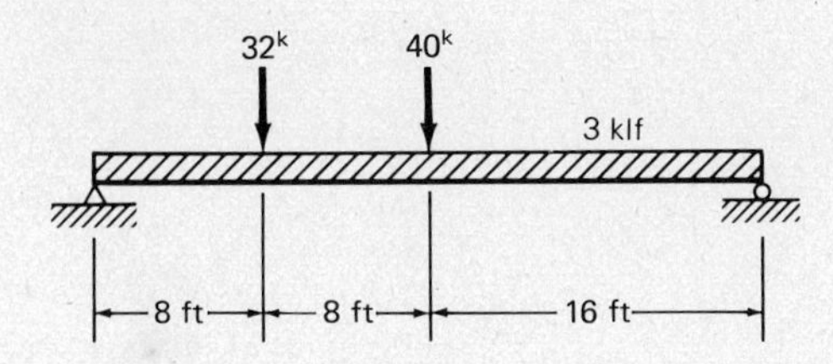

Problem 4.6 *Ans.* max $V = 34.17^k$, max $M = 225'^k$)

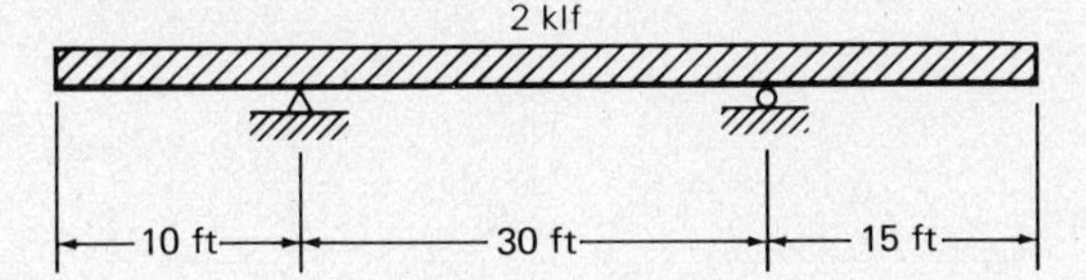

Problem 4.7

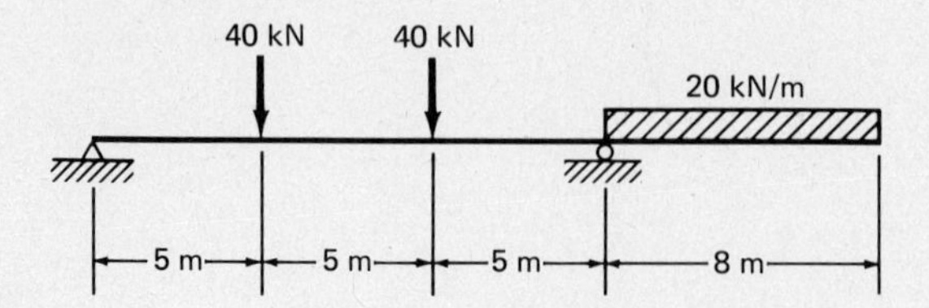

Problem 4.8 (*Ans.* max $V = 100.25^k$, max $M = 697.5'^k$)

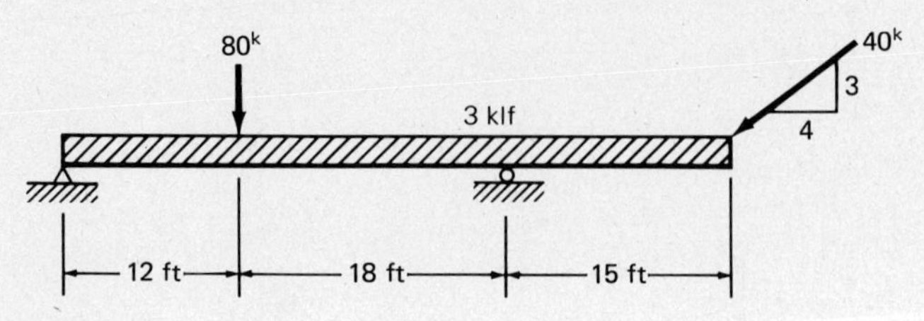

Problem 4.9

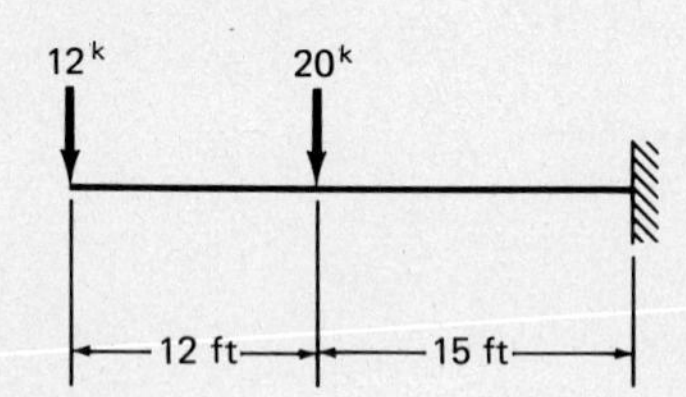

Problem 4.10 (*Ans.* max $V = 72^k$, max $M = 960'^k$)

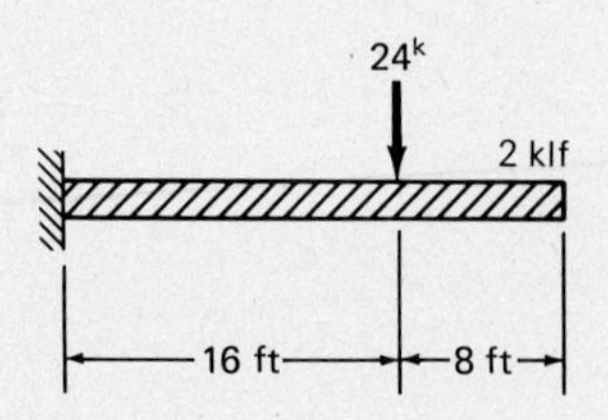

Problem 4.11

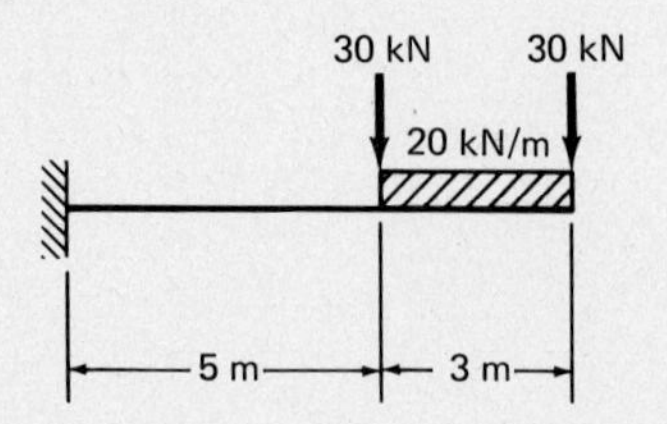

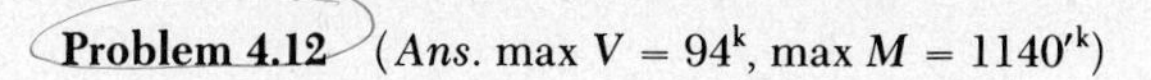

Problem 4.12 (*Ans.* max $V = 94^k$, max $M = 1140'^k$)

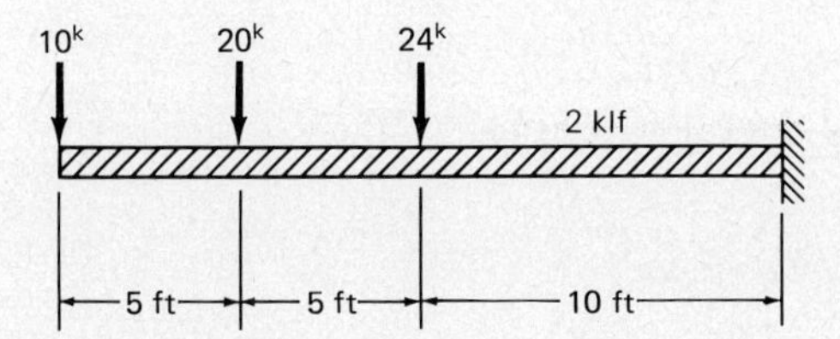

Problem 4.13

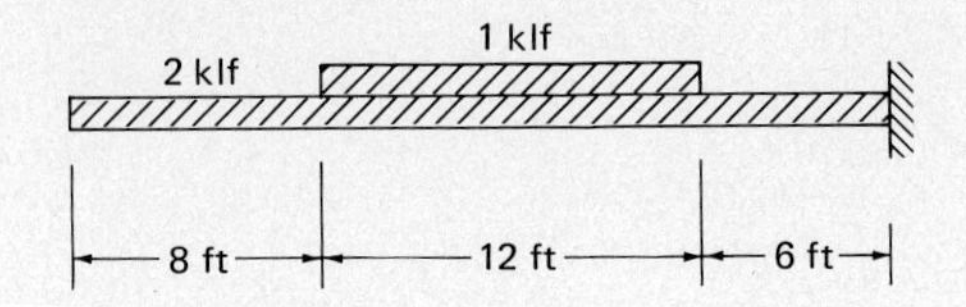

Problem 4.14 (*Ans.* max $V = 30^k$, max $M = 200'^k$)

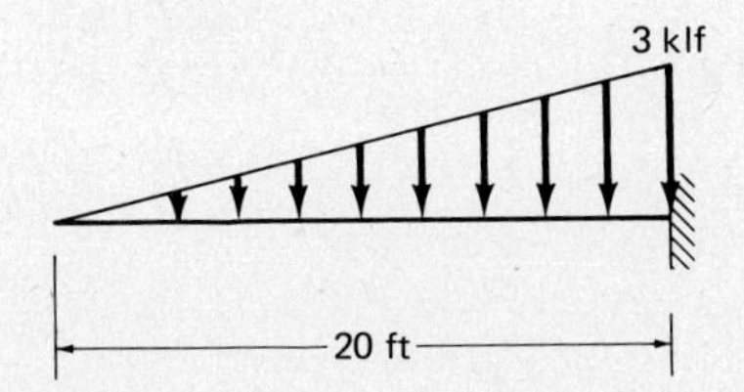

Problem 4.15

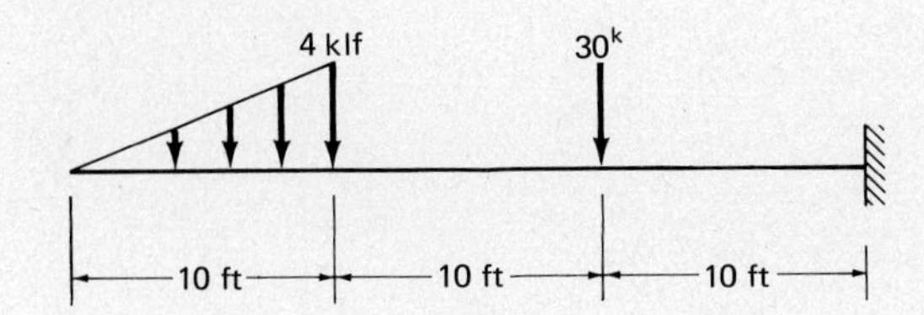

Problem 4.16 (*Ans.* max $V = 87.5^k$, max $M = 1125'^k$)

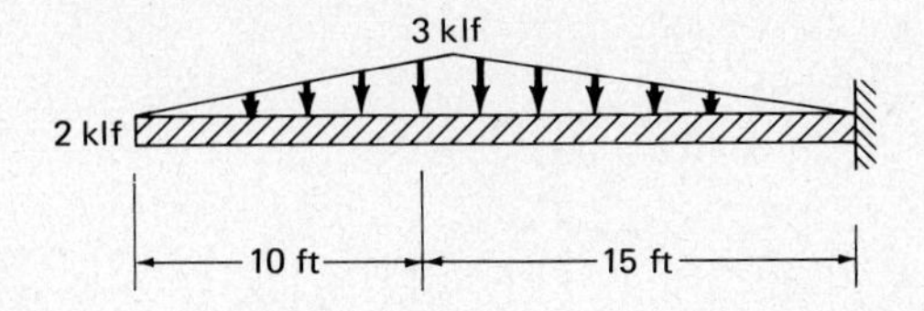

Problem 4.17

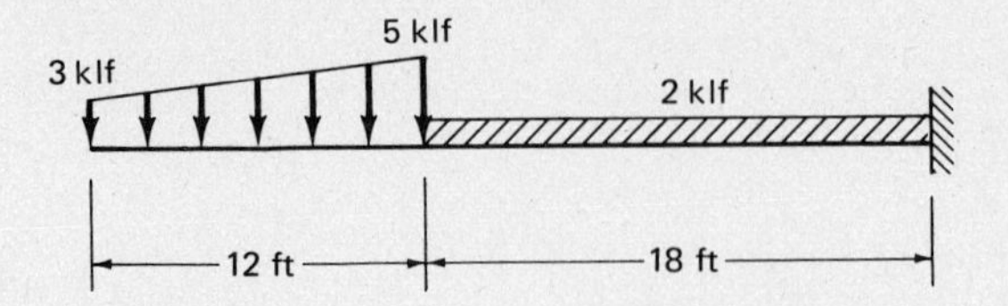

Problem 4.18 (*Ans.* max $V = 17.98^k$, max $M = 160.7'^k$)

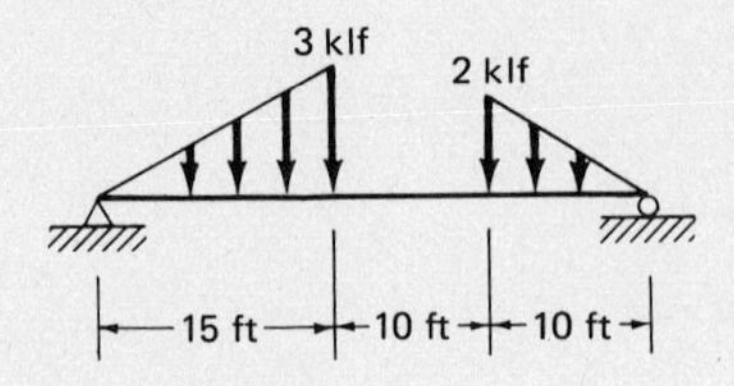

Problem 4.19

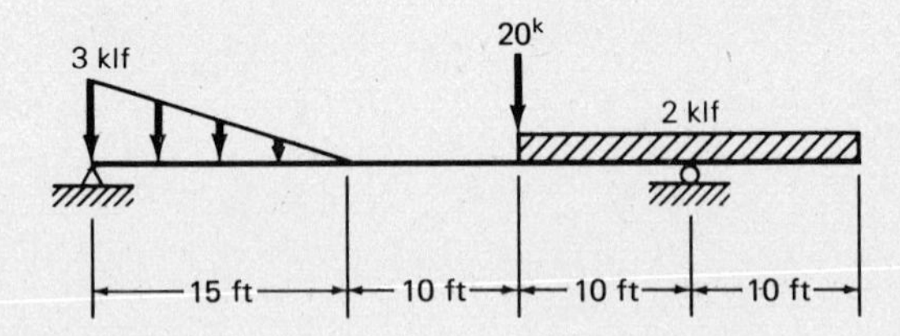

Problem 4.20 (*Ans.* max $V = 70^k$, max $M = 453'^k$)

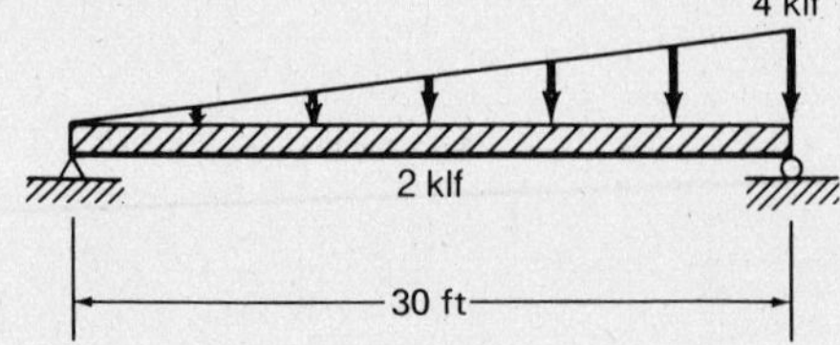

Problem 4.21

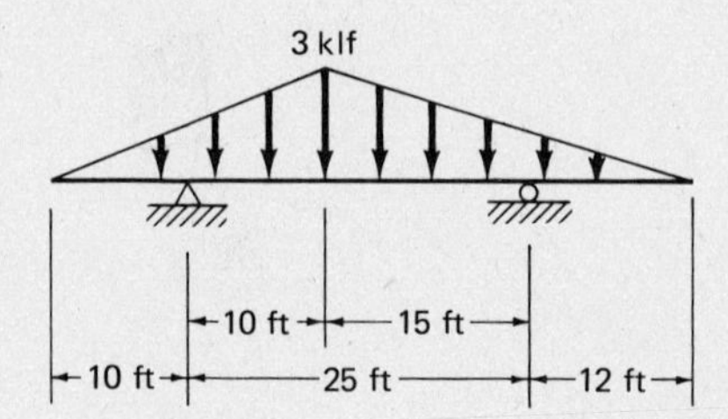

Problem 4.22 (*Ans.* max $V = 43.1^k$, max $M = 188'^k$)

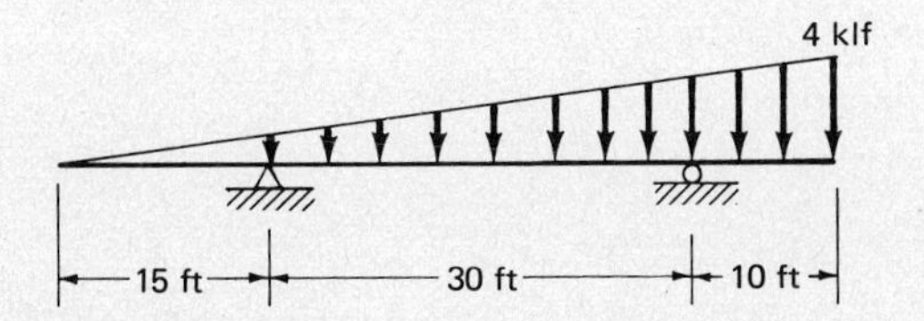

Problem 4.23

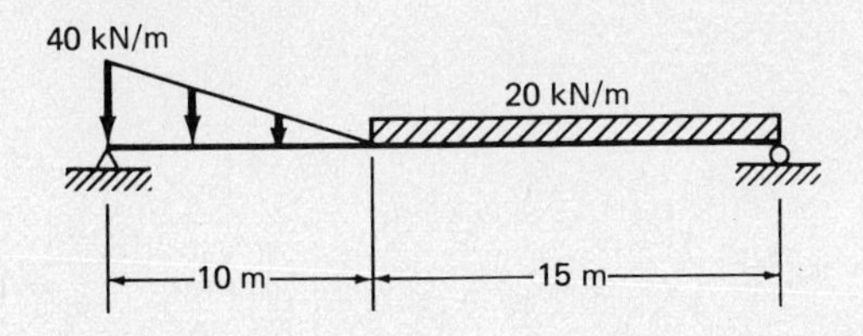

Problem 4.24 (*Ans.* max $V = 58.13^k$, max $M = 440.6'^k$)

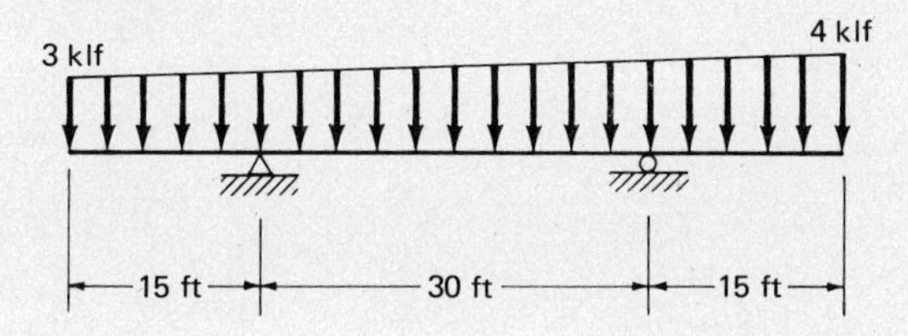

Problem 4.25

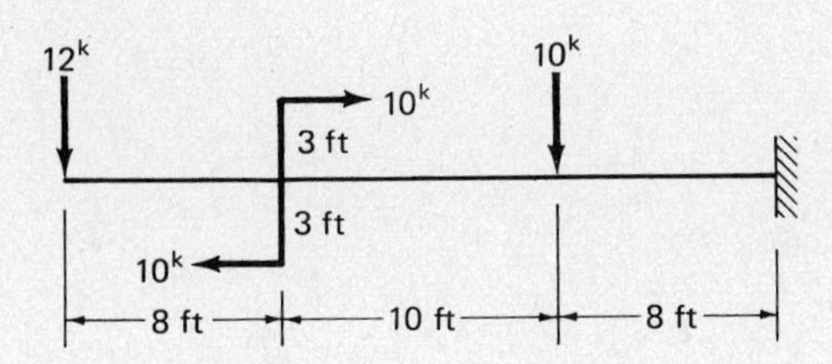

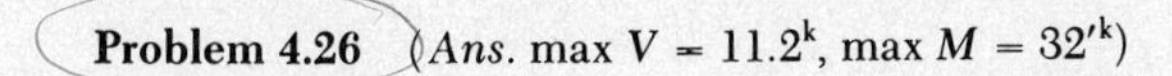

Problem 4.26 (*Ans.* max $V = 11.2^k$, max $M = 32'^k$)

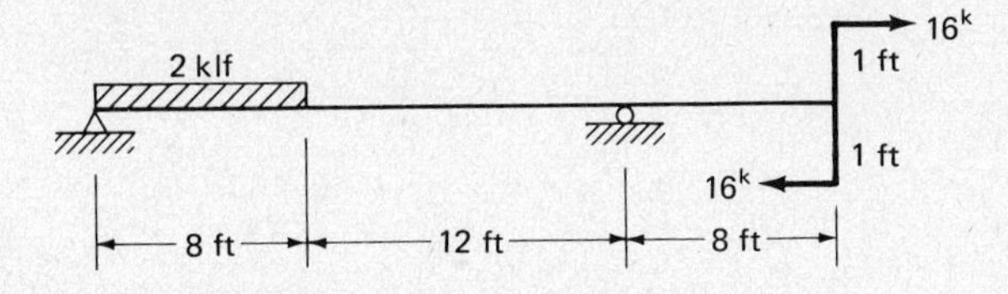

Problem 4.27

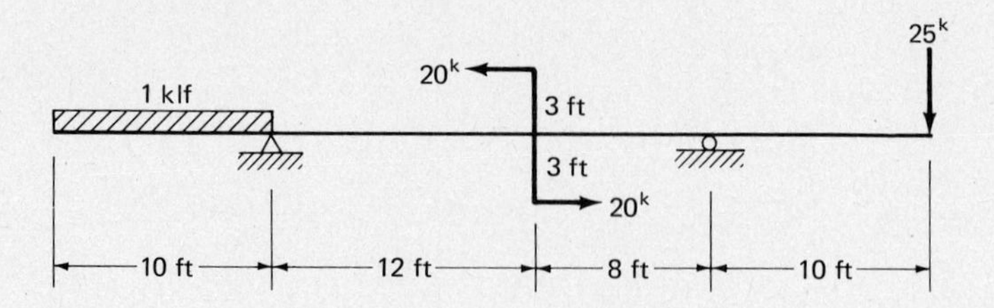

Problem 4.28 (*Ans.* max $V = 119.4^k$, max $M = 1138'^k$)

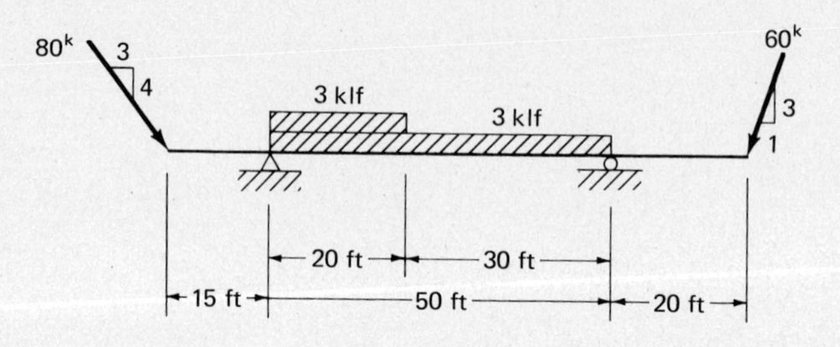

Problem 4.29

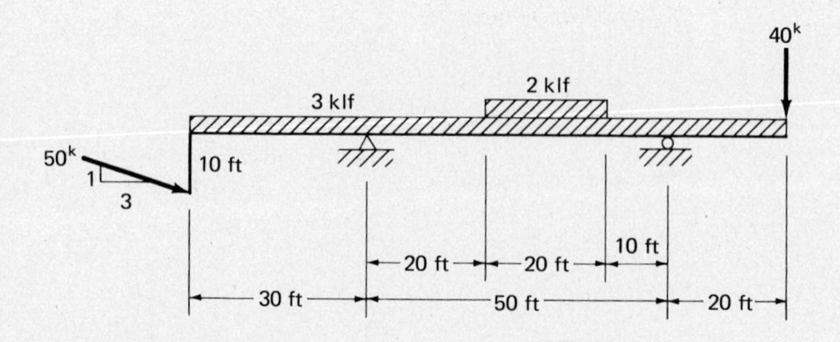

Problem 4.30 (*Ans.* max $V = 73.3^k$, max $M = 616'^k$)

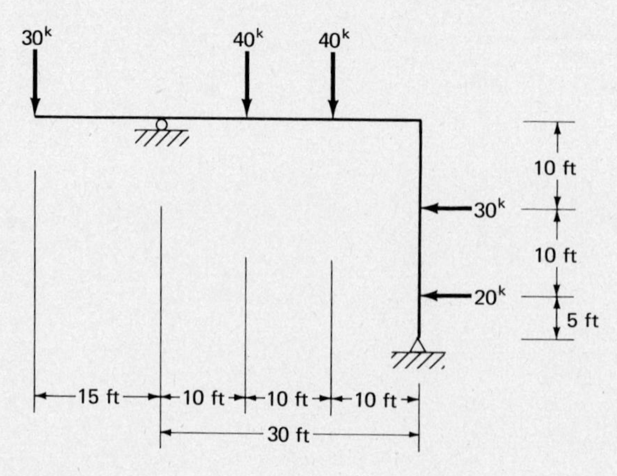

Problem 4.31

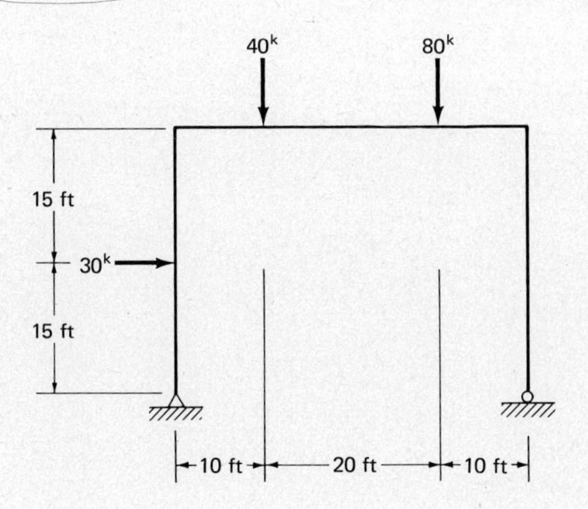

Problem 4.32 (*Ans.* max $V = 75^k$, max $M = 937'^k$)

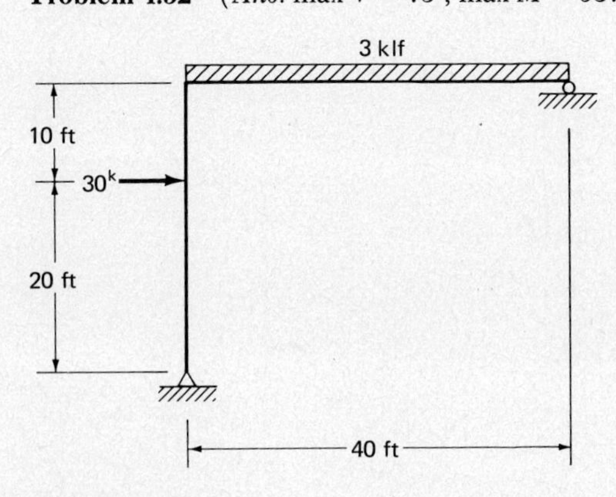

Problem 4.33

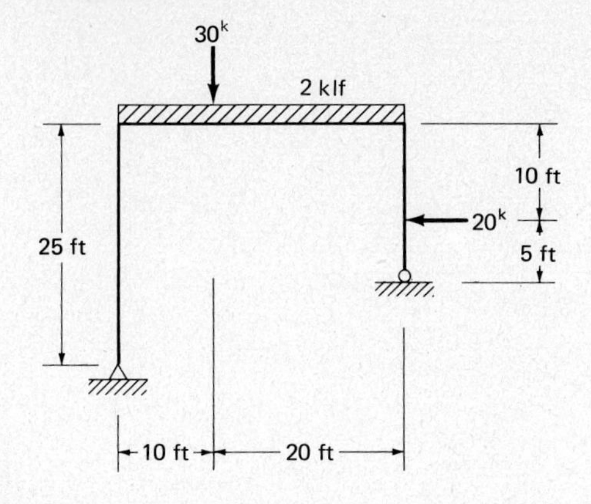

Problem 4.34 (*Ans.* max $V = 70.4^k$, max $M = 676.7'^k$)

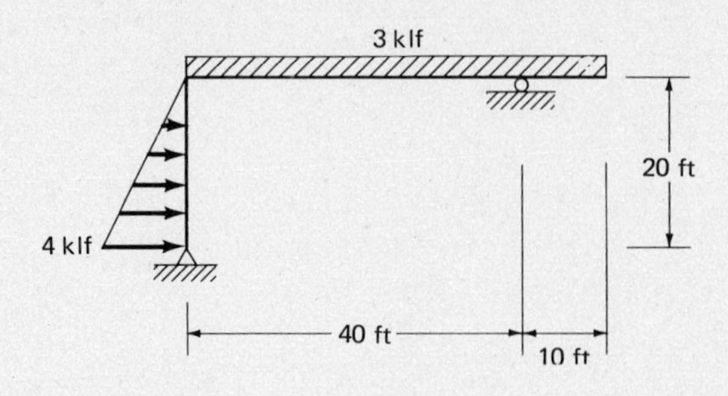

Problem 4.35

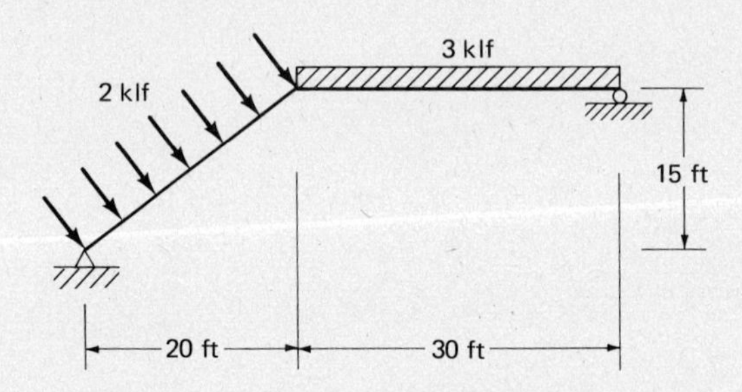

Problem 4.36 (*Ans.* max $V = 325^k$, max $M = 1200'^k$)

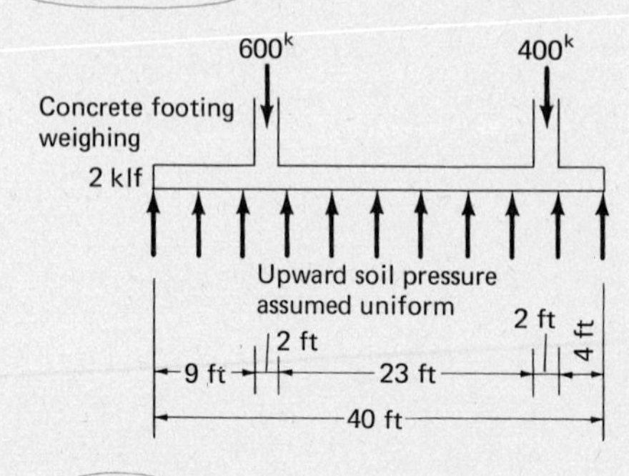

Problem 4.37

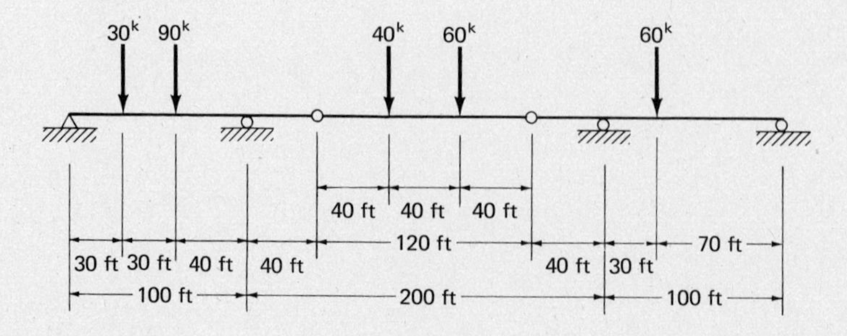

Problem 4.38 (*Ans.* max $V = 215^k$, max $M = 7000'^k$)

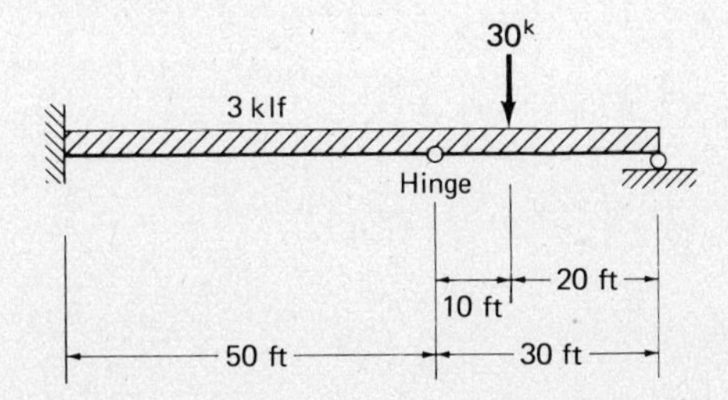

Problem 4.39

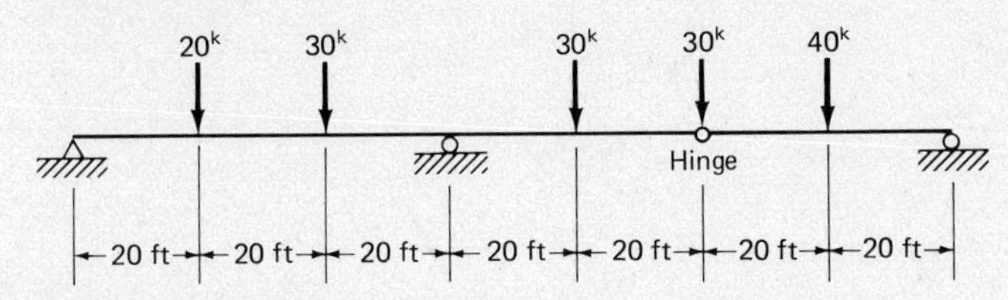

Problem 4.40 Given: Moment at interior support = $-1274'^k$ (*Ans.* max $V = 144.35^k$, max $M = 1274'^k$)

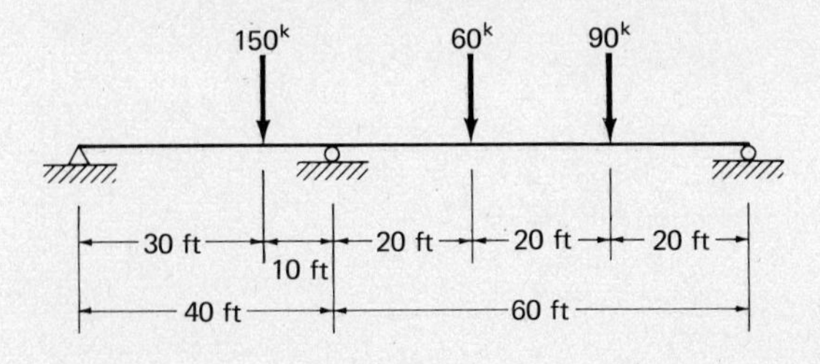

Problem 4.41 Given: Moment at fixed end = $-147.7'^k$; other reactions as shown.

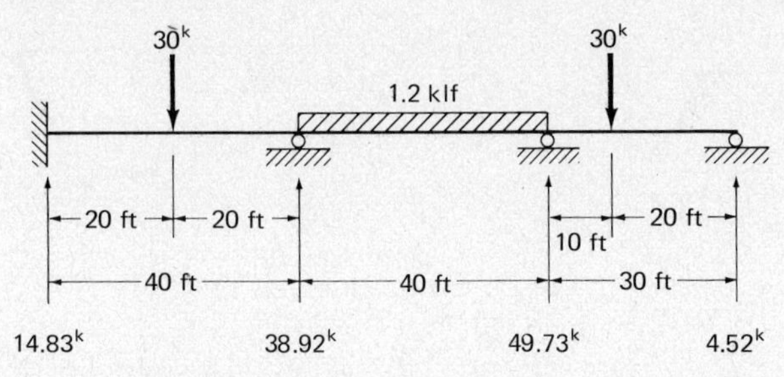

Problem 4.42 Given: Moment at $B = -238'^{k}$, moment at $C = -217'^{k}$. (*Ans.* max $V = 40.56^{k}$, max $M = 331'^{k}$)

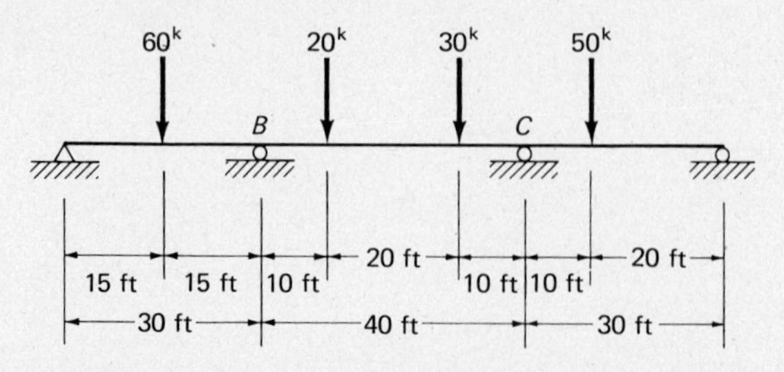

For Probs. 4.43 to 4.45, for the moment diagrams and dimensions given, draw the shear diagrams and load diagrams. Assume upward forces are reactions.

Problem 4.43

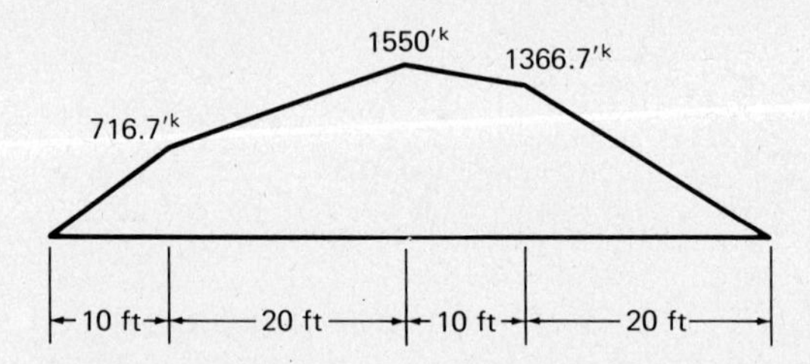

Problem 4.44 (*Ans.* Reactions and loads left to right: 80 kN, 100 kN, 216 kN, 216 kN, 232 kN)

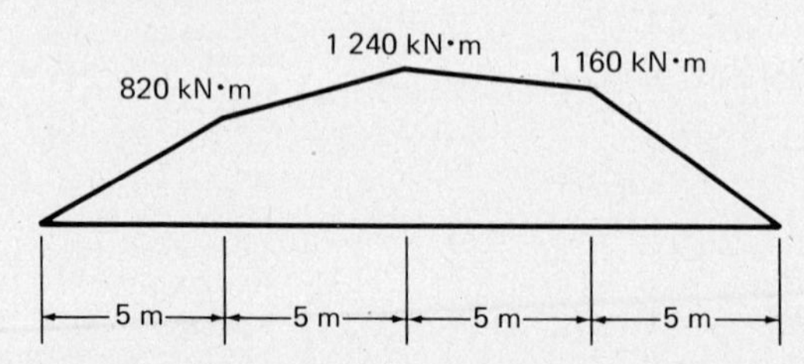

Problem 4.45

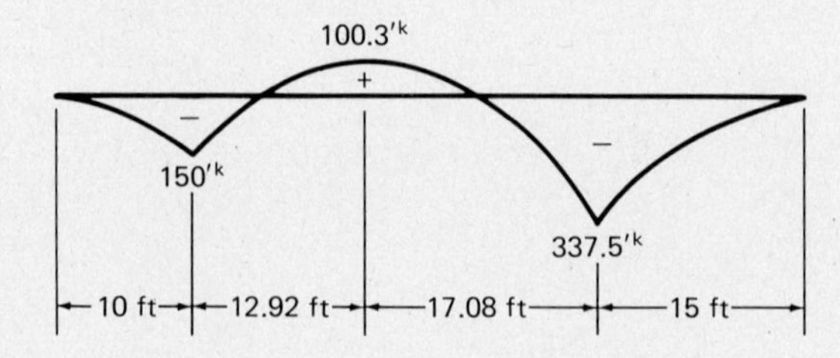

5 Introduction to Trusses

5.1 GENERAL

An Italian architect, Andrea Palladio (1518–1580), is believed to have developed the first trusses. His extensive writings on architecture include detailed descriptions and drawings of several wooden trusses quite similar to those in use today [1].

A truss is defined in Chap. 1 as a structure formed by a group of members arranged in the shape of one or more triangles. Because the members are assumed to be connected with frictionless pins, the triangle is the only stable shape. Study of the truss of Fig. 5.1(a) shows that it is impossible for the triangle to change shape under load unless one or more of the sides is bent or broken. Figures of four or more sides are not stable and may collapse under load, as seen in Fig. 5.1(b) and (c). These structures may be deformed without a change in length of any of their members. It will be seen in Chap. 9 that there are many stable trusses that include one or more figures which are not triangles. A careful study, however, will show they consist of separate groups of triangles that are connected together according to definite rules, forming nontriangular but stable figures in between.

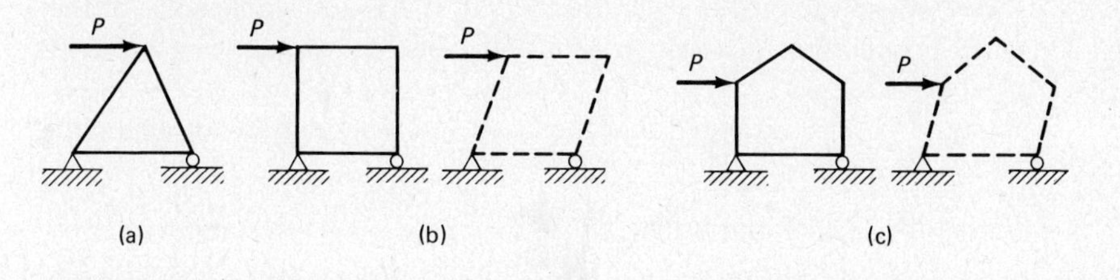

Figure 5.1

Truss for conveyor of West Virginia quarry of Pennsylvania Glass Sand Corporation. (Courtesy of Bethlehem Steel Corporation.)

5.2 ASSUMPTIONS FOR TRUSS ANALYSIS

The following assumptions are made in order to simplify the analysis of trusses:

1. Truss members are connected together with frictionless pins. (Pin connections are used for very few trusses erected today, and no pins are frictionless. A heavy bolted or welded joint is a far cry from a frictionless pin.)
2. Truss members are straight. (If they were not straight, the axial forces would cause them to have bending moments.)
3. The deformations of a truss under load, caused by the changes in lengths of the individual members, are not of sufficient magnitude to cause appreciable changes in the overall shape and dimensions of the truss. Special consideration may have to be given to some very long and flexible trusses.
4. Members are so arranged that the loads and reactions are applied only at the truss joints.

Examination of roof and bridge trusses will prove this last statement to be generally true. In buildings with roof trusses the beams, columns, and bracing frame directly into the truss joints. Roof loads are transferred to trusses by horizontal beams, called *purlins,* that span the distance between the trusses. The roof is supported directly by the purlins or it is supported by rafters, or subpurlins, which run parallel to the trusses and are supported by the purlins. The purlins are placed at the truss joints unless the top-chord panel lengths become exceptionally long, in which case it is sometimes economical to place purlins in between the joints, although some bending will be developed in the top chords. (Some types of roofing, such as corrugated steel, gypsum slabs, and others may be laid directly on the purlins. The purlins then have to be spaced at intermediate points along the top chord so as to provide a proper span for the roofing they directly support.) The loads supported by a highway bridge are transferred to the trusses at the joints by beams running underneath the roadway as shown in Fig. 12.1 and described in Sec. 12.2.

5.3 EFFECT OF ASSUMPTIONS

The effect of the foregoing assumptions is to produce an ideal truss, whose members have only axial forces. A member with axial force only is subject to a push or pull with no bending present, as illustrated in Fig. 5.2. (Even if all of the assumptions were perfectly true, there would be some bending in a member caused by its own weight.)

Forces obtained on the basis of these simplifying assumptions are very satisfactory in most cases and are referred to as ***primary forces.*** Structures are sometimes analyzed without the aid of some or all of these assumptions. Forces caused by conditions not considered in the primary force analysis are said to be ***secondary forces.***

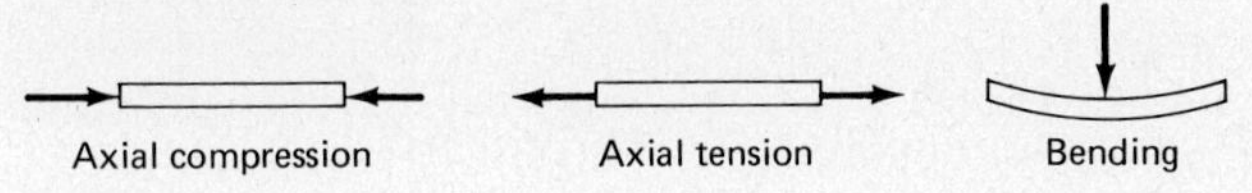

Figure 5.2

5.4 TRUSS NOTATION

A common system of denoting the members of a truss is shown in Fig. 5.3. The joints are numbered from left to right, the bottom joints having L (for lower) prefixes and the top joints having U (for upper) prefixes. Should there be, in more complicated trusses, joints in between the lower and upper joints, they may be given M (for middle) prefixes.

The various members of a truss are often referred to by the following names, reference being made to Fig. 5.3.

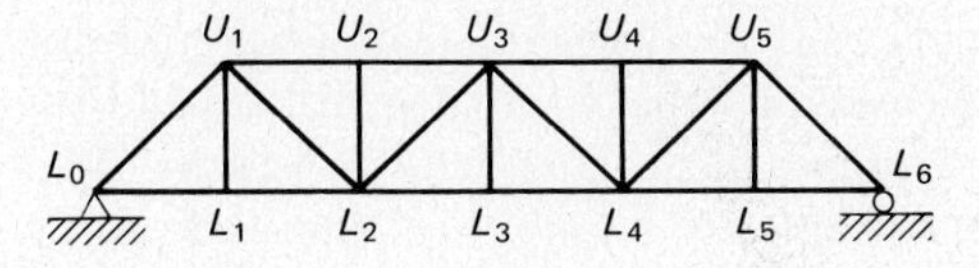

Figure 5.3

1. *Chords* are those members forming the outline of the truss, such as members U_1U_2 and L_4L_5.
2. *Verticals* are named on the basis of their direction in the truss, such as members U_1L_1 and U_3L_3.
3. *Diagonals* also are named on the basis of their direction in the truss, such as members U_1L_2 and L_4U_5.
4. *Web members* include the verticals and diagonals of a truss, and most engineers consider them to include the end diagonals, or *end posts*, such as L_0U_1 and U_5L_6.

5.5 DISCUSSION OF ROOF TRUSSES

The purposes of roof trusses are to support the roofs which keep the elements out (rain, snow, wind) and to support the loads connected underneath (ducts, piping, ceiling) as well as their own weights.

The designer is often concerned with the problem of selecting a truss or a beam to span a given opening. Should no other factors be present, the decision would probably be based on consideration of economy. The smallest amount of material will nearly always be used if a truss is selected for spanning a certain opening; however, the cost of fabrication and erection of trusses will probably be appreciably higher than required for beams. For shorter spans the overall cost of beams (material plus fabrication and erection) will definitely be less but as the spans become greater, the higher fabrication and erection costs of trusses will be more than canceled by their weight saving. A further advantage of trusses is that for the same amounts of material they have greater stiffnesses than do beams.

It is impossible to give a lower economical span for trusses. They may be used for spans as small as 30 or 40 ft and as large as 300 to 400 ft. Beams may be economical for some applications for spans much greater than the lower limits mentioned for trusses.

Roof trusses can be flat or peaked. In the past the peaked roof trusses have probably been used more for short-span buildings and the flatter trusses for the longer spans. The trend today for both long and short spans, however, seems to be away from the peaked trusses and toward the flatter ones, the change being due to the appearance desired and perhaps a more economical construction of roof decks.

In the pages to follow the terms *pitch* and *slope* are often used. The pitch of a symmetrical truss is referred to as the rise of the top chord of the truss divided by the span. Should the truss be unsymmetrical, the numerical value of its pitch is not of much use. For such cases the slope of the truss on each side can be given. The slope is the rise of the top chord to its horizontal projection, often given as so many inches per horizontal foot. For symmetrical trusses the slope will equal twice the pitch.

5.6 SPACING AND SUPPORT OF ROOF TRUSSES

The center-to-center spacing of roof trusses depends on the type of roof construction, truss spans, and foundation conditions. The usual spacings vary from 12 to 30 ft, the lower spacings being used for the shorter spans and the larger spacings for the longer spans. For the 50- and 60-ft spans, spacings of roughly 12 to 20 ft on centers are common, whereas values of from 15 to 24 ft are common for 90- to 100-ft spans. For very long-span roof trusses, say those of more than 140 or 150 ft, the center-to-center spacings may be as great as 50 or 60 ft. The purlins for such large spacings are probably trusses themselves framing into the sides of the main trusses.

It is desirable to space the trusses uniformly for a given section of a building or, even better, for the entire building so that as many trusses as possible will be identical.

The usual truss is supported on brick, block, or concrete walls or on steel or reinforced concrete columns. To provide for temperature expansion and contraction it is usually necessary to have the anchor bolts at one end set in a slotted hole in the bearing plate to permit the required shortening or lengthening.

5.7 COMMON TYPES OF ROOF TRUSSES

Several of the more commonly used types of roof trusses are shown in Fig. 5.4. A good many of these trusses have been named for the engineers or architects who first developed them. A remark or two is made about each of these trusses in the paragraphs to follow. The letter at the beginning of each of these paragraphs corresponds to the letter in Fig. 5.4 beneath the type of truss being considered.

(a) The Warren and Pratt trusses have probably been used more for the flatter roofs (slopes of from $\frac{3}{4}$ to $1\frac{1}{4}$ in/ft) where built-up roofing can be satisfactorily applied than have the other types of trusses. These trusses can be economically used for flat roofs for spans of roughly 40 to 125 ft, although they have been used for spans as great as 200 ft. The Warren truss is usually a little more satisfactory than the Pratt. The roofs may be completely flat for spans not exceeding 30 or 40 ft, but for longer spans the slopes mentioned are used for drainage purposes.

(b) The pitched Pratt and Howe trusses are probably the most common types of medium-rise trusses. The slopes intended here fall in between those given for (a) and (c). They have maximum economical spans of about 90 or 100 ft.

(c) For steep roofs (with slopes of 5 or 6 in/ft) the Fink truss is very popular. The Pratt and Howe trusses may also be used for steep slopes, but they are usually not as economical. The Fink truss has been used for spans as great as 120 ft. A feature that makes it more economical is that most of the members are in tension whereas those that are in compression are fairly short. The panel

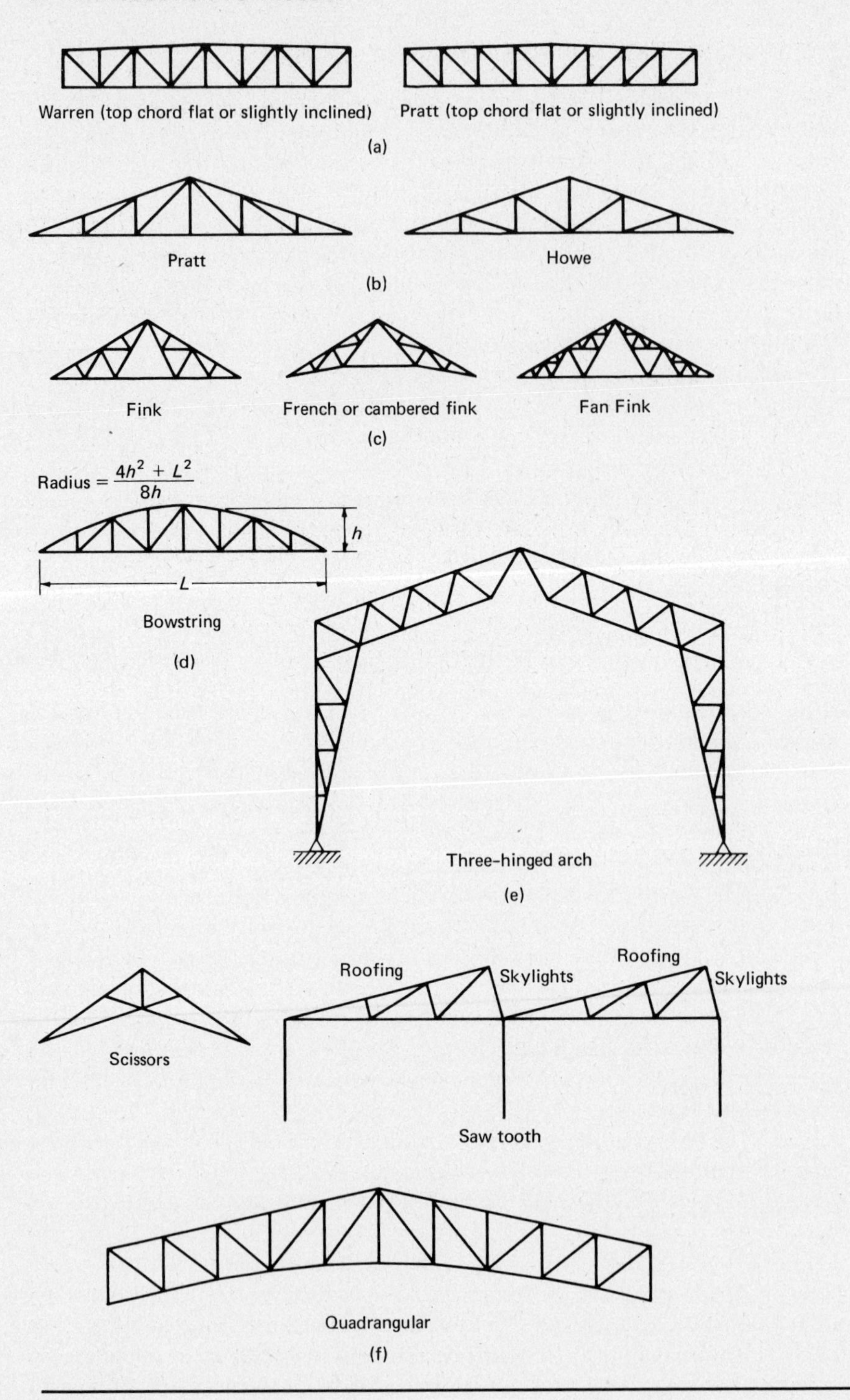
Warren (top chord flat or slightly inclined)
Pratt (top chord flat or slightly inclined)
(a)
Pratt
Howe
(b)
Fink
French or cambered fink
Fan Fink
(c)
Radius = $\frac{4h^2 + L^2}{8h}$
h
L
Bowstring
(d)
Three-hinged arch
(e)
Scissors
Roofing
Skylights
Roofing
Skylights
Saw tooth
Quadrangular
(f)

Figure 5.4

Fairview-St. Mary's Skyway System, Minneapolis, Minnesota. (Courtesy of the American Institute of Steel Construction, Inc.)

layout of a truss may be controlled by the purlin spacings. As it is usually desirable to have purlins placed at panel points only, the main panels may be subdivided. Fink trusses can be divided up into a large number of panels to suit almost any span or purlin spacing. The fan Fink shown illustrates subdivision very well.

(d) If a curved roof is acceptable, the bowstring truss can be used economically for spans of up to 120 ft, although on occasions it has satisfactorily been used for much longer spans. When properly designed, this truss has the unusual feature of having very small forces in the web members. Despite the fact that there is some expense in bending the top chord, the bowstring has proved quite popular for warehouses, supermarkets, garages, and small industrial buildings. A recommended radius of curvature for the top chord is given in the figure [2].

(e) When spans appreciably above 100 ft are planned, consideration can be given to using steel arches as they may provide the most economical solutions. The three-hinged arch is the only one shown here.

(f) In this part of the figure several miscellaneous types of trusses are shown. The scissors truss (which is so named because of its resemblance to a pair of scissors) may be satisfactory for supporting short-span churches and other buildings with steep roofs. Sawtooth trusses (not used much today) may be used when adequate natural lighting is desired from skylights in wide buildings. Their steep faces support skylights which usually face the north for more evenly diffused light. Sawtooth trusses are used when their numerous columns are not objectionable. A long-span roof truss that has been used for spans well over 100 ft is the quadrangular truss. Near the centerline of this truss the diagonals are reversed for the purpose of keeping as many of them in tension as possible.

5.8 DISCUSSION OF BRIDGE TRUSSES

As bridge spans become longer and loads heavier, trusses begin to be competitive economically. Early bridge trusses were constructed with wood but they had several disadvantages. First, they were subject to deterioration from wind and water. As a result, covered bridges were introduced and such structures would often last for quite a few decades. Nevertheless, wooden truss bridges were quite subject to destruction by fire and this was particularly true for railroad bridges. In addition, there were some problems with the loosening of the fasteners under moving loads as the years went by and as the loads became heavier.

As a result of the several preceding disadvantages, wooden truss bridges faded from use toward the end of the nineteenth century and structural steel bridges, which did not require extensive protection from the elements and whose joints had higher fatigue resistance, took over the market. (There were also some earlier iron truss bridges.) Section 5.10 presents a discussion of some steel truss bridges.

Today existing steel truss bridges are being steadily replaced with steel, precast-concrete, or prestressed-concrete beam bridges. It seems that the age of steel truss bridges is over except for spans of more than several hundred feet (a very small percentage of the total).

5.9 THROUGH, DECK, AND HALF-THROUGH BRIDGES

The reader has often seen highway bridges in which the trusses were on the sides. As he rode across the bridge he could see overhead lateral bracing between the trusses. This type of bridge is said to be a ***through bridge.*** The floor system is supported by floor beams that run under the roadway and between the bottom chord joints of the trusses.

In the ***deck bridge*** the roadway is placed on top of the trusses or girders. Deck construction has every advantage over through construction except for underclearance. There is unlimited overhead horizontal and vertical clearance, and future expansion is more feasible. Another very important advantage is that supporting trusses or girders can be moved closer together, reducing lateral moments in the floor system. Other advantages of the deck truss are simplified floor systems and possible reduction in the sizes of piers and abutments due to reductions in their heights. Finally, the very pleasing appearance of deck structures is another reason for their popularity.

Cornish-Windsor covered bridge over Connecticut River between Vermont and New Hampshire. (Courtesy of the New Hampshire Department of Public Works and Highways.)

Today's bridge designer tries to prevent any sense of confinement for the users of his or her bridges. In trying to achieve this goal the designer attempts to eliminate any overhead bracing or truss members that will protrude above the roadway level. The result is that the deck structure is again desirable unless underclearance requirements prevent its use or the spans are so large as to make it impractical.

Sometimes short-span through bridge trusses were so shallow that adequate depth was not available to provide overhead bracing and at the same time leave sufficient vertical clearance above the roadway for the traffic. As a result, the bracing was placed underneath the roadway. Bridges of the through type without overhead bracing are referred to as *half-through* or *pony bridges*. One major problem with pony truss bridges was the difficulty of providing adequate lateral bracing for the top-chord compression members of the trusses. It would be very unlikely for a pony truss to be economical today in the age where beams have captured the short-span bridge market.

5.10 COMMON TYPES OF BRIDGE TRUSSES

A few types of bridge trusses that can be used for medium spans are shown in Fig. 5.5. The Pratt truss, which is reasonably economical for spans of 150 to 200 ft, has one advantage in that the diagonals are all in tension under dead loads. The end posts, however, are always in compression and the movement of live loads back and forth across the span may cause force reversals in some of the other diagonals.

The Warren truss is thought by many to have a little more attractive appearance than the Pratt and may be a little more popular for the same spans. It is probably used more as a deck truss than as a through truss because it is particularly economical when so used.

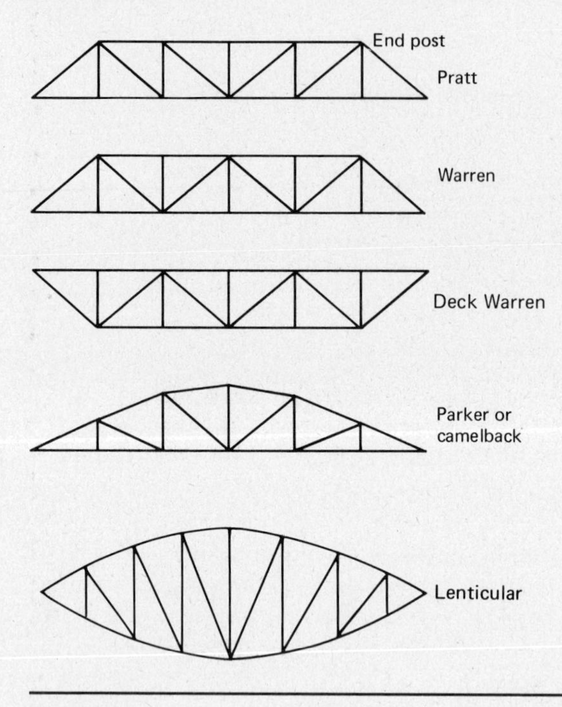

Figure 5.5

The deeper a truss is for the same size chord members, the greater is its resisting moment. For longer spans it is economical to increase truss depths where the moments are larger. These varying depth trusses are definitely lighter than corresponding parallel chord trusses, but their fabrication costs are higher. Above approximately 180 to 200 ft the weight saving will more than cancel the extra fabrication costs and the so-called curved-chord trusses become economical. The Parker truss is quite satisfactory for spans from roughly 180 to 360 ft. (If the top chord of the Parker truss has exactly five slopes it is sometimes called a camelback truss.) For long continuous and cantilever span trusses the depths are almost always varied with the moments. The greatest depths in these trusses will occur at the supports where the moments are largest.

A variation of the Pratt truss is the lenticular truss. This structure, which may be the most striking-looking type of bridge truss, has parabolic upper and lower chords. It was generally used for spans from 150 to 400 ft but the Smithfield Street bridge in Pittsburgh is of the lenticular type and has a 720-ft span. A major disadvantage of this truss is its high fabrication cost.

Willard Bridge over Kansas River north of Willard, Kansas. (Courtesy of the American Institute of Steel Construction, Inc.)

Various economic studies for truss bridges have shown that the diagonals should be kept at angles of approximately 45° with the horizontal, and the depth-to-span ratios should vary from about one-fifth to one-eighth with the smaller ratios used for the larger spans. Should these recommendations be followed for spans much greater than 300 ft, the panel lengths will become excessive.

When panels are very long, the compression member sizes get out of hand because of their excessive unsupported lengths. Furthermore, the floor system between panel points becomes unreasonably heavy and expensive for large panels. In fact floor-beam spacings greater than 25 ft are not recommended.

The subdivided trusses shown in Fig. 5.6(a) and (b) can be used to keep the panel lengths within the desired values for longer trusses while at the same

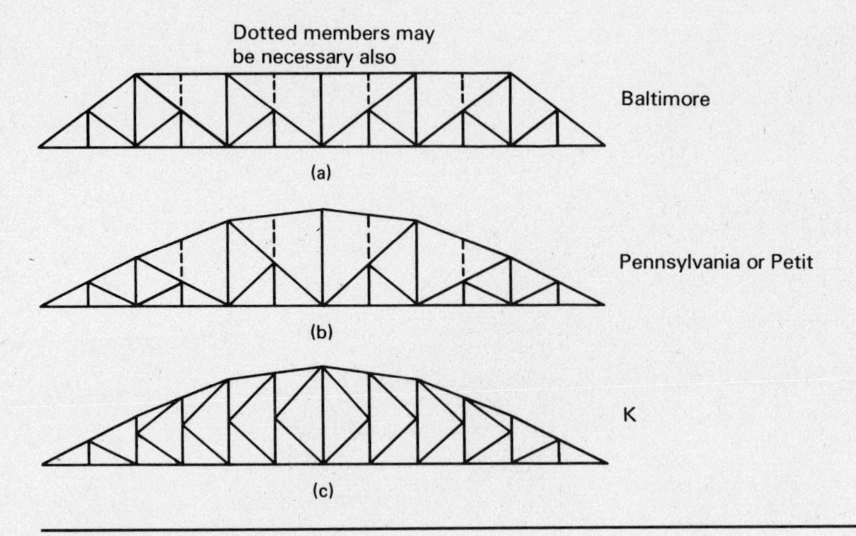

Figure 5.6

time following economic requirements previously mentioned (diagonal slope and depth-to-span ratio). In part (a) of the figure the parallel-chord Baltimore or subdivided Pratt truss is shown. Again it is usually more economical to use trusses that vary in depth with the moments and a curved-chord Baltimore is shown in part (b) of the figure. This truss is usually called a *Pennsylvania* (because of its extensive use on the Pennsylvania Railroad) or *Petit truss*. The K truss shown in (c) accomplishes the same purposes as the other subdivided trusses, has a more attractive appearance, and probably has smaller secondary forces.

REFERENCES

[1] Linton E. Grinter, *Theory of Modern Steel Structures* (New York: Macmillan, 1962), p. 7.

[2] John E. Lothers, *Design in Structural Steel,* 2nd ed. (Englewood Cliffs, N.J.: Prentice-Hall, 1965), p. 348.

6 Truss Analysis by Method of Joints

6.1 USE OF SECTIONS

An indispensable part of truss analysis, as in beam analysis, is the separation of the truss into two parts with an imaginary section. The part of the truss on one side of the section is removed and studied independently. The loads applied to this free body include the axial forces of the members that have been cut by the section and any loads and reactions that may be applied externally.

Application of the equations of statics to isolated free bodies enables one to determine the forces in the cut members if the free bodies are carefully selected so that the sections do not pass through too many members whose forces are unknown. There are only three equations of statics, and no more than three unknowns may be determined from any one section.

After he or she has analyzed a few trusses, the student will have little difficulty in most cases in selecting satisfactory locations for the sections. The student is not encouraged to remember specific sections for specific trusses, although he or she will probably unconsciously fall into such a habit as time goes by. At this stage the student needs to consider each case individually without reference to other, similar trusses.

6.2 HORIZONTAL AND VERTICAL COMPONENTS

It is convenient to work with horizontal and vertical components in the computation of forces of truss members, as in the computation of reactions. The $\Sigma H = 0$ and $\Sigma V = 0$ equations of statics are generally written on the basis of a pair of axes that are horizontal and vertical. Because the forces in truss members are determined successively across a truss, much time will be saved if the vertical and horizontal components of forces in inclined members are recorded for use in applying the equations to other members. The use of components is clearly illustrated in the example problems of the sections that follow.

6.3 ARROW CONVENTION

The sign convention for tensile and compressive forces (+ and −, respectively) has already been mentioned. Arrows are also used throughout the text to represent the character of forces. The arrows indicate what members are doing to resist the axial forces applied to them by the remainder of the truss. For example, if a truss is compressing a certain member from each end (→ ——— ←), the member will push back against the compressive forces (←———→). This arrow convention is used for members in compression. The arrow convention for a member in tension is just the opposite, because a member that is being pulled or stretched from the ends (← ——— →) will resist by pulling back (→———←).

After some practice in the analysis of trusses, it is possible to determine by examination the character of the forces in many of the members of a truss. The reader should try to picture whether a member is in tension or compression

Ford Automobile Assembly Plant, Milpitas, California. (Courtesy of the American Institute of Steel Construction.)

before making the actual calculations. In this way a better understanding of the action of trusses under load will be obtained. The following paragraphs will show it is possible to determine entirely by mathematical means the character as well as the numerical value of forces.

6.4 METHOD OF JOINTS

An imaginary section may be completely passed around a joint in a truss, regardless of its location, completely isolating it from the remainder of the truss. The joint has become a free body that is in equilibrium under the forces applied to it. The equations $\Sigma H = 0$ and $\Sigma V = 0$ may be applied to the joint to determine the unknown forces in members meeting there. It is evident that no more than two unknowns can be determined at a joint with these two equations.

A person learning the method of joints may initially find it necessary to draw a free-body sketch for every joint in a truss he is analyzing. After he has computed the forces in two or three trusses, he will find it necessary to draw the diagrams for very few joints, because he will be able to visualize with ease the free bodies involved. The most important thing for the beginner to remember is that he is interested in only one joint at a time. He must keep his mind away from the loads and forces at other joints. His only concern is with the forces at the one joint on which he is working. Another very helpful suggestion for the reader is *draw large sketches*. The method of joints is illustrated by Example 6.1.

Example 6.1

By using the method of joints, find all forces in the truss shown in Fig. 6.1.

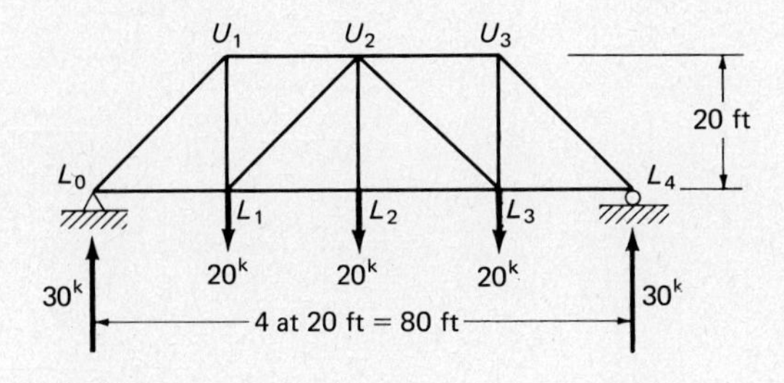

Figure 6.1

SOLUTION

Considering joint L_0,

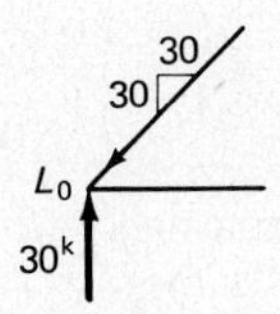

$$\Sigma V = 0$$

$$30 - V_{L_0U_1} = 0$$

$$V_{L_0U_1} = 30^k \text{ compression}$$

An examination of the joint shows a vertical reaction of 30^k acting upward. The equation $\Sigma V = 0$ indicates the members meeting there must supply 30^k downward. A member that is horizontal, such as L_0L_1, can have no vertical component of force; therefore L_0U_1 must supply the entire amount and 30^k will be its vertical component. The arrow convention shows that L_0U_1 is in compression. From its slope (20:20, or 1:1) the horizontal component can be seen to be 30^k also.

$$\Sigma H = 0$$
$$-30 + F_{L_0L_1} = 0$$
$$F_{L_0L_1} = 30^k \text{ tension}$$

The application of the $\Sigma H = 0$ equation shows L_0U_1 to be pushing horizontally to the left against the joint with a force of 30^k. For equilibrium, L_0L_1 must pull to the right away from the joint with the same force. The arrow convention shows the force is tensile.

Considering joint U_1,

$$\Sigma V = 0$$
$$30 - F_{U_1L_1} = 0$$
$$F_{U_1L_1} = 30^k \text{ tension}$$

The force in L_0U_1 has previously been found to be compressive with vertical and horizontal components of 30^k each. Since it is pushing upward at joint U_1 with a force of 30^k, U_1L_1 (the only other member at the joint that has a vertical component) must pull down with a force of 30^k in order to satisfy the $\Sigma V = 0$ equation.

$$\Sigma H = 0$$
$$30 - F_{U_1U_2} = 0$$
$$F_{U_1U_2} = 30^k \text{ compression}$$

Member L_0U_1 is pushing to the right horizontally with a force of 30^k. For equilibrium U_1U_2 is pushing back to the left with 30^k.

Considering joint L_1,

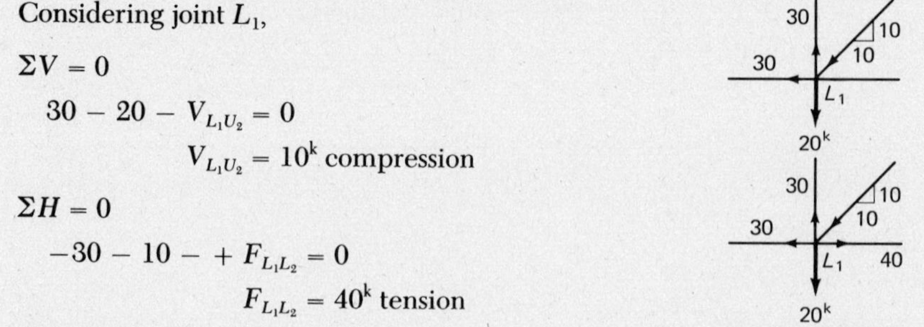

$$\Sigma V = 0$$
$$30 - 20 - V_{L_1U_2} = 0$$
$$V_{L_1U_2} = 10^k \text{ compression}$$

$$\Sigma H = 0$$
$$-30 - 10 - + F_{L_1L_2} = 0$$
$$F_{L_1L_2} = 40^k \text{ tension}$$

The forces in all of the truss members may be calculated in a similar manner with the following results:

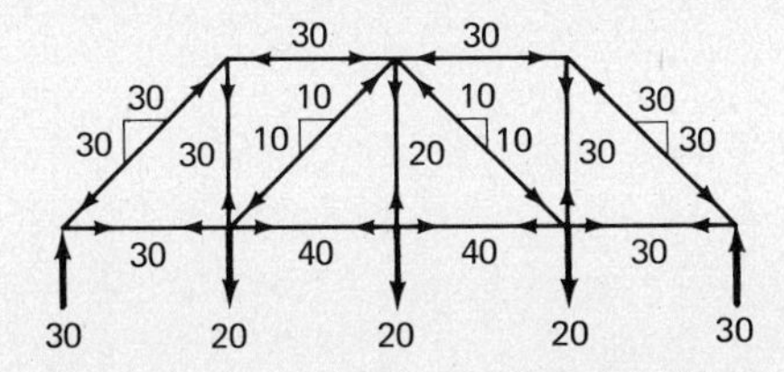

The resultant forces for inclined members may be determined from the square root of the sum of the squares of the vertical and horizontal components of force. An easier method is to write ratios comparing the resultant axial force of a member and its horizontal or vertical component with the true length of the member and its horizontal or vertical component. By letting F, H, and V represent the force and its components and l, h, and v the length and its components, the ratios of Fig. 6.2 are developed.

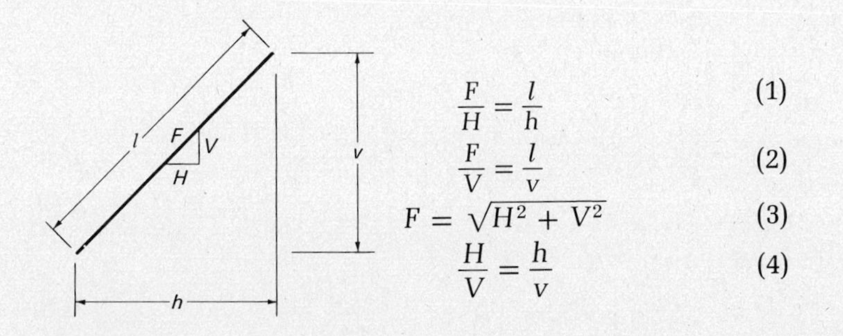

$$\frac{F}{H} = \frac{l}{h} \tag{1}$$

$$\frac{F}{V} = \frac{l}{v} \tag{2}$$

$$F = \sqrt{H^2 + V^2} \tag{3}$$

$$\frac{H}{V} = \frac{h}{v} \tag{4}$$

Figure 6.2

The method of joints may be used conveniently to compute the forces in all of the members of many trusses. The trusses of Examples 6.2 and 6.3 and the problems at the end of the chapter fall into this category. There are, however, a large number of trusses that need to be analyzed by a combination of the method of joints with the methods discussed in the following chapter. The author likes to calculate as many forces as possible in a truss by using the method of joints. At joints where he has a little difficulty, he takes moments, as described in the next chapter, and then continues his calculations as far as possible by joints until he reaches another point of difficulty, where he takes moments again, and so on.

Example 6.2

Find all forces in the truss of Fig. 6.3.

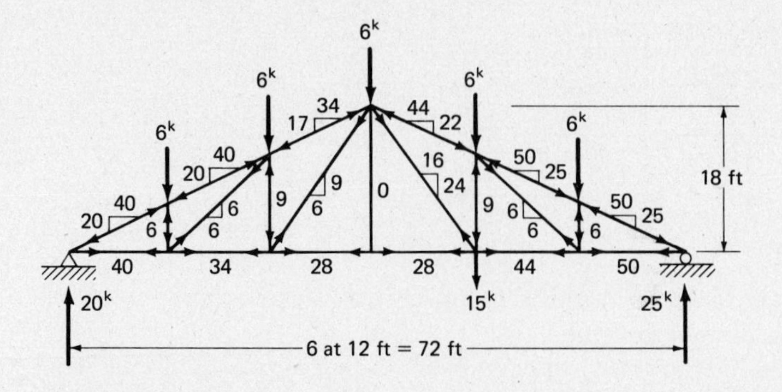

Figure 6.3

Example 6.3

Find all forces in the truss of Fig. 6.4.

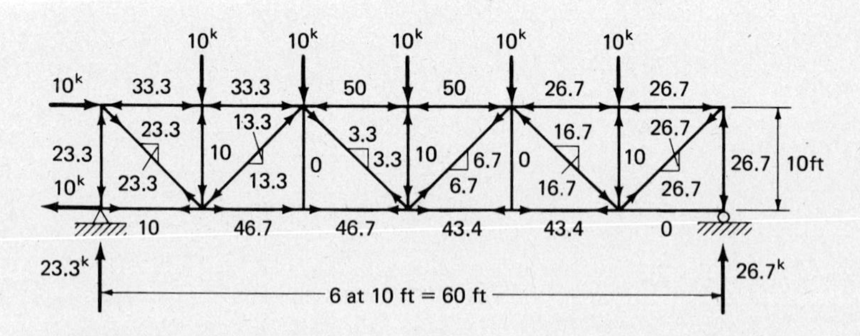

Figure 6.4

PROBLEMS

For Probs. 6.1 to 6.36 compute the forces in all the members of the trusses by using the method of joints.

Problem 6.1

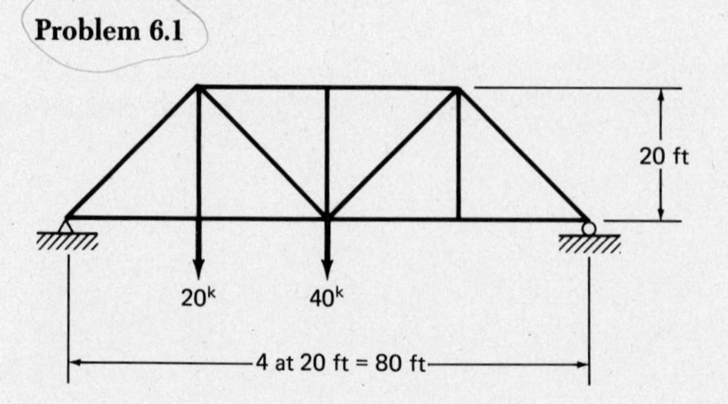

Problem 6.2 (*Ans.* $L_0L_1 = +85^k$, $U_1U_2 = -95^k$, $L_1L_2 = +55^k$)

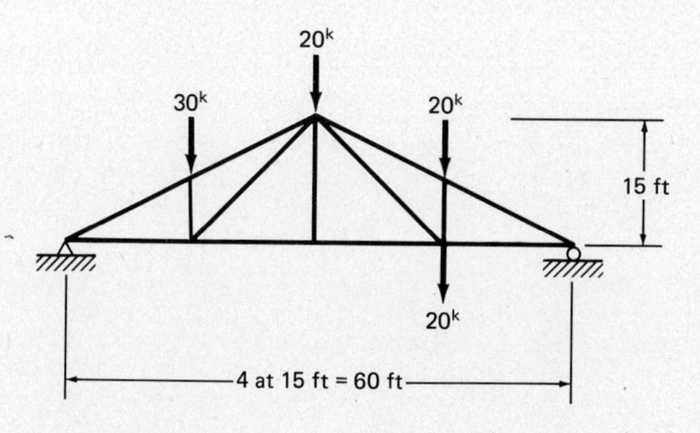

Problem 6.3

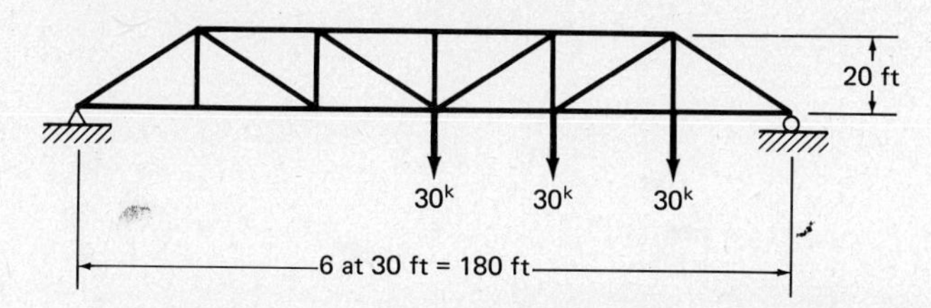

Problem 6.4 (*Ans.* $U_0U_1 = -82.5^k$, $L_1L_2 = +115^k$, $U_2L_3 = -38.9^k$)

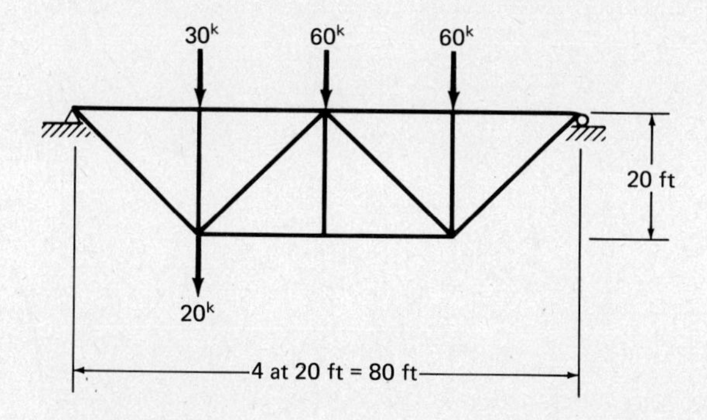

6.5 Rework Prob. 6.1 if the truss depth is reduced to 15 ft and the loads doubled.

6.6 Rework Prob. 6.2 if the 30-kip load is removed. (*Ans.* $L_1U_2 = 0$, $L_1L_2 = +40^k$, $U_2L_3 = +56.6^k$)

6.7 Rework Prob. 6.3 with the panels changed from 6 at 30 ft to 6 at 20 ft.

6.8 Rework Prob. 6.4 if the 20-kip load is removed and the panels are changed from 4 at 20 ft to 4 at 30 ft. (*Ans.* $L_0L_1 = +121.7^k$, $U_2U_3 = -123.8^k$, $U_2L_3 = -40.6^k$)

Problem 6.9

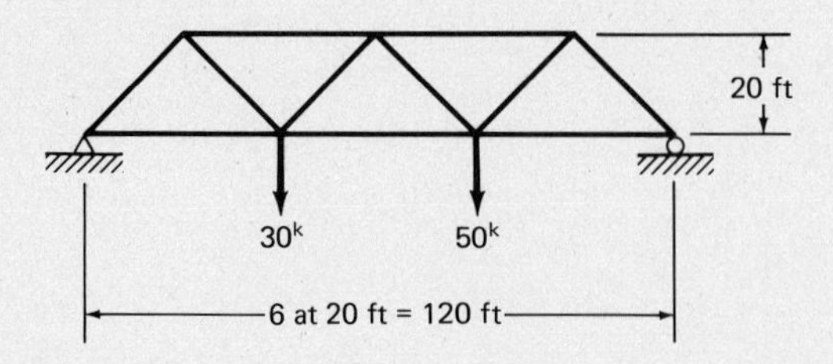

Problem 6.10 (*Ans.* $U_0L_0 = +11.25^k$, $L_0U_1 = -85.44^k$, $L_1L_2 = +110^k$)

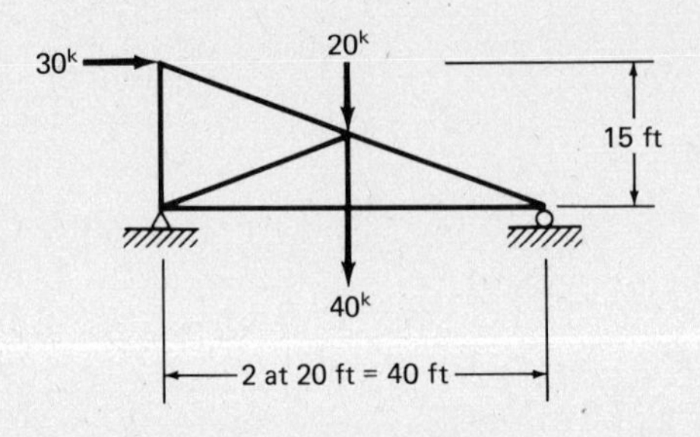

6.11 Repeat Prob. 6.10 if the roller support (due to friction, corrosion, etc.) is assumed to supply one-third of the total horizontal force resistance needed with the other two-thirds supplied by the pin support.

Problem 6.12 (*Ans.* $L_2U_3 = -9^k$, $U_3U_4 = -105^k$, $L_3L_4 = +127.5^k$)

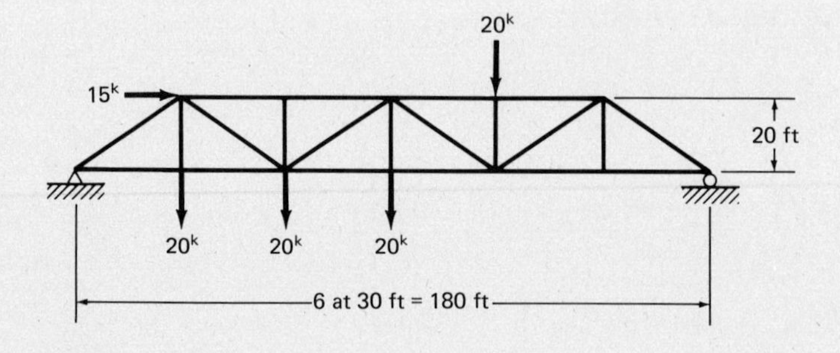

Problem 6.13

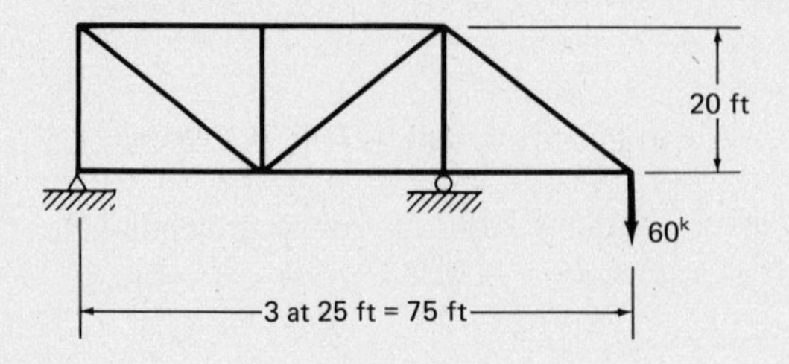

Problem 6.14 (*Ans.* $L_0L_1 = -12.5^k$, $U_1U_2 = +12.5^k$, $U_2L_2 = -81.7^k$)

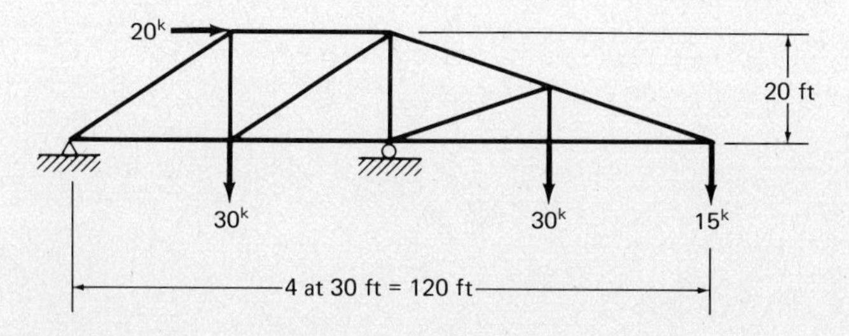

Problem 6.15

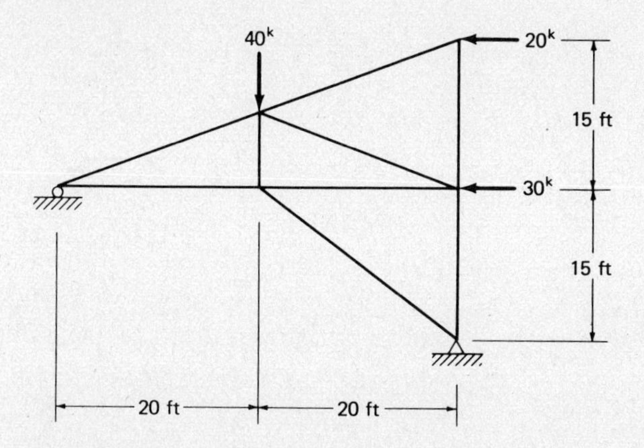

Problem 6.16 (*Ans.* $L_2L_4 = +40^k$, $U_2U_3 = -20^k$, $U_2L_4 = -28.3^k$)

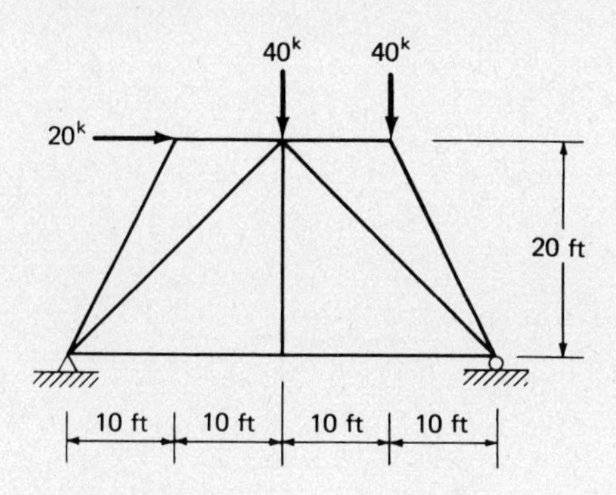

Problem 6.17

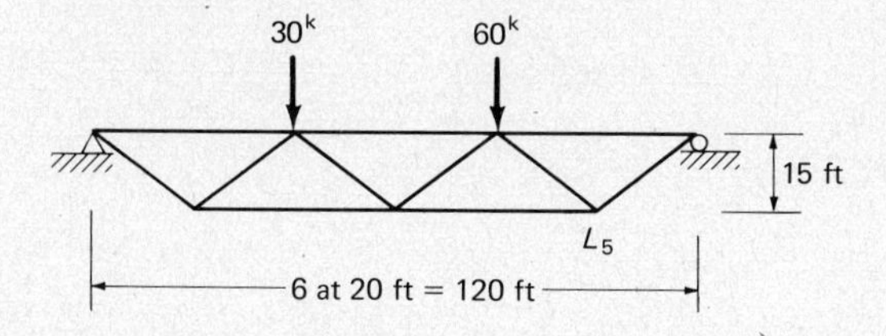

6.18 Repeat Prob. 6.17 if the roller support is placed beneath joint L_5. (*Ans.* $U_2U_4 = -80^k$, $U_2L_3 = 0$, $L_3L_5 = +80^k$)

Problem 6.19

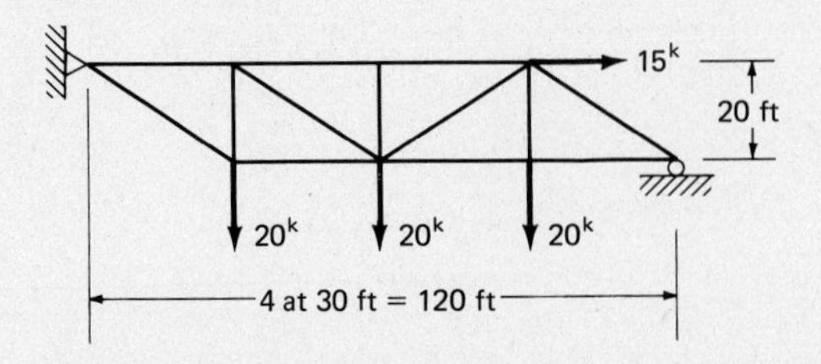

6.20 Rework Prob. 6.19 with the panels changed from 4 at 30 ft to 4 at 25 ft. (*Ans.* $U_1U_2 = -35^k$, $U_1L_2 = +16.0^k$, $L_2L_3 = +37.5^k$)

Problem 6.21

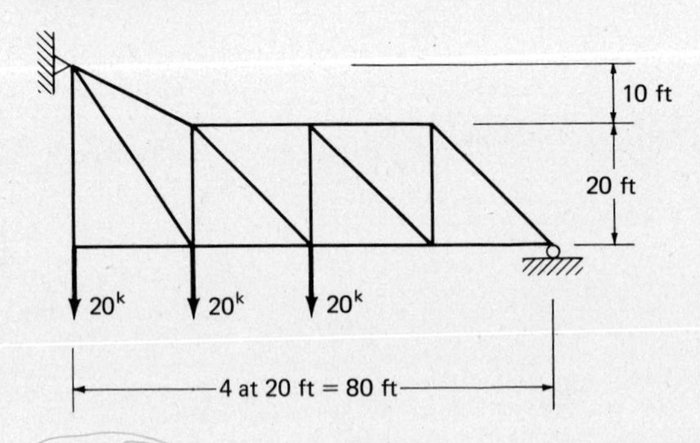

Problem 6.22 (*Ans.* $AB = 0$, $DF = -37.5^k$, $EG = +37.5^k$)

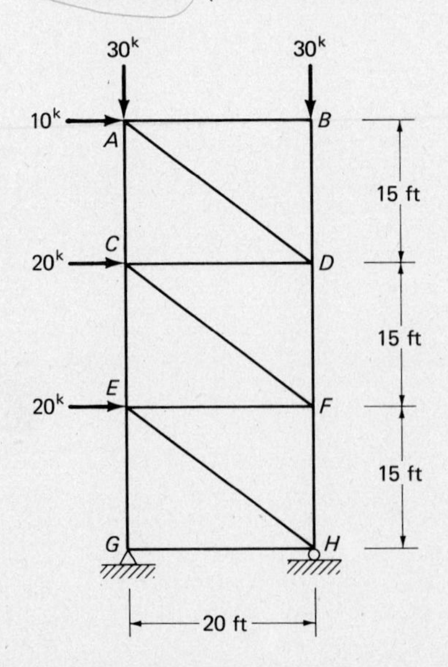

Problem 6.23

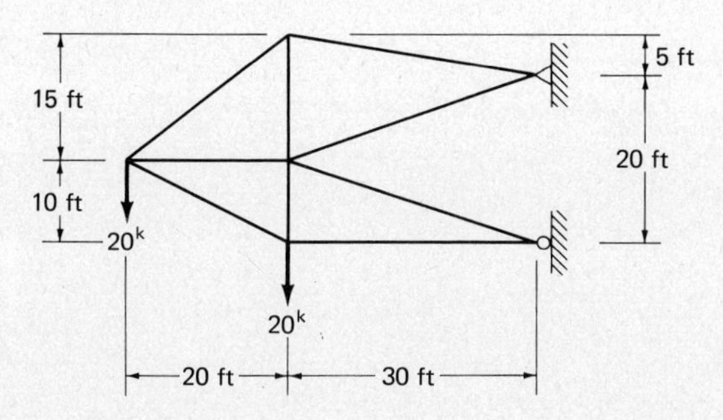

Problem 6.24 (*Ans.* $AB = +47.5^k$, $CD = -31.7^k$, $DF = -142.3^k$)

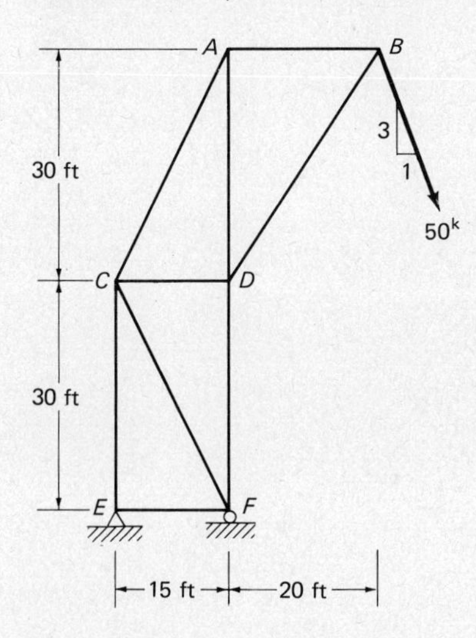

Problem 6.25

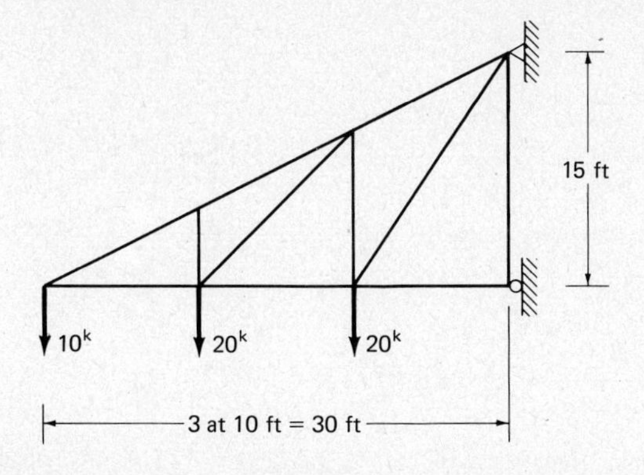

Problem 6.26 (*Ans.* $U_1U_2 = +65.6^k$, $L_0U_1 = -143^k$)

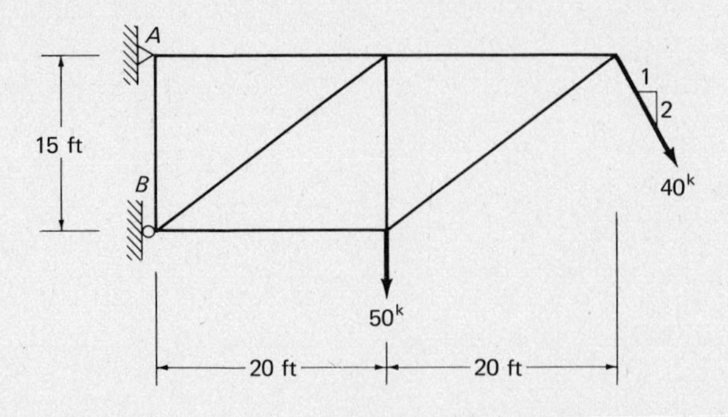

Problem 6.27

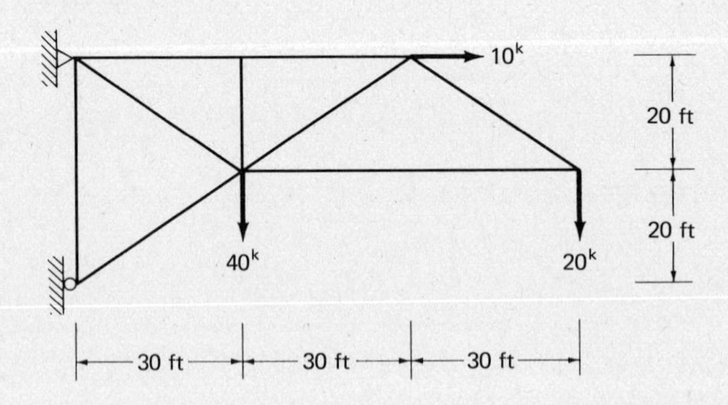

Problem 6.28 (*Ans.* $U_0L_0 = +35^k$, $U_1L_1 = +34^k$, $L_1L_2 = -13.33^k$)

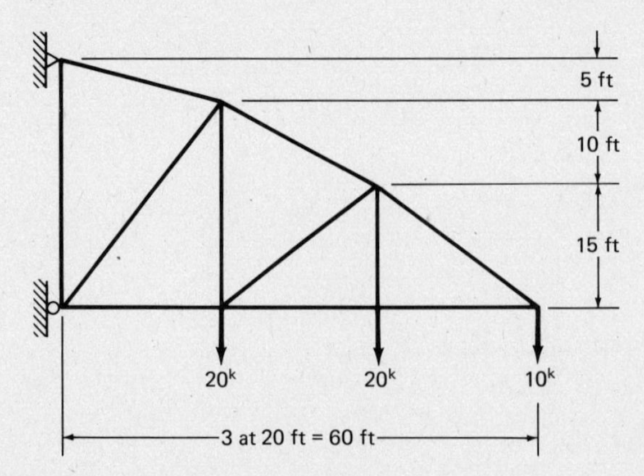

Problem 6.29

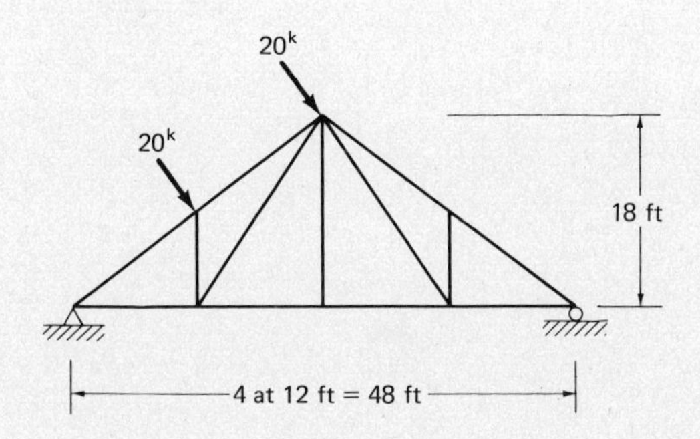

6.30. Rework Prob. 6.29 if the horizontal components of the external loads are assumed to be equally resisted by the pin and the roller due to friction, corrosion, and so on, at the latter. (*Ans.* $L_0U_1 = -22.1^k$, $L_1L_2 = +13^k$, $U_2U_3 = -31.25^k$)

6.31. Repeat Prob. 6.17 if the supporting surface beneath the roller is changed to a 45° slope (slope 1 : 1).

6.32. Rework Prob. 6.16 if the supporting surface beneath the roller is changed as follows (slope 3 : 4). (*Ans.* $L_0U_1 \doteq 0$, $L_0U_2 = -28.3^k$, $L_2L_4 = -5^k$)

Problem 6.33

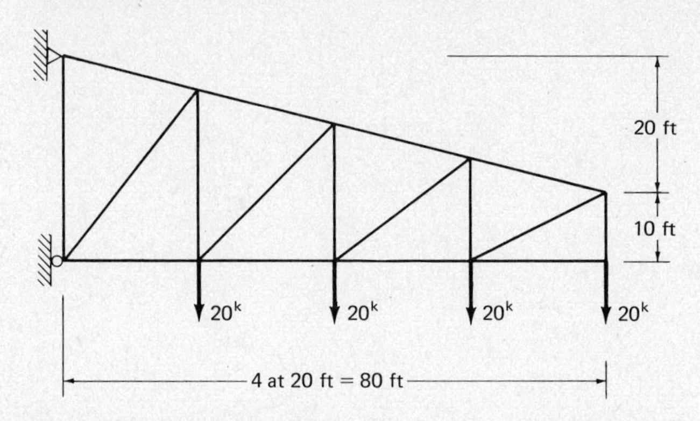

Problem 6.34 (*Ans.* $U_0U_1 = -45^k$, $U_1L_1 = -90^k$, $U_2L_3 = +63.6^k$)

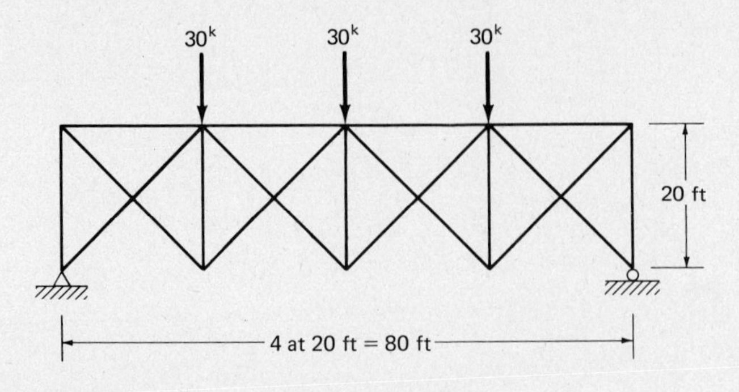

Problem 6.35

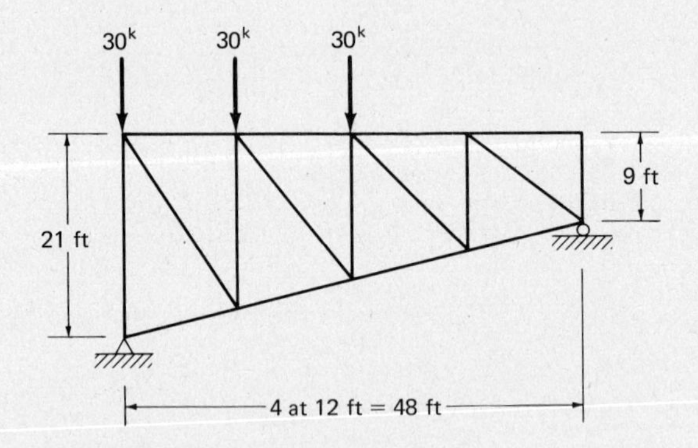

Problem 6.36 (*Ans.* $U_1L_1 = -54.3^k$, $U_2U_3 = -100.1^k$, $L_2L_3 = +69.7^k$)

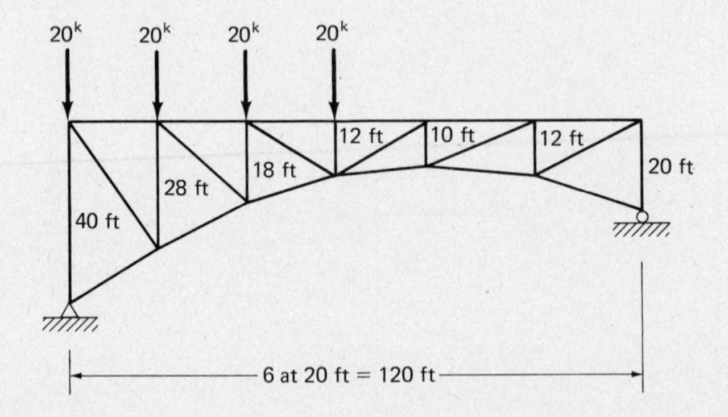

7 Truss Analysis by Moments and Shears

7.1 METHOD OF MOMENTS

The equilibrium of free bodies is the basis of force computation by the method of moments as it is by the method of joints. To obtain the value of the force in a particular member, an imaginary section is passed completely through the truss to divide it into two free bodies. A section is placed to cut the member whose force is desired and as few other members as possible.

The moment of all the forces applied to the free body under consideration about any point in the plane of the truss is zero. If it is possible to take moments of the forces about a point so that only one unknown force appears in the equation, the value of the force can be obtained. This objective can usually be attained by selecting a point along the line of action of one or more of the forces of the other members. Some familiar trusses have special locations for placing sections that greatly simplify the work involved. These cases will be discussed in the chapters to follow.

One advantage of the method of moments is that if the force in only one member of a truss is desired and the member is not near the end of the truss, it may be obtained directly in most cases without first determining the forces in other members. If the method of joints were used, it would be necessary to calculate the forces in the members joint by joint from the end of the truss until the member in question was reached.

7.2 FORCES IN MEMBERS CUT BY SECTIONS

Should tension and compression members actually be cut, the results would be as described in the following paragraphs and as pictured in Fig. 7.1

1. A tension member is being stretched, and should it be cut would tend to resume its original length, leaving a gap at the section. A tension member pulls away from the free body in Fig. 7.1(a).

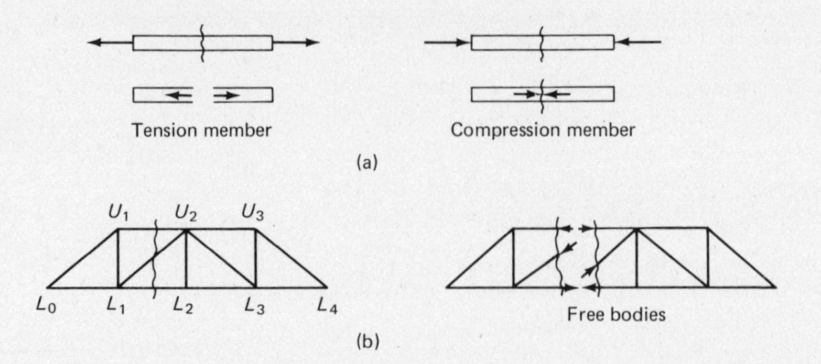

Figure 7.1

2. A compression member has been shortened, and should it be cut in half would tend to resume its original length, that is, try to expand. A compression member pushes against the free body in Fig. 7.1(a).

A truss is divided into two free bodies in Fig. 7.1(b). Members U_1U_2 and L_1U_2 are assumed to be in compression, and member L_1L_2 is assumed to be in tension. On the basis of these assumptions the directions of the forces on the two free bodies are shown.

Final truss slipped into place for Newport Bridge linking Jamestown and Newport, Rhode Island. (Courtesy of the Bethlehem Steel Corporation.)

7.3 APPLICATION OF THE METHOD OF MOMENTS

Examples 7.1 to 7.5 illustrate in detail the computation of forces with the $\Sigma M = 0$ equation. In writing the moment equation it is to be noted that the unknown force may be assumed to be tension or compression. If the mathematical solution yields a negative number, the character of force is the opposite of that which was assumed. The numerical answer is correct regardless of the sign.

It is probably simpler always to assume the unknown force to be in tension, that is, pulling away from the free body. If the solution yields a positive number, the force is tensile; if a negative number, the force is compressive. Therefore the sign always agrees with the normal sign convention of + for tension and − for compression. This practice is followed in the illustrative problems throughout this book.

Example 7.1

Find the forces in members L_1L_2 and U_2U_3 in Fig. 7.2 by moments.

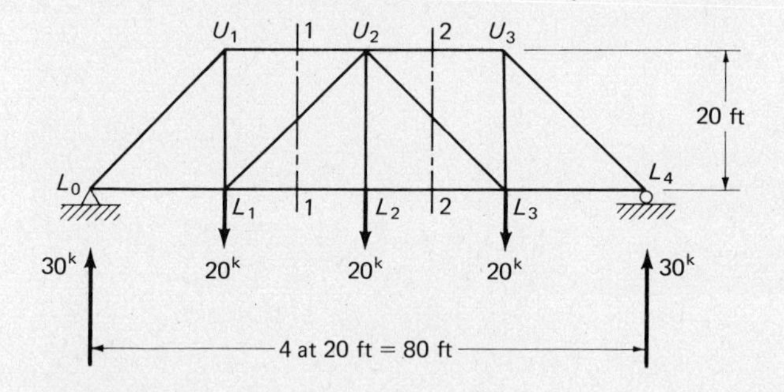

Figure 7.2

SOLUTION

Member L_1L_2. Section 1–1 is passed through the truss, and the part of the truss to the left of the section is considered to be the free body. The forces acting on the free body are the 30-kip reaction, the 20-kip load at L_1, and the axial forces in the members cut by the section (U_1U_2, L_1U_2, and L_1L_2). Moments of these forces are taken about U_2, which is the point of intersection of L_1U_2 and U_1U_2. The moment equation contains one unknown force, L_1L_2, and its value may be found by solving the equation.

$$\Sigma M_{U_2} = 0$$

$$(30)(40) - (20)(20) - 20F_{L_1L_2} = 0$$

$$F_{L_1L_2} = +40^k \text{ tension}$$

Timber trusses for tannery, South Paris, Maine. (Courtesy of the American Wood Preservers Institute.)

Member U_2U_3. Section 2–2 is passed through the truss, and the portion of the truss to the right of the section is considered to be the free body. Members U_2U_3, U_2L_3, and L_2L_3 are cut by the section. Taking moments at the intersection of L_2L_3 and U_2L_3 at L_3 eliminates those two members from the equation because the lines of action of their forces pass through the center of moments. The force in U_2U_3 is the only unknown appearing in the equation, and its value may be determined.

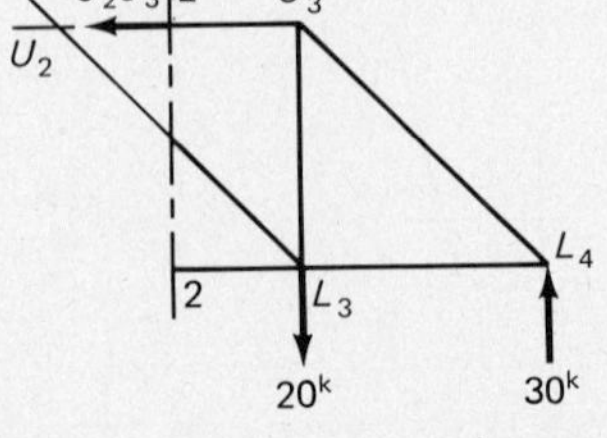

$$\Sigma M_{L_3} = 0$$

$$-(30)(20) - 20F_{U_2U_3} = 0$$

$$F_{U_2U_3} = -30^{k} \text{ compression}$$

Example 7.2

Find the forces in all of the members of the truss shown in Fig. 7.3. Use both the method of joints and the method of moments as convenient.

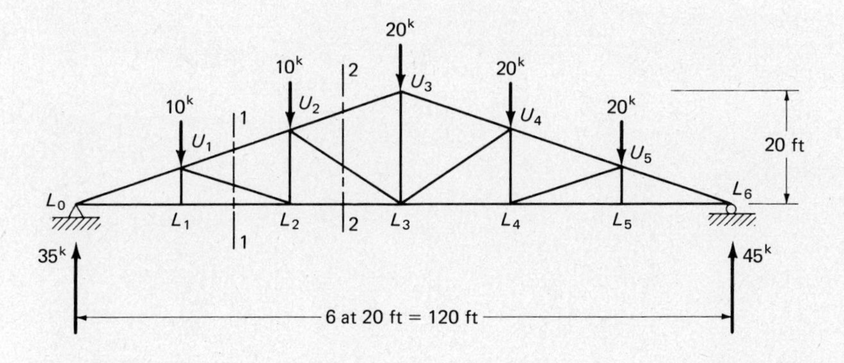

Figure 7.3

SOLUTION

The forces of members meeting at L_0 and L_1 are quickly determined by the method of joints. To calculate the force in U_1U_2, section 1–1 is passed and moments are taken at L_2. Because U_1U_2 is an inclined member, the force is resolved into its vertical and horizontal components. The components of a force may be assumed to act anywhere along its line of action. It is convenient in this case to break the force down into its components at joint U_2, because the vertical component will pass through the center of moments and the moment equation may be solved for the horizontal component of force.

$$\Sigma M_{L2} = 0$$

$$(35)(40) - (10)(20) + 13.33 H_{U_1U_2} = 0$$

$$H_{U_1U_2} = -90^k \text{ compression}$$

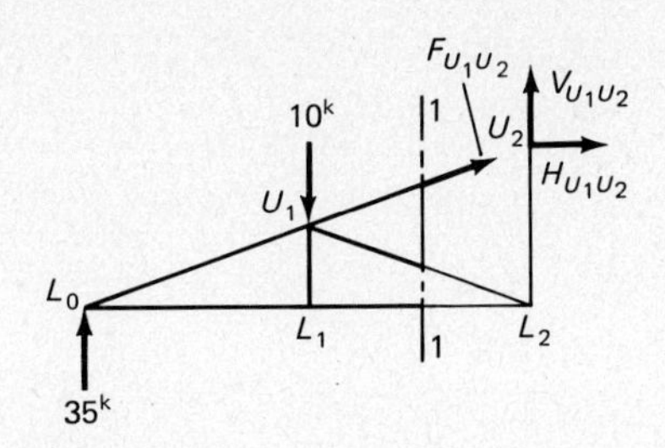

By joints, the unknown forces in members meeting at U_1 and L_2 may now be obtained. Section 2–2 is passed through the truss, and moments are taken at L_3 to find the force in U_2U_3. By knowing this force, the remaining forces in the truss can be found by joints with the results shown.

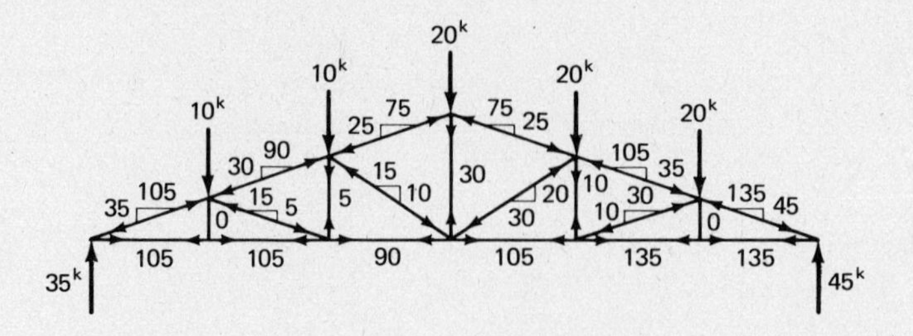

Example 7.3

Determine the forces in all of the members of the truss shown in Fig. 7.4.

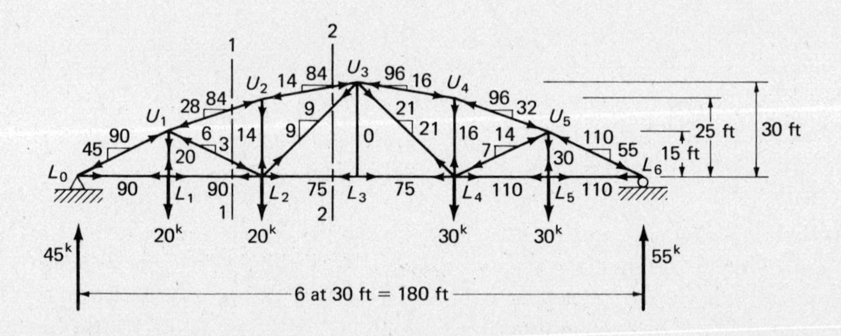

Figure 7.4

SOLUTION

$\Sigma M_{L_2} = 0$; free body to left of section 1–1

$$(45)(60) - (20)(30) + 25H_{U_1U_2} = 0$$

$$H_{U_1U_2} = -84^k \text{ compression}$$

$\Sigma M_{U_3} = 0$; free body to left of section 2–2

$$(45)(90) - (20)(30) - (20)(60) - 30F_{L_2L_3} = 0$$

$$F_{L_2L_3} = +75^k \text{ tension}$$

Example 7.4

Determine the forces in all of the members of the Fink truss shown in Fig. 7.5.

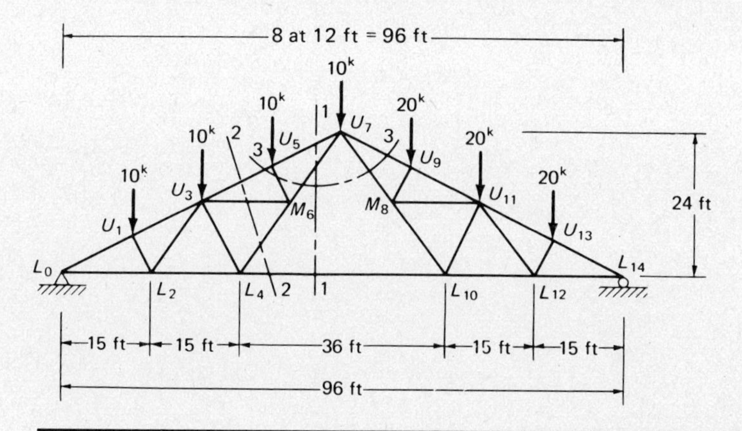

Figure 7.5

SOLUTION

The forces in members meeting at joints L_0, L_2, and U_1 can be found by the methods of joints and moments without any difficulty. At each of the next two joints, U_3 and L_4, there are three unknown forces that cannot be determined directly with sections. It is convenient to compute the forces in some members further over in the truss and then work back to these joints. The sections numbered 1–1, 2–2, and 3–3 may be used to advantage. From the first of these sections the forces in any of the three members cut may be obtained by moments. By using section 2–2 and taking moments at U_3, the force in L_4M_6 may be found. It is important to note that four members have been cut by the section and only two of them pass through the point where moments are being

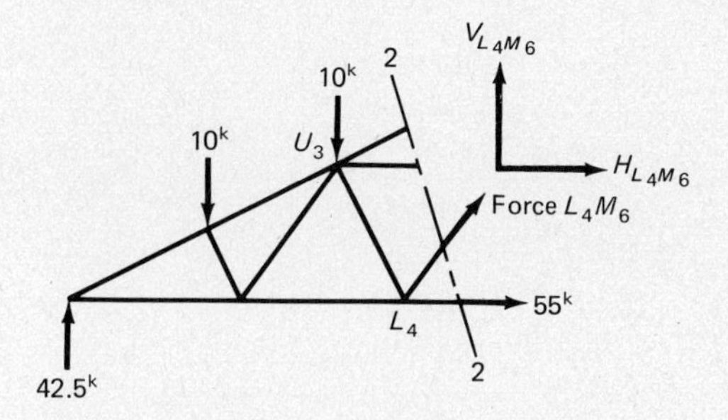

taken; however, the force in one of these members, L_4L_{10}, was previously found with section 1–1, and only one unknown is left in the equation. The remaining forces in the truss may be calculated by the usual methods. These two sections are sufficient for analyzing the truss, but should another approach be desired, a section such as 3–3 may be considered. From this section the force in U_5M_6 can be found by taking moments at U_7 because all of the other members cut by the section pass through U_7. The forces in all of the truss members are as shown.

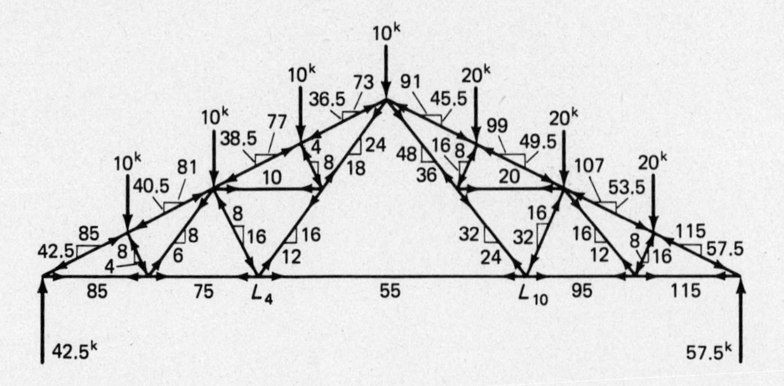

Example 7.5

Calculate the force in member *cg* of the truss of Fig. 7.6.

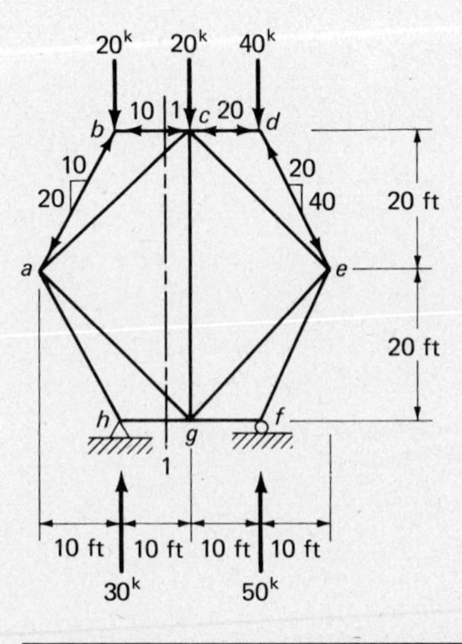

Figure 7.6

SOLUTION

The force in the member in question cannot be determined immediately by joints or moments. It is necessary to know the force values for several other members before the value for *cg* can be found. The forces in members *ba*, *bc*, *dc*, and *de* may be found by joints as shown, and the force in member *ac* can be found by moments. Considering section 1–1 and the free body to the left, moments may be taken about *g*. By noting that the force in *bc* is in compression and pushes against the free body from the outside and by assuming member *ac* to be in tension, the following equation may be written. The force is broken into its vertical and horizontal components at *c*.

$$\Sigma M_g = 0$$

$$(30)(10) - (20)(10) - (10)(40) + (H_{ac})(40) = 0$$

$$H_{ac} = +7.5^{k} \text{ tension}$$

By having the force in *ac*, the forces in *ce* and *cg* can be determined by joints as shown.

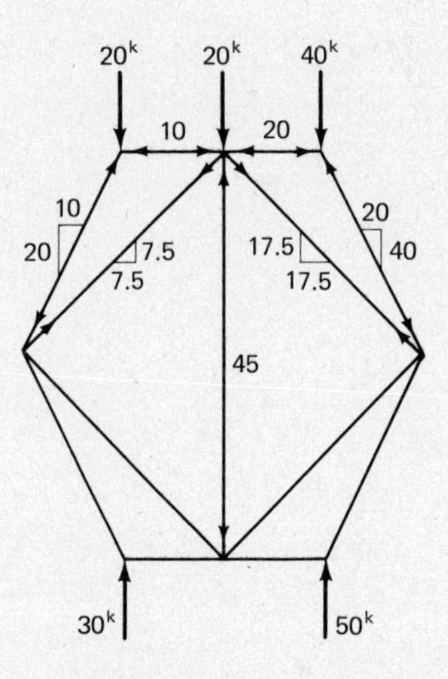

7.4 METHOD OF SHEARS

It should be obvious by this time that if a vertical section is passed through a truss and divides it into two separate free bodies, the sum of the vertical forces to the left of the section must be equal and opposite in direction to the sum of the vertical forces to the right of the section. The summation of these forces to the left or to the right of a section has been defined as the *shear*.

The inclined members cut by a section must have vertical components of force equal and opposite to the shear along the section, because the horizontal members can have no vertical components of force. For most parallel-chord trusses there is only one inclined member in each panel, and the vertical component of force in that inclined member must be equal and opposite to the shear in the panel. The vertical components of force are computed by shears for the diagonals of the parallel-chorded truss of Example 7.6.

Example 7.6

Determine the vertical components of force in the diagonals of the truss shown in Fig. 7.7. Use the method of shears.

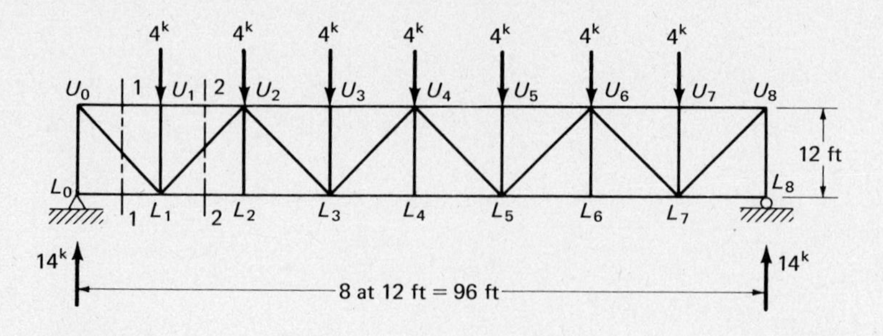

Figure 7.7

SOLUTION

By considering section 1–1 and free body to the left,

shear to the left = $14^k \uparrow$

$V_{U_0L_1} = 14^k \downarrow$ tension (pulling away from free body)

By considering section 2–2 and free body to the left,

shear to left = $10^k \uparrow$

$V_{L_1U_2} = 10^k \downarrow$ compression (pushing against free body)

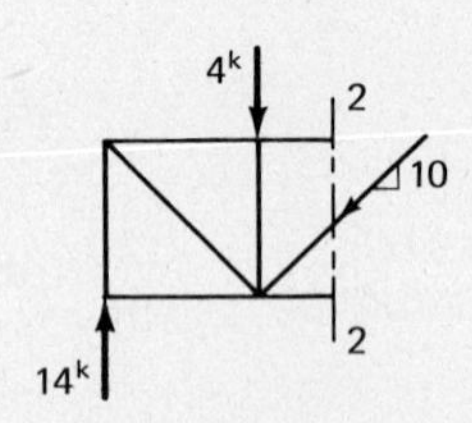

The vertical components of force in all the diagonals are as follows:

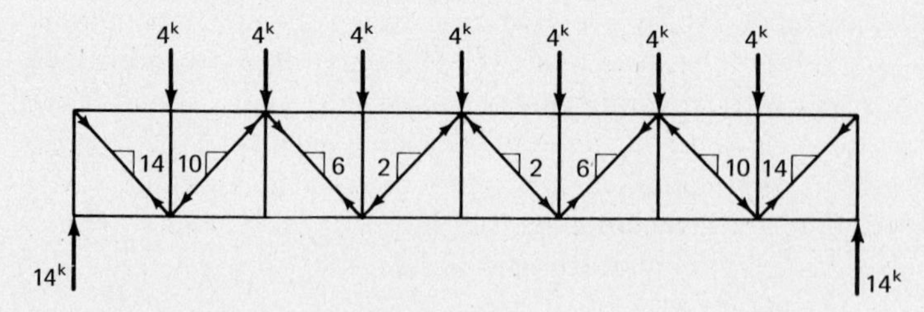

Nonparallel-chord trusses have two or more diagonals in each panel, and they all may have vertical components of force; however, their sum must be equal and opposite to the shear in the panel. If all but one of the diagonal forces in a panel are known, the remaining one may be determined by shears, as illustrated in Example 7.7.

Example 7.7

Referring to the truss of Example 7.3, sections 1–1 and 2–2, and assuming that the forces in the chords U_1U_2 and U_2U_3 are known, find the vertical components of force in U_1L_2 and L_2U_3 by the method of shears.

SOLUTION

By considering section 1–1 and free body to the left,

$$\text{shear to left} = 25^k \uparrow$$
$$V_{U_1U_2} = 28^k \downarrow$$
$$V_{U_1}L_2 = 3^k \uparrow \text{ compression}$$

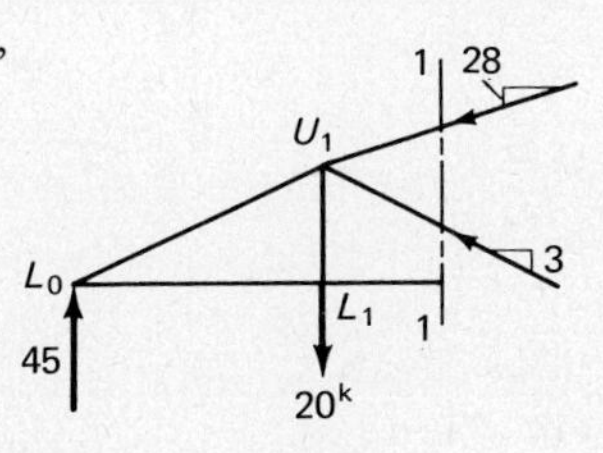

By considering section 2–2 and free body to the right,

$$\text{shear to right} = 5^k \downarrow$$
$$V_{U_2U_3} = 14^k \uparrow$$
$$V_{L_2U_3} = 9^k \downarrow \text{ tension}$$

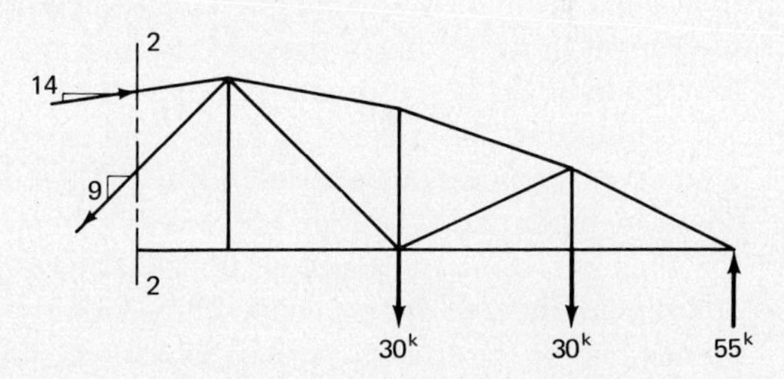

7.5 WHEN ASSUMPTIONS ARE NOT CORRECT

The designer should realize that often his or her assumptions regarding the behavior of a structure (pinned joints, loads applied at joints only, frictionless rollers, or whatever) may not be entirely valid. As a result he or she should give consideration to what might happen to a structure if the assumptions made in analysis were appreciably in error. Perhaps an expansion device or roller will resist (due to friction) a large proportion of any horizontal forces present. How would this affect the member forces of a particular truss? For this very reason the truss of Prob. 6.30 was included at the end of Chap. 6 where it was assumed that half of the horizontal load was resisted by the roller.

The actual types of end supports can have an appreciable effect on the magnitude of forces in truss members caused by lateral loads.

For fairly short roof trusses generally no provisions are made for temperature expansion and contraction and both ends of the trusses are bolted down to their supports. These trusses are actually statically indeterminate, but the usual practice is to assume that the horizontal loads split equally between the supports.

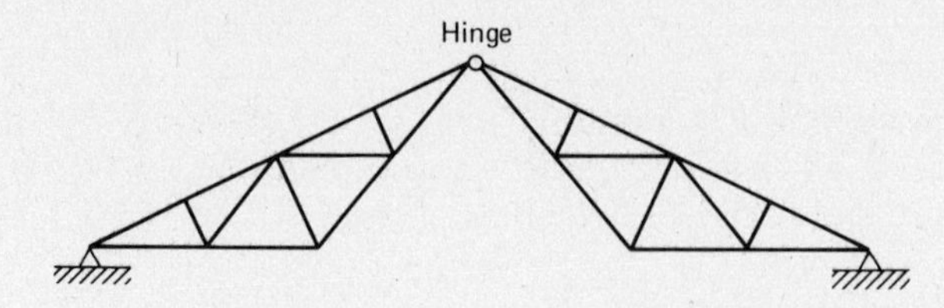

Figure 7.8

For longer roof trusses provisions for expansion and contraction are considered necessary. Usually the bolts at one end are set in a slotted hole so as to provide space for the anticipated length changes. A base plate is provided at this end on which the truss can slide.

Actually it is impossible to provide a support that has no friction. From a practical standpoint the maximum value of the horizontal reaction at the expansion end equals the vertical reaction times the coefficient of friction (one-third being a reasonable guess). Should corrosion occur, thus preventing movement (a rather likely prospect), a half-and-half split of the lateral loads may again be the best estimate.

The author once read of an interesting case where the owners of a building with a roof supported by a series of Fink trusses decided that the middle lower chord member (L_4L_{10} in Fig. 7.5) was in their way. They therefore removed the member from quite a few of the trusses and much to the designer's amazement the roof did not collapse. Apparently the roller or expansion device on one end of the trusses (perhaps bolts in a slot) permitted very little or no movement. As a result each of the trusses apparently behaved as a three-hinged arch as shown in Fig. 7.8. In many situations where the assumptions do not prove to be correct the results are more unpleasant than they were for these Fink trusses.

PROBLEMS

For Probs. 7.1 to 7.47 determine the forces in all members of the trusses.

Problem 7.1

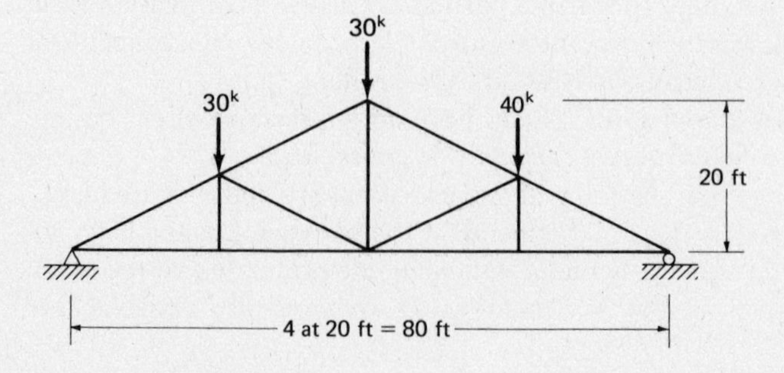

Problem 7.2 (*Ans.* $U_1U_2 = -94.9^k$, $U_3L_3 = +20^k$, $L_3L_4 = +60^k$)

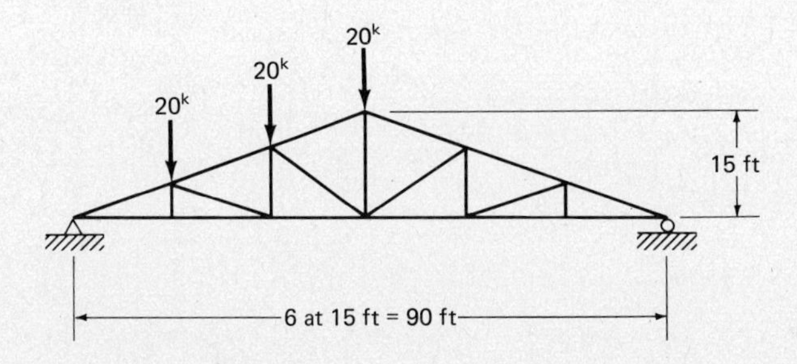

7.3. Rework Prob. 7.1 if a uniform load of 2 klf. is applied for the entire span in addition to the loads shown. This additional load is to be applied to the bottom chord joints.

7.4. Rework Prob. 7.2 if the panels are changed from 6 at 15 ft to 6 at 12 ft. (*Ans.* $L_2L_3 = +72^k$, $U_3L_3 = +20^k$, $U_4U_5 = -52^k$)

Problem 7.5

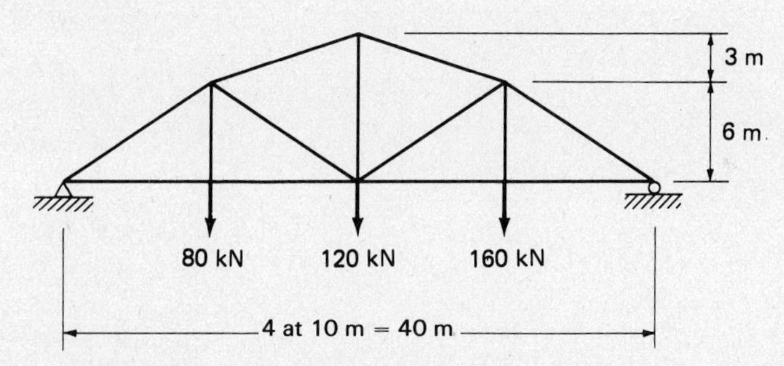

Problem 7.6 (*Ans.* $U_2L_2 = +72^k$, $L_3L_4 = +144^k$, $U_4U_5 = -152^k$)

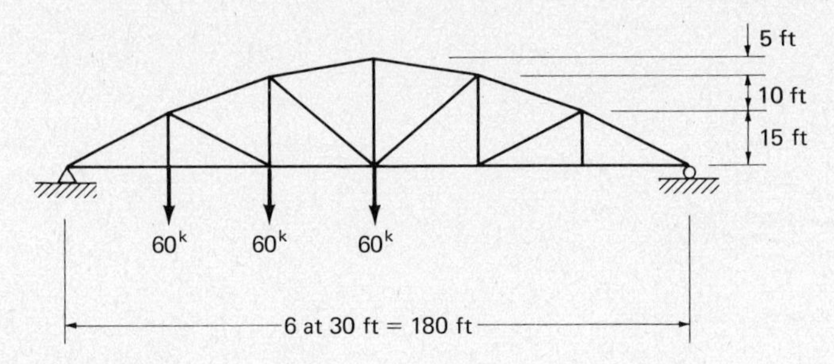

Problem 7.7

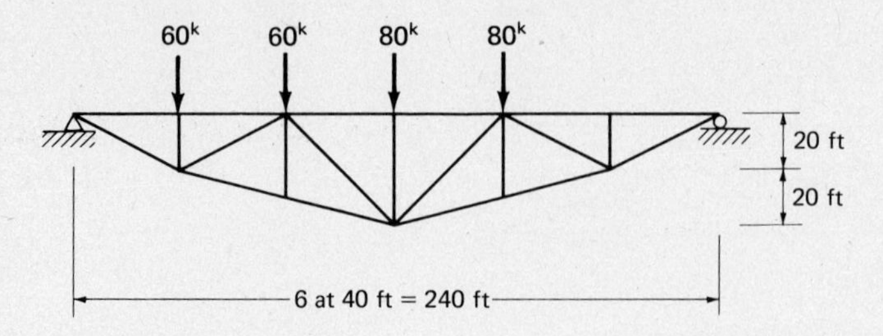

7.8. Repeat Prob. 7.7 if a 20-kip load is added at each of the lower joints. (*Ans.* $L_1L_2 = +458^k$, $U_2U_3 = -380^k$, $L_5L_6 = +388^k$)

Problem 7.9

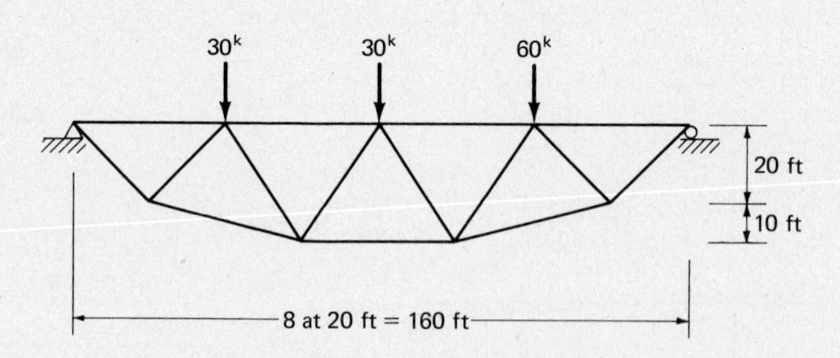

Problem 7.10 (*Ans.* $AB = -3.5^k$, $BF = -54.1^k$, $CF = -37.5^k$)

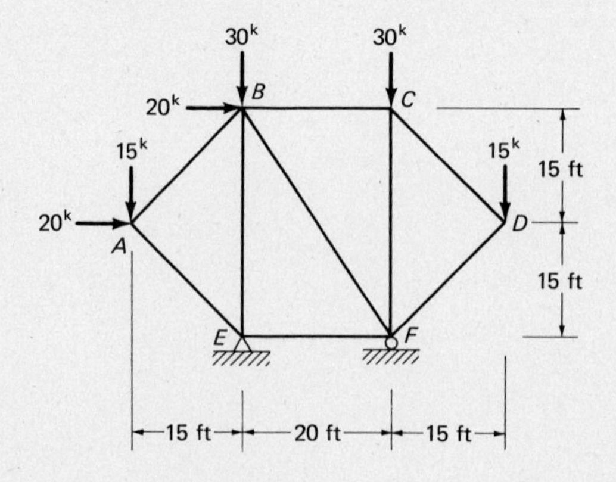

Problem 7.11

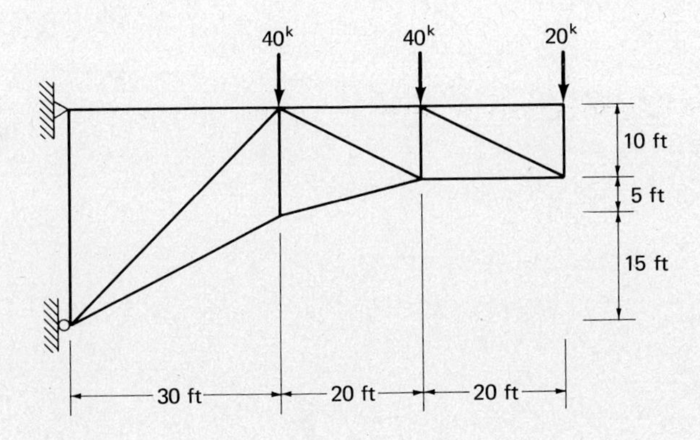

Problem 7.12 (*Ans.* $L_0L_1 = +17.9^k$, $U_1L_2 = -62.6^k$, $U_2L_3 = +44.7^k$)

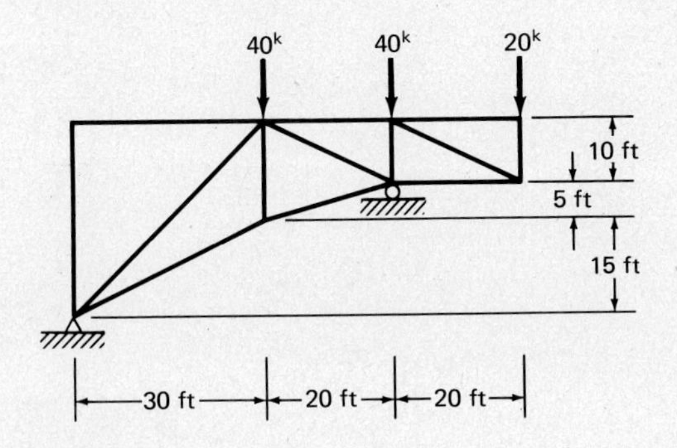

Problem 7.13

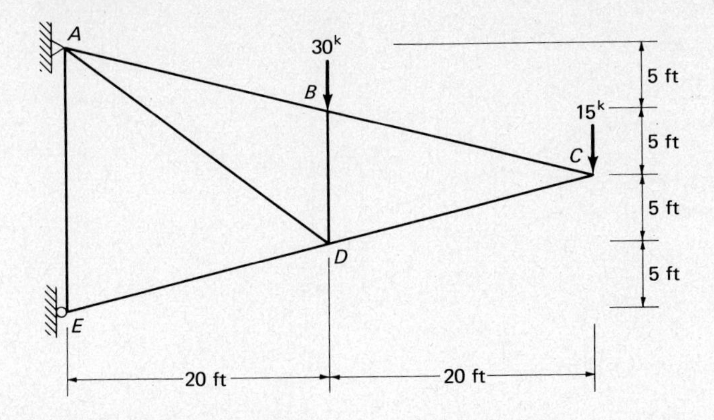

Problem 7.14 (*Ans.* $L_2L_3 = -70^k$, $U_4U_5 = +21.3^k$, $U_6L_6 = -102.5^k$)

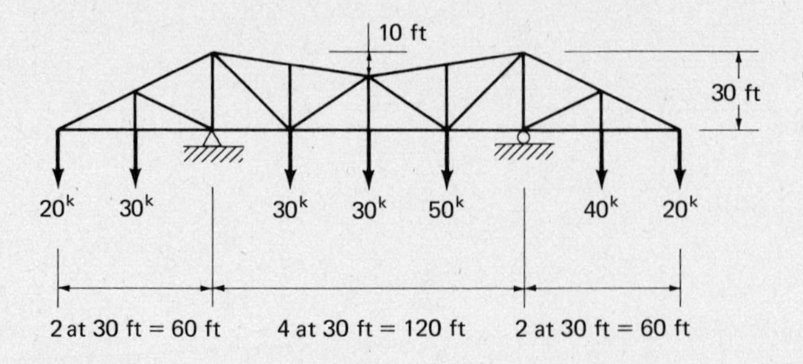

Problem 7.15

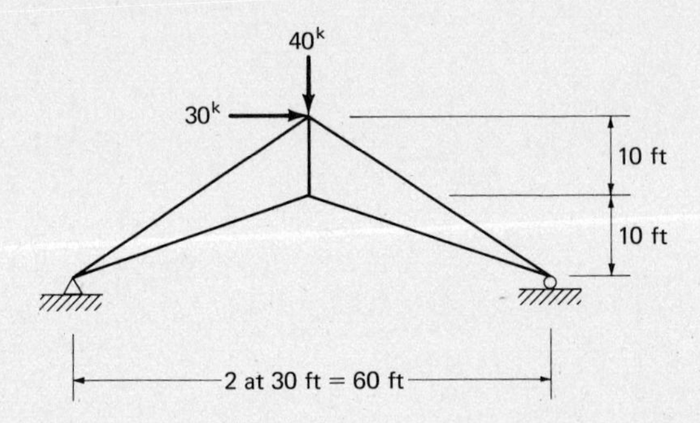

Problem 7.16 (*Ans.* $U_0L_1 = +36.9^k$, $U_1U_2 = -15.8^k$, $U_2L_2 = +3.3^k$)

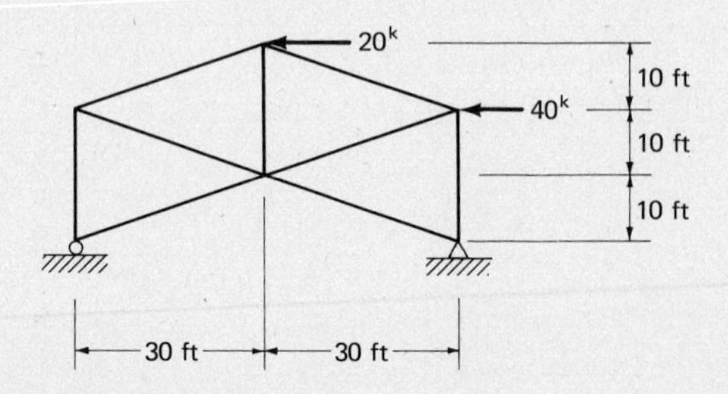

Problem 7.17

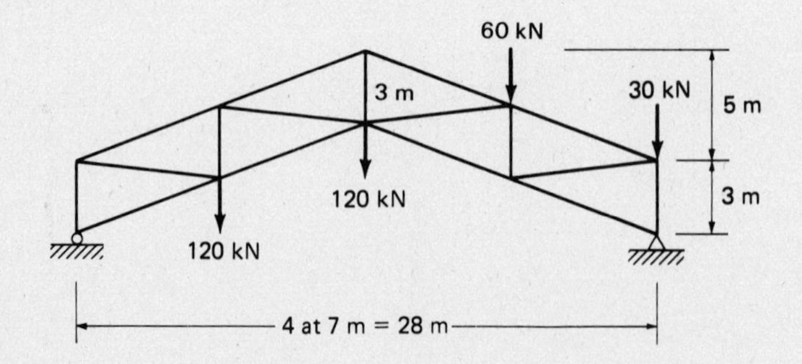

Problem 7.18 (*Ans.* $U_0L_1 = +74.8^k$, $U_2L_2 = +41.6^k$, $U_3U_4 = -102^k$)

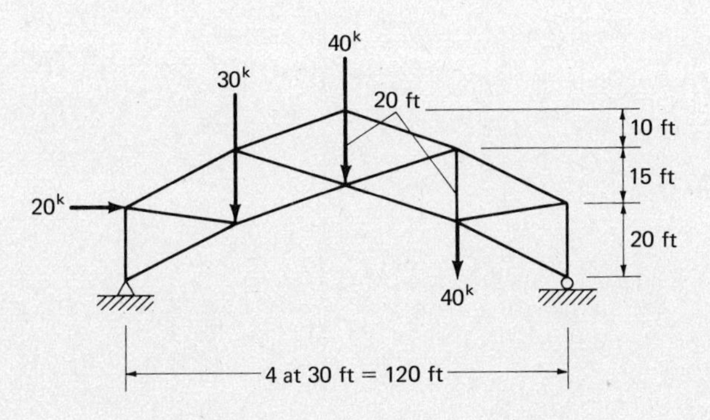

Problem 7.19

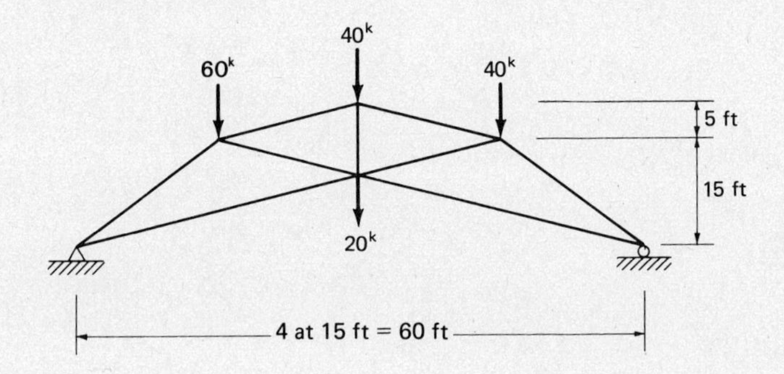

Problem 7.20 (*Ans.* $U_1L_2 = -20^k$, $L_2U_3 = +145.3^k$, $L_4L_6 = +144.2^k$)

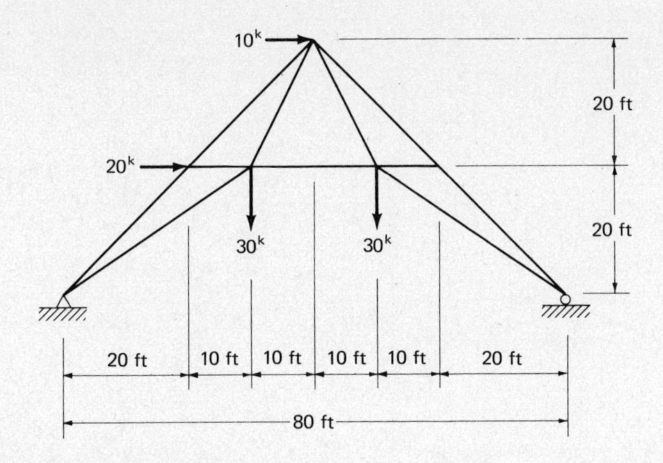

Problem 7.21

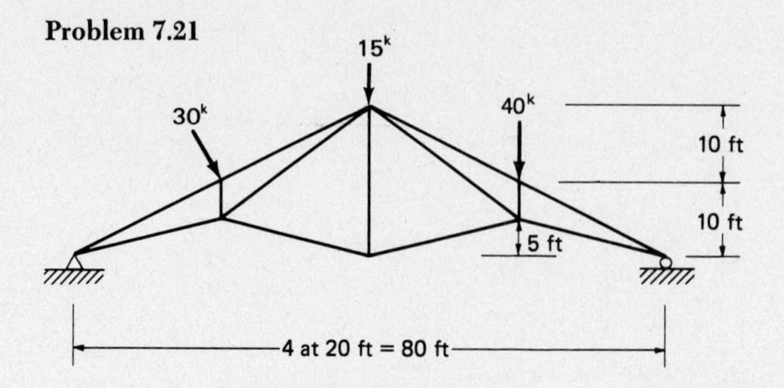

Problem 7.22 (*Ans.* $L_0L_2 = +40$ kN, $U_1L_2 = -16.0$ kN, $L_2U_4 = +72.1$ kN)

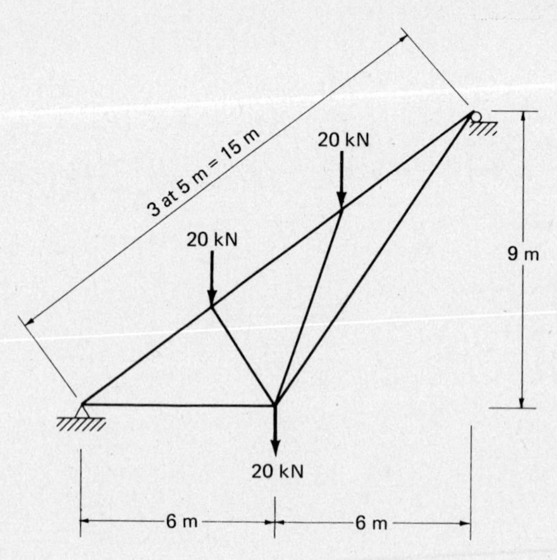

Problem 7.23

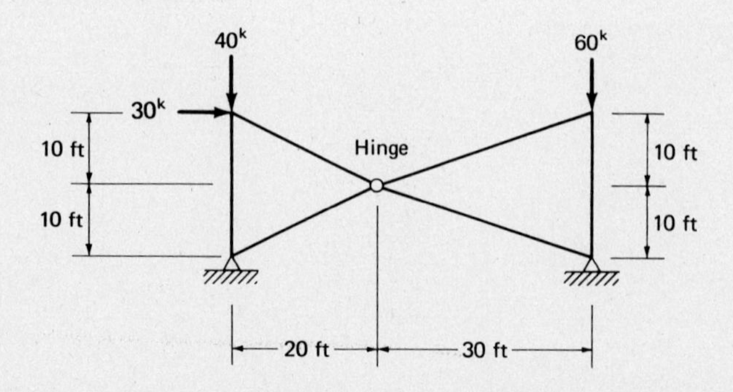

Problem 7.24 (*Ans.* $U_0U_1 = -30.4^k$, $U_2L_3 = +41.2^k$, $U_3U_4 = -63.2^k$)

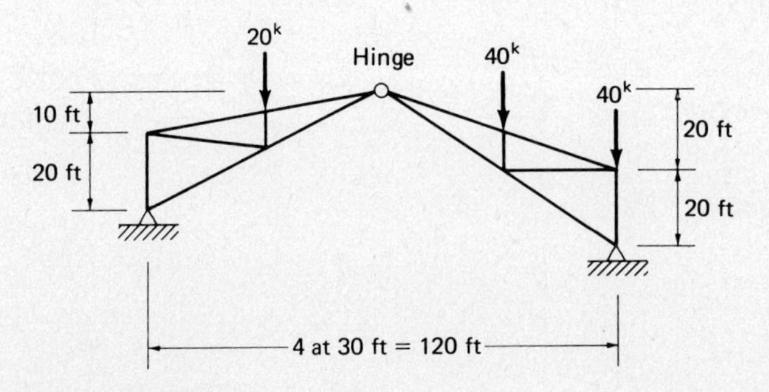

Problem 7.25

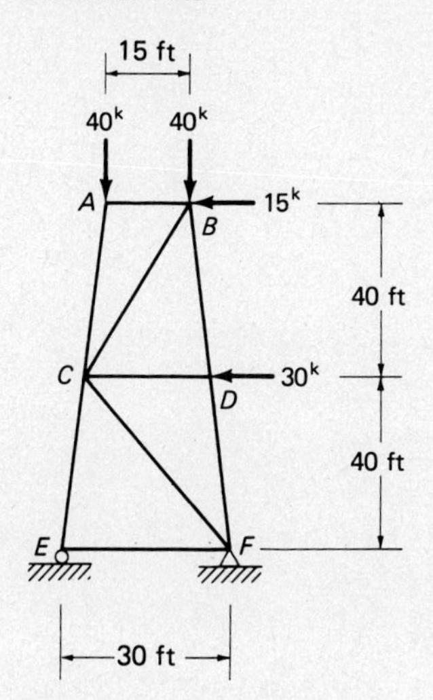

Problem 7.26 (*Ans.* $AB = +185.2^k$, $BC = -132.9^k$, $DE = -14.2^k$)

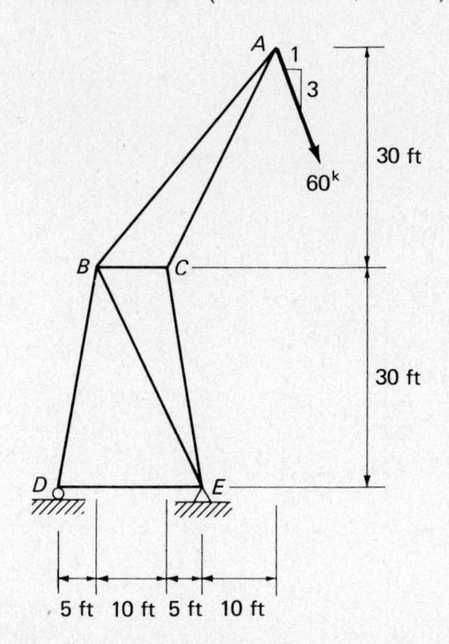

Problem 7.27

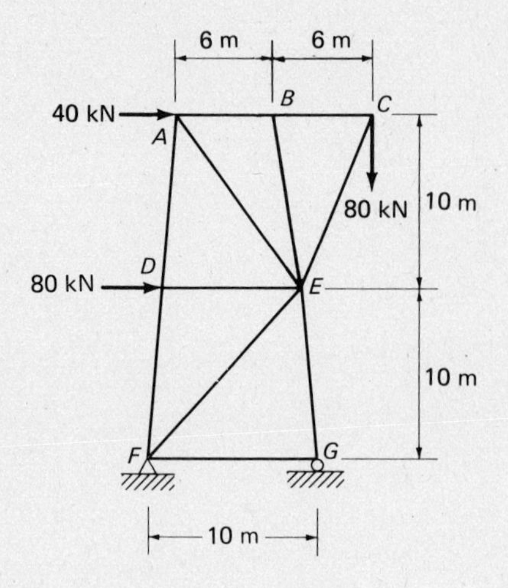

Problem 7.28 (*Ans.* $U_1L_2 = -23.6^k$, $U_2L_2 = +10^k$, $L_2L_3 = +144.3^k$)

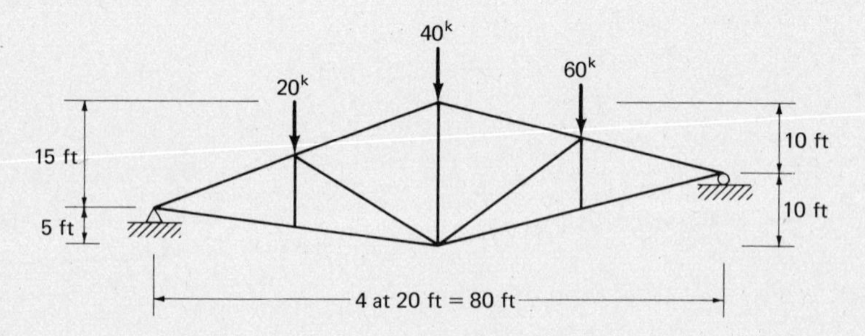

Problem 7.29

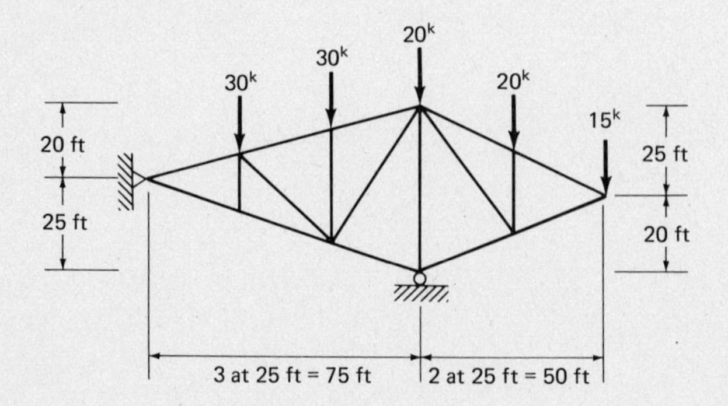

Problem 7.30 (*Ans.* $L_0U_1 = -48^k$, $U_2L_2 = -16.7^k$)

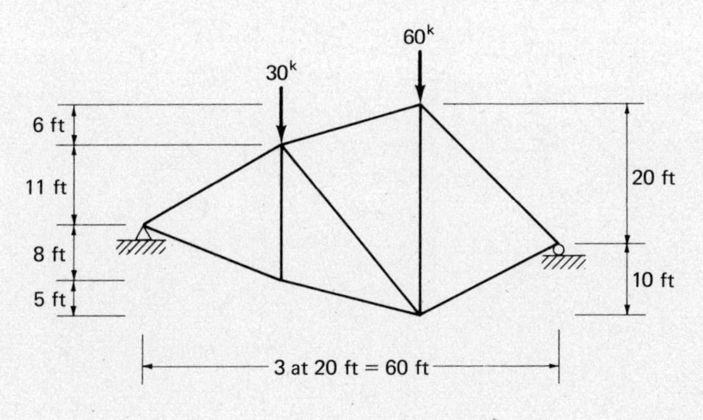

Problem 7.31

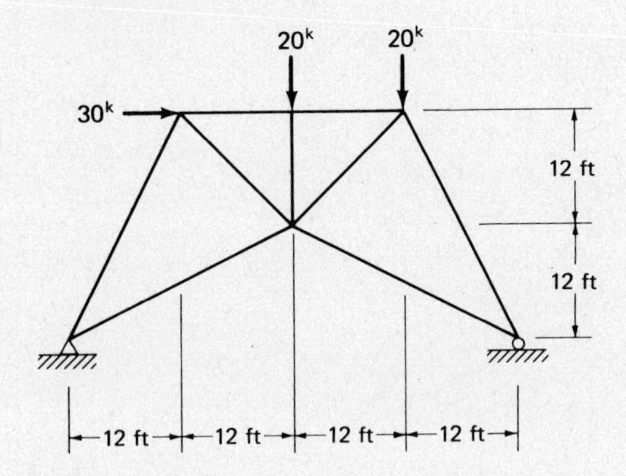

7.32. Repeat Prob. 7.31 with the roller support inclined as shown (slope 1:2). (*Ans.* $U_1U_2 = -40^k$, $L_2U_3 = +28.3^k$)

Problem 7.33

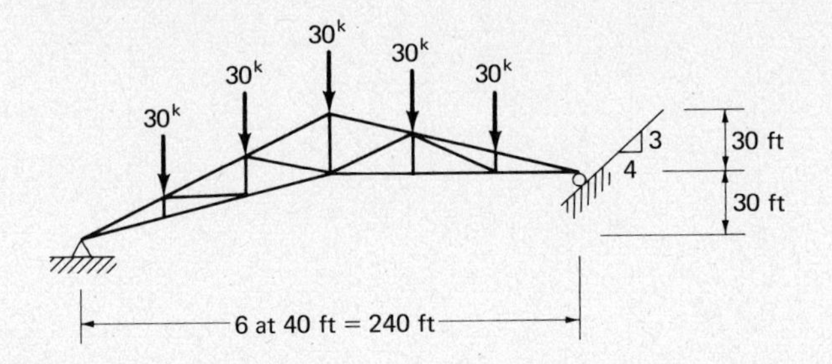

Problem 7.34 (*Ans.* $L_0L_2 = +74.5^k$, $L_2U_3 = +53^k$, $U_3L_4 = +43.7^k$)

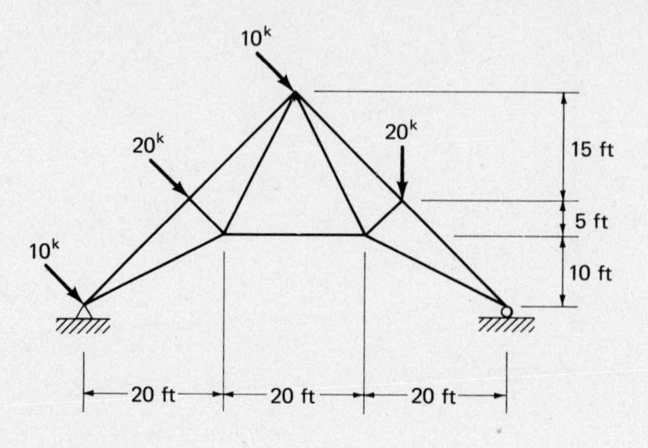

Problem 7.35

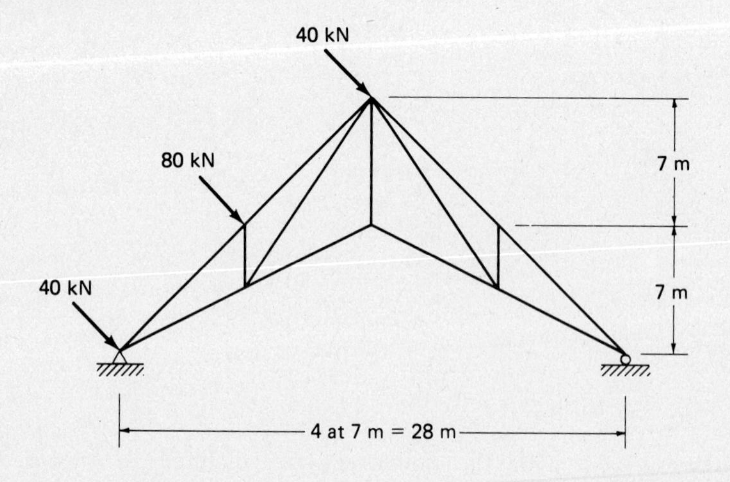

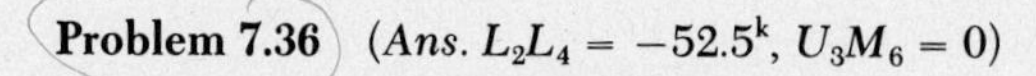

Problem 7.36 (*Ans.* $L_2L_4 = -52.5^k$, $U_3M_6 = 0$)

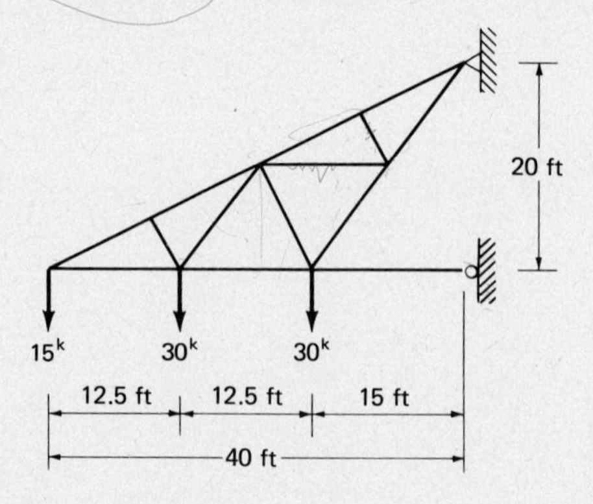

Problem 7.37

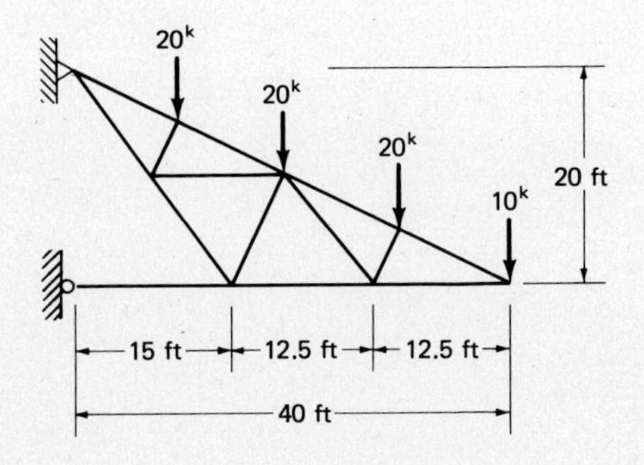

Problem 7.38 (*Ans.* $U_0U_1 = +13.3^k$, $L_1L_2 = +3.7^k$, $L_5U_6 = -36^k$)

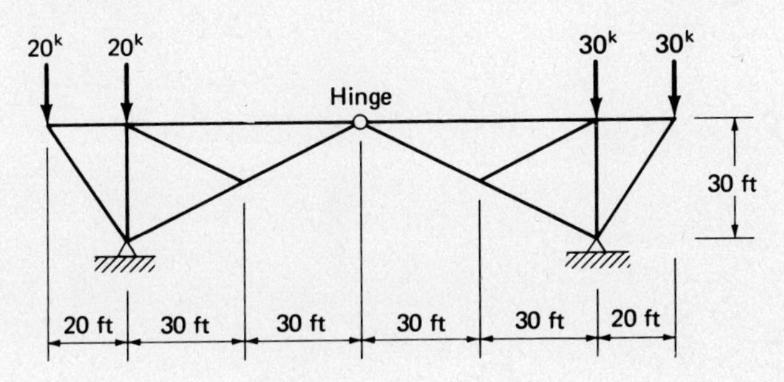

Problem 7.39

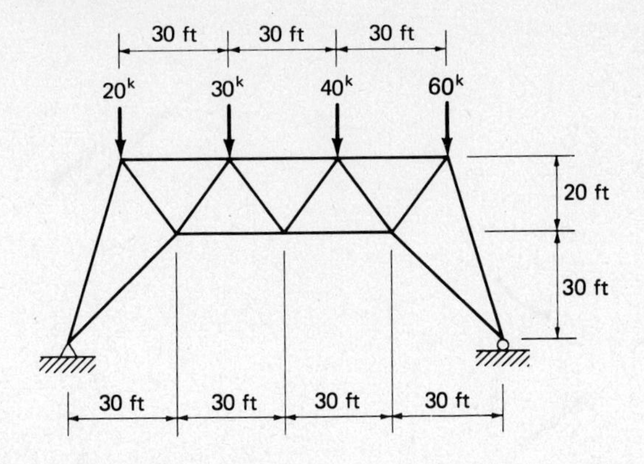

Problem 7.40 (*Ans.* $L_2L_4 = +85^k$, $U_3M_6 = +10^k$, $U_7M_8 = +60^k$)

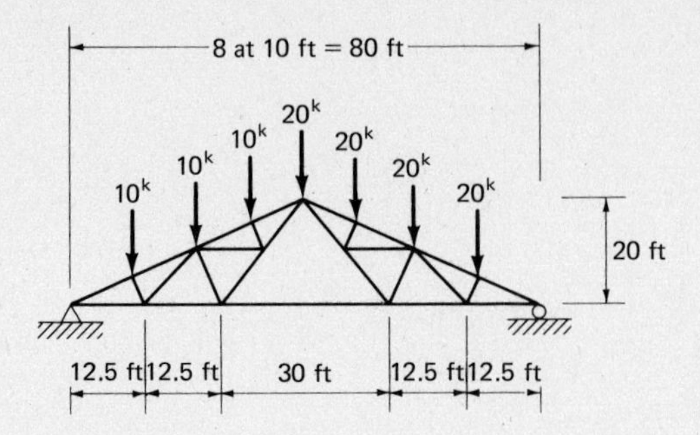

Problem 7.41

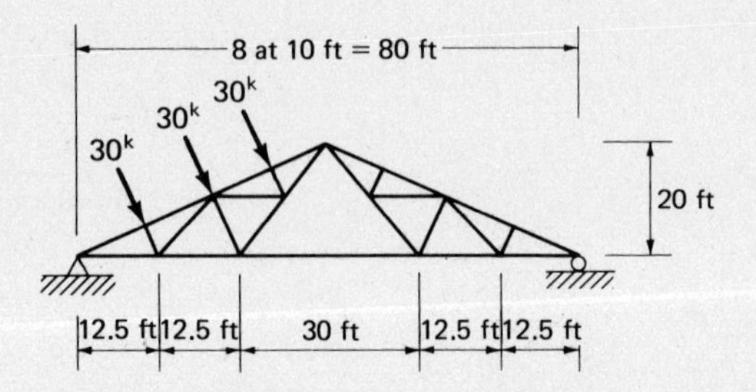

Problem 7.42 (*Ans.* $AB = -1.25^k$, $DE = -13.75^k$, $DG = +51.5^k$)

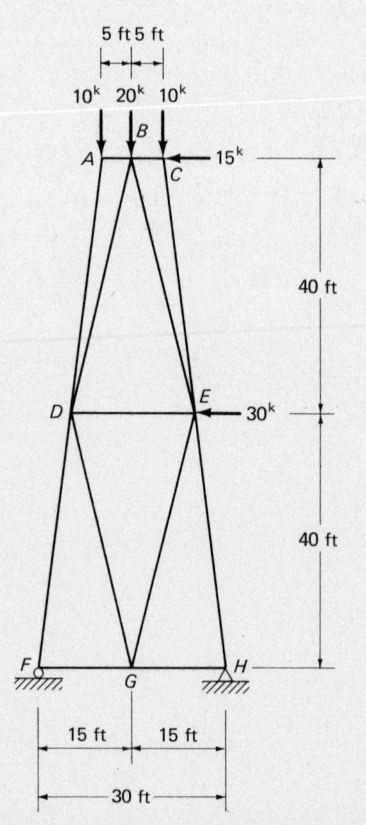

Problem 7.43

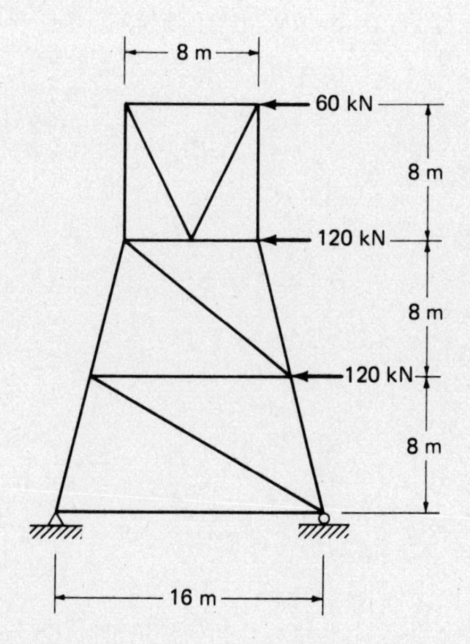

Problem 7.44 (*Ans.* $M_0L_1 = +44.7^k$, $U_2L_2 = -16^k$, $L_2M_4 = +80.8^k$)

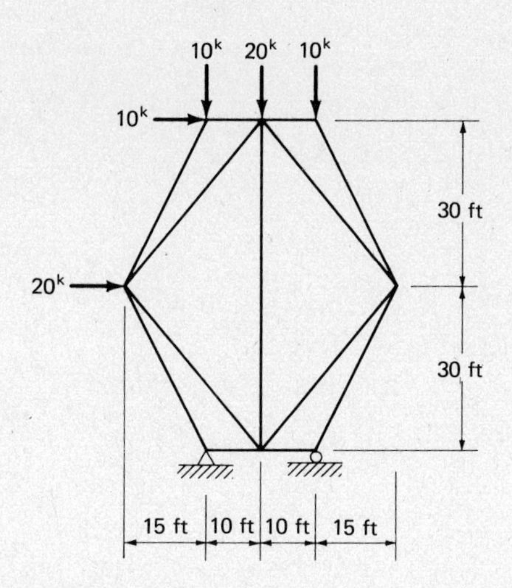

Problem 7.45

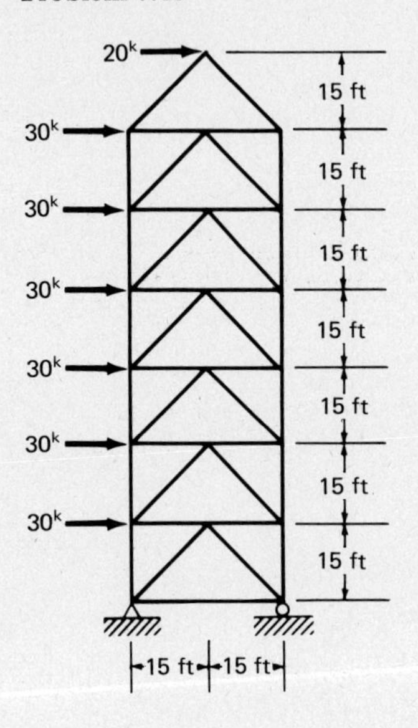

Problem 7.46 (*Ans.* $AE = -13.2^k$, $BE = -49.5^k$, $CF = -8.75^k$)

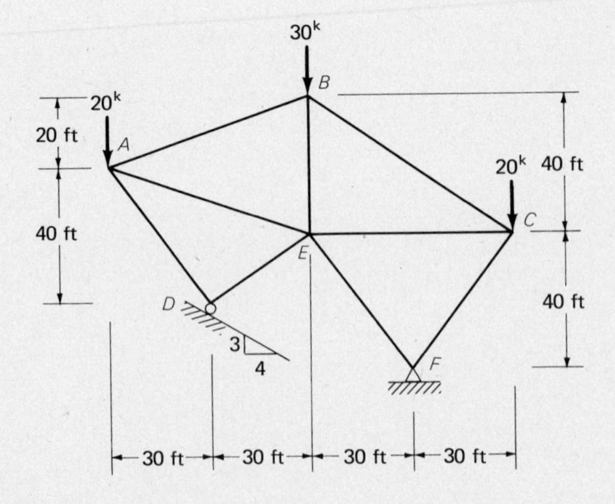

Problem 7.47

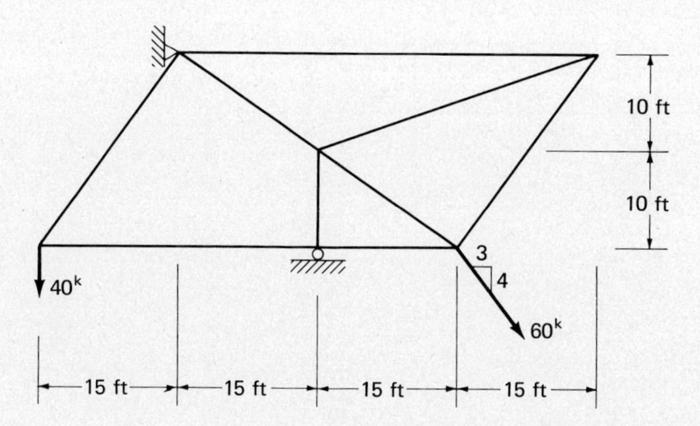

For Probs. 7.48 to 7.51 determine directly the forces in the designated members using the method of moments.

7.48. Members U_2U_3, L_2L_3, and U_2L_2 of the truss of Prob. 7.2 (*Ans.* $U_2U_3 = -63.2^k$, $L_2L_3 = +90^k$, $U_2L_2 = +10^k$)

7.49. Members L_2L_3, U_3U_4, and U_2L_3 of the truss of Prob. 7.6.

7.50. Members L_1U_2, L_1L_2, and L_3U_4 of the truss of Prob. 7.7. (*Ans.* $L_1U_2 = -27.4^k$, $L_1L_2 = 348^k$, $L_3U_4 = -55.0^k$)

7.51. Members U_0U_1, U_1U_2, and U_1L_2 of the truss of Prob. 7.30.

For Probs. 7.52 to 7.54 determine the forces in the designated members.

Problem 7.52 Members L_2M_3, M_1M_3, and L_6M_7. (*Ans.* $L_2M_3 = +10^k$, $M_1M_3 = -16^k$, $L_6M_7 = +10^k$)

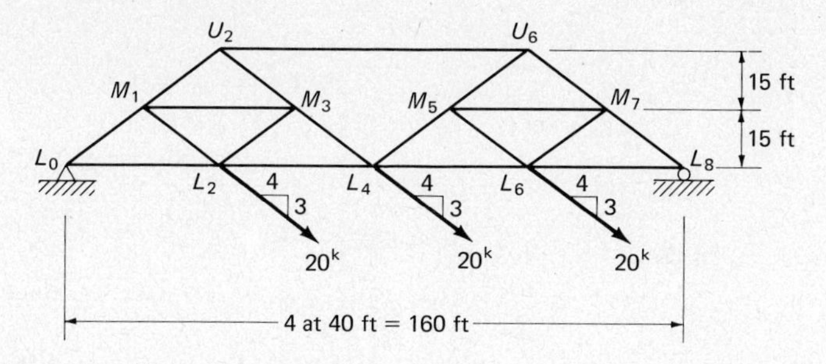

Problem 7.53 Members *ab*, *bc*, and *ad*.

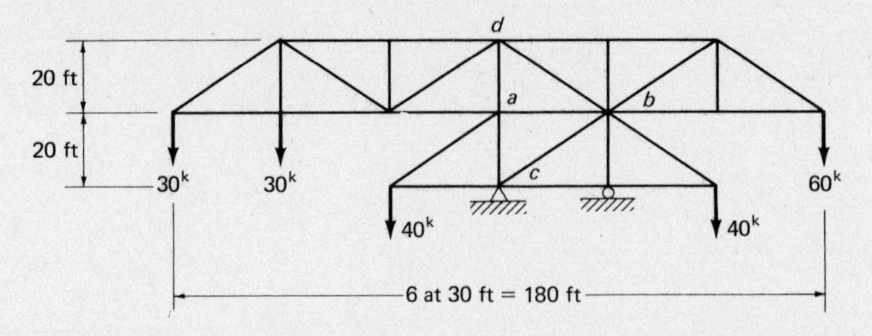

Problem 7.54 Members *ab*, *bc*, and *ad*. (*Ans.* $ab = -300^k$, $bc = -116.6^k$, $ad = 0$)

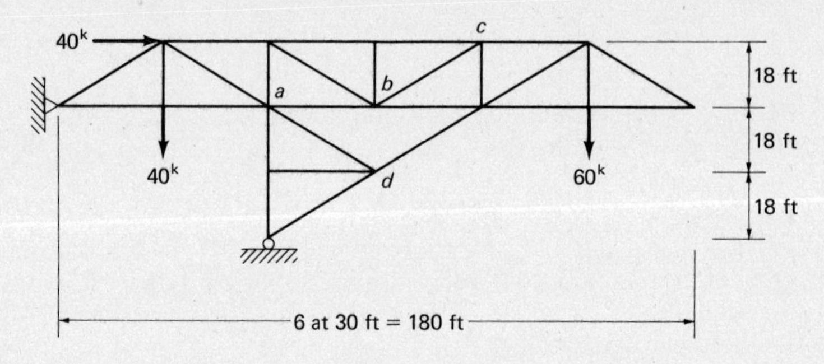

8
Subdivided Trusses

8.1 Reasons for Subdivision

In Chap. 5 the statement was made that it is desirable for reasons of economy to keep bridge truss diagonals at angles of approximately 45° with the horizontal and to vary truss depths from one-fifth to one-eighth of their spans (the flatter trusses being used for the longer spans). To maintain these relationships for long-span steel trusses, it becomes necessary to have very wide and deep panels. With large panels the unsupported lengths of chords and diagonals become excessive and the weight of the floor system increases greatly. The Pratt truss of Fig. 8.1 illustrates these disadvantages. For this truss the following facts are evident:

1. Ratio of depth to span is one to eight.
2. Diagonals are at 45° with horizontal.
3. Unsupported length of top chord members is 50 ft.
4. Unsupported length of diagonals is 70.7 ft.
5. Unsupported length of floor system is 50 ft.

The moments in bridge floor systems longer than 20 to 25 ft are excessive. A floor system would have to be exceptionally heavy, thus expensive, to span a

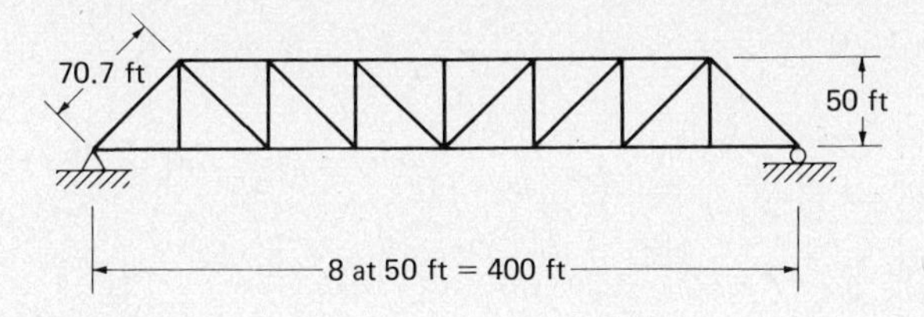

Figure 8.1

Brown's Bridge, Forsyth and Hall Counties, Gainesville, Georgia. (Courtesy of the American Institute of Steel Construction, Inc.)

distance of 50 ft between panel points from which it receives its support. The increased weight of the floor system would produce corresponding increases in the forces, sizes, and cost of the supporting trusses. Long diagonals and chords subject to compression must be made quite heavy because of the danger of buckling. A large compressive load is not required to buckle a 70-ft column even though the cross-sectional area of the column may be quite large.

8.2 Trusses with Multiple Web Systems

Toward the end of the nineteenth century, these problems were solved with multiple-web system trusses such as the ones shown in Fig. 8.2. These trusses reduced lengths of members and floor systems, but they introduced other disadvantages: they are inconvenient to analyze, because they are statically

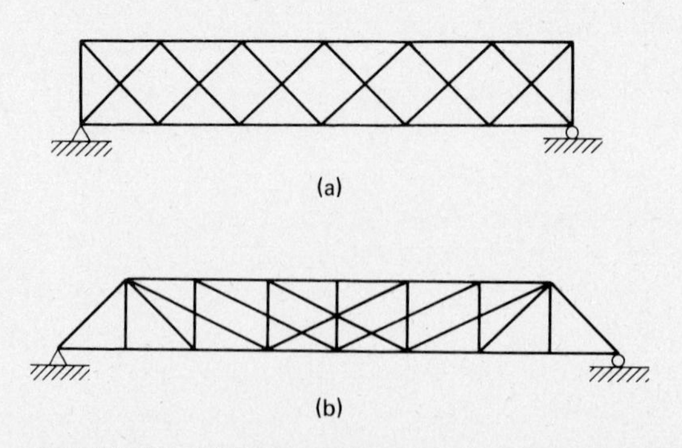

Figure 8.2 (a) Double Warren system truss. (b) Whipple truss.

indeterminate, and they are somewhat expensive to construct. For these reasons they are almost extinct today. The reader who is interested in their analysis by an approximate method may refer to *Stresses in Framed Structures* by Hool and Kinne [1].

8.3 Subdivided Trusses

Subdivided trusses are more satisfactory for long-span trusses than are the multiple-web system trusses. The Pratt truss of Fig. 8.1 may be subdivided by the addition of the members shown in Fig. 8.3.

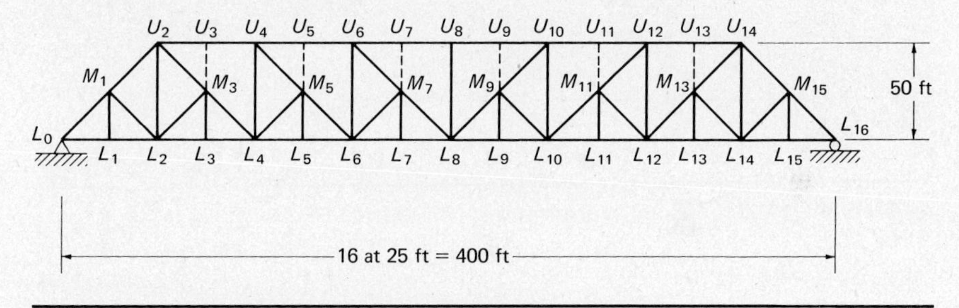

Figure 8.3 Subdivided Pratt or Baltimore truss.

The resulting truss is a subdivided Pratt, or Baltimore, truss. The newly added members are referred to as *subdiagonals* and *subverticals* and may also be referred to as *substruts* (if they are compression members) or *subties* (if they are tension members).

The subverticals (M_3L_3, $M_{11}L_{11}$, and so forth) transfer the loads at their panel points (L_3 and L_{11}) by tension to the main diagonals. The subdiagonals (L_2M_3, $M_{11}L_{12}$, and so forth) prevent the diagonals from being bent laterally by the tensile loads applied from the subverticals.

Subdivision reduces the unsupported length of the floor system between panel points by one-half; therefore the bending moment, which varies approximately as the square of the span length, becomes approximately one-fourth of its former value. The unsupported lengths of the main diagonals are one-half as long as they were, which in design will allow them to be considerably smaller, particularly if they are subject to compression. Members U_3M_3, U_5M_5, and so forth (shown with dotted lines), are sometimes used to reduce the unsupported lengths of top-chord members.

Other types of subdivided trusses are shown in Fig. 8.4(a) and (b). As spans become longer it becomes economical to vary truss depths approximately in proportion to bending moments as illustrated by the Pettit or Pennsylvania truss of part (b) of the figure. Figure 8.4(c) shows a K truss, with its unusual arrangement of members, which accomplishes the same purpose as ordinary subdivided trusses and probably has smaller secondary forces.

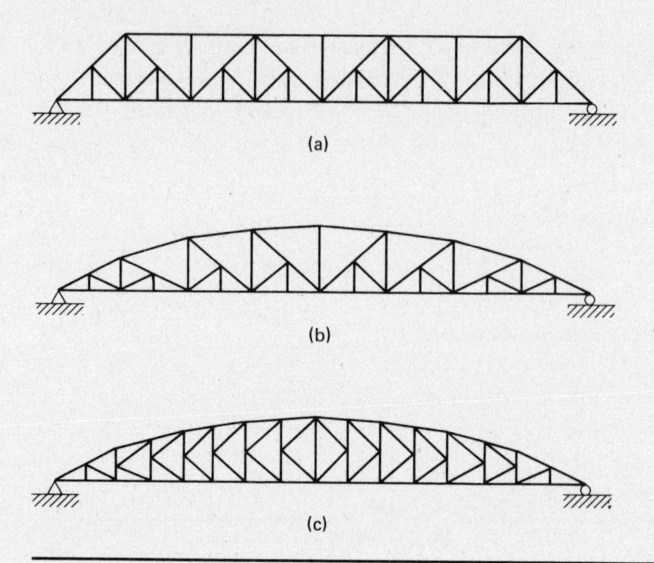

Figure 8.4 (a) Subdivided Warren truss. (b) Pettit or Pennsylvania truss. (c) K truss.

8.4 Disadvantages of Subdivision

Several advantages of subdivision have been discussed in the preceding paragraphs. However, subdivision introduces three main disadvantages that accompany the advantages. They are (1) higher pound prices of steel, (2) less attractive trusses, and (3) higher secondary forces in many cases.

8.5 Analysis of Baltimore Truss

The Baltimore truss of Fig. 8.5 may be analyzed with little difficulty and with no new information presented by the method of joints and moments; however, one short cut is available that will appreciably expedite the analysis.

The force in each subvertical obviously equals the load applied to the panel point below. If no load is applied at the panel point, the subvertical can have no force.

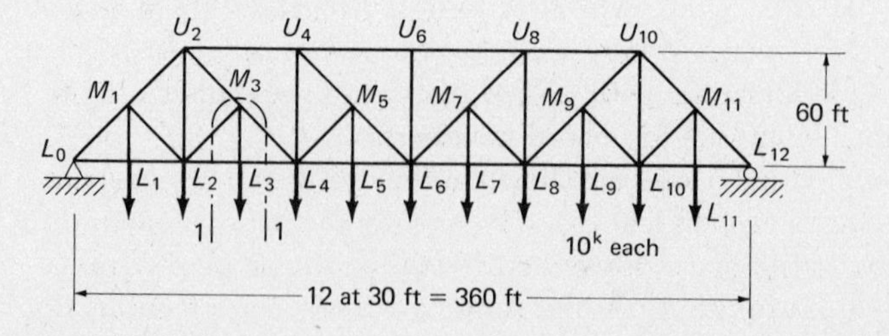

Figure 8.5

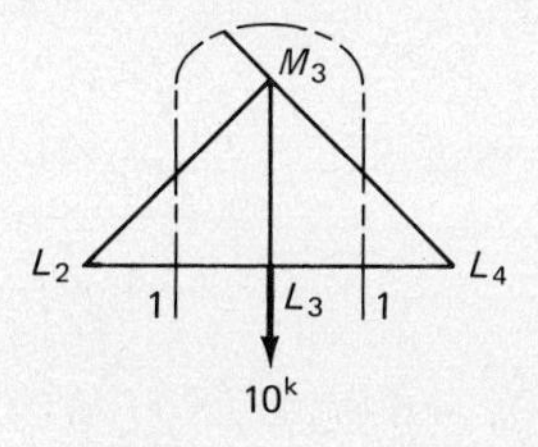

Figure 8.6

It is desired to find a simple method of calculating the force in the subdiagonal L_2M_3, so the horseshoe-shaped section 1–1, Fig. 8.5 is passed around joints M_3 and L_3 as shown in Fig. 8.6.

Considering the free body enclosed by the horseshoe, the imaginary section can be seen to pass through five different members (L_2L_3, L_2M_3, U_2M_3, M_3L_4, and L_3L_4). Moments are taken about L_4 to find the force in L_2M_3 because the lines of action of the forces in all of the cut members, except L_2M_3, pass through this joint.

$$\Sigma M_{L_4} = 0$$

$$-(10)(30) - (V_{L_2M_3})(60) = 0$$

$$V_{L_2M_3} = -5^{\text{k}} \text{ compression}$$

The moment equation shows that the vertical component of force in the subdiagonal equals one-half of the subpanel load and is in compression. This statement is always true if the subpanel point lies midway between the main panel points. Should the subpanel point not lie at the midpoint (quite unusual because the purpose of the subvertical is to reduce as much as possible the unsupported length of the floor system), the value of the vertical component may be determined by moments as described for this truss.

With this simplification the forces for the subverticals and subdiagonals of the truss can be quickly written in. The remaining forces can be computed by the method of joints. Should the subdivided truss under consideration have nonparallel chords, it may be necessary to use the method of moments for a few chord members to complete the solution.

8.6 Analysis of K Truss

The short panels of the K truss eliminate the necessity for subverticals to help support the floor system. The forces may be found by joints and moments with little difficulty, but as with the Baltimore truss some simplifications of the analysis are available.

For the discussion to follow, the K truss of Fig. 8.7 is considered. The force in chord member U_2U_3 may be found by passing section 1–1 and taking moments at L_2 because the lines of action of the forces in all of the members cut by the section except U_2U_3 pass through that joint.

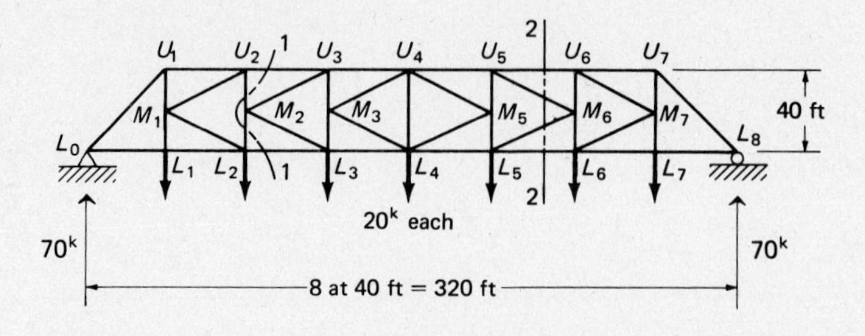

Figure 8.7

$$\Sigma M_{L_2} = 0 \text{ (free body to left of section)}$$

$$(70)(80) - (20)(40) + (F_{U_2U_3})(40) = 0$$

$$F_{U_2U_3} = -\ 120\text{k compression}$$

A complete analysis may be made by using sections of the same nature as 1–1 in each panel for finding chord forces and determining the forces in the other members by joints. Another simplification, however, can be found by studying the middle joints. By passing a section completely around joint M_1, the free body of Fig. 8.8(a) is obtained. Unless an inclined external load is applied at M_1 (highly improbable), the horizontal components of force in M_1U_2 and M_1L_2 must be equal numerically and opposite in character. If one diagonal is in tension, the other is in compression. The numerical relationship between their vertical components depends on their respective slopes. If they have the same slopes, as is usually the case, the components are equal. The two possible force situations in these members are presented in Figs. 8.8(b) and (c). The vertical components of force in these diagonals act in the same direction in opposing the shear in the panel, regardless of which diagonal is in tension and which is in compression. The vertical components are both acting down or are both acting up; there is no other possible combination.

The slopes of the diagonals of the truss of Fig. 8.7 are equal and their vertical components will be equal. Passing section 2–2, as shown in Fig. 8.9, will permit the calculation of the diagonal forces in the panel by considering the free body to the right of the section.

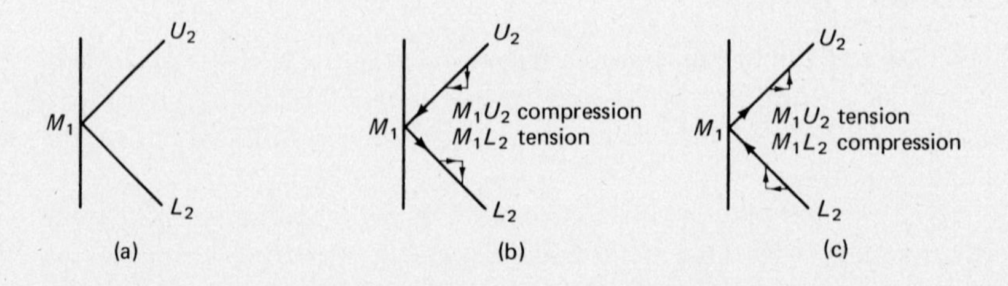

Figure 8.8

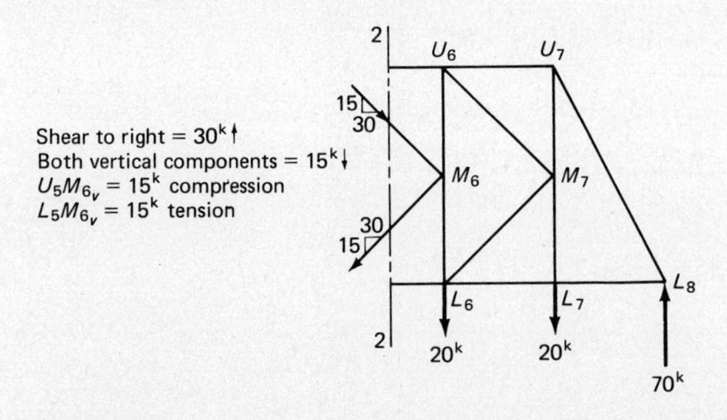

Figure 8.9

The K trusses often have nonparallel chords, and the inclined chord has a vertical component of force that must be included in considering the forces resisting shear in a panel. Forces in sloping chords may be determined by taking moments.

Example 8.1

Find the forces in all of the members of the K truss shown in Fig. 8.7.

SOLUTION

The final forces are as shown.

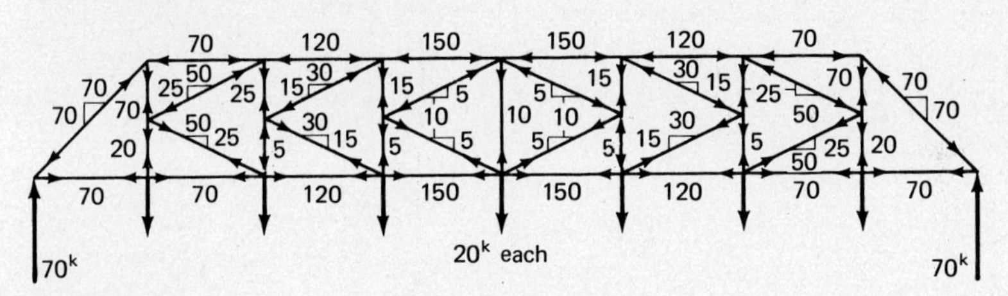

PROBLEMS

For Probs. 8.1 to 8.24 analyze the subdivided trusses.

Problem 8.1

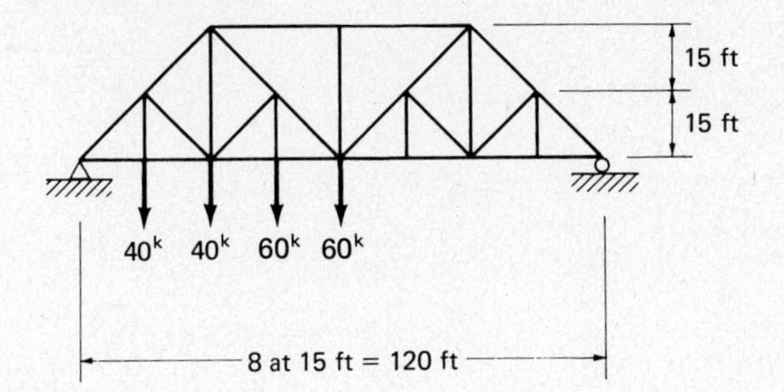

Problem 8.2 (*Ans.* $U_1U_2 = -100^k$, $M_2L_3 = +33.3^k$, $L_4L_5 = +60^k$)

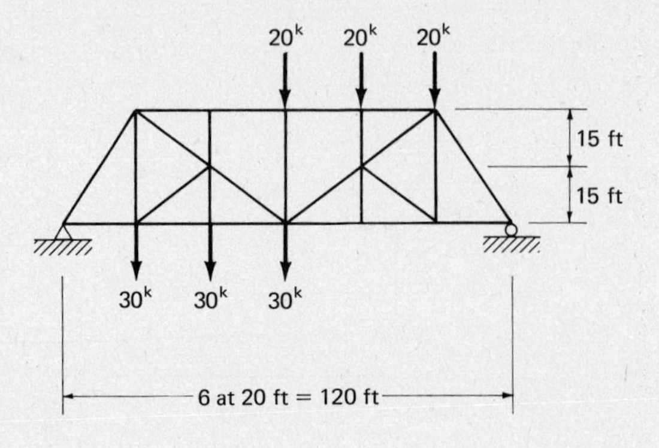

Problem 8.3

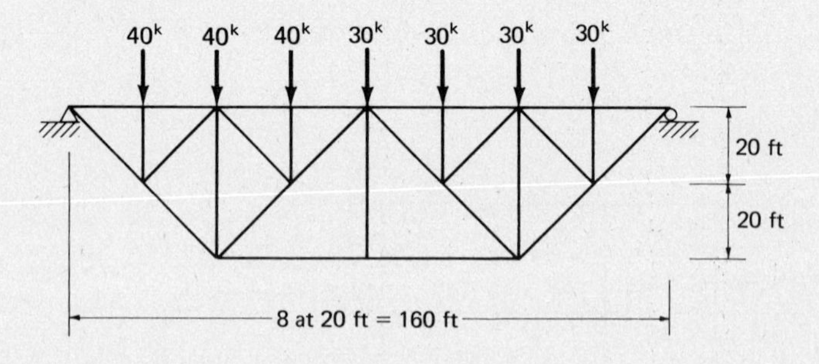

Problem 8.4 (*Ans.* $L_2L_3 = +75^k$, $U_4L_4 = -20^k$, $U_6M_7 = -92^k$

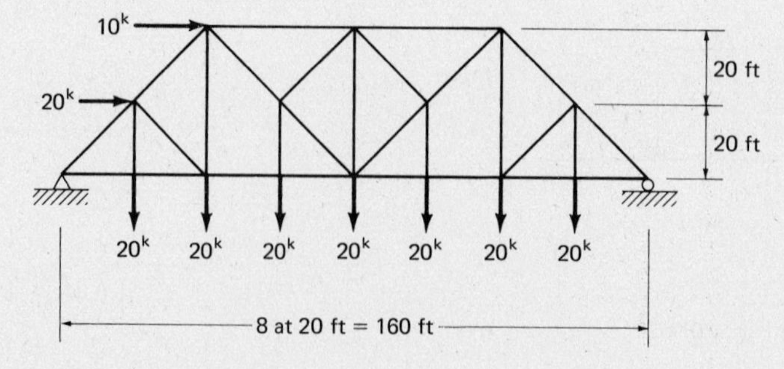

Problem 8.5

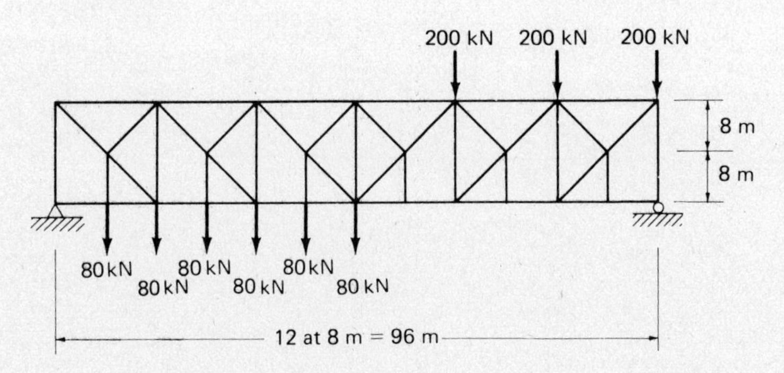

Problem 8.6 (*Ans.* $M_1U_2 = -235^k$, $U_4L_4 = +86.6^k$, $M_5U_6 = +26.1^k$)

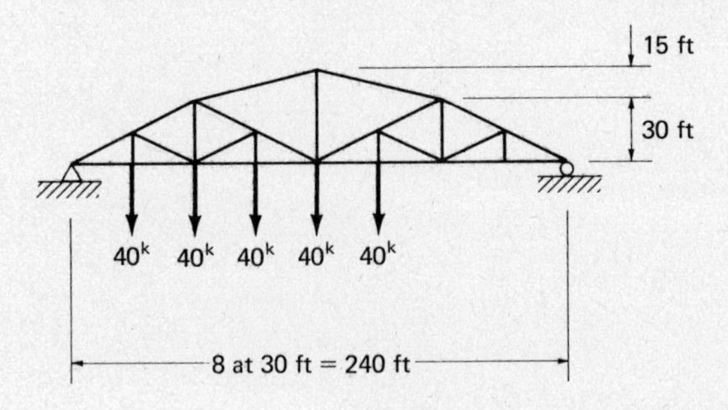

Problem 8.7

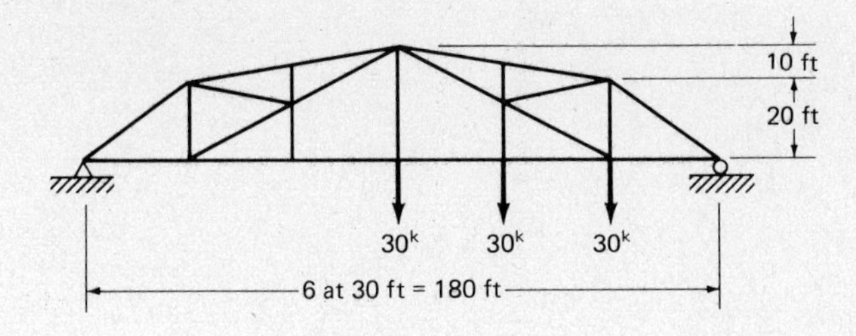

8.8. Rework Prob. 8.7 if the loads are doubled and the panels changed from 6 at 30 ft to 6 at 20 ft. (*Ans.* $L_1L_2 = +120^k$, $M_2U_3 = -75^k$, $M_4L_5 = 0$)

Problem 8.9

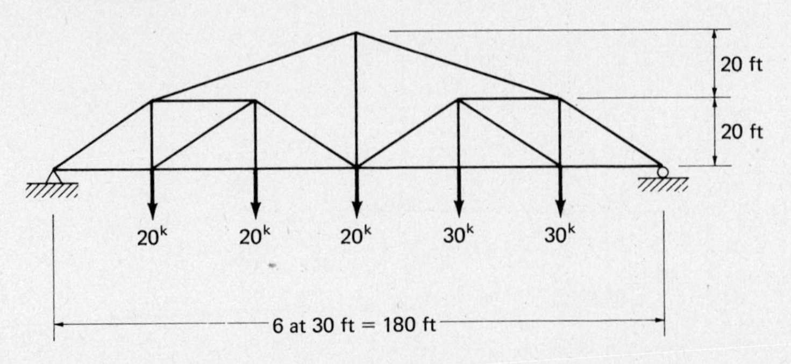

Problem 8.10 (*Ans.* $U_4L_4 = -16.7^k$, $U_6U_7 = -125^k$, $L_8L_{10} = +134.8^k$)

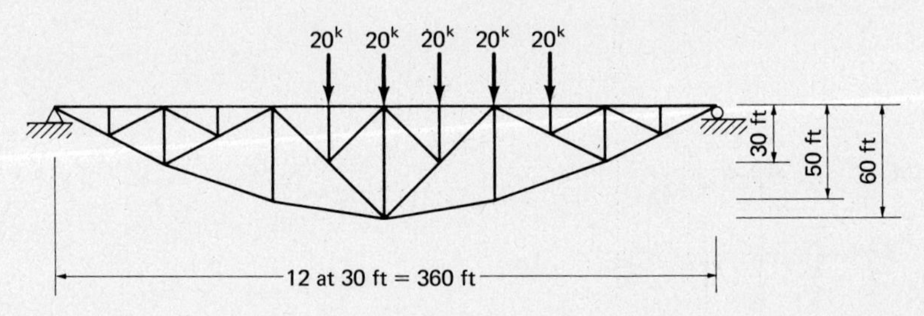

Problem 8.11

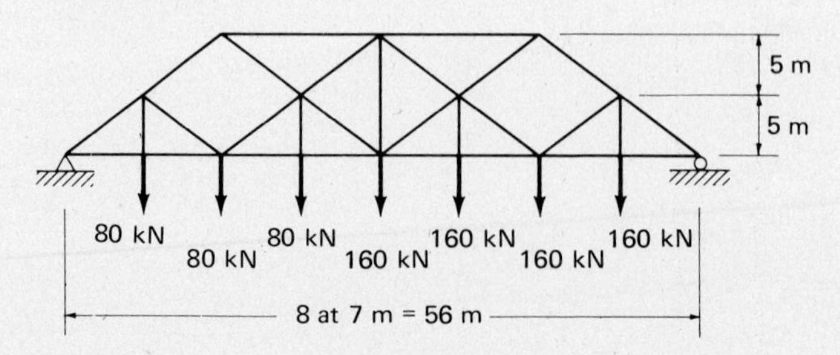

Problem 8.12 (*Ans.* $M_2L_2 = +30^k$, $U_4L_4 = +20^k$, $U_5U_6 = -210^k$)

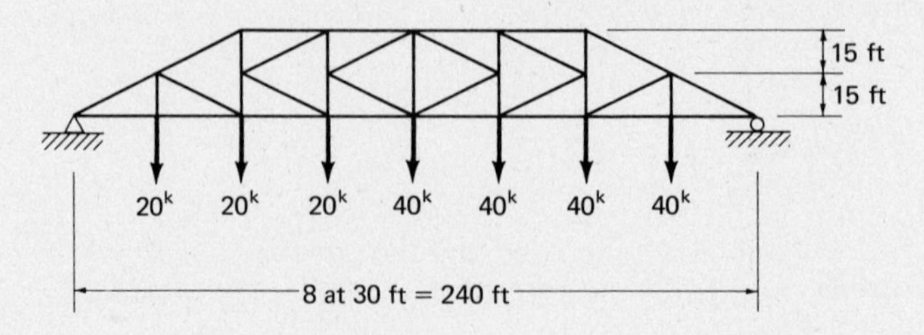

Problem 8.13

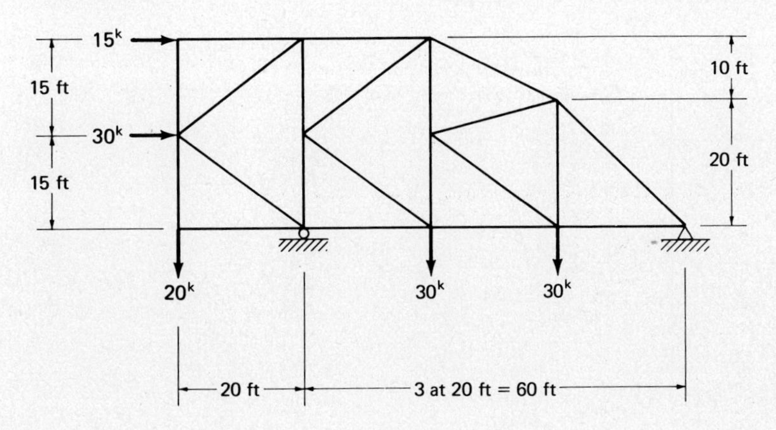

Problem 8.14 (*Ans.* $L_2L_3 = +171^k$, $U_3L_3 = +17.5^k$, $L_3M_4 = +31^k$)

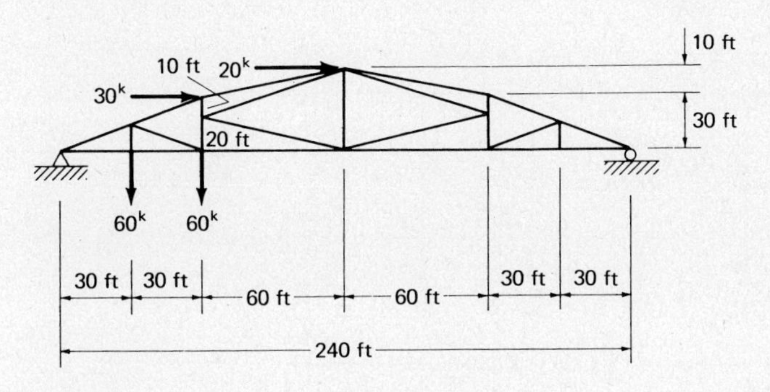

Problem 8.15

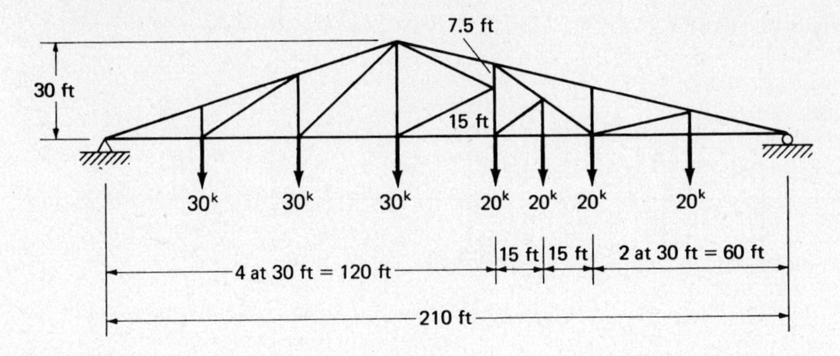

Problem 8.16 (*Ans.* $AB = -7.5^k$, $EH = -45^k$, $FJ = -67.5^k$)

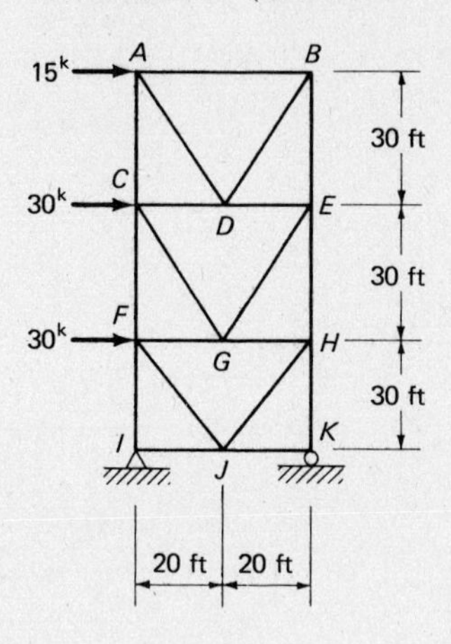

Problem 8.17

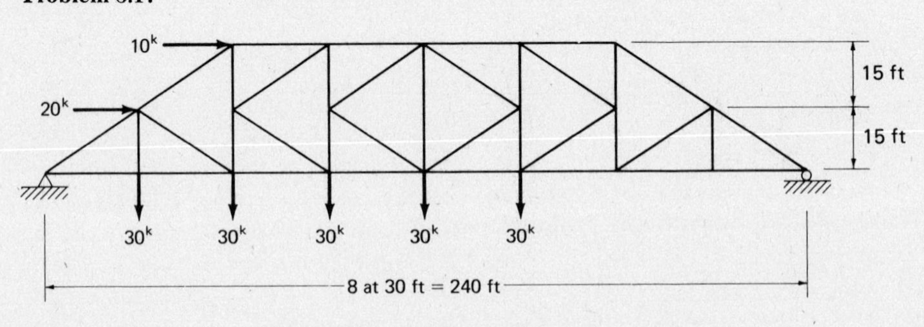

Problem 8.18 (*Ans.* $L_2L_3 = +1\,125$ kN, $U_3U_4 = -1\,062$ kN, $M_5L_5 = -46.9$ kN)

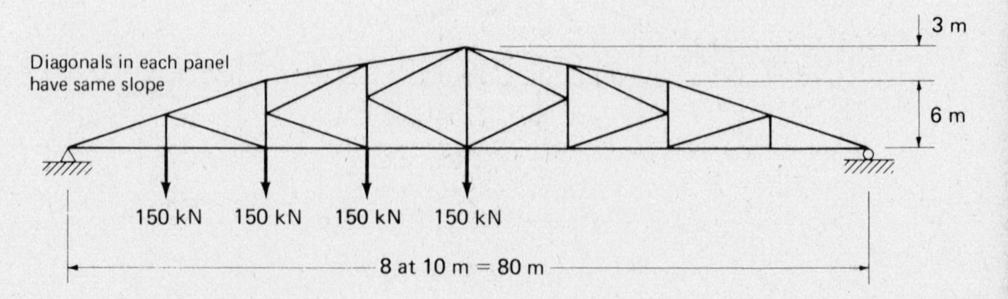

Problem 8.19

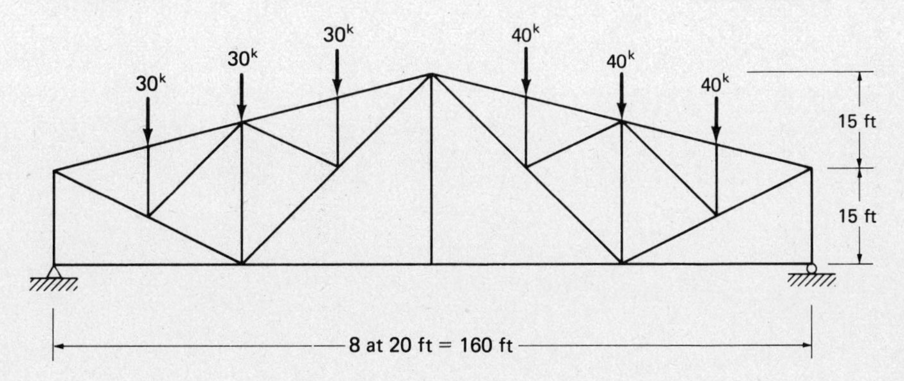

Problem 8.20 (*Ans.* $DE = -128.4^k$, $HI = +145.1^k$, $FI = +103.8^k$)

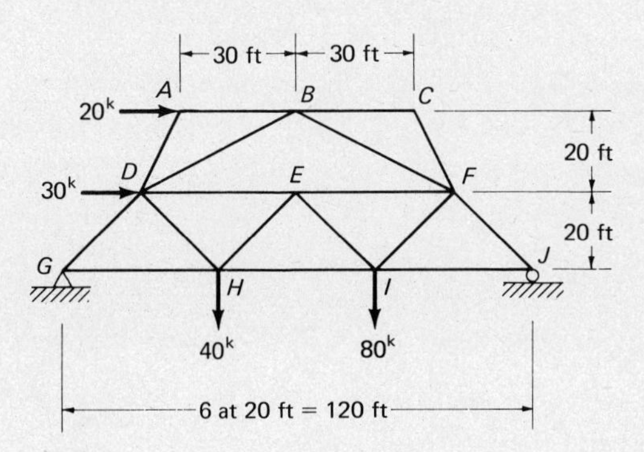

Problem 8.21

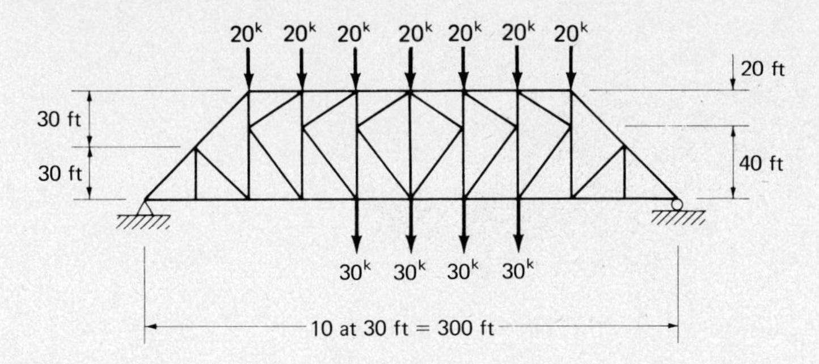

Problem 8.22 (*Ans.* $AB = -61.2^k$, $CD = -210^k$, $EF = +49.3^k$, $GH = +162^k$)

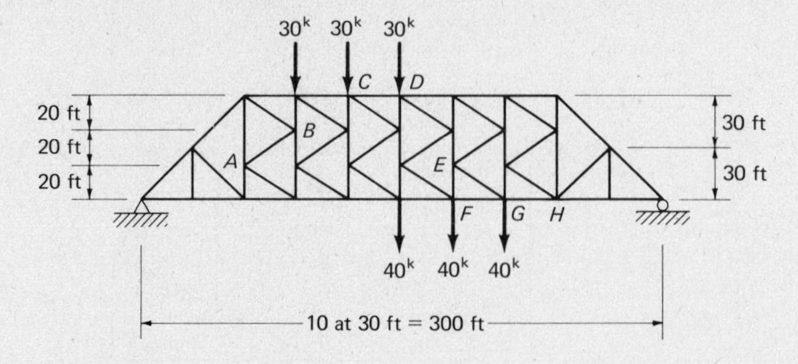

Problem 8.23

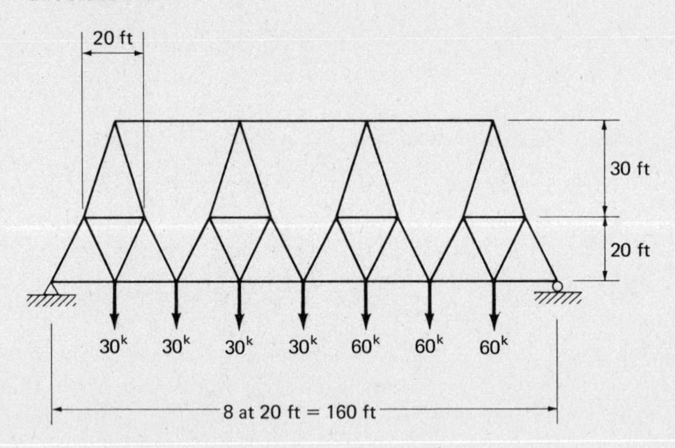

Problem 8.24 (*Ans.* $AC = +22.9^k$, $AD = -13.3^k$, $DE = -14.1^k$)

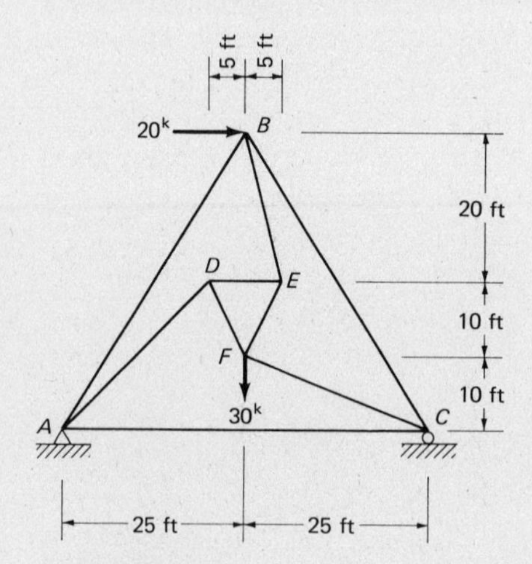

REFERENCES

[1] G. A. Hool and W. S. Kinne, Eds. *Stresses in Framed Structures* (New York: McGraw-Hill, 1923), pp. 279–286.

9 Truss Types and Stability

9.1 ARRANGEMENT OF TRUSS MEMBERS

A detailed discussion of the assembly of trusses has been delayed until this chapter so that the reader will have had some contact with the elementary types. The background should enable him or her to understand the material to follow more easily.

The triangle has been shown to be the basic shape from which trusses are developed because it is the only stable shape. Other shapes such as the ones shown in Figs. 9.1(a) and (b) are obviously unstable and may possibly collapse under load. Structures such as these can, however, be made stable by one of the following methods.

1. Addition of members so that the shapes are made to consist of triangles. The structures of Figs. 9.1(a) and (b) are stabilized in this manner in (c) and (d), respectively.
2. Using a member to tie the unstable structure to a stable support. Member AB performs this function in Fig. 9.1(e).

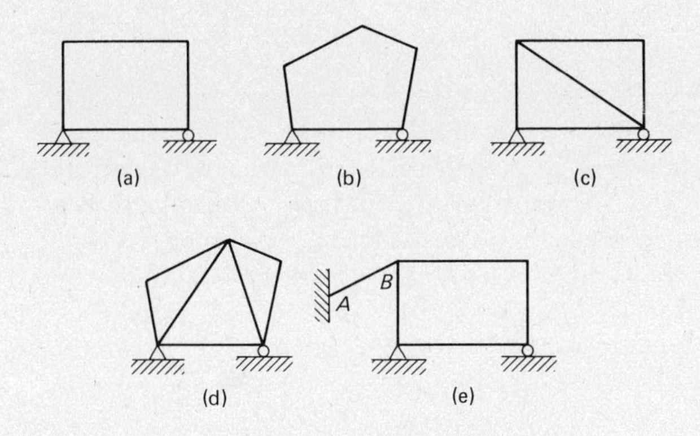

Figure 9.1

3. Making some or all of the joints of an unstable structure rigid, so they become moment resisting. A figure with moment-resisting joints, however, does not coincide with the definition of a truss (that is, members connected with frictionless pins, and so on).

9.2 STATICAL DETERMINANCY OF TRUSSES

The simplest form of truss, a single triangle, is illustrated in Fig. 9.2(a). To determine the unknown forces and reaction components for this truss, it is possible to isolate the joints and write two equations, $\Sigma H = 0$ and $\Sigma V = 0$, for each. From experience obtained in the last few chapters, there should be little difficulty in making the necessary calculations.

The single-triangle truss may be expanded into a two-triangle one by the addition of two new members and one new joint. In Fig. 9.2(b), triangle *ABD* is added by installing new members *AD* and *BD* and the new joint *D*. A further expansion with a third triangle is made in part (c) of the figure by the addition of members *BE* and *DE* and joint *E*. For each of the new joints, *D* and *E*, a new pair of equations is available for calculating the two new-member forces. As long as this procedure of expanding the truss is followed, the truss will be statically determinate internally. Should new members be installed without adding new joints, such as member *CE* in Fig. 9.2(d), the truss will become statically indeterminate because no new joint equations are made available to find the new-member forces.

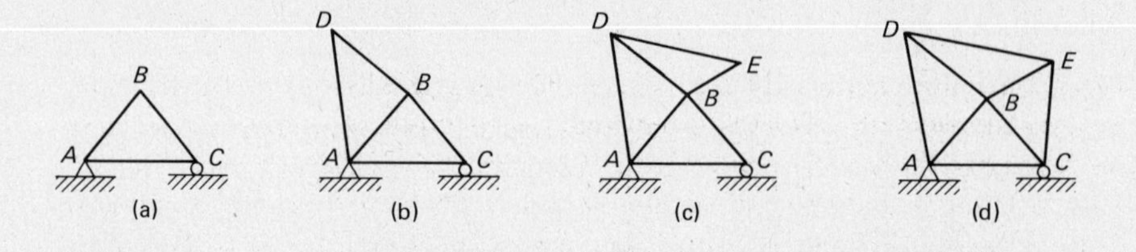

Figure 9.2

From the information above an expression can be written for the relationship that must exist between the number of joints and the number of members and reaction components for a particular truss if it is to be statically determinate internally. (The identification of externally determinate structures has previously been discussed.) In the following discussion, m is the number of members, j is the number of joints, and r is the number of reaction components.

If the number of equations available ($2j$) is sufficient to obtain the unknowns, the structure is statically determinate, from which the following relation may be written:

$$2j = m + r$$

Or as more commonly written,

$$m = 2j - r$$

Before an attempt is made to apply this equation, it is necessary to have a structure that is stable externally or the results are meaningless; therefore r is the least number of reaction components required for external stability. Should the structure have more external reaction components than necessary for stability (and thus be statically indeterminate externally), the value of r remains the least number of reaction components required to make it stable externally. This statement means that r will equal three for the usual statics equations plus the number of any additional condition equations that may be available.

It is possible to build trusses that have too many members to be analyzed by statics, in which case they are statically indeterminate internally, and m will exceed $2j - r$ because there are more members present than are absolutely necessary for stability. The extra members are said to be redundant members. If m is three greater than $2j - r$, there are three redundant members, and the truss is internally statically indeterminate to the third degree. Should m be less than $2j - r$, there are not enough members present for stability.

A brief glance at a truss will usually show if it is statically indeterminate. Trusses having members that cross over each other or members that serve as the sides for more than two triangles may quite possibly be indeterminate. The $2j - r$ expression should be used, however, if there is any doubt about the determinancy of a truss, because it is not difficult to be mistaken. Figure 9.3 shows several trusses and the application of the expression to each. The small circles on the trusses indicate the joints.

Little explanation is necessary for most of the structures shown, but some remarks may be helpful for a few. The truss of Fig. 9.3(e) has five reaction components and is statically indeterminate externally to the second degree; however, two of the reaction components could be removed and leave a structure with sufficient reactions for stability. The least number of reaction components for stability is 3, m is 21, and j is 12; applying the equation $m = 2j - r$ yields.

$$21 = 24 - 3 = 21 \qquad \text{statically determinate internally}$$

The truss of Fig. 9.3(j) is externally indeterminate because there are five reaction components and only four equations available. With r equal to 4 the structure is shown to be statically determinate internally. The three-hinged arch of Fig. 9.3(k) has four reaction components, which is the least number of reaction components required for stability; so r equals 4. Application of the equation shows the arch to be statically determinate internally.

In the chapters pertaining to the analysis of statically indeterminate structures it will be seen that the values of the redundants may be obtained by applying certain simultaneous equations. The number of simultaneous equations equals the total number of redundants, whether internal, external, or

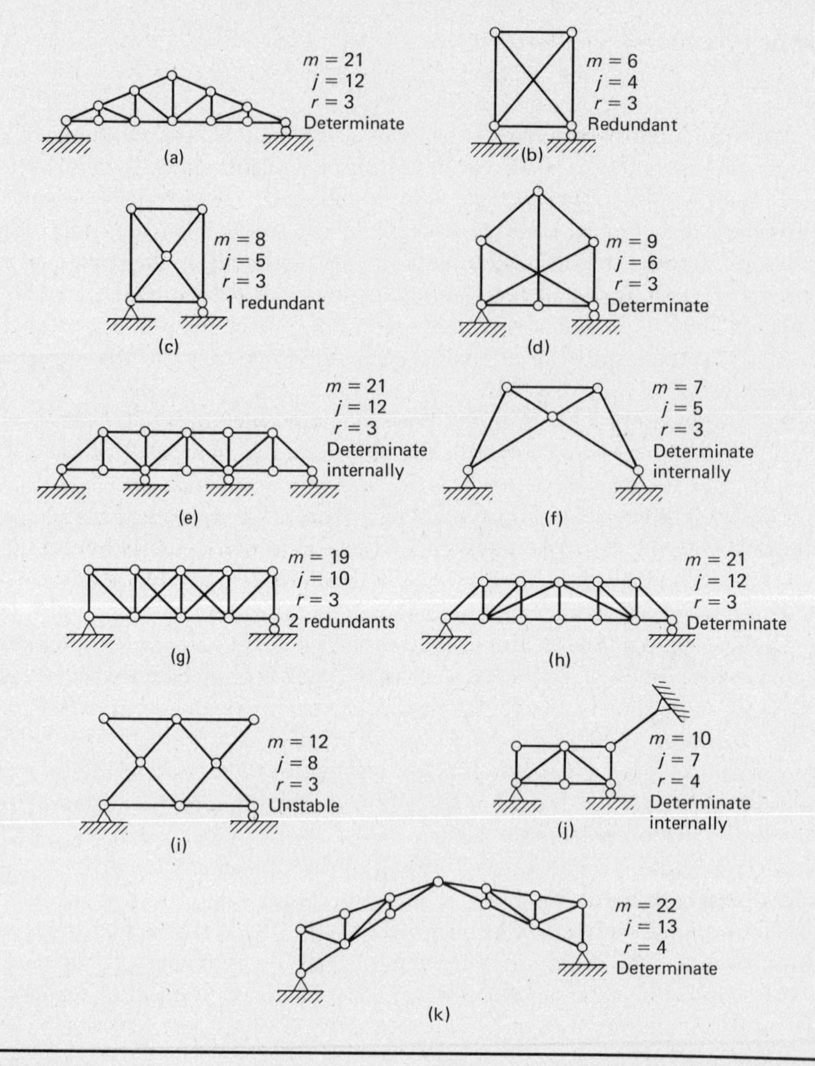

Figure 9.3

both. It therefore may seem a little foolish to distinguish between internal and external determinancy. The separation is particularly questionable for some types of internally and externally redundant trusses where no solution of the reactions is possible independently of the member forces, and vice versa.

If a truss is externally determinate and internally indeterminate, the reactions may be obtained by statics. If the truss is externally indeterminate and internally determinate, the reactions are dependent on the internal-member forces and may not be determined by a method independent of those forces. If the truss is externally and internally indeterminate, the solution of the forces and reactions will be performed simultaneously. (For any of these situations, it may be possible to obtain a few forces here and there by joints without going through the indeterminate procedure necessary for complete analysis.) This entire subject is discussed in detail in later chapters.

9.3 SIMPLE TRUSSES

The first step in forming a truss has been shown to be the connecting of three members at their ends to form a triangle. Subsequent figures are formed by adding two members and one joint; the new members meet at the new joint and each is pinned at its opposite ends into one of the existing joints. Trusses formed in this way are said to be ***simple trusses.*** (Some of these trusses, however, are not very "simple" to analyze.)

9.4 COMPOUND TRUSSES

A ***compound truss*** is a truss made by connecting two or more simple trusses. The simple trusses may be connected by three nonparallel nonconcurrent links, by one joint and one link, by a connecting truss, by two or more joints, and so on. An almost unlimited number of trusses may be formed in this way. The Fink truss shown in Fig. 9.4(a), consisting of the two crosshatched trusses connected by one joint and one link, is one example. Other compound trusses are shown in Figs. 9.4(b) and (c). The $2j - r$ equation applies equally well to compound trusses and simple trusses.

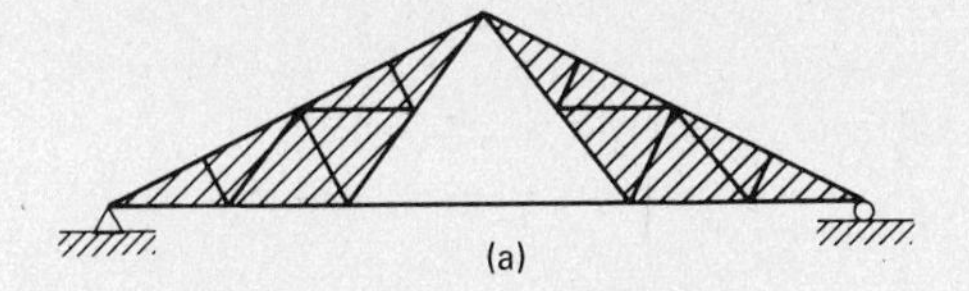

(a)

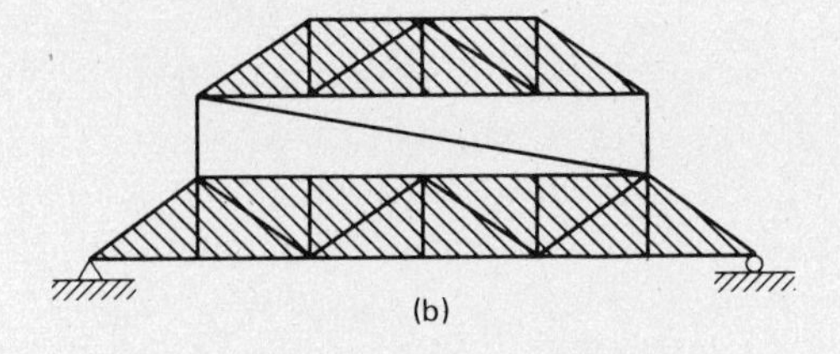

(b)

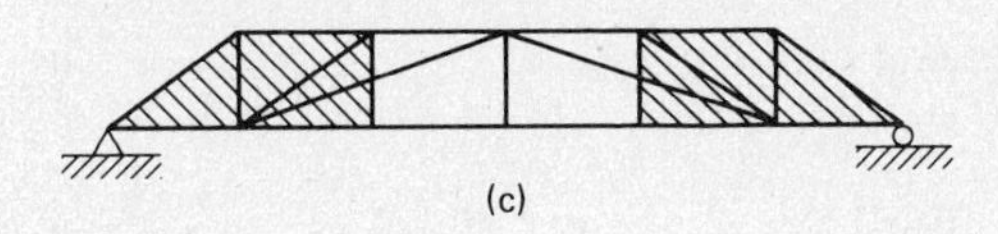

(c)

Figure 9.4

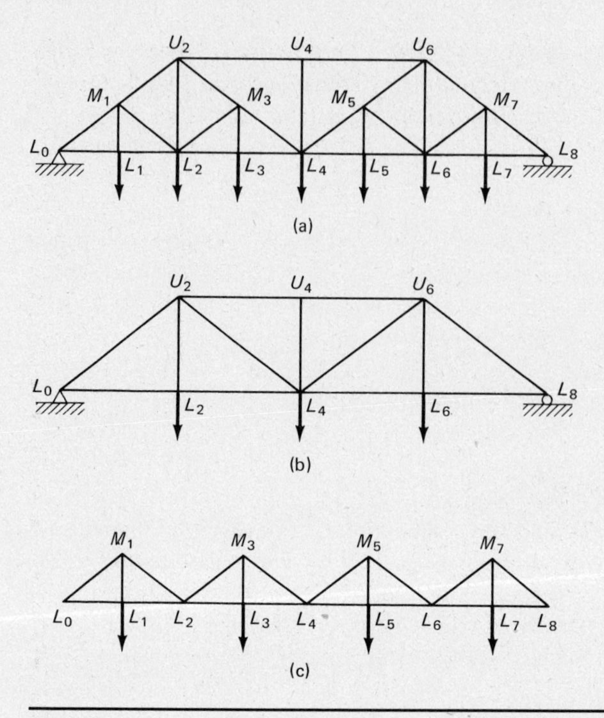

Figure 9.5

Another type of compound truss is the subdivided truss; an example is shown in Fig. 9.5(a). Analysis of subdivided trusses was discussed in Chap. 8. The subdivided truss of Fig. 9.5(a) may be considered to consist of the large truss of (b) with the small trusses of (c) superimposed on it. In fact, the truss analysis may be handled in exactly this manner, although such a procedure is not usually very practical.

9.5 COMPLEX TRUSSES

There are a few trusses that are statically determinate but do not have the requirements necessary to fall within the classification of either simple or compound trusses. These are referred to as ***complex trusses.*** The members of simple and compound trusses are usually arranged so that sections may be passed through three members at a time, moments taken at the intersection of two of them, and the force found in the third.

Complex trusses may not be analyzed in this manner. Not only does the method of moments result in failure, but the methods of shears and joints are also of no avail. The difficulty lies in the fact that there are three members meeting at almost every joint, and therefore there are too many unknowns at every location in the truss to pass a section and obtain the force in any member directly by the equations of statics. Two complex trusses are shown in Fig. 9.6.

College fieldhouse, Largo, Maryland. (Courtesy of the Bethlehem Steel Corporation.)

The number of joints and members is sufficient for them to be statically determinate.

One method of computing the forces in complex trusses is to write the equations $\Sigma H = 0$ and $\Sigma V = 0$ at each joint giving a total of $2j$ simultaneous equations. These equations may be solved simultaneously for the member forces and external reactions. (It is often possible to calculate the external reactions initially and their values may be used as a check against the results obtained from the solution of the simultaneous equations.) This method will work for any complex truss but the solution of the equations is very tedious unless a digital computer is available and usually other methods are more desirable.

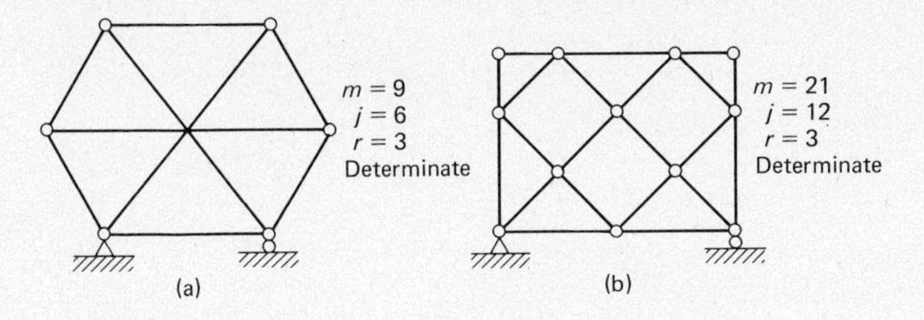

Figure 9.6

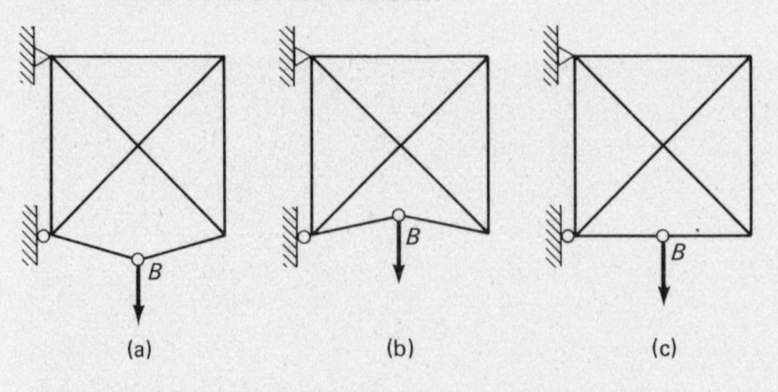

Figure 9.7

One precedure, applicable to many complex trusses, involves the assumption of the force in one of the truss members. A convenient member is selected and given a force of X. Forces in the surrounding members are computed in terms of X. The process is continued until it is possible to pass a section completely through the truss and write one of the equations of statics for the free body so the only members appearing in the equation have forces that have been calculated in terms of X. Solution of the resulting equation will give the value of X, and the other forces may be determined by statics. This method, however, is not easy to apply for many complex trusses.

The equation $m = 2j - r$ may be used to determine whether a complex truss is statically determinate or indeterminate; but to determine whether it is stable or unstable may not be so simple. Complex trusses often consist of not only triangles but also other shaped figures with the result that they are particularly susceptible to geometrical instability. They may be completely unstable and yet their instability may not be obvious until a solution is attempted. *If an analysis is attempted on an unstable truss, the results will always be inconsistent or noncompatible.*

Should the truss be geometrically unstable it is said to have ***critical form.*** The critical form of a truss may or may not be obvious, with the result that an analysis may have to be attempted before its stability is known. One arrangement or configuration of a truss may be stable and a slightly varying configuration may be unstable. The $m = 2j - r$ equation will be satisfied whether the truss is stable or whether it has critical form.

The trusses of Figs. 9.7(a) and (b) are stable under the action of a vertical load at joint B. In between these two arrangements lies the truss of part (c) of the same figure. This truss is unstable because it will not hold its position under the action of the load at B. Joint B will begin to deflect downward without causing calculable forces in the members.

9.6 THE ZERO-LOAD TEST

A structure that is stable has only one possible set of forces for a given loading condition and it is said to have a ***unique solution.*** Therefore if it is possible to show that more than one solution can be obtained for a structure for a given set of conditions, the structure is unstable.

This discussion leads into the idea of the so-called ***zero-load test.*** If no external loads are applied to a truss, it is logical to assume that all of the members will have zero forces. Should an assumed force (not zero) be given to one of the members of a truss that has no external loads and the forces be computed in the other members, the results must be incompatible if the truss is stable. If the calculated forces are compatible, the truss is unstable.

To illustrate this procedure the top horizontal member of each of the three trusses of Fig. 9.8 is assumed to have tension of X and the other member forces are computed working down from the top by joints. It will be seen that the lower three joints of the trusses of parts (a) and (b) of the figure cannot be balanced whereas they can for the truss of part (c). A structure that has zero external loads should also have zero internal forces and truss (c) must therefore be unstable or it has *critical form.* For the first two arrangements, (a) and (b), it is impossible to assume a set of member forces other than zero for which the joints will balance, and they must therefore be stable [1].

For another illustration of the zero-load test the reader might work with the truss of Fig. 9.6(a). He or she will find that it also has critical form.

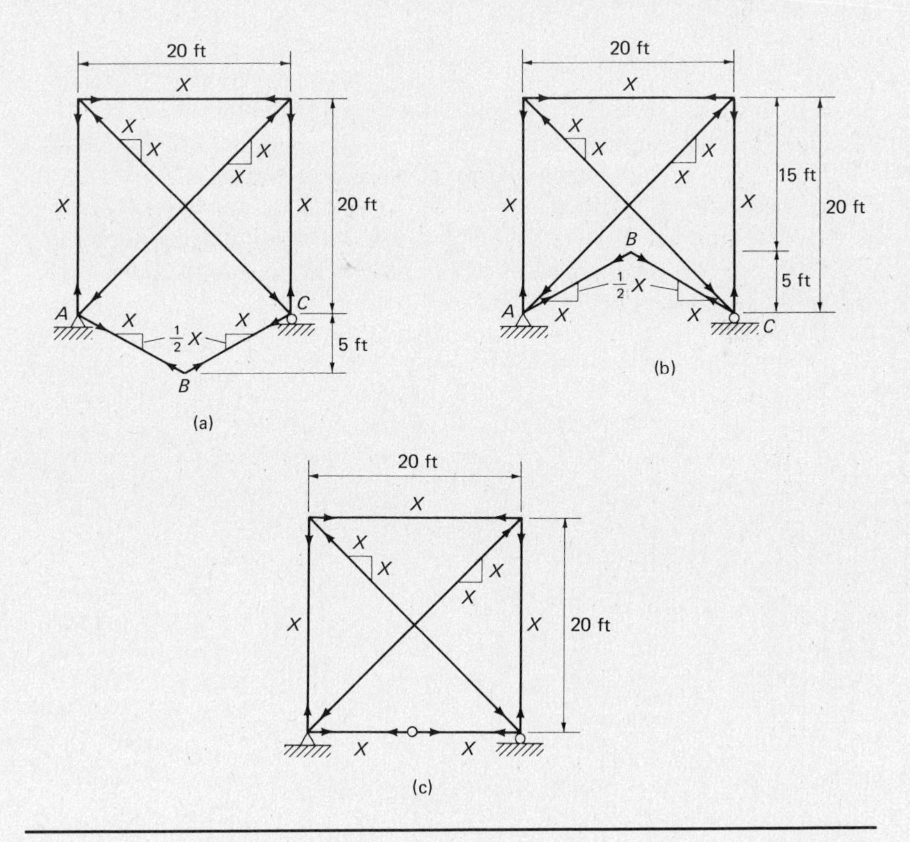

Figure 9.8 (a) Joints A, B, and C do not balance. Truss is stable. (b) Joints A, B, and C do not balance. Truss is stable. (c) All joints balance. Truss is unstable.

It is possible to determine by matrix algebra (a subject presented in Chap. 27) if a structure is stable by setting up the simultaneous equations $\Sigma H = 0$ and $\Sigma V = 0$ at each joint and finding the determinant of the equations. If the determinant is not zero, the structure has a unique solution and is stable. Should, however, the determinant be zero, the structure is unstable, as there are an infinite set of answers that will satisfy the simultaneous equations.

For a more comprehensive discussion of complex trusses the reader may refer to the method of substitute members described in *Theory of Structures* by S. Timoshenko and D. H. Young [2].

Generally speaking there is little need for complex trusses because it is possible to select simple or compound trusses that will serve the desired purpose equally well.

9.7 STABILITY

The following paragraphs discuss several situations that may cause a structure to be unstable.

Less than $2j - r$ Members

A truss that has less than $2j - r$ members is obviously unstable internally, but a truss may have as many or more than $2j - r$ members and still be unstable. The truss of Fig. 9.9(a) satisfies the $2j - r$ relationship and is statically determinate and stable; however, if the diagonal in the second panel is removed and added to the first panel as shown in Fig. 9.9(b), the truss is unstable even though the number of members remains equal to $2j - r$. The part of the truss to the left of panel 2 can move with respect to the part of the

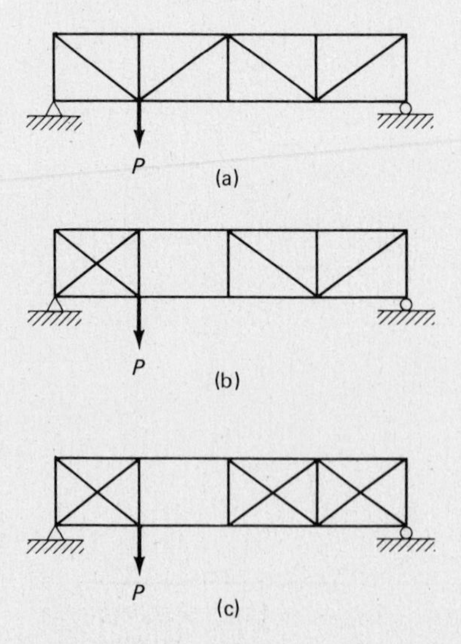

Figure 9.9

truss to the right of panel 2, because panel 2 is a rectangle. (As previously indicated, a rectangular shape is unstable unless restrained in some way.)

Similarly, addition of diagonals to panels 3 and 4, as shown in Fig. 9.9(c), will not prevent the truss from being unstable. There are two more than $2j - r$ members, and the truss is seemingly statically indeterminate to the second degree; but it is unstable because panel 2 is unstable.

Trusses Consisting of Figures That Are Not All Triangles

As the reader becomes more familiar with trusses, he or she will be able in most cases to tell with a brief glance if a truss is stable or unstable. For the present, though, it may be a good idea for the reader to study trusses in detail if he or she thinks there is a possibility of instability. One indication that instability may be present is when a truss does not consist entirely of triangles. The trusses of Figs. 9.9(b) and (c) fall in this category.

The fallacy of this idea is that the number of perfectly stable trusses that can be assembled not consisting entirely of triangles is endless. As an example consider the truss of Fig. 9.10(a). The basic triangle *ABC* has been extended by the addition of joint *D* and members *AD* and *CD* and a stable truss maintained, even though figure *ABCD* is not a triangle. Joint *D* is firmly held in position and cannot move without changing the length of one or more members. Compound trusses such as the ones of Figs. 9.3(h) and (i) and the subdivided types of Chap. 8 often present a nontriangular but stable situation. Another example is presented in Fig. 9.10(b). Structures of these types have the joints around the nontriangular figures tied to the rest of the truss so that they are not free to move. If a truss consists of figures that are not all triangles, it should be carefully examined to see if any of the joints can possibly move in any direction without causing changes in length of one or more of the truss members.

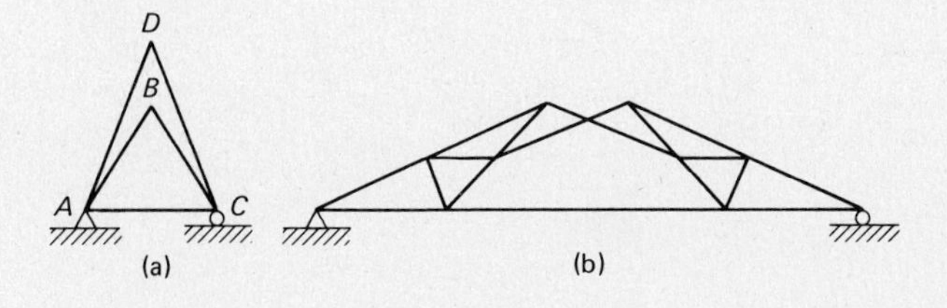

Figure 9.10

Analysis as Means of Finding Instability

The members of a truss must be arranged to support the external loads. What will satisfactorily support the external loads is a rather difficult question to answer with only a glance at the truss under consideration, but an analysis of the structure will always provide the answer. If the structure is stable, the analysis will yield reasonable results, but if it is unstable, the analysis will never balance.

Unstable Supports

A structure cannot be stable if its supports are unstable. To be stable, it must be supported by at least three nonparallel, nonconcurrent forces. This subject was discussed in Chap. 2.

9.8 EQUATIONS OF CONDITION

On some occasions two or more separate structures are connected together so that only one type of force can be transmitted through the connection. The three-hinged arch and cantilever types of structures of Chap. 3 have been shown to fall into this class because they are connected with interior hinges unable to transmit rotation.

Perhaps the simplest way to produce a hinge in a truss is by omitting a chord member in one of the panels, as shown in Fig. 9.11(a). It is obvious that the moment of all the external forces on the part of the structure to the left or the right of the pin connection at joint L_3 must be zero. The truss is statically determinate because there are three statics equations and one condition equation available for calculating the four reaction components.

The omission of members in some other situations may produce equations of condition. A diagonal of the truss of Fig. 9.11(b) has been omitted between the two interior supports. With no members in the panel able to have a vertical component of force, no shear can be transmitted through the panel, and an equation of condition is available. The supports on each side of the usually unstable rectangular shape prevent it from collapsing.

Practically speaking, the bars mentioned as being omitted are probably not omitted at all, because they would detract from the appearance of the structure and might be useful in its erection. They are frequently assembled so

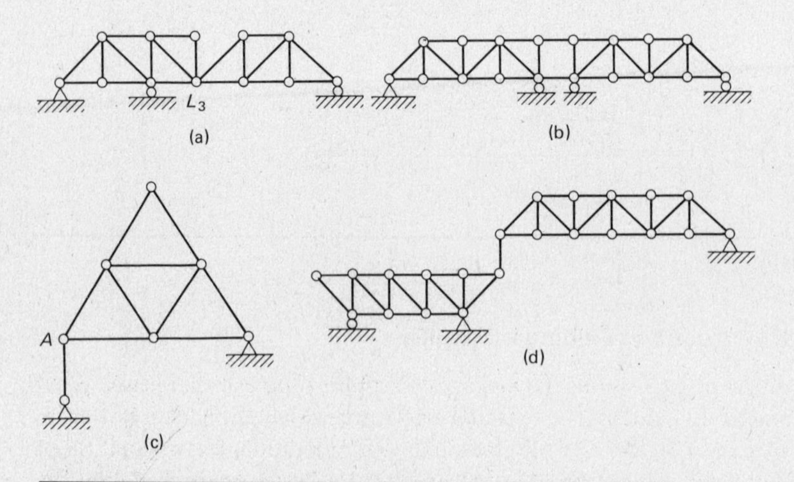

Figure 9.11

they can be adjusted to be inactive in the completed truss. The Wichert truss, discussed in Chap. 15, is a continuous truss that has omitted verticals (actually omitted) over the interior supports, each omission providing an equation of condition.

Figures 9.11(c) and (d) present two more situations in which equations of condition are produced. In the first of these there are four reaction components, and the structure may appear to be statically indeterminate externally; however, the joint at A, Fig. 9.11(c), is pin connected and cannot transmit rotation. This equation of condition makes the structure statically determinate externally. Figure 9.11(d) shows two separate trusses that are connected by a link. The link makes available two equations of condition, because rotation may not be transmitted at either end.

PROBLEMS

For Probs. 9.1 to 9.22 classify the structures as to their internal and external stability and determinancy. For statically indeterminate structures include the degree of redundancy internally or externally.

Problem 9.1

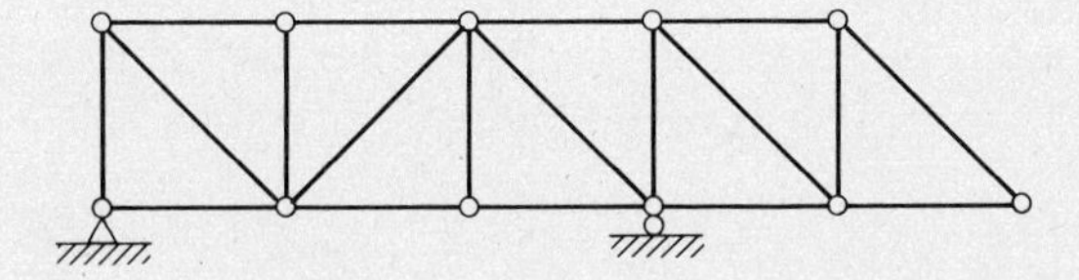

Problem 9.2 (*Ans.* Statically indeterminate internally to first degree)

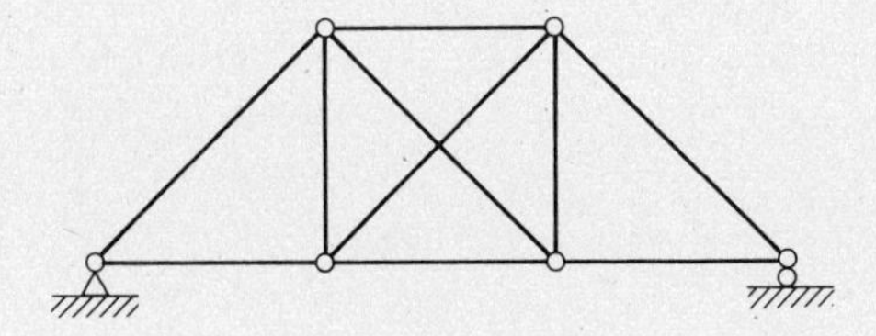

Problem 9.3

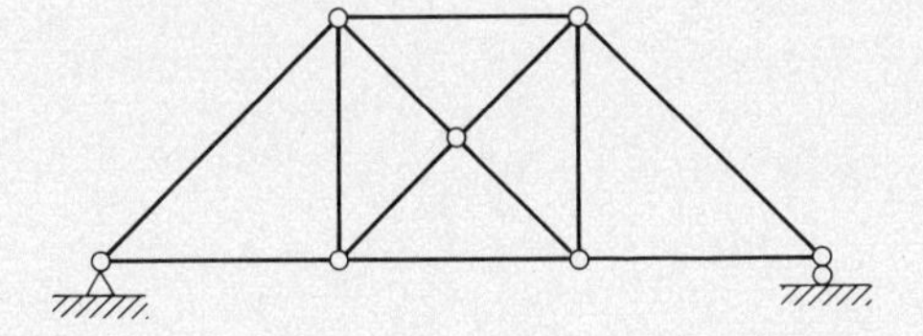

Problem 9.4 (*Ans.* Statically indeterminate externally to second degree)

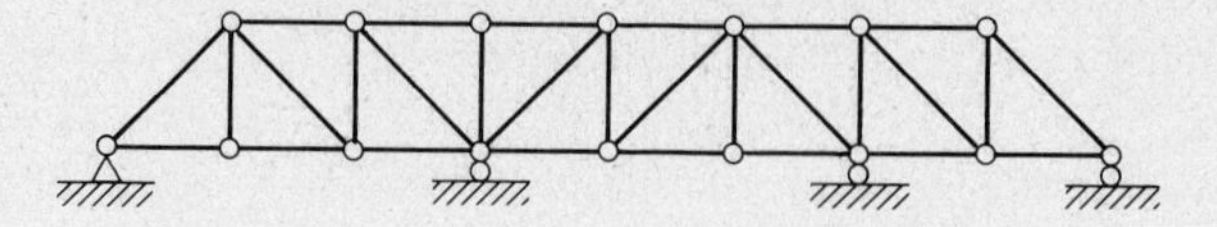

Problem 9.5

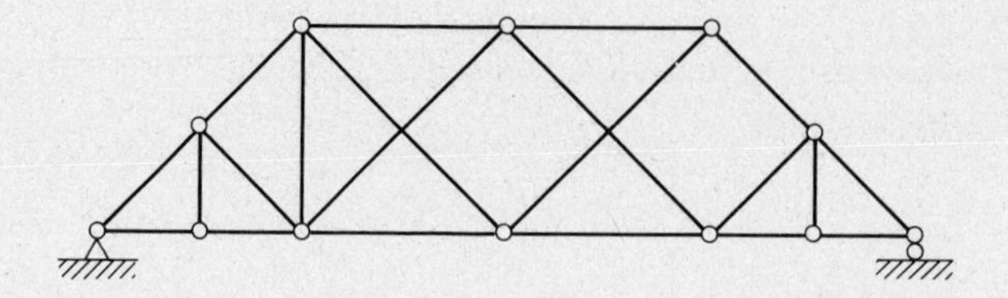

Problem 9.6 (*Ans.* Statically determinate externally and internally)

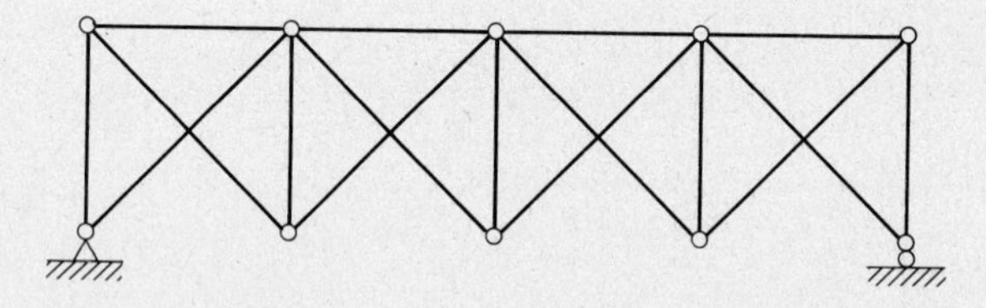

Problem 9.7

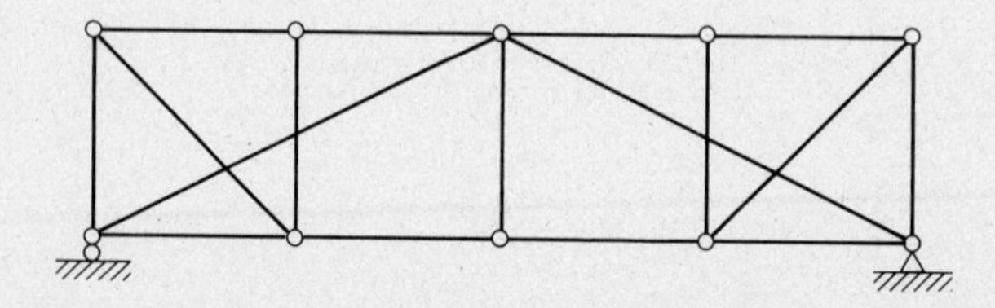

Problem 9.8 (*Ans.* Statically determinate externally and internally)

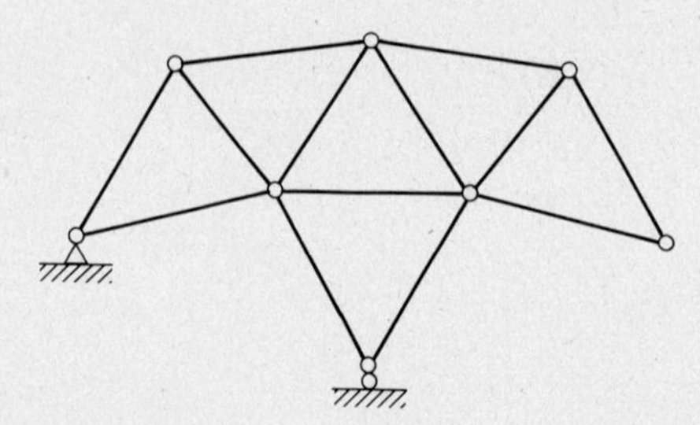

Problem 9.9

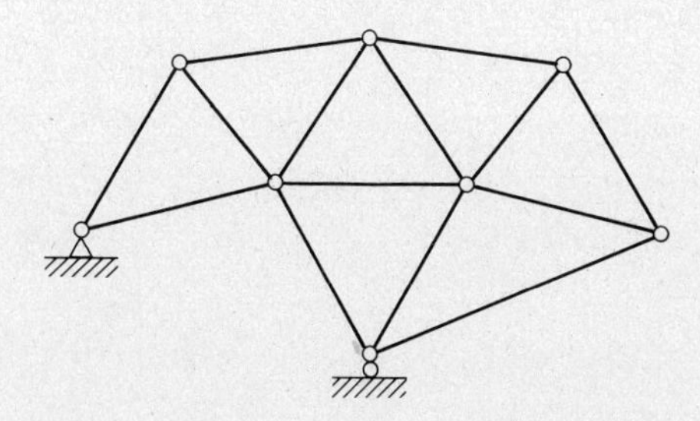

Problem 9.10 (*Ans.* Statically indeterminate externally to second degree and internally to first degree)

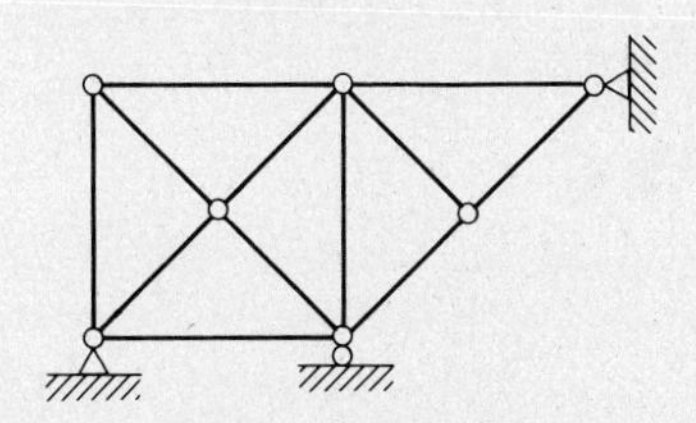

Problem 9.11

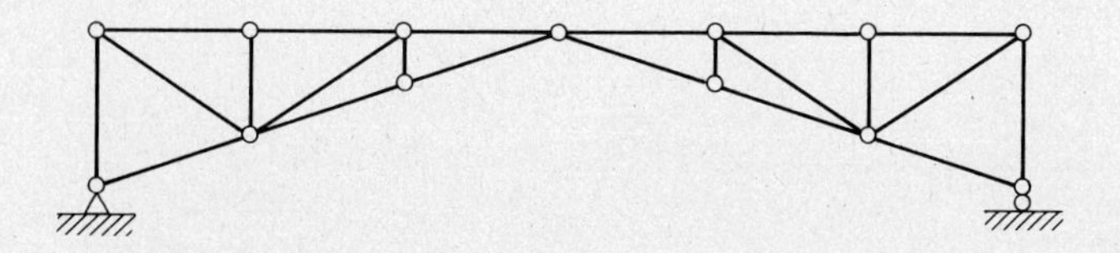

Problem 9.12 (*Ans.* Statically indeterminate externally to second degree)

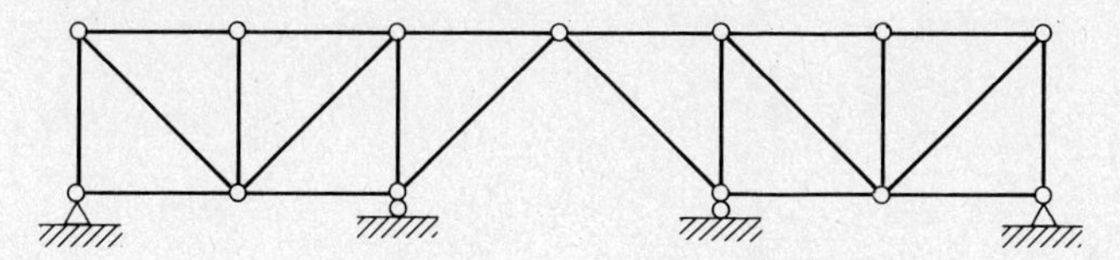

Problem 9.13

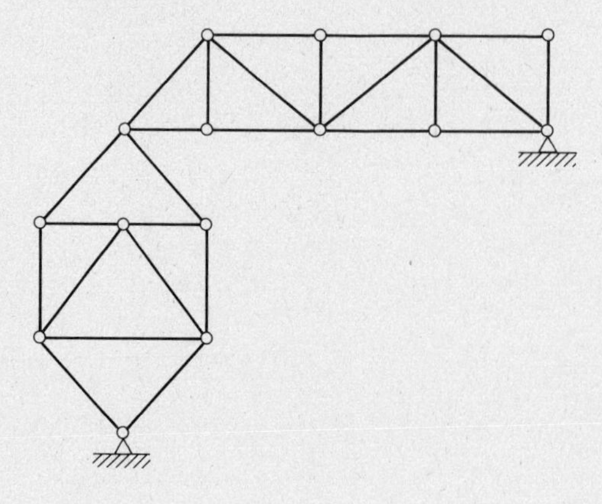

Problem 9.14 (*Ans.* Statically determinate externally and internally)

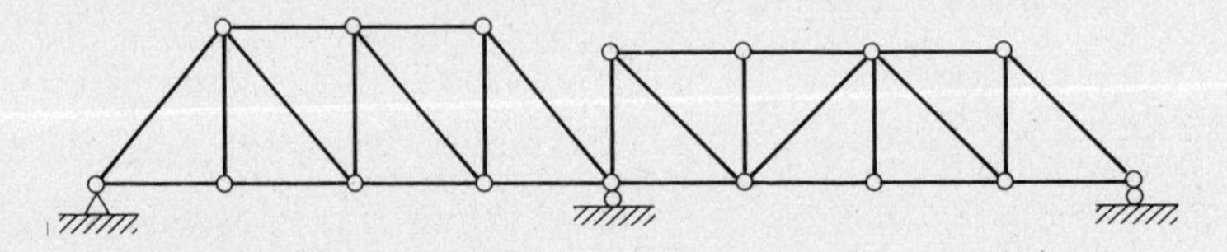

Problem 9.15

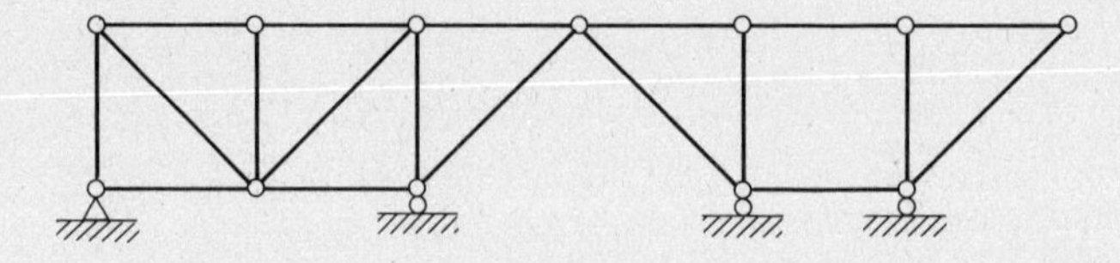

Problem 9.16 (*Ans.* Statically determinate externally and internally)

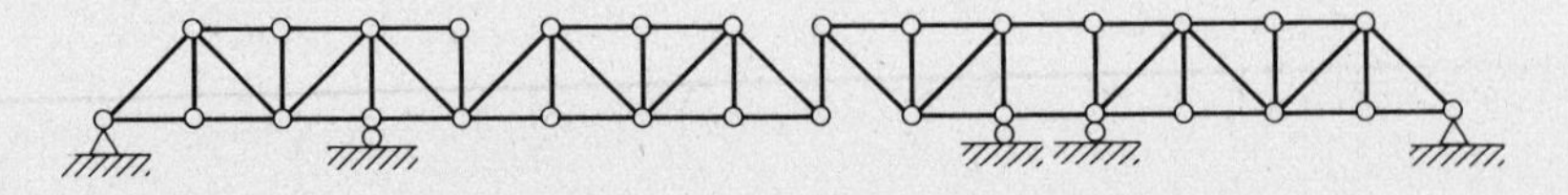

Problem 9.17

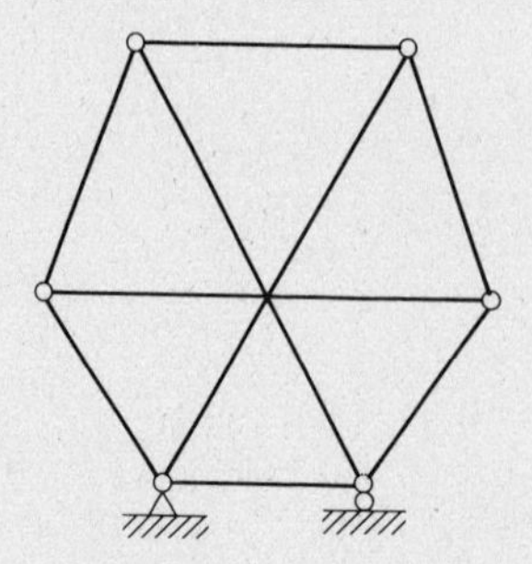

Problem 9.18 (*Ans*. Statically indeterminate internally to first degree)

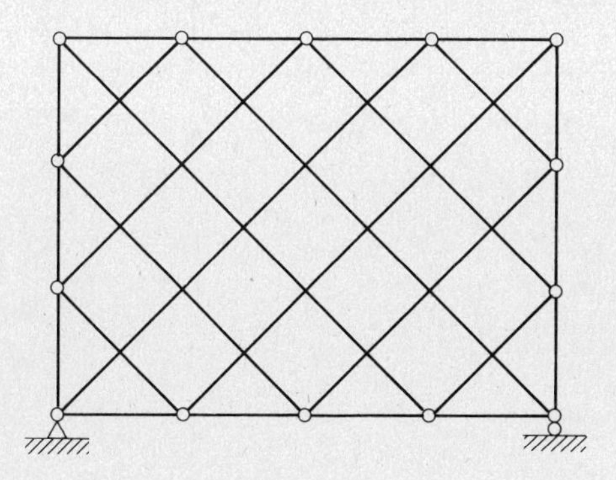

Problem 9.19

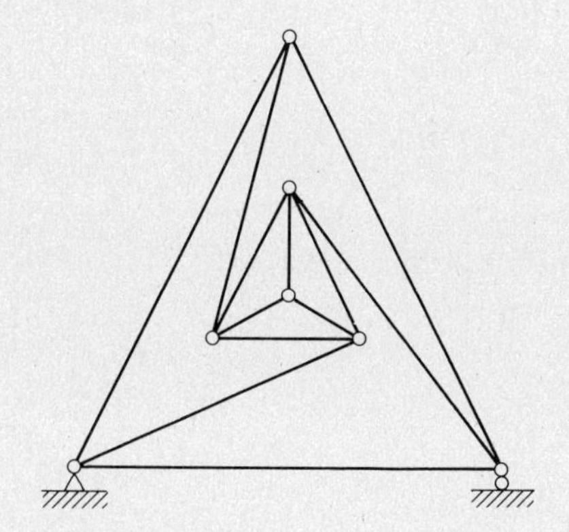

Problem 9.20 (*Ans*. Statically indeterminate internally to second degree)

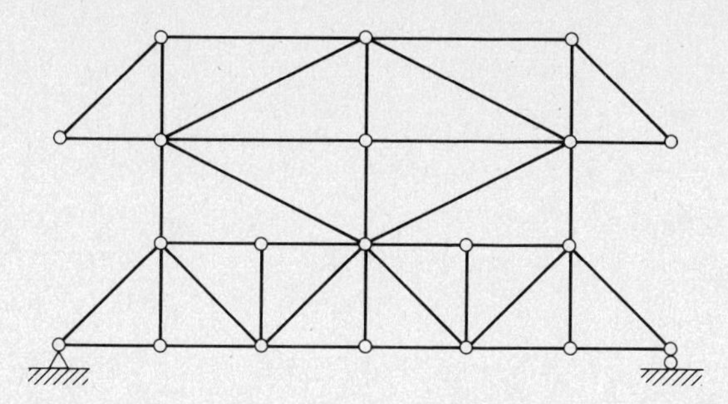

Problem 9.21

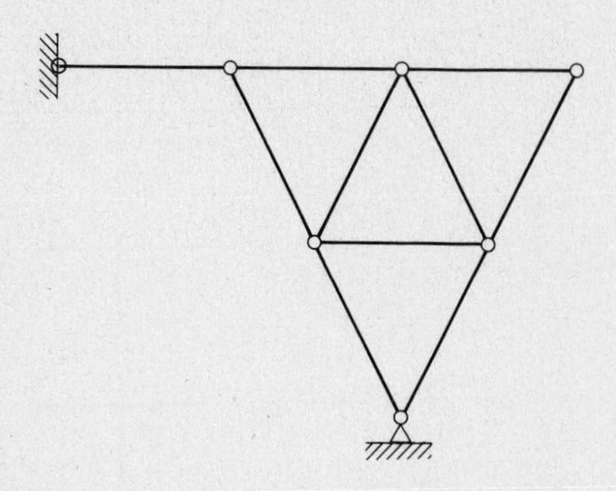

Problem 9.22 (*Ans.* Statically determinate externally and internally)

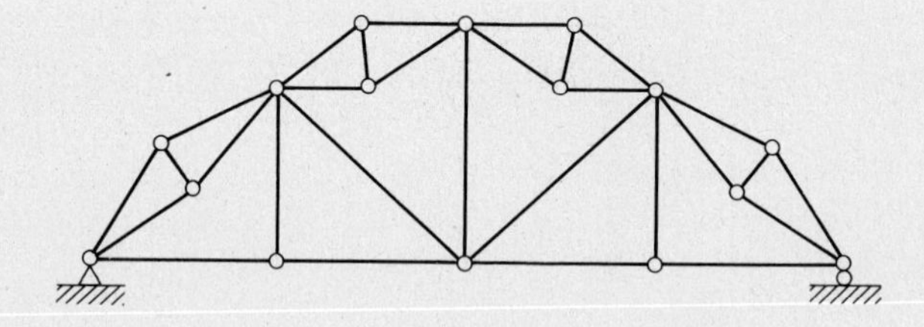

For Probs. 9.23 to 9.29 use the zero-load test to determine whether the structures have critical form.

Problem 9.23

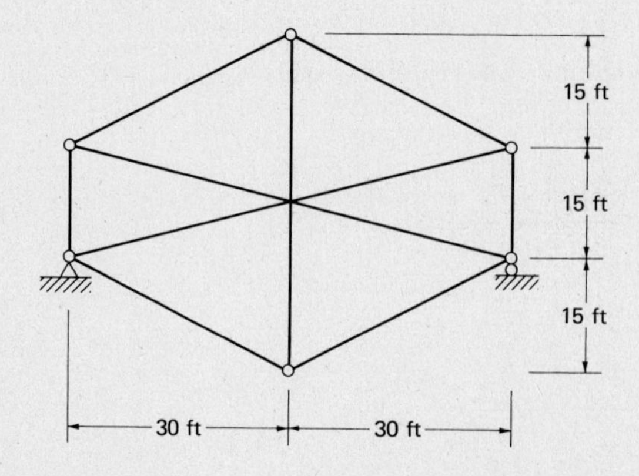

Problem 9.24 (*Ans.* Stable)

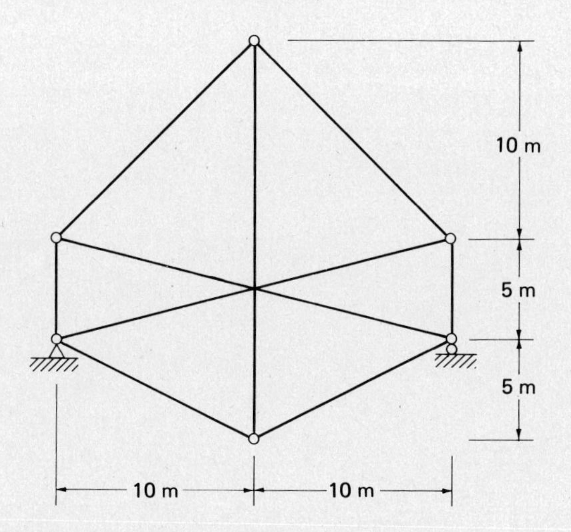

Problem 9.25

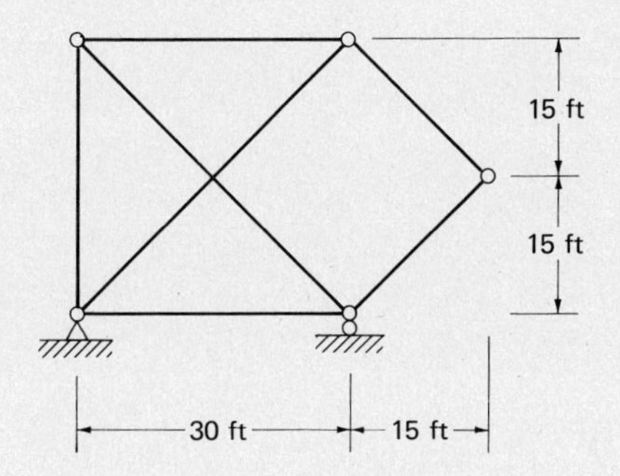

Problem 9.26 (*Ans.* Unstable)

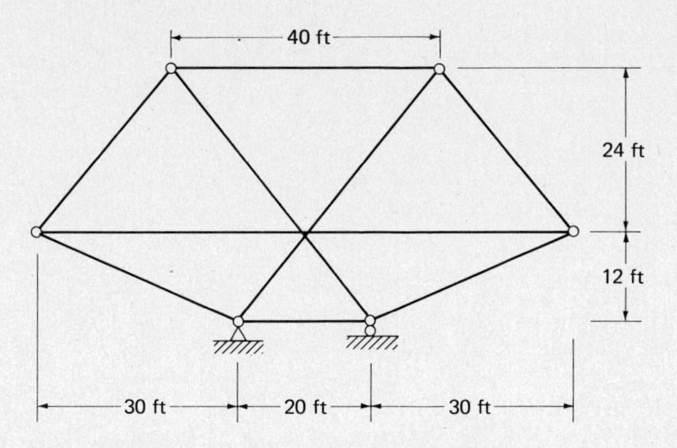

Problem 9.27

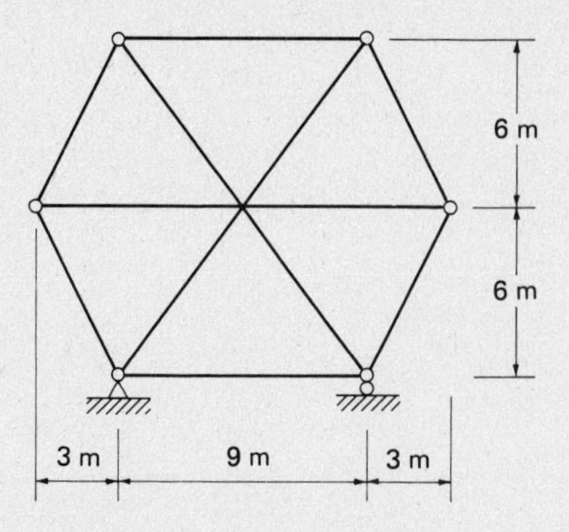

Problem 9.28 (*Ans*. Stable)

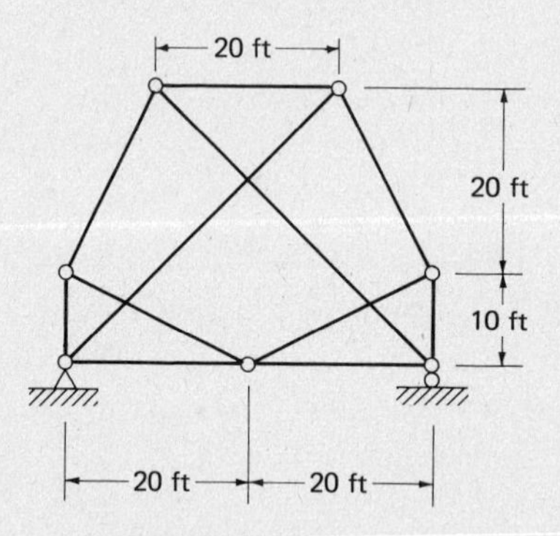

Problem 9.29

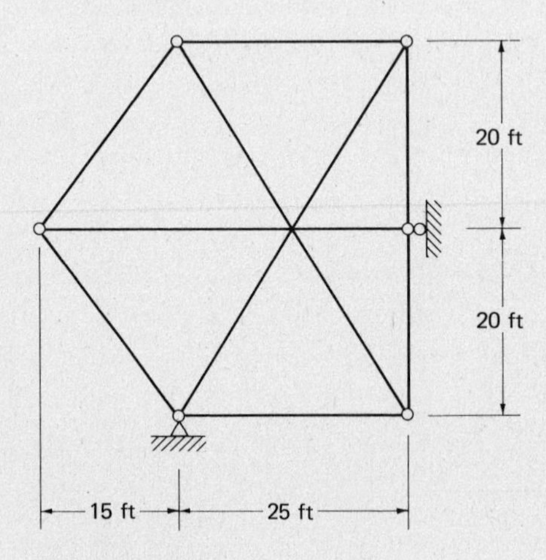

REFERENCES

[1] G. L. Rogers and M. L. Causey, *Mechanics of Engineering Structures* (New York: Wiley, 1962), pp. 19–20.

[2] S. Timoshenko and D. H. Young, *Theory of Structures* (New York: McGraw-Hill, 1965), pp. 92–103.

10
Graphic Statics

10.1 GENERAL

Reactions and forces for all types of structural frameworks may be determined from carefully constructed geometric figures as well as by the algebraic methods discussed in earlier chapters. The solution of these and other statics problems by graphical methods is known as *graphic statics*.

There are several situations in which graphical solutions are of advantage. Most nonparallel-chord trusses may be analyzed by graphics as quickly, if not more quickly, than by algebraic methods. The advantage becomes even greater for very complicated types of trusses, such as towers, which are often quite difficult to analyze by the former methods. The author believes that the "brain work" necessary for most graphical analysis is less than that required for algebraic analysis of the same structures. The simplicity of the method permits companies, such as steel fabricators, to use draftsmen for much of their analysis. The draftsmen may be completely unfamiliar with the theory of graphic statics and of algebraic solutions and yet be able to make the analyses because they have memorized the few simple steps involved. The situation pictured here is not altogether desirable, but it is occasionally encountered.

Some steel design and fabricating offices use graphics for the analysis of roof trusses such as the bowstring and the Pratt. In addition the method is occasionally used by companies that make prefabricated timber trusses. This application of graphics makes it desirable for structural engineers to be familiar with the process.

10.2 BASIC CONCEPTS

Complete mastery of the elementary concepts of graphics can be obtained immediately, because only a few basic mechanics principles are involved. These principles are reviewed in the following paragraphs.

Installation of aluminum platforms in Lake Maracaibo, Venezuela. (Courtesy of Reynolds Metals Company.)

Force Representation

A force is completely identified when its magnitude, direction, and point of application are known. A fully identified force may be represented with a line or vector. The line is drawn parallel to the force, with an arrow representing its direction, and to a scaled length representing its magnitude. (Representation of forces with fine penciled lines seems to indicate that they are concentrated at fine points. The loads are actually distributed over relatively large areas, however, and the lines merely represent their centers of gravity.)

Combination of Forces

It may be convenient to replace a group of forces with a single force, referred to as the *resultant of the forces*. Two or more nonparallel forces intersecting at one point, such as the forces in a group of truss members meeting at a joint, may be graphically combined into one resultant. The forces P_1 and P_2 of Fig. 10.1 are combined with a force triangle in (b) and with a force parallelogram in (c). The magnitude of the resultant force R can be obtained by scaling it with the same scale used initially to draw P_1 and P_2.

A group of four forces is combined into one resultant in Fig. 10.2. An arbitrary point was selected as the starting position, and successive lines were

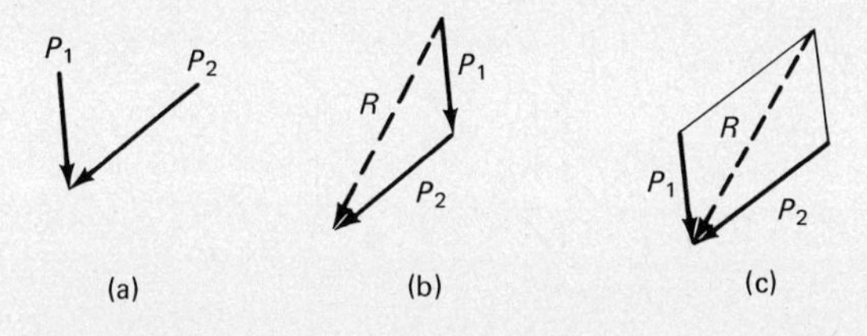

Figure 10.1

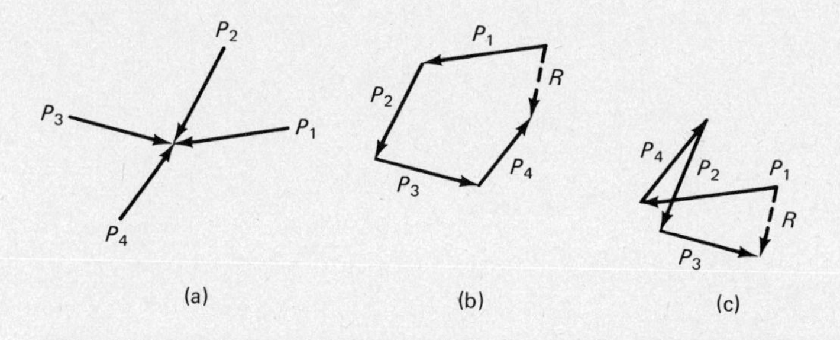

Figure 10.2

drawn representing each of the forces, the lines being parallel to the actual forces and scaled to the proper magnitudes. The order in which the forces are considered is immaterial. They are drawn in completely different orders in Figs. 10.2(b) and (c), but the resultant is the same in magnitude and direction in each case. The resultant was found by drawing a line (dashed in the figure) from the starting to the ending point. The resultant of all of the forces applied to a body in equilibrium is zero, and the starting and ending points of the polygon of forces coincide. Each of the joints of a stable truss is in equilibrium, and the resultant of all of the loads and member forces at a joint is zero.

Resolution of Forces

Not only can two or more forces that meet at a point be combined into one resultant force, but any single force may itself be broken down arbitrarily into two or more components the resultant of which is the force. Force P_1 is resolved into components a and b in Fig. 10.3.

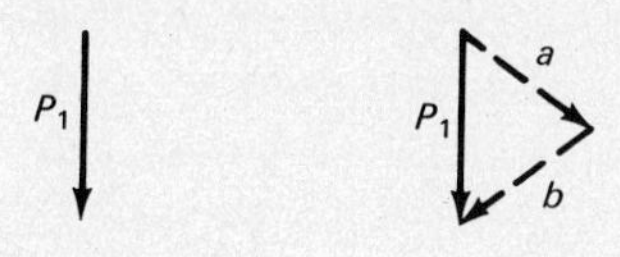

Figure 10.3

Rigid Bodies in Equilibrium

The structures considered in this book are assumed to be in equilibrium or at rest, and the summation of all the forces applied to any one of them is zero. Furthermore they are assumed to be rigid bodies, which will not deform under load. The application of load to any body causes some deformation, but it is usually so small as to be negligible in its effect on the stresses of the body.

10.3 BOW'S NOTATION

A convenient system for numbering the members, loads, and reactions of a truss to be analyzed by graphics is shown in Fig. 10.4. This system is known as *Bow's notation* and consists in placing a number in each of the triangles of a truss and a letter in the space between each of the external loads and reactions. By this method each of the members can be designated by the letters or numbers on each side of it; examples are members *B*–1, *E*–6, 6–7, and 9–10 in the truss shown. Similarly, each of the external forces is designated by a pair of letters, such as loads *B*–*C* and *H*–*A* and reaction *G*–*H*. Bow's notation is more satisfactory for graphics than the joint-numbering system used in the earlier chapters, where the joints were numbered L_0, L_1, U_1, U_2, and so forth. To use the joint system it would be necessary to give additional letters or numbers to the external loads and reactions on the structure.

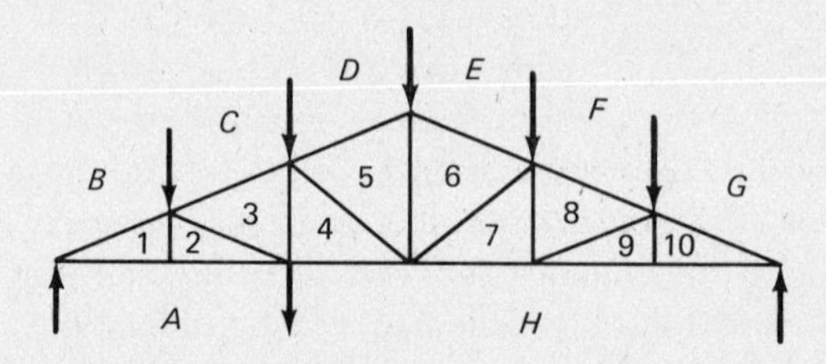

Figure 10.4

10.4 THE FORCE POLYGON

Three forces (P_1, P_2, and P_3) acting on a rigid, nondeforming body are represented in Fig. 10.5(a). For the body to be in equilibrium there must be another force applied to it: a reaction equal and opposite to the resultant of the three loads. A force polygon is drawn in Fig. 10.5(b) for all four forces. This polygon should be carefully examined, because it illustrates the following two important facts: (1) The polygon closes because the forces are in equilibrium; and (2) the arrows on the forces follow each other successively around the polygon.

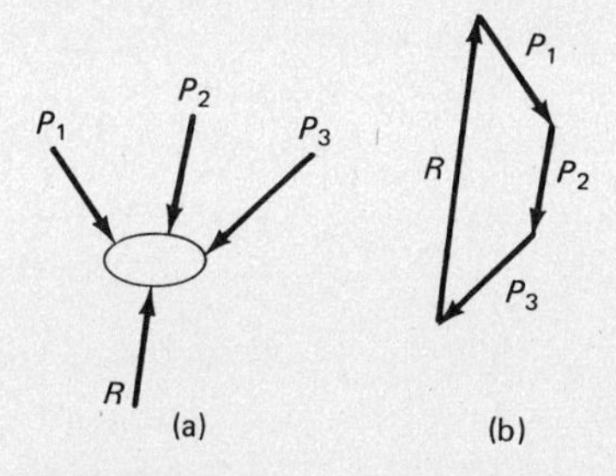

Figure 10.5

10.5 FORCE POLYGONS FOR INDIVIDUAL JOINTS OF A TRUSS

Two or more forces meet or intersect at each joint of a truss. If the truss is in equilibrium, each and every joint of the truss is in equilibrium, and the resultant of all the forces at any joint is zero; therefore a force polygon drawn for any joint must close. It will be seen in the following paragraphs that if one or two, but not more than two, forces meeting at a joint are unknown, they may be determined by drawing a force polygon.

Figure 10.6 presents a truss for which the forces are to be determined graphically. Truss reactions may be determined algebraically or graphically

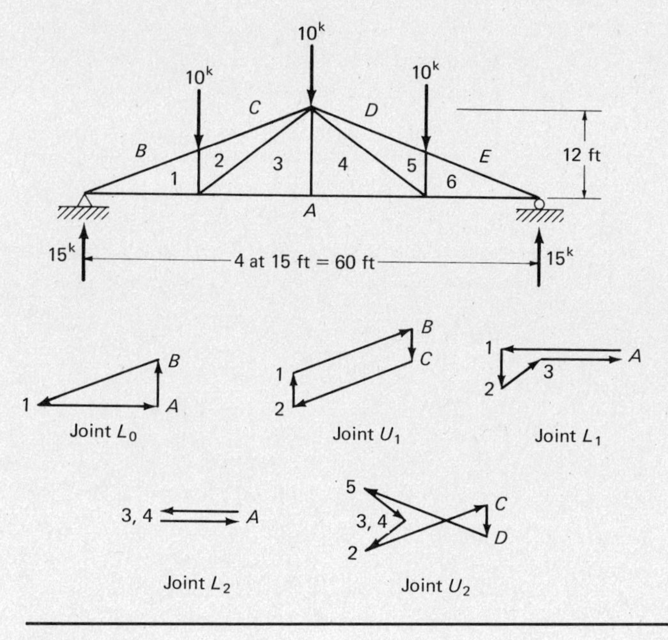

Figure 10.6

(as described later in the chapter), but for the usual loading conditions they are determined so easily by algebraic methods that a graphical solution would waste time. A procedure closely related to the algebraic method of joints is used to calculate the forces in truss members by graphic statics. Joints are taken out one by one and a diagram (or force polygon) is drawn for each, which in effect applies the equations of statics that set the horizontal forces equal to zero and the vertical forces equal to zero. Force polygons are shown in Fig. 10.6 for each of the joints on the left-hand side of the truss. Because the truss and loading are symmetrical on each side, it is unnecessary to consider the right-hand side of the truss. The polygons are prepared as described in the following paragraphs.

Joint L_0

There are three forces at the joint—a 15-kip reaction and two unknown member forces. The magnitude and character of the two forces may be determined by drawing a force polygon for the joint. The order in which the forces are considered in drawing a separate polygon for each joint is of little importance, but in drawing a combined polygon for an entire truss (to be described) it is essential for all the forces to be taken successively in a clockwise order or in a counterclockwise order. To form the proper habit, the individual polygons are drawn by taking the forces one by one as they are encountered in a clockwise order. The forces at joint L_0 in a clockwise order are A–B, B–1, and 1–A. If they were taken in a counterclockwise order, they would be B–A, A–1, and 1–B.

A scaled line is drawn to represent the reaction A–B. In going clockwise around the joint from A to B, the line must go up 15 kips. Force B–1 is a force applied at the joint acting to the right or left and parallel to the member, so a light construction line is drawn through B in this direction. Force 1–A acts either to the right or left parallel to the member and, because the polygon must close for equilibrium, ends up at A, the starting point of the polygon. A construction line is drawn through point A parallel to force 1–A. As point 1 must fall on each of the two construction lines it can only be at one point: the intersection of the two lines. The force polygon for joint L_0 is complete and may be darkened with the pencil.

The arrow on force A–B was up, and since the arrows on a force polygon must follow each other successively around the polygon, the directions of forces B–1 and 1–A are known. The character of the two forces is now obvious from the arrows on the polygon. Force B–1 is acting to the left and pushing against L_0 (compression), and 1–A is to the right pulling away from L_0 (tension). Their magnitudes are available by scaling.

Joint U_1

There are three unknown forces at joint L_1 and they cannot be determined conveniently at the present time. Joint U_1 is therefore considered next; it has

Aluminum trusses, Reynolds Hangar, Byrd Field, Richmond, Virginia. (Courtesy of Reynolds Metals Company.)

only two unknowns. The members are again taken in a clockwise order around the joint, and the member previously denoted as B–1 at joint L_0 has become 1–B.

The force in 1–B, which was determined by the force polygon for L_0 is compressive and is pushing against the joints at each of its two ends. Its value is scaled off by going to the right and parallel to the member. A line representing load B–C is scaled downward 10 kips from B. Through point C a line is drawn parallel to member C–2; and through point 1 a line is drawn parallel to member 2–1, their intersection being point 2. The directions and magnitudes of all the forces at joint U_1 are now known. Force C–2 is to the left pushing against the joint and is compressive, whereas force 2–1 is pushing up against the joint and also is compressive.

Joint L_1

The determination of force 2–1 at joint U_1 leaves only two unknown forces at joint L_1, which permits the construction of a force polygon there. Member A–1 is taken first; its magnitude and direction are available from the polygon for joint L_0. The force is tensile and pulls away from the joint to the left. Member 1–2 is in compression and pushes down against the joint. After the known forces are drawn in, a line is drawn through point 2 parallel to member 2–3, and another line is drawn through A parallel to member 3–A. Their intersection is point 3, which in this case lies along the previously drawn line, A–1. The force polygon starts at A, goes left to 1, down to 2, up to 3, and right to A again.

Other Joints

The other joints for which polygons have been prepared, L_2 and U_2, are handled in a similar manner. Notice that when the diagram for joint L_2 was drawn, points 3 and 4 were found to be at the same location, thus indicating that the force in member 3–4 is zero.

10.6 THE MAXWELL DIAGRAM

The preparation of separate force polygons for each of the joints in a truss wastes considerable time and space, because it is necessary to repeat lines used in previous polygons. For instance, the force 1–A in the polygon for L_0 is redrawn as A–1 in the polygon for L_1; the force C–2 at U_1 is repeated as 2–C at U_2; and so forth. The force polygons for all of the joints of a truss may be combined in one large diagram in which each force is represented with only one line. The combined drawing is the ***Maxwell diagram*** (sometimes called the ***reciprocal polygon diagram***), which is drawn by the identical methods used for the individual-joint polygons. A Maxwell diagram is drawn in Fig. 10.7 for the same roof truss considered in Fig. 10.6, and its preparation is described in the following paragraphs. The forces and loads at each of the joints are taken in a clockwise order, although the same numerical results would be obtained by taking them in a counterclockwise order. A step-by-step construction of the diagram is presented in Fig. 10.7.

Joint L_0

The diagram for joint L_0 is drawn in Fig. 10.7 exactly as in Fig. 10.6 except that no attempt is made to place arrows on the lines; they would be confusing and are unnecessary in determining the character of the forces. The forces have been considered successively around the joint in a clockwise order; therefore it is known that in going from B to 1 on the Maxwell diagram, the line goes to the left. Force B–1 acts to the left and pushes against the joint (compression). Continuing around the joint, the next force is 1–A. In going from 1 to A on the

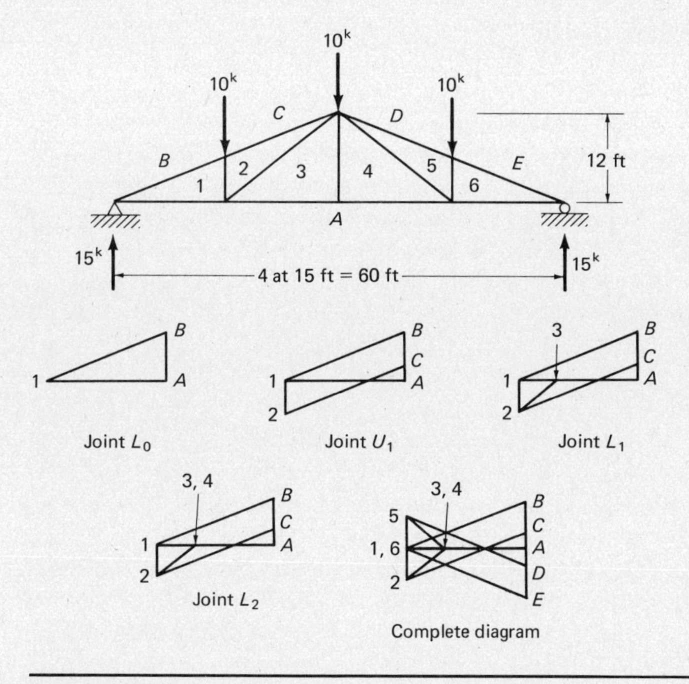

Figure 10.7

diagram the line goes to the right and the force is pulling away from the joint (tension).

Joint U_1

The polygon for joint U_1 is drawn as a continuation of the one for L_0. On starting around the joint clockwise, the first force encountered is 1–*B*, which was previously developed on the diagram. The next force is the load *B*–*C*, and it is represented by a 10-kip line downward. Finally, a line is drawn through *C* parallel to *C*–2 and another line parallel to 2–1 is drawn through 1, their intersection being point 2. The diagram shows *C*–2 is to the left, pushes against the joint, and is compressive. Similarly, the direction from 2 to 1 on the diagram is up, the force pushes against the joint, and member 2–1 is in compression.

Joint L_1

By using the figure prepared for joints L_0 and U_1 and going clockwise around L_1, the first forces encountered are *A*–1 and 1–2. The lines representing these forces are already on the diagram. A line is drawn through point 2 parallel to 2–3, and another line is drawn through *A* parallel to 3–*A*. The intersection of the two is point 3.

Other Joints

The other joints are handled in the same manner; each is an extension of the previous diagram. If the Maxwell diagram is accurately drawn, results of a high degree of accuracy, satisfactory for any design work, will be obtained. Sharp pencils, careful scaling of distances and great care in drawing parallel lines are essential. Most drafting offices have equipment to facilitate the construction of parallel lines, but in the absence of special equipment a T square and a pair of large triangles (or, better, a pair of adjustable triangles) will produce satisfactory results.

10.7 SAMPLE PROBLEMS

Figures 10.8 to 10.10 illustrate the analysis of three different types of trusses by graphic statics. The procedure is identical with the one used for analyzing the truss of Fig. 10.7, but a few explanatory remarks may be helpful.

The hinge reactions for the trusses of Figs. 10.9 and 10.10 have been broken down into their vertical and horizontal components as shown. If resultant hinge reactions were used, the results would be the same. In Fig. 10.9, *A–B* is the vertical reaction component and *B–C* is the horizontal reaction component. Should the resultant of the two components be used, it would

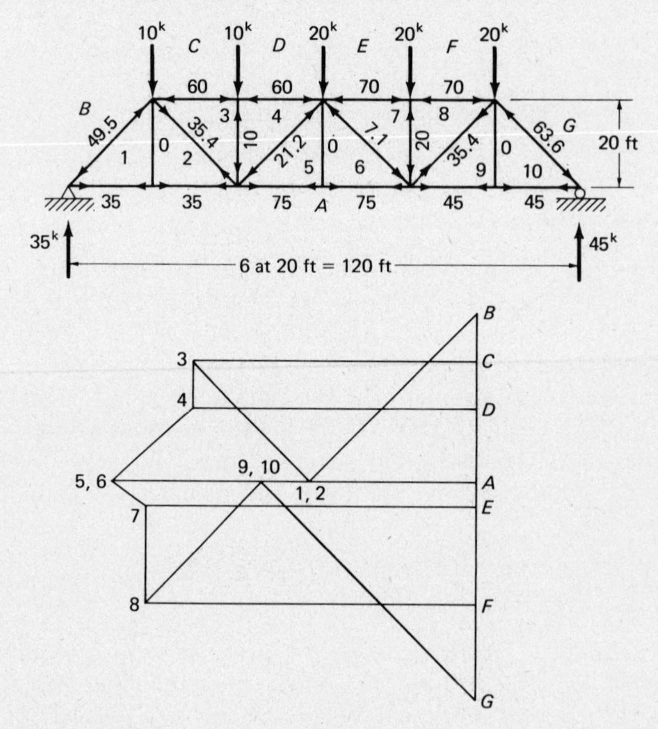

Figure 10.8

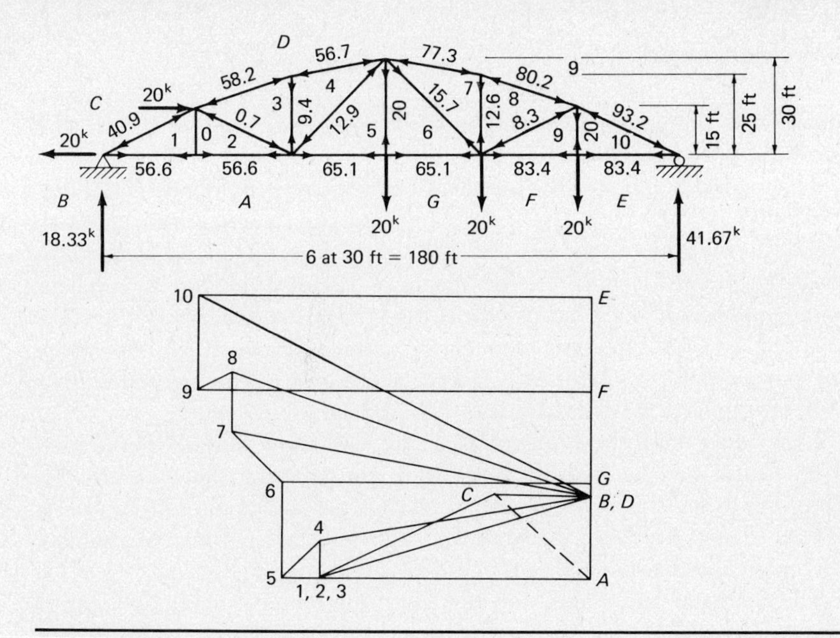

Figure 10.9

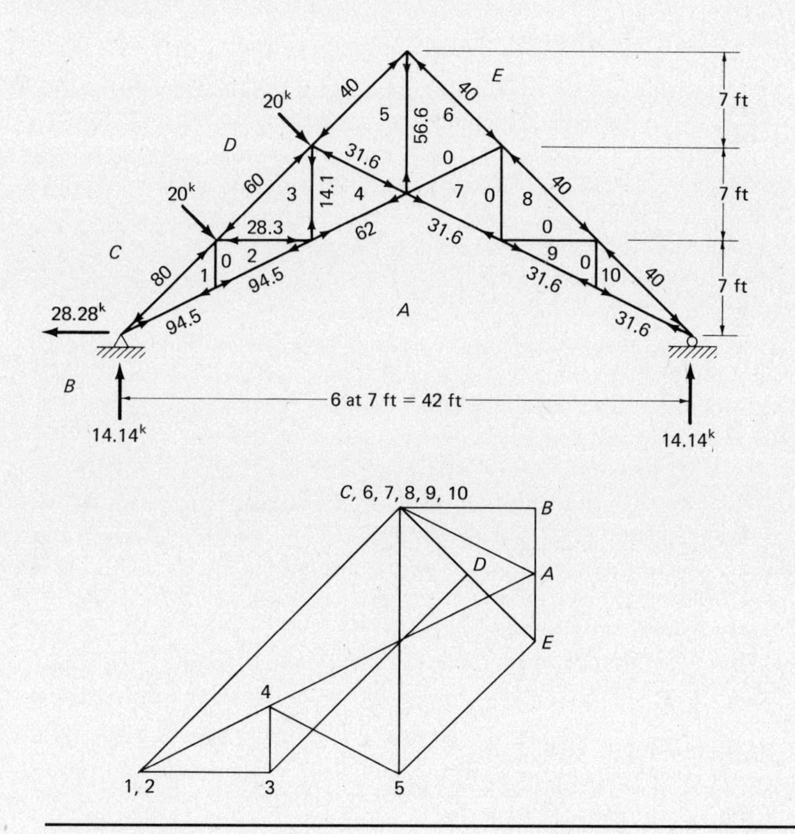

Figure 10.10

coincide with the dotted line A–C in the figure. The reactions for all three trusses have been obtained algebraically.

The reader should note the manner in which the bottom panel point loads are handled in Fig. 10.9. Upon reaching joint L_3, having constructed the polygons for the joints on the left-hand side of the truss up to L_3, it is noted that point A is already located on the diagram. Because the forces are being considered as encountered in a clockwise direction around the joints, the 20-kip load is lettered G–A. On going from G to A, a line is drawn 20 kips downward to the existing point A; therefore a bottom-chord load is seemingly handled in exactly the opposite manner from a top-chord load.

Experience with the Maxwell diagram will show that all of the loads and reactions may be laid off initially in constructing the diagram. The student, however, is not encouraged to try this procedure until he or she has practiced several problems, because the loads may be confused, particularly when top and bottom panel points of a truss are loaded.

The character and magnitude of each of the forces as obtained from the Maxwell diagram are indicated on the truss members in Figs. 10.8 to 10.10.

10.8 SUBSTITUTE MEMBERS

The Maxwell diagram can be drawn quite easily for simple trusses, but the work is a little more difficult for compound trusses. It may be possible to handle several joints of a compound truss by the usual methods, but there will come a time when each of the subsequent joints will have three unknown member forces. The Fink truss of Fig. 10.11(a) is an excellent example. The lines representing the forces in members meeting at joints L_0, U_1, and L_2 are quickly drawn, but there are three unknown forces meeting at each of the next two joints, U_3 and L_4. These forces may not be determined by the graphical methods discussed so far unless one of the unknowns is otherwise determined.

One solution to the problem is to compute algebraically the force in one of the unknown members and draw its proper position on the diagram. In Fig. 10.11(b) the diagram has been developed for joints L_0, U_1, and L_2, and it is assumed that the force in member D–5 has been calculated algebraically to be 69.8 kips in compression. Starting around U_3 in a clockwise direction, 3–2 is known, 2–C is known, and C–D is drawn downward 10 kips from C. The force in member D–5 is compressive and pushing against the joint, so 69.8 kips is scaled off to the left parallel to the member to point 5 as shown in Fig. 10.11(b). With D–5 represented on the diagram the forces in 5–4 and 4–3 can be obtained by the usual graphical methods. The diagram is completed for the left-half of the truss, which is symmetrical as to both dimensions and loading about the centerline. The completed diagram is the same as Fig. 10.11(d) with point X and the dotted line X–6 omitted.

Another method of handling joints in a compound truss having three unknown forces is the method of substitute members. By this method two of the members of the truss at the difficult joints are assumed to be replaced with

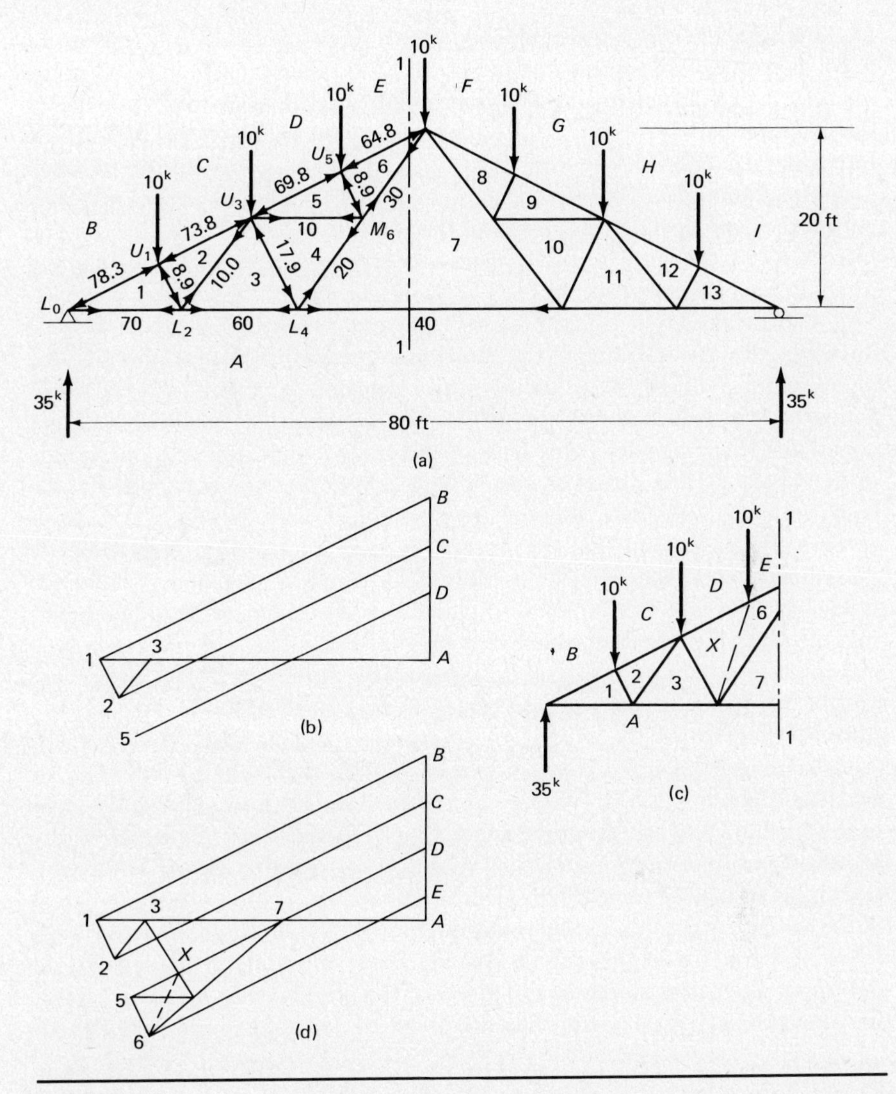

Figure 10.11

a single member, leaving only two unknowns at one of the joints. This substitution permits the continuation of the diagram, although several fictitious forces will be obtained in the undisturbed members of the truss as well as in the substitute member. Eventually, however, a force that is correct for the original truss will be obtained. Having obtained a correct force, it will be possible to work back through the joints of the original truss and find the correct forces in all of the members.

The truss of Fig. 10.11(a) is considered in the following discussion, and it is assumed that the Maxwell diagram has been constructed for joints L_0, L_2, and

U_1. Members 4–5 and 5–6 are replaced with a single member running from L_4 to U_5, as shown by the dotted line in Fig. 10.11(c). The triangle to the left of the member is given the letter X. The insertion of a fictitious member in place of the two members leaves only two unknowns at joint U_3, D–X, and X–3. These forces, correct only for the fictitious truss, can be obtained with the Maxwell diagram. Although not correct for the actual truss, they are useful because they permit the graphical determination of the forces E–6 and 6–X at joint U_5. The force in E–6 is the same for both trusses and is the key to finding the remaining forces in the original truss.

A study of section 1–1 in Figs. 10.11(a) and (c) shows the force E–6 to be unchanged by the insertion of the substitute member. To determine the force in this member algebraically, moments can be taken about joint L_4 of the forces acting on the free body to the left of the section. The resulting moment equation is the same with either truss; therefore the graphically obtained value of E–6 in the fictitious truss is correct. Now that E–6 has been established, the truss with the substitute member is abandoned and the original truss is reconsidered. At joint U_5 the forces in 6–5 and 5–D are graphically determined; then the diagram is drawn for joint U_3, which has only two unknowns, 5–4 and 4–3. No other difficulties will be encountered in completing the drawing for the remainder of the truss. The point X is only a construction point and is not used subsequently; it may be erased to prevent possible confusion in scaling force values. Figure 10.11(d) shows the complete Maxwell diagram for the left-half of the truss.

There are several other compound trusses that require algebraic or substitute-member assistance for complete analysis. The procedure is the same as the one used for the Fink truss, although some trusses may require substitute members (or algebraic computations) in more than one location. Two precautions must be kept in mind when a substitute member is used: (1) The removal of the actual truss members and replacement with the substitute member must leave a stable truss; and (2) the arrangement must allow the analyst to determine the correct force in a member of the original truss. In some cases it may be necessary to go two or three joints before a true force can be obtained.

10.9 GRAPHICAL DETERMINATION OF REACTIONS; NONPARALLEL LOADS

For a structure to be in equilibrium, the resultant of the applied external loads must coincide with the resultant of the reactions and be equal in magnitude and opposite in direction. Similarly, the resultant of any group of reactions and loads on a structure must be equal and opposite to the resultant of the remaining reactions and loads.

The load P on the simple beam of Fig. 10.12(a) produces the reactions R_1 and R_2. The direction of the left-hand reaction R_1 at the hinge is not known, but the right-hand reaction R_2 is vertical since it is perpendicular to the horizontal supporting surface beneath the roller. If the lines of action of P and R_2 are

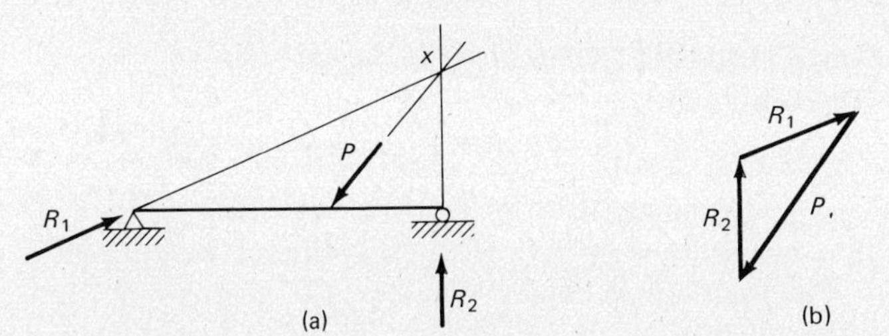

Figure 10.12

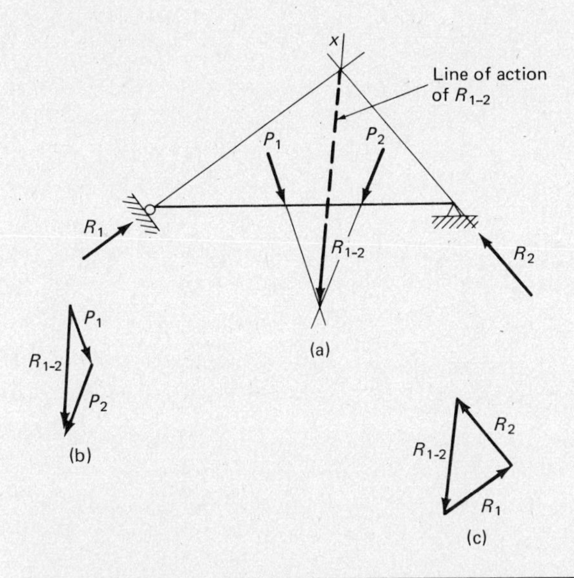

Figure 10.13

extended, they will intersect at some point x. The only remaining force on the beam is R_1, and it must pass through x and be equal and opposite to the resultant of P and R_2. If R_1 did not pass through the point, the statics equation $\Sigma M = 0$ of the forces acting on the beam taken about x could not be satisfied.

The direction of R_1 is known, since the line of action passes through the left hinge and point x. Having the directions of all three forces and the magnitude of one, it is possible to determine the magnitude of the other two by drawing a force polygon, Fig. 10.12(b).

The beam of Fig. 10.13(a) has two loads, P_1 and P_2. These loads are combined into one resultant $R_{1\text{-}2}$ with a force triangle in Fig. 10.13(b). The line of action of the resultant is indicated by the dotted line in (a). The problem is now the same as the one handled in Fig. 10.12. The resultant of the loads is extended until it intersects the line of action of the roller reaction R_1 at x. The line of action of the right-hand reaction R_2 must also pass through x. A force polygon is drawn in Fig. 10.13(c), and the magnitudes of R_1 and R_2 can be determined by scaling.

10.10 GRAPHICAL DETERMINATION OF RESULTANT OF SERIES OF PARALLEL LOADS

The beams of Figs. 10.12 and 10.13 had conveniently supported loads whose lines of action made fairly large angles with each other. This relationship made it possible to determine the resultants of the loads at the points of intersection of the lines of action. The resultants, having considerable inclination, were intersected with the lines of action of the roller reactions within the confines of an ordinary size sheet of paper, which permitted the determination of the reactions.

When a structure is loaded with a series of forces that are parallel, the method discussed in Sec. 10.9 cannot be applied successfully. Although the magnitude of the resultant is easily found from a force polygon, the fact that the lines of action of the forces do not intersect makes the location of the resultant something of a problem. Even if the forces are not parallel but nearly so, they will not intersect on an ordinary size sheet of paper. When the supporting surface underneath the roller is horizontal, the reaction is vertical, and it will not intersect the resultant of a set of gravity loads (except possibly at infinity). The first of these problems, the one of locating the resultant of a group of parallel loads, is discussed in this section. Section 10.11 explains how the reactions of a beam loaded with parallel loads may be determined graphically.

The magnitude and direction of the resultant of the three parallel loads acting on the beam in Fig. 10.14(a) are determined by drawing a force polygon, shown in 10.14(b). The resultant is represented by a line connecting the starting and ending points of the diagram, its direction being parallel to the line.

It is often convenient to resolve a force into two components, an infinite number of pairs of components being available. The forces P_1, P_2, and P_3 of Fig. 10.14 are each resolved into a pair of components making a considerable angle with each other, as shown in Fig. 10.15. This procedure is followed to permit the lines of action of the assumed components to intersect within the extent of an ordinary size sheet of drawing paper.

A convenient method of assuming components is illustrated in Fig. 10.15. The loads are plotted end to end, and an arbitrary point is selected either to the right or to the left of the loads. From this point, labeled 0 in the figure, lines are drawn to the starting and ending points of each of the plotted loads. Examination of the force polygon shows each load has been broken down into a

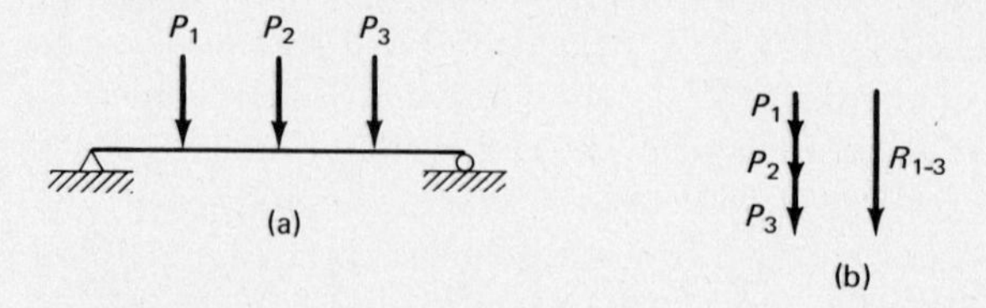

Figure 10.14

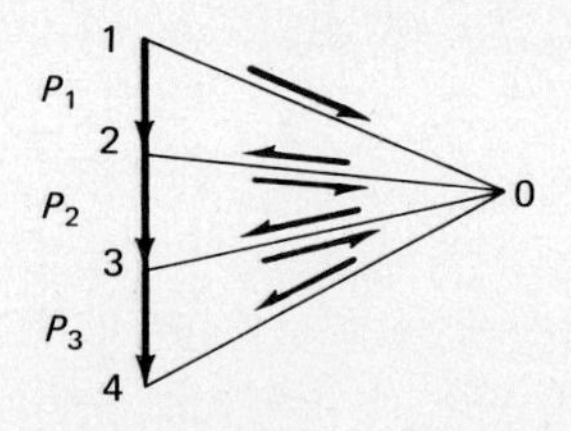

Figure 10.15

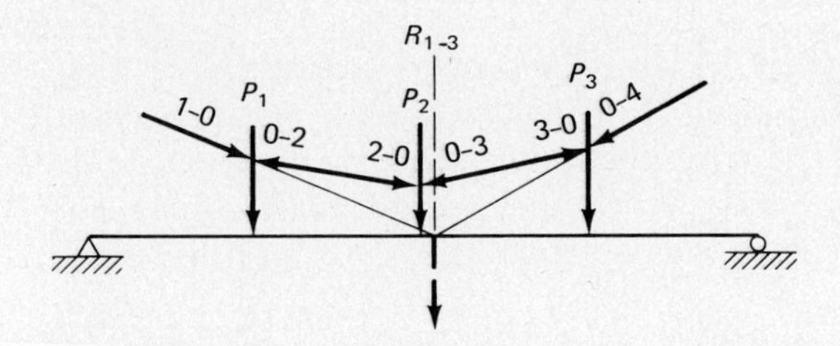

Figure 10.16

pair of components. Load P_1 has the components 1–0 and 0–2; load P_2 has the components 2–0 and 0–3; and load P_3 has the components 3–0 and 0–4.

Each of the loads having been replaced with a pair of components, the components may be combined as shown in Fig. 10.16. At an arbitrary point along its line of action, P_1 is replaced with its components 1–0 and 0–2. The line of action of component 0–2 is extended until it intersects the line of action of P_2. At this point P_2 is replaced with components 2–0 and 0–3. Finally, the line of action of 0–3 is extended until it intersects the line of action of P_3, at which point P_3 is replaced with components 3–0 and 0–4.

By examining this diagram, called the *funicular polygon* or *string polygon*, it can be seen that components 2–0 and 0–2 cancel each other (because they are equal and opposite along the same line of action) as do components 3–0 and 0–3. The entire force system has been reduced to two components 1–0 and 0–4. On referring to the force polygon of Fig. 10.15, it is obvious that the resultant of these two components is the same as the resultant of the three loads. The direction and magnitude of the resultant can be obtained by drawing a line through the starting and ending points of the polygon. The intersection of the two remaining components, 1–0 and 0–4, is a point along the line of action of the resultant of the three loads. Any number of loads can be reduced to two components with the string polygon.

10.11 GRAPHICAL DETERMINATION OF REACTIONS; PARALLEL LOADS

The same general approach is used for determining the reactions of a beam loaded with parallel loads as was used for determining the resultant of a set of parallel loads. The loads P_1 to P_4 applied to the beam of Fig. 10.17 are each

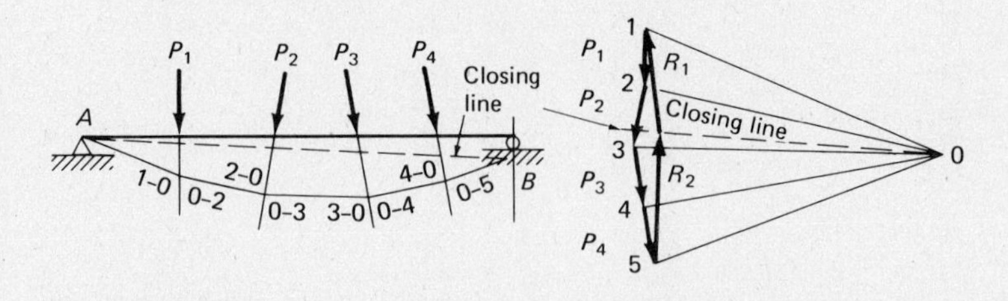

Figure 10.17

resolved into a pair of components. The components reduce to a single pair, 1–0 and 0–5, as described in Sec. 10.10. The resultant of the two components is the resultant of the four loads and, for equilibrium, is equal and opposite to the resultant of the two reactions. Similarly, the resultant of the left reaction and component 1–0 is equal and opposite to the resultant of the right reaction and component 0–5.

To handle the problem, each of the loads and reactions can be resolved into a pair of components somewhere along its line of action. The lines of action of the loads and the roller reaction are known, but that of the hinge reaction is unknown. One point along the latter line of action is known: the hinge itself, as indicated by point A in Fig. 10.17. At this point the hinge reaction will be broken down into components.

By drawing the funicular polygon for the loads, P_1 is resolved into two components, 1–0 and 0–2. Reaction R_1 is to be resolved into component 1–0 and some other component, as yet unknown. Because the resolving must be done along the lines of action of each of the forces, 1–0 is passed through the hinge at point A and continued until it intersects the line of action of P_1. Nothing more is done with R_1 at the present time. Each of the external loads is broken down into its components as its line of action is encountered, until the line of action of component 0–5 is extended to intersect the line of action of R_2, the roller reaction.

The entire force system has now been reduced to four forces: 1–0, 0–5, R_1, and R_2. The string polygon shows that forces R_2 and 0–5 intersect at point B and forces R_1 and 1–0 intersect at point A. If equilibrium is present, the resultant of R_1 and 1–0 must be equal and opposite to the resultant of R_2 and 0–5. A *closing line* is therefore drawn between points A and B because it must be the line of action of the two resultants.

The values of R_1 and R_2 can now be obtained from the force polygon by drawing a line from point 0 parallel to the closing line. The resultant of 0–5 and R_2 must be along the closing line and R_2 can be drawn from 5 up to the closing line. Since the resultant of R_1, R_2, 1–0, and 0–5 is zero, R_1 must run from the point where R_2 intersects the closing line to point 1. The values of R_1 and R_2 may be scaled from the polygon.

Two other reaction problems are solved graphically in Figs. 10.18 and 10.19.

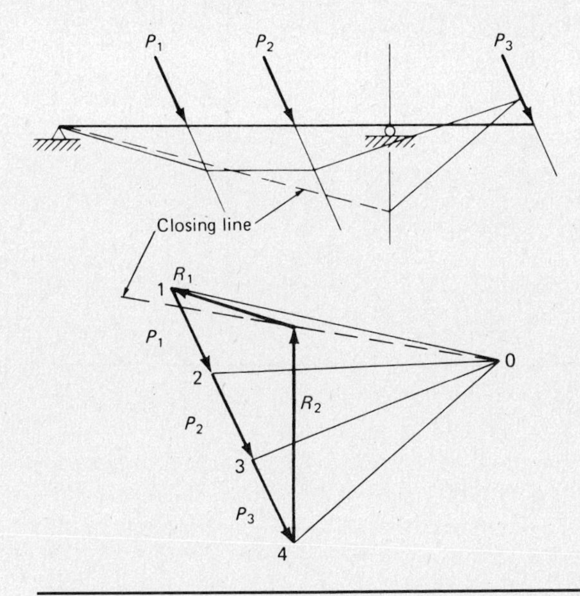

Figure 10.18

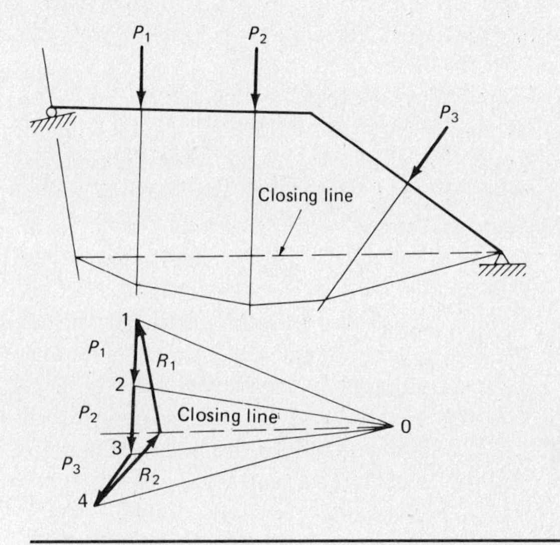

Figure 10.19

10.12 REACTIONS FOR A THREE-HINGED ARCH

Several methods are available for determining the reactions of a three-hinged arch graphically, but only one is discussed here. At each of the hinges of the arch of Fig. 10.20(a) the point of application of the resultant reaction is known but the magnitudes and lines of action are not known; therefore the lines of the

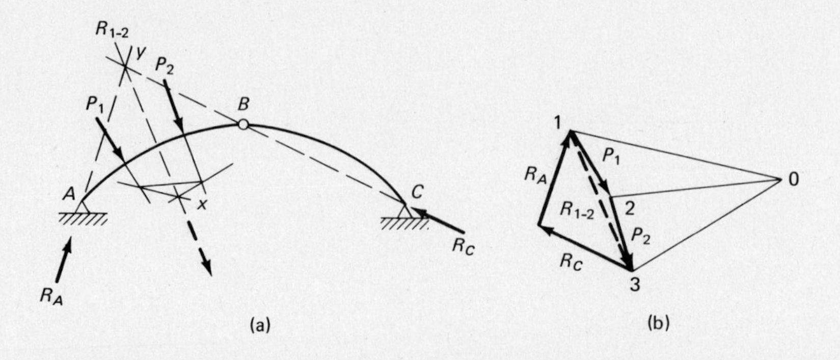

Figure 10.20

string polygon must pass through all of the hinges for a solution to be made. The resultant of all the forces on either side of the crown hinge must pass through that hinge because the moment of those forces about the hinge must be zero.

The arch shown in Fig. 10.20 is loaded only to the left of the crown hinge. The resultant of the loads R_{1-2} and a point along its line of action x are determined with the force polygon in Fig. 10.20(b) and the string polygon of Fig. 10.20(a). The sum of the moments of the forces to the right of hinge B is zero, and since the only force to the right of this hinge is the right-hand resultant reaction R_C, its line of action necessarily passes through B.

The line of action R_C is extended through B until it intersects the line of action of R_{1-2} at y. All of the external forces of the arch have been considered except R_A, and for equilibrium its line of action also passes through y. The resultant of R_A and R_C must be equal and opposite to the resultant of the loads, R_{1-2}, and since their lines of action are now available, their magnitudes may be obtained as shown in Fig. 10.20(b).

Should an arch be loaded on both sides of the crown hinge, such as the one of Fig. 10.21(a), it is possible to handle the problem in exactly the same manner as was used for the arch loaded on one side only. Considering only the loads acting to the left of hinge B, the reaction R_{A_1} and R_{C_1} can be determined. Similarly, their values R_{A_2} and R_{C_2} can be determined for the loads acting to the right of the crown hinge. The total values of the reactions for R_A and R_C for all loads can be obtained by combining the two values previously found.

The loads are plotted on the force polygon as shown in Fig. 10.21(b). Considering only the loads to the left of the hinge, the intersection point of R_{A_1}, R_{C_1}, and R_{1-2} is found at y. In the same manner, considering only the loads on the right side of the structure, the intersection of R_{A_2}, R_{C_2}, and R_{3-4} is found at point z.

Returning to the force polygon, the values of the reactions for loads on the left side only, R_{A_1} and R_{C_1}, are drawn to be equal and opposite to the resultant of R_{1-2}. Similarly, R_{A_2} and R_{C_2} are drawn to be equal and opposite to R_{3-4}. The total values of R_A and R_C can be found, as shown by the dotted lines in the figure.

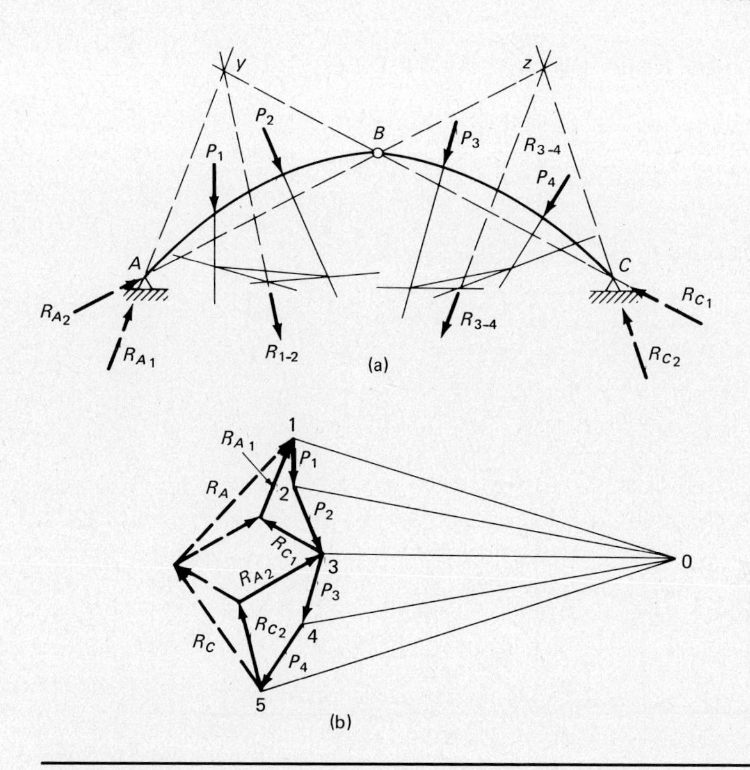

Figure 10.21

The reaction R_{C_1} is moved down to the upper end of R_{C_2} and R_{A_2} is moved to the upper end of R_{C_1}. Then R_{A_2} will come into the lower end of R_{A_1}. The resultants are found by drawing the lines shown.

Should a three-hinged trussed arch be encountered, the reactions can be determined as described and the forces determined by the usual Maxwell diagram.

PROBLEMS

For Probs. 10.1 to 10.22 compute graphically the values of the forces for all of the members of the trusses. Reactions may be determined algebraically.

Problem 10.1

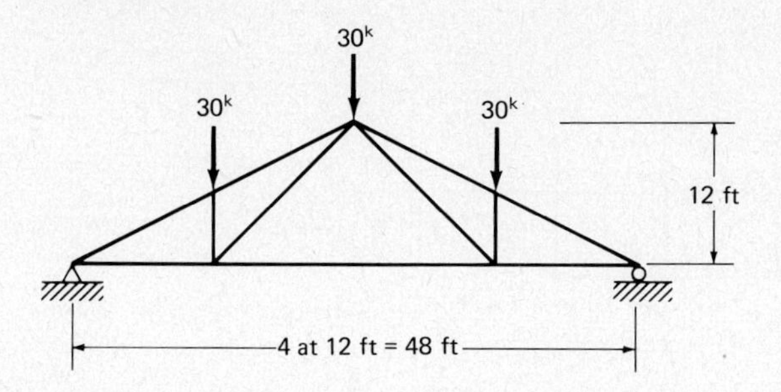

Problem 10.2 (*Ans.* $L_0L_2 = +25^k$, $U_1U_3 = -30^k$, $U_3L_4 = -49.5^k$)

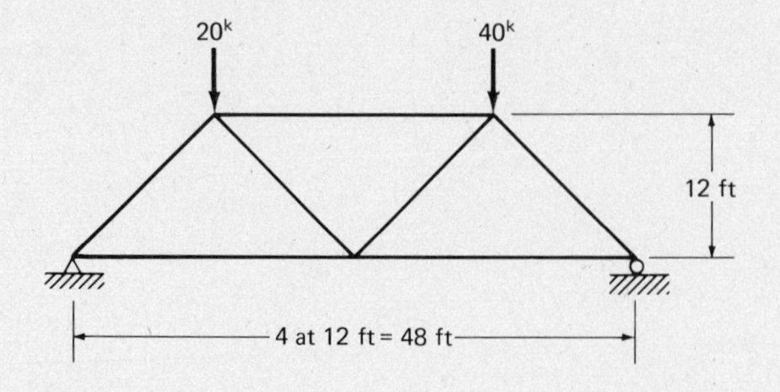

Problem 10.3

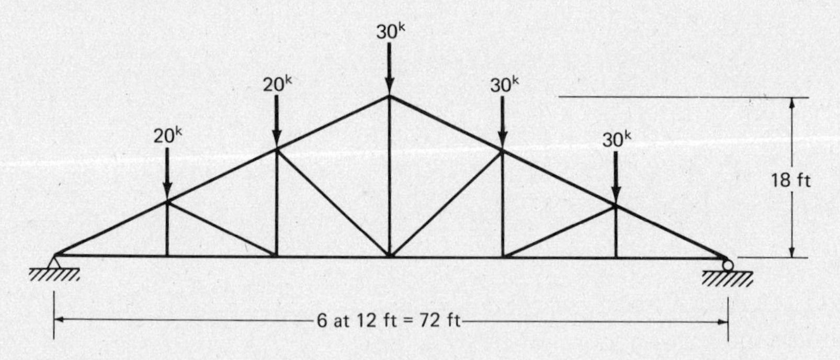

Problem 10.4 (*Ans.* $U_1U_2 = -83.9,^k$ $L_2L_3 = +135,^k$ $U_3L_4 = -150.9^k$)

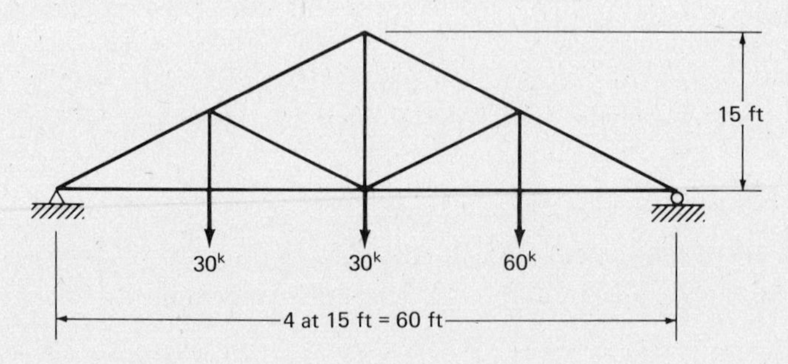

Problem 10.5

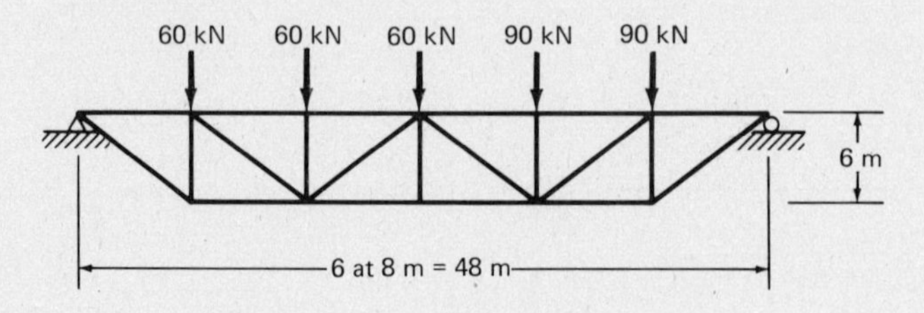

Problem 10.6 (*Ans.* $U_1U_2 = -36.8^k$, $U_3L_3 = +15^k$, $U_4L_5 = -33.4^k$)

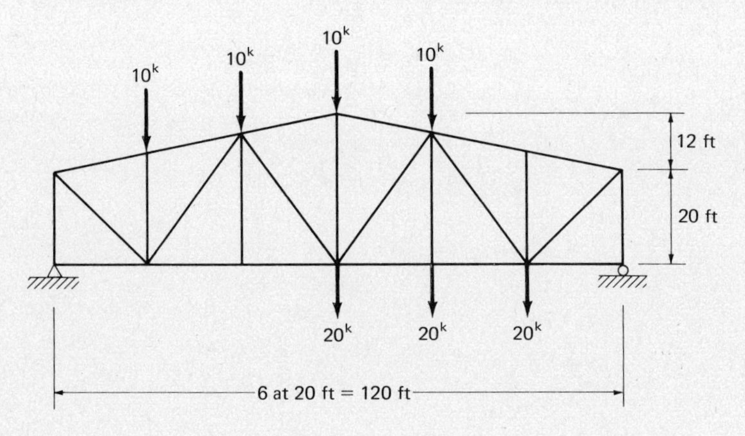

Problem 10.7

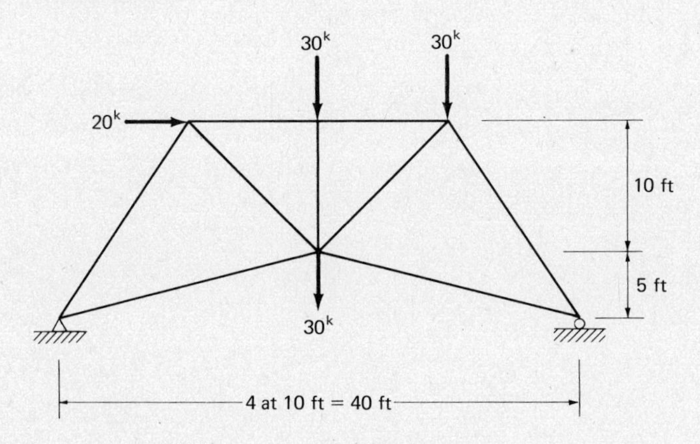

Problem 10.8 (*Ans.* $L_1L_2 = +92.2^k$, $U_2U_3 = -116^k$, $L_3L_4 = -70.8^k$)

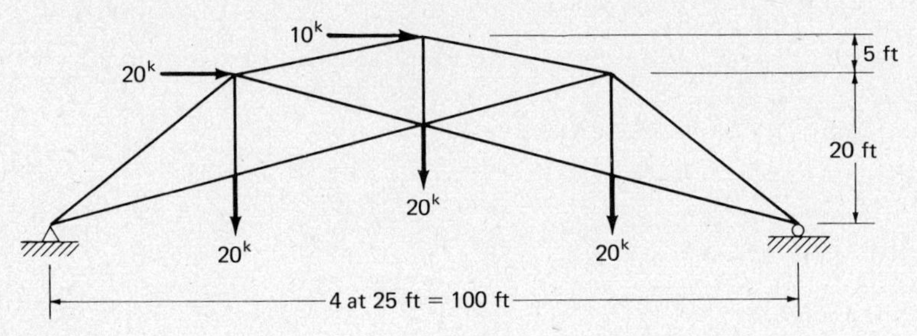

Problem 10.9

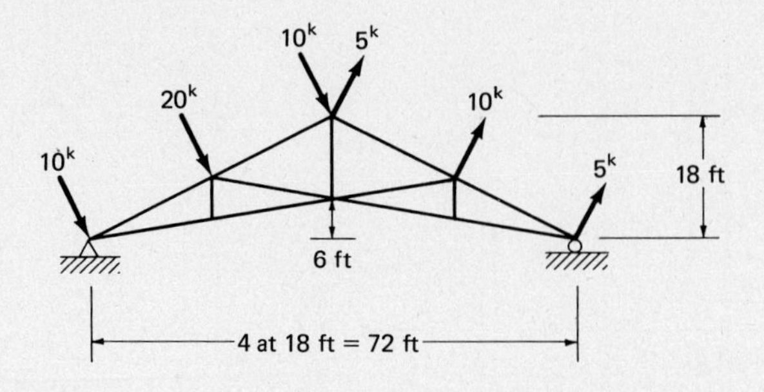

Problem 10.10 (*Ans.* $U_0L_0 = +35^k$, $U_1L_1 = +20^k$, $L_1L_2 = -68^k$)

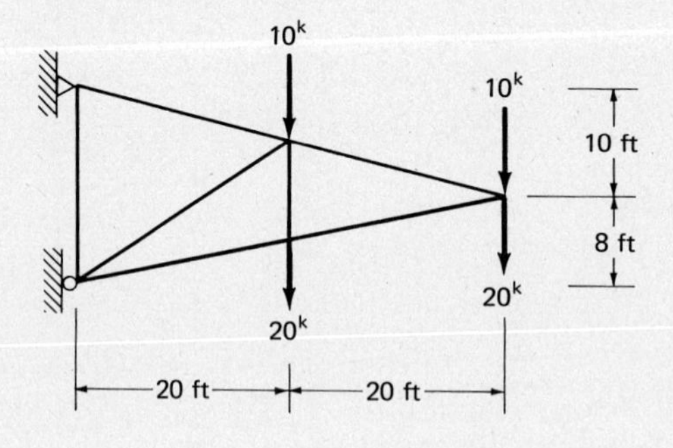

Problem 10.11

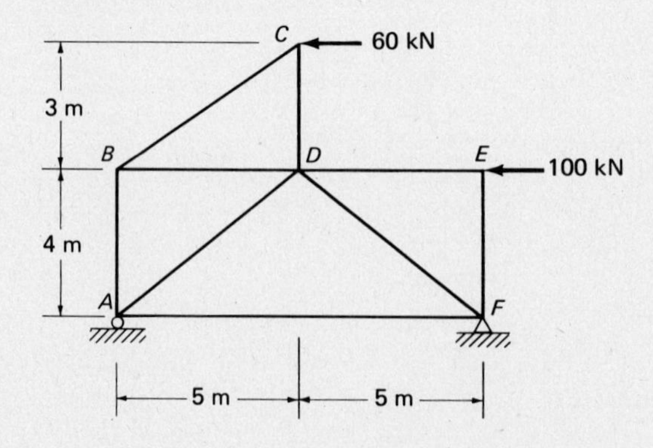

Problem 10.12 (*Ans.* $L_0L_1 = +33.8^k$, $U_3L_3 = -75.6^k$, $U_4U_5 = +13.7^k$)

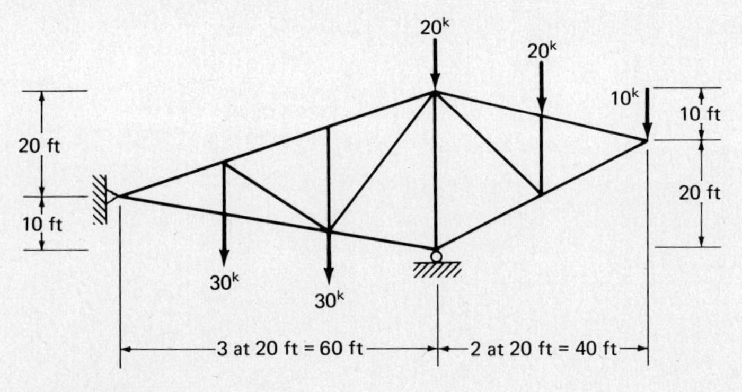

Problem 10.13

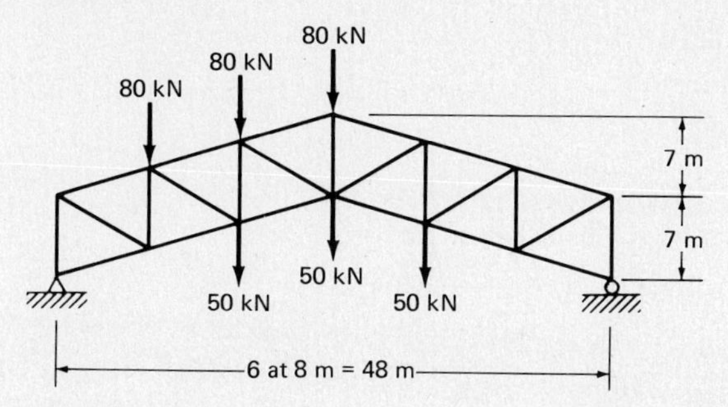

Problem 10.14 (*Ans.* $L_0M_1 = -110.4^k$, $M_1M_4 = -40^k$, $U_3M_4 = -20.1^k$)

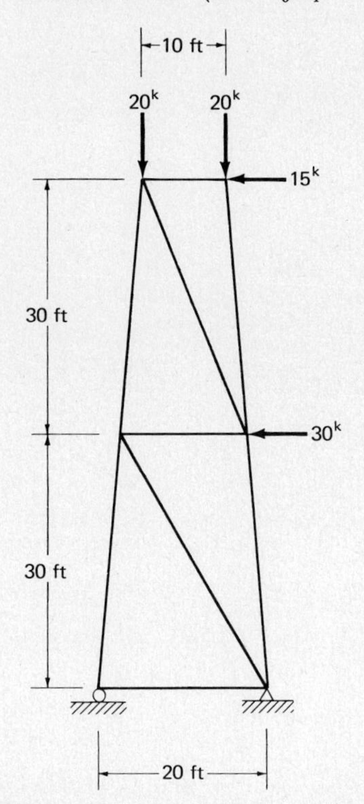

Problem 10.15

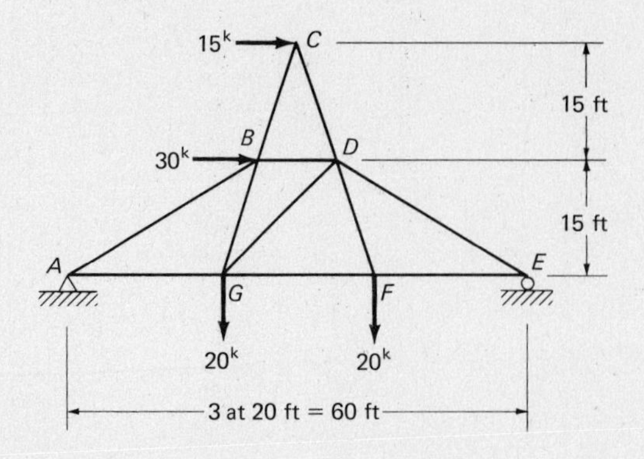

Problem 10.16 (*Ans.* $U_1U_3 = +31.3^k$, $U_3L_4 = -35.8^k$, $M_6U_7 = +60^k$)

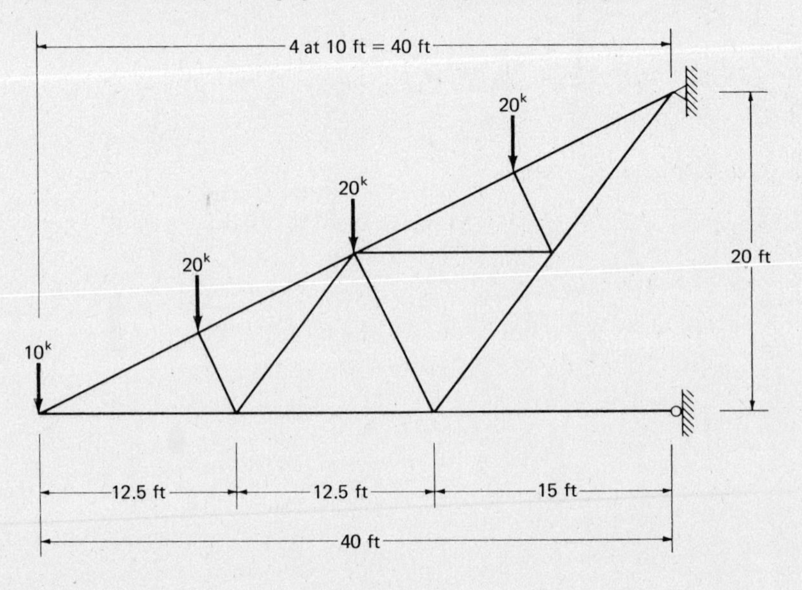

10.17. The truss of Prob. 7.26.

10.18. The truss of Prob. 7.30. (*Ans.* $L_0U_1 = -48^k$, $U_2L_2 = -16.7^k$)

10.19. The truss of Prob. 7.35.

10.20. The truss of Prob. 7.46. (*Ans.* $AE = -13.2^k$, $BE = -49.5^k$, $CE = -14.2^k$)

10.21. The truss of Prob. 7.47.

Problem 10.22 (*Ans.* $L_0L_1 = +63.3^k$, $L_1L_3 = +75.5^k$, $U_2L_3 = -11.2^k$)

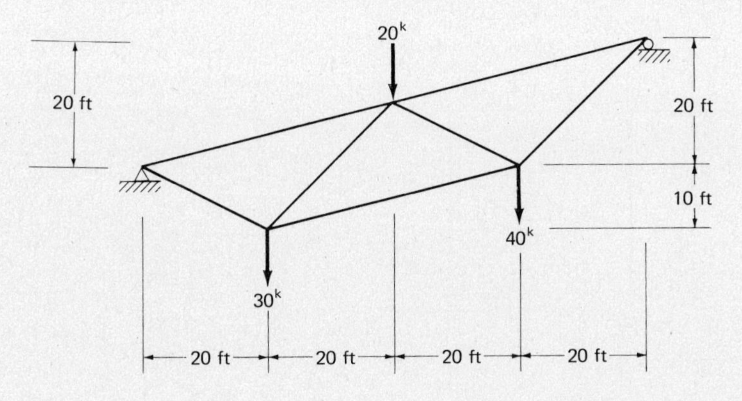

For Probs. 10.23 to 10.34 graphically determine the reactions for the structures.

Problem 10.23

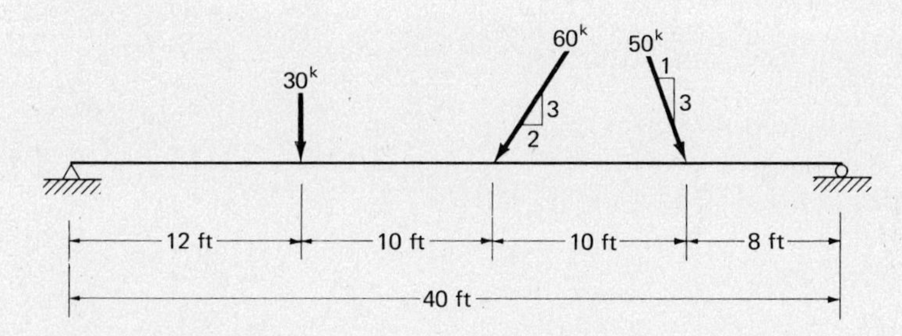

Problem 10.24 (*Ans.* $R_L = 42.8^k$ ↖, $R_R = 57.3^k$ ↑)

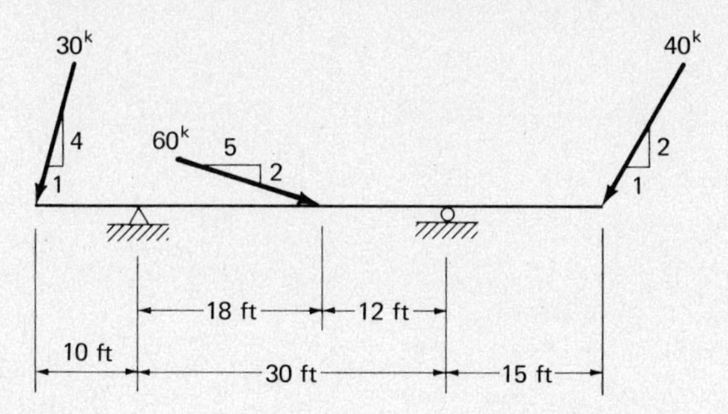

Problem 10.25

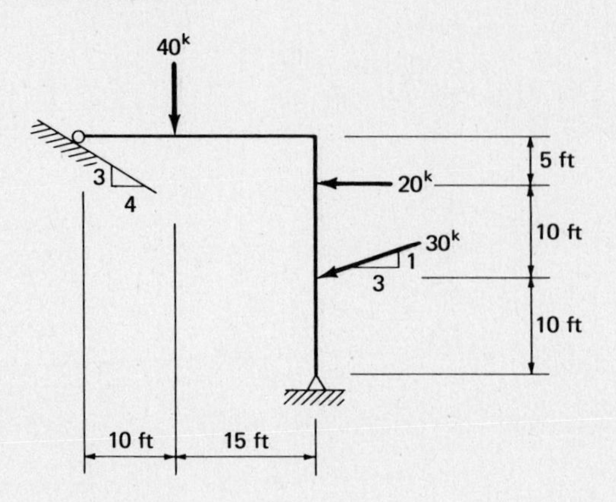

10.26. The truss of Prob. 2.9. (*Ans.* $R_L = 45.3^k$ ↖, $V_R = 15.8^k$ ↑)

10.27. The truss of Prob. 10.8.

10.28. The beam of Prob. 2.18. (*Ans.* $R_L = 28.2^k$ ↗, $R_R =$ ↖ 20.8^k)

10.29. The frame of Prob. 2.12.

10.30. The truss of Prob. 10.11. (*Ans.* $V_L = 82$ kN ↑, $R_R = 179.8$ kN ↘)

10.31. The truss of Prob. 10.10.

10.32. The truss of Prob. 10.14. (*Ans.* $V_L = 110^k$ ↑, $R_R = 83.2^k$ ↘)

10.33. The arch of Prob. 3.9.

10.34. The arch of Prob. 3.18 (*Ans.* $R_L = 67.9^k$ ↗, $R_R = 78.7^k$ ↗)

For Probs. 10.35 to 10.37 graphically determine the reactions and member forces for the structures.

10.35. The truss of Prob. 6.26

10.36. The truss of Prob. 7.34. (*Ans.* $V_R = 29.1^k$↑, $U_1U_3 = -64^k$, $L_3L_5 = +28.7^k$)

10.37. The arch of Prob. 7.38.

11 Influence Lines for Beams

11.1. INTRODUCTION

Structures supporting groups of loads fixed in one position have been discussed in the foregoing chapters. Whether beams, frames, or trusses were being considered and whether the functions sought were shears, reactions, or forces, the loads were stationary. The engineer in practice, however, rarely deals with structures supporting only fixed loads. Nearly all structures are subject to loads moving back and forth across their spans. Perhaps bridges with their vehicular traffic are the most noticeable examples, but industrial buildings with traveling cranes, office buildings with furniture and human loads, frames supporting conveyor belts, and so forth, are in the same category.

Each member of a structure must be designed for the most severe conditions that can possibly develop in that member. The designer places the live loads at the positions where they will produce these conditions. The critical positions for placing live loads will not be the same for every member. For example, the maximum force in one member of a bridge truss may occur when there is a line of trucks from end to end of the bridge, whereas the maximum force in some other member may occur when the trucks extend only from that member to one end of the bridge. The maximum forces in certain beams and columns of a building will occur when the live loads are concentrated in certain portions of the building, whereas the maximum forces in other beams and columns will occur when the loads are placed elsewhere.

On some occasions it is possible by inspection to determine where to place the loads to give the most critical forces, but on many other occasions it is necessary to resort to certain criteria or diagrams to find the locations. The most useful of these devices is the influence line.

Tennessee-Tombigbee Waterway Bridge in Mississippi. (Courtesy of the Mississippi State Highway Department.)

11.2 THE INFLUENCE LINE DEFINED

The influence line, which was first used by Professor E. Winkler, of Berlin, in 1867 [1], shows graphically how the movement of a unit load across a structure influences some function of the structure. The functions that may be represented include reactions, shears, moments, forces, and deflections.

An influence line may be defined as a diagram whose ordinates show the magnitude and character of some function of a structure as a load of unity moves across the structure. Each ordinate of the diagram gives the value of the function when the load is at that ordinate.

Influence lines are used primarily for calculating forces and for determining positions for live loads to cause maximum forces. The procedure for drawing the diagrams is simply the plotting of values of the function under study as ordinates for various positions of the unit load along the span and the connecting of these ordinates. The reader should mentally picture the load moving across the span and try to imagine what is happening to the function in question during the movement. The study of influence lines can immeasurably increase his or her knowledge of what happens to a structure under different loading conditions.

Study of the following sections should fix clearly in his or her mind what an influence line is. The actual mechanics of developing the diagrams are elementary, once the definition is completely understood. No new fundamentals are introduced here; rather, a method recording information in a convenient and useful form is given.

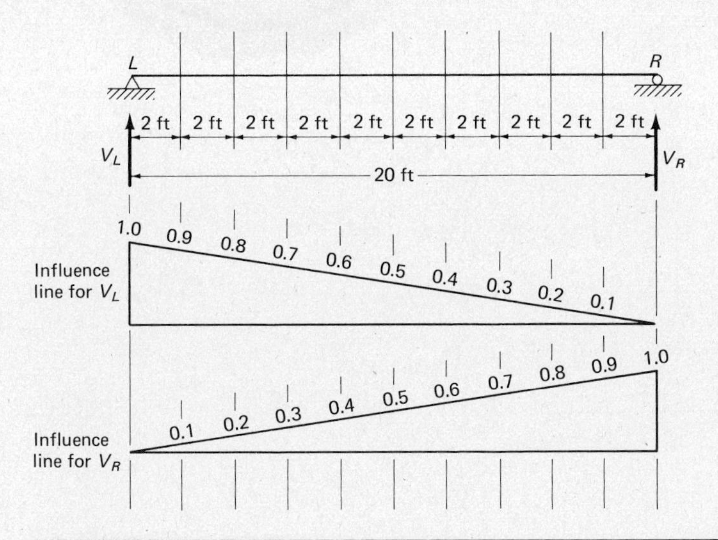

Figure 11.1

11.3 INFLUENCE LINES FOR SIMPLE BEAM REACTIONS

Influence lines for the reactions of a simple beam are given in Fig. 11.1. The variation of the left-hand reaction V_L as a unit load moves from left to right across the beam is considered initially. When the load is directly over the left support, V_L equals 1; when it is 2 ft to the right of the left support, V_L equals 18/20, or 0.9; when it is 4 ft to the right, V_L equals 16/20, or 0.8; and so on.

Values of V_L are shown at 2-ft intervals as the unit load moves across the span. These values lie in a straight line because they change uniformly for equal intervals of the load. For every 2-ft interval the ordinate changes 0.1. The values of V_R, the right-hand reaction, are plotted similarly for successive 2-ft intervals of the unit load. For each position of the unit load the sum of the ordinates of the two diagrams at any point equals (and for equilibrium certainly must equal) the unit load.

11.4 INFLUENCE LINES FOR SIMPLE BEAM SHEARS

Influence lines are plotted in Fig. 11.2 for the shear at two sections in a simple beam. The usual sign convention for shear is used: Positive shear occurs when the sum of the transverse forces to the left of a section is up or when the sum of the forces to the right of the section is down.

Placing the unit load over the left support causes no shear at either of the two sections. Moving the unit load 2 ft to the right of the left support results in a left-hand reaction of 0.9 and the sum of the forces to the left of section 1–1 is 0.1 down, or a shear of -0.1. When the load is 4 ft to the right of the left support and an infinitesimal distance to the left of section 1–1, the shear to the left is -0.2. If the load is moved a very slight distance to the right of section

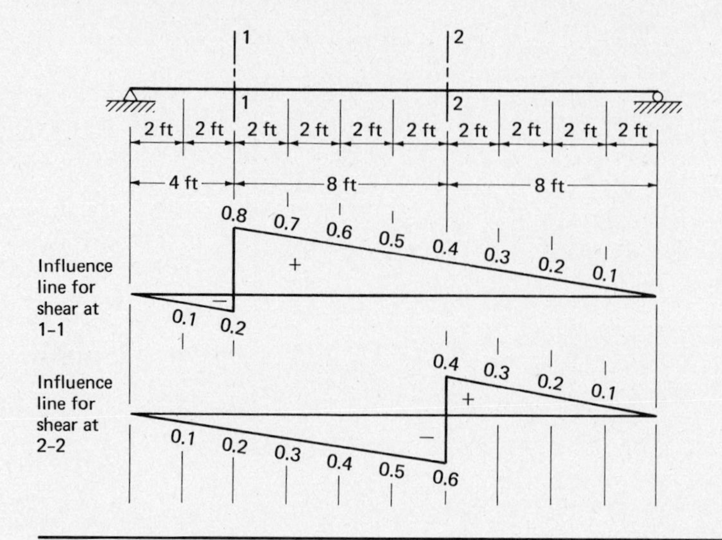

Figure 11.2

1–1, the sum of the forces to the left of the section becomes 0.8 up, or a +0.8 shear. Continuing to move the load across the span toward the right support results in changing values of the shear at section 1–1. These values are plotted for 2-ft intervals of the unit load. The influence line for shear at section 2–2 is developed in the same manner.

Precast concrete Kalihiwai Bridge near Kilauea, Kauai, Hawaii. (Courtesy of the Hawaii Department of Transportation.)

11.5 INFLUENCE LINES FOR SIMPLE BEAM MOMENTS

Influence lines are plotted in Fig. 11.3 for the moment at the same sections of the beam used in Fig. 11.2 for the shear illustrations. To review, a positive moment causes tension in the bottom fibers of a beam and occurs at a particular section when the sum of the moments of all the forces to the left is clockwise or when the sum to the right is counterclockwise. Moments are taken at each of the sections for 2-ft intervals of the unit load.

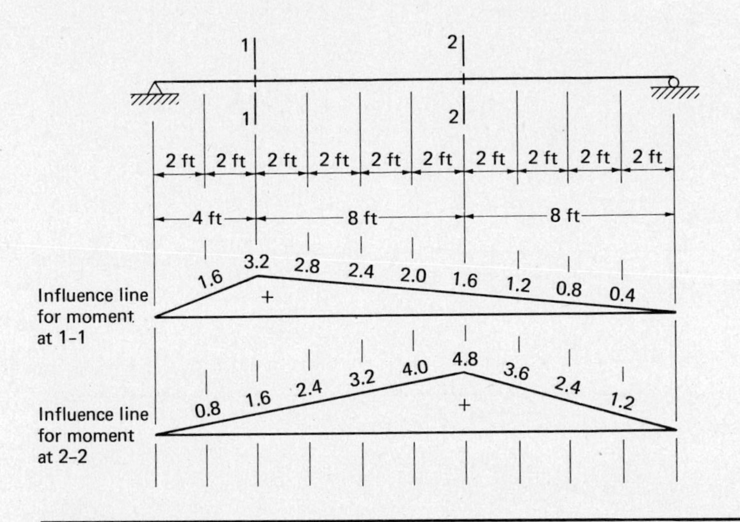

Figure 11.3

The major difference between shear and moment diagrams as compared with influence lines should now be clear. A shear or moment diagram shows the variation of shear or moment across an entire structure for loads fixed in one position. An influence line for shear or moment shows the variation of that function at one section in the structure caused by the movement of a unit load from one end of the structure to the other.

Influence lines for functions of statically determinate structures consist of a set of straight lines. An experienced analyst will be able to compute values of the function under study at a few critical positions and connect the plotted values with straight lines. A person beginning his or her study, however, must be very careful to compute the value of the function for enough positions of the unit load. The shapes of influence lines for forces in truss members (Chap. 12) are often deceptive in their seeming simplicity. It is obviously better to plot ordinates for several extra positions of the load than to fail to plot one essential value.

Several influence lines for moment, shear, and reactions for an overhanging beam are plotted in Fig. 11.4.

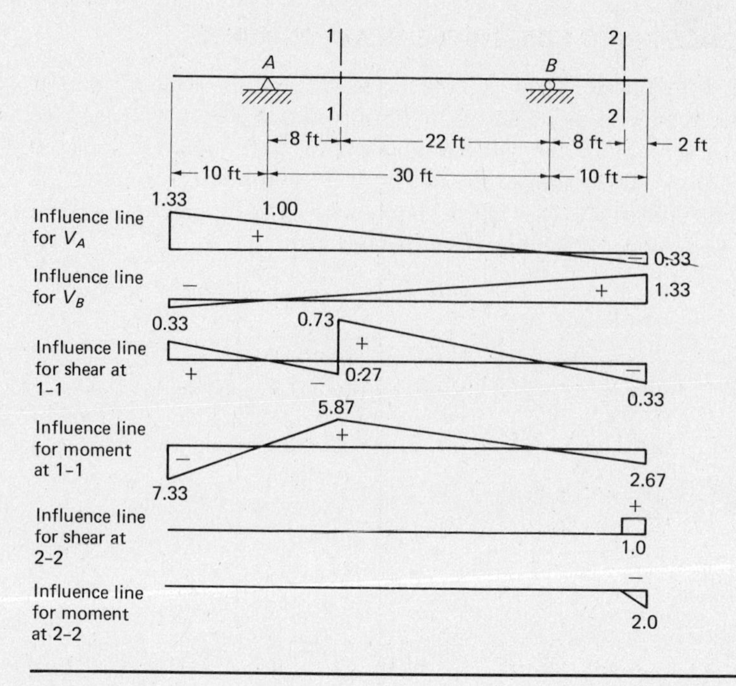

Figure 11.4

11.6 QUALITATIVE INFLUENCE LINES

The average reader initially has a great deal of difficulty in drawing influence lines. For this reason qualitative influence lines are introduced at this time as they will enable the reader to obtain immediately the correct shape of the desired figures and hopefully will give him or her a better understanding of these useful devices.

The influence lines drawn in the preceding sections for which numerical values were computed are referred to as ***quantitative influence lines.*** It is possible, however, to make rough sketches of these diagrams with sufficient accuracy for many practical purposes without computing any numerical values. These latter diagrams are referred to as ***qualitative influence lines.***

A detailed discussion of the principle on which these sketches are made is given in Chap. 19 together with a consideration of their usefulness. Such a discussion is delayed until the student has had some exposure to deflections. Qualitative influence lines are based on a principle introduced by the German Professor Heinrich Müller-Breslau. This principle follows: **The deflected shape of a structure represents to some scale the influence line for a function such as reaction, shear, or moment if the function in question is allowed to act through a small distance.** In other words the structure draws its own influence line when the proper displacement is applied.

As a first example the qualitative influence line for the left reaction of the beam of Fig. 11.5(a) is considered. The constraint at the left support is removed

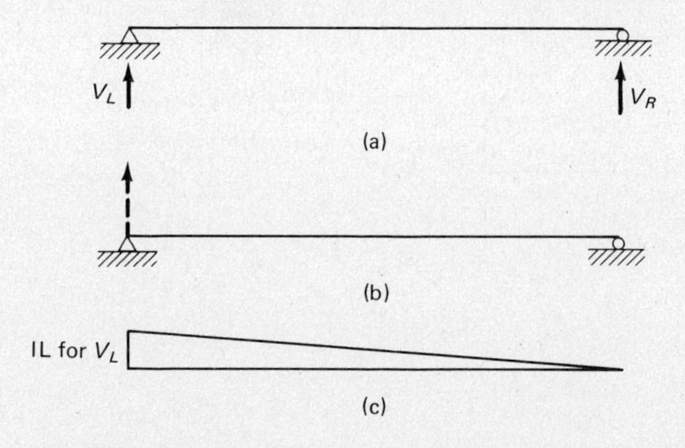

Figure 11.5

and a unit displacement introduced there in the direction of the reaction as shown in part (b) of the figure. When the left end of the beam is pushed up, the area between the original and final position of the beam is the influence line for V_L to some scale.

In a similar manner the influence lines for the left and right reactions of the beam of Fig. 11.6 are sketched.

As a third example the influence line for moment at section 1–1 in the beam of Fig. 11.7 is considered. This diagram can be obtained by cutting the beam at the point in question and applying moments just to the left and just to the right of the cut section as shown. It can be seen in the figure that the moment on each side of the section is positive with respect to the segment of

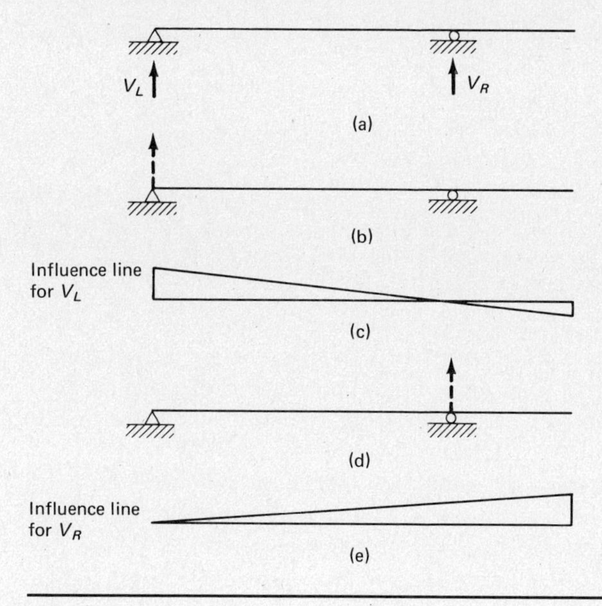

Figure 11.6

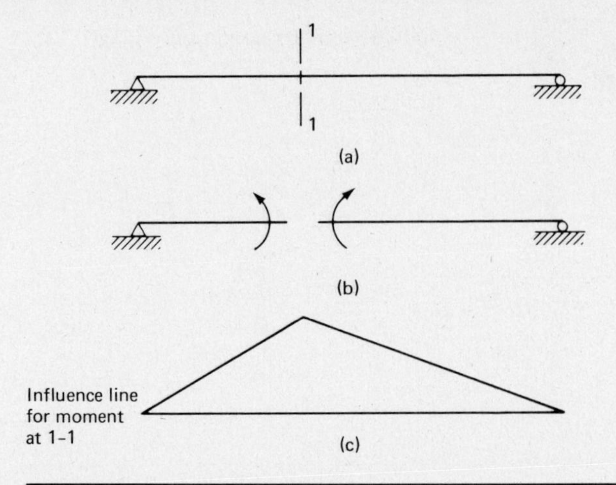

Figure 11.7

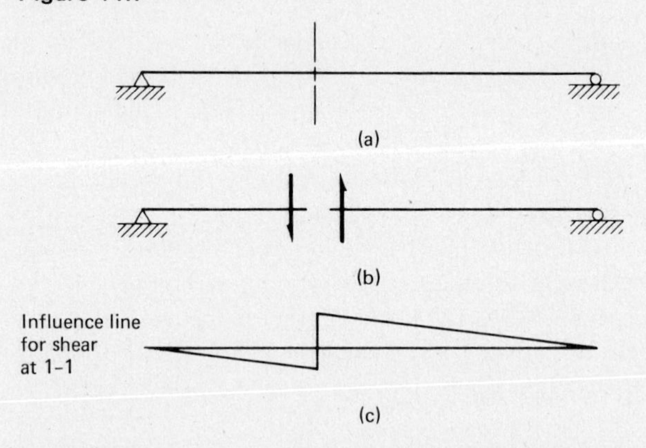

Figure 11.8

the beam on that side of the section. The resulting deflected shape of the beam is the qualitative influence line for moment at section 1–1.

To draw a qualitative influence line for shear the beam is assumed to be cut at the point in question and a vertical force of the nature required to give positive shear applied to the beam on each side of the section [see Fig. 11.8(b)]. To understand the direction used for these forces the reader should notice that they are applied to the left and to the right of the cut section so as to produce a positive shear for each segment. In other words the force on the left segment is in the direction of a positive shear force applied from the right side ($\downarrow$) and vice versa. Additional examples are presented for qualitative influence lines in Fig. 11.9.

Müller-Breslau's principle is useful for sketching influence lines for statically determinate structures but its greatest value is for statically indeterminate structures. Though the diagrams are drawn exactly as before, the reader should notice that they consist of curved lines instead of straight lines as was the case for statically determinate structures. Figure 11.10 shows several such examples.

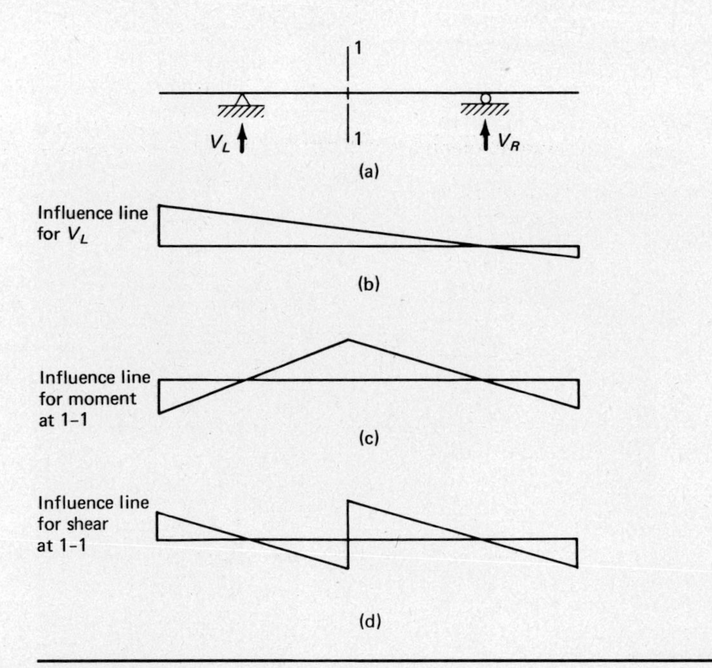

Figure 11.9

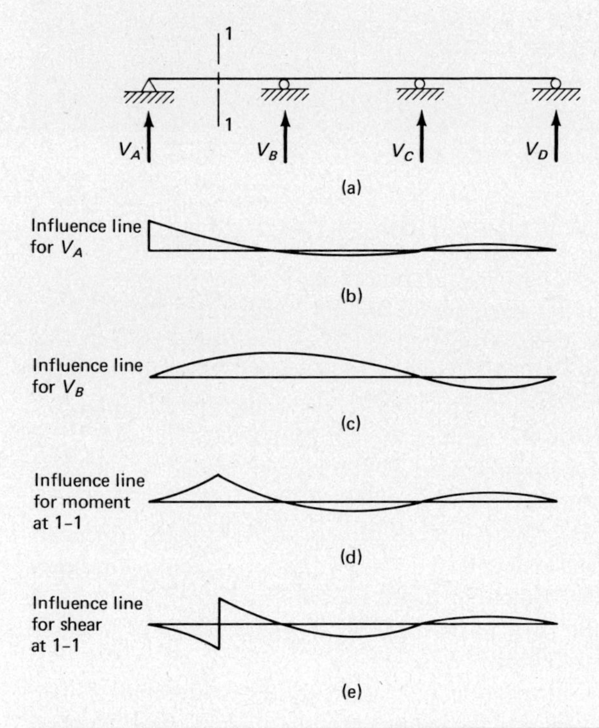

Figure 11.10

Overpass, Boise, Idaho. (Courtesy of the American Concrete Institute.)

11.7 USES OF INFLUENCE LINES; CONCENTRATED LOADS

Influence lines are the plotted values of functions of structures for various positions of a unit load. Having an influence line for a particular function of a structure makes the value of the function for a concentrated load at any position on the structure immediately available. The beam of Fig. 11.1 and the influence line for the left reaction are used to illustrate this statement. A concentrated 1-kip load 4 ft to the right of the left support would cause V_L to equal 0.8 kip. Should a concentrated load of 175 kips be placed in the same position, V_L would be 175 times as great, or 140 kips.

The value of a function due to a series of concentrated loads is quickly obtained by multiplying each concentrated load by the corresponding ordinate of the influence line for that function. Loads of 150 kips 6 ft to the right of L in Fig. 11.1 and 200 kips 16 ft to the right of L would cause V_L to equal $(150)(0.7) + (200)(0.2)$, or 145 kips.

Influence lines for the left reaction and the centerline moment are shown for a simple beam in Fig. 11.11, and the values of these functions are calculated for the several loads supported by the beam.

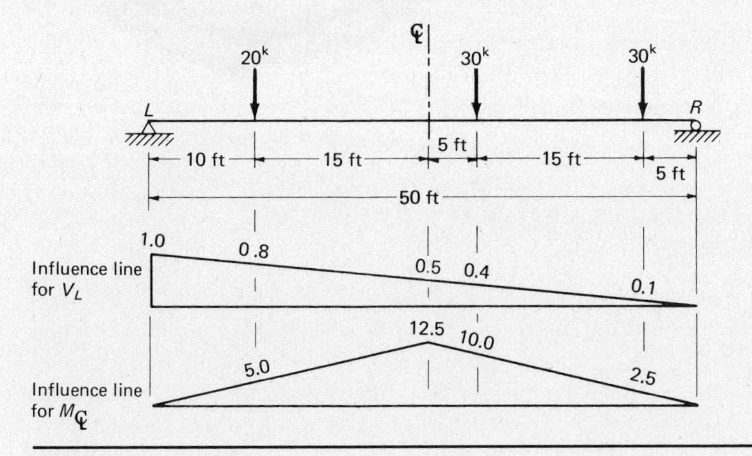

Figure 11.11

$$V_L = (20)(0.8) + (30)(0.4) + (30)(0.1) = 31^k$$

$$M_{℄} = (20)(5.0) + (30)(10.0) + (30)(2.5) = 475'^k$$

11.8 USES OF INFLUENCE LINES; UNIFORM LOADS

The value of a certain function of a structure may be obtained from the influence line, when the structure is loaded with a uniform load, by multiplying the area of the influence line by the intensity of the uniform load. The following discussion proves this statement to be correct.

A uniform load of intensity w lb/ft (pound per foot) is equivalent to a continuous series of smaller loads of $(w)(\frac{1}{10})$ lb on each $\frac{1}{10}$ ft, or $w\,dx$ lb on each dx distance. Considering each dx distance to be loaded with a concentrated load of $w\,dx$, the value of the function under study for one of these small loads is $(w\,dx)(y)$, or $wy\,dx$, where y is the ordinate of the influence line at that point. The effect of all of these concentrated loads is equal to $\int wy\,dx$. This expression shows that the effect of a uniform load on some function of a structure equals the intensity of the uniform load (w) times the area of the influence line $(\int y\,dx)$ along the section of the structure covered by the uniform load.

Assuming the beam of Fig. 11.11 to be loaded with a uniform load of 3 klf for the entire span, the values of V_L and $M_{℄}$ would be as follows:

$$V_L = (3)(\tfrac{1}{2} \times 1.0 \times 50) = 75^k$$

$$M_{℄} = (3)(\tfrac{1}{2} \times 12.5 \times 50) = 937.5'^k$$

If the uniform load extended only from the left end to the centerline of the beam, the values of V_L and $M_{℄}$ would be

$$V_L = (3)\left(\frac{1.0 + 0.5}{2} \times 25\right) = 56.25^k$$

$$M_{℄} = (3)(\tfrac{1}{2} \times 12.5 \times 25) = 468.75'^k$$

Should a structure support uniform and concentrated loads, the value of the function under study can be found by multiplying each concentrated load by its respective ordinate on the influence line and the uniform load by the area of the influence line opposite the section covered by the uniform load.

11.9 COMMON SIMPLE BEAM FORMULAS FROM INFLUENCE LINES

Several useful expressions for moments in simple beams can be determined with influence lines. Formulas are developed for moment at the centerline of a simple beam in Fig. 11.12(a), the beam being loaded first with a uniform load and second with a concentrated load at the centerline. In Fig. 11.12(b) formulas are developed for moment at any point in a simple beam loaded with a uniform load and for moment at any point where a concentrated load is located.

(a) LOADED WITH A UNIFORM LOAD	(b) LOADED WITH A UNIFORM LOAD
$M_{℄} = (w)\left(\frac{1}{2} \times l \times \frac{l}{4}\right) = \frac{wl^2}{8}$	$M_{1\text{-}1} = (w)\left(\frac{1}{2} \times \frac{ab}{l} \times l\right) = \frac{wab}{2}$
Loaded with a concentrated load P at centerline:	Loaded with a concentrated load P at section 1–1:
$M_{℄} = \frac{Pl}{4}$	$M_{1\text{-}1} = \frac{Pab}{l}$

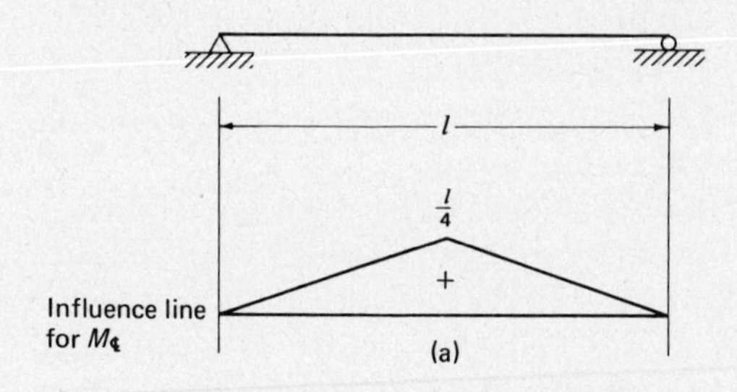

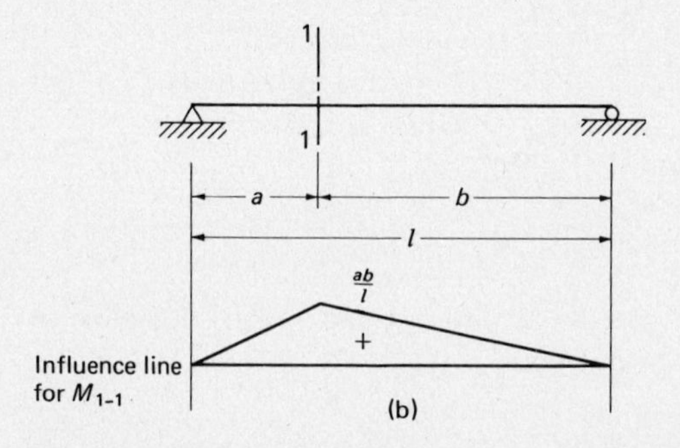

Figure 11.12

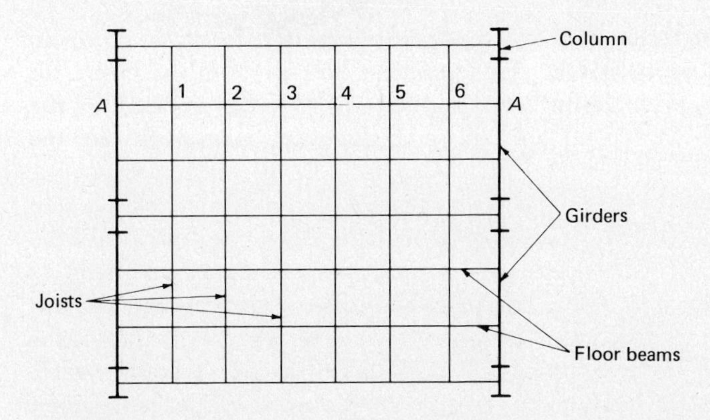

Figure 11.13

11.10 INFLUENCE LINES FOR BEAMS WITH FLOOR JOISTS

Frequently the loads supported by a beam are not applied directly to the beam but are transmitted to the beam from a system of other members that are supported by the beam. Perhaps the most obvious cases of this type of framing exist in building floors and bridge floors. A common building floor framing system is shown in Fig. 11.13.

Loads from the floor slab are supported directly by the joists, which are supported at their ends by the floor beams, which are supported at their ends by the girders, which receive their support from the columns. Bridge floor arrangements and building floor arrangements are closely related; a discussion of the former is presented in Sec. 12.2.

Floor beam A–A is removed from the floor system of Fig. 11.13 and considered in Fig. 11.14 for the purpose of preparing influence lines. Live

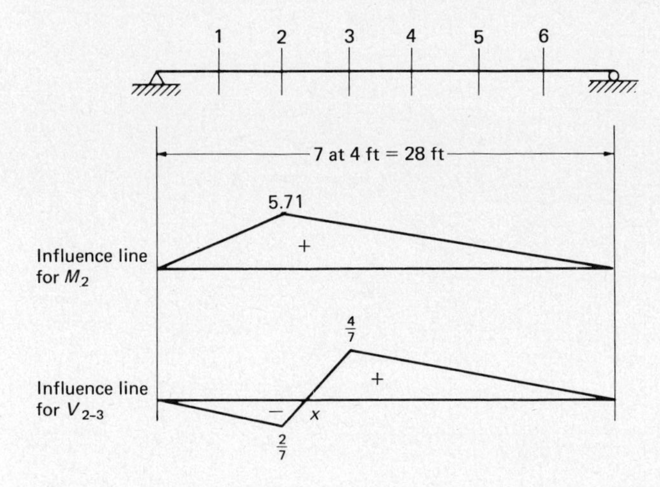

Figure 11.14

loads can be applied to this floor beam only at points 1, 2, 3, and so on, where joists frame into it on each side. Influence lines are shown in the figure for moment at point 2 and for shear between points 2 and 3. Development of the first diagram needs no explanation, but some discussion is necessary for the second.

For each position of the unit load the shear is constant from 2 to 3 because the load can be applied to the beam only through the joists. The ordinates of the influence lines at each of the points can be calculated by the usual methods. As the load moves from 2 to 3, it is being applied to the beam through the two joists in proportion to its position between them. For instance, if the load is three-fourths of the distance from 2 to 3 (Fig. 11.14), the effect is $\frac{1}{4}$ down at 2 and $\frac{3}{4}$ down at 3. As the load moves from 2 to 3, the value of the left reaction varies linearly, as does the amount of load actually applied at point 2. The difference between these two values is the shear between 2 and 3, and it varies linearly, which permits the construction of a straight line between the two points. This type of influence line is a forerunner of those encountered in trusses in the next chapter. There is a section between the two points where a unit load will cause no shear in the panel; this point is designated with the letter x on the influence line. Points of this type may occur in influence lines drawn for any functions.

PROBLEMS

For Probs. 11.1 to 11.6 draw qualitative influence lines for all of the reactions and for shear and moment at section 1–1 for each of the beams.

Problem 11.1

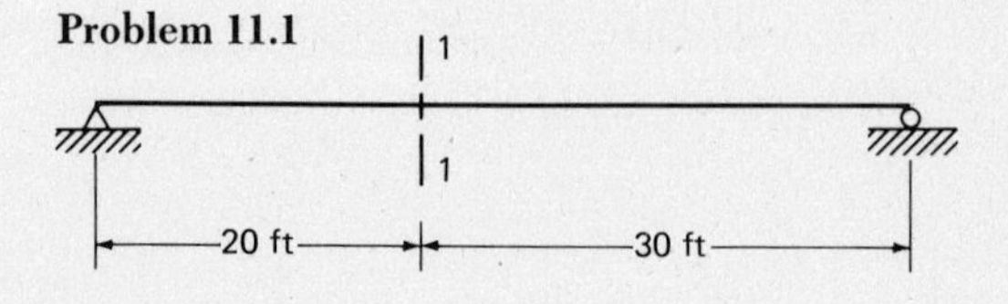

Problem 11.2

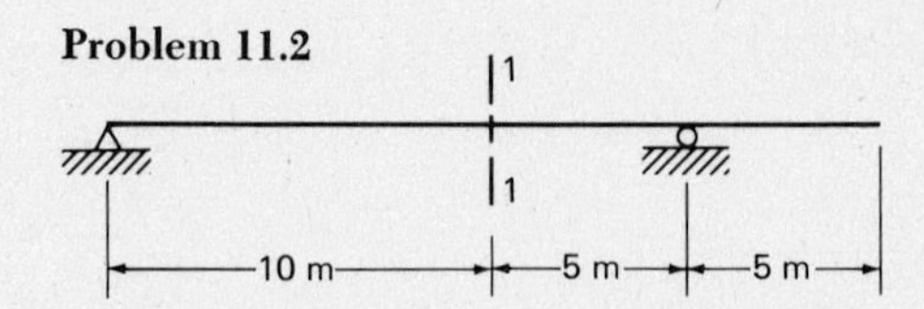

Problem 11.3

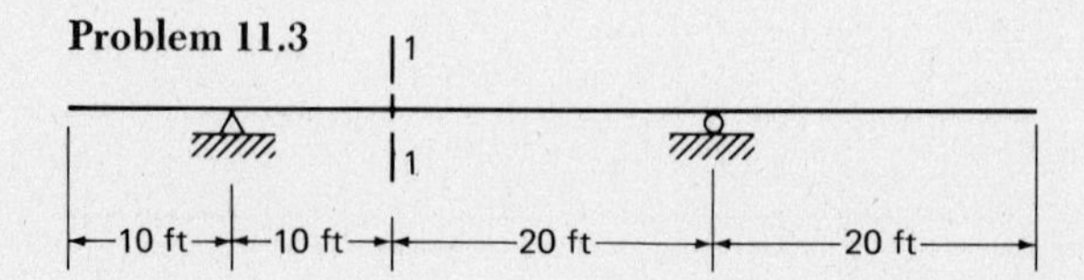

Problem 11.4

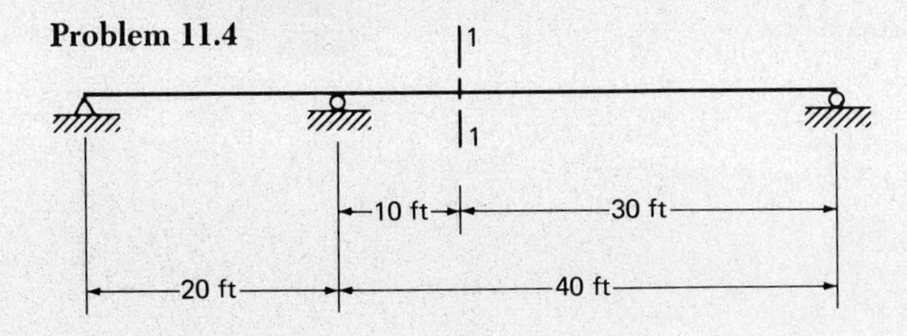

Problem 11.5

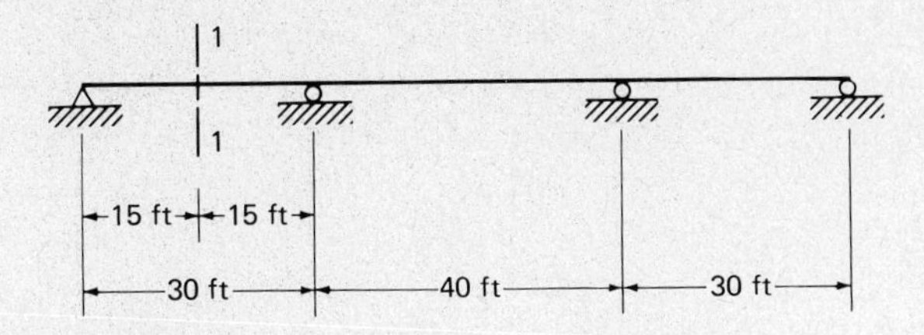

Problem 11.6

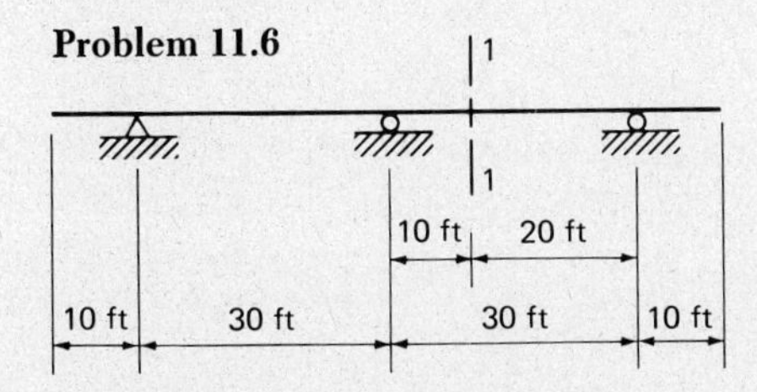

For Probs. 11.7 to 11.18, draw quantitative influence lines for the situations listed.

Problem 11.7 Both reactions and shear and moment at section 1–1

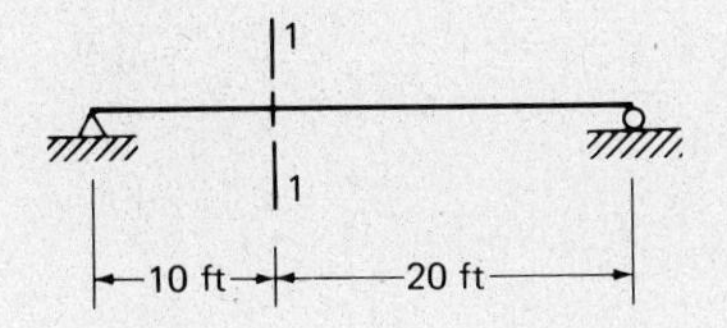

Problem 11.8 Both reactions and shear and moment at section 1–1. (*Ans.* Load at free end; $V_L = 0.33 \downarrow$, $V_R = 1.33 \uparrow$, $V_{1\text{-}1} = -0.33$, $M_{1\text{-}1} = -6.67$)

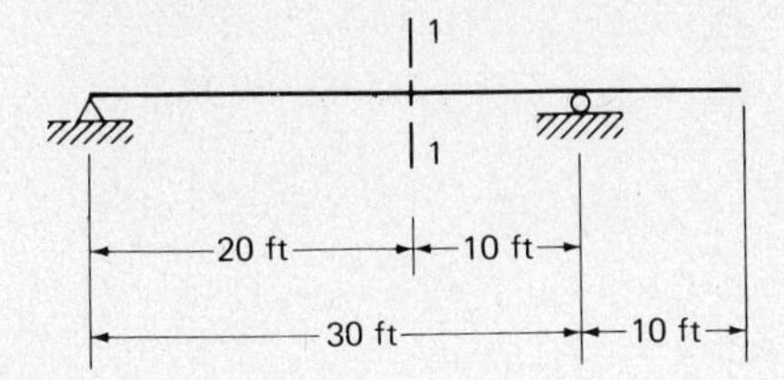

Problem 11.9 Both reactions and shear and moment at section 1–1

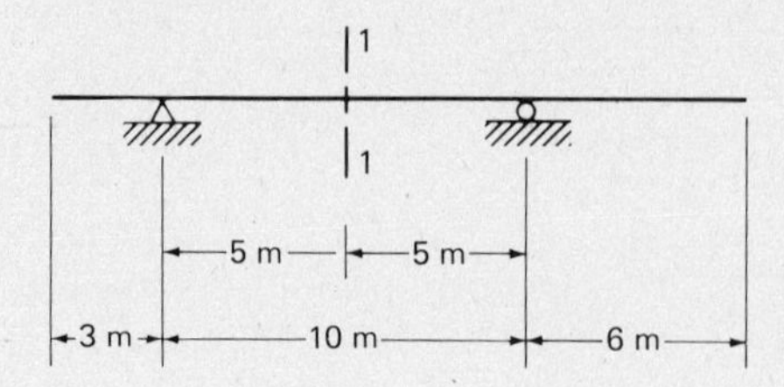

Problem 11.10 Both reactions and shear at sections 1–1, 2–2 (just to left and right of left support) and section 3–3 (*Ans.* Load at left end; $V_L = 1.67 \uparrow$, $V_R = 0.67 \downarrow$, $V_{1\text{-}1} = -1.00$, $V_{2\text{-}2} = +0.67$, $V_{3\text{-}3} = +0.67$)

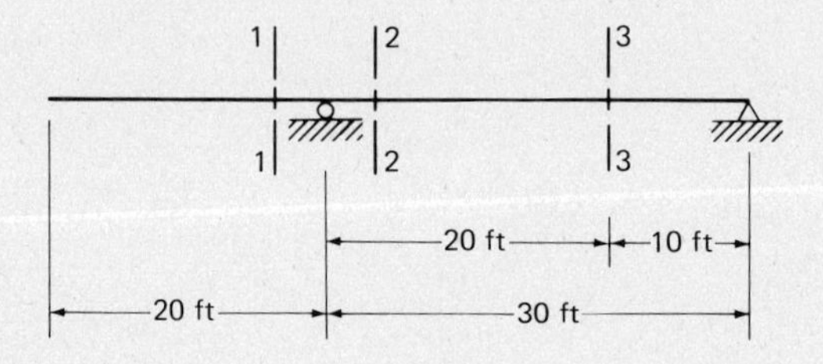

Problem 11.11 Both reactions, shear at sections 1–1 and 2–2 and moment at section 2–2

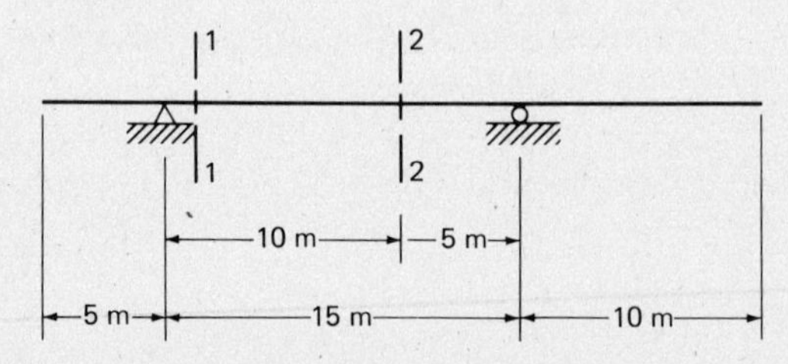

Problem 11.12 Vertical reaction and moment reaction at fixed end, shear and moment at section 1–1 (*Ans.* Load at free end; $V_L = 1.00 \uparrow$, $M_L = -25$, $V_{1\text{-}1} = +1.00$, $M_{1\text{-}1} = -15$)

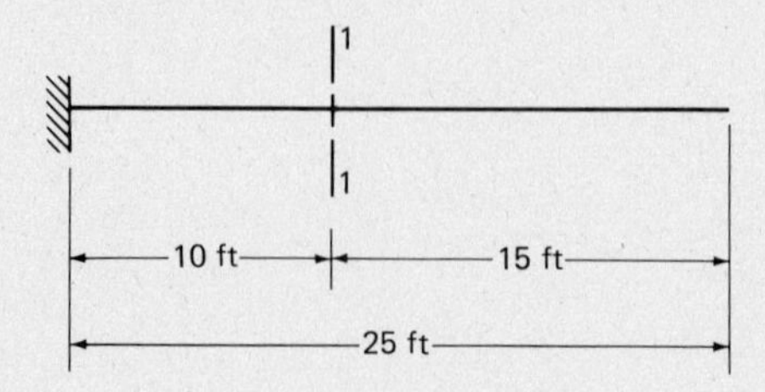

Problem 11.13 Shear and moment at sections 1–1 and 2–2

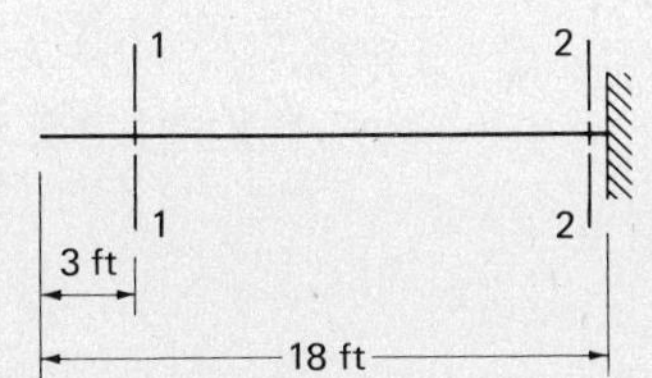

Problem 11.14 Both reactions as load moves from A to B (*Ans.* Load at B; V_L = 0.75 ↓, V_R = 1.75 ↑)

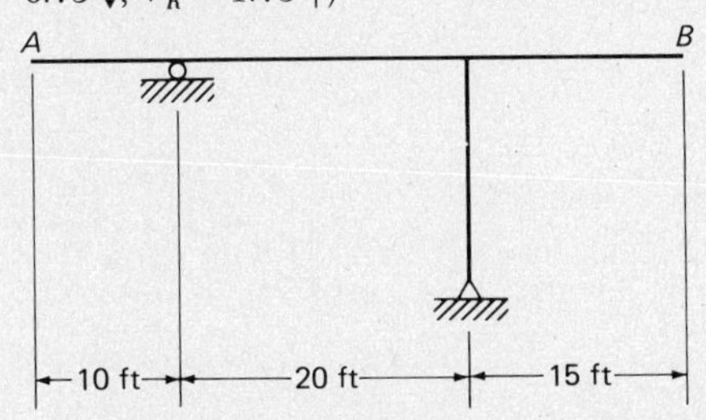

Problem 11.15 All reactions

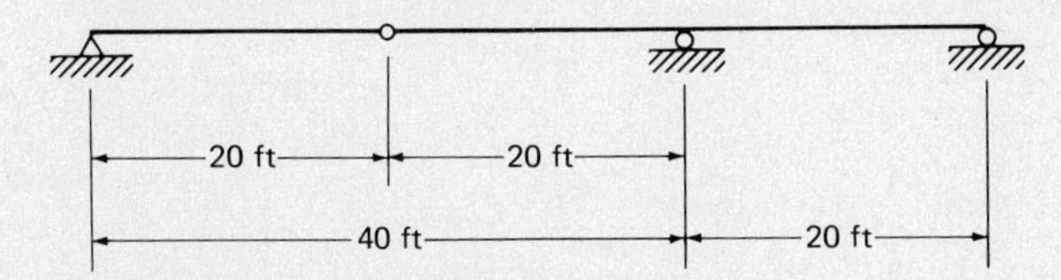

Problem 11.16 All vertical reactions, moment and shear at section 1–1 (*Ans.* Load just to left of section 1–1; V_L = 0.33 ↑, V_R = 0.67 ↑, $V_{1\text{-}1} = -0.67$, $M_{1\text{-}1} = +6.67$)

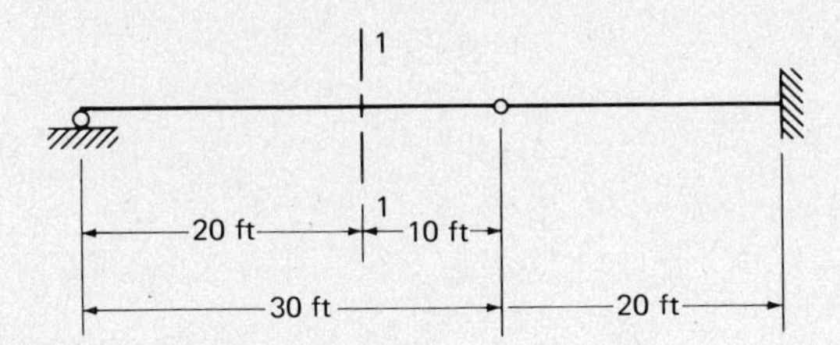

Problem 11.17 Reactions at supports A and B

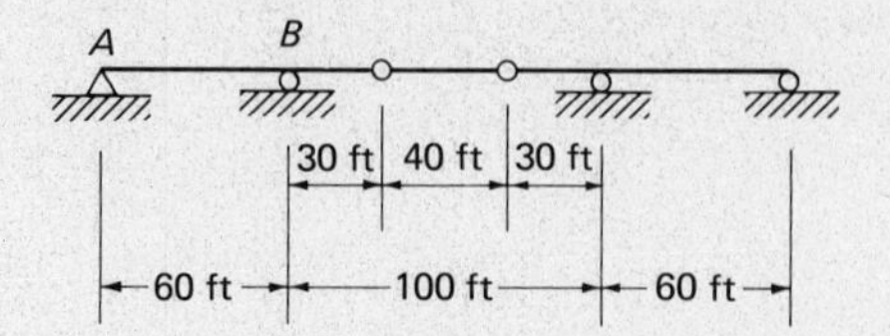

Problem 11.18 Vertical reactions at supports A and B (*Ans.* Load at free end; V_A = 1.50 ↑, V_B = 0.50 ↓)

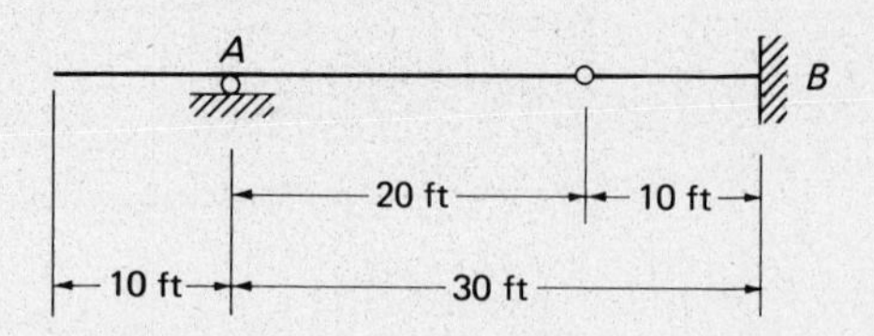

11.19 Draw influence lines for both reactions, and for shear, just to the left of the 16-kip load and for moment at the 16-kip load. Determine the magnitude of each of these functions using the influence lines for the loads fixed in the positions shown in the accompanying illustration.

Problem 11.19

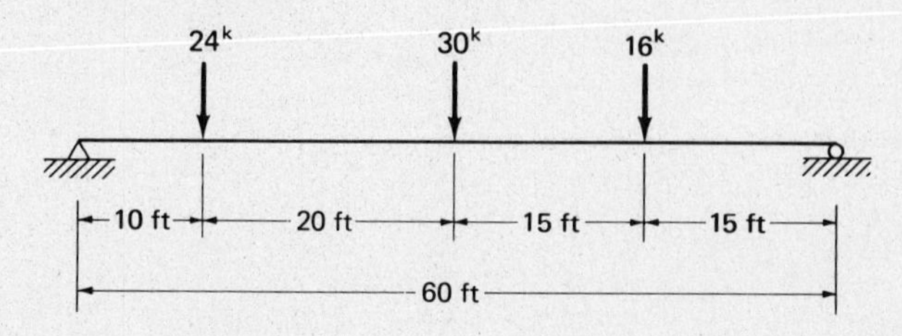

For Probs. 11.20 to 11.26 using influence lines determine the quantities asked for in Probs. 11.20 to 11.26 for a uniform dead load of 2 klf, a moving uniform load of 3 klf, and a floating concentrated load of 20 kips.

Problem 11.20 Maximum left reaction and maximum plus shear and moment at section 1–1 (*Ans.* $V_L = +95^k$, $V_{1\text{-}1} = +21.25^k$, $M_{1\text{-}1} = +712.5'^k$)

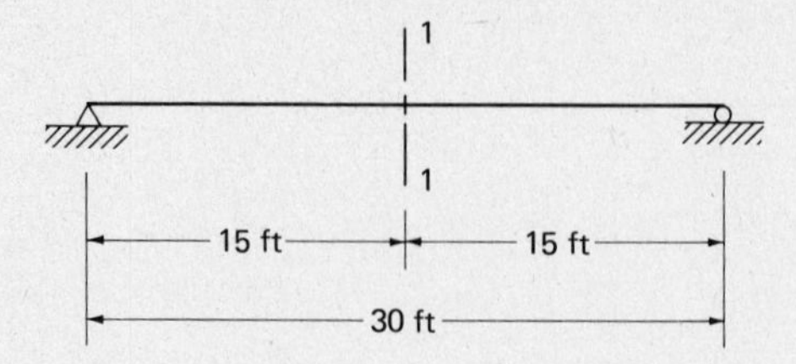

Problem 11.21 Maximum positive values of left reaction and shear and moment at section 1–1

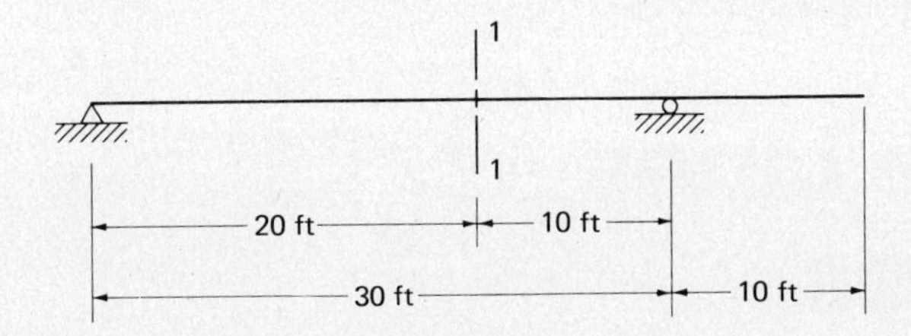

11.22. Maximum negative values of shear and moment at section 1–1, shear just to left of support and moment at the support for the beam of Prob. 11.13. (*Ans.* $V_{1\text{-}1} = -35.0^k$, $M_{1\text{-}1} = -82.5'^k$, $V_{2\text{-}2} = -110.0'^k$, $M_R = -1170'^k$)

Problem 11.23 Maximum negative shear and moment at sections 1–1 and 2–2. Section 2–2 is just to left of support

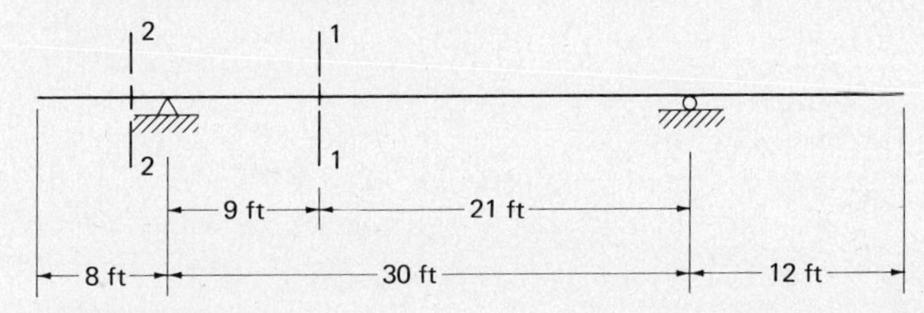

Problem 11.24 Maximum upward value of reactions at *A* and *B* and maximum negative shear and moment at section 1–1 (*Ans.* $V_A = 75^k \uparrow$, $V_B = 472.5^k$, $V_{1\text{-}1} = -142.5^k$, $M_{1\text{-}1} = -2150'^k$)

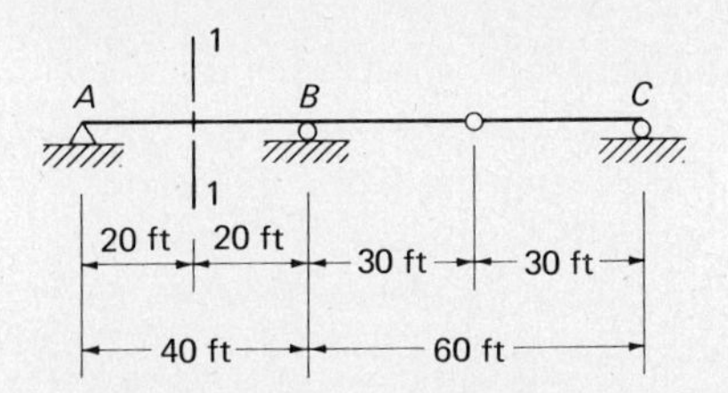

11.25. Maximum positive and negative values of shear and moment at section 1-1 in the beam of Prob. 11.9. Change loads to SI units.

11.26. Maximum positive shear at unsupported hinge, maximum negative moment at support *B*, and maximum downward value of reaction at support *A* for the beam of Prob. 11.24. (*Ans.* $+V$ at unsupported hinge $= +95^k$, $M_B = -5100'^k$, $V_A = -87.5^k$)

REFERENCE

[1] J. S. Kinney, *Indeterminate Structural Analysis* (Reading, Mass.: Addison-Wesley, 1957), Chap. 1.

12 Influence Lines for Trusses

12.1 GENERAL

The variation of forces in truss members due to moving loads is of great importance. Influence lines may be drawn and used for making force calculations, or they may be sketched roughly without computing the values of the ordinates and used only for placing the moving loads to cause maximum or minimum forces.

The procedure used for preparing influence lines for trusses is closely related to the one used for beams, particularly those that have loads applied to them from joists, as described in Sec. 11.10. The exact manner of application of loads to a bridge truss from the bridge floor is described in the following section. A similar discussion could be made for the application of loads to roof trusses.

12.2 ARRANGEMENT OF BRIDGE FLOOR SYSTEMS

The arrangement of the members of a bridge floor system should be carefully studied so that the manner of application of loads to the truss will be fully understood. Probably the most common type of floor system consists of a concrete slab supported by steel stringers running parallel to the trusses. The stringers run the length of each panel and are supported at their ends by floor beams that run transverse to the roadway and frame into the panel points or joints of the truss (Fig. 12.1).

The foregoing discussion apparently indicates that the stringers rest on the floor beams and the floor beams on the trusses. This method of explanation has

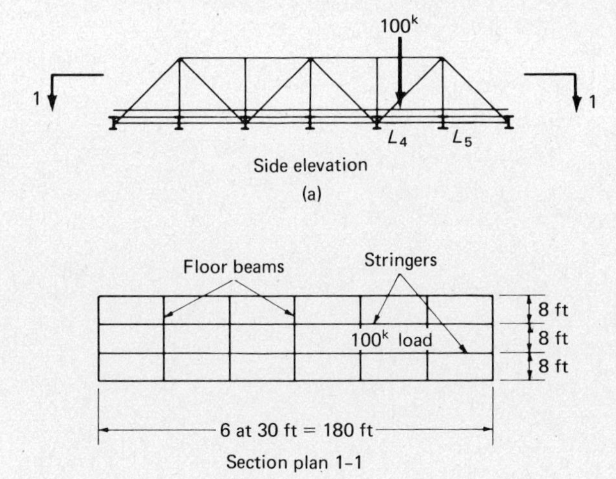

Figure 12.1

been used to emphasize the manner in which loads are transferred from the pavement to the trusses, but the members are usually connected directly to each other. Stringers are conservatively assumed to be simply supported, but actually there is some continuity in their construction.

A 100-kip load is applied to the floor slab in the fifth panel of the truss of Fig. 12.1. The load is transferred from the floor slab to the stringers, thence to the floor beams, and finally to joints L_4 and L_5 of the supporting trusses. The amount of load going to each stringer depends on the position of the load between the stringers; if halfway, each stringer would carry half. Similarly, the amount of load transferred from the stringers to the floor beams depends on the longitudinal position of the load.

Figure 12.2 illustrates the calculations involved in figuring the transfer of the 100-kip load to the trusses. The final reactions shown for the floor beams represent the downward loads applied at the truss panel points. The computation of truss loads is usually a much simpler process than the one described here.

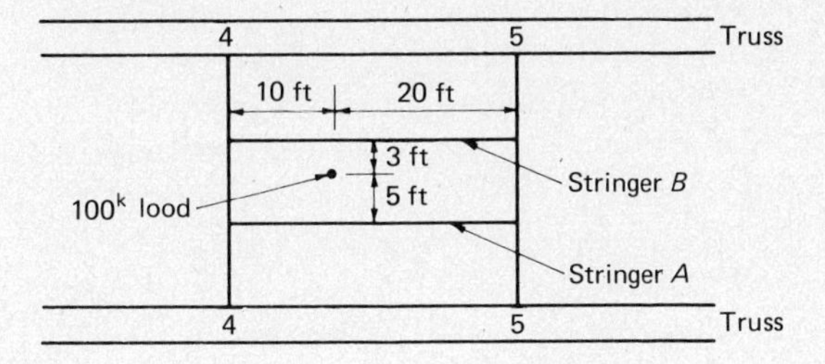

Load transferred to each stringer,

$$A = \tfrac{3}{8} \times 100 = 37.5^k$$

$$B = \tfrac{5}{8} \times 100 = 62.5^k$$

Load transferred from stringer *A* to floor beams,

$$4\text{–}4 = \tfrac{20}{30} \times 37.5 = 25^k$$

$$5\text{–}5 = \tfrac{10}{30} \times 37.5 = 12.5^k$$

Load transferred from stringer *B* to floor beams,

$$4\text{–}4 = \tfrac{20}{30} \times 62.5 = 41.67^k$$

$$5\text{–}5 = \tfrac{10}{30} \times 62.5 = 20.83^k$$

Floor beams loaded as follows:

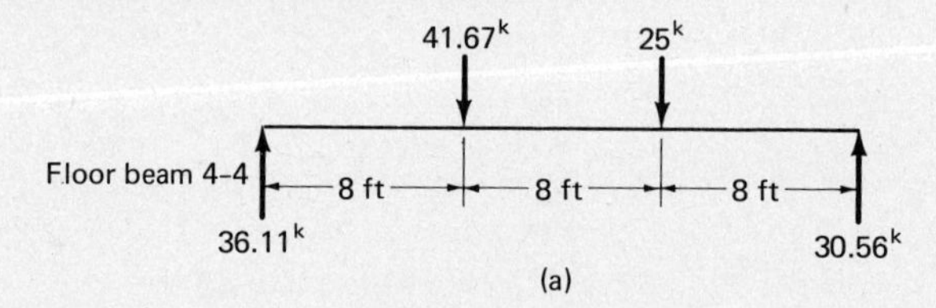

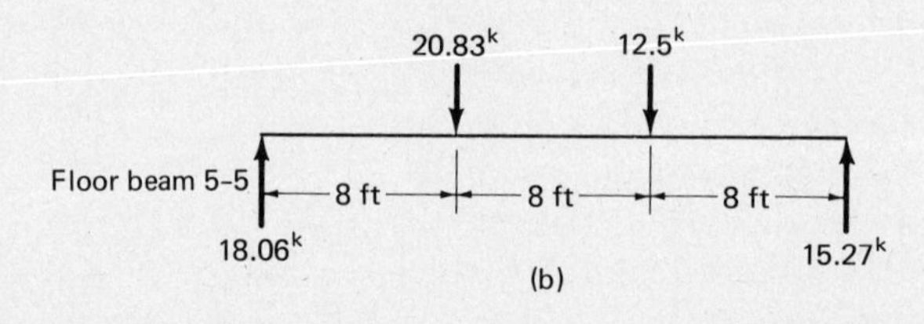

Figure 12.2

12.3 INFLUENCE LINES FOR TRUSS REACTIONS

Influence lines for reactions of simply supported trusses are used to determine the maximum loads that may be applied to the supports. Although their preparation is elementary, they offer a good introductory problem in learning the construction of influence lines for truss members.

Influence lines for the reactions at both supports of an overhanging truss are given in Fig. 12.3. Loads can be applied to the truss only by the floor beams at the panel points, and floor beams are assumed to be present at each of the panel points including the end ones. A load applied at the very end of the truss opposite the end panel point will be transferred to that panel point by the end floor beam.

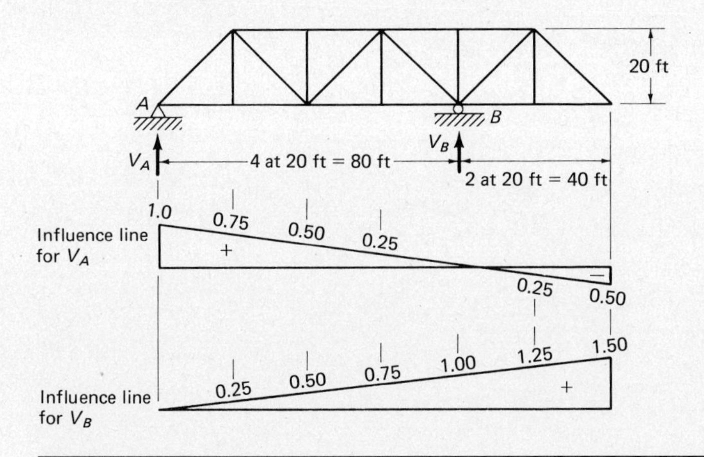

Figure 12.3

12.4 INFLUENCE LINES FOR MEMBER FORCES OF PARALLEL-CHORD TRUSSES

Influence lines for forces in truss members may be constructed in the same manner as those for various beam functions. The unit load moves across the truss, and the ordinates for the force in the member under consideration may be computed for the load at each panel point. In most cases it is unnecessary to place the load at every panel point and calculate the resulting value of the force, because certain portions of influence lines can readily be seen to consist of straight lines for several panels.

One method used for calculating the forces in a chord member of a truss consists in passing an imaginary section through the truss cutting the member in question and taking moments at the intersection of the other members cut by the section. The resulting force in the member is equal to the moment divided by the lever arm; therefore the influence line for a chord member is the same shape as the influence line for moment at its moment center.

The truss of Fig. 12.4 is used to illustrate this point. The force in member L_1L_2 is determined by passing section 1–1 and taking moments at U_1. An influence line is shown for the moment at U_1 and for the force in L_1L_2, the ordinates of the latter figure being those of the former divided by the lever arm. Similarly, section 2–2 is passed to compute the force in U_4U_5, and influence lines are shown for the moment at L_4 and for the force in U_4U_5.

The forces in the diagonals of parallel-chord trusses may be calculated from the shear in each panel. The influence line for the shear in a panel is of the same shape as the influence line for the force in the diagonal, because the vertical component of force in the diagonal is equal numerically to the shear in the panel. Figure 12.5 illustrates this fact for two of the diagonals of the same truss considered in Fig. 12.4. For some positions of the unit loads the diagonals are in compression, and for others they are in tension.

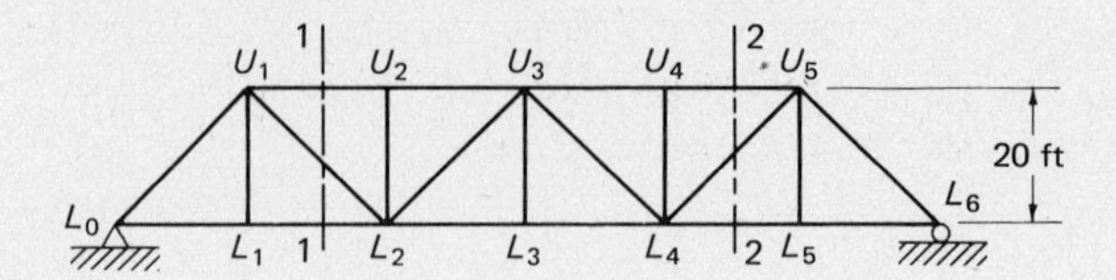

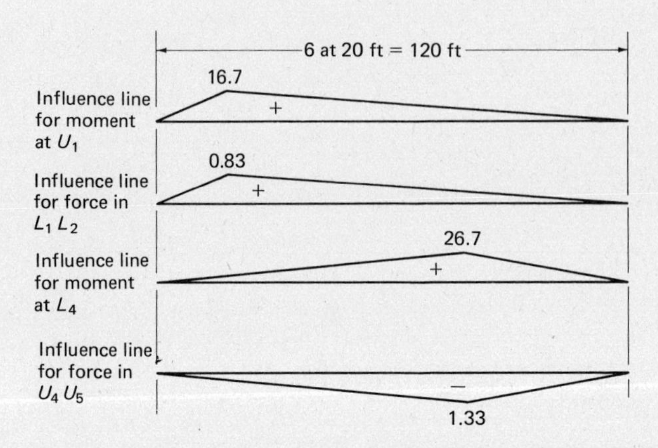

Figure 12.4

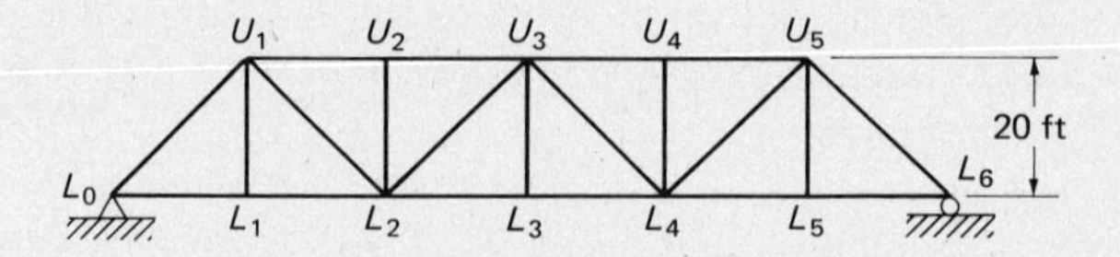

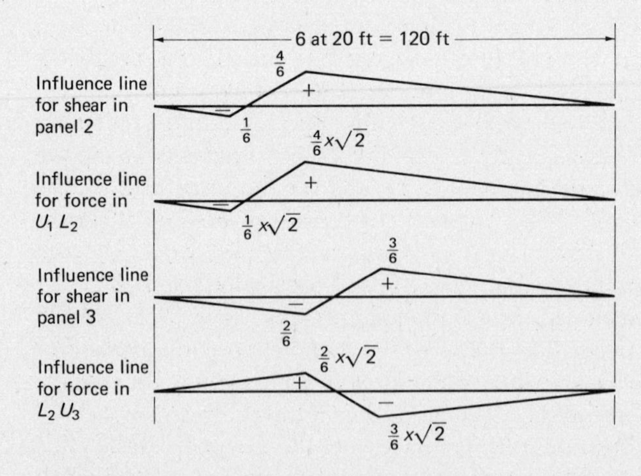

Figure 12.5

The vertical components of force in the diagonals can be converted into the actual forces from their slopes. The sign convention for positive and negative shears is the same as the one used previously.

12.5 INFLUENCE LINES FOR MEMBERS OF NONPARALLEL-CHORD TRUSSES

Influence-line ordinates for the force in a chord member of a "curved-chord" truss may be determined by passing a vertical section through the panel and taking moments at the intersection of the diagonal and the other chord. Several such influence lines are drawn for chords of a Parker truss in Fig. 12.6.

The ordinates for the influence line for force in a diagonal may be obtained by passing a vertical section through the panel and taking moments at the intersection of the two chord members, as illustrated in Fig. 12.6, where the force in U_1L_2 is obtained by passing section 1–1 and taking moments at the intersection of chords U_1U_2 and L_1L_2 at point x. The influence line is drawn for the vertical component of force in the inclined member. In the following pages

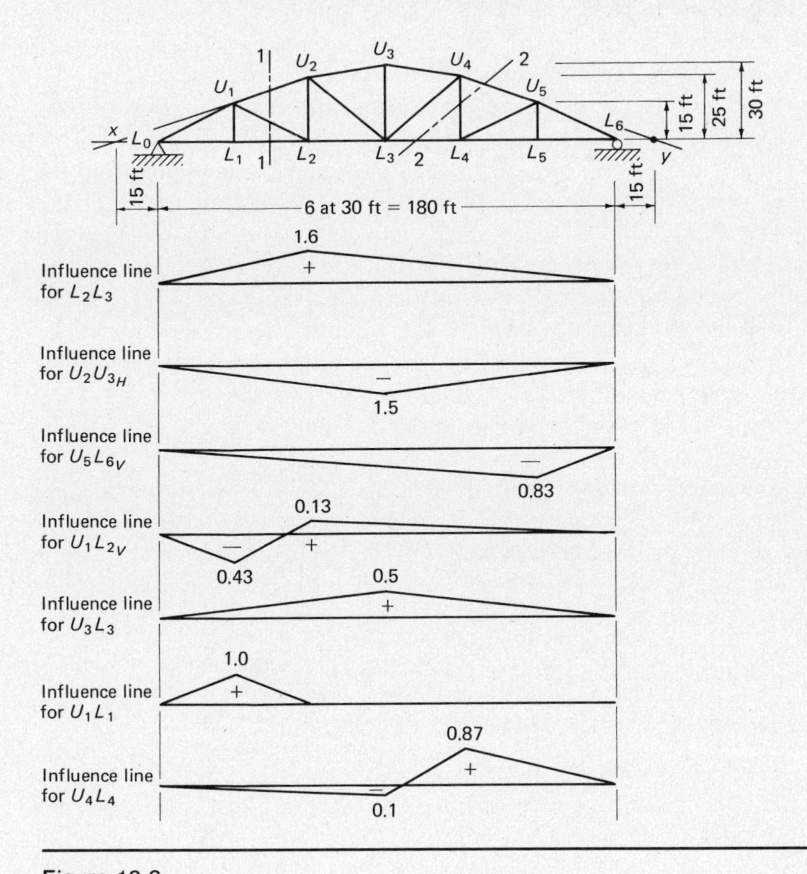

Figure 12.6

many influence lines are drawn for either the vertical or horizontal components of force for inclined members. Force components obtained from the diagrams may be quickly adjusted to resultant forces from slopes of the members.

The influence line for the midvertical U_3L_3 is obtained indirectly by computing the vertical components of force in U_2U_3 and U_3U_4. The ordinates for U_3L_3 are found by summing up these components. The influence lines for the other verticals are more easily drawn. Member U_1L_1 can have a force only when the unit load lies between L_0 and L_2. It has no force if the load is at either of these joints, but a tension of unity occurs when the load is at L_1. Influence lines for verticals such as U_4L_4 can be drawn by two methods. A section such as 2–2 may be passed through the truss and moments taken at the intersection of the chords at point y, or if the influence diagram for L_4U_5 is available, its vertical components may be used to calculate the ordinates for U_4L_4.

12.6 INFLUENCE LINES FOR K TRUSS

Figure 12.7 shows influence lines for several members of a K truss. The calculations necessary for preparing the diagrams for the chord members are equivalent to those used for the chords of trusses previously considered. The values needed to plot the diagrams for vertical and diagonal members are slightly more difficult to obtain.

The forces in the two diagonals of each panel may be obtained from the shear in the panel. By knowing that the horizontal components are equal and opposite, the relationship between their vertical components can be found from their slopes. If the slopes are equal, the shear to be carried divides equally between the two. The influence lines for the verticals, such as U_5M_5, may be determined from the influence lines for the adjoining diagonals if available. On the other hand the ordinates may be computed independently for various positions of the unit load. The reader should make a careful comparison of the influence lines for the upper and lower verticals, such as those given for M_3L_3 and U_3M_3 in the figure.

The influence line for the midvertical U_4L_4 can be developed by computing the vertical components of force in M_3L_4 and L_4M_5, or in M_3U_4 and U_4M_5, for each position of the unit load. The vertical components of force in each of these pairs of members will cancel each other unless the shear in panel 4 is unequal to the shear in panel 5, which is possible only when the unit load is at L_4.

12.7 DETERMINATION OF MAXIMUM FORCES

Truss members are designed to resist maximum forces that may be caused by any combination of the dead, live, and impact loads to which the truss may be subjected. The live load probably consists of a series of moving concentrated loads representing the wheel loads of the vehicles using the structure, but for

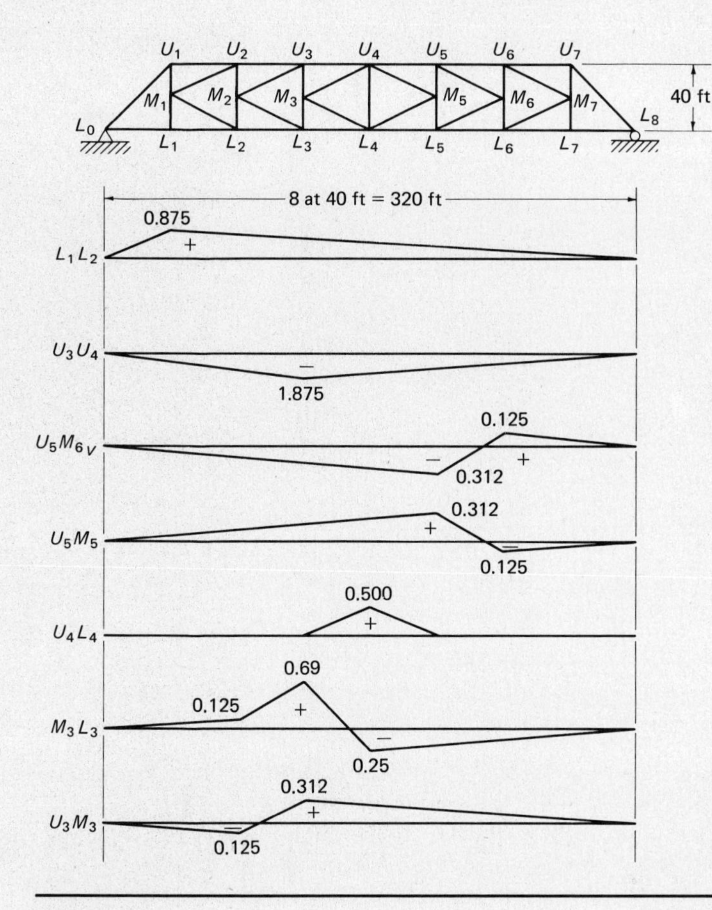

Figure 12.7

convenience in force analysis an approximately equivalent uniform live load with only one or two concentrated loads is often used in their place. Live loads for which highway and railroad bridges are designed and common impact expressions are discussed in detail in Chap. 13.

Two methods are available for calculating the maximum forces in the members of a truss. These are the exact and panel-load methods.

Exact Method

The exact method, previously illustrated for beams in Secs. 11.7 and 11.8, requires the preparation of an influence line for force in the member under consideration.

The dead load, representing the weight of the structure and permanent attachments, extends for the entire length of the truss, but the uniform and concentrated live loads are placed at the points on the influence line that cause

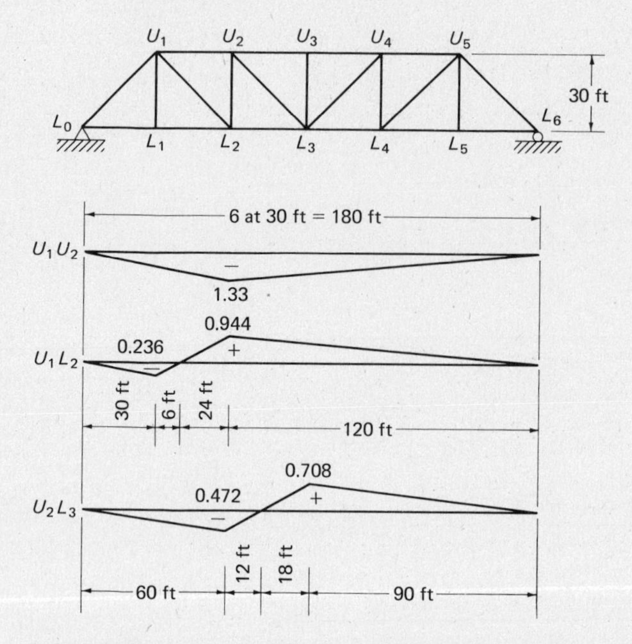

Figure 12.8

maximum force of the character being studied. If tension is being studied, the live uniform load is placed along the section of the truss corresponding to the positive or tensile section of the influence line, and the live concentrated loads are placed at the maximum positive tensile ordinates on the diagram.

Members whose influence lines have both positive and negative ordinates may possibly be in tension for one combination of loads and in compression for another. A member subject to ***force reversal*** must be designed to resist both the maximum compressive and maximum tensile forces.

In members U_1U_2, U_1L_2, and U_2L_3 of the truss of Fig. 12.8, the maximum possible forces due to the following loads are desired.

1. Dead uniform load of 1.5 klf
2. Live uniform load of 2 klf
3. Moving concentrated load of 20 kips
4. Impact of 24.4%

The influence lines are drawn, and the forces are computed by the exact method as described in the following paragraphs.

U_1U_2

The member is in compression for every position of the unit load; therefore the dead uniform load and the live uniform load are placed over the entire span. The moving concentrated load of 20 kips is placed at the maximum compression ordinate on the influence line. The impact factor is multiplied by the live load forces and added to the total.

$$\begin{aligned}
DL &= (1.5)(180)(-1.33)(\tfrac{1}{2}) &&= -180.0 \\
LL &= (2)(180)(-1.33)(\tfrac{1}{2}) &&= -240.0 \\
&\quad +(20)(-1.33) &&= -\ \ 26.7 \\
I &= (0.244)(-240.0 - 26.7) &&= \underline{-\ \ 65.1} \\
&\quad \text{total force} &&= -511.8^k \text{ compression}
\end{aligned}$$

U_1L_2

Examination of the influence line for U_1L_2 shows that for some positions of the unit load the member is in compression, whereas for others it is in tension. The live loads should be placed over the positive portion of the diagram and the dead loads across the entire structure to obtain the largest possible tensile force. Similarly, the live loads should be placed over the negative portion of the diagram and the dead loads over the entire structure to obtain the largest possible compressive force.

Maximum tension:

$$\begin{aligned}
DL &= (1.5)(144)(+0.944)(\tfrac{1}{2}) &&= +102.0 \\
&\quad +(1.5)(36)(-0.236)(\tfrac{1}{2}) &&= -\ \ \ 6.4 \\
LL &= (2)(144)(+0.944)(\tfrac{1}{2}) &&= +136.0 \\
&\quad +(20)(+0.944) &&= +\ \ 18.9 \\
I &= (0.244)(+136.0 + 18.9) &&= \underline{+\ \ 37.8} \\
&\quad \text{total force} &&= +288.3^k \text{ tension}
\end{aligned}$$

Maximum compression:

$$\begin{aligned}
DL &= (1.5)(144)(+0.944)(\tfrac{1}{2}) &&= +102.0 \\
&\quad +(1.5)(36)(-0.236)(\tfrac{1}{2}) &&= -\ \ \ 6.4 \\
LL &= (2)(36)(-0.236)(\tfrac{1}{2}) &&= -\ \ \ 8.5 \\
&\quad +(20)(-0.236) &&= -\ \ \ 4.7 \\
I &= (0.244)(-8.5 - 4.7) &&= \underline{-\ \ \ 3.2} \\
&\quad \text{total force} &&= +\ 79.2^k \text{ tension}
\end{aligned}$$

U_2L_3

The calculations for U_1L_2 proved it could have only tensile forces regardless of the positioning of the live loads given. The following calculations show force reversal may occur in member U_2L_3.

Maximum tension:

$$DL = (1.5)(108)(+0.708)(\tfrac{1}{2}) = +\ 57.3$$

$$+(1.5)(72)(-0.472)(\tfrac{1}{2}) = -\ 25.5$$

$$LL = (2)(108)(+0.708)(\tfrac{1}{2}) = +\ 76.4$$

$$+(20)(+0.708) = +\ 14.2$$

$$I = (0.244)(+76.4 + 14.2) = \underline{+\ 22.1}$$

$$\text{total force} = +144.5^{k} \text{ tension}$$

Maximum compression:

$$DL = (1.5)(108)(+0.708)(\tfrac{1}{2}) = +57.3$$

$$+(1.5)(72)(-0.472)(\tfrac{1}{2}) = -25.5$$

$$LL = (2.0)(72)(-0.472)(\tfrac{1}{2}) = -34.0$$

$$+(20)(-0.472) = -\ 9.4$$

$$I = (0.244)(-34.0 - 9.4) = \underline{-10.6}$$

$$\text{total force} = -22.2^{k} \text{ compression}$$

Panel-Load Method

The first part of this section considered the exact method of calculating maximum tensile and compressive forces due to a set of moving loads. For maximum tension the load was placed on the exact portion of the truss corresponding to the positive section of the influence line, and the tension caused equaled the intensity of the uniform load times the positive area of the diagram. The results were exact or mathematically correct, but an approximate method called the *panel-load method* is sometimes more convenient to use.

One basis for using an approximate method can be found in Sec. 13.4, where common empirical formulas for impact are presented. The values obtained from such expressions are only estimates of the effect of impact and are given as percentages of the live-load forces. Since the total force in a truss member is the sum of the dead load plus live load plus impact forces and since the impact is only approximately correct, it seems unnecessary to use great precision in computing the live-load forces. It is therefore satisfactory to use a reasonable estimate for the live-load portion, if the calculations are appreciably expedited thereby. It will be seen that the panel-load method cannot give forces smaller than those obtained by the exact method and will give the same values in some cases.

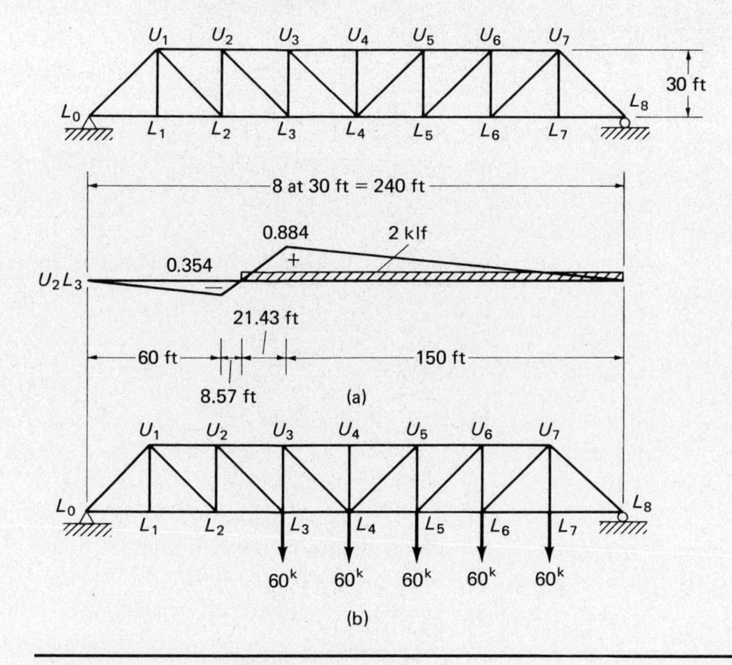

Figure 12.9

For this discussion the influence line for member U_2L_3 of the truss of Fig. 12.9 and a moving uniform load of 2 klf are used. By the exact method the largest possible tension in U_2L_3 occurs when the uniform load is placed along the positive portion of the diagram, as shown in Fig. 12.9(a). The force would be as follows:

$$(2)(171.43)(+0.884)(\tfrac{1}{2}) = 151.5^{k} \text{ tension}$$

In the panel-load method, full panel loads are assumed to be placed at each of the panel points on the side of the influence line being loaded. This placement makes it unncessary to compute the numerical values of the ordinates for the influence line, because a rough sketch of the diagram will reveal which panel points should be loaded for maximum tension and which ones should be loaded for maximum compression. The force in the member may be computed by some other method.

For maximum tension in U_2L_3, full panel loads ($2 \times 30 = 60$ kips) are placed at each panel point inside the positive section of the influence line. To have full panel loads, the uniform load must extend from L_2 to the right end, whereas the positive section of the influence line extends only from the zero point between L_2 and L_3 to the right end. A larger panel load is caused at L_3 than is used by the exact method; and since the effect of the half-panel load at L_2 having a negative influence-line ordinate is neglected, the resulting positive force will be larger than the value obtained by the exact method. For the full panel loads the force in U_2L_3 obtained by statics is $+159.0$ kips, or 5% higher than the exact value.

The results cannot possibly be smaller than the exact values because the panel load next to the point where the influence line changes from tension to compression is made larger. Should the influence line consist of a single triangle, identical results will be obtained by the two methods because the uniform load will be assumed to extend over the same length in both cases, that is, the entire span.

12.8 COUNTERS IN BRIDGE TRUSSES

The fact that a member in compression is in danger of bending or buckling reduces its strength and makes its design something of a problem. The design of a 20-ft member for a tensile force of 100 kips will result in a much smaller section than is required for a member of the same length subject to a compressive force of the same magnitude. The ability of a member to resist compressive loads depends on its stiffness, which is measured by the ***slenderness ratio.*** The slenderness ratio is the ratio of the length of a member to its least radius of gyration. As a section becomes longer, or as its slenderness ratio increases, the danger of buckling increases, and a larger section is required to withstand the same load.

This discussion shows there is a considerable advantage in keeping the diagonals of a truss in tension if possible. If a truss supported only dead load, it would be a simple matter to arrange the diagonals so that they were all in tension. All of the diagonals of the Pratt truss of Fig. 12.10(a) would be in tension for a uniform dead load extending over the entire span. The calculations in Sec. 12.7, however, have shown that live loads may cause the forces in some of the diagonals of a bridge truss to alternate between tension and compression. The constant passage of trains or trucks back and forth across a bridge will probably cause the forces in some of the diagonals to change continually from tension to compression and back to tension.

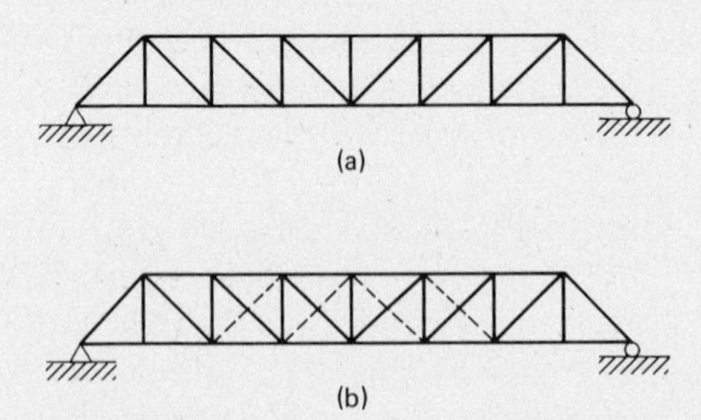

Figure 12.10

The possibilities of force reversal are much greater in the diagonals near the center of a truss. The reason for this situation can be seen by referring to the truss of Fig. 12.9, where a positive shear obviously causes tension in members U_1L_2 and U_2L_3. The positive dead-load shear is much smaller in panel 3 than in panel 2, and it is more likely for the live load to be in a position to cause a negative shear large enough to overcome the positive shear and produce compression in the diagonal.

A few decades ago, when it was common for truss members to be pin connected, the diagonals were actually eyebars which were capable of resisting little compression. The same force condition exists in trusses erected today with diagonals consisting of a pair of small steel angles or other shapes of little stiffness. It was formerly common to add another tension-resisting diagonal to the panels where force reversal could occur, the new diagonal running across the first one and into the previously unconnected corners of the panel. These members, called ***counters*** or ***counter diagonals***, can be seen in hundreds of older bridges across the country but rarely in new ones.

Figure 12.10(b) shows a Pratt truss to which counters have been added in the middle four panels, the counters being represented by dotted lines. When counters have been added in a panel, both diagonals may consist of relatively slender and light members, neither being able to resist appreciable compression. With light and slender diagonals the entire shear in the panel is assumed to be resisted by the diagonal which would be in tension for that type of shear, whereas the other diagonal is relaxed or without stress. The two diagonals in a panel may be thought of as cables that can resist no compression whatsoever. If compression were applied to one of the cables, it would become limp, whereas the other one would be stretched. A truss with counters is actually statically indeterminate unless the counter is adjusted to have zero force under dead load.

Today's bridges are designed with diagonals capable of resisting force reversals. In fact all bridge truss members, whether subject to force reversal or not, must be capable of withstanding the large force changes that occur when vehicles move back and forth across those structures. A member that is subject to frequent force changes even though the character of the force does not change (as $+50$ to $+10$ to $+50$ kips, etc.) is in danger of a fatigue failure unless it is specifically designed for that situation.

Modern steel specifications provide a maximum permissible stress range (from high to low) for each truss member. The stress range is defined as the algebraic difference between the maximum and minimum stresses. For this calculation tensile stress is given an algebraic sign that is opposite to that of compression stress. The AASHTO and AISC specifications provide a permissible stress range that is dependent on the estimated number of cycles of stress, on the type and location of a particular member, and on its type of connection. Obviously the more critical the situation the smaller is the permissible stress range.

PROBLEMS

For Probs. 12.1 to 12.18 draw influence lines for the members indicated.

Problem 12.1 L_0U_1, L_0L_1, U_1L_1, U_1U_2

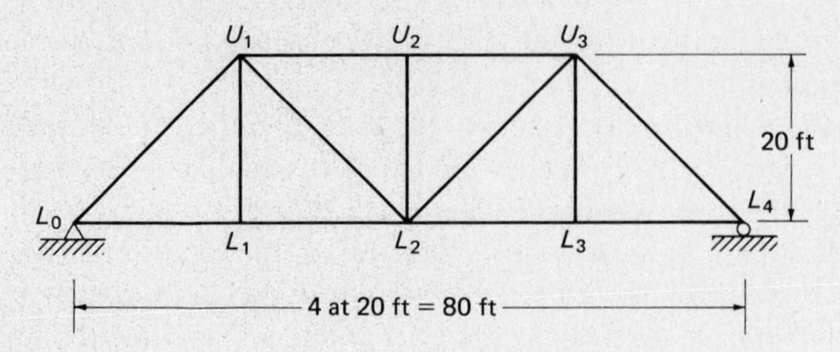

Problem 12.2 L_0L_1, U_1U_2, L_1U_2 (*Ans.* L_0L_1 + 1.50 at L_1; $U_1U_{2_H}$ − 1.50 at L_1; $L_1U_{2_V}$ + 1.00 at L_1, 0 at L_2)

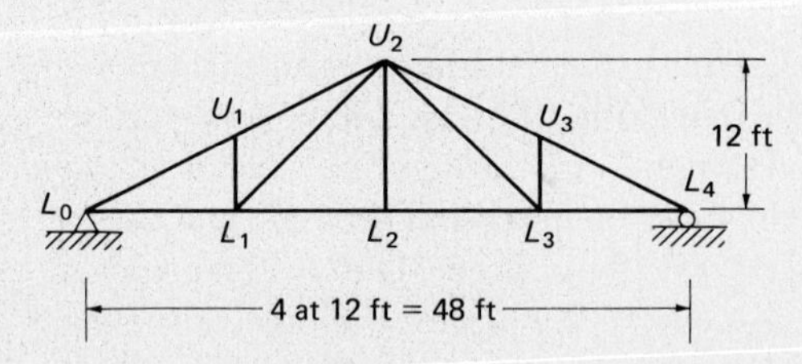

Problem 12.3 U_1U_2, L_1L_2, U_1L_2, L_2U_3

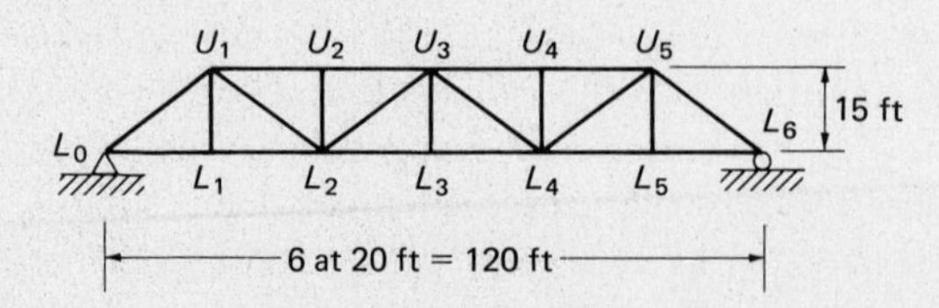

Problem 12.4 L_1L_2, U_2L_2, U_2L_3, L_3U_4 as unit load moves across top of truss (*Ans.* $L_1L_{2_H}$ + 1.00 at U_2; U_2L_2 − 1.00 at U_2; $U_2L_{3_V}$0 at U_2, +0.50 at U_3; $L_3U_{4_V}$ + 0.75 at U_3)

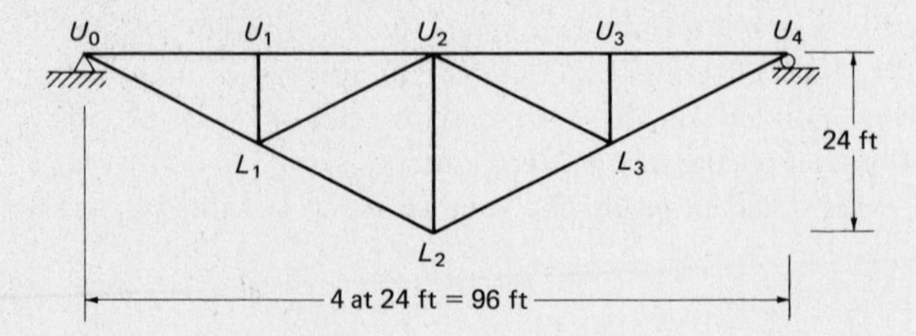

Problem 12.5 U_1U_2, U_2L_3, U_3L_3, L_4L_5

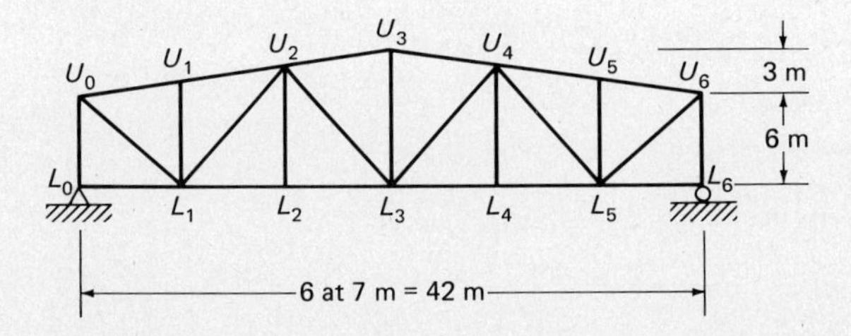

Problem 12.6 L_0U_1, L_1L_2, U_1U_2 (*Ans.* $L_0U_{1_V}$ $-$ 0.75 at L_1; L_1L_2 + 1.125 at L_1; $U_1U_{2_H}$ $-$ 1.00 at L_2)

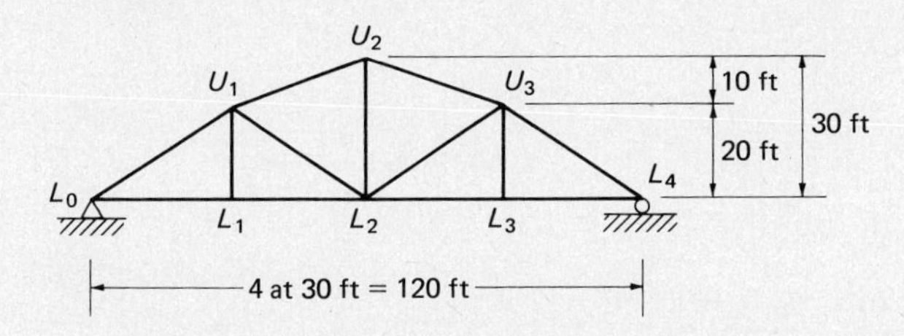

12.7. Members L_3L_4, U_1L_2, and U_2L_2 of the truss of Prob. 12.6.

Problem 12.8 U_1U_3, L_2L_4, L_2U_3, L_4U_5 (*Ans.* U_1U_3 $-$ 1.78 at L_2; L_2L_4 + 1.33 at L_2 and L_4; $L_2U_{3_V}$ + 0.33 at L_2; $L_4U_{5_V}$ + 0.67 at L_4)

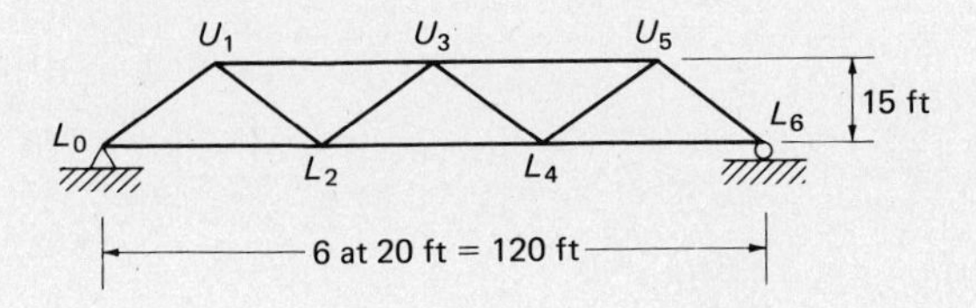

Problem 12.9 U_0L_1, U_1U_2, U_1L_2, L_3U_4 as unit load moves across top of truss

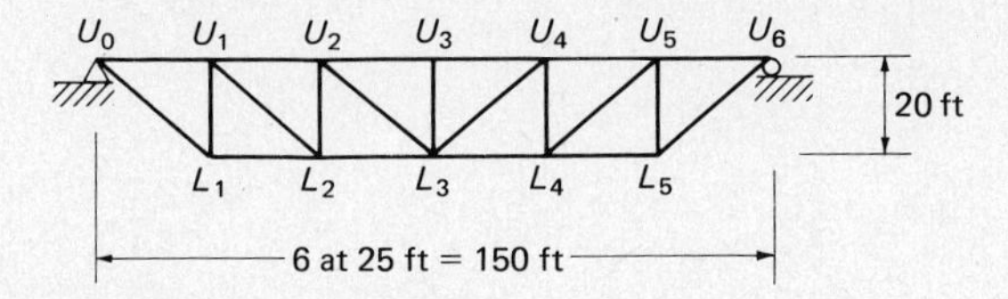

Problem 12.10 L_1L_2, U_1L_1, U_1L_2, U_3L_4 as unit load moves across top of truss (*Ans.* $L_1L_{2_H}$ + 1.67 at U_1; U_1L_1 − 0.28 at U_1; $U_1L_{2_V}$ + 0.222 at U_2; $U_3L_{4_V}$ − 0.25 at U_3)

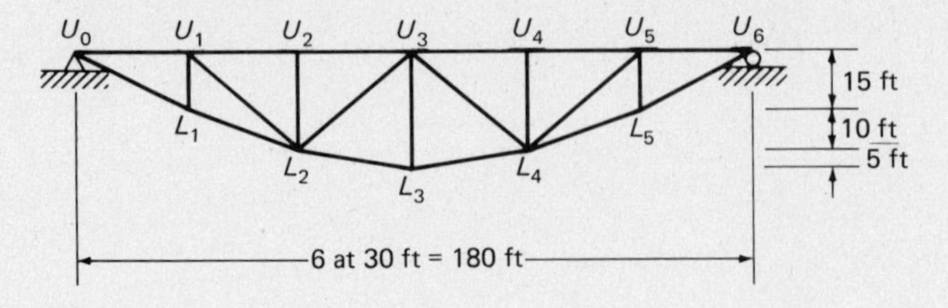

Problem 12.11 L_0L_1, U_1L_2, U_2U_3

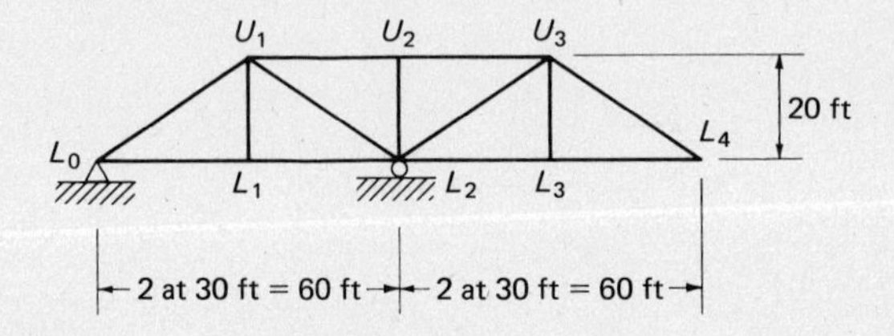

Problem 12.12 L_0U_1, U_1L_2, U_3L_4, U_5U_6 (*Ans.* $L_0U_{1_V}$ + 1.0 at L_0, 0 at L_1; $U_1L_{2_V}$ − 1.0 at L_0 and L_1, 0 at L_2; $U_3L_{4_V}$ + 0.667 at L_0, − 0.333 at L_3, +0.333 at L_4, −0.667 at L_7; U_5U_6 0 at L_6, +1.33 at L_7)

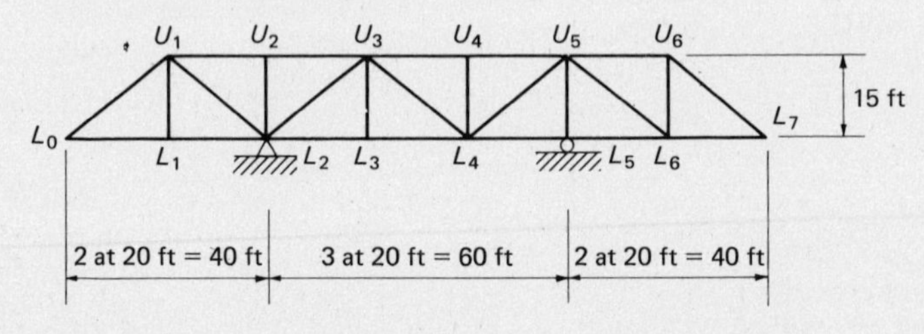

Problem 12.13 L_0U_1, U_1L_2, U_3L_4, U_5U_6

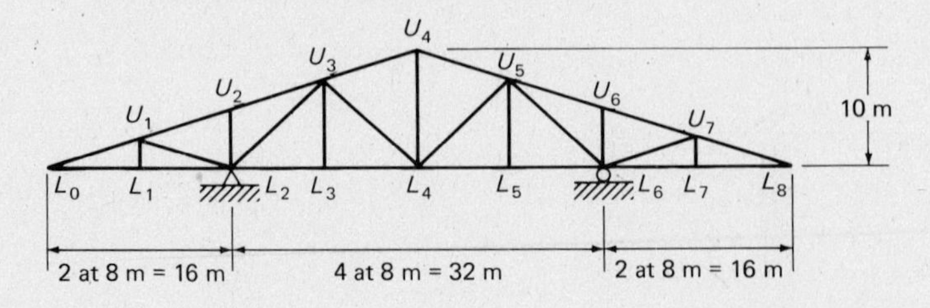

Problem 12.14 U_1L_2, L_2L_3, U_3L_4 (*Ans.* $U_1L_{2_v}$ − 0.467 at L_1; L_2L_3 − 2.00 at L_5; $U_3L_{4_v}$ + 0.5 at L_5)

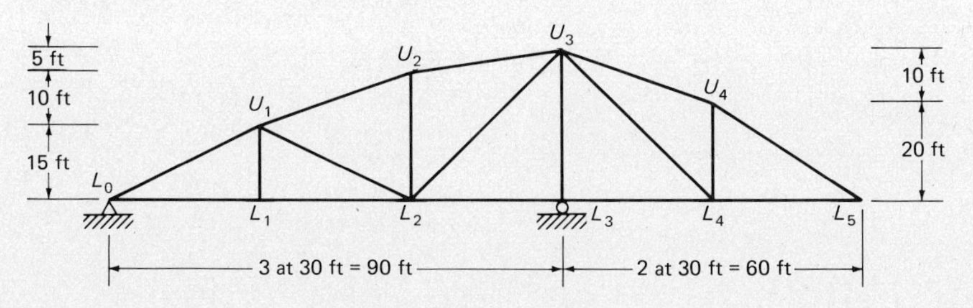

Problem 12.15 U_1U_2, U_2L_3, U_3L_3

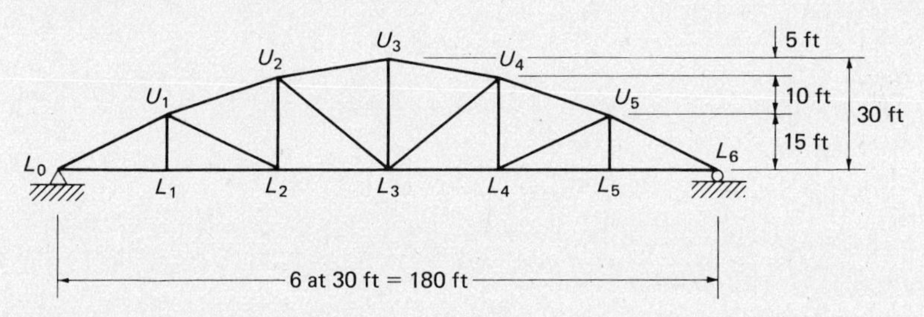

Problem 12.16 U_2U_4, L_3L_5, U_4L_5, L_5U_6 as unit load moves across top of truss (*Ans.* U_2U_4 − 1.25 at U_2, −1.50 at U_4; L_3L_5 + 2.00 at U_4; $U_4L_{5_v}$ − 0.50 at U_4, +0.25 at U_6; $L_5U_{6_v}$ + 0.167 at U_4; −0.75 at U_6)

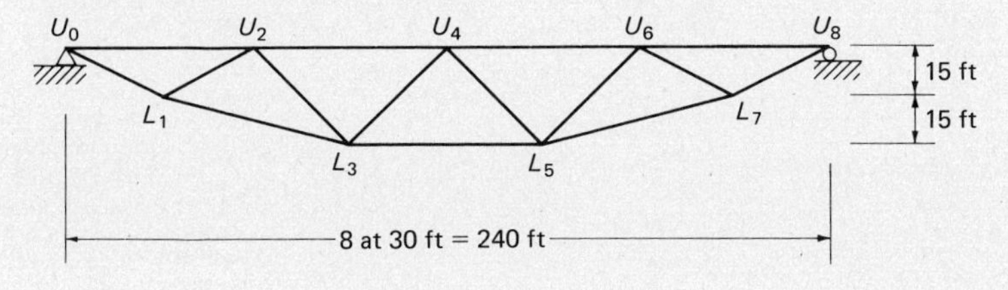

Problem 12.17 U_1U_2, U_1L_1, L_2U_3. Assume unit load moves across top of truss

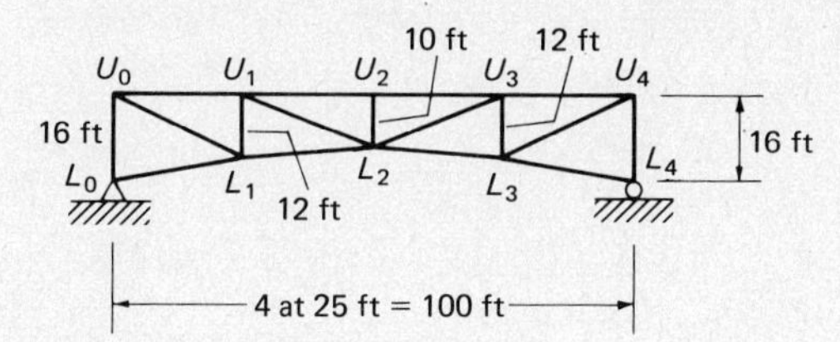

Problem 12.18 U_2U_3, M_1L_2, M_2L_2, U_3L_3 (*Ans.* U_2U_3 − 1.33 at L_2; $M_1L_{2_V}$ − 0.083 at L_1, +0.33 at L_2; M_2L_2 + 0.083 at L_1, +0.67 at L_2, −0.25 at L_3; U_3L_3 + 0.50 at L_3, 0 at L_2 and L_4)

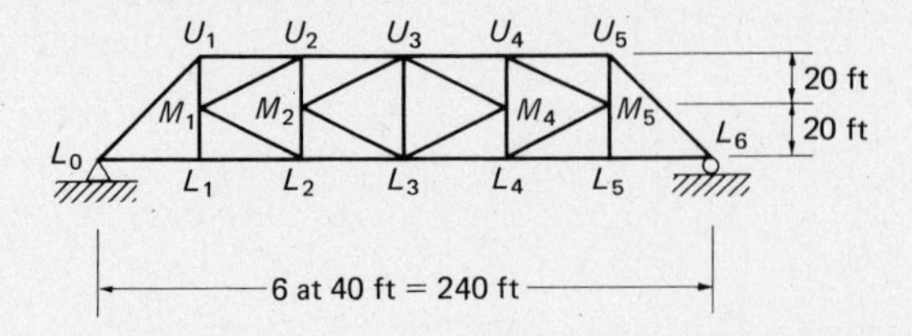

12.19. By the exact method compute the maximum and minimum forces in L_2U_3 of the truss of Prob. 12.3 for a uniform dead load of 1 klf, a moving uniform load of 2 klf, a moving concentrated load of 20 kips and an impact factor of 27%.

12.20. By the exact method determine if force reversal is possible in member U_1L_2 of the truss of Prob. 12.15 for the loads and impact factor used in Prob. 12.19. (*Ans.* Yes, -96.5^k, $+49.0^k$ resultant forces)

12.21. Solve Prob. 12.19 using the panel-load method.

12.22. Solve Prob. 12.20 using the panel-load method. (*Ans.* Yes, -105^k, $+57.7^k$ resultant forces)

For Probs. 12.23 to 12.29 draw influence lines for the members indicated.

Problem 12.23 U_1U_2, M_3L_4, U_4L_4, U_6L_6

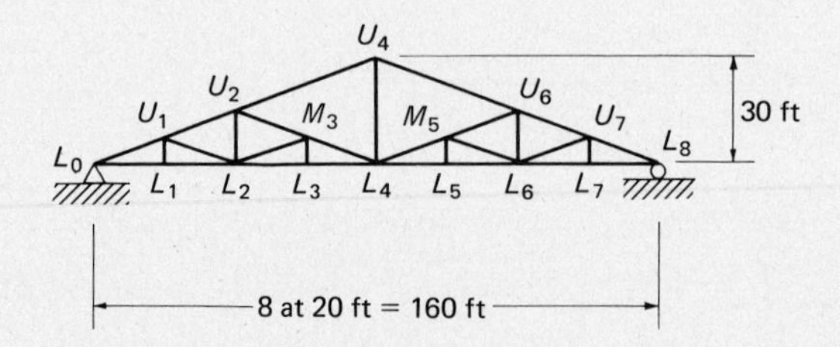

12.24. Members L_2L_3, U_3U_4, U_4L_5, and U_6L_6 of the truss of Prob. 7.14. (*Ans.* L_2L_3 − 2.0 at L_0, 0 at L_2; $U_3U_{4_H}$ + 1.8 at L_0, −0.9 at L_3, +0.6 at L_8; $U_4L_{5_V}$ + 0.6 at L_0, −0.6 at L_4, +0.1 at L_5, 0 at L_6, −0.2 at L_8; U_6L_6 + 0.5 at L_0, −0.75 at L_5, 0 at L_6, −1.5 at L_8)

12.25. Members U_1U_3, L_3L_4, and M_4U_5 of the truss of Prob. 8.9.

12.26. Members U_0U_1, U_2L_3, and L_5L_6 of the truss of Prob. 3.10. (*Ans.* U_0U_1 − 1.00 at U_1, 0 at U_4; $U_2L_{3_V}$0 at U_2, +0.5 at U_3, 0 at U_4; $L_5L_{6_H}$ − 2.10 at U_4, +0.79 at U_6)

Problem 12.27 Members *a*, *b*, and *c*

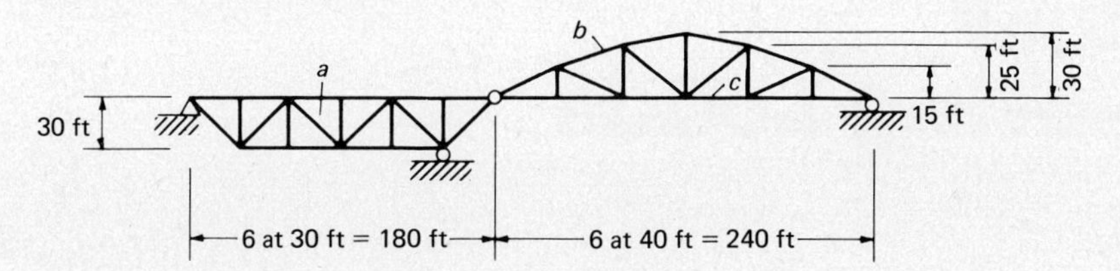

Problem 12.28 U_1U_2, L_2L_3, L_4U_5 as unit load moves across top of span (*Ans.* U_1U_2 − 1.5 at U_1; $L_2L_{3_H}$ + 1.0 at U_1; $L_4U_{5_H}$ = −1.0 at U_3)

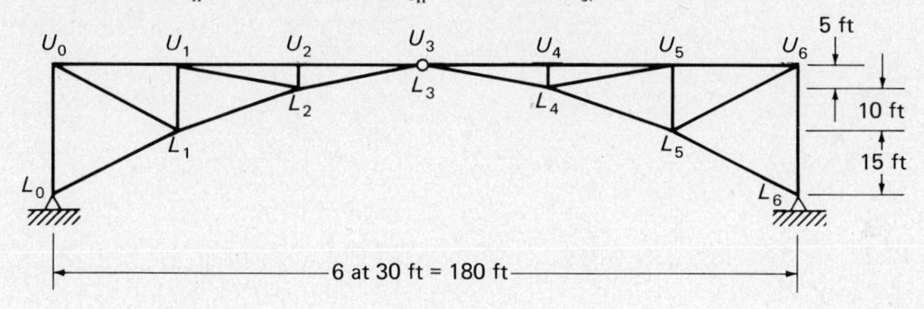

Problem 12.29 U_2U_3, U_2M_2, M_2L_2, U_3L_3

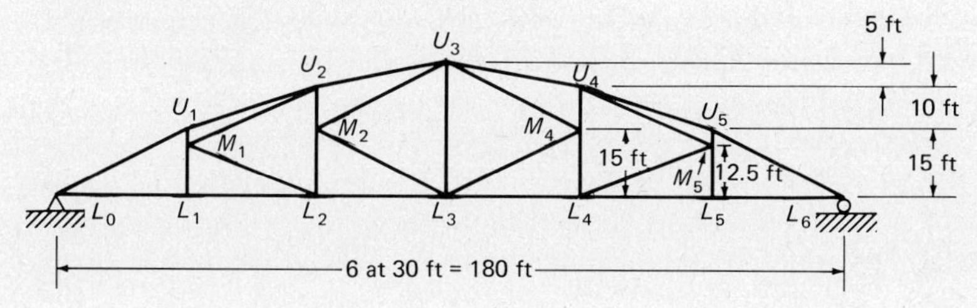

13 Moving Loads

13.1 GENERAL

Chapters 11 and 12 have repeatedly indicated that to design a beam, girder, truss, or any other structure supporting moving loads, the designer must be able to determine which positions of these loads cause maximum shear, moment, and so on, at various points in the structure. If one can place the loads at the positions causing maximums, he or she need not worry about any other positions the loads might take on the structure. Should a structure be loaded with a uniform live load and not more than one or two moving concentrated loads, the critical positions for placing the loads will be obvious from the influence lines.

If, however, the structure is to support a series of concentrated loads of varying magnitudes, such as groups of truck or train wheels, the problem is not as simple. The influence line will, of course, indicate the approximate positions for placing the loads, because it is reasonable to assume that the heaviest loads should be grouped in the vicinity of the largest ordinates of the diagram. The procedure for finding exactly the critical positions of the loads is substantially a trial-and-error method for which the influence line will provide a good initial estimate.

13.2 LIVE LOADS FOR HIGHWAY BRIDGES

Although highway bridges must support several different types of vehicles, the heaviest possible loads are caused by a series of trucks. The AASHTO specifies that highway bridges shall be designed for lines of motor trucks occupying 10-ft-wide lanes. Only one truck is placed in each span for each lane. The truck loads specified are designated with an H prefix followed by a number indicating the total weight of the truck, in tons. The weight may be followed by

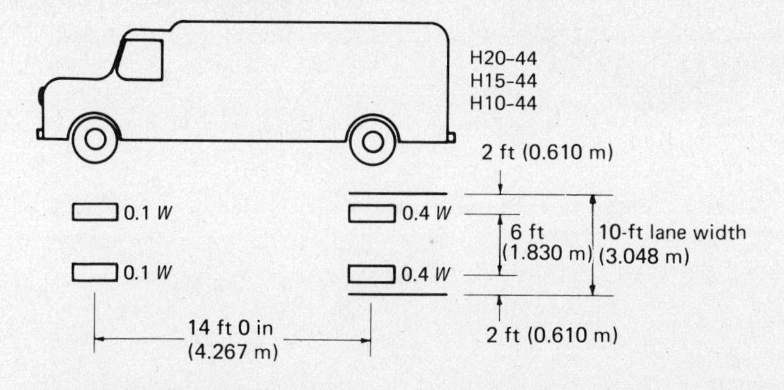

Figure 13.1

another number indicating the year of the specifications. For example, an H20–44 loading indicates a 20-ton truck and the 1944 specifications. A sketch of the truck and the distances between axle centers, wheel centers, and so on, is shown in Fig. 13.1.

The selection of the particular truck loading to be used in design depends on the bridge location, anticipated traffic, and so on. These loadings may be broken down into three groups as follows.

Two-Axle Trucks: H20, H15, and H10

The weight of an H truck is assumed to be distributed two-tenths to the front axle (for example, 4 tons, or 8 kips, for an H20 loading) and eight-tenths to the rear axle. The axles are spaced 14 ft, 0 in on center, and the center-to-center lateral spacing of the wheels is 6 ft, 0 in. Should a truck loading varying in weight from these be desired, one that has axle loads in direct proportion to the standard ones listed here may be used. A loading as small as the H10 may be used only for bridges supporting the lightest traffic.

Two-Axle Trucks Plus One-Axle Semitrailer; HS15-44 and HS20-44

For today's highway bridges carrying a great amount of truck traffic, the two-axle truck loading with a one-axle semitrailer weighing 80% of the main truck load is commonly specified for design. The HS20-44 truck has 4 tons on the front axle, 16 tons on the rear axle, and 16 tons on the trailer axle. The distance from the rear truck axle to the semitrailer axle is varied from 14 to 30 ft, depending on which spacing will cause the most critical conditions.

Uniform Lane Loadings

Computation of forces caused by a series of concentrated loads, whether they represent two-axle trucks or two-axle trucks with semitrailers, is a tedious job with a hand calculator; therefore a lane loading that will produce approximately the same forces is frequently used. The lane loading consists of a

John F. Fitzgerald Expressway, Mystic River Bridge to Haymarket Square, Boston, Massachusetts. (Courtesy of the American Institute of Steel Construction, Inc.)

uniform load plus a single moving concentrated load. This load system represents a line of medium-weight traffic with a heavy truck somewhere in the line. The uniform load per foot is equal to 0.016 times the total weight of the truck to which the load is to be roughly equivalent. The concentrated load equals 0.45 times the truck weight for moment calculations and 0.65 times the truck weight for shear calculations. These values for an H20 loading would be as follows: 0.016×20 tons equals 640 lb/ft of lane; concentrated load for moment 0.45×20 tons equals 18 kips; and concentrated load for shear 0.65×20 tons equals 26 kips.

For continuous spans another concentrated load of equal weight is to be placed in one of the other spans in such a position as to cause maximum negative moment. For positive moment only one concentrated load is to be used per lane with the uniform load placed in as many spans as necessary to produce the maximum positive value.

The lane loading is more convenient to handle, but it should not be used unless it produces moments or shears equal to or greater than those produced by the corresponding H loading. Based on the information presented later in this chapter, calculations can be made which will show that the equivalent lane loading for the HS20-44 will produce greater moments in simple spans of 145 ft and above and greater shears for simple spans of 128 ft and above. Appendix A of the AASHTO specifications contains tables that give the maximum shears and moments in simple spans for the various H loadings or for their equivalent lane loadings, whichever controls.

The possibility of having a continuous series of heavily loaded trucks in every lane of a bridge that has more than two lanes does not seem as great as that for a bridge that has only two lanes. The AASHTO therefore permits the values caused by full loads in every lane to be reduced by a certain factor if the bridge has more than two lanes.

Interstate Highway System Loading

Another loading system can be used instead of the HS20-44 in the design of structures for the Interstate Highway System. This alternate system, which consists of a pair of 24-kip axle loads spaced 4 ft on center, is critical for short spans only. From the theory presented later in this chapter it is possible to show that this loading will produce maximum moments for simple spans from 11.5 to 37 ft and maximum shears for spans from 6 to 22 ft. For other spans the HS20-44 loading or its equivalent lane loading will be critical.

13.3 LIVE LOADS FOR RAILWAY BRIDGES

Railway bridges are commonly analyzed for a series of loads devised by Theodore Cooper. His loads, referred to as E loadings, represent two locomotives followed by a line of freight cars. A series of concentrated loads is used for the locomotives, and a uniform load represents the freight cars. Mr. Cooper introduced his loading system in 1894; it was the so-called E-40 load, which is pictured in Fig. 13.2. The train is assumed to have a 40-kip load on the driving axle of the engine. Since his system was introduced, the weights of trains have been increased considerably, until at the present time bridges are designed on the basis of loads in the vicinity of an E–72 loading, and the use of E–80 and E–90 loadings is not uncommon.

Various tables that are obtainable present detailed information pertaining to Cooper's loadings such as axle loads, moments, and shears. If information is

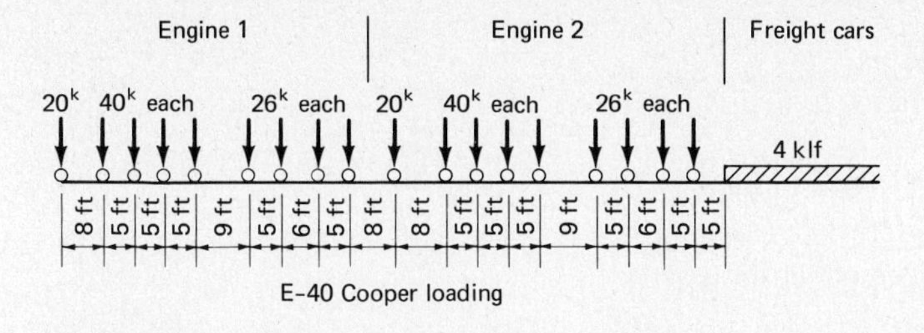

Figure 13.2

available for one E loading, the information for any other E loading can be obtained by direct proportion. The axle loads of an E–75 are 75/40 those for an E–40; those for an E–60 are 60/72 of those for an E–72; and so on. Tables used in conjunction with the maximum criteria presented later in this chapter greatly reduce the computations.

Cooper's loadings do not accurately picture today's trains, but they are still in general use despite the availability of several more modern and more realistic loadings such as Dr. D. B. Steinman's M–60 loading [1].

13.4 IMPACT LOADINGS

The truck and train loads applied to highway and railroad bridges are applied not gently and gradually but rather violently, which causes forces to increase. Additional loads, called *impact loads*, must be considered, and they are taken into account by increasing the live-load forces by some percentage, the percentage being obtained from purely empirical expressions. Numerous formulas have been presented for estimating impact. One example is the following AASHTO formula for highway bridges, in which I is the percent of impact and **L is the length of the span, in feet, over which live load is placed to obtain a maximum stress.** The AASHTO says that it is unnecessary to use an impact percentage greater than 30%, regardless of the value given by the formula. Notice that the longer the span length becomes, the smaller becomes the impact.

$$I = \frac{50}{L + 125} \quad \text{or} \quad \frac{15.24}{L + 38} \text{ if } L \text{ is in meters}$$

Impact factors or percentages for railroad bridges are higher than those for highway bridges because of the much greater vibrations caused by the wheels of a train as compared to the relatively soft rubber-tired vehicles on a highway bridge. A person need only stand near a railroad bridge for a few seconds while a fast-moving and heavily loaded freight train passes over to see the difference. Tests have shown the impact on railroad bridges will often run as high as 100% or more. Not only does a train have a direct vertical impact, or bouncing up and down, but it also has a lurching or swaying back-and-forth type of motion. Some AREA impact formulas are as follows:

Direct vertical effect for beams, girders, floor beams, and so on:

$$I = 60 - \frac{L^2}{500} \qquad \text{for } L < 100 \text{ ft}$$

$$I = \frac{1800}{L - 40} + 10 \qquad \text{for } L = 100 \text{ ft or more}$$

Direct vertical effect for trusses:

$$I = \frac{4000}{L + 25} + 15$$

The AISC specification states that unless otherwise specified live loads shall be increased by certain percentages. Some of these values are: 100% for elevators, 33% for hangers supporting floors and balconies, not less than 50% for supports of reciprocating machinery or power driven units, and so on.

13.5 MAXIMUM SHEAR IN A BEAM SUPPORTING UNIFORM LIVE LOADS

The simple beam of Fig. 13.3 and the influence lines for shear at section 1–1 and at the beam ends are considered here. It is apparent from these diagrams that maximum shear is caused when the moving uniform load completely covers the beam. Under this loading condition, the reactions are at their maximums, as are the shears an infinitesimal distance from the supports. A uniform load of 6 klf would cause shears at each end of the beam equal to $6 \times \frac{1}{2} \times 20 \times 1.0 = 60$ kips.

Examination of the influence line for shear at section 1–1 shows that maximum shear at the section would be developed if the uniform load were placed over the positive or negative portion of the diagram having the largest area. For section 1–1 the 6-klf load is placed across the beam from the section to the right support, which causes a positive shear of $6 \times \frac{1}{2} \times 14 \times 0.7 = +29.4$ kips. The maximum negative shear at the section is developed when the uniform load extends from the left support to the section and equals $6 \times \frac{1}{2} \times 6 \times 0.3 = -5.4$ kips.

To review, the problem of determining the maximum shear in a simple beam loaded with a moving uniform load is elementary. The beam is loaded for the entire span and the reactions equal the maximum shear that can occur. If the maximum shear is desired at a certain section in the beam, the uniform load is so placed that it extends from the section to the support that is the

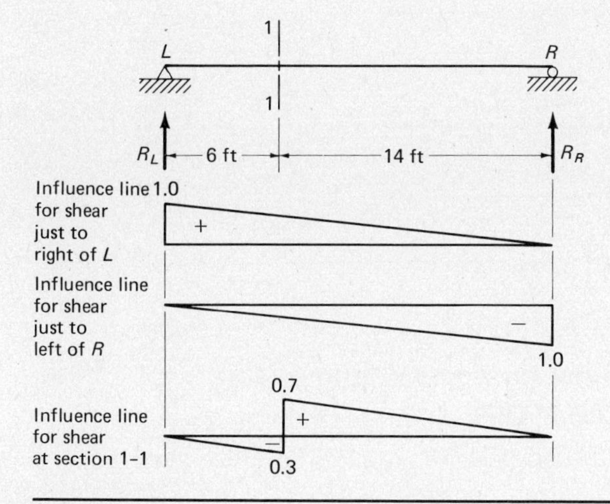

Figure 13.3

greatest distance away. The value of the shear so developed is computed from the area of the influence line corresponding to the portion of the beam along which the load is placed. The maximum shears at points in beams other than simple ones due to a moving uniform load can be determined from the influence lines in a similar manner.

13.6 MAXIMUM BENDING MOMENTS AT POINTS IN BEAMS SUPPORTING UNIFORM LIVE LOADS

Influence lines for bending moment at two sections in a simple beam are presented in Fig. 13.4. Since these influence lines have positive ordinates for their entire lengths, it is readily apparent that maximum moments will occur at both sections when the uniform load extends across the entire span. For a uniform load of 4 klf the maximum moments at sections 1–1 and 2–2 are as follows:

$$M_{1\text{-}1} = 4 \times \tfrac{1}{2} \times 30 \times 6.67 = 400'^{k}$$

$$M_{2\text{-}2} = 4 \times \tfrac{1}{2} \times 30 \times 4.17 = 250'^{k}$$

The ordinates of the influence lines for moments at sections in other types of beams will rarely be all positive or all negative. Maximum positive moment for a moving uniform load can be found by placing the load over the positive portion of the diagram and maximum negative moment by placing the load over the negative portion. The intensity of the load is multiplied by the corresponding influence-line area.

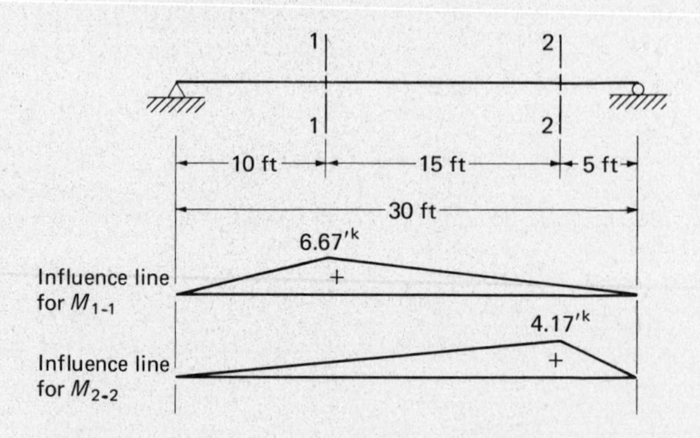

Figure 13.4

13.7 MAXIMUM END SHEAR IN A BEAM SUPPORTING MOVING CONCENTRATED LOADS

The largest shear that can be developed in a simple beam has been shown to occur an infinitesimal distance from one of the supports. At this point the shear equals the reaction, and the influence line for the reaction is the same as the

Fort Henry Bridge, Wheeling, West Virginia. (Courtesy of the American Bridge Division, U.S. Steel Corporation.)

influence line for the shear next to it. Influence lines are plotted for the left reaction and the shear an infinitesimal distance to the right of the reaction for a simple beam in Fig. 13.5.

If a simple beam is loaded with a series of moving concentrated loads, the maximum shear occurs at the supports, and the problem becomes a question of which position of the loads will cause the greatest end reaction and thus the greatest shear. For example, the beam of Fig. 13.5 is to be loaded with the series of loads shown, and it is desired to find the maximum shear that can occur at the left support. As the series of loads moves onto the span from the

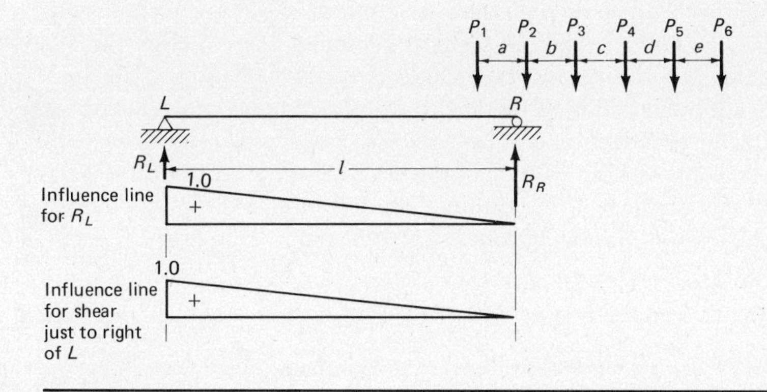

Figure 13.5

right side, the left reaction begins to build up from zero to a maximum when P_1 comes over the support. Immediately after P_1 leaves the span, the reaction decreases by the amount of the load. As the loads continue to the left, the left reaction begins to increase again until it reaches another maximum when P_2 is over the support. This process is repeated as each of the loads moves toward the left support. The left reaction reaches a maximum for each and then falls off as the load moves off the span.

One method of determining the maximum value of the left reaction is trial and error. The reaction may be computed when P_1 is over the support, when P_2 is over the support, and so forth for all of the loads. The largest value of the left reaction obtained is the largest shear possible in the beam for that loading condition. The method is tedious, and a simpler procedure based on a consideration of the change in the reaction as each load passes off the span is available.

For the following discussion ΣP is considered to be the sum of the loads remaining on the beam at any time. As P_1 passes off the span and P_2 moves over the support, the shear changes as follows:

$$dV = \frac{\Sigma Pa}{l} - P_1$$

As P_2 passes off the span and P_3 moves over the left support, the shear change is

$$dV = \frac{\Sigma Pb}{l} - P_2$$

If the resulting change is positive, the shear has increased; if negative, it has decreased. The loads may be considered one by one as they leave the span and the next load moves over the support. The first load leaving the span that causes a decrease should be the one placed over the support for computation of maximum shear. For normal loading conditions, one of the first two or three loads will be found to be the critical one.

Should another load come onto the span from the right end, as the loads are moved successively over the left support, its increasing effect on the left reaction, and thus the shear, must be included in the calculations. The shear increase equals the magnitude of the load times the distance moved onto the span divided by the span length, or simply the magnitude times the ordinate of the influence line opposite the position of the load. Example 13.1 illustrates the application of this method of determining maximum end shear in a simple beam caused by a series of moving concentrated loads.

Example 13.1

Determine the maximum shear developed at the left end of the beam shown in Fig. 13.6 as the concentrated loads shown move across the span from right to left.

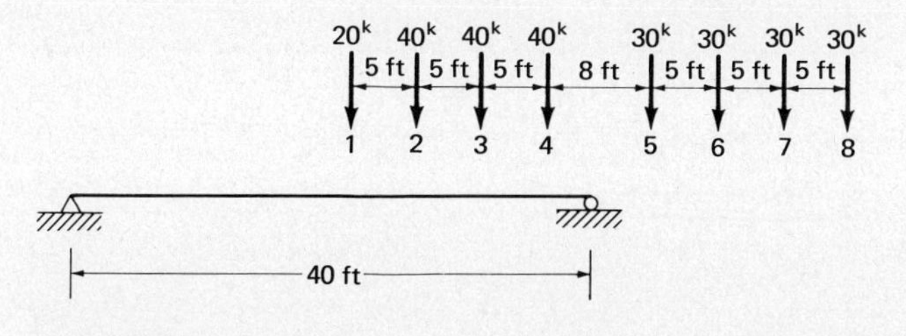

Figure 13.6

SOLUTION

Move load 1 off span to left and load 2 over support:

$$dV = \frac{(240)(5)}{40} - 20 = +10 \qquad \text{an increase}$$

Move load 2 off span to left and load 3 over support:

$$dV = \frac{(200)(5)}{40} - 40 = -15 \qquad \text{a decrease}$$

Place load 2 over left support for maximum end shear:

$$V = \frac{(30)(7 + 12 + 17 + 22) + (40)(30 + 35 + 40)}{40} = 148.5^k$$

13.8 MAXIMUM SHEAR AT INTERIOR POINTS OF BEAMS SUPPORTING MOVING CONCENTRATED LOADS

The maximum possible shear in a beam at some point away from the supports caused by the passage of a series of concentrated loads is often needed for design. Its value may be determined by a method closely related to the one used for computing maximum end shears. The beam and loads of Fig. 13.7 will be used in the following discussion.

The problem may be solved by trial and error, as could the end-shear problem. The loads may be moved across the span from right to left, and the shear then computed when the first load is over the section, when the second load is over the section, and so on. Eventually the maximum shear will be determined. The work is not as lengthy as this discussion may seem to indicate because the maximum will probably occur when one of the first two or three loads is above the section. The influence line for shear at section 1–1 in the beam of Fig. 13.7 shows that a good estimate for positioning the loads can be made by placing as many loads as possible over the positive portion of the influence line and as few as possible over the negative portion. If the loads are moved onto the span from the right side until the first 20-kip load is at the section, the shear developed is probably close to its absolute maximum possible value.

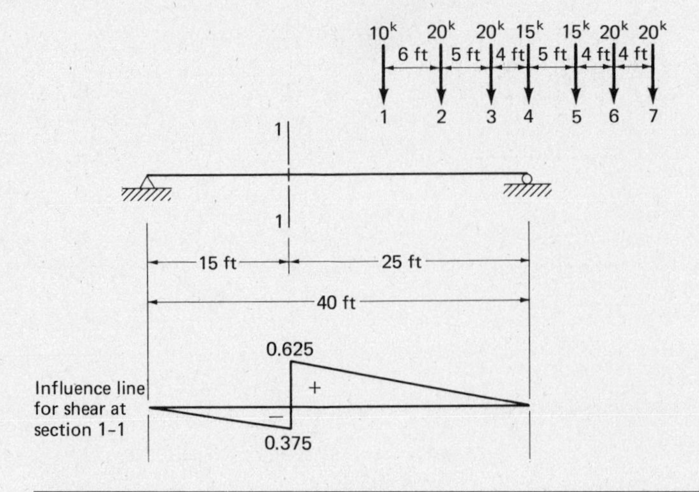

Figure 13.7

A consideration of the shear change as each load passes over the section presents a much quicker solution than the trial-and-error method discussed in the preceding paragraph. As the loads move across the span, the shear builds up as each load approaches the section. When the 10-kip load is at the section, a maximum shear occurs, but after it moves an infinitesimal distance to the left, the shear falls off 10-kips. Continuing to move the loads to the left causes the shear to increase gradually until the first 20-kip load reaches the section. At that time another maximum occurs, but as soon as the load moves slightly beyond the section, the shear is reduced by the amount of the load. The absolute maximum shear occurs when one of the concentrated loads is at the section.

As the loads move across the span, the maximum shears that occur for each load as it passes over the section will increase until the load causing the absolute maximum reaches the section. After it passes the section, the maximums occurring for the successive loads will decrease. If it were possible to find which load caused the absolute maximum, it would be necessary to compute the shear for only one position of the loads. This load can be determined from a consideration of the changes in shear as each of the loads passes the section.

As the loads are moved one after another to the section the shear change equals the increase in the left reaction due to the movement of the loads to the left, plus the increase in the left reaction due to any additional loads that have come onto the span from the right, less the load that has just passed over the section. Should the sum of these values be positive, the shear has increased. The first load that in moving past the section causes a decrease is the one which will cause absolute maximum shear, and the computations are made with that load over the section. Example 13.2 illustrates the computation of maximum shear at an interior point of a beam.

Cold Springs Interchange Bridge (U.S. 395), North of Reno, Nevada. (Courtesy of the Nevada Department of Transportation.)

Example 13.2

Compute the maximum shear at section 1–1 in the beam of Fig. 13.7 for the loads given in the figure.

SOLUTION

Move load 1 past section and load 2 up to section, noting that load 7 moves onto span:

$$dV = \frac{(100)(6)}{40} + \frac{(20)(3)}{40} - 10 = +6.5 \qquad \text{an increase}$$

Move load 2 past section and load 3 up to section:

$$dV = \frac{(120)(5)}{40} - 20 = -5 \qquad \text{a decrease}$$

By using load 2 to compute maximum shear,

$$R_L = \frac{(20)(3 + 7 + 20 + 25) + (15)(11 + 16) + (10)(31)}{40} = 45.4^{k}$$

$$V = 45.4 - 10 = 35.4^{k}$$

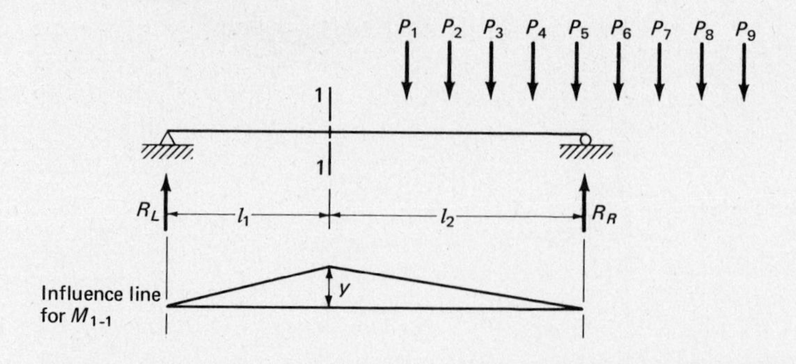

Figure 13.8

13.9 MAXIMUM MOMENT AT A POINT IN A BEAM SUPPORTING CONCENTRATED LIVE LOADS

The design of a beam may require the calculation of maximum moments at several points in the beam caused by the movement of a series of concentrated loads across the span. The trial-and-error method of moving one load after another up to the point in question and computing the moment in each case is one way of handling the problem. The influence line presents a method of estimating the value of the maximum moment. Examination of the simple beam of Fig. 13.8 and the influence line for moment at section 1–1 indicates that maximum moment will occur when as many of the large loads as possible are located near the peak ordinates of the diagram. The loads could be positioned on the diagram in this manner and the moment calculated. The result should be a fairly good estimate of the maximum possible moment at the point.

A study of the moment change as each load moves up to and past the section provides a method of quickly obtaining the exact maximum moment. As the group of loads in Fig. 13.8 is moved onto the span and up to the section, the moment at section 1–1 equals $R_L \times l_1$ minus any loads that move past the section times their lever arms back to the section. As a load moves toward the section from the right, the moment increases until the load is at the section, because its movement to the left causes an increase in R_L and a corresponding increase in $R_L \times l_1$. When the load passes the section, the moment begins to decrease, because the increase in $R_L \times l_1$ is not as great as the increase in the load times the distance back to the section. A maximum moment occurs when each load is at the section, and the problem is to determine which load causes the absolute maximum.

The movement of the loads from right to left on the beam of Fig. 13.8 or up the influence line causes a moment increase. The movement of the first loads past the section or down the influence line may cause a decrease in total moment. The change that occurs in the moment is equal to the increase due to the loads on the right of the section times their increase in influence-line ordinates less the loads on the left of the section times their ordinate decrease.

The total moment at section 1–1 is greater if the increases to the right are greater than the decreases to the left. A movement of the loads on the right of the section to the left a distance of 1 ft causes each of the ordinates of the loads to increase by $1 \times y/l_2$. The total increase in moment would be as follows:

$$\text{moment increase} = \text{total load to right} \times \frac{y}{l_2}$$

The movement of the loads on the left of the section to the left a distance of 1 ft causes each of the ordinates of the loads to decrease by $1 \times y/l_1$. The total decrease in moment may be written as

$$\text{moment decrease} = \text{total load to left} \times \frac{y}{l_1}$$

The rate of change of moment at the section is constant until one of the loads passes the section. The new rate of change of moment will remain constant until another load passes the section. Although the rate of change of moment changes as each load passes the section, it will be positive until the load causing absolute maximum moment reaches the section; thereafter the moment will decrease. As this particular load reaches the section, the rate of change of moment at the section becomes zero, because the rate of increase equals the rate of decrease.

$$\text{moment decrease} = \text{moment increase}$$

$$\text{total load to left} \times \frac{y}{l_1} = \text{total load to right} \times \frac{y}{l_2}$$

By canceling y from the equation,

$$\frac{\text{total load to left}}{l_1} = \frac{\text{total load to right}}{l_2}$$

Absolute maximum moment at any point in a beam due to a moving series of concentrated loads occurs when the average load to the left of the point is equal to the average load to the right of the point. To determine the absolute maximum moment at a point in a beam, each load is moved up until it is an infinitesimal distance to the right of the point, and the average load on each side of the point is determined. The load is moved just to the left of the point, and the averages are computed again. If the movement of the load from the right of the point to the left of the point causes the average load to the right to change from larger than the average load to the left to smaller, the critical load has been found. The moment is computed with that load over the point because during its passage the criterion is satisfied. Example 13.3 illustrates the application of the average-load method. A convenient table is given in Fig. 13.9 for recording the average-load calculations.

Try load 6

	Average load to left	Average load to right
With load to right of point		
With load to left of point		

Figure 13.9

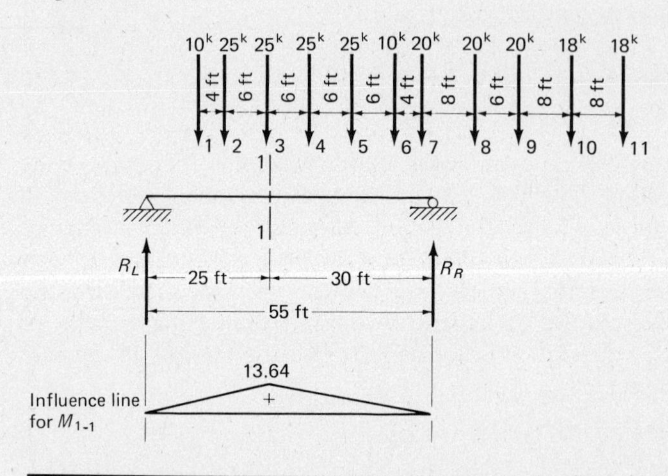

Figure 13.10

Example 13.3

Calculate the absolute maximum moment at a point 25 ft from the left end of the simple beam shown in Fig. 13.10 as the series of concentrated loads shown moves across the span.

SOLUTION

The influence line indicates that maximum moment will occur when as many loads as possible are on the span.

TRY 3

$\frac{35}{25} = 1.40$	$\frac{105}{30} = 3.50$
$\frac{60}{25} = 2.40$	$\frac{80}{30} = 2.67$

TRY 4

$\frac{60}{25} = 2.40$	$\frac{100}{30} = 3.33$
$\frac{85}{25} = 3.40$	$\frac{75}{30} = 2.50$

By using load 4 at section 1–1.

$$R_L = \frac{(20)(6 + 14) + (10)(18) + (25)(24 + 30 + 36 + 42) + (10)(46)}{55} = 78.9^{k}$$

$$M = (78.9)(25) - (25)(6 + 12) - (10)(16) = 1362'^{k}$$

13.10 ABSOLUTE MAXIMUM MOMENT DEVELOPED BY A SERIES OF CONCENTRATED LIVE LOADS

The absolute maximum moment in a simple beam is usually thought of as occurring at the beam centerline. Maximum moment does occur at the centerline if the beam is loaded with a uniform load or a single concentrated load. A beam, however, may be required to support a moving series of varying concentrated loads such as the wheels of a train, and the absolute maximum moment will in all probability occur at some position other than the centerline.

The largest possible moment should be determined, because the beam must be capable of withstanding the worst possible conditions. To calculate the moment, it is necessary to find the point where it occurs and the position of the loads causing it. Assuming the largest moment to be developed at the centerline of long span beams is reasonable, but for short-span beams this assumption may be considerably in error. It is therefore necessary to have a definite procedure for determining absolute maximum moment.

The moment diagram for a simple beam loaded with a group of concentrated loads will consist of a set of straight lines regardless of the position of the loads; therefore the absolute maximum moment occurring during the movement of these loads across the span will occur at one of the loads, usually the one nearest the center of gravity of the group. The beam in Fig. 13.11 and the series of loads P_1, P_2, P_3, and so on, are studied in the following paragraphs. The load P_3 is assumed to be the one nearest the center of gravity of the loads on the span, and it is located a distance l_1 from P_R (the resultant of all the loads on the span) and a distance l_2 from $P_{1\text{-}2}$ (the resultant of loads P_1 and P_2). The left reaction R_L is located a distance x from P_R. In the following paragraphs maximum moment is assumed to occur at P_3, and a definite method is developed for placing this load to cause the maximum.

The moment at P_3 may be written as follows:

$$M = R_R(l - x - l_1) - (P_{1\text{-}2})(l_2)$$

Substituting the value of R_R, P_Rx/l, gives

$$M = \left(\frac{P_Rx}{l}\right)(l - x - l_1) - (P_{1\text{-}2})(l_2)$$

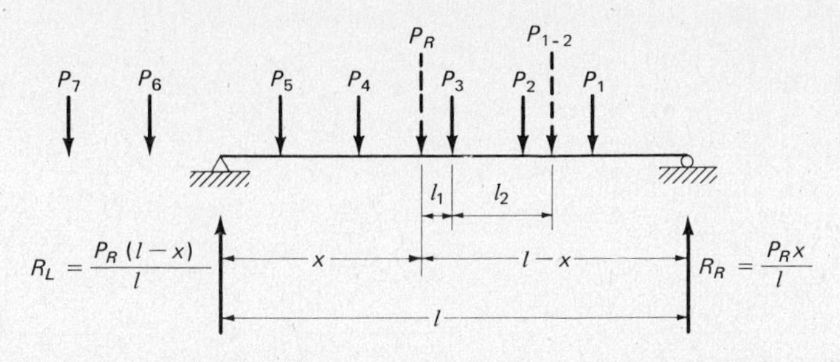

Figure 13.11

It is desired to find the value of x for which the moment at P_3 will be a maximum. Maximum moment at P_3, which occurs when the shear is zero, may be found by differentiating the moment expression with respect to x, equating the result to zero, and solving for x.

$$\frac{dM}{dx} = l - 2x - l_1 = 0$$

$$x = \frac{l}{2} - \frac{l_1}{2}$$

From the preceding derivation a general rule for absolute maximum moment may be stated as follows: **Maximum moment in a beam loaded with a moving series of concentrated loads will usually occur at the load nearest the center of gravity of the loads on the beam when the center of gravity is the same distance on one side of the centerline of the beam as the load nearest the center of gravity of the loads is on the other side.**

Should the load nearest the center of gravity of the loads be a relatively small one, the absolute maximum moment may occur at some other load nearby. Occasionally two or three loads have to be considered to find the greatest value; however, the problem is not a difficult one because the other moment criteria—average load to left equals average load to right—must be satisfied, and there will be little trouble in determining which of the nearby loads will govern. (Actually it can be shown that the absolute maximum moment occurs under the load that would be placed at the centerline of the beam to cause maximum moment there, when that wheel is placed as far on one side of the beam centerline as the center of gravity of all the loads is on the other [2].)

Example 13.4 illustrates the calculation of absolute maximum moment in a beam.

Example 13.4

Determine the absolute maximum moment in the 50-ft simple beam of Fig. 13.12 caused by the moving concentrated load system shown.

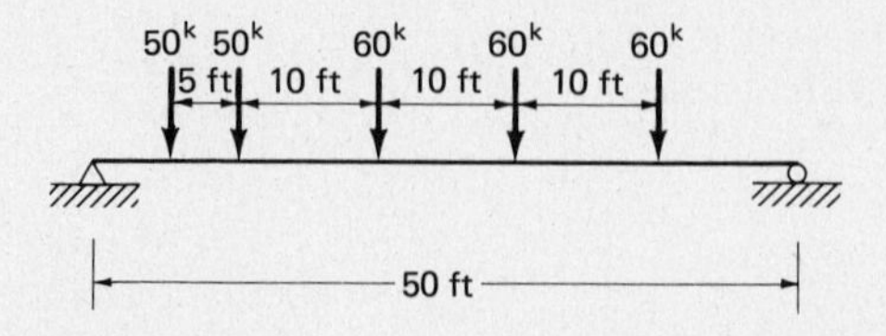

Figure 13.12

SOLUTION

Center of gravity of loads:

$$\frac{(50)(5) + (60)(15 + 25 + 35)}{280} = 16.96 \text{ ft}$$

Place loads as follows:

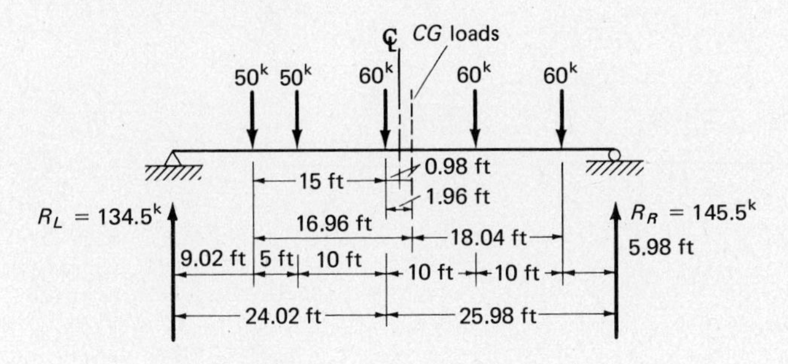

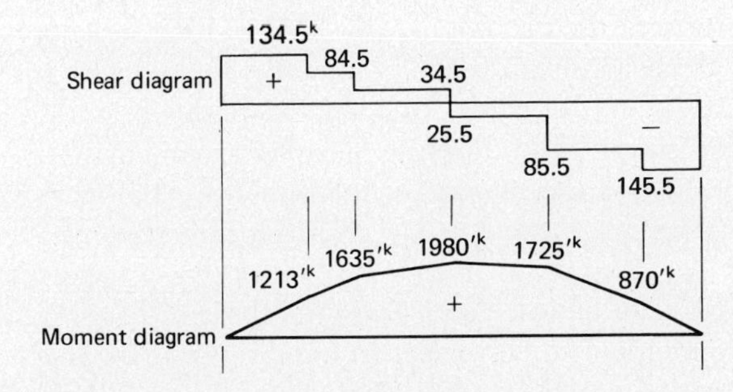

13.11 DISCUSSION OF METHODS

Two methods were introduced for determining maximum shears and moments at special points in a beam. These methods are the so-called increase-decrease method and the average-load criterion method. The first of the two is the process of determining if the function increases or decreases as the loads are successively moved up to the point, and the second consists in computing the average load to the left and to the right of the point for each position of the loads. It is important to realize that the increase-decrease method is perfectly applicable for all types of influence lines, whereas the average-load criterion was derived only for a triangular-shaped influence line. For another type of influence line another criterion would have to be developed; however, once developed it would be more easily applied than the first method.

The chapter has included methods for calculating maximum shears and moments in simple beams only. The discussion was not expanded to include trusses and other types of beams because it was felt that if the reader could handle the simple-beam problems, he or she could handle the others, since their solutions are so closely related to the simple-beam solutions.

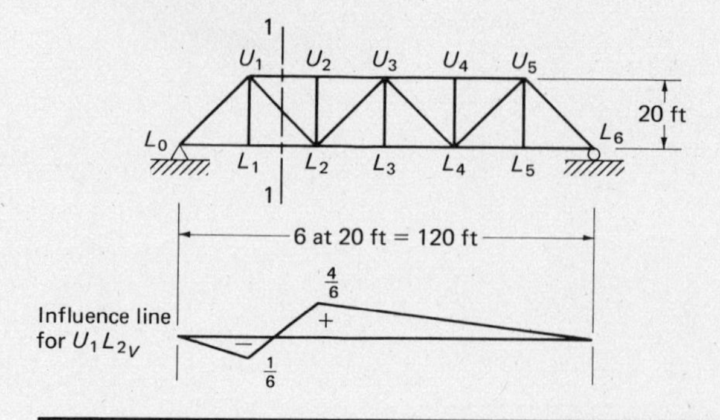

Figure 13.13

As an illustration of the relationship between simple-beam problems and truss problems, the truss of Fig. 13.13 is considered. For a set of loads in a given position, the force in member U_1U_2 may be found by passing section 1–1 and taking moments at L_2. To determine the maximum force in U_1U_2 for a series of moving loads, the first part of the problem is the positioning of the loads to cause maximum moment at L_2 in the truss. An identical problem is the positioning of loads on a 120-ft simple beam to cause maximum moment 40 ft from the left end. Similarly, to find the maximum force in L_3L_4, the loads would be placed to cause maximum moment at U_3, or at the centerline of a 120-ft simple beam.

The force is member U_1L_2 can be calculated from the shear in panel 2. The influence line shows the member to be subject to force reversal for live loads; therefore maximum tension would be developed when the loads were placed to cause maximum positive shear at L_2, or 40 ft from the left end of a simple beam, and maximum compression is developed when maximum negative shear is developed at L_1.

PROBLEMS

13.1 (a) Determine the maximum shear at the left end of the beam shown in the accompanying illustration due to the right-to-left movement of the load system given.
(b) Repeat if the span is 15 ft.

Problem 13.1

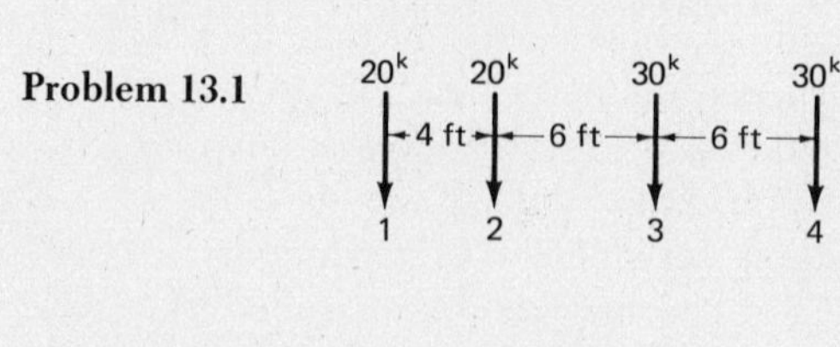

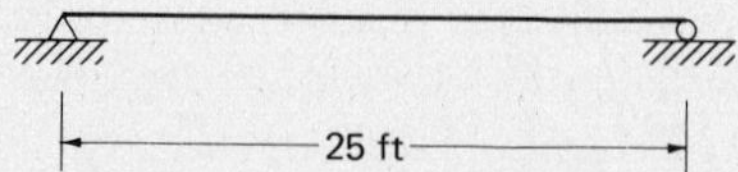

13.2. Determine the maximum shear at sections 1–1 and 2–2 of the beam shown in the accompanying illustration caused by the right-to-left movement of the load system of Prob. 13.1 (*Ans.* 38.0^k, 22.3^k)

Problem 13.2

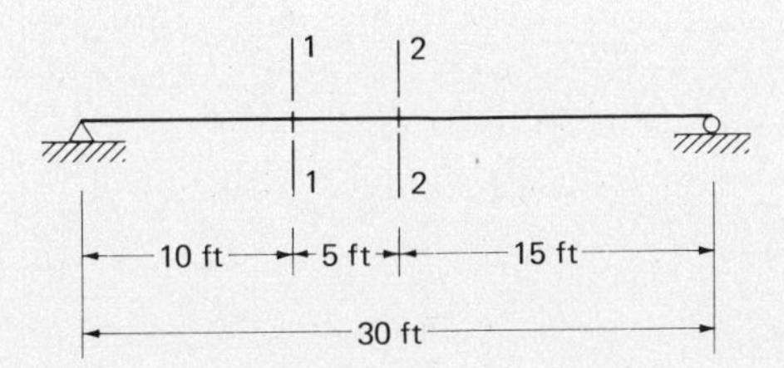

13.3. For a simply supported 40-ft span determine the critical positions and the following values for the load system of Prob. 13.1: (a) maximum shear at support; (b) maximum moment 10 ft from left support; (c) maximum shear at centerline; (d) maximum moment at centerline.

13.4. Determine the absolute maximum shear and moment possible in a 50-ft simple beam due to the load system of Prob. 13.1. (*Ans.* $V_{max} = 85.2^k$, $M_{max} = 1001'^k$)

13.5. Compute the maximum possible shear at the following points on a 60-ft simple beam caused by the right-to-left movement of the loads shown in the accompanying illustration: (a) left end; (b) 10 ft from left end; (c) centerline.

Problem 13.5

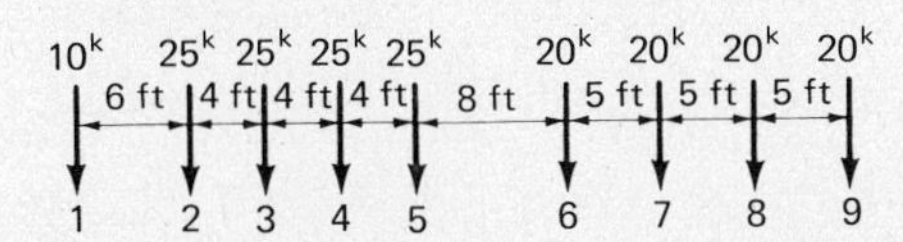

13.6. Compute the maximum possible moments at points 10, 20, and 30 ft from the left end of the beam of Prob. 13.5 (*Ans.* $1067'^k$, $1687'^k$, $1840'^k$)

13.7. What is the absolute maximum moment that can occur in the beam of Prob. 13.5?

13.8. A simple beam of 16-m span supports a pair of 60 kN moving concentrated loads 4-m apart. Compute the maximum possible moment at the centerline of the beam and the absolute maximum moment in the beam. (*Ans.* 360 kN · m, 367.5 kN · m)

13.9. Compute the maximum shear that can occur just to the left and just to the right of the left support of the beam shown in the accompanying illustration for the load system of Prob. 13.5.

Problem 13.9

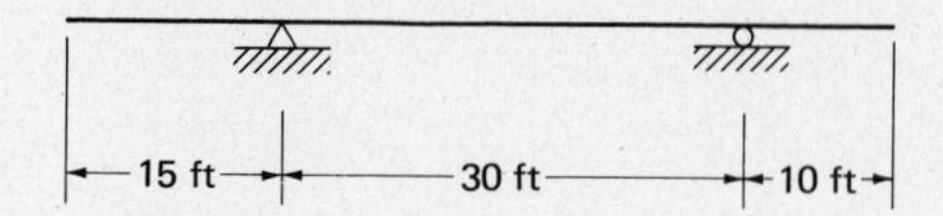

13.10. For a 50-ft simple beam supporting a uniform dead load of 2 klf, a 3-klf uniform live load, and the moving concentrated load system of Prob. 13.5, find the maximum shear at the left support and at a section 15 ft from the left support. (*Ans.* 249^k, 125^k)

13.11. Find the maximum possible moment at the centerline of the beam of Prob. 13.10.

13.12. A simply supported 20-m crane runway girder is to be designed to support two cranes represented by the loading systems shown in the accompanying illustration. Determine the maximum possible shears at the end and quarter points, the maximum centerline moment and the absolutely maximum moment possible in the girder due to these loads. (*Ans.* 256 kN, 173 kN, 1164 kN · m, 1165.4 kN · m)

Problem 13.12

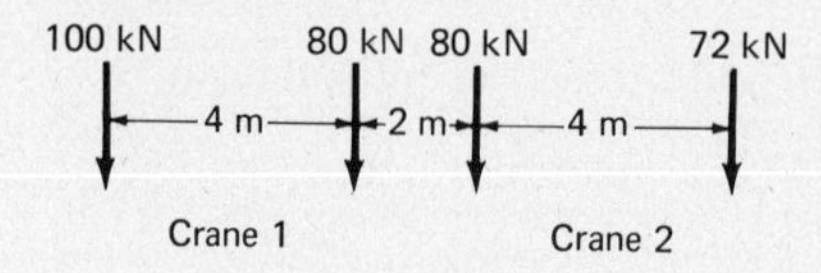

13.13. Calculate the maximum possible moment at joints L_2 and L_3 of the Pratt truss shown in the accompanying illustration for the load system of Prob. 13.5.

Problem 13.13

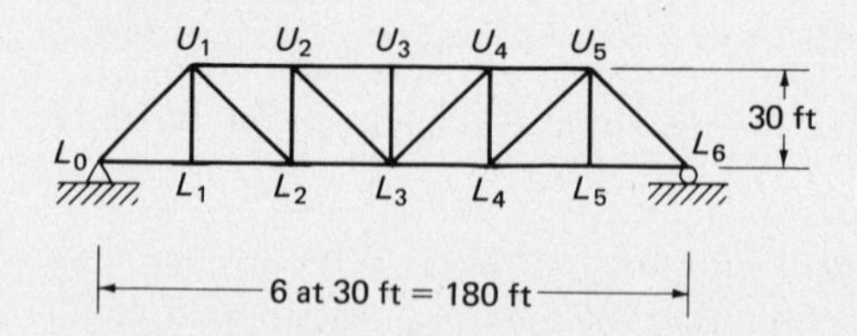

13.14. Determine the maximum possible shears at joints L_2 and L_3 of the truss of Prob. 13.13 using the load system of Prob. 13.5. (*Ans.* 105.1^k, 73.4^k)

13.15. For the truss of Prob. 13.13 determine the maximum possible forces in members L_1L_2, U_2U_3, U_2L_3, and L_4U_5.

13.16 By using the load system of Prob. 13.5 compute the maximum forces in members L_1L_2 and U_2U_3 of the Parker truss shown in the accompanying illustration. (*Ans.* $+282^k$, -255^k)

Problem 13.16

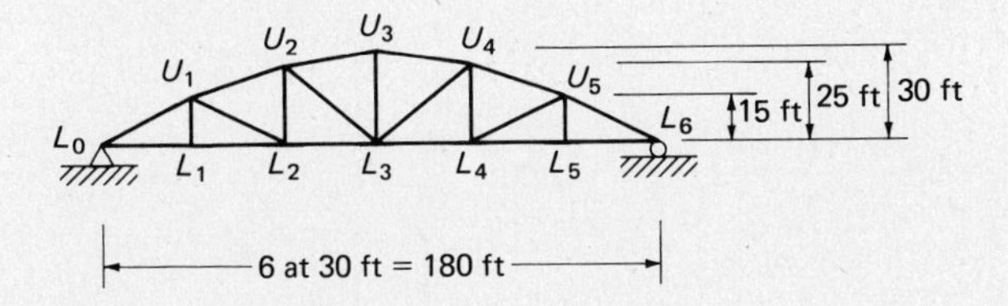

13.17. As the load system of Prob. 13.5 moves across the deck truss shown in the accompanying illustration compute the maximum forces produced in members U_0L_1, L_1L_2, and U_2U_3.

Problem 13.17

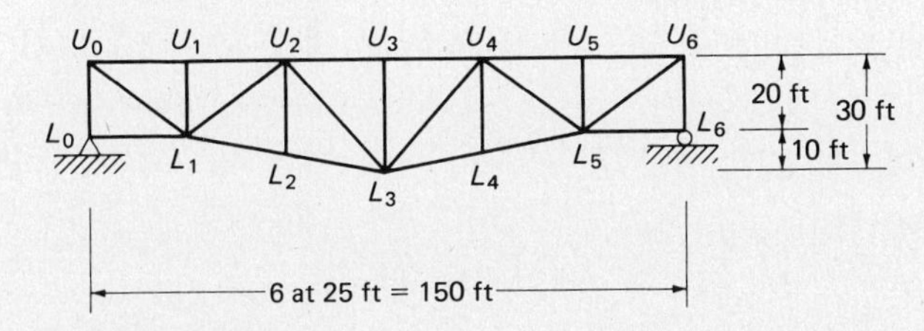

REFERENCES

[1] "Locomotive Loadings for Railway Bridges," *Transactions of the American Society of Civil Engineers* 86 (1923):606–636.

[2] A. A. Jakkula, and H. K. Stephenson, *Fundamentals of Structural Analysis* (New York: Van Nostrand, 1953), pp. 241–242.

14 Space Frames

14.1 GENERAL

A structure whose members do not all lie in the same plane is a space frame. Nearly all engineering structures fall into this classification. Bridges and buildings are two leading examples. They, however, may usually be broken down into separate systems, each lying in a single plane at right angles to the other. Figure 14.1(a) shows that an independent analysis of each of the systems is permissible. Two members, *AB* and *BC*, at right angles to each other, are shown in the figure. It is evident that a force in *AB* has no effect on the force in *BC*, because the component of a force at 90° is zero. Similarly, the forces in one truss have no effect on the forces in another truss framed into it at a right angle.

The members joining two systems together serve as members of both systems, and their total force is obtained by combining the forces developed as a part of each of the systems. The end posts of bridge trusses having end portals (see Figs. 16.7 and 16.8) are one illustration. They serve as the end posts of the bridge trusses and as the columns of the portal.

Many towers, domes, and derricks are three-dimensional structures made up of members so arranged that it is impossible to divide them into different

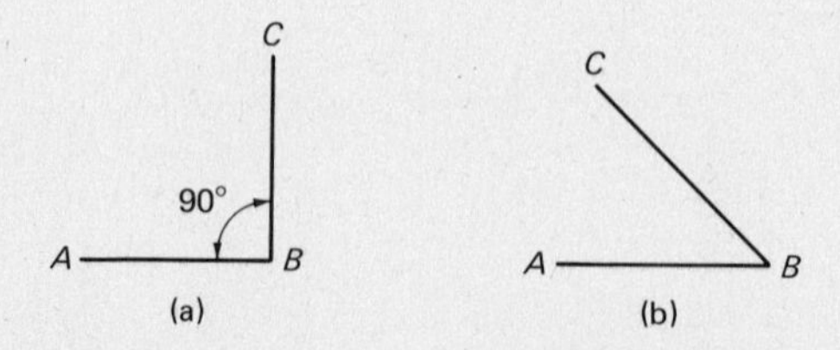

Figure 14.1

systems, each lying in a single plane, which may be handled individually. The difference is that the truss systems lie in planes that are not at right angles to each other. The forces in one truss framed into another at an angle other than 90° affect the forces in that truss, just as a force in member *AB* of Fig. 14.1(b) causes a force in member *BC*. For trusses of this type it is necessary to analyze the whole structure as a unit, rather than consider the systems in various planes individually. This chapter is devoted to these types of space frames.

The average structural designer is so accustomed to visualizing structures in one plane that when he or she encounters space frames, he or she frequently makes mistakes because his or her mind is still operating on a single-plane basis. If the layout of a space frame is not completely clear, the construction of a small model will probably clarify the situation. Even the most simple models of paper, cardboard, or wire are helpful.

14.2 BASIC PRINCIPLES

Prior to introducing a method of analyzing space frames, a few of the basic principles pertaining to such structures need to be considered. Three-dimensional structures, as were two-dimensional structures, are assumed to be made up of members subject to axial force only. In other words, the frames are assumed to have members that are straight between joints, to have loads applied at joints only, to have members whose ends are free to rotate (note that for this situation to be true the members would have to be connected with universal joints or at least with several frictionless pins), and so forth. Analyses based on these assumptions are usually quite satisfactory despite the welded and bolted connections used in actual practice.

A system of forces coming together at a single point, though not all in the same plane, may be combined into one resultant force. Similarly, it can be seen that an inclined force may have three coordinate components, and three reference planes will be used in handling them. The three planes used here are one horizontal and two verticals, each being perpendicular to the other. The intersections of these planes form the three coordinate axes used, *X*, *Y*, and *Z*. The force in any member inclined to these axes may be broken down into components along them, their magnitudes being proportional to their length projections on the axes. The inclined force *S* in Fig. 14.2 is graphically broken down into components S_x, S_y, and S_z.

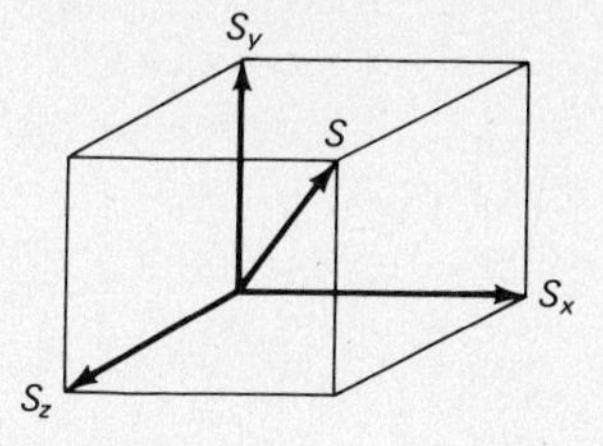

Figure 14.2

Transmission towers for the country's first 345,000-volt transmission line, Chief Joseph-Snohomish Dam, Washington State. (Courtesy of the Bethlehem Steel Corporation.)

The same values may be computed algebraically from the following relationship: The force in a member is to the length of the member as the *X*, *Y*, or *Z* component of force is to the corresponding *X*, *Y*, or *Z* component of length.

$$\frac{S}{l} = \frac{S_x}{l_x} = \frac{S_y}{l_y} = \frac{S_z}{l_z}$$

$$l^2 = l_x^2 + l_y^2 + l_z^2$$

$$S^2 = S_x^2 + S_y^2 + S_z^2$$

Space frames may be either statically determinate or statically indeterminate; consideration is given here only to those that are statically determinate. The methods developed in later chapters for statically indeterminate structures apply equally to three-dimensional and two-dimensional structures.

14.3 STATICS EQUATIONS

There are more statics equations available for determining the reactions of three-dimensional structures because there are two more axes to take moments about and one new axis along which to sum up forces. For equilibrium the sum of the forces along each of the three reference axes equals zero, as does the sum of the moments of all the forces about each of the axes. A total of six equations is available ($\Sigma X = 0$, $\Sigma Y = 0$, $\Sigma Z = 0$, $\Sigma M_x = 0$, $\Sigma M_y = 0$, and $\Sigma M_z = 0$), and six reaction components may be determined directly from them.

Should a structure have more than six reaction components, it is statically indeterminate externally; if less than six, it is unstable; and if equal to six, it is statically determinate externally. Many space frames, however, have more than six reaction components and yet are statically determinate internally. Example 14.2 shows that reactions for this type of structure may be determined by solving them concurrently with the member forces.

The basic figure of the space frame is the triangle. A triangle can be extended into a space frame by adding three members and one joint. Each of the new members frames into one of the joints of the basic triangle, the other ends coming together to form a new joint. The elementary space frame formed has six members and four joints. It may be enlarged by the addition of three members and one joint. For each of the joints of a space frame three equations ($\Sigma X = 0$, $\Sigma Y = 0$, and $\Sigma Z = 0$) are available to calculate the unknowns. Letting j be the number of joints, m the number of members, and r the number of reaction components, it can be seen that for a space frame to be statically determinate the following relation must hold:

$$3j = m + r$$

Should there be joints in the frame where the members are all in one plane, only two equations are available at each, and it is necessary to subtract one from the left-hand side of the equation for each such joint. The omission of one member for each reaction component in excess of six will cause this equation to be satisfied, and the structure will be statically determinate internally. When this situation occurs, it is possible to compute by three-dimensional statics the forces and reactions for the frame, although it is statically indeterminate externally.

The general rule for stability as regards outer forces is that the projection of the structure on any one of the three planes must itself be stable; therefore, as with two-dimensional structures, there must be at least three nonconcurrent reaction components in any one plane. The results of reaction computations will be inconsistent for any other case.

In the preceding paragraphs external stability and determinateness and internal stability and determinateness have been treated as though they were completely independent subjects. The two have been divorced for clarity for the reader who has not previously encountered space frames. The reader will learn, however, from the example problems in the pages that follow, that it is impossible in a majority of cases to consider the two separately. For instance, many frames are statically indeterminate externally and statically determinate internally and can be completely analyzed by statics. Few two-dimensional structures fall into this class.

14.4 SPECIAL THEOREMS APPLYING TO SPACE FRAMES

From the principles of elementary statics there may be developed two theorems that are useful in analyzing space frames. These are discussed in the following paragraphs.

1. The component of a force at 90° is zero, because no matter how large the force may be it equals zero when multiplied by the cosine of 90°. A force in one plane cannot have components in a plane normal to the original plane. Furthermore a force in one plane cannot cause moment about any axis in its plane, because it will either intersect the axis or be parallel to it.

From the foregoing it is evident that if several members of a frame come together at a joint, all but one lying in the same plane, the component of force

Northern Arizona University ensphere, Flagstaff, Arizona. (Courtesy of the American Institute of Timber Construction.)

in the member normal to the plane of the other members must equal the sum of the components of the external forces at the joint normal to the same plane. If no external forces are present, the member has a force of zero.

2. The equations of statics clearly show that if there is a joint in a truss where no external loads are applied and where it has been proved that all but two of the members coming into the joint have no force, these two members must have zero force unless they happen to lie in a straight line.

14.5 TYPES OF SUPPORT

Trusses in one plane have been assumed to be supported with rollers or hinges that could supply one or two reaction components. For three-dimensional structures the same types of support are used, but the number of reaction components may vary from one to three.

1. The *hinge* may have three reaction components because it can resist forces in X, Y, and Z directions.

2. The *slotted roller* is free to move in one direction parallel to the supporting surface. Movement is prevented in the other direction parallel to the surface as well as perpendicular to it, giving a total of two reaction components.

3. *Plane rollers, flat plates*, or *steel balls* have resistance only to movement perpendicular to the supporting surface, or one reaction component.

This discussion indicates it is possible to select a type of support having three reaction components or one that may have one or two of the components eliminated. A little thought on the subject shows that the possibility of limiting the number of reaction components of a space frame is very advantageous. A frame that is statically indeterminate externally may have its total reaction components limited to six, making it statically determinate. (Advantages of statically determinate and statically indeterminate structures are discussed in Chap. 15.) For some structures it is desirable to eliminate the reaction components in certain directions. The most obvious example occurs when a space frame is supported on walls where a reaction or thrust perpendicular to the wall is undersirable.

The directions in which reaction components are possible are indicated herein by dark heavy lines at the support points, as shown in the diagrams of the frames analyzed in Examples 14.1 to 14.3.

14.6 ILLUSTRATIVE EXAMPLES

Examples 14.1 and 14.2 illustrate the application of the foregoing principles to elementary space frames. Example 14.1 considers a structure supported at three points with six reaction components, which can be computed directly. The second example presents a space frame supported at four points with seven reaction components, which cannot be solved directly.

Example 14.1

Determine the reactions and member forces in the structure shown in Fig. 14.3.

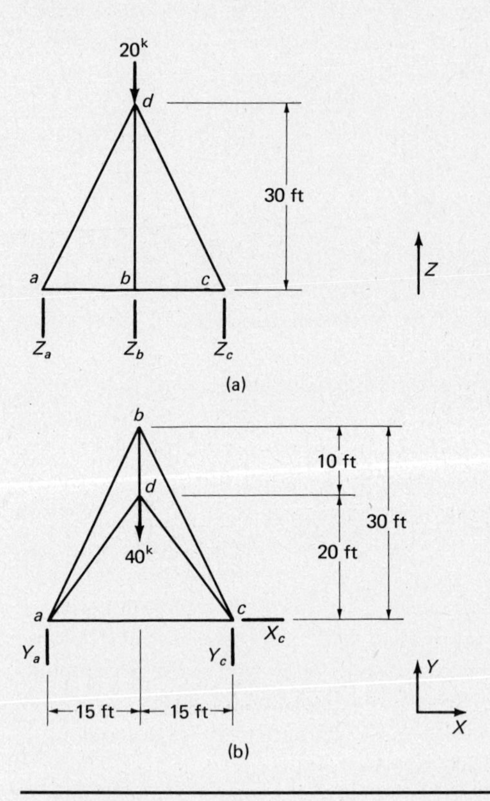

Figure 14.3

SOLUTION

The frame is statically determinate and stable externally because there is a total of six reaction components, three nonconcurrent ones in each plane. Internally it is statically determinate as proved with the joint equation.

$$3j = m + r$$
$$12 = 6 + 6$$
$$12 = 12$$

For a frame with three vertical reaction components, moments may be taken about an axis through any two of them to find the third.

$\Sigma M_x = 0$ about ac

$$(40)(30) - (20)(20) + 30Z_b = 0$$
$$Z_b = -26.7^k \downarrow$$

$\Sigma M_y = 0$ about line of action of Y_a

$$(20)(15) + (26.7)(15) - 30Z_c = 0$$
$$Z_c = +23.3^k \uparrow$$

$\Sigma Z = 0$

$$-20 - 26.7 + 23.3 + Z_a = 0$$
$$Z_a = +23.4^k \uparrow$$

Similarly, where there are three unknown horizontal reaction components, moments may be taken about a vertical axis passing through the point of intersection of two of the components.

$\Sigma M_z = 0$ about line of action of Z_c

$$-(40)(15) + 30Y_a = 0$$
$$Y_a = +20^k \uparrow$$

$\Sigma Y = 0$

$$20 - 40 + Y_c = 0$$
$$Y_c = +20^k \uparrow$$

$\Sigma X = 0$

$$0 + X_c = 0$$
$$X_c = 0$$

When the reactions have been found, the member forces can readily be computed by the method of joints. At joint *a*, member *ad* is the only member having a Z component of length; therefore its component must be equal and opposite to Z_a, or 23.4-kip compression. The *X* and *Y* components of *ad* are proportional to its components of length in those directions. Setting up a table similar to the one shown simplifies the computation of components and resultant forces.

Considering joint *a*, the *Y* component of force in member *ab* can be determined by joints now that the *Y* component of *ad* is known.

ΣY at joint $a = 0$

$$20 - 15.6 - Y_{ab} = 0$$
$$Y_{ab} = -4.4 \text{ compression}$$

The other member forces are computed by joints and shown in Table 14.1.

Table 14.1

MEMBER	PROJECTION X	Y	Z	LENGTH	COMPONENT OF FORCE X	Y	Z	FORCE
ab	15	30	0	33.5	−2.2	−4.4	0	−4.92
ad	15	20	30	39.1	−11.7	−15.6	−23.4	−30.5
ac	30	0	0	30.0	+13.9	0	0	+13.9
bc	15	30	0	33.5	−2.2	−4.5	0	−5.02
bd	0	10	30	31.6	0	+8.9	+26.7	+28.1
cd	15	20	30	39.1	−11.7	−15.5	−23.3	−30.3

Example 14.2

Find all reactions and member forces of the space frame shown in Fig. 14.4.

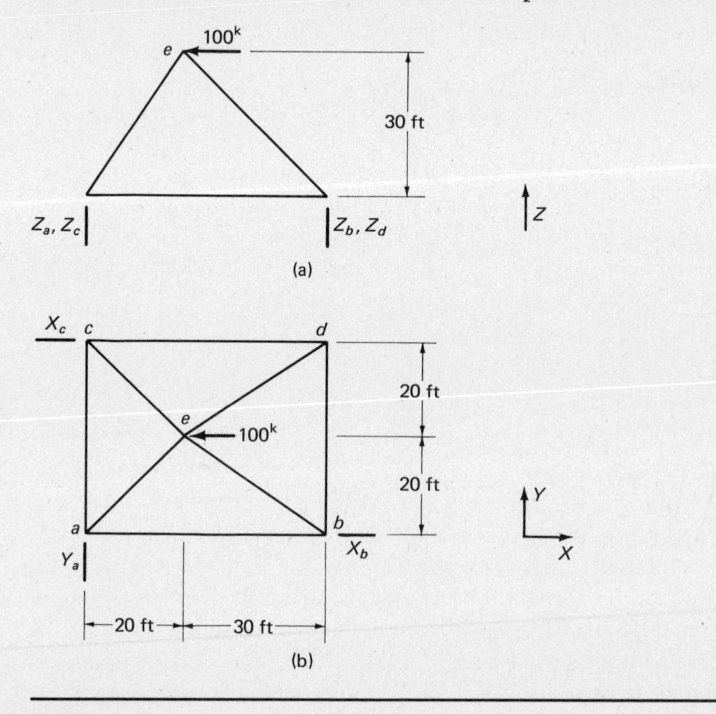

Figure 14.4

SOLUTION

Examination of the frame shows it to be statically indeterminate externally because there are seven definite reaction components and only six equations of statics. Internally, however, it is statically determinate as shown, and the analysis may be handled by statics.

$$3j = m + r$$

$$15 = 8 + 7$$

$$15 = 15$$

Although the frame is statically indeterminate externally, there are only three unknown reaction components in the XY plane, and these may be determined immediately. The other four components will be solved in conjunction with the member forces.

$\Sigma M_z = 0$ about line of action of Z_a

$$-(100)(20) + 40X_c = 0$$

$$X_c = +50^k \rightarrow$$

$\Sigma X = 0$

$$+50 - 100 - X_b = 0$$

$$X_b = +50^k \rightarrow$$

$\Sigma Y = 0$

$$0 + Y_a = 0$$

$$Y_a = 0$$

If the value of one of the Z reaction components should be known, the values of the other three could be determined by statics. It is assumed that Z_d has a value of S downward, and the reaction components are computed in terms of S.

$\Sigma M_y = 0$ about ac

$$-(100)(30) + 50S + 50Z_b = 0$$

$$Z_b = 60 - S$$

$\Sigma M_x = 0$ about ab

$$+40S - 40Z_c = 0$$

$$Z_c = +S$$

$\Sigma Z = 0$

$$+S - S - (60 - S) + Z_a = 0$$

$$Z_a = 60 - S$$

checking by $\Sigma M_x = 0$ about cd

$$(60 - S)(40) - 40Z_a = 0$$

$$Z_a = 60 - S$$

The calculation of member forces may now be started from the reaction components in terms of S. These computations are continued until the forces at both ends of one bar are determined in terms of S. The two values must be equal, and they are equated to give the correct value of S.

The Z component of force in de equal S and is in tension, whereas the Z component of force in be equals $60 - S$ and is also in tension. The Y component of force in de equals $(20/30)(S) = 2/3\ S$, and the Y component of force in be is

$(20/30)(60 - S) = 40 - 2/3\ S$. By $\Sigma Y = 0$ at joint d, member bd is in compression with a force of $\frac{2}{3}S$. Similarly, by $\Sigma Y = 0$ at joint b, member bd is seen to have a compressive force of $40 - \frac{2}{3}S$. Equating the two expressions yields the value of S.

$$\frac{2}{3}S = 40 - \frac{2}{3}S$$
$$\frac{4}{3}S = 40$$
$$S = 30^k$$

The numerical values of the Z reaction components can now be found from S, and the forces in the frame can be determined by joints. The use of a table to work with length and force components is again convenient. The results are shown in Table 14.2.

Table 14.2

	PROJECTION				COMPONENT OF FORCE			
MEMBER	X	Y	Z	LENGTH	X	Y	Z	FORCE
ab	50	0	0	50	+20	0	0	+20
ae	20	20	30	41.2	−20	−20	−30	−41.2
ac	0	40	0	40	0	+20	0	+20
be	30	20	30	46.9	+30	+20	+30	+46.9
bd	0	40	0	40	0	−20	0	−20
de	30	20	30	46.9	+30	+20	+30	+46.9
cd	50	0	0	50	−30	0	0	−30
ce	20	20	30	41.2	−20	−20	−30	−41.2

14.7 MORE COMPLICATED FRAMES

The method of joints was used to determine the forces in the elementary space frames analyzed in Sec. 14.6. For more complicated space frames, even if the reactions are available, the analyst may have difficulty calculating any of the forces by the method of joints. The frame of Fig. 14.5 falls into this class.

It is possible to compute all reaction components immediately, but no forces may be determined by the method of joints as presented in Examples 14.1 and 14.2. The application of the zero-member principle of paragraph (1) of Sec. 14.4, however, will quickly prove that several members have no force. For example, at joint f, members fc, fb, and fe lie in the same plane; member fd does not, and it has a component of force perpendicular to the plane of the other three members. This component must be equal and opposite to all components perpendicular to the plane from external loads applied at the joint. Since there are no loads present, member fd has no force. A similar analysis can be made at joint e to show the force in member fe is zero.

Two of the four members meeting at joint f have been proved to have no force. The remaining two members are not in a straight line and must have zero forces, since there are no external loads applied at the joint [paragraph (2)

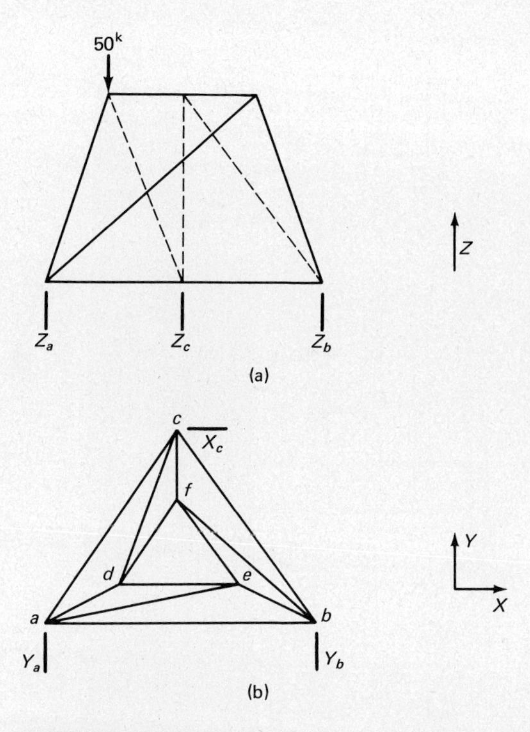

Figure 14.5

of Sec. 14.4]. It is now possible to compute the forces in the remaining members by joints for this particular truss.

For other structures the use of the zero-member principle will not be sufficient (if it is of any value at all) in determining forces. In this latter type of frame the use of moments will probably enable the reader to make the analysis. The frame of Fig. 14.5 could be analyzed by taking moments. If an imaginary section were passed completely around joint *e* isolating it as a free body, moments could be taken about line *ab* to find the force in member *ef*. Forces in *ae* and *eb* intersect line *ab*, and the line of action of *de* is parallel to *ab*. Several moment equations of this type are used in Example 14.3.

As space frames become more complicated, it may not be possible to find a single force by the methods introduced to this point. If a member is arbitrarily given a force, say S, it may be possible to compute the forces in several members in terms of S. Once the force at both ends of a member is known in terms of S, the two values may be equated to find S, and the force analysis for the frame may be carried out.

Example 14.3

Analyze the space frame of Fig. 14.6.

Fieldhouse, University of Wichita, Wichita, Kansas. (Courtesy of the American Institute of Steel Construction.)

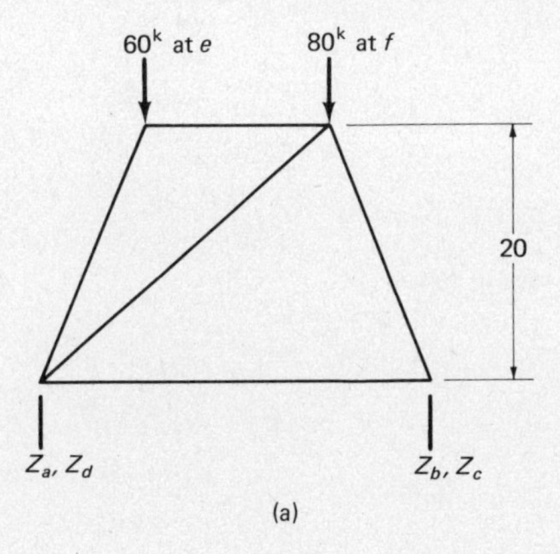

(a)

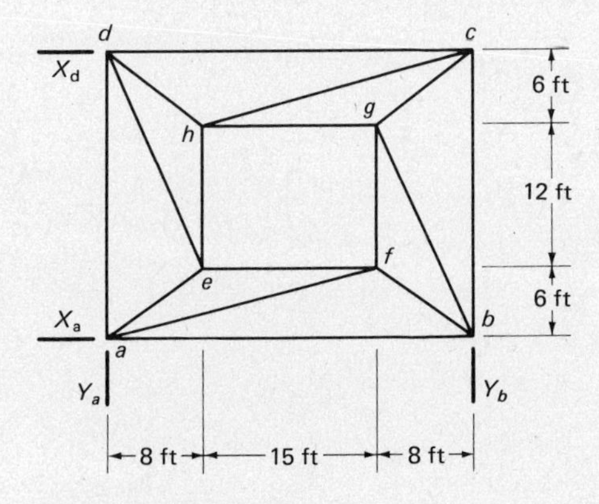

Figure 14.6

SOLUTION

Although there are eight reaction components, the structure is statically determinate.

$$3j = m + r$$
$$(3)(8) = 16 + 8$$
$$24 = 24$$

A brief examination of the structure will reveal four zero members. At joint h member eh has no force, and at joint g the same can be said for member gh; therefore members hd and hc meeting at joint h have zero forces.

By removing joint e as a free body and taking moments about line ad, the force in ef is determined, and by taking moments about line ab, force de is found. The method of joints is then used to find the forces in ae and da.

$\Sigma M_y = 0$ about line ad

$$(60)(8) - 20S_{ef} = 0$$

$$S_{ef} = 24^{\text{k}} \text{ compression}$$

$\Sigma M_x = 0$ about line ab

$$(60)(6) - 20Y_{de} - 6Z_{de} = 0$$

$$360 - 20Y_{de} - (6)(20/18)(Y_{de}) = 0$$

$$Y_{de} = 13.5^{\text{k}} \text{ compression}$$

$\Sigma Y = 0$ at joint e

$Y_{ae} = 13.5^{\text{k}}$ compression

$\Sigma Y = 0$ at joint d

$Y_{da} = 13.5^{\text{k}}$ tension

An imaginary section is passed around joint f and moments are taken about line ab to find the force in fg. By using the same free body, moments are taken about line bc to find the force in af.

$\Sigma M_X = 0$ about line ab

$$(80)(6) - 20S_{fg} = 0$$

$$S_{fg} = 24^{\text{k}} \text{ compression}$$

$\Sigma M_y = 0$ about line bc

$$-(80)(8) + (24)(20) + 20X_{af} + 8Z_{af} = 0$$

$$X_{af} = 5.9^{\text{k}} \text{ compression}$$

A similar moment procedure may be used, with joint g as the free body, to determine the force in member bg, and the values of all remaining forces and reactions can be determined by joints. A summation of the results is given in Table 14.3.

Reactions:

$Za = 50\uparrow \qquad X_a = 2\rightarrow$

$Zb = 55\uparrow \qquad X_d = 2\leftarrow$

$Zc = 20\uparrow \qquad Y_a = 1.5\uparrow$

$Zd = 15\uparrow \qquad Y_b = 1.5\downarrow$

14.8 SIMULTANEOUS EQUATION ANALYSIS

An analyst who has access to a digital computer will often find the easiest analysis can be made by writing simultaneous equations and solving them with the computer. As a matter of fact, standard computer programs for space frames are generally available throughout the country. The simultaneous equations ($\Sigma X = 0$, $\Sigma Y = 0$, and $\Sigma Z = 0$ at each joint) can be used to obtain both reactions and member forces. Although the subject of matrix formulation for solving simultaneous equations is discussed in Chap. 27, the solution of the

Table 14.3

MEMBER	PROJECTION X	Y	Z	LENGTH	COMPONENT OF FORCE X	Y	Z	FORCE
ab	31	0	0	31	+22	0	0	+22
bc	0	24	0	24	0	+ 6	0	+ 6
cd	31	0	0	31	+ 8	0	0	+ 8
da	0	24	0	24	0	+13.5	0	+13.5
ae	8	6	20	22.4	−18	−13.5	−45	−50.4
af	23	6	20	31.1	− 5.9	− 1.5	− 5.1	− 7.9
bf	8	6	20	22.4	−30	−22.5	−75	−84
bg	8	18	20	28.1	+ 8	+18	+20	+28.1
cg	8	6	20	22.4	− 8	− 6	−20	−22.4
ch	23	6	20	31.1	0	0	0	0
dh	8	6	20	22.4	0	0	0	0
de	8	18	20	28.1	− 6	−13.5	−15	−21.1
ef	15	0	0	15	−24	0	0	−24
fg	0	12	0	12	0	−24	0	−24
gh	15	0	0	15	0	0	0	0
he	0	12	0	12	0	0	0	0

equations by the process of elimination is used in this chapter. Actually, this type of solution may be rather quick for small space frames, despite the large number of equations, because of the small number of unknowns that appear in each equation.

Before considering the preparation of simultaneous equations for an entire structure a single joint is considered. Should there only be three unknown forces at a joint, it may be possible to obtain their values by writing the three statics equations for that joint. (Frequently this procedure will not work because the equations may be linearly dependent.)

The preparation and solution of simultaneous equations for space frames can be simplified by making use of ***tension coefficients*** [1]. The tension coefficient for a member equals its force divided by its length. In each of the following expressions for force components the value S/l is replaced by T, the tension coefficient.

$$S_x = \frac{l_x}{l} S = \frac{S}{l} l_x = T l_x$$

$$S_y = \frac{l_y}{l} S = \frac{S}{l} l_y = T l_y$$

$$S_z = \frac{l_z}{l} S = \frac{S}{l} l_z = T l_z$$

The space frame of Fig. 14.3 is redrawn in Fig. 14.7 and for convenience in preparing the equations the members are numbered as shown. The following equations are written for joint d of this frame using tension coefficients and assuming all members are in tension. The reactions are assumed to be in the directions shown.

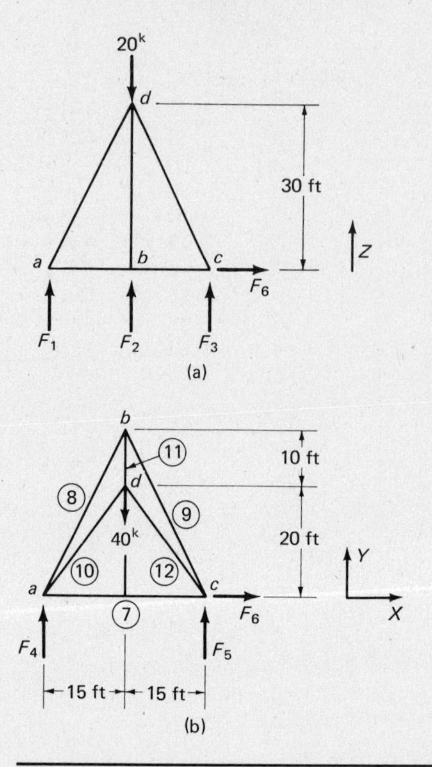

Figure 14.7

$$\Sigma X = -15T_{10} + 15T_{12} = 0$$
$$\Sigma Y = -20T_{10} + 10T_{11} - 20T_{12} = 40$$
$$\Sigma Z = -30T_{10} - 30T_{11} - 30T_{12} = 20$$

There should be no difficulty in solving these equations by the usual form of elimination; however, the elimination procedure described here [2] for solving the equations is of advantage when there are a large number of equations. First, the coefficients of both sides of the equations are written as follows:

−15	0	15	0
−20	10	−20	40
−30	−30	−30	20

Each equation is divided through by its leading nonzero constant with the results:

1	0	−1	0
1	−0.5	1	−2
1	1	1	−0.667

The first equation is left as is and each of the other equations not starting with a zero value is subtracted from the first.

$$\begin{matrix} 1 & 0 & -1 & 0 \\ 0 & 0.5 & -2 & 2 \\ 0 & -1 & -2 & 0.667 \end{matrix}$$

These equations are each divided through by their leading nonzero constants with the following results:

$$\begin{matrix} 1 & 0 & -1 & 0 \\ 0 & 1 & -4 & 4 \\ 0 & 1 & 2 & -0.667 \end{matrix}$$

The process is repeated with all rows after the second one whose second element is not zero being subtracted from row 2 with the resulting rows being divided through by their leading nonzero element. The final result is

$$\begin{matrix} 1 & 0 & -1 & 0 \\ & 1 & -4 & 4 \\ & & 1 & -0.777 \end{matrix}$$

These rows now have the form of a triangular matrix (see Chap. 27) and correspond to the following equations:

$$\begin{aligned} T_{10} \qquad\qquad - T_{12} &= 0 \\ T_{11} \quad -4T_{12} &= 4 \\ T_{12} &= -0.777 \end{aligned}$$

The value of T_{12} is calculated from the last equation; T_{11}, from the next to last equation; and so on. The results are $T_{12} = -0.777$; $T_{11} = 0.892$; and $T_{10} = -0.777$. The member forces are formed by multiplying the tension coefficients by the member lengths.

$$\begin{aligned} S_{10} &= (39.1)(-0.777) = -30.4^{k} \\ S_{11} &= (31.6)(\ 0.892) = +28.2^{k} \\ S_{12} &= (39.1)(-0.777) = -30.4^{k} \end{aligned}$$

In a similar manner the equations for the entire space frame of Fig. 14.7 are written at the end of this paragraph. When a computer is not available and the simultaneous equation procedure is used, it may be more practical to start at joints having only three unknowns and determine their values (if feasible) and use the values so computed in later equations, rather than considering all the equations for an entire structure at one time. Nevertheless all the equations for the frame of Fig. 14.7 are considered as a group here.

$$
\text{joint } a \begin{cases} 30T_7 + 15T_8 + 15T_{10} = 0 & (7) \\ F_4 + 30T_8 + 20T_{10} = 0 & (4) \\ F_1 + 30T_{10} = 0 & (1) \end{cases}
$$

$$
\text{joint } b \begin{cases} -15T_8 + 15T_9 = 0 & (8) \\ -30T_8 - 30T_9 - 10T_{11} = 0 & (9) \\ F_2 + 30T_{11} = 0 & (2) \end{cases}
$$

$$
\text{joint } c \begin{cases} F_6 - 30T_7 - 15T_9 - 15T_{12} = 0 & (6) \\ F_5 + 30T_9 + 20T_{12} = 0 & (5) \\ F_3 + 30T_{12} = 0 & (3) \end{cases}
$$

$$
\text{joint } d \begin{cases} -15T_{10} + 15T_{12} = 0 & (12) \\ -20T_{10} + 10T_{11} - 20T_{12} = 40 & (11) \\ -30T_{10} - 30T_{11} - 30T_{12} = 20 & (10) \end{cases}
$$

To facilitate the solution of the equations it is very important to arrange them properly. Should they not be arranged properly the work involved in solving them can be enormous. They should be arranged as nearly as possible to the desired final triangular matrix. In Table 14.4 the equations are listed in the order given by the numbers to the right of each of the preceding equations. Each row is also divided by its leading nonzero constant.

Table 14.4

F_1	F_2	F_3	F_4	F_5	F_6	T_7	T_8	T_9	T_{10}	T_{11}	T_{12}	LOAD
1									30			
	1									30		
		1									30	
			1				30		20			
				1				30			20	
					1	−30		−15			−15	
						1	0.5		0.5			
							1	−1				
							1	−1		0.33		
									1	1	1	−0.67
									1	−0.5	1	− 2
									1		−1	

These equations are already almost in the desired final form. Following the procedure previously described, the results shown in Table 14.5 can be obtained. These equations can be solved with the same results shown in Example 14.1.

Table 14.5

F_1	F_2	F_3	F_4	F_5	F_6	T_7	T_8	T_9	T_{10}	T_{11}	T_{12}	LOAD
1									30			
	1									30		
		1									30	
			1				30		20			
				1				30			20	
					1	−30		−15			−15	
						1	0.5		0.5			
							1	−1				
								1		0.17		
									1	1	1	−0.67
										1		0.89
											1	−0.78

PROBLEMS

For Probs. 14.1 to 14.8 compute the reaction components and the member forces for the space frames.

Problem 14.1

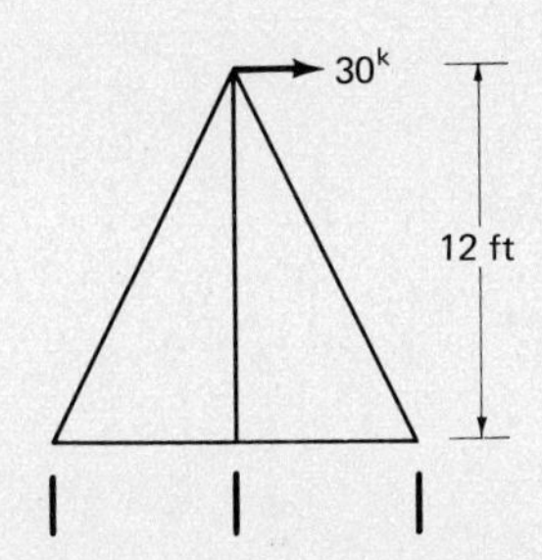

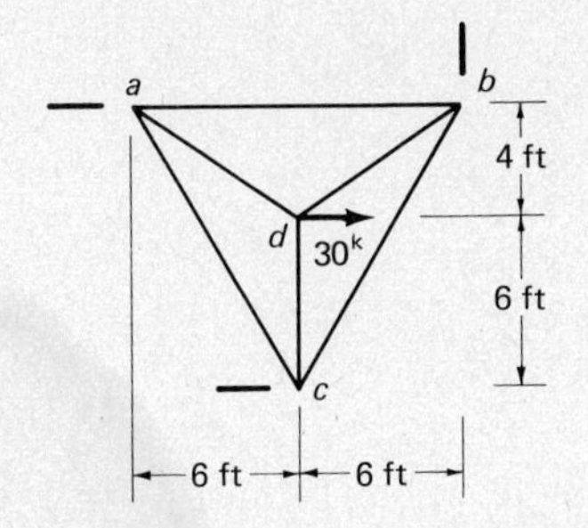

Problem 14.2 (*Ans.* $Z_b = 36^k \downarrow$, $Y_c = 37.5^k \downarrow$, $ad = -123.8^k$, $bc = -10.1^k$, $bd = +43.2^k$)

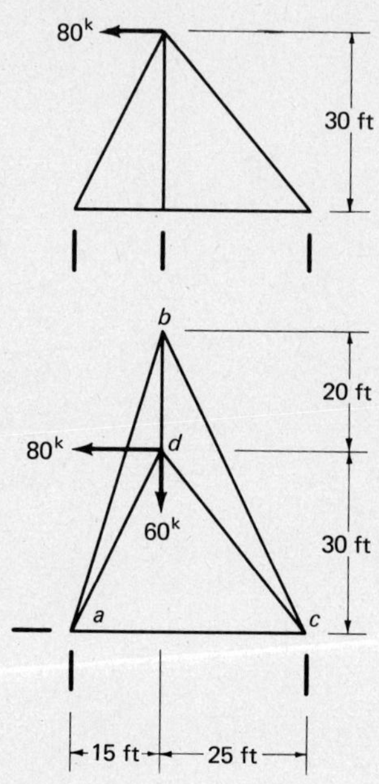

Problem 14.3

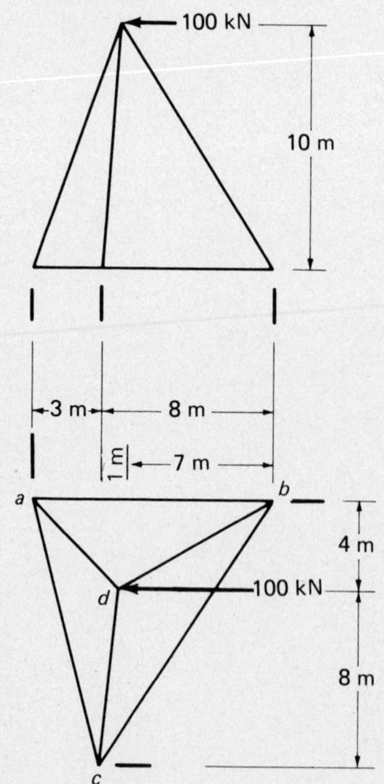

Problem 14.4 (*Ans.* $X_a = 15^k \leftarrow$, $Z_b = 30^k \uparrow$, $ab = 15^k$, $ce = 0$, $be = -36.8^k$).

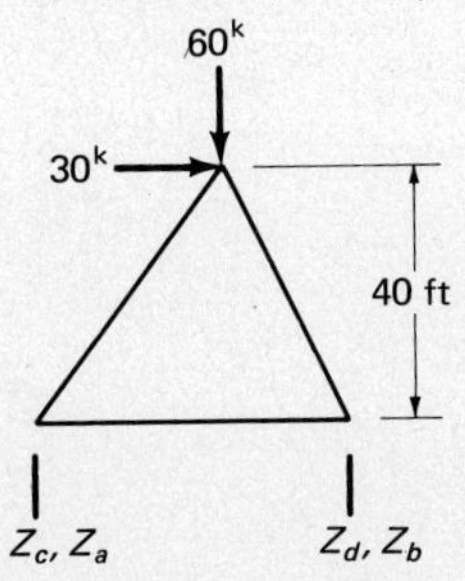

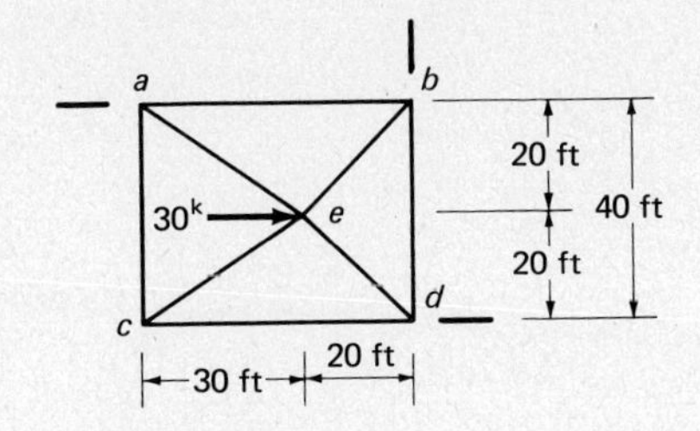

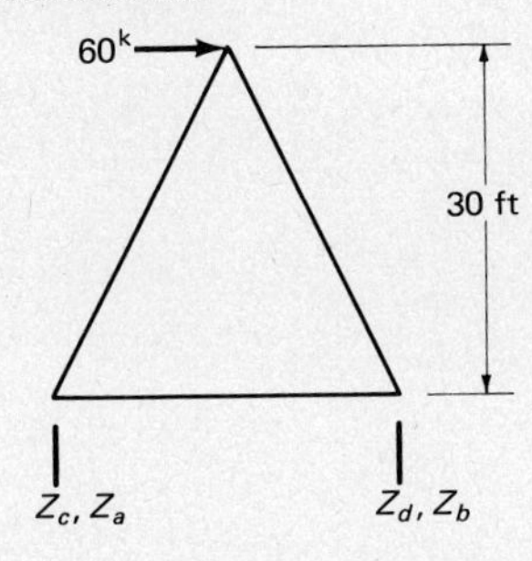

Problem 14.5

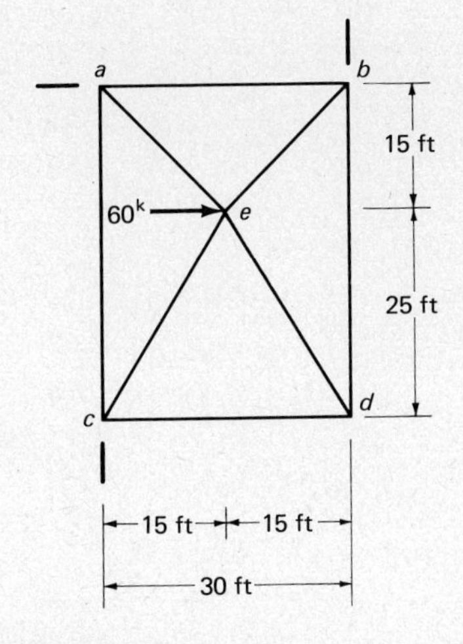

Problem 14.6 (*Ans.* $cd = +12.5^k$, $gh = -50^k$, $bf = 0$, $ch = +63.7^k$, $Z_c = 50^k \downarrow$, $Y_d = 12.5^k \uparrow$)

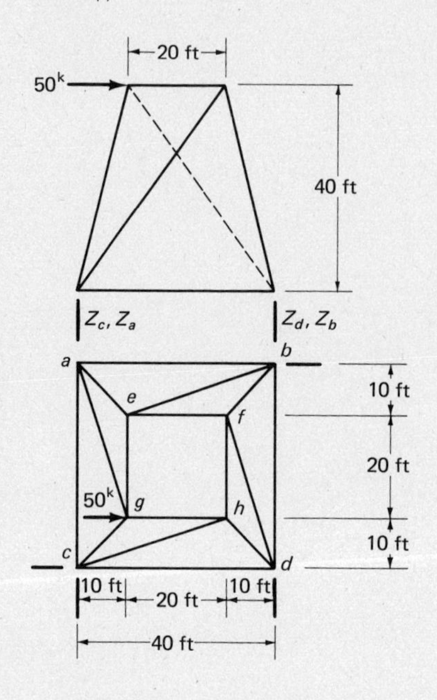

Problem 14.7

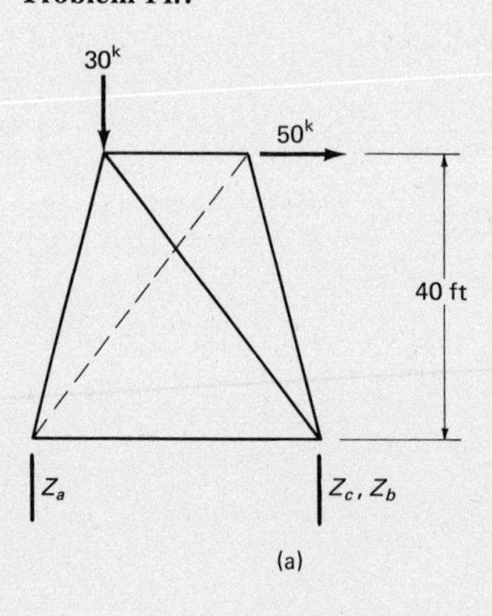

(a)

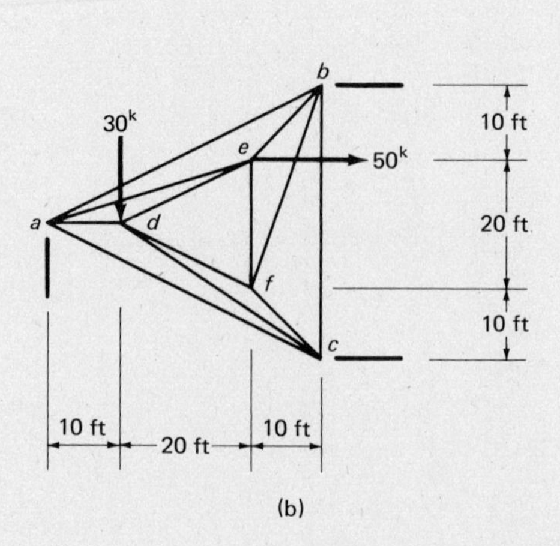

(b)

Problem 14.8 (*Ans.* $de = -14.8^k$, $eh = +15.0^k$, $fj = -8.3^k$, $hi = -32.7^k$, $Xb = 3.33^k \rightarrow$, $Y_a = 6.67^k \downarrow$)

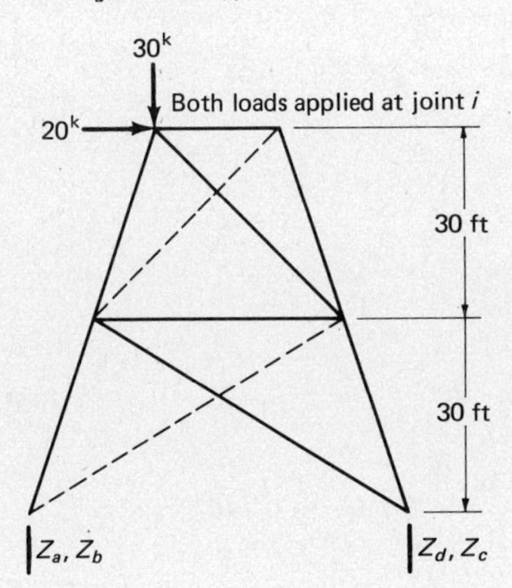

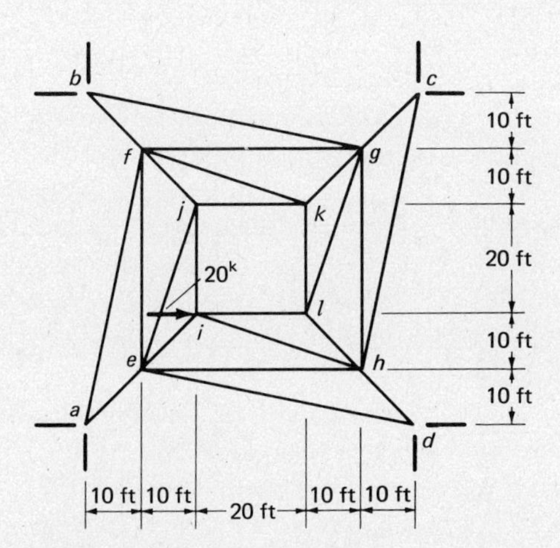

14.9. Solve Prob. 14.1 using simultaneous equations for the entire frame.

14.10. Solve Prob. 14.3 using simultaneous equations for the entire frame. (*Ans.* $X_b = 66.7$ kN $\rightarrow$, $Z_a = 90.9$ kN $\uparrow$, $ad = -104.5$ kN, $bc = -43.7$ kN, $cd = 0$)

REFERENCES

[1] R. V. Southwell, "Primary Stress Determination in Space Frames," *Engineering* 109 (1920) 165.

[2] G. L. Rogers and M. L. Causey, *Mechanics of Engineering Structures* (New York: Wiley, 1962), pp. 27–31.

15 Discussion of Statically Indeterminate Structures

15.1 INTRODUCTION

The use of statically indeterminate structures is becoming more extensive each year; in the past their use had been more common in Europe, where the resulting savings in material were of primary importance. Up until the early part of the twentieth century statically indeterminate structures were avoided by most American engineers, if possible, but three great developments completely changed the picture. These were (1) monolithic reinforced concrete structures, (2) arc welding of steel structures, and (3) modern methods of analysis.

The preceding chapters may have led the reader to believe that statically determinate beams and trusses are the rule in modern structures. The truth is that it is a difficult task to find an ideal simply supported beam. Probably the best place to look for one would be in a structures textbook; for bolted or welded beam to column connections do not produce ideal simple supports with zero moments.

The same situation holds for statically determinate trusses. Those rare trusses that are pin connected are not blessed with frictionless pins, and if pins are used, they frequently rust in place and then permit little rotation. The other assumptions made about trusses in the earlier chapters are not altogether true, and in a strict sense all trusses are statically indeterminate because they have some bending and secondary forces.

15.2 CONTINUOUS STRUCTURES

As the spans of simple structures become longer, their bending moments increase rapidly. If the weight of a structure per foot remained constant, regardless of the span, the dead-load moment would vary in proportion to the

Colorado River arch bridge Utah Route 95. (Courtesy of the Utah Department of Transportation.)

square of the span length ($M = wl^2/8$). This proportion, however, is not correct, because the weight of structures must increase with longer spans to be strong enough to resist the increased bending moments; therefore the dead-load moment increases at a greater rate than does the square of the span.

For economy, it pays in long spans to introduce types of structures that have smaller moments than the tremendous ones that occur in long-span, simply supported structures. Chapter 3 introduced one type of structure that considerably reduced bending moments: the cantilever-type construction. Two other moment-reducing structures are discussed in the following paragraphs.

In some locations it may be possible to have a beam with fixed ends, rather than one with simple supports. A comparison of the moments developed in a uniformly loaded simple beam with those in a uniformly loaded fixed-ended beam is made in Fig. 15.1.

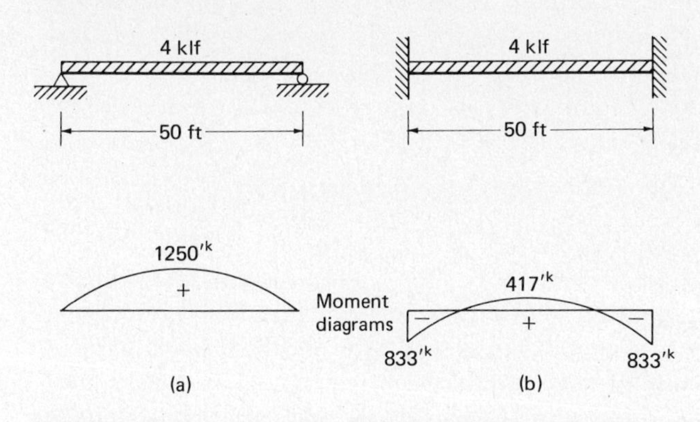

Figure 15.1

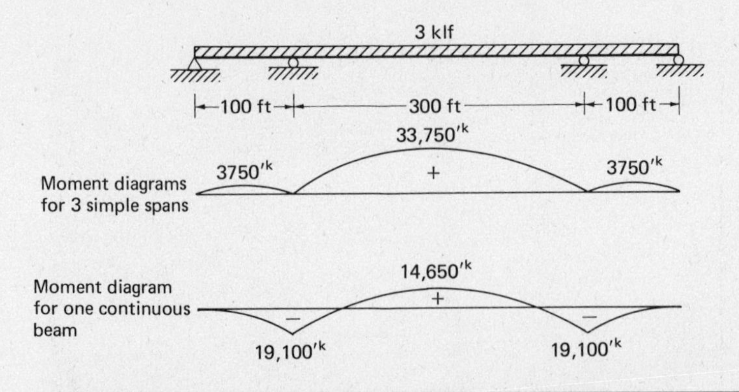

Figure 15.2

The maximum bending moment for the fixed-ended beam is only two-thirds of that for the simply supported beam. Usually it is difficult to fix the ends, particularly in the case of bridges; for this reason flanking spans are often used, as illustrated in Fig. 15.2. These spans will partially fix the interior supports, thus tending to reduce the moment in the center span. This figure presents a comparison of the bending moments that occur in three uniformly loaded simple beams (spans 100, 300, and 100 ft) with the moments of a uniformly loaded beam continuous over the same three spans.

The maximum bending moment for the continuous beam is approximately 43% less than that for the simple beams. Unfortunately, there will not be a corresponding 43% reduction in total cost. The cost-reduction figure probably is only 2 or 3% of the total structure cost because such items as foundations, connections, and floor systems are not reduced a great deal by the moment reductions.

In the foregoing discussion, the moments developed in beams have been shown to be reduced appreciably by continuity. This reduction occurs where beams are rigidly fastened to each other or where beams and columns are rigidly connected. There is a continuity of action in resisting a load applied to any part of a continuous structure because the load is resisted by the combined efforts of all of the members of the frame.

15.3 ADVANTAGES OF STATICALLY INDETERMINATE STRUCTURES

In comparing statically indeterminate structures with statically determinate structures, the first consideration, to most, would pertain to cost. It is, however, impossible to make a statement favoring one type, economically, without reservation. Each structure presents a different situation, and all factors must be considered, economic or otherwise. In general, statically indeterminate structures have the advantages discussed in the following paragraphs.

Broadway Street Bridge showing cantilever erection, Kansas City, Missouri. (Courtesy of American Bridge.)

Savings in Materials

The smaller moments developed permit the use of smaller members, the material saving possibly running as high as 10 to 20% of the steel used in bridges. The large number of force reversals occurring in railroad bridges keeps their maximum saving nearer the 10% value.

A structural member of a given size can support more load if it is part of a continuous structure than if it is simply supported. The continuity permits the use of smaller members for the same loads and spans or increased spacing of supports for the same size members. The possibility of fewer columns in buildings or piers in bridges may permit a reduction in overall costs.

Continuous structures of concrete or steel are cheaper without the joints, pins, and so on, required to make them statically determinate, as was frequently the practice in past years. Monolithic reinforced-concrete structures are erected so that they are naturally continuous and statically indeterminate. To install the hinges and other devices necessary to make them statically determinate would not only be a difficult construction problem but also be very expensive. Furthermore, if a building frame consisted of columns and simple beams, it would be necessary to have objectionable diagonal bracing between the joints to make the frame stable and rigid.

More Rigid Structures

Statically indeterminate structures are more rigid than statically determinate ones. Such structures are of particular importance where there are many moving loads and considerable vibration.

More Attractive Structures

It is difficult to imagine statically determinate structures having the gracefulness and beauty of many statically indeterminate arches and rigid frames being erected today.

Adaptation to Cantilever Erection

The cantilever method of erecting bridges is of particular value where conditions underneath (probably naval traffic or deep water) hinder the erection of falsework. Continuous statically indeterminate bridges and cantilever-type bridges are conveniently erected by the cantilever method.

15.4 DISADVANTAGES OF STATICALLY INDETERMINATE STRUCTURES

A comparison of statically determinate and statically indeterminate structures shows the latter have several disadvantages which may make their use undesirable on many occasions. They are discussed in the following paragraphs.

Support Settlement

Statically indeterminate structures are not desirable where foundation conditions are poor, because seemingly minor support settlements or rotations may cause major changes in the moments, shears, reactions, and bar forces. Where statically indeterminate bridges are used despite the presence of poor foundation conditions, it is often felt necessary to weigh the dead-load reactions. The supports of the bridge are jacked up or down until the calculated reaction is obtained, after which the support is built to that elevation.

Development of Other Stresses

Support settlement is not the only condition that causes stress variations in statically indeterminate structures. Variation in the relative positions of members caused by temperature changes, poor fabrication, or internal deformation of members of the structure under load may cause serious force changes throughout the structure.

Difficulty of Analysis and Design

The forces of statically indeterminate structures depend not only on their dimensions but also on their elastic properties (moduli of elasticity, moments of inertia, and cross-sectional areas). This situation presents a major design difficulty: The forces cannot be determined until the member sizes are known, and the member sizes cannot be determined until their forces are known. The problem is handled by assuming member sizes and computing the forces, designing the members for these forces and computing the forces for the new sizes, and so on, until the final design is obtained. Design by this method—the method of successive approximations—takes more time than the design of a comparable statically determinate structure, but the extra cost is only a small part of the total structure cost. Such a design is best done by interaction between the designer and the computer. Interactive computing is now used extensively in the aircraft and auto industries.

Force Reversals

Generally more force reversals occur in statically indeterminate structures than in statically determinate structures. Additional material may be required at certain sections to resist the different force conditions.

15.5 THE WICHERT TRUSS

Continuous trusses are thought of as being statically indeterminate, but there are a few exceptions. Two of these have been discussed in preceding chapters: the cantilever-type trusses and trusses with certain members omitted (shown in Fig. 9.11). Mr. E. M. Wichert patented another type of statically determinate continuous truss in 1932. His truss falls into the omitted-member class, because the verticals over interior supports are left out. (Müller-Breslau, a professor in Berlin, had discussed this form of truss as early as 1887 [1].) The Wichert truss of Fig. 15.3, which has 32 members, 18 joints, and 4 reaction components, is shown to be statically determinate as follows:

$$m = 2j - r$$

$$32 = 36 - 4$$

$$32 = 32$$

The reader is reminded that the satisfying of the $m = 2j - r$ criterion is a *necessary* but not a *sufficient* condition for statical determinancy. Unless the

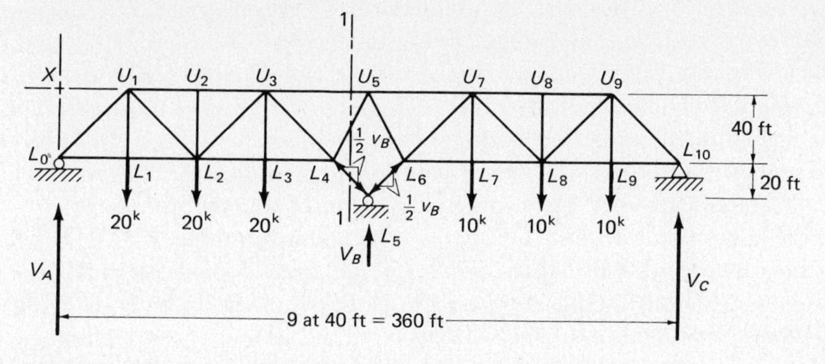

Figure 15.3

criterion is satisfied, the structure cannot be statically determinate. If it is satisfied, the structure may be statically determinate or it may be unstable. The structure will indeed be statically determinate and stable if the equations have a unique solution (see Sec. 9.6).

A Wichert truss has all of the advantages but none of the disadvantages of continuity. Being statically determinate, it is easy to analyze and design and is not appreciably affected by support settlements, minor fabrication errors, and so on. The method of analysis is based on the conditions at the interior support. The support at L_5 in the truss of Fig. 15.3 is a roller, and the reaction there is vertical with no horizontal components; therefore the horizontal components of force in the members L_4L_5 and L_5L_6 must be equal and opposite for equilibrium (that is, both tension or both compression). Members having the same slope and the same horizontal force components must have the same forces and the same vertical components of force. Assuming the reaction V_B at the roller to be up, the two forces are in compression, and the sum of their vertical components is equal and opposite to the reaction. This relation may be stated as follows:

$$V_{L_4L_5} = V_{L_5L_6} = \tfrac{1}{2}V_B$$

The horizontal components of the two forces can be expressed in terms of the vertical components, which are in terms of V_B. For this particular truss the members have an inclination of 45°, and the vertical and horizontal components are equal, as shown on the members in the figure. Moments may be taken about joint U_5 of all the forces to the left of section 1–1. Since the components of force in L_4L_5 are expressed in terms of V_B, the equation will contain two unknowns, V_A and V_B. Moments may then be taken about L_{10} of all the external forces acting on the truss. The resulting equation has the same two unknowns, V_A and V_B, and their values may be determined by solving the two equations simultaneously. Once two of the reactions are obtained, the analysis may be completed by the usual method of statics. Example 15.1 presents a complete analysis of this truss.

Analysis of several Wichert trusses of different arrangements will show that instability is possible under some circumstances. This situation will occur when the lower-chord members meeting at the interior supports become very flat, and instability will be indicated by the tremendous values of those forces. There is a definite slope of the bottom-chord members at which the truss becomes unstable, and this slope can be seen for the truss in Fig. 15.3. Should L_4L_5 be so flat that its line of action intersects the line of action of the top chord as far to the left as point X in the figure, the truss will be unstable.

Wichert trusses of more than two spans are quite tedious to analyze, although they are statically determinate. Dr. D. B. Steinman, in his book *The Wichert Truss* [2] presents a detailed discussion of the various types, including methods of analysis for multispan trusses, design, and economy.

Example 15.1

Calculate the reactions and member forces of the Wichert truss of Fig. 15.3.

SOLUTION

$\Sigma M_{U_5} = 0$ (to left of section 1–1)

$$180V_A - (20)(60 + 100 + 140) + 60H_{L_4L_5} = 0$$

$$180V_A - 6000 + (60)(\tfrac{1}{2}V_B) = 0$$

$$180V_A + 30V_B = 6000 \tag{1}$$

$\Sigma M_{L_{10}} = 0$ (entire structure)

$$-(10)(40 + 80 + 120) - (20)(240 + 280 + 320) + 360V_A + 180V_B = 0$$

$$360V_A + 180V_B = 19{,}200 \tag{2}$$

Solving Eqs. (1) and (2) simultaneously gives

$$V_A = 23.3^k$$

$$V_B = 60.0^k$$

by $\Sigma V = 0$

$$V_c = 6.7^k$$

By statics the following forces are obtained:

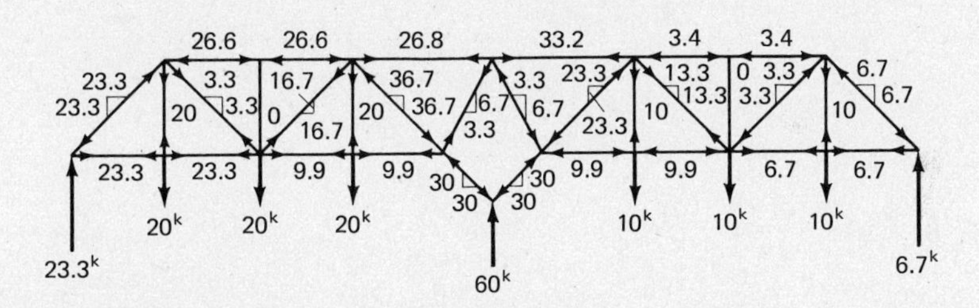

15.6 ANALYSIS OF STATICALLY INDETERMINATE STRUCTURES

There are two general approaches used for the analysis of statically indeterminate structures. These are the *force methods* and the *displacement methods*. In the first of these methods, assumed redundants are removed from the structure such that static equilibrium is not affected. The resulting inconsistencies in geometry are determined and are corrected in such a manner that equilibrium is not disturbed. In the second method, a solution which satisfies the geometrical requirements is obtained, and it is corrected for the inconsistencies that result in static equilibrium. The following paragraphs are presented to elaborate on these confusing definitions.

Force or Flexibility or Compatability Method

In this method of analysis, the redundants (reactions and/or member forces) are assumed and removed from the structure so that a stable and statically determinate structure remains. One equation is written for the deflection condition at and in the direction of each redundant that has been removed. These are written in terms of the redundants and the resulting equations are solved for the numerical values of these redundants.

As a simple example the propped beam of Fig. 15.4(a), which has one redundant reaction, is considered. The vertical reaction component at B is considered to be the redundant. It is assumed to be removed from the structure leaving a statically determinate and stable beam. As shown in part (b) of the figure the right end deflects an amount δ_B.

Next the external or real loads are assumed to be removed from the structure and a unit load of 1 kip applied at B. The resulting deflection is indicated in part (c) of Fig. 15.4 by δ_{bb}. Should that load be increased to V_B the total deflection at B will be $V_B\delta_{bb}$.

As there is actually no settlement at B the total deflection there is zero and the following equation can be written:

$$\delta_B + V_B\delta_{bb} = 0$$

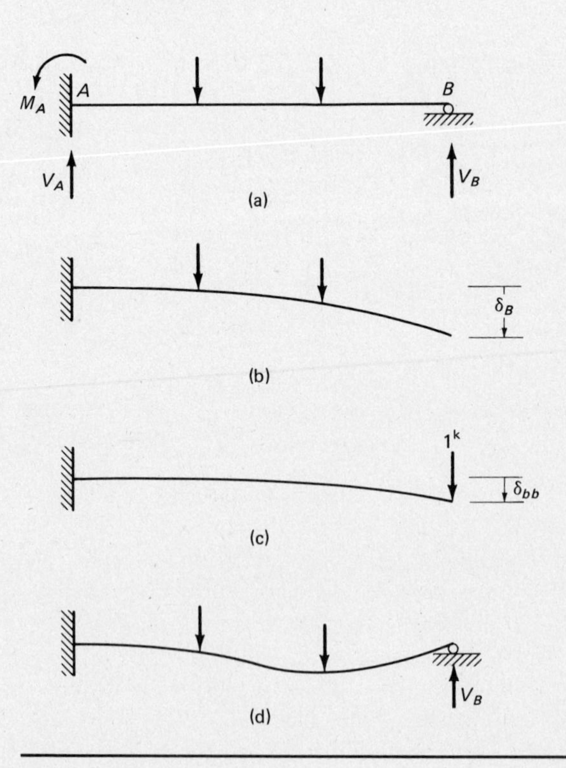

Figure 15.4

In other words if V_B were acting upward, it would be sufficient to push point B back to its original nondeflected position as shown in part (d) of Fig. 15.4. Solution of the equation will yield the value of V_B and the other reaction components can be computed by statics. Similar expressions can be written for multiredundant structures. Chapters 17, 18, and part of 21 are devoted to the determination of slopes and deflections, and Chaps. 19, 20, and part of 21 present in detail the force methods of analysis.

Displacement or Stiffness Methods

In the displacement methods of analysis the displacements of the joints (rotations and translations) necessary to describe fully the deformed shape of the structure are used in the equations instead of the redundant actions as used in the force methods. When the simultaneous equations are solved, these displacements are determined and substituted into the original equations to determine the various internal forces. Slope deflection (Chap. 22) is a displacement method and moment distribution (Chaps. 23 and 24) is a successive approximation method based on the same general theory.

PROBLEMS

15.1 Determine the forces in all of the members of the truss shown in the accompanying illustration.

Problem 15.1

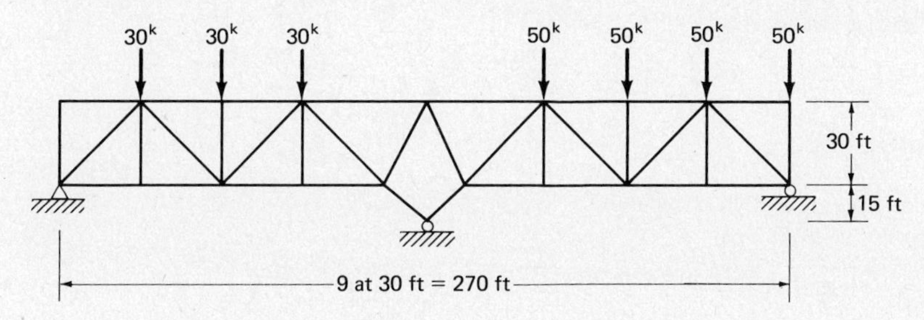

15.2. Repeat Prob. 15.1 if the 50-kip loads are doubled. (*Ans.* $U_1L_2 = -33.0^k$, $U_3U_5 = +153.4^k$, $U_5L_6 = +52.2^k$, $L_7L_8 = +70.0^k$)

15.3. Draw influence lines for the vertical reaction components for the left and center supports of the truss of Prob. 15.1. Assume that unit load moves across the top span.

15.4. Compute the forces in all of the members of the truss shown in the accompanying illustration.

Problem 15.4 (*Ans.* $L_2U_3 = +71.6^k$, $L_4L_5 = -124.8^k$, $U_7U_8 = -47.8^k$, $L_8U_9 = +33.8^k$)

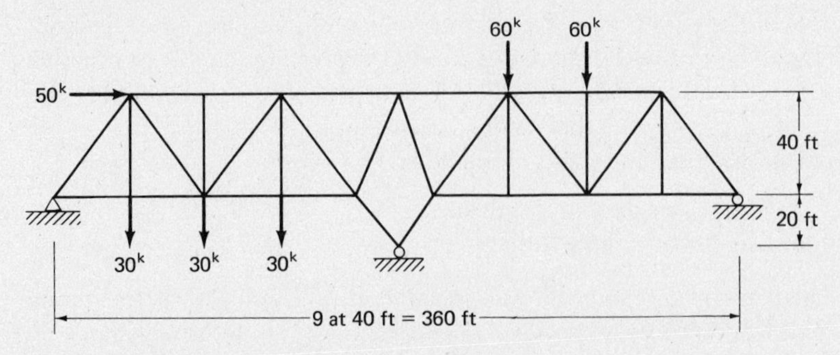

15.5. Compute the forces in all the members of the truss shown in the accompanying illustration.

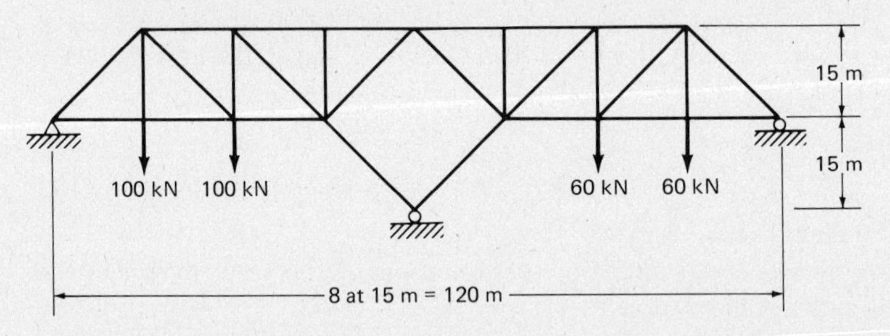

15.6. For the truss of Prob. 15.4 draw influence lines for the left vertical reaction component and for the forces in members U_3L_4 and U_3U_5. (*Ans.* Unit load at L_4 : $V_L = -0.111$, $U_3L_4 = -0.157$, $U_3U_5 = +0.444$)

REFERENCES

[1] H. Sutherland and H. L. Bowman, *Structural Theory* (New York: Wiley, 1954), Chap. 5.

[2] D. B. Steinman, *The Wichert Truss* (New York: D. Van Nostrand, 1932).

16 Approximate Analysis of Statically Indeterminate Structures

16.1 GENERAL

Statically indeterminate structures may be analyzed "exactly" or "approximately." Several "exact" methods, which are based on elastic distortions, are discussed in Chap. 19 to 25. Approximate methods, involving the use of simplifying assumptions, are presented in this chapter. These latter methods have many practical applications such as the following:

1. An "exact" analysis may be so tedious and cumbersome that time is not available to perform the necessary computations.
2. The structure may be so complicated that no one who has the knowledge to make an "exact" analysis is available.
3. For some structures either method may be subject to so many errors and imperfections that approximate methods may yield forces as accurate as an exact analysis. A specific example is the analysis of a building frame for wind loads where the walls, partitions, and floors contribute an indeterminate amount to wind resistance. Wind forces calculated in the frame by either method are not accurate.
4. To design the members of a statically indeterminate structure, it is necessary to make an estimate of their sizes before structural analysis can begin by an "exact" method. Approximate analysis of the structure will yield forces from which reasonably good initial estimates can be made as to member sizes.
5. Today computers are available with which "exact" analyses and designs of highly indeterminate structures can be quickly and economically made. To make use of such a computer program it is prudent or economically advisable to make some preliminary estimates as to member sizes. If an approximate analysis of the structure has been

made it will be possible to make very reasonable member size estimates. The result will be appreciable saving of both computer time and money.

6. Approximate analyses are quite useful in rough checking "exact" solutions.

Many different methods are available for making approximate analyses. A few of the more common ones are presented here, with consideration being given to trusses and building frames.

To be able to analyze a structure by statics, there must be no more unknowns than there are equations of statics available. If a truss or frame has 10 more unknowns than equations, it is statically indeterminate to the tenth degree. To analyze it by an approximate method, one assumption must be made for each degree of indeterminancy, or a total of 10 assumptions. It will be seen that each assumption presents another equation to use in the calculations.

16.2 TRUSSES WITH TWO DIAGONALS IN EACH PANEL

Diagonals Having Little Stiffness

The truss of Fig. 16.1 has two diagonals in each panel. If one of these diagonals were to be removed from each of the six panels, the truss would become statically determinate. The structure is statically indeterminate to the sixth degree.

If the diagonals are relatively long and slender, such as those made of a pair of small steel angles, they will be able to carry reasonably large tensile forces but negligible compression. For this situation it is logical to assume that the shear in each panel is carried entirely by the diagonal that would be in tension for that type of shear. The other diagonal has no force. Making this assumption in each panel makes a total six assumptions for six redundants, and the equations of statics may be used to complete the analysis. The forces in Fig. 16.1 were obtained on this basis.

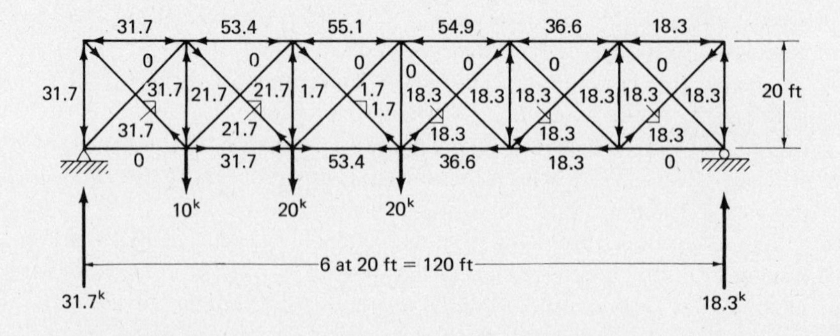

Figure 16.1

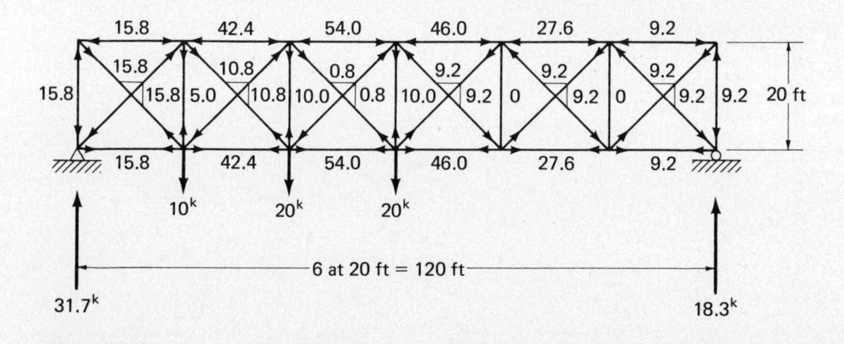

Figure 16.2

Diagonals Having Considerable Stiffness

In some trusses the diagonals are constructed of sufficient stiffness to resist compressive loads. For panels having two diagonals the shear may be considered to be taken by both of them. The division of shear causes one diagonal to be in tension and the other to be in compression. The usual approximation made is that each diagonal takes 50% of the shear. It is possible to assume some other division, such as one-third of shear to compression diagonal and two-thirds to tension diagonal.

The forces calculated for the truss in Fig. 16.2 are based on a 50% division of the shear in each panel.

16.3 ANALYSIS OF MILL BUILDINGS

The building trusses analyzed in previous chapters have been assumed to rest on top of masonry walls or on top of columns on the sides of buildings. A different, but common, type of industrial construction is the mill bent or mill building, the trusses of which are rigidly fastened to the columns so that they act together. Figure 16.3 shows two types of framing commonly used for mill buildings.

For gravity loads these trusses are analyzed as though they were simply supported on walls instead of being rigidly fastened to the columns, but for

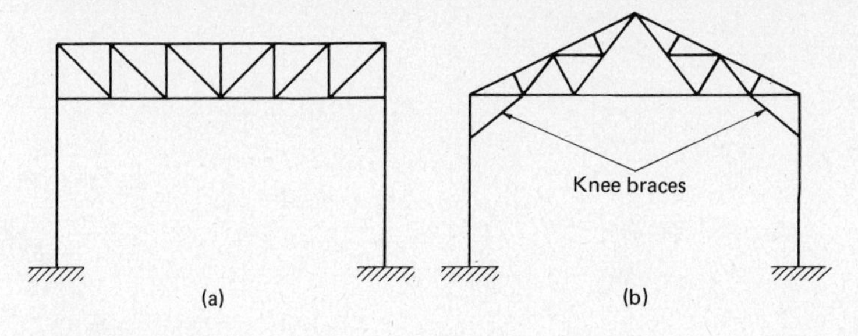

Figure 16.3

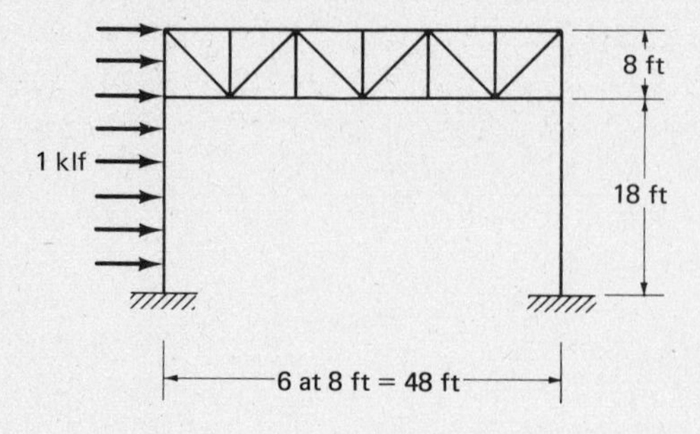

Figure 16.4

lateral loads the columns and trusses should be analyzed as a unit. In office buildings of corresponding heights, the inside walls and partitions offer considerable resistance to wind, in many cases supplying sufficient resistance. Mill bents, however, must be analyzed and designed for wind loads, because there are no interior walls to help resist the wind forces. These buildings may have traveling cranes whose operation will cause additional lateral loads which need to be considered in design. The mill building of Fig. 16.4 is analyzed for a wind load of 1 kip per foot of height. Should there be cranes loads, they would be handled in exactly the same manner.

If the column bases are fixed, there will be three unknown reactions at each support, giving a total of six unknowns. The structure is statically indeterminate to the third degree, and to analyze it by an approximate method, three assumptions must be made.

When a column is rigidly attached to the foundation, there can be no column rotation at the base. Even though the building is subjected to wind loads causing the columns to bend laterally, a tangent to the column at the base will remain vertical. If the truss at the tops of the columns is very stiff and rigidly fastened to them, a tangent to the column at the junction will remain vertical. A column rigidly fixed top and bottom will assume the shape of an S curve when it is subjected to lateral loads (Fig. 16.5).

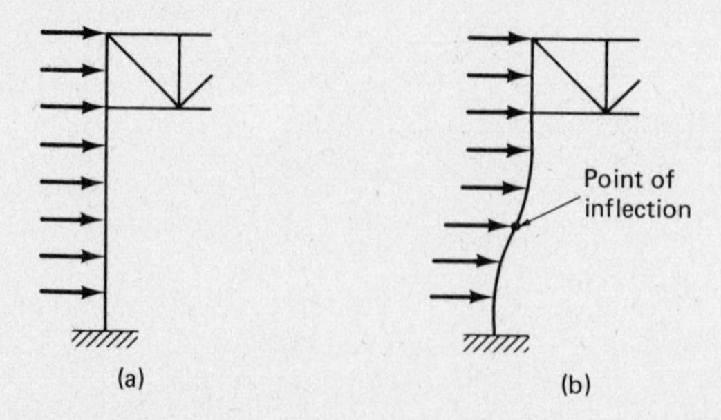

Figure 16.5

At a point midway from the column base to the bottom of the truss or knee brace the moment is zero because it changes from a moment causing tension on one side of the column to a moment causing tension on the other side. The point of zero moment is commonly called a ***point of inflection*** (PI) or a ***point of contraflexure.*** If points of inflection are assumed in each of the columns, two of the necessary three assumptions have been made (two $\Sigma M = 0$ equations are made available).

The discussion of the location of points of inflection has been based on the assumption that the column bases are completely fixed. If the columns are anchored in a deep concrete foundation or a concrete foundation wall, the assumption is good. Frequently, though, the columns are supported by small concrete footings which offer little resistance to rotation, and the column bases act as hinges, in which case the points of inflection are at the bases. The usual situation probably lies in between the two extremes, with the column bases only partially fixed. The points of inflection are commonly assumed to lie about one-fourth to one-third of the distance from the base up to the bottom of the truss, or to the knee brace if one is used.

The third assumption made is that the horizontal shear divides equally between the two columns at the plane of contraflexure. An exact analysis proves this to be a very reasonable assumption if the columns are approximately the same size. If they are not similar in size, an assumption may be made that the shear splits between them in a little different proportion, the stiffest column carrying the largest amount of shear. The methods of analysis discussed in later chapters show that distributing the shear in proportion to the I/l^3 values of the columns is a very good assumption.

The mill building of Fig. 16.4 is analyzed in Fig. 16.6 on the basis of the following assumptions: The plane of contraflexure is assumed to be at the one-third point, or 6 ft above the column bases, and the total shear above the plane of contraflexure of 20 kip divides equally between the columns.

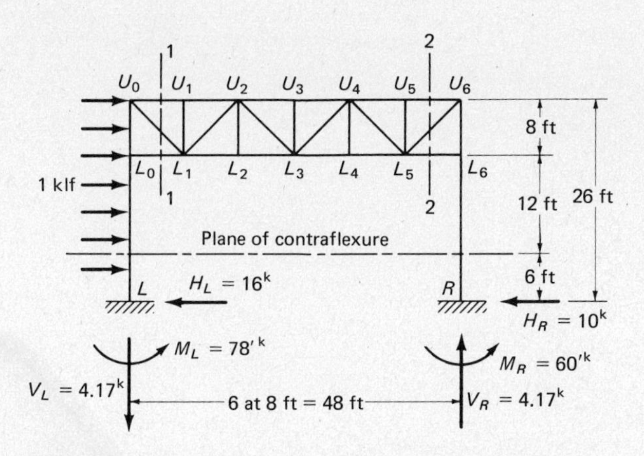

Figure 16.6

Moments are taken of the forces above the plane of contraflexure about the point of inflection in the left column to find the vertical force in the right-hand column. By $\Sigma V = 0$ the vertical force in the left-hand column is obtained. With the wind blowing from left to right, the right-hand (or leeward) column is in compression and the left-hand (or windward) column is in tension.

Moments at each column base are determined by taking moments about the base of the forces applied to the column at and below the point of inflection.

Finally, forces in the truss are computed by statics. In computing the forces of members of this truss which are connected to a column, all of the forces acting on the column must be taken into account because the columns are subject to both axial force and bending moment. Section 1–1 is passed through the truss, and moments are taken about joint U_0 to obtain the force in L_0L_1. By using the same section, moments are taken about joint L_1 to find the force in U_0U_1. Similarly, section 2–2 is passed through the truss and moments are taken about U_6 to find the force in L_5L_6 and about L_5 to obtain force in U_5U_6. The remaining forces can be obtained by the method of joints.

$\Sigma M_{\text{left PI}} = 0$ of forces above plane of contraflexure

$$(1)(20)(10) - 48V_R = 0$$

$$V_R = 4.17^k \uparrow = \text{axial force in right column}$$

By $\Sigma V = 0$,

$$V_L = 4.17^k \downarrow = \text{axial force in left column}$$

Assuming total horizontal shear above plane of contraflexure of (1)(20) divides equally between the two columns (10^k each),

$$H_L = 10 + (1)(6) = 16^k \leftarrow$$

$$H_R = 10^k \leftarrow$$

Moment reactions for each column are found by taking moments at base of forces on column up to points of inflection,

$$M_L = (10)(6) + (1)(6)(3) = 78'^k \circlearrowleft$$

$$M_R = (10)(6) = 60'^k$$

Force in L_0L_1

$$\Sigma M_{U_0} = 0$$

$$(10)(20) - (20)(1)(10) - 8L_0L_1 = 0$$

$$L_0L_1 = 0$$

Force in U_0U_1

$$\Sigma M_{L_1} = 0$$

$$(10)(12) - (20)(1)(2) - (4.17)(8) + 8U_0U_1 = 0$$

$$U_0U_1 - 5.83^k$$

Bowaters Southern Corporation Plant, Calhoun, Tennessee. (Courtesy of the American Institute of Steel Construction.)

16.4 LATERAL BRACING FOR BRIDGES

Bridge trusses may be braced laterally by bracing systems in the planes of the top and bottom chords as well as by vertical or inclined planes of bracing. The planes of bracing tie the main trusses together and cause the entire structure to act as a rigid framework. They prevent excessive vibrations and resist the lateral loads of wind, earthquake, nosing of locomotives, and the centrifugal effect of traffic on curved bridges. Figure 16.7 shows a Warren bridge truss with the bracing systems that might be used in the plane of the top and bottom chords.

The loads are applied as concentrated loads at the joints of the lateral trusses, which permits analysis by the approximate methods discussed in Sec. 16.2. The top lateral bracing is usually subject to light loads, and the diagonals

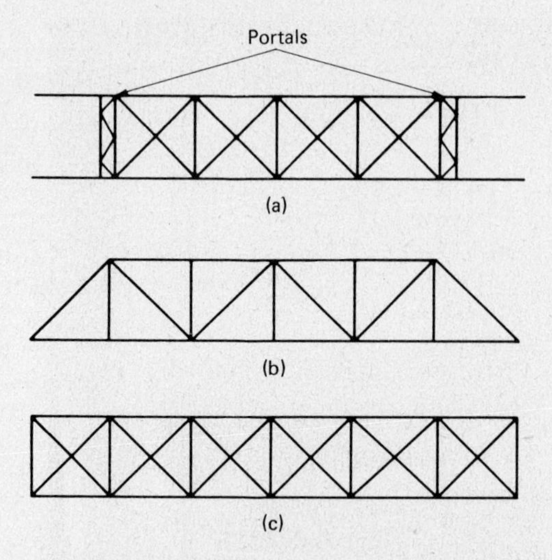

Figure 16.7 (a) Top lateral truss. (b) Plan. (c) Bottom lateral truss.

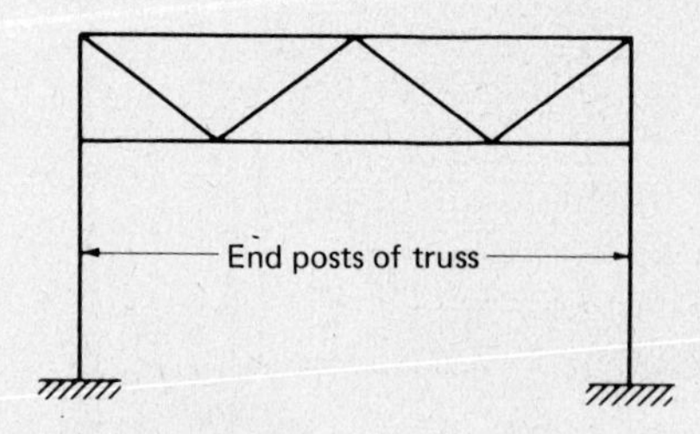

Figure 16.8

will probably be slender and able to resist only tensile loads. The design lateral loads are probably larger on the bottom of the truss, and the diagonals of the bottom-chord lateral system may be large enough to carry some compressive forces. The chord members of the lateral trusses are the chord members of the main trusses, but the AASHTO specifications do not require a strengthening of these members unless their forces, as a part of the lateral system, are greater than 25% of their normal forces as parts of the main trusses.

A system of bracing is frequently used in the plane of the end posts, similar to the one shown in Figs. 16.7 and 16.8. This type of bracing is commonly referred to as "portal bracing." The portal has the purpose of furnishing the end reaction to the top lateral system and transferring it down to the supports. The arrangement of the members of portals is quite similar to that of mill buildings, and the portals may be analyzed exactly as were the mill buildings.

Portals for girder bridges may be of the types shown in Fig. 16.9, in which the horizontal beam is rigidly connected to the columns (actually the girders). Similar structures are an essential part of steel building frames, and they also

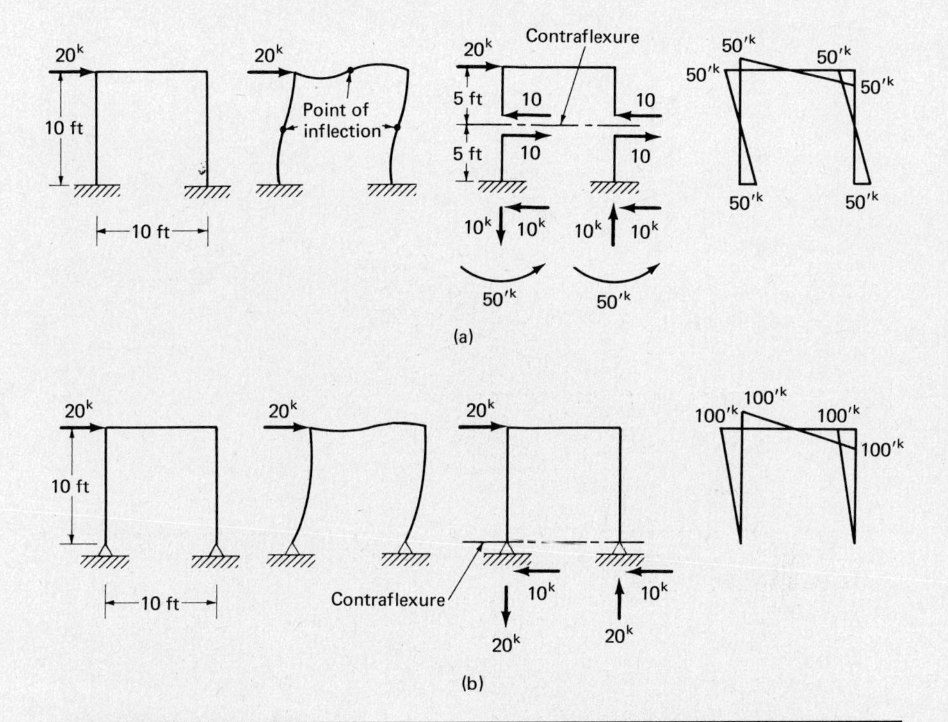

Figure 16.9

may be analyzed on the basis of the same assumptions used for mill buildings. A portal or bent that is fixed at the column bases is analyzed in Fig. 16.9(a). Figure 16.9(b) shows the analysis of a portal that is hinged at the column bases. In each case the deformed shape of the bent, the reactions, and the moment diagrams are shown.

16.5 ANALYSIS OF BUILDING FRAMES FOR VERTICAL LOADS

One approximate method for analyzing building frames for vertical loads involves the estimation of the location of points of inflection in the girders. A common practice is to assume the PI's are located at the one-tenth points from each girder end. In addition the axial forces in the girders are assumed to be zero [1].

These assumptions have the effect of creating a simple beam between the points of inflection, and the positive moments in the beam can be determined by statics. Negative moments occur in the girders between their ends and the points of inflection. They may be computed by considering the portion of the beam out to the point of inflection to be a cantilever.

The shear at the end of each of the girders contributes to the axial forces in the columns. Similarly, the negative moments at the ends of the girders are

Multistory steel frame building, showing structural skeleton; Hotel Sheraton, Philadelphia, Pennsylvania. (Courtesy of the Bethlehem Steel Corporation.)

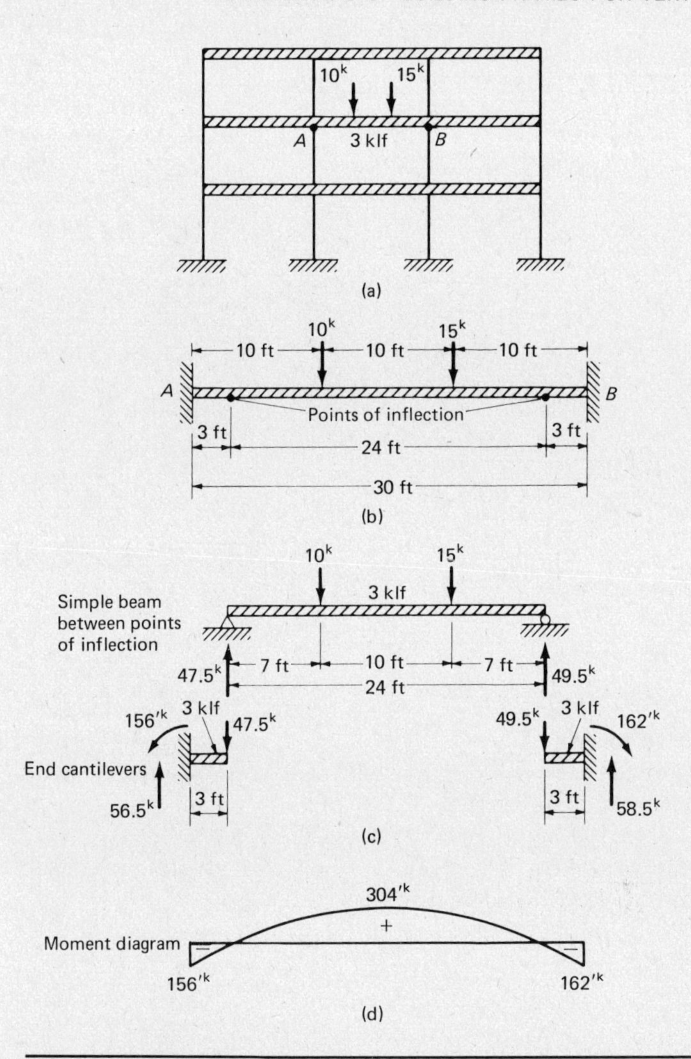

Figure 16.10

transferred to the columns. For interior columns the girder moments on each side oppose each other and may cancel. Exterior columns have moments on only one side, caused by the girders framing into them, and these need to be considered in design.

In Fig. 16.10, beam *AB* of the building frame shown is analyzed by assuming points of inflection at one-tenth points and fixed supports at beam ends.

To make reasonable estimates for PI locations the reader may find the sketching of the approximate deflected shape of the structure in question very helpful. For illustration a continuous beam is drawn to scale in Fig. 16.11(a) and a sketch of its estimated deflected shape for the loads shown is given in part

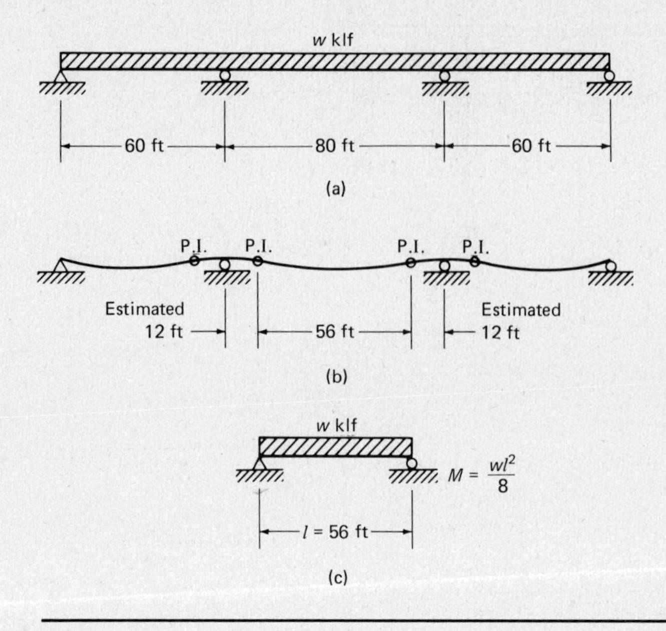

Figure 16.11

(b). From the sketch an approximate location of the PI's is estimated. Finally in part (c) of the figure the part of the beam between the PI's in the center span is taken out as an assumed simple span.

It might be useful for the reader to see where PI's occur for a few types of statically indeterminate beams. These may be helpful in estimating where PI's will occur in other structures. Moment diagrams are shown for several beams in Fig. 16.12. The PI's obviously occur where the moment diagrams change from + to − or vice versa.

16.6 ANALYSIS OF BUILDING FRAMES FOR LATERAL LOADS

Building frames are subjected to lateral loads as well as to vertical loads. The necessity for careful attention to these forces increases as buildings become taller. Not only must a building have sufficient lateral resistance to prevent failure but it also must have sufficient resistance to deflecting to prevent injury to its various parts.

Another item of importance is the provision of sufficient lateral rigidity to give the occupants a feeling of safety. They might not have this feeling in tall buildings that have a great deal of lateral movement in times of high winds. There have been actual instances of occupants of the upper floors of tall buildings complaining of seasickness on very windy days.

Lateral loads can be taken care of by means of X or other types of bracing, by shear walls, or by moment resisting wind connections. Only the last of these three methods is considered in this chapter.

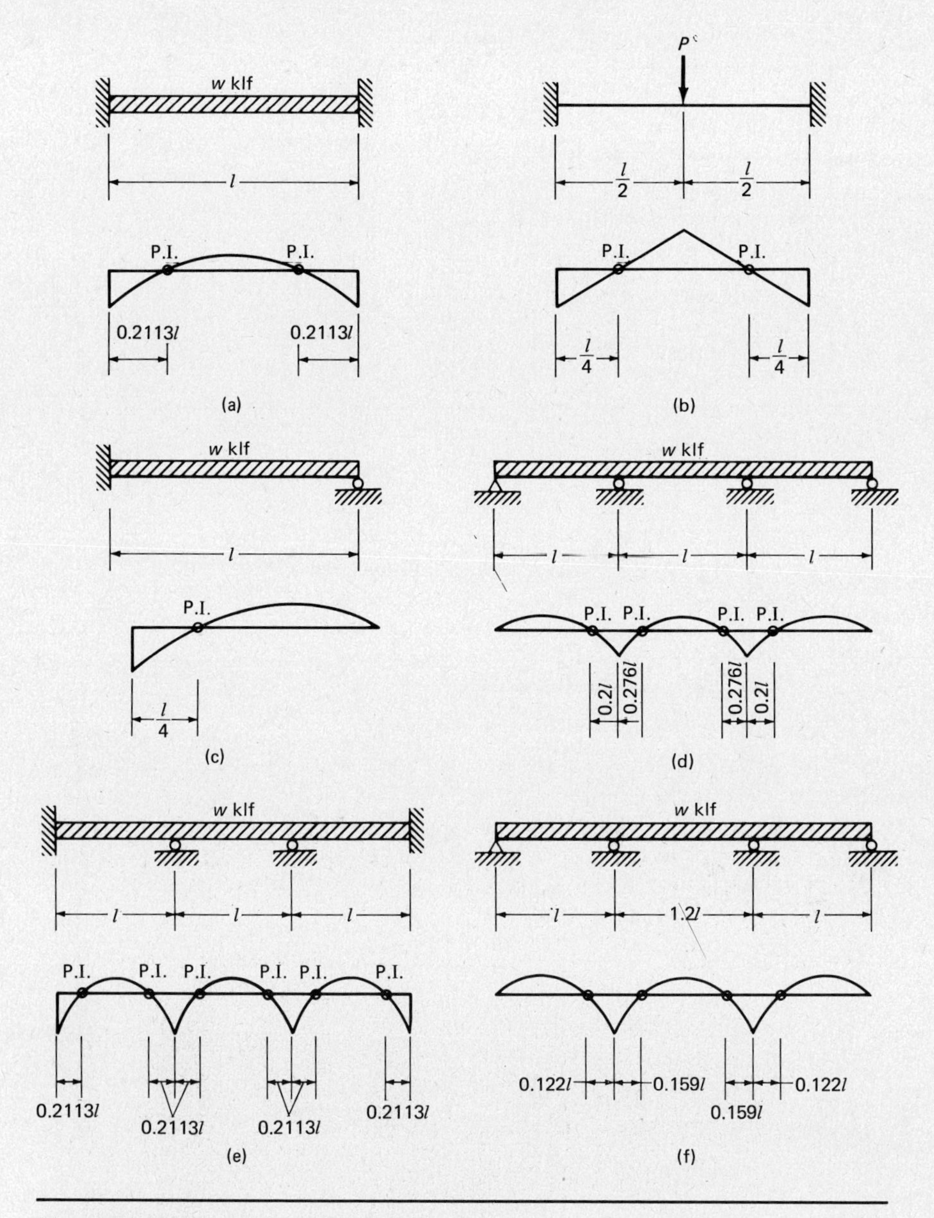

Figure 16.12

Rigid frame buildings are highly indeterminate, and their analysis by the usual "exact" methods is so lengthy as to make the approximate methods very popular. The total degree of indeterminancy of a building frame (internal and external) can be determined by considering it to consist of separate portals. One level of the rigid frame of Fig. 16.13 is broken down into a set of portals in Fig. 16.14. Each of the portals is statically indeterminate to the third degree, and the total degree of indeterminancy of a building equals three times the number of individual portals in the frame.

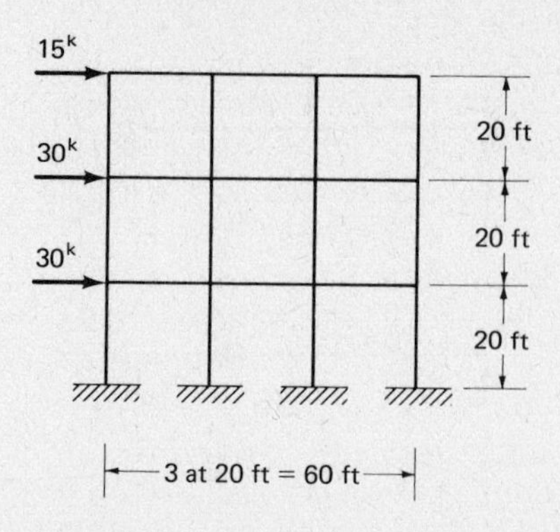

Figure 16.13

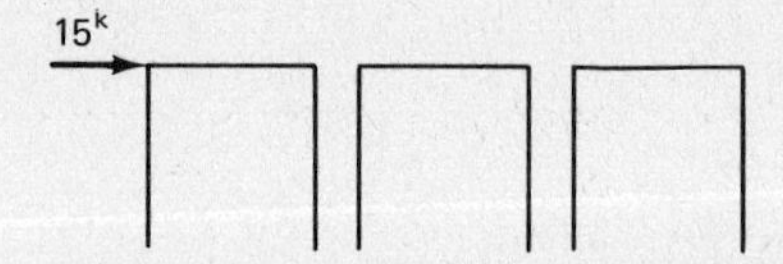

Figure 16.14 One level of frame of Fig. 16.13.

Syracuse University Library, Syracuse, New York. (Courtesy of the American Concrete Institute.)

Another method of obtaining the degree of indeterminancy is to assume that each of the girders is cut by an imaginary section. If these values—shear, axial force, and moment—are known in each girder, the free bodies produced can be analyzed by statics. The total degree of indeterminancy equals three times the number of girders.

The building frame of Fig. 16.13 is analyzed by two approximate methods in the pages to follow, and the results are compared with those obtained by one of the "exact" methods considered in a later chapter. The dimensions and loading of the frame are selected to illustrate the methods involved while keeping the computations as simple as possible. There are nine girders in the frame, giving a total degree of indeterminancy of 27, and at least 27 assumptions will be needed to permit an approximate solution.

The reader should be aware that today with the availability of digital computers it is feasible to make "exact" analyses in appreciably less time than that required to make approximate analyses (without the use of computers). The more accurate values obtained permit the use of smaller members. The results of computer usage are money saving in analysis time and in the use of smaller members. Today it is possible to analyze structures (such as tall buildings) that are statically indeterminate in minutes with hundreds or even thousands of simultaneous equations via the displacement method. The results for these highly indeterminate structures are far more accurate and more economically obtained than from approximate analyses.

The two methods considered here are the portal and cantilever methods. These methods have been used in so many successful building designs that they were almost the unofficial standard procedure for the design profession before the advent of modern computers. No consideration is given in either of these methods to the elastic properties of the members. These omissions can be very serious in unsymmetrical frames and in very tall buildings. To illustrate the seriousness of the matter, the changes in member sizes are considered in a very tall building. In such a building there will probably not be a great deal of change in beam sizes from the top floor to the bottom floor. For the same loadings and spans the changed sizes would be due to the large wind moments in the lower floors. The change, however, in column sizes from top to bottom would be tremendous. The result is that the relative sizes of columns and beams on the top floors are entirely different from the relative sizes on the lower floors. When this fact is not considered, it causes large errors in the analysis.

In both the portal and cantilever methods, the entire wind loads are assumed to be resisted by the building frames, with no stiffening assistance from the floors, walls, and partitions. Changes in length of girders and columns are assumed to be negligible. They, however, are not negligible in tall slender buildings the height of which is five or more times the least horizontal dimension.

If the height of a building is roughly five or more times its least lateral dimension, it is generally felt that a more precise method of analysis should be used than the portal or cantilever methods. There are several excellent approximate methods which make use of the elastic properties of the structures and which give values closely approaching the results of the "exact" methods. These include the Factor method [2], the Witmer method of K percentages [3], and the Spurr method [4]. Should an "exact" method be desired, the slope-deflection method (Chap. 22) and the moment-distribution method (Chaps. 23 and 24) are available. If the slope-deflection procedure is used, the designer will have the problem of solving a large number of simultaneous equations. The problem is not so serious, however, if one has access to a digital computer.

The Portal Method

The most common approximate method of analyzing building frames for lateral loads is the portal method. Because of its simplicity, it has probably been used more than any other approximate method for determining wind forces in building frames. This method, which was presented by Albert Smith in the *Journal of the Western Society of Engineers* in April, 1915, is said to be satisfactory for most buildings up to 25 stories in height [5].

At least three assumptions must be made for each individual portal or for each girder. In the portal method, the frame is theoretically divided into independent portals (Fig. 16.14) and the following three assumptions are made:

1. The columns bend in such a manner that there is a point of inflection at middepth, Fig. 16.9(a).
2. The girders bend in such a manner that there is a point of inflection at their centerlines.
3. The horizontal shears on each level are arbitrarily distributed between the columns. One commonly used distribution (and the one illustrated here) is to assume the shear divides among the columns in the ratio of one part to exterior columns and two parts to interior columns. The reason for this ratio can be seen in Fig. 16.14. Each of the interior columns is serving two bents, whereas the exterior columns are serving only one. Another common distribution is to assume that the shear V taken by each column is in proportion to the floor area it supports. The shear distribution by the two procedures would be the same for a building with equal bays, but for one with unequal bays the results would differ with the floor area method, probably giving more realistic results.

For this frame there are 27 redundants; to obtain their values, one assumption as to the location of the point of inflection has been made for each of the 21 columns and girders. Three assumptions are made on each level as to the shear split in each individual portal, or the number of shear assumptions equals one less than the number of columns on each level. For the frame, 9 shear assumptions are made, giving a total of 30 assumptions and only 27 redundants. More assumptions are made than necessary, but they are consistent with the solution (that is, if only 27 of the assumptions were used and the remaining values were obtained by statics, the results would be identical).

Frame Analysis

The frame is analyzed in Fig. 16.15 on the basis of these assumptions. The arrows shown on the figure give the direction of the girder shears and the column axial forces. The reader can visualize the stress condition of the frame if he or she assumes the wind is tending to push it over from left to right, stretching the left exterior columns and compressing the right exterior columns. Briefly the calculations were made as follows.

1. COLUMN SHEARS

The shears in each column on the various levels were first obtained. The total shear on the top level is 15 kip. Because there are two exterior and two interior columns, the following expression may be written:

$$x + 2x + 2x + x = 15^k$$
$$x = 2.5^k$$
$$2x = 5.0^k$$

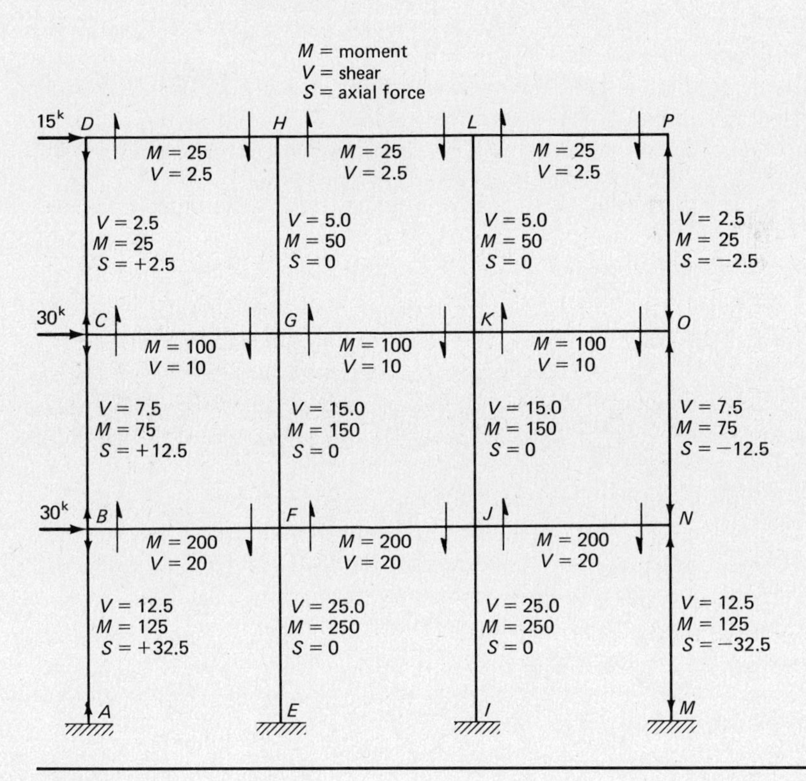

Figure 16.15

The shear in column *CD* is 2.5 kip; in *GH* it is 5.0 kip; and so on. Similarly, the shears were determined for the columns on the first and second levels, where the total shears are 75 and 45 kip, respectively.

2. COLUMN MOMENTS

The columns are assumed to have points of inflection at their middepths; therefore their moments, top and bottom, equal the column shears times half the column heights.

3. GIRDER MOMENTS AND SHEARS

At any joint in the frame the sum of the moments in the girders equals the sum of the moments in the columns. The column moments have been previously determined. By beginning at the upper left-hand corner of the frame and working across from left to right, adding or subtracting the moments as the case may be, the girder moments were found in this order: *DH*, *HL*, *LP*, *CG*, *GK*, and so on. It follows that with points of inflection at girder centerlines, the girder shears equal the girder moments divided by half-girder lengths.

4. COLUMN AXIAL FORCES

The axial forces in the columns may be directly obtained from the girder shears. Starting at the upper left-hand corner, the column axial force in *CD* is numerically equal to the shear in girder *DH*. The axial force in column *GH* is equal to the difference between the two girder shears *DH* and *HL*, which equals zero in this case. (If the width of each of the portals is the same, the shears in the girder on one level will be equal, and the interior columns will have no axial force, since only lateral loads are considered.)

The Cantilever Method

Another simple method of analyzing building frames for lateral forces is the cantilever method presented by A. C. Wilson in *Engineering Record*, September 5, 1908. This method is said to be a little more desirable for high narrow buildings than the portal method and may be used satisfactorily for buildings with heights not in excess of 25 to 35 stories [6]. It is not as popular as the portal method.

Mr. Wilson's method makes use of the assumptions that the portal method uses as to locations of points of inflection in columns and girders, but the third assumption differs somewhat. Rather than assume the shear on a particular level to divide between the columns in some ratio, the axial force in each column is considered to be proportional to its distance from the center of gravity of all the columns on that level. The wind loads are tending to overturn the building, and the columns on the leeward side will be compressed, whereas those on the windward side will be put in tension. The greater the distance a column is from the center of gravity of its group of columns, the greater will be its axial force.

The new assumption is equivalent to making a number of axial-force assumptions equal to one less than the number of columns on each level. Again, the structure has 27 redundants and 30 assumptions are made (21 column and girder point-of-inflection assumptions and 9 column axial-force assumptions), but the extra assumptions are consistent with the solution.

Frame Analysis

The frame previously analyzed by the portal method is analyzed by the cantilever method in Fig. 16.16. Briefly the calculations are made as follows.

1. COLUMN AXIAL FORCES

Considering first the top level, moments are taken about the point of contraflexure in column *CD* of the forces above the plane of contraflexure through the columns on that level. According to the third assumption, the axial force in *GH* will only be one-third of that in *CD*, and these forces in *GH* and *CD* will be tensile, whereas those in *KL* and *OP* will be compressive. The following expression is written, with respect to Fig. 16.17, to determine the values of the column axial forces on the top level.

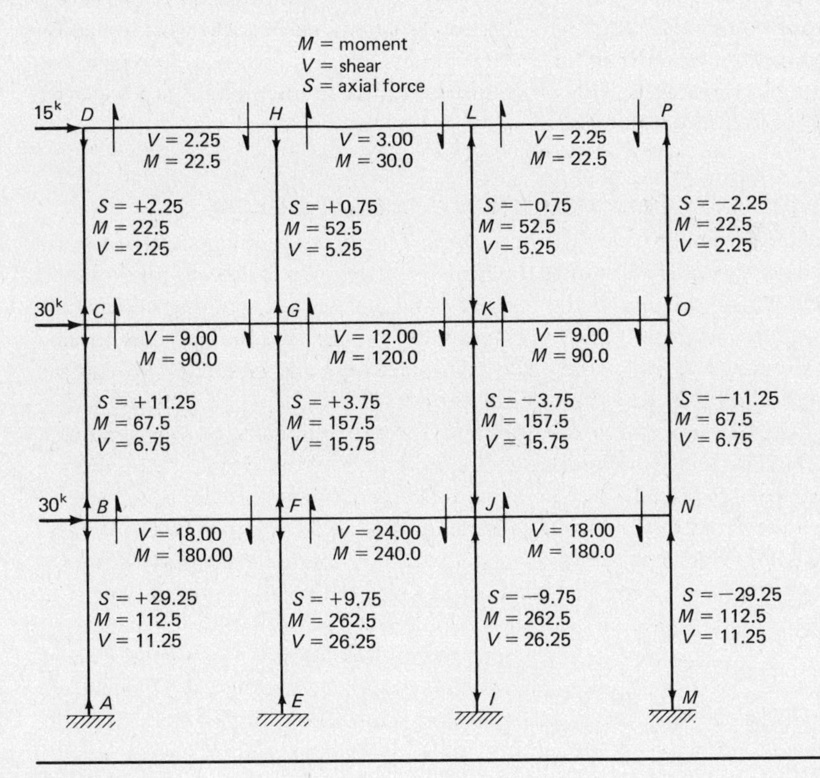

Figure 16.16

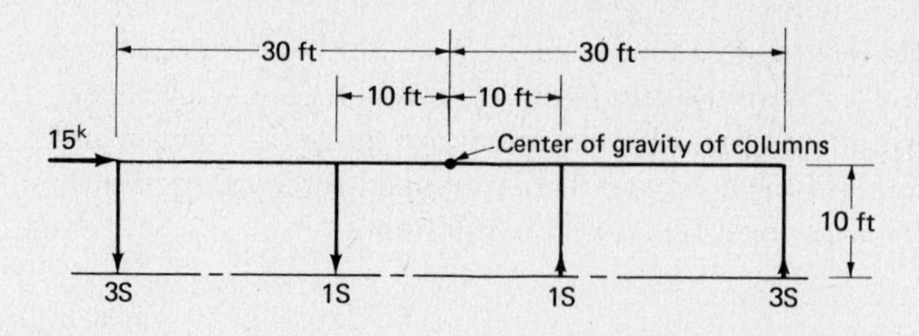

Figure 16.17

$$(15)(10) + (1S)(20) - (1S)(40) - (3S)(60) = 0$$

$$S = 0.75^k$$

$$3S = 2.25^k$$

The axial force in CD is 2.25 kip and that in GH is 0.75 kip, and so on. Similar moment calculations are made for each level to obtain the column axial forces.

2. GIRDER SHEARS

The next step is to obtain the girder shears from the column axial forces. These shears are obtained by starting at the top left-hand corner and working across the top level and adding or subtracting the axial forces in the columns according to their signs. This procedure is similar to the method of joints used for finding truss forces.

3. COLUMN AND GIRDER MOMENTS AND COLUMN SHEARS

The final steps can be quickly summarized. The girder moments, as before, are equal to the girder shears times the girder half-lengths. The column moments are obtained by starting at the top left-hand corner and working across each level in succession, adding or subtracting the previously obtained column and girder moments as indicated. The column shears are equal to the column moments divided by half the column heights.

Table 16.1 compares the moments in the members of this frame as determined by the two approximate methods and by the moment-distribution method described in Chaps. 23 and 24. It is noted that for several members the approximate results vary considerably from the results obtained by the "exact" method. Experience with the "exact" methods for handling indeterminate building frames will show that the points of inflection will not occur exactly at the midpoints. Using more realistic locations for the assumed inflection points will greatly improve results. The Bowman method [6] involves the location of the points of inflection in the columns and girders according to a specified set of rules depending on the number of stories in the building. In addition the shear is divided between the columns on each level according to a set of rules that are based on the moments of inertia of the columns as well as the bay

Table 16.1 MEMBER MOMENTS

MEMBER	PORTAL	CANTILEVER	MOMENT DISTRIBUTION	MEMBER	PORTAL	CANTILEVER	MOMENT DISTRIBUTION
AB	125	112.5	199	*IJ*	250	262.5	244
BA	125	112.5	127	*JI*	250	262.5	177
BC	75	67.5	62	*JF*	200	240	139
BF	200	180	189	JK	150	157.5	122
CB	75	67.5	92	*JN*	200	180	161
CD	25	22.5	10	*KJ*	150	157.5	141
CG	100	90	102	*KG*	100	120	83
DC	25	22.5	36	*KL*	50	52.5	34
DH	25	22.5	36	*KO*	100	90	93
EF	250	262.5	224	*LK*	50	52.5	56
FE	250	262.5	177	*LH*	25	30	26
FB	200	180	161	*LP*	25	22.5	30
FG	150	157.5	122	*MN*	125	112.5	199
FJ	200	240	139	*NM*	125	112.5	127
GF	150	157.5	141	*NJ*	200	180	189
GC	100	90	93	*NO*	75	67.5	62
GH	50	52.5	34	*ON*	75	67.5	92
GK	100	120	83	*OK*	100	90	102
HG	50	52.5	56	*OP*	25	22.5	10
HD	25	22.5	30	*PO*	25	22.5	36
HL	25	30	26	*PL*	25	22.5	36

widths. Application of the Bowman method gives much better results than the portal and cantilever procedures.

16.7 ANALYSIS OF VIERENDEEL TRUSS

A type of truss that is frequently used in Europe but only occasionally in the United States is the Vierendeel truss, developed by M. Vierendeel in 1896. This type of truss, illustrated in Fig. 16.18, is actually not a truss by the usual definition and requires the use of moment-resisting joints. Loads are supported by means of the bending resistance of its short heavy members. Although analysis and design are quite difficult, it is a fairly efficient structure.

These highly indeterminate structures can be approximately analyzed by the portal and cantilever methods described in the preceding section. Figure 16.19 shows the results obtained by applying the portal method to a Vierendeel truss. (For this symmetrical case the cantilever method will yield the same results.) To follow the calculations the reader may find it convenient to turn the structure sideways because the shear being considered for the Vierendeel is vertical, whereas it was horizontal for the building frames previously considered.

An "exact" analysis of this truss by the computer program STRUDL yields the values shown in Fig. 16.20. For this computer program the members were assumed to have constant areas and moments of inertia. As the points of inflection are not exactly at the midpoints of the members, the moments at the member ends vary somewhat and both moments are given in the figure. Though the results obtained with the portal method seem quite reasonable for

Vierendeel truss, Clinic Inn Pedestrian Bridge, Cleveland, Ohio. (Courtesy of the American Institute of Steel Construction, Inc.)

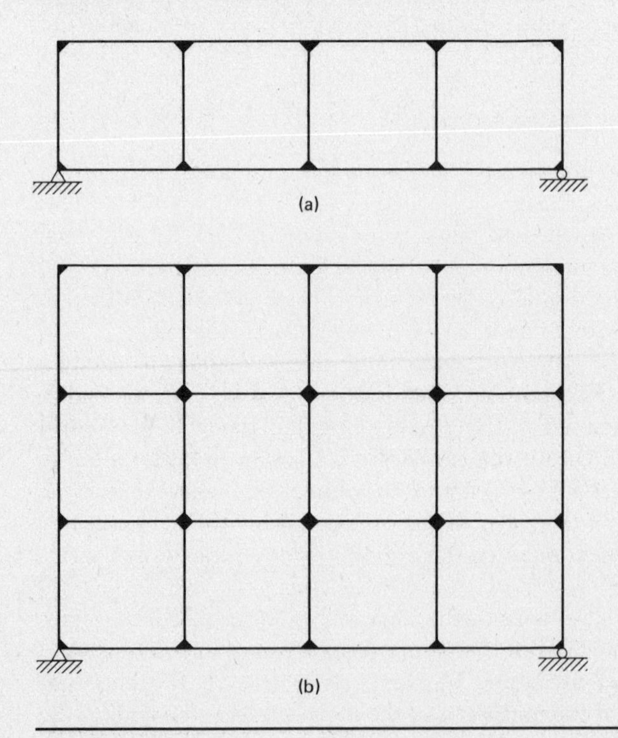

Figure 16.18

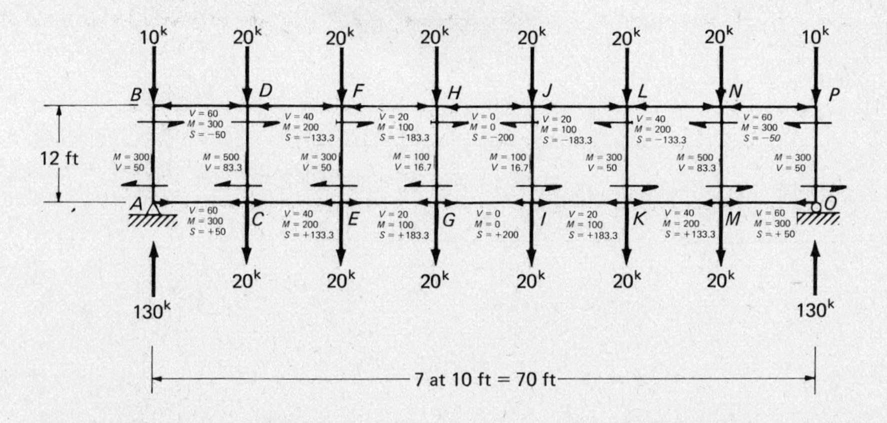

Figure 16.19

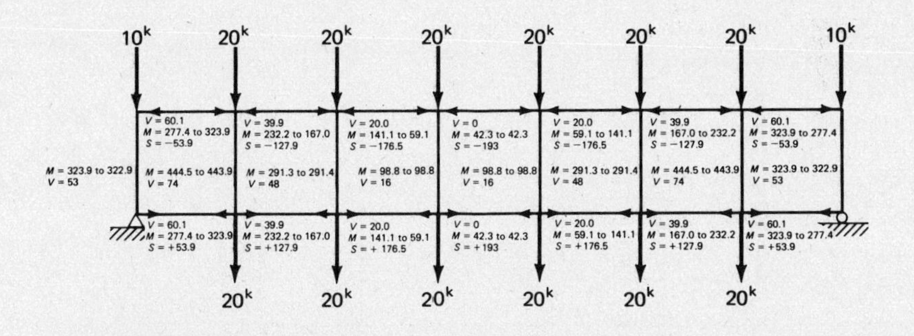

Figure 16.20

this particular truss, they could have been improved by adjusting the assumed locations of the points of inflection a little.

For many Vierendeels, particularly those of several stories, the lower horizontal members may be much larger and stiffer than the other horizontal members. To obtain better results with the portal or cantilever methods the nonuniformity of sizes should be taken into account by assuming more of the shear is carried by the stiffer members.

PROBLEMS

16.1. Compute the forces in the members of the truss shown in the accompanying illustration for each of the following conditions: (a) diagonals unable to carry compression; (b) diagonal that would be in compression can resist half of shear in panel; (c) diagonal that would be in compression can resist one-third of shear in panel.

Problem 16.1

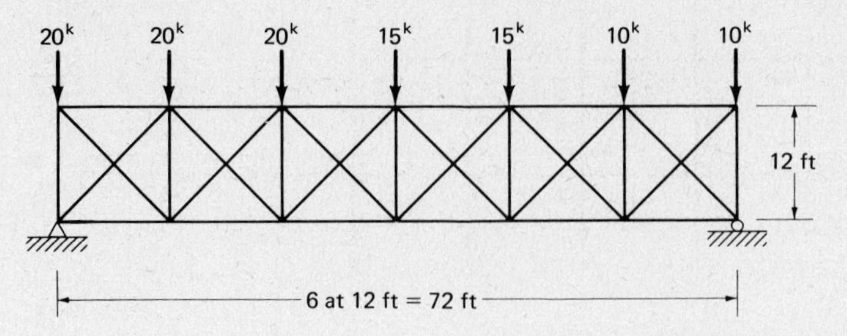

16.2. Determine the forces for all of the members of the mill building truss shown in the accompanying illustration if points of inflection are assumed to be 12 ft above the column bases.

Problem 16.2 (*Ans.* $V_R = 13.07^k \uparrow$, $U_1U_2 = -14.95^k$, $L_2L_3 = -26.14^k$, $U_4L_5 = -18.5^k$)

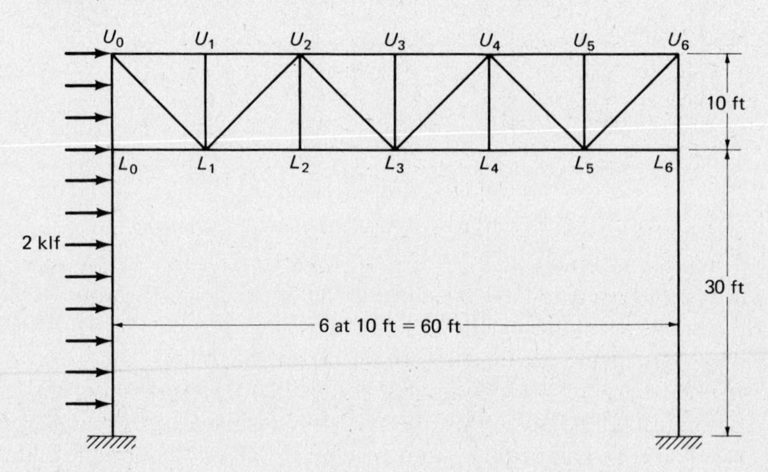

16.3. Repeat Prob. 16.2 if points of inflection are assumed to be 15 ft above column bases.

16.4. Assuming points of inflection to be 10 ft from the column bases, determine the forces in all members of the structure shown in the accompanying illustration.

Problem 16.4 (*Ans.* $L_0L_2 = -10.05^k$, $U_1U_3 = -28.0^k$, $L_4L_{10} = -13.76^k$)

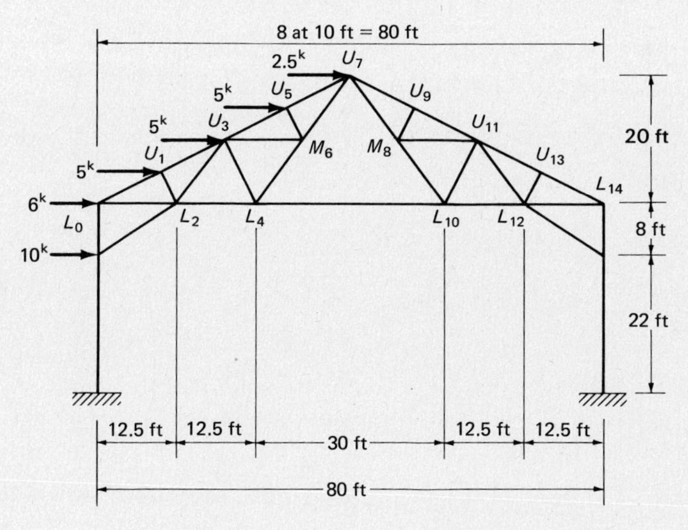

16.5. Rework Prob. 16.4 with points of inflection assumed to be 6 ft from the column bases.

For Probs. 16.6 to 16.13 compute moments, shears and axial forces for all of the members of the frames shown (a) by using the portal method and (b) by using the cantilever method.

Problem 16.6 (*Ans.* Portal method for BC: $V = 5^k$, $M = 37.5'^k$, $S = +4.16^k$; cantilever method for BC; $V = 4.5^k$, $M = 33.75'^k$, $S = +3.75^k$)

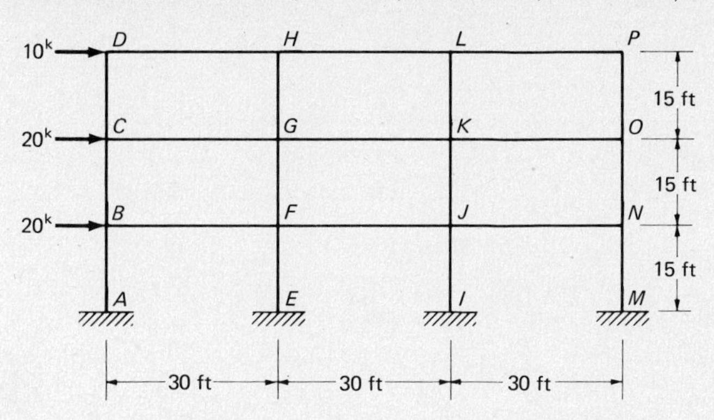

Problem 16.7

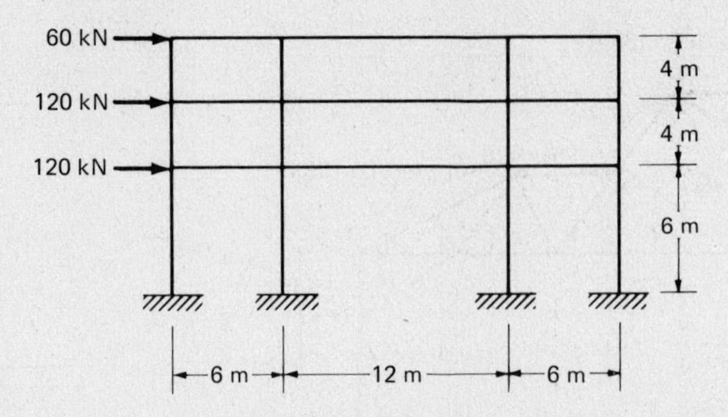

Problem 16.8 (*Ans.* Portal method for FG: $V = 12.5^k$, $M = 125'^k$, $S = +22.5^k$; cantilever method for FG: $V = 7.5'^k$, $M = 75'^k$, $S = +17.5^k$)

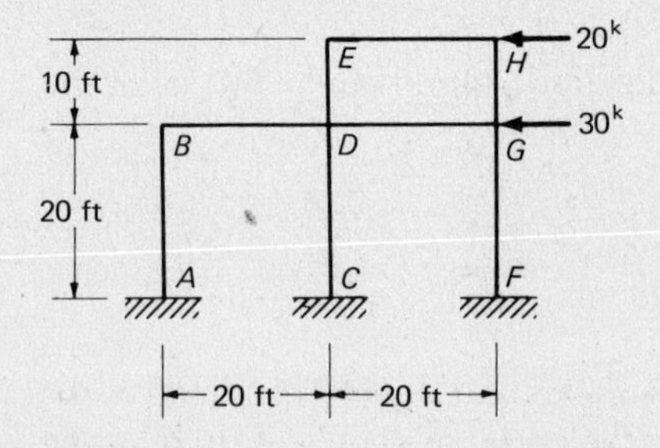

Problem 16.9

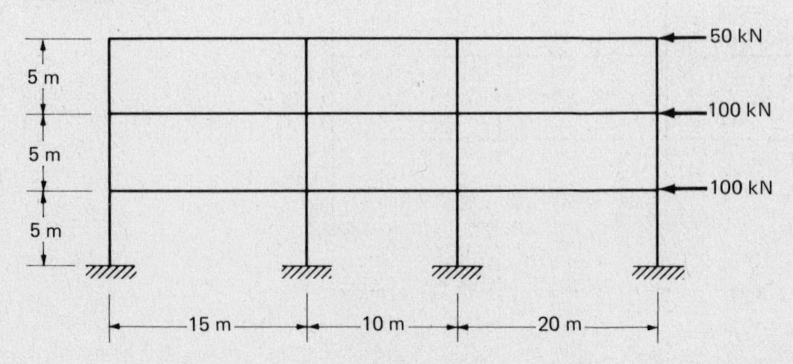

Problem 16.10 (*Ans.* Portal method for FG: $V = 12.0^k$, $M = 72'^k$, $S = +2.0^k$; cantilever method for FG: $V = 11.65^k$, $M = 69.9'^k$, $S = +1.06^k$)

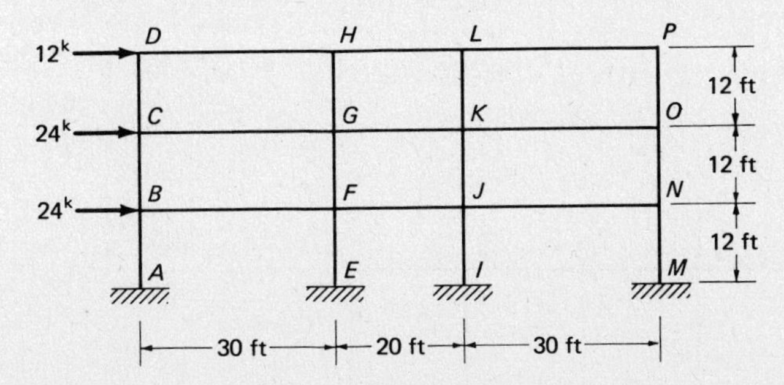

16.11. Rework Prob. 16.10 if the column bases are assumed to be pinned.

Problem 16.12 (*Ans.* Portal method for OP: $V = 12.5^k$, $M = 125'^k$, $S = -22.08^k$; cantilever method for OP: $V = 12.48^k$, $M = 124.8'^k$, $S = -6.28^k$)

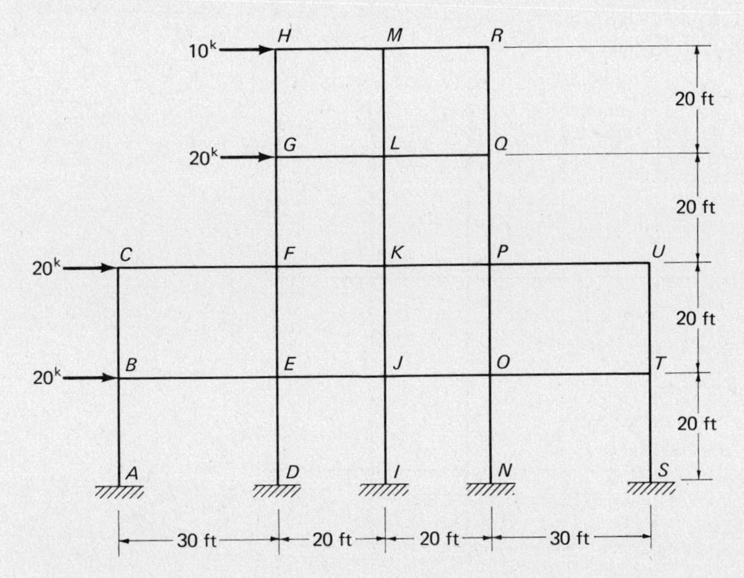

Problem 16.13

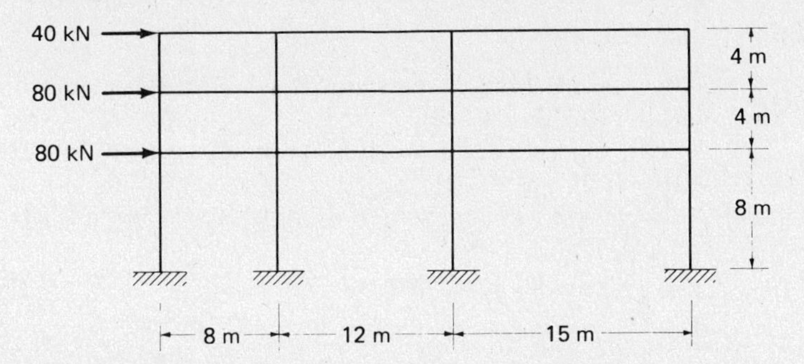

For Probs. 16.14 and 16.15, compute moments, shears, and axial forces for all of the members of the Vierendeel trusses using the portal method.

Problem 16.14 (*Ans.* For *CD*: $V = 60^k$, $M = 360'^k$; for *FH*: $V = 15^k$, $M = 90'^k$, $S = -135^k$)

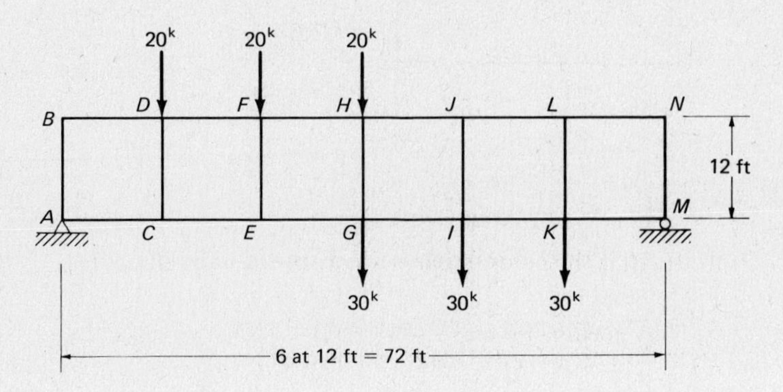

Problem 16.15

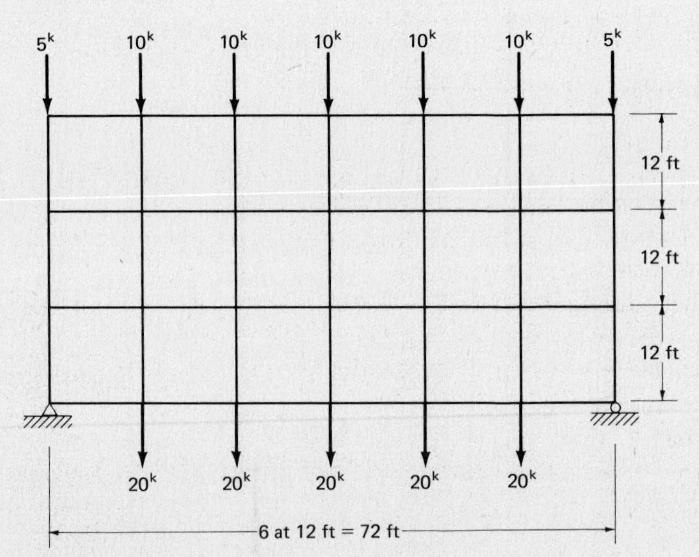

REFERENCES

[1] C. H. Norris, J. B. Wilbur, and S. Utku, *Elementary Structural Analysis*, 3rd ed. (New York: McGraw-Hill, 1976), pp. 200–201.

[2] C. H. Norris, J. B. Wilbur, and S. Utku, *Elementary Structural Analysis*, 3rd ed. (New York: McGraw-Hill, 1976), pp. 207–212.

[3] "Wind Bracing in Steel Buildings," *Transactions of the American Society of Civil Engineers* 105 (1940): 1725–1727.

[4] Ibid, pp. 1723–1725.

[5] Ibid, p. 1723.

[6] H. Sutherland and H. L. Bowman, *Structural Theory* (New York: Wiley, 1950), pp. 295–301.

17 Deflections

17.1 REASONS FOR COMPUTING DEFLECTIONS

The members of all structures are made up of materials that deflect when loaded. If deflections exceed allowable values, they may detract from the appearance of the structures and the materials attached to the members may be damaged. For example, a floor joist that deflects too much may cause cracks in the ceiling below, or if it supports concrete or tile floors, it may cause cracks in the floors. In addition, the use of a floor supported by beams that "give" appreciably does not inspire confidence, although the beams may be perfectly safe. Excessive vibration may occur in a floor of this type, particularly if it supports machinery.

Standard American practice is to limit deflections caused by live load to 1/360 of the spans. This figure was probably originated for beams supporting plastered ceilings and was thought to be sufficient to prevent plaster cracks. (Although most of the deflections in a building are due to dead load, they will have substantially taken place before plaster is applied.)

The 1/360 deflection is only one of many maximum deflection values in use because of different loading situations, different designers, and different specifications. For situations in which precise and delicate machinery is supported, maximum deflections may be limited to 1/1500 or 1/2000 of the span lengths. The 1977 AASHTO specifications limit deflections in steel beams and girders due to live load and impact to 1/800 of the span. This value which is applicable to both simple and continuous spans is preferably reduced to 1/1000 for bridges in urban areas which are used in part by pedestrians. Corresponding AASHTO values for cantilever arms are 1/300 and 1/375.

The deflections of members may be controlled by cambering. The members are constructed of such a shape that they will assume their theoretical shape under some loading condition (usually dead load). A simple beam would be constructed with a slight convex bend so that under gravity loads it would

Houston Ship Canal Bridge, Houston, Texas. (Courtesy of the Texas State Department of Highways and Public Transportation.)

become straight as assumed in the calculations. Some designers take into account both dead and live loads in figuring the amount of camber.

Deflection computations may be used for computing the reactions for statically indeterminate beams and trusses as well as the forces in the members of redundant trusses. In fact, the importance of deflections in the analysis of statically indeterminate structures is nearly as great as the importance of the equations of statics is in the analysis of statically determinate structures.

Despite the importance of deflections, it is rarely necessary, even for statically indeterminate structures, to compute structure deformations for the purpose of correcting the original structure dimensions on which computations are based. The deformations of the materials used in ordinary work are quite small as compared to the overall dimensions. For example, the strain that occurs in a steel section that has a modulus of elasticity of 29×10^6 psi (pounds per square inch) when the stress is 20,000 psi is only

$$\epsilon = \frac{f}{E} = \frac{20 \times 10^3}{29 \times 10^6} = 0.000690$$

or 0.0690% of the member length.

Quite a few methods are available for determining deflections. It is desirable for the structural designer to be familiar with several of these. For some structures one method may be easier to apply; for others another method is more satisfactory. In addition, the ability to handle any structural problem by more than one method is of great importance for checking results.

In this book the following methods of calculating slopes and deflections are presented. In parentheses are listed the locations of the presentations.

1. Moment-area theorems (Chap. 17)
2. Conjugate-beam method (Chap. 17)
3. Virtual work (Chap. 18)
4. Williot-Mohr diagrams (Chap. 18)
5. Castigliano's first theorem (Chap. 21)

17.2 THE MOMENT-AREA THEOREMS

The first method presented for the calculation of deflections is the very interesting and valuable moment-area method presented by Charles E. Greene of the University of Michigan about 1873. Under changing loads the neutral axis of a member changes in shape according to the positions and magnitudes of the loads. The elastic curve of a member is the shape the neutral axis takes under temporary loads. Professor Greene's theorems are based on the shape of the elastic curve of a member and the relationship between bending moment and the rate of change of slope at a point on the curve.

To develop the theorems, the simple beam of Fig. 17.1 is considered. Under the loads P_1 to P_4 it deflects downward as indicated in the figure.

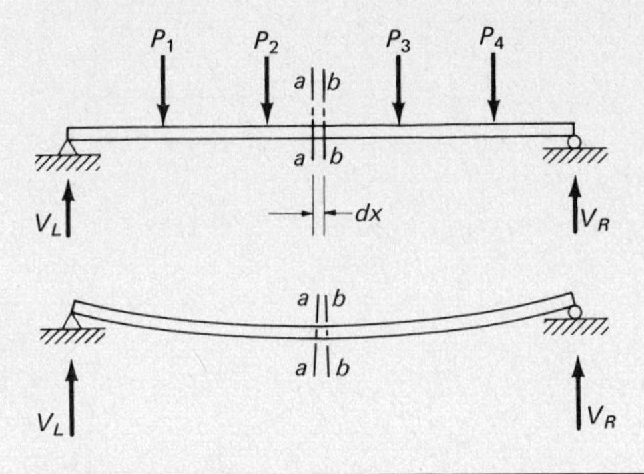

Figure 17.1

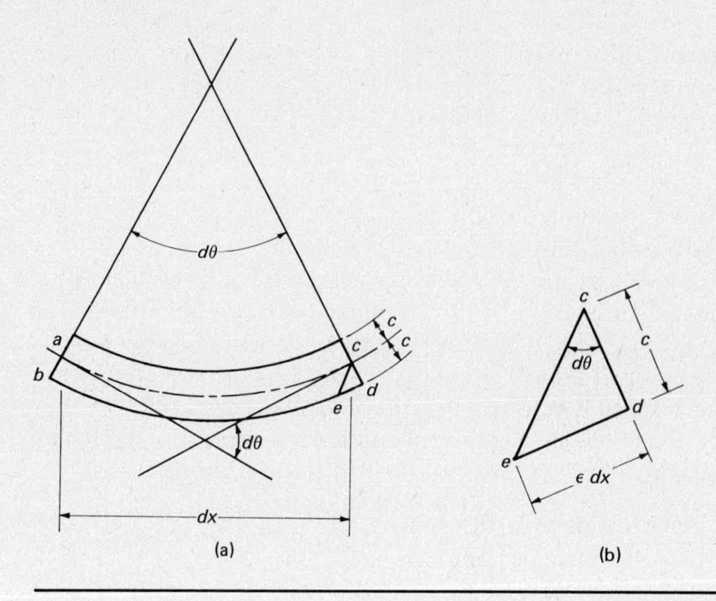

Figure 17.2

The dx section bounded on its ends by sections a–a and b–b is shown in Fig. 17.2. The size, degree of curvature, and distortion of the segment are tremendously exaggerated so that the slopes and deflections to be discussed can be seen easily. Line ac lies along the neutral axis of the beam and is unchanged in length. Line ce is drawn parallel to ab; therefore, be equals ac and de represents the lengthening of the bottom fiber of the dx section. Figure 17.2(b) shows an enlarged view of triangle cde and the angle $d\theta$, which is the change in slope of the tangent to the elastic curve at the left end of the section from the tangent at the right end. Sufficient information is now available to determine $d\theta$. In the derivation to follow it is to be remembered that the $d\theta$ angle being considered is minute, and for a very small angle the sine, tangent, and the angle in radians are identical, which permits their values to be used interchangeably. It is worthwhile to check a set of natural trigonometry tables to see the range of angles for which the functions coincide.

The bending moments developed by the external loads are positive and cause shortening of the upper beam fibers and lengthening of the lower fibers. The changes in fiber dimensions have caused the change in slope $d\theta$. The modulus of elasticity is known, and the stress at any point can be determined by the flexure formula; therefore the strain in any fiber can be found because it equals the stress divided by the modulus of elasticity. The value of $d\theta$ may be expressed as follows:

$$\tan d\theta = d\theta \text{ in radians (rad)} = \frac{\text{strain}}{c} = \frac{ed}{cd}$$

$$d\theta = \frac{\epsilon\, dx}{c}$$

By substituting the value of ϵ,

$$d\theta = \frac{(f/E)\,dx}{c}$$

But f is equal to Mc/I, and

$$d\theta = \frac{(Mc/EI)\,dx}{c} = \frac{M\,dx}{EI}$$

The change in slope in a dx distance is equal to $M\,dx/EI$, and the total change in slope from one point A in the beam to another point B can be expressed as the summation of all the $d\theta$ changes in the dx distances between the two points.

$$\theta_{AB} = \int_A^B \frac{M\,dx}{EI}$$

If the M/EI diagram is drawn for the beam, the above expression will be seen to equal each dx distance, from A to B, multiplied by its respective M/EI ordinate. The summation of these multiplications is the area of the diagram between the two points. From this discussion the first moment-area theorem

Steel framed dome for Ford Motor Company, Detroit, Michigan. (Courtesy of the Bethlehem Steel Corporation.)

may be expressed as follows: **The change in slope between the tangents to the elastic curve at two points is equal to the area of the M/EI diagram between the two points.**

Once we have a method by which changes in slopes between tangents to the elastic curve at various points may be determined, it is only a brief step to a method for computing deflections between the tangents. In a dx distance the neutral axis changes in direction by an amount $d\theta$. The deflection of one point on the beam with respect to the tangent at another point due to this angle change is equal to x (the distance from the point at which deflection is desired to the particular differential distance) times $d\theta$.

$$d\delta = x\,d\theta$$

The value of $d\theta$ from the first theorem is substituted in this expression:

$$d\delta = x\frac{M\,dx}{EI} = \frac{Mx\,dx}{EI}$$

To determine the total deflection from the tangent at one point A to the tangent at another point B on the beam, it is necessary to obtain a summation of the products of each $d\theta$ angle (from A to B) times the distance to the point where deflection is desired. The preceding sentence is a statement of the second moment-area theorem.

$$\delta_{AB} = \int_A^B \frac{Mx\,dx}{EI}$$

The deflection of a tangent to the elastic curve of a beam with respect to a tangent at another point is equal to the moment of the M/EI diagram between the two points, taken about the point at which deflection is desired.

17.3 APPLICATION OF THE MOMENT-AREA THEOREMS

Moment area is most conveniently used for beams in which the direction of the tangent to the elastic curve at one or more points is known, such as cantilever beams where the tangent at the fixed end does not change in slope. The method is applied quite easily to beams loaded with concentrated loads, because the moment diagrams consist of straight lines. These diagrams can be broken down into single triangles and rectangles, which facilitates the mathematics. Beams supporting uniform loads or uniformly varying loads may be handled, but the mathematics is slightly more difficult.

The properties of several figures given in Fig. 17.3 are useful in handling the M/EI diagrams.

Examples 17.1 to 17.6 illustrate the application of the moment-area theorems. It may occasionally be possible to simplify the mathematics by drawing the moment diagram and making the calculations in terms of symbols, such as P for a concentrated load, w for a uniform load, or l for span

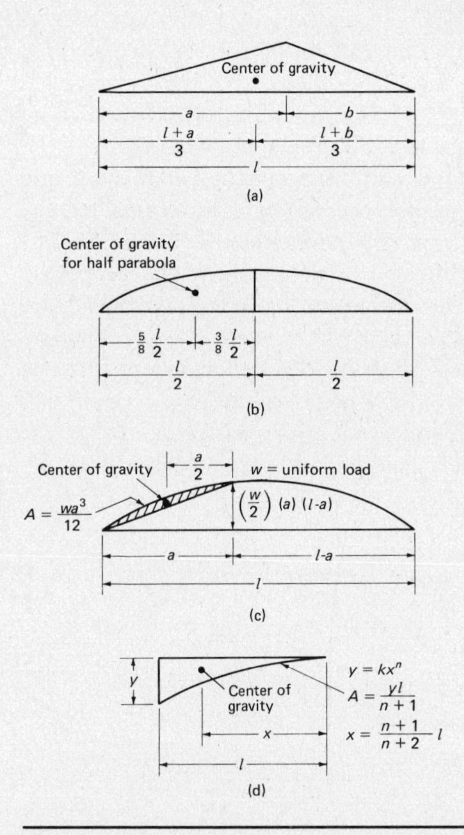

Figure 17.3

length, as illustrated by Examples 17.1 and 17.3. The numerical values of each of the symbols are substituted in the final step to obtain the slope or deflection desired.

Care must be taken to use consistent units in the calculations. The procedure here is to use all of the distances in feet and all of the loads and reactions in kilopounds. At the end of each problem the kilopounds are changed to pounds and the feet to inches. The resulting deflections will be in inches and the slopes in radians. (There are 2π rad in 360°.)

To prevent mistakes in the application of moment-area theory, it is emphasized that the slopes and deflections that are obtained are with respect to tangents to the elastic curve at the points being considered. The theorems do not directly give the slope or deflection at a point in the beam as compared to the horizontal (except in one or two special cases); they give the change in slope of the elastic curve from one point to another or the deflection of the tangent at one point with respect to the tangent at another point.

If a beam or frame has several loads applied, the M/EI diagram may be inconvenient to handle. The calculations may be simplified by drawing a separate diagram for each of the loads and determining the slopes and deflections for each diagram separately. The final results for a particular point can be found by adding the results for all of the loads. The principle of superposition applies to area moment and to any of the other methods of determining slopes and deflections discussed in subsequent pages.

Should a beam or frame be loaded with both concentrated loads and uniform loads it is advisable to separate the moment diagrams for the uniform loads from the diagrams for the concentrated loads. Such a separation is desirable because of difficulty in determining the properties (areas and centers of gravities) of the resulting combined diagrams which are necessary in applying the moment-area theorems as well as the conjugate-beam procedure presented later in this chapter. The deflections needed for solution of Example 19.1 were obtained in this manner.

Example 17.6 shows that moment area is one method that may be used to determine the moments at the ends of a fixed-ended beam, which is statically indeterminate to the third degree.

Example 17.1

Determine the slope and deflection of the right end of the cantilever beam shown in Fig. 17.4.

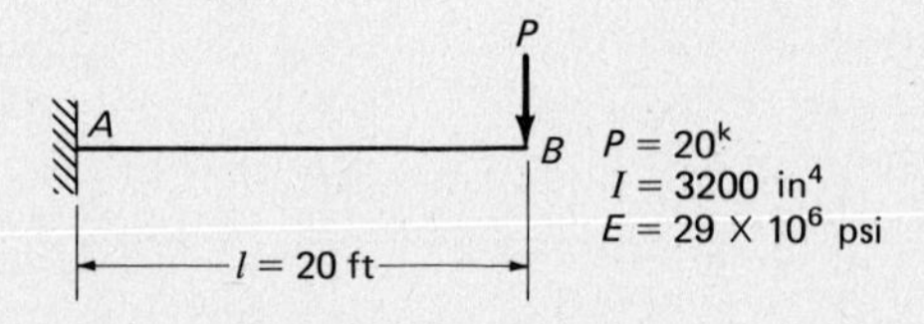

Figure 17.4

SOLUTION

A tangent to the elastic curve at the fixed end is horizontal; therefore the changes in slope and deflection of a tangent at the free end with respect to a tangent at the fixed end are the slope and deflection of that point.

M diagram Pl

$\frac{M}{EI}$ diagram $\frac{Pl}{EI}$

Slope at B equals the area of the M/EI diagram from A to B

$$\theta_B = \left(\frac{1}{2}\right)(l)\left(\frac{Pl}{EI}\right) = \frac{Pl^2}{2EI}$$

$$= \frac{(20 \times 1000)(20 \times 12)^2}{(2)(29 \times 10^6)(3200)} = 0.00621 \text{ rad} = 0.36°$$

Deflection at B equals the moment of the M/EI diagram from A to B about B

$$\delta_B = \left(\frac{1}{2}\right)(l)\left(\frac{Pl}{EI}\right)\left(\frac{2}{3}l\right) = \frac{Pl^3}{3EI}$$

$$= \frac{(20 \times 1000)(20 \times 12)^3}{(3)(29 \times 10^6)(3200)} = 0.99 \text{ in}$$

Example 17.2

Determine the slope and deflection of the beam at point B, 10 ft from the left end of the structure shown in Fig. 17.5.

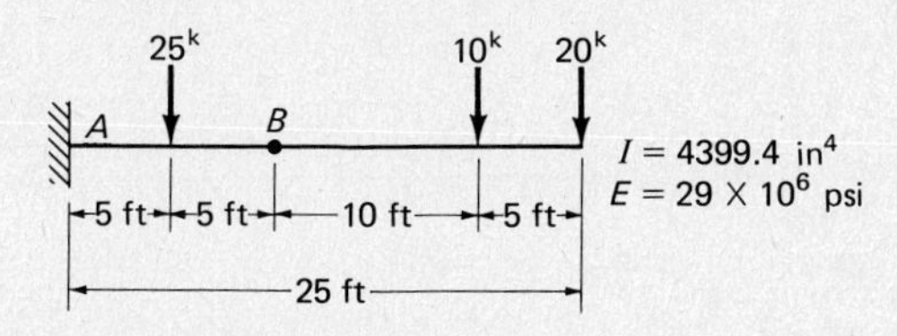

Figure 17.5

SOLUTION

The left end is again fixed; the slope at B equals the area of the M/EI diagram from A to B; and the deflection at B equals the moment of the M/EI diagram from A to B taken about B. The diagram is broken down into convenient triangles as shown for making the calculations.

10 ft

$\frac{M}{EI}$ diagram

$\frac{825'^k}{EI}$ $\frac{550'^k}{EI}$ $\frac{400'^k}{EI}$ $\frac{100'^k}{EI}$

Slope:

$$\theta_B = \frac{(\frac{1}{2})(825)(5) + (\frac{1}{2})(550)(5) + (\frac{1}{2})(550)(5) + (\frac{1}{2})(400)(5)}{EI} = \frac{5812.5 \text{ ft}^2\text{-k}}{EI}$$

$$= \frac{(5812.5)(12 \times 12)(1000)}{(29 \times 10^6)(4399.4)} = 0.00656 \text{ rad} = 0.38°$$

Deflection:

$$\delta_B = \frac{(\frac{1}{2})(825)(5)(8.33) + (\frac{1}{2})(550)(5)(6.67) + (\frac{1}{2})(550)(5)(3.33) + (\frac{1}{2})(400)(5)(1.67)}{EI}$$

$$= \frac{32{,}600 \text{ ft}^3\text{-k}}{EI} = \frac{(32{,}600)(12 \times 12 \times 12)(1000)}{(29 \times 10^6)(4399.4)} = 0.442 \text{ in}$$

Example 17.3

Determine the slope and deflection at the free end of the cantilever beam shown in Fig. 17.6.

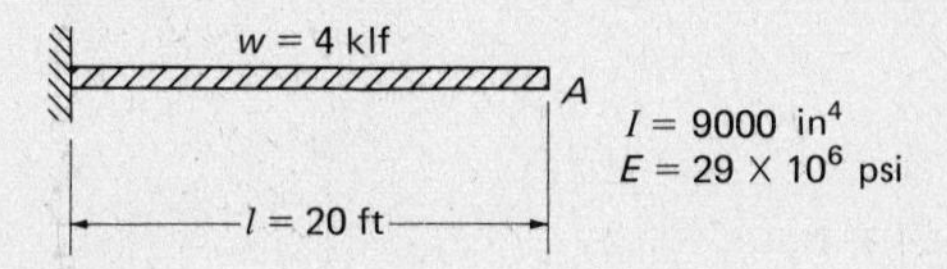

Figure 17.6

SOLUTION

Notice the units used in solving the slope and deflection equations developed. To substitute in the equations the units inches and pounds are used herein. The value of w, the uniform load, is then 4000/12 lb/in and not just 4000 lb/ft. If the reader is not careful with the units for w he or she can easily miss the slope or deflection for a particular problem by a multiple of 12.

$\frac{M}{EI}$ diagram

$\frac{wl^2}{2EI}$

Slope:

$$\theta_A = \left(\frac{1}{3}\right)(l)\left(\frac{wl^2}{2EI}\right) = \frac{wl^3}{6EI}$$

$$= \frac{(4000/12)(20 \times 12)^3}{(6)(29 \times 10^6)(9000)} = 0.00294 \text{ rad} = 0.17°$$

Deflection:

$$\delta_A = \left(\frac{1}{3}\right)(l)\left(\frac{wl^2}{2EI}\right)\left(\frac{3}{4}l\right) = \frac{wl^4}{8EI}$$

$$= \frac{(4000/12)(20 \times 12)^4}{(8)(29 \times 10^6)(9000)} = 0.530 \text{ in}$$

Example 17.4

Compute the slope and deflection at the free end of the cantilever beam shown in Fig. 17.7.

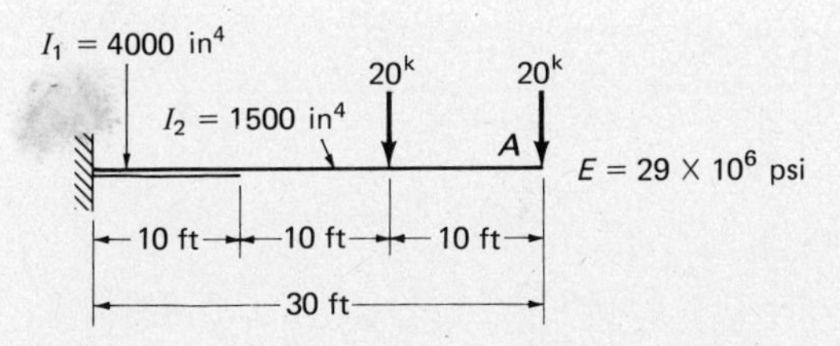

Figure 17.7

SOLUTION

The moment of inertia of the beam has been increased near the support where bending moment is greatest. The M/EI diagram is drawn by keeping the constant E as a symbol but dividing the ordinates by the proper moments of inertia. The resulting figure is conveniently divided into triangles and the computations made as before.

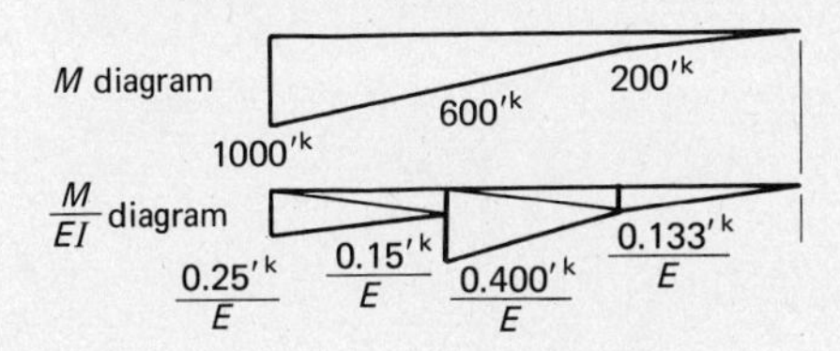

Slope:

$$\theta_A = \frac{(\frac{1}{2})(10)(0.25) + (\frac{1}{2})(10)(0.15) + (\frac{1}{2})(10)(0.400) + (\frac{1}{2})(10)(0.133)(2)}{E}$$

$$= \frac{5.333 \text{ ft}^2\text{-k}}{E} = \frac{(5.333)(144)(1000)}{29 \times 10^6} = 0.0265 \text{ rad} = 1.52°$$

Deflection:

$$\delta_A = [(\tfrac{1}{2})(10)(0.25)(26.67) + (\tfrac{1}{2})(10)(0.15)(23.33) + (\tfrac{1}{2})(10)(0.400)(16.67) + (\tfrac{1}{2})(10)(0.133)(13.33) + (\tfrac{1}{2})(10)(0.133)(6.67)]/E$$

$$= \frac{97.47 \text{ ft}^3\text{-k}}{E} = \frac{(97.47)(1728)(1000)}{29 \times 10^6} = 5.81 \text{ in}$$

Example 17.5

Compute the deflection at the centerline of the uniformly loaded simple beam shown in Fig. 17.8.

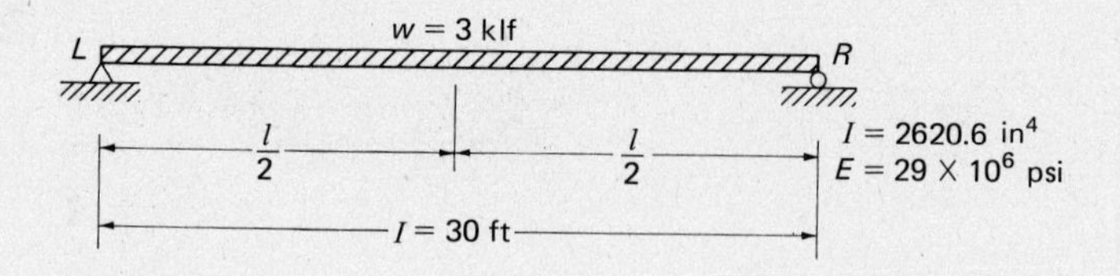

Figure 17.8

SOLUTION

The tangents to the elastic curve at each end of the beam are inclined. It is a simple matter to determine the deflection between a tangent at the centerline and one of the end tangents, but the result is not the actual deflection at the centerline of the beam. To obtain the correct deflection, it is necessary to work in a somewhat roundabout manner as follows:

1. The deflection δ_1 of the tangent at the right end R from the tangent at the left end L is found.
2. The deflection of a tangent at the centerline from a tangent at L, δ_2, is found.
3. By proportions the distance from the original chord between L and R and the tangent at L, δ_3, can be computed. The difference between δ_3 and δ_2 is the centerline deflection.

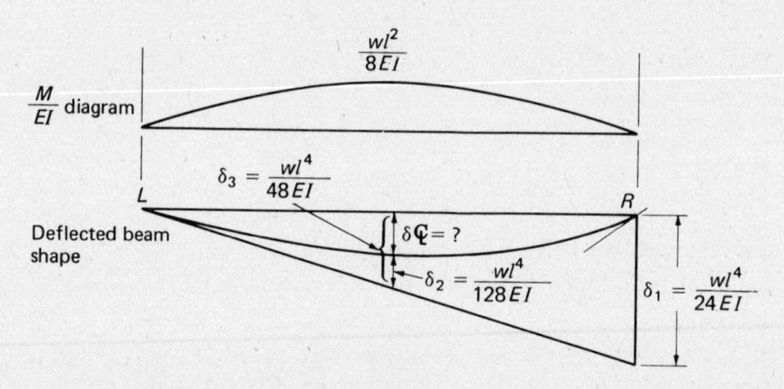

$$\delta_1 = \left(\frac{2}{3}\right)(l)\left(\frac{wl^2}{8EI}\right)\left(\frac{l}{2}\right) = \frac{wl^4}{24EI}$$

$$\delta_2 = \left(\frac{2}{3}\right)\left(\frac{l}{2}\right)\left(\frac{wl^2}{8EI}\right)\left(\frac{3}{8}\right)\left(\frac{l}{2}\right) = \frac{wl^4}{128EI}$$

$$\delta_3 = \frac{1}{2} \times \delta_1 = \frac{1}{2} \times \frac{wl^4}{24EI} = \frac{wl^4}{48EI}$$

$$\delta_{℄} = \frac{wl^4}{48EI} - \frac{wl^4}{128EI} = \frac{5wl^4}{384EI}$$

$$\delta_{℄} = \frac{(5)(3000/12)(30 \times 12)^4}{(384)(29 \times 10^6)(2620.6)} = 0.719 \text{ in}$$

NOTE: The procedure followed in this example was a general one which is applicable to many problems; however, the calculations of this particular problem could have been appreciably shortened by computing the deflection from the tangent at the centerline of the beam (which is horizontal due to symmetry) to the tangent at one of the supports by taking moments at that support as follows:

$$\delta_{\mathfrak{L}} = \left(\frac{wl^2}{8EI}\right)\left(\frac{l}{2}\right)\left(\frac{2}{3}\right)\left(\frac{5}{8}\frac{l}{2}\right) = \frac{5wl^4}{384EI}$$

Example 17.6

Determine the moments at the ends of the fixed-ended beam shown in Fig. 17.9 for which E and I are constant.

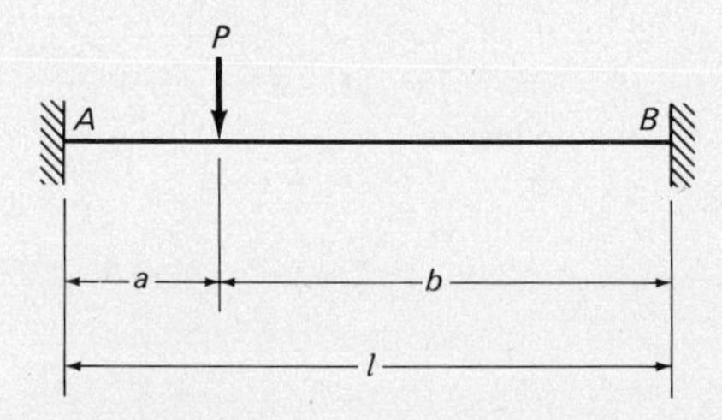

Figure 17.9

SOLUTION

Examination of the beam reveals no change of slope and no deflection of the tangent at A from the tangent at B; therefore the total area of the M/EI diagram from A to B is zero, and the moment of the M/EI diagram about either end is zero.

The M/EI diagram may be drawn in two parts: the simple beam moment diagram the ordinates of which are known, and the moment diagram due to the unknown end moments, M_A and M_B. For each of the latter moments, a triangular-shaped diagram may be drawn and the two combined into one trapezoid. The moment-area theorems are written to express the change in slope and deflection from B to A. Each of the two equations contains the two unknowns M_A and M_B, and the equations are solved simultaneously.

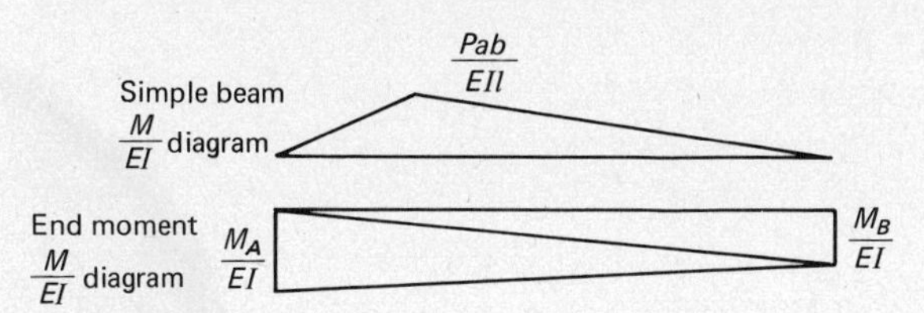

Theorem 1:

$$\left(\frac{1}{2}\right)\left(\frac{Pab}{EIl}\right)(l) + \left(\frac{1}{2}\right)\left(\frac{M_A}{EI}\right)(l) + \left(\frac{1}{2}\right)\left(\frac{M_b}{EI}\right)(l) = 0$$

$$\frac{Pab}{2EI} + \frac{M_A l}{2EI} + \frac{M_B l}{2EI} = 0 \qquad (1)$$

Theorem 2 (taking moments about left end A):

$$\left(\frac{Pab}{2EI}\right)\left(\frac{l+a}{3}\right) + \left(\frac{M_A l}{2EI}\right)\left(\frac{1}{3}l\right) + \left(\frac{M_B l}{2EI}\right)\left(\frac{2}{3}l\right) = 0$$

$$\frac{Pabl + Pa^2b}{6EI} + \frac{M_A l^2}{6EI} + \frac{M_B l^2}{3EI} = 0 \qquad (2)$$

By solving equations (1) and (2) simultaneously for M_A and M_B,

$$M_A = -\frac{Pab^2}{l^2} \qquad M_B = -\frac{Pa^2b}{l^2}$$

17.4 METHOD OF ELASTIC WEIGHTS

A careful study of the procedure used in applying the area-moment theorems will reveal a simpler and more practical method of computing slopes and deflections for most beams. In reviewing this procedure the beam and M/EI diagram of Fig. 17.10 are considered.

By letting A equal the area of the M/EI diagram, the deflection of the tangent at R from the tangent at L equals Ay, and the change in slope between the two tangents is A. An imaginary beam is loaded with the M/EI diagram, as shown in Fig. 17.11, and the reactions R_L and R_R are determined. They equal Ay/l and Ax/l, respectively.

In Fig. 17.10 the slopes of the tangents to the elastic curve at each end of the beam (θ_L and θ_R) are equal to the deflections between the tangents at each end divided by the span length, as follows:

$$\theta_L = \frac{\delta_R}{l} \qquad \theta_R = \frac{\delta_L}{l}$$

The values of δ_L and δ_R have previously been found to equal Ax and Ay, respectively, and may be substituted in these expressions.

$$\theta_L = \frac{Ay}{l} \qquad \theta_R = \frac{Ax}{l}$$

The end slopes are exactly the same as the reactions for the beam in Fig. 17.11. At either end of the fictitious beam the shear equals the reaction and thus the slope in the actual beam. Further experiments will show that the shear at any point in the beam loaded with the M/EI diagram equals the slope at that point in the actual beam.

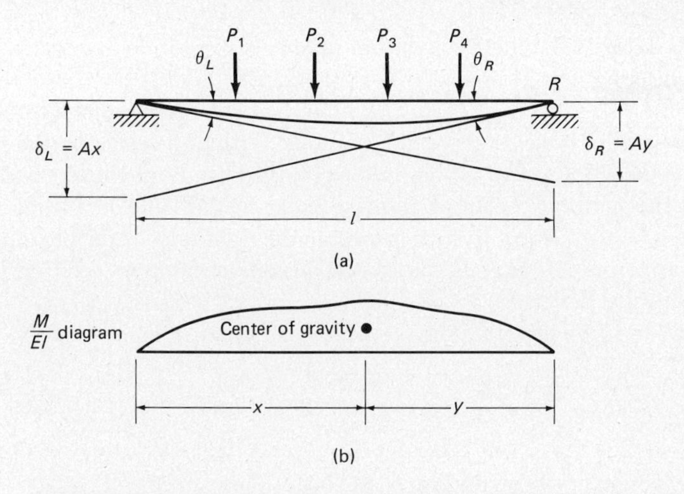

Figure 17.10

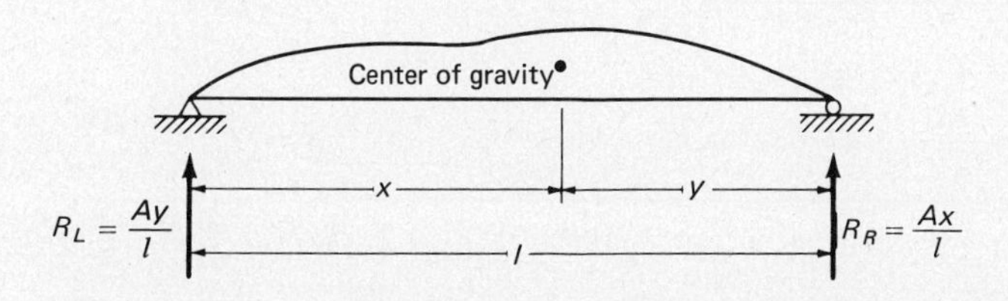

Figure 17.11

A similar argument can be made concerning the computation of deflections, and it will be found that the deflection at any point in the actual beam equals the moment at that point in the fictitious beam. In detail, the two theorems of elastic weights may be stated as follows:

1. The slope of the elastic curve of a simple beam at a point, measured with respect to a chord between the supports, equals the shear at that point if the beam is loaded with the M/EI diagram.

2. The deflection of the elastic curve of a simple beam at a point, measured with respect to a chord between the supports, equals the moment at that point if the beam is loaded with the M/EI diagram.

17.5 APPLICATION OF THE METHOD OF ELASTIC WEIGHTS

The method of elastic weights in its present form is applicable only to beams simply supported at each end. It will be found in using the method that maximum deflections in the actual beam occur at points of zero shear in the imaginary beam. The reasoning is the same as that presented for shear and moment diagrams in Sec. 4.5, where maximum moments were found to occur at points of zero shear.

Consideration has not been given to the subject of sign conventions for either the area-moment or the elastic-weight methods. With little difficulty the reader can see the directions of slopes and deflections by study of the shears and moments on the fictitious beam. A positive shear in the fictitious beam shows the left side is being pushed up with respect to the right side, or the beam is sloping downward from left to right. Similarly, a positive moment (see Fig. 4.7) indicates downward deflection.

Examples 17.7 to 17.10 illustrate the application of elastic weights.

Example 17.7

Determine the deflection at the centerline of the beam shown in Fig. 17.12.

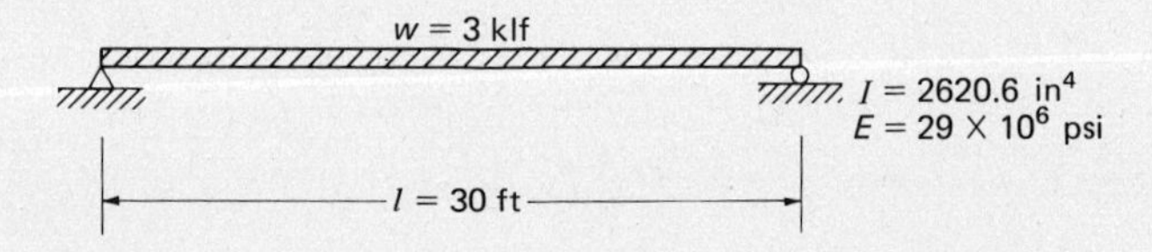

Figure 17.12

SOLUTION

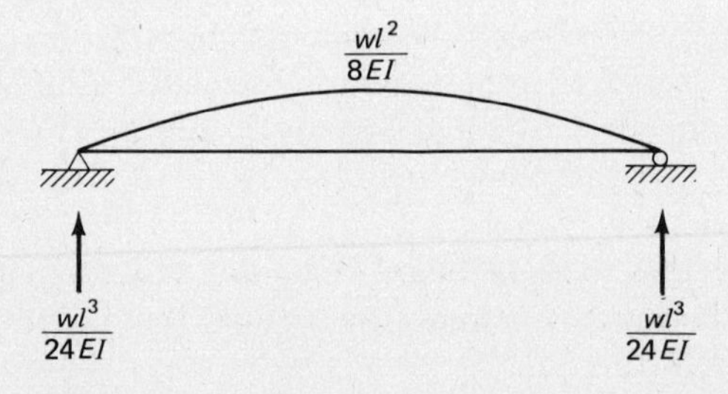

Deflection at centerline:

$$\delta_{\mathrm{CL}} = \text{moment}_{\mathrm{CL}} = \left(\frac{wl^3}{24EI}\right)\left(\frac{l}{2}\right) - \left(\frac{2}{3}\right)\left(\frac{l}{2}\right)\left(\frac{wl^2}{8EI}\right)\left(\frac{3}{8}\frac{l}{2}\right)$$

$$= \frac{5wl^4}{384EI} = \frac{(5)(3000/12)(30 \times 12)^4}{(384)(29 \times 10^6)(2620.6)} = 0.719 \text{ in}$$

Example 17.8

Determine the slope and deflection at the centerline of the beam shown in Fig. 17.13.

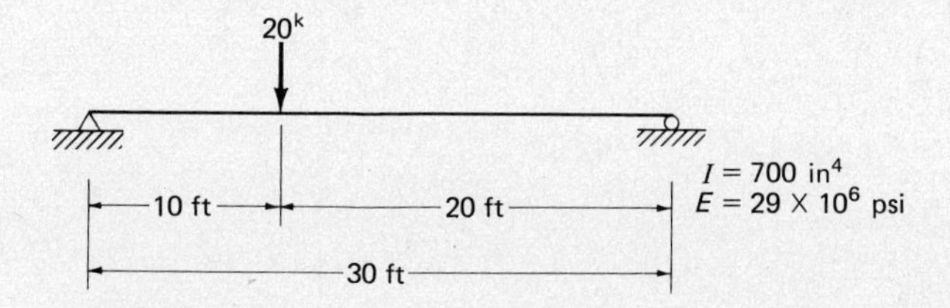

Figure 17.13

SOLUTION

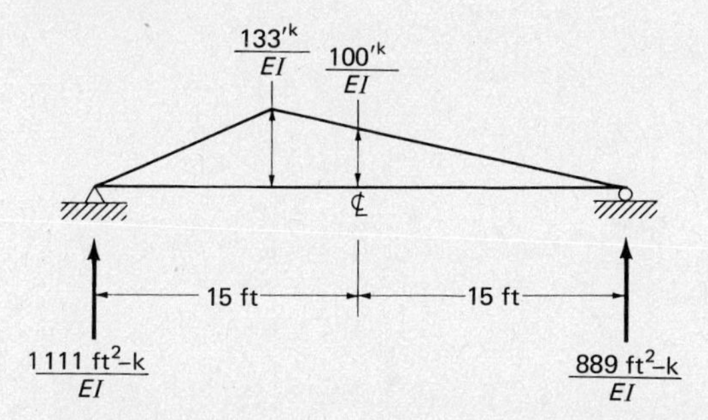

Deflection

$$\delta_{\mathrm{C\!L}} = \frac{(889)(15) - (\frac{1}{3})(100)(15)(5)}{EI} = \frac{9585\ \text{ft}^3\text{-k}}{EI}$$

$$= \frac{(9585)(1728)(1000)}{(29 \times 10^6)(700)} = 0.816\ \text{in}$$

Slope:

$$\theta_{\mathrm{C\!L}} = \frac{-889 + (\frac{1}{2})(100)(15)}{EI} = -\frac{139\ \text{ft}^2\text{-k}}{EI}$$

$$= -\frac{(139)(144)(1000)}{(29 \times 10^6)(700)} = -0.000986\ \text{rad}$$

$$= -0.056^\circ \qquad \text{(negative slope /)}$$

Example 17.9

Compute the maximim deflection for the beam shown in Fig. 17.14.

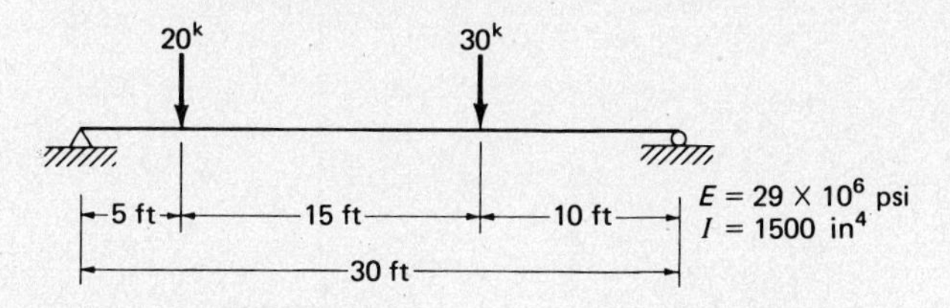

Figure 17.14

SOLUTION

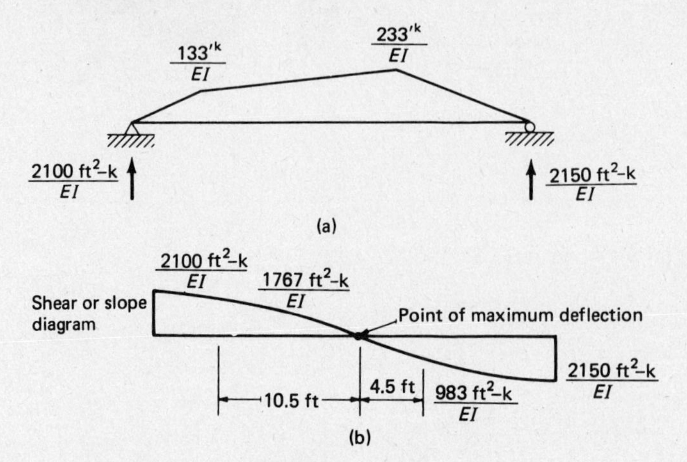

$$\delta_{\max} = \frac{(2100)(15.5) - (\frac{1}{2})(5)(133)(12.17) - (10.5)(133)(5.25) - (\frac{1}{2})(10.5)(70)(3.5)}{EI}$$

$$= \frac{19{,}890 \text{ ft}^3\text{-k}}{EI} = \frac{(19{,}890)(1728)(1000)}{(29 \times 10^6)(1500)} = 0.790 \text{ in}$$

Example 17.10

Compute the centerline deflection for the simple beam shown in Fig. 17.15.

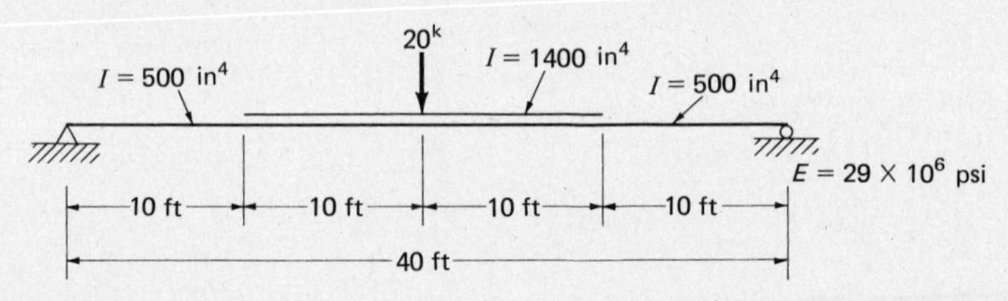

Figure 17.15

SOLUTION

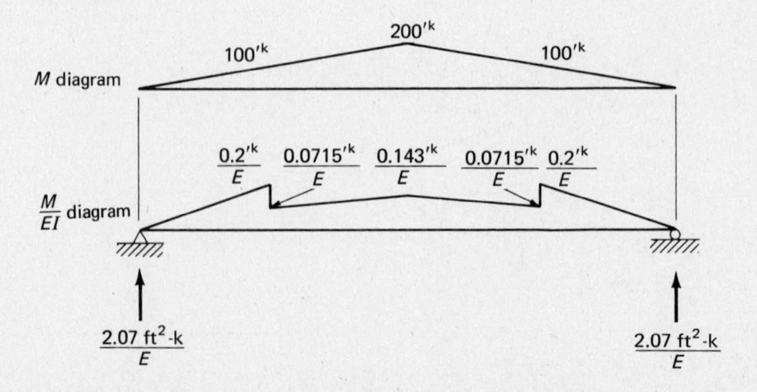

$$\delta_{\mathbb{C}} = \left(\frac{2.07}{E}\right)(20) - \left(\frac{1}{2}\right)(10)\left(\frac{0.2}{E}\right)(13.33) - (10)\left(\frac{0.0715}{E}\right)(5) - \left(\frac{1}{2}\right)(10)\left(\frac{0.0715}{E}\right)(3.33)$$

$$= \frac{23.3}{E} = \frac{(23.3)(1728)(1000)}{29 \times 10^6} = 1.39 \text{ in}$$

17.6 LIMITATIONS OF ELASTIC-WEIGHT METHOD

The method of elastic weights was developed for simple beams and in its present form will not work for cantilever beams, overhanging beams, fixed-ended beams, and continuous beams. The moment-area theorems are used to determine the correct slope and deflection at the free end of the uniformly loaded cantilever beam of Fig. 17.16.

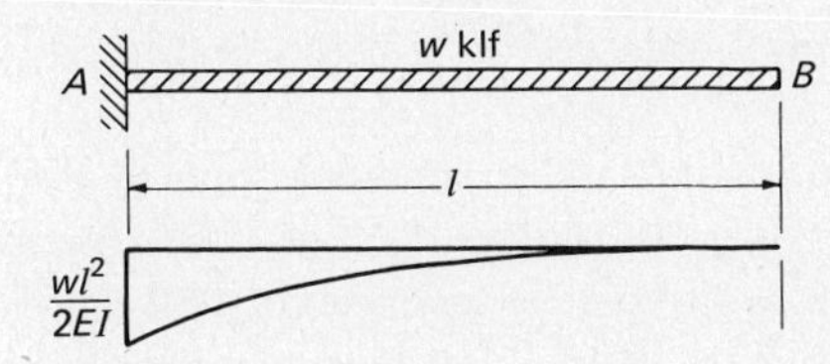

Figure 17.16

$$\theta_B = \left(\frac{1}{3}\right)(l)\left(\frac{wl^2}{2EI}\right) = \frac{wl^3}{6EI}$$

$$\delta_B = \left(\frac{1}{3}\right)(l)\left(\frac{wl^2}{2EI}\right)\left(\frac{3}{4}l\right) = \frac{wl^4}{8EI}$$

If the elastic-weight method were used in an attempt to find the slope and deflection at the ends of the same beam, the result would be slopes and deflections of zero at the free end and $wl^3/6EI$ and $wl^4/24EI$ at the fixed end, as shown in Fig. 17.17.

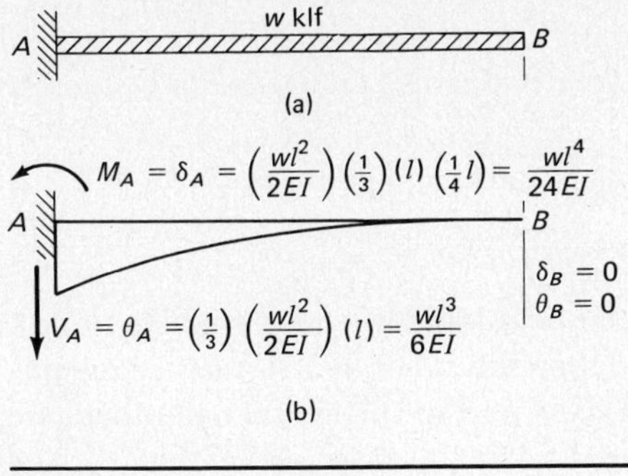

w klf

A B

(a)

$M_A = \delta_A = \left(\frac{wl^2}{2EI}\right)\left(\frac{1}{3}\right)(l)\left(\frac{1}{4}l\right) = \frac{wl^4}{24EI}$

A B

$\delta_B = 0$

$\theta_B = 0$

$V_A = \theta_A = \left(\frac{1}{3}\right)\left(\frac{wl^2}{2EI}\right)(l) = \frac{wl^3}{6EI}$

(b)

Figure 17.17

The slope and deflection at the fixed end A must be zero; but application of elastic weights to the beam results in both shear and moment, falsely indicating slope and deflection.

If the fixed end of the beam were moved to the free end and the resulting beam loaded with the M/EI diagram, the shears and moments would correspond exactly to the slopes and deflections on the actual beam as found by the moment-area method.

17.7 CONJUGATE-BEAM METHOD

The conjugate-beam method makes use of an "analogous" or "conjugate" beam to be handled by elastic weights in place of the actual beam to which it cannot be correctly applied. The shear and moment in the imaginary beam, loaded with the M/EI diagram, must correspond exactly with the slope and deflection of the actual beam.

The correct mathematical relationship is obtained for a beam simply end supported if it is loaded "as is" with the M/EI diagram. If the elastic-weight method is applied to other types of beams, the largest moments due to the M/EI loading occur at the supports, incorrectly indicating that the largest deflections occur at those points. For elastic weights to be applied correctly, use must be made of substitute beams or conjugate beams that have the supports changed so the correct relationships are obtained.

The loads and properties of the true beam have no effect on the manner in which the conjugate beam is supported. The only factors affecting the supports of the imaginary beam are the supports of the actual beam. The lengths of the two beams are equal. In the following paragraphs is a discussion of what the various types of beam support must become in the conjugate beam so that the elastic-weight method will apply. The mathematical proof of these relationships is explained in detail in texts on strength of materials.

Free End

The free end of a beam slopes and deflects when the beam is loaded. The conjugate beam must have both shear and moment at that end when it is loaded with the M/EI diagram. The only type of end support having both shear and moment is the fixed end. **A free end in the actual beam becomes a fixed end in the conjugate beam.**

Fixed End

A similar discussion in reverse order can be made for a fixed end. No slope or deflection can occur at a fixed end, and there must not be any shear or moment in the conjugate beam at that point. **A fixed end in the actual beam becomes a free end in the conjugate beam.**

Simple End Support

A simple end slopes but does not deflect when the beam is loaded. The imaginary beam will have shear but no moment at that point, a situation that can occur only at a simple support. **A simple end support in the actual beam remains a simple end support in the conjugate beam.**

Simple Interior Support

There is no deflection at either a simple interior or a simple end support. Both types may slope when the beam is loaded, but the situations are somewhat different. The slope at a simple interior support is continuous across the support; that is, no sudden change of slope occurs. This condition is not present at a simple end support where the slope suddenly begins. (See the deflection curve for the beam of Fig. 17.18.) If there is no change of slope at a simple interior support, there can be no change of shear at the corresponding support in the conjugate beam. Any type of external support at this point would cause a change in the shear; therefore an internal pin (or unsupported hinge) is required. **A simple interior support in the actual beam becomes an unsupported internal hinge in the conjugate beam.**

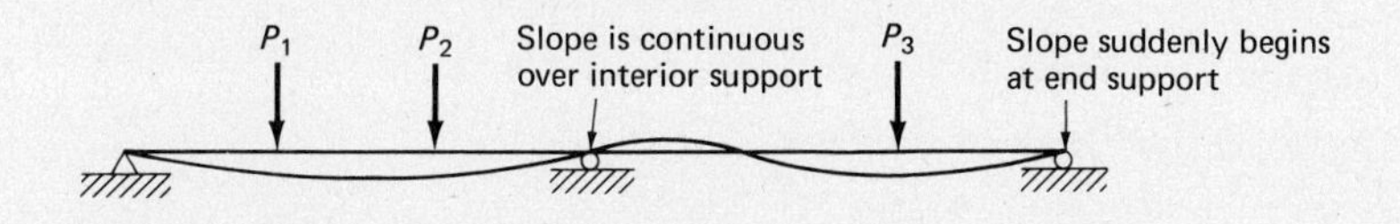

Figure 17.18

Internal Hinge

At an unsupported internal hinge there is both slope and deflection, which means that the corresponding support in the conjugate beam must have shear and moment. **An internal hinge in the actual beam becomes a simple support in the conjugate beam.**

Summary

Figure 17.19 shows several types of common beams and their corresponding conjugates.

Equilibrium

The reactions, moments, and shears of the conjugate beam are easily computed by statics because the conjugate beam is always statically determinate even though the real beam may be statically indeterminate. Sometimes the conjugate beam may appear to be completely unstable. The most conspicuous

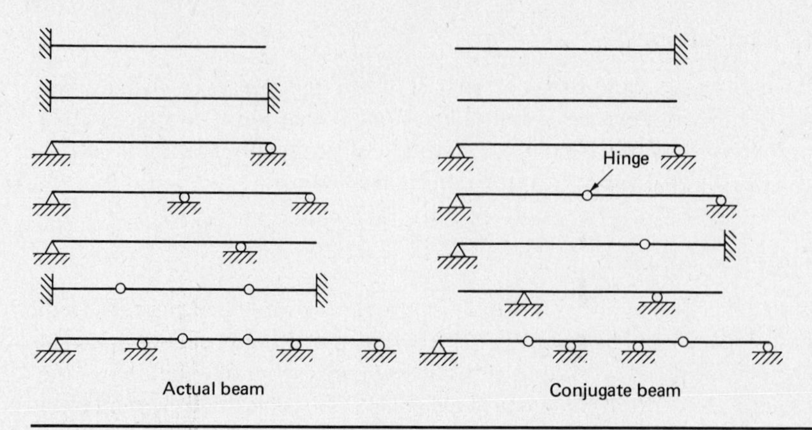

Figure 17.19

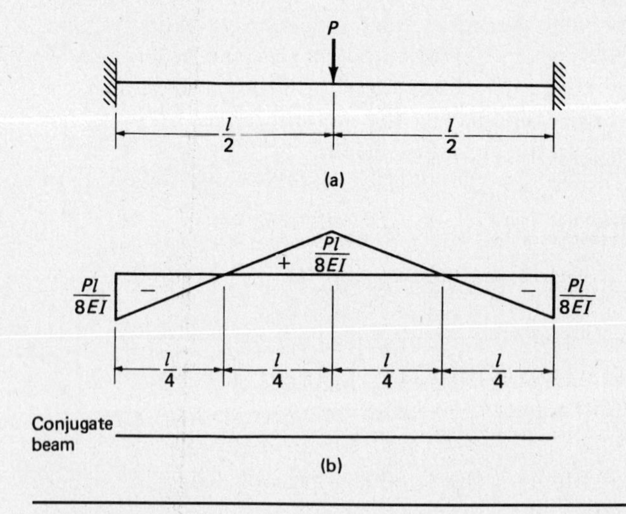

Figure 17.20

example is the conjugate beam for the fixed-end beam (Fig. 17.20), which has no supports whatsoever. On second glance the areas of the M/EI diagram are seen to be so precisely balanced between downward and upward loads (positive and negative areas of the diagram, respectively) as to require no supports. Any supports, seemingly required, would have zero reactions, and the proper shears and moments are supplied to coincide with the true slopes and deflections. Even a real beam continuous over several simple supports has a conjugate that is simply end supported.

17.8 SUMMARY OF BEAM RELATIONS

A brief summary of the relations that exist between loads, shears, moments, slope changes, slopes, and deflections is presented in Fig. 17.21. The relations

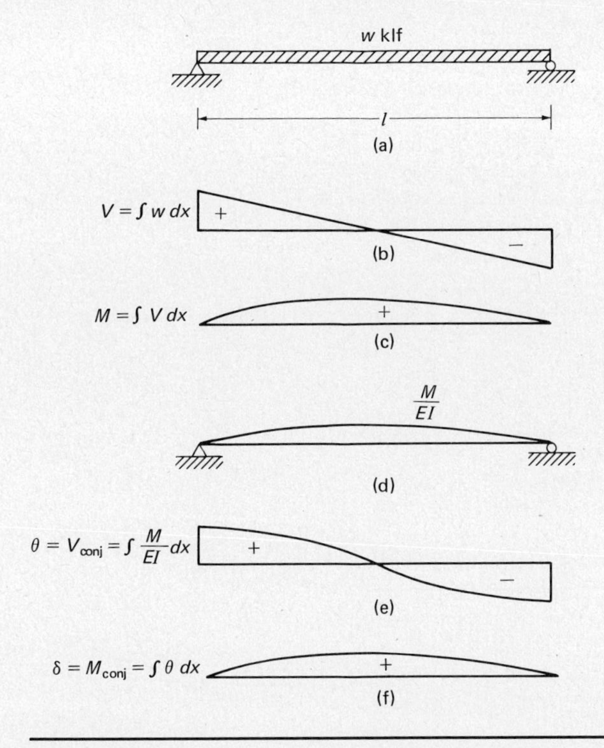

Figure 17.21

are shown for a uniformly loaded beam but are applicable to any type of loading. For the two sets of curves shown, the ordinate on one curve equals the slope at that point on the following curve. It is obvious from these figures that the same mathematical relations that exist between load, shear, and moment hold for M/EI loading, slope, and deflection.

17.9 APPLICATION OF CONJUGATE METHOD TO BEAMS

Examples 17.11 and 17.12 illustrate the conjugate method of calculating slopes and deflections for beams. The procedure as to symbols and units used in applying the method is in general the same as that used for the moment-area and elastic-weight methods. Maximum deflections occur at points of zero shear on the conjugate structure. For example, the point of zero shear in the beam of Fig. 17.20 is the centerline. The deflection is as follows:

$$\delta_{\mathbb{C}} = \text{moment}_{\mathbb{C}}$$
$$= \left(\frac{1}{2}\right)\left(\frac{l}{4}\right)\left(\frac{Pl}{8EI}\right)\left(\frac{5}{12}l\right) - \left(\frac{1}{2}\right)\left(\frac{l}{4}\right)\left(\frac{Pl}{8EI}\right)\left(\frac{l}{12}\right)$$
$$= \frac{Pl^3}{192EI}$$

Example 17.11

Determine the slope and deflection of point A, Fig. 17.22.

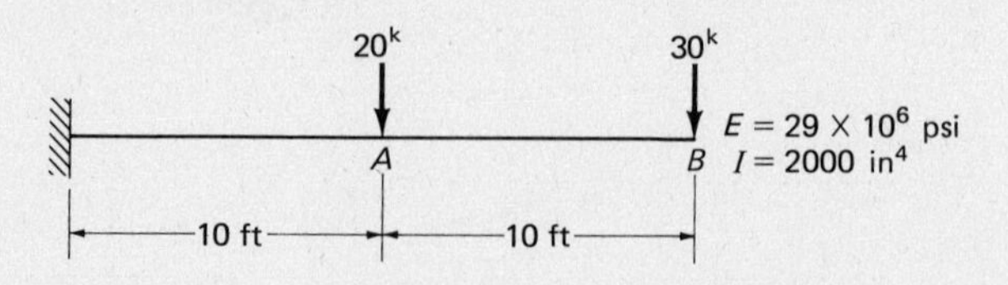

Figure 17.22

SOLUTION

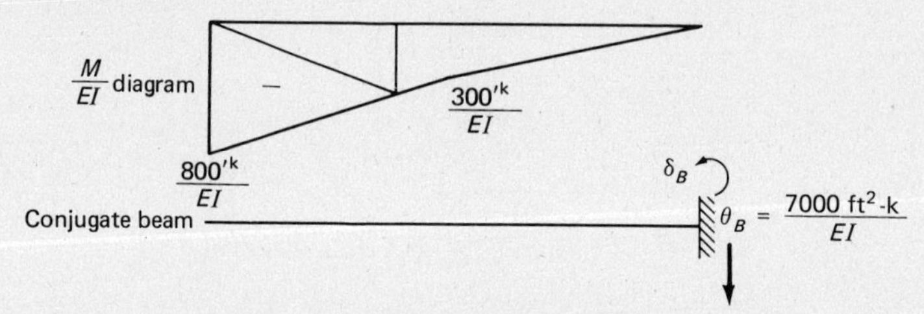

Slope:

$$\theta_A = \frac{(\frac{1}{2})(300)(10) + (\frac{1}{2})(800)(10)}{EI} = \frac{5500 \text{ ft}^2\text{-k}}{EI}$$

$$= \frac{(5500)(144)(1000)}{(29 \times 10^6)(2000)} = 0.0136 \text{ rad} = 0.78° \backslash$$

Deflection:

$$\delta_A = \frac{(\frac{1}{2})(300)(10)(3.33) + (\frac{1}{2})(800)(10)(6.67)}{EI} = \frac{31{,}667 \text{ ft}^3\text{-k}}{EI}$$

$$= \frac{(31{,}667)(1728)(1000)}{(29 \times 10^6)(2000)} = 0.943 \text{ in} \downarrow$$

Example 17.12

Determine deflections at points A and B in the overhanging beam of Fig. 17.23.

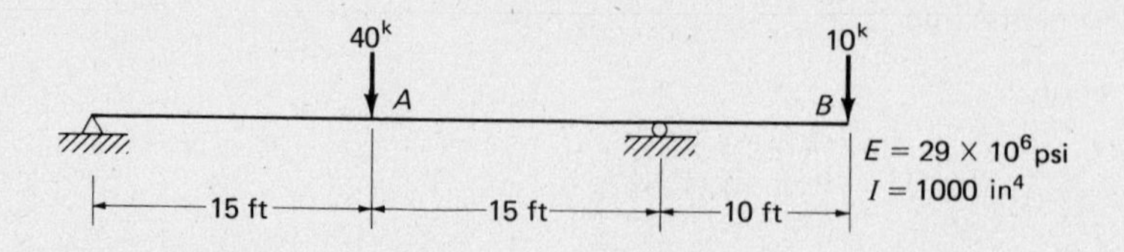

Figure 17.23

SOLUTION

The M/EI diagram is drawn and placed on the conjugate beam, which has an interior hinge. The reactions are determined as they were for the cantilever-type structures of Chap. 3. The portion of the beam to the left of the hinge is considered as a simple beam, and its reactions are determined. The reaction at the hinge is applied as a concentrated load acting at the end of the cantilever to the right of the hinge in the opposite direction, and the reactions at the fixed end are determined. To simplify the mathematics, a separate moment diagram is drawn for each of the concentrated loads.

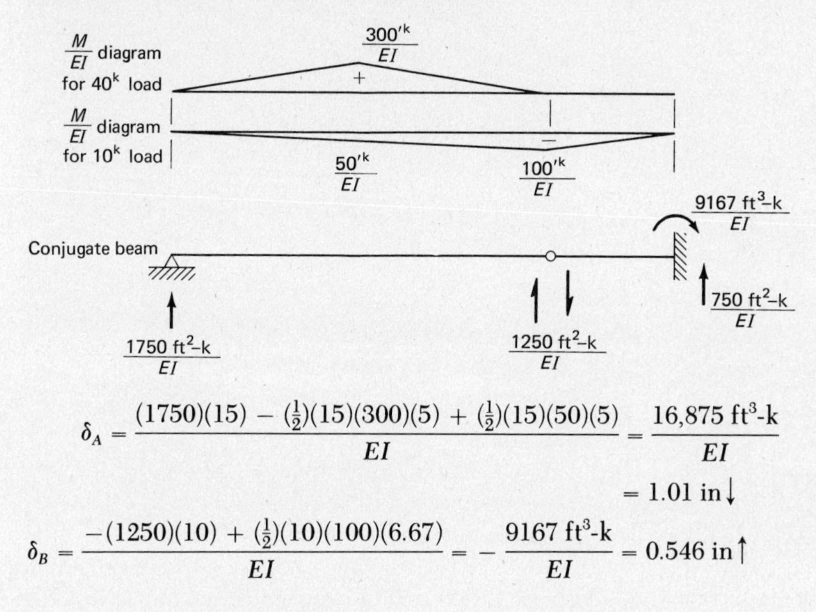

$$\delta_A = \frac{(1750)(15) - (\frac{1}{2})(15)(300)(5) + (\frac{1}{2})(15)(50)(5)}{EI} = \frac{16{,}875 \text{ ft}^3\text{-k}}{EI}$$

$$= 1.01 \text{ in} \downarrow$$

$$\delta_B = \frac{-(1250)(10) + (\frac{1}{2})(10)(100)(6.67)}{EI} = -\frac{9167 \text{ ft}^3\text{-k}}{EI} = 0.546 \text{ in} \uparrow$$

17.10 COMMENTS ON MODULI OF ELASTICITY

In computing deflections the analyst needs to pay careful attention to the values of the moduli of elasticity. For some materials, such as structural steel, its value is practically constant whereas for other materials, particularly concrete, its value may vary appreciably depending on several factors. For instance the modulus of elasticity for concrete varies with its strength, age, type of loading, cement, and aggregate characteristics, as well as with the definition of the modulus (tangent, initial, or secant). The result is that it is almost impossible to predict accurately the modulus of elasticity for a given concrete [1].

When a concrete member is loaded in compression it will have an immediate or elastic shortening. Should the load be left in place for an appreciable length of time the shortening will increase a great deal, perhaps several times the initial shortening. To obtain realistic deflections for reinforced concrete structures it is therefore necessary to separate the sustained or long-term loads from the short-term loads and use one value of the modulus for

Precast concrete, Kitchens of Sara Lee, Inc., Deerfield, Illinois. (Courtesy of the Portland Cement Association.)

the short-term loads and a much smaller value for the sustained loads. The result of the two deflection calculations can then be combined. The sustained loads for a building would include the dead load plus some percentage of the live load. For an apartment house or for an office building perhaps only 20 to 25% of the live load should be considered as being sustained, while as much as 70 to 80% of the live load of a warehouse might fall into this category.

A similar discussion can be made for timber structures. Seasoned timber members subjected to long-term loads develop a permanent deformation or sag approximately equal to twice the deflection computed for short-term loads of the same magnitude [2].

17.11 APPLICATION OF CONJUGATE METHOD TO FRAMES

The rigid frames of Fig. 17.24 consist of members rigidly connected at their joints. The joints are moment resisting and prevent the frame members from having freedom of rotation.

The slopes and deflections of frames such as these can be computed by the conjugate method as they were for beams. The procedure, however, is confusing and the ***author does not recommend its use for the usual frame.*** The virtual-work method (Chap. 18) and Castigliano's first theorem (Chap. 21) provide simpler and more logical methods for frames.

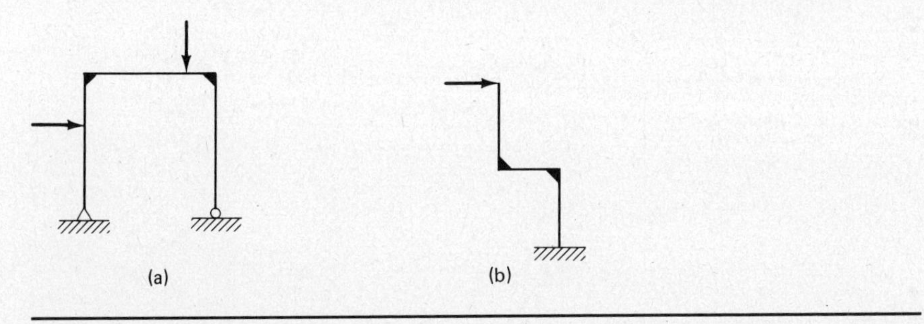

Figure 17.24

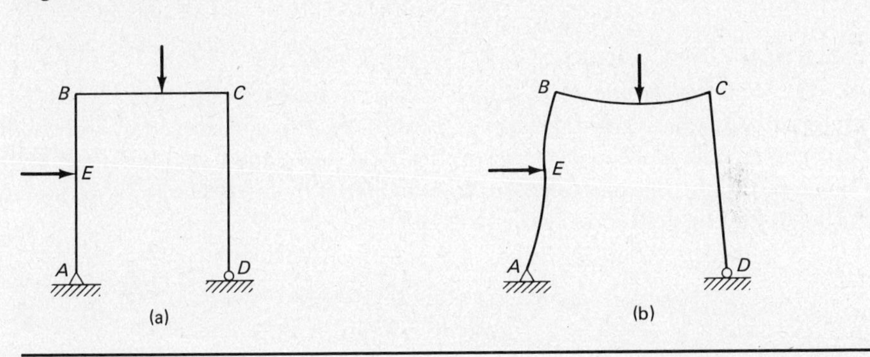

Figure 17.25

Should the conjugate method be applied to frames there will be additional factors (from those encountered in beams) which must be included in the calculations.

Consider the frame and loads of Fig. 17.25(a) and let the deflection of point E be desired. Figure 17.25(b) shows a sketch of the estimated deformed shape of the frame. It is seen that there are two factors affecting the total deflection at E, these being the deflection at E if joint B were prevented from moving laterally plus the effect of the actual movement to the right of joint B. The frame sways to the right, which causes E to deflect an additional amount. Similarly, to find the total change in slope or the rotation of a particular joint in a complicated frame, it may be necessary to consider the rotations of several joints.

The foregoing discussion has shown that the calculation of deflections for a rigid frame by the usual conjugate-beam procedure may involve the taking of moments, the structure being loaded with the M/EI diagram, plus a complicated consideration of joint rotations. Slope and deflection are determined for an elementary frame in Example 17.13 in this manner.

Example 17.13

Determine the deflections at points B and D of the frame shown in Fig. 17.26.

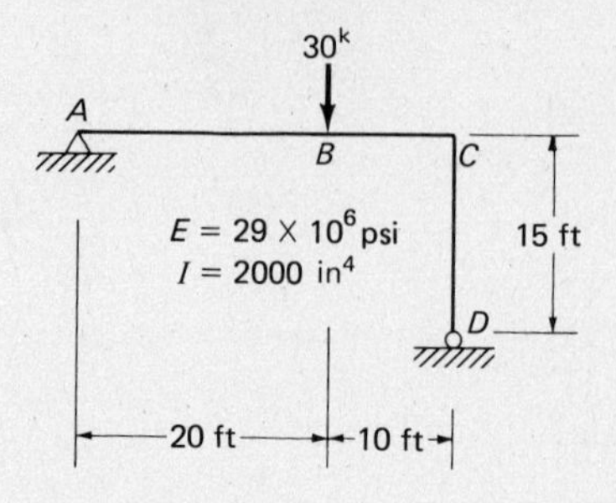

Figure 17.26

SOLUTION

A sketch of the estimated deflected shape of the structure and the M/EI diagram are drawn. The deflection of point B is easily determined by the usual procedure. The deflected shape of the structure shows that member CD remains straight, because it has no moment, and it inclines outward at an angle equal to θ_c. The deflection at D equals the length of CD times θ_c.

30k

θ_C

Deflected shape

θ_C

$\frac{1333\ \text{ft}^2\text{-k}}{EI}$

δ_D

$\frac{200'^{k}}{EI}$

$\frac{M}{EI}$ diagram

$\frac{1667\ \text{ft}^2\text{-k}}{EI}$

$$\delta_B = \frac{(1333)(20) - (\frac{1}{2})(20)(200)(6.67)}{EI} = \frac{13{,}333\ \text{ft}^3\text{-k}}{EI} = 0.397\ \text{in}\downarrow$$

$$\theta_C = \frac{1667\ \text{ft}^2\text{-k}}{EI} = 0.00414\ \text{rad}$$

$$\delta_D = (15 \times 12)(0.00414) = 0.745\ \text{in}\rightarrow$$

The conjugate method for frames is quite popular with some engineers. Readers who are interested in the topic should refer to Kinney [3] where it is fully described.

PROBLEMS

Using the moment-area method determine the quantities asked for each of Probs. 17.1 to 17.13.

Problem 17.1 θ_A, δ_A: $E = 29 \times 10^6$ psi, $I = 1000$ in^4

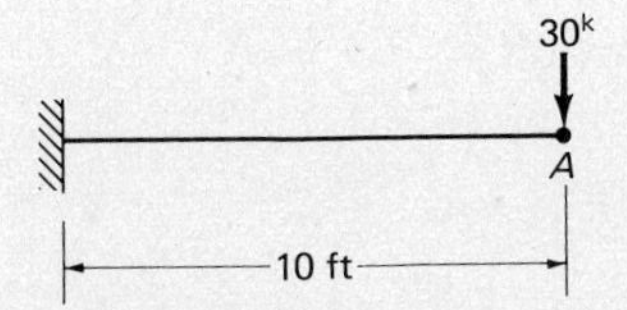

Problem 17.2 $\theta_A, \theta_B, \delta_A, \delta_B$: $E = 1.5 \times 10^6$ psi, $I = 1500$ in^4 (*Ans.* $\theta_A = \theta_B = 0.0384$ rad/, $\delta_A = 5.36$ in↓, $\delta_B = 3.07$ in↓)

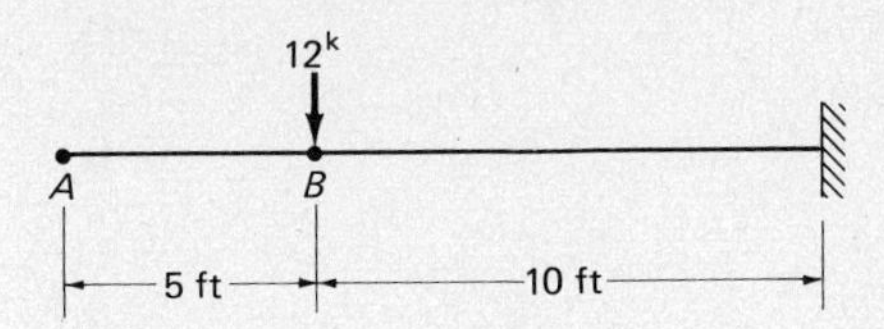

Problem 17.3 $\theta_A, \theta_B, \delta_A, \delta_B$: $E = 29 \times 10^6$ psi, $I = 4100$ in^4

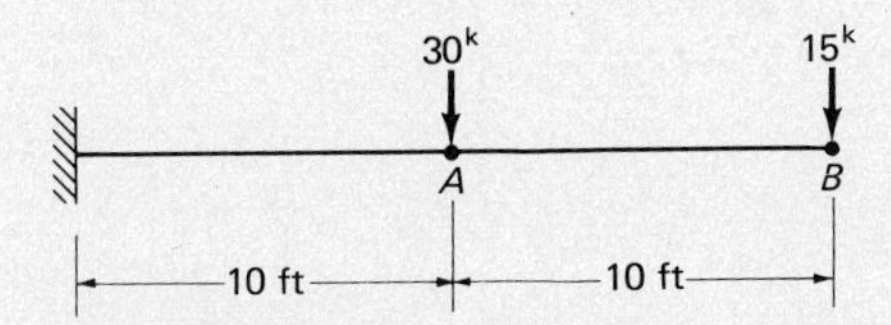

Problem 17.4 θ_A, δ_A: $E = 29 \times 10^6$ psi, $I = 843$ in^4 (*Ans.* $\theta_A = 0.00338$ rad/, $\delta_A = 0.366$ in↓)

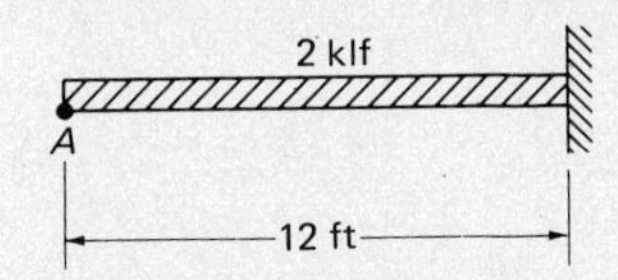

Problem 17.5 $\theta_A, \theta_B, \delta_A, \delta_B$: $E = 29 \times 10^6$ psi, $I = 1140$ in^4

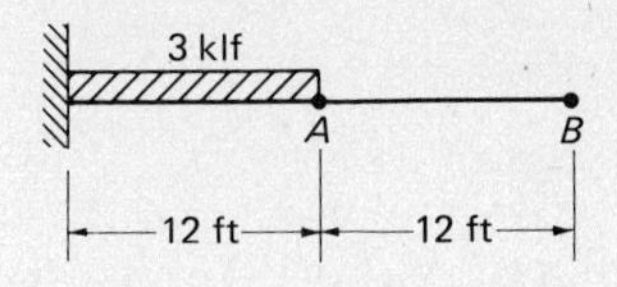

Problem 17.6 $\theta_A, \theta_B, \delta_A, \delta_B$: $E = 29 \times 10^6$ psi, $I = 4000$ in^4 (*Ans.* $\theta_A = 0.00435$ rad\, $\theta_B = 0.00538$ rad\, $\delta_A = 0.296$ in↓, $\delta_B = 0.906$ in↓)

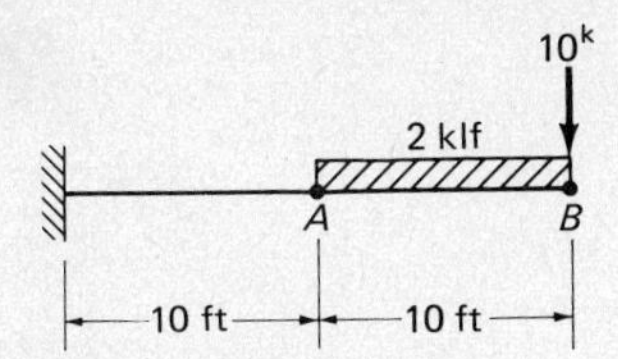

Problem 17.7 θ_A, θ_B, θ_C, δ_A, δ_B, δ_C: $E = 29 \times 10^6$ psi, $I = 1330$ in^4

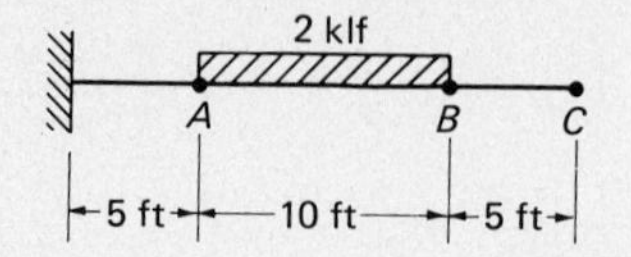

Problem 17.8 θ_A, θ_B, δ_A, δ_B: $E = 29 \times 10^6$ psi (*Ans.* $\theta_A = 0.00443$ rad\, $\theta_B = 0.00643$ rad\, $\delta_A = 0.389$ in $\downarrow$, $\delta_B = 1.08$ in $\downarrow$)

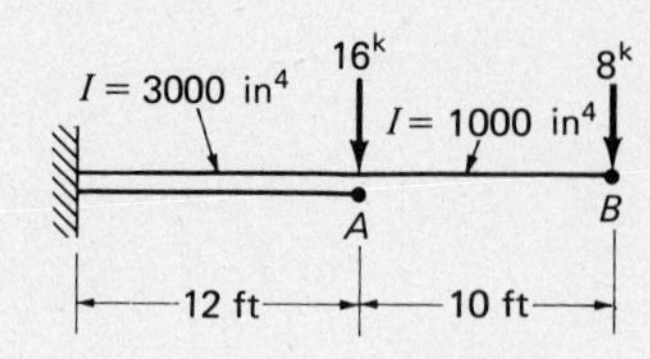

Problem 17.9 θ_A, θ_B, δ_A, δ_B: $E = 29 \times 10^6$ psi

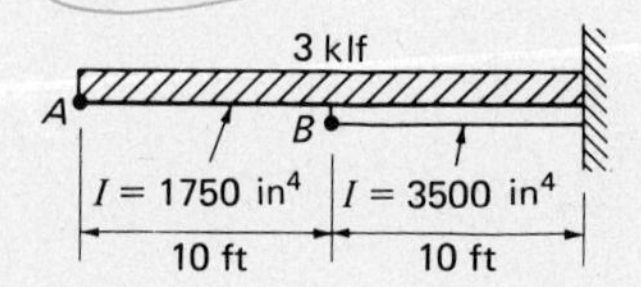

Problem 17.10 θ_A, δ_A: $E = 200\ 000$ MPa, $I = 3.0 \times 10^8$ mm^4 (*Ans.* $\theta_A = 0.01$ rad\, $\delta_A = 11.67$ mm$\downarrow$)

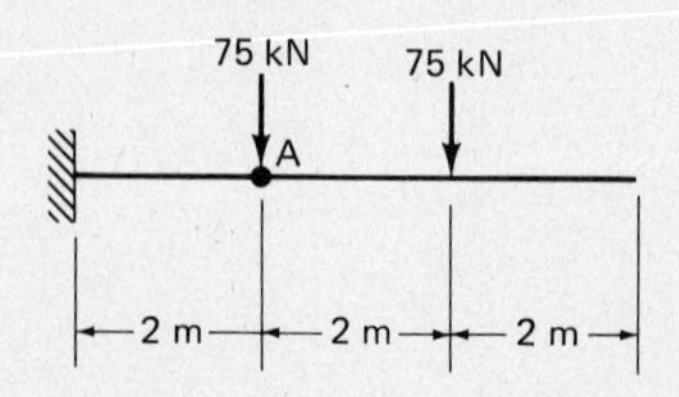

Problem 17.11 θ_A, δ_A: $E = 200\ 000$ MPa

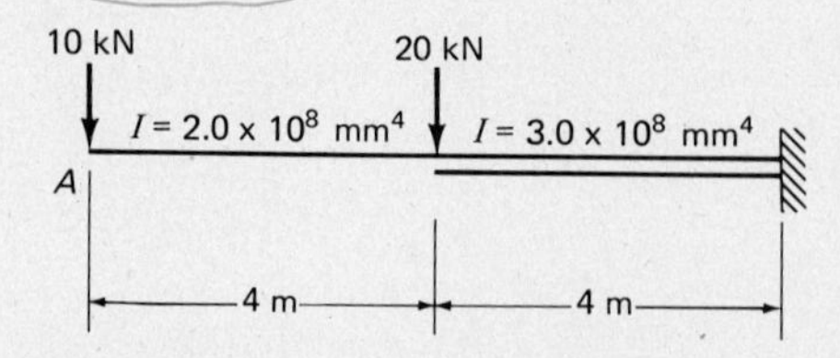

Problem 17.12 θ_A, δ_A: $E = 29 \times 10^6$ psi, $I = 1500$ in^4 (*Ans.* $\theta_A = 0.00331$ rad/, $\delta_A = 0.594$ in$\downarrow$)

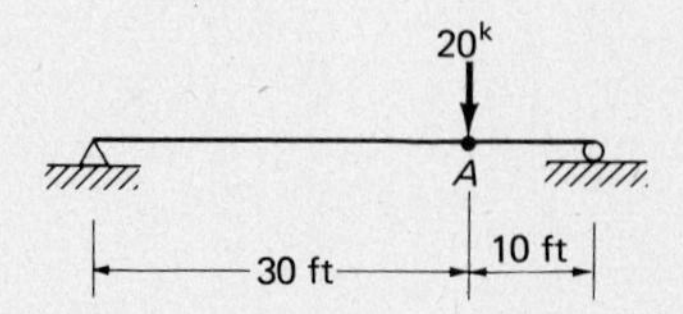

Problem 17.13 δ_A: $E = 29 \times 10^6$ psi, $I = 3200$ in^4

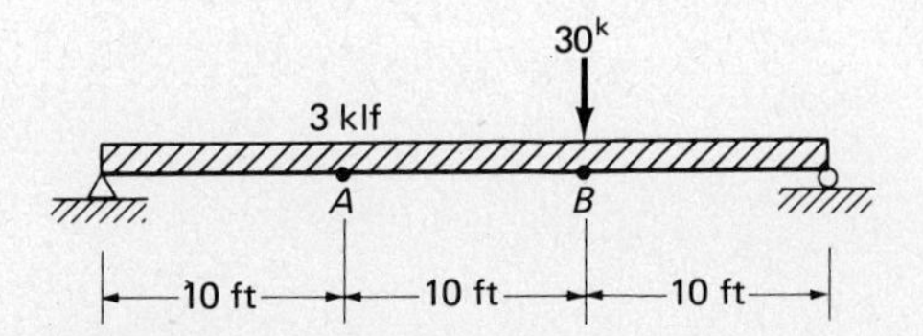

For Probs. 17.14 to 17.18 compute the fixed-end moments for the beams; E and I constant except as shown.

Problem 17.14 (*Ans.* $M_A = -225'^{k}$, $M_B = -75'^{k}$)

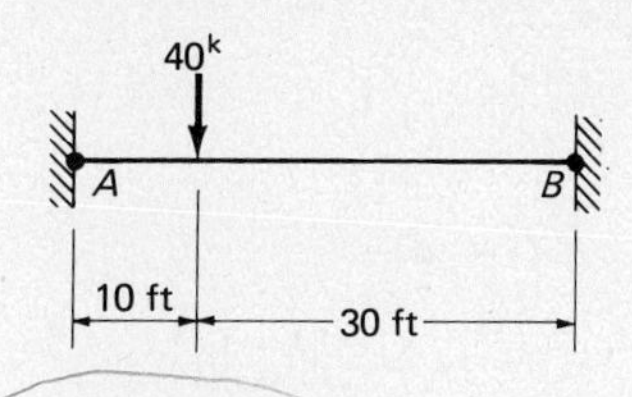

Problem 17.15

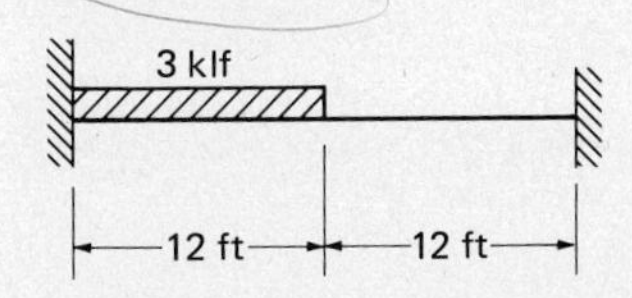

Problem 17.16 (*Ans.* $M_A = M_B = -366.6'^{k}$)

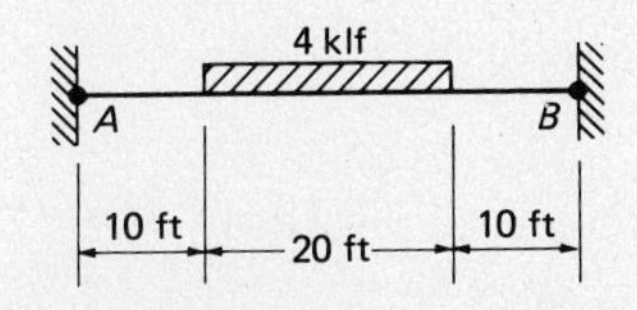

Problem 17.17

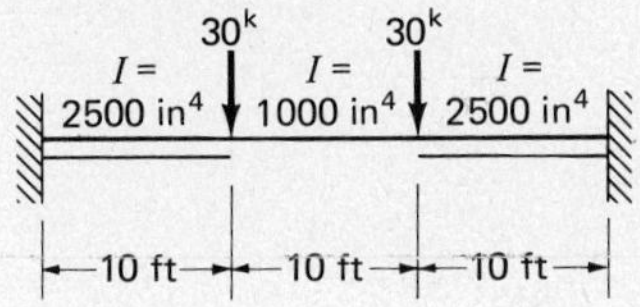

Problem 17.18 (*Ans.* $M_A = -325'^{k}$, $M_B = -105'^{k}$)

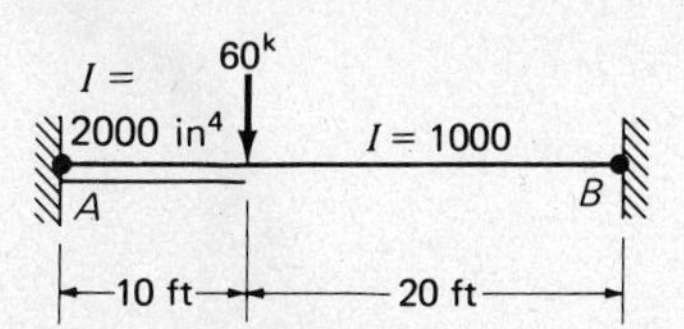

Use the conjugate-beam method for solving Probs. 17.19 to 17.44.

Problem 17.19 θ_A, δ_A: $E = 29 \times 10^6$ psi, $I = 1800$ in^4

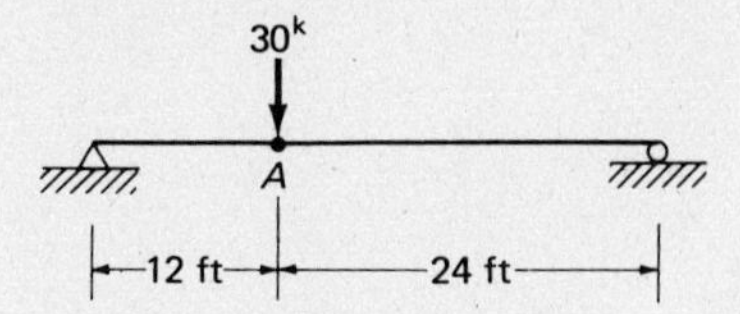

17.20. Determine the maximum deflection in the beam of Prob. 17.19. (*Ans.* 0.832 in↓)

Problem 17.21 θ_A, θ_B, δ_A, δ_B: $E = 29 \times 10^6$ psi, $I = 4000$ in^4

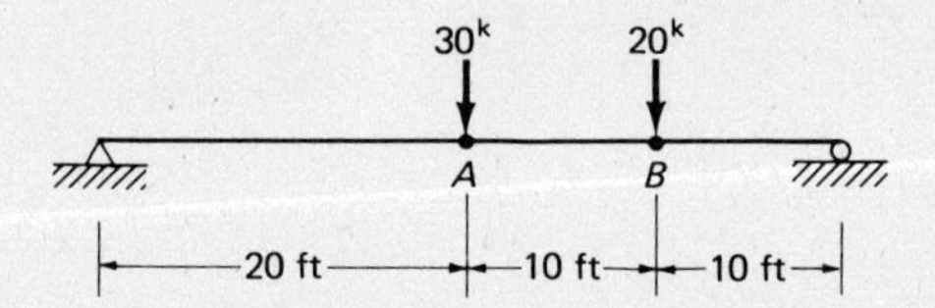

17.22. Determine the maximum deflection in the beam of Prob. 17.21. (*Ans.* 0.869 in↓)

17.23. Determine the slope and deflection at points A and B for the beam of Prob. 17.13.

Problem 17.24 θ_A, δ_A: $E = 29 \times 10^6$ psi, $I = 2370$ in^4 (*Ans.* $\theta_A = 0.00130$ rad/, $\delta_A =$ 1.08 in↓)

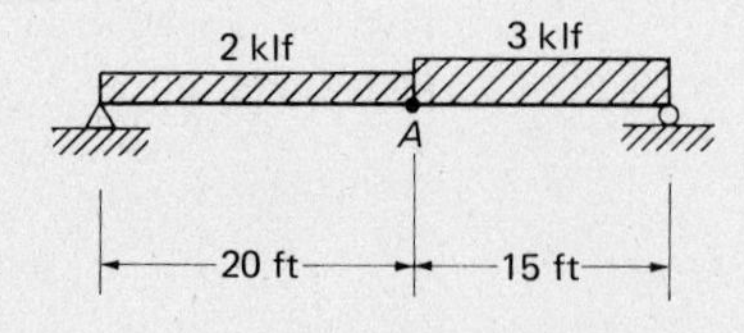

Problem 17.25 θ_A, θ_B, δ_A, δ_B: $E = 29 \times 10^6$ psi, $I = 1820$ in^4

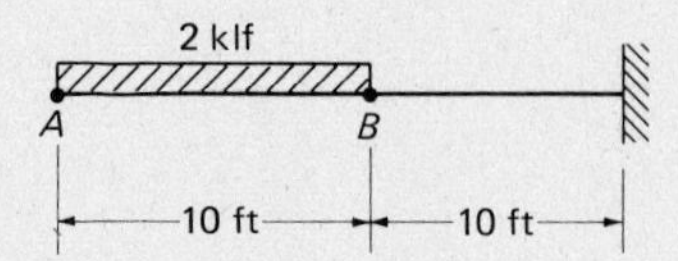

Problem 17.26 θ_A, δ_A: $E = 200\ 000$ MPa, $I = 6.0 \times 10^8$ mm^4 (*Ans.* $\theta_A = 0.002$ 25 rad\, $\delta_A = 11.25$ mm↓)

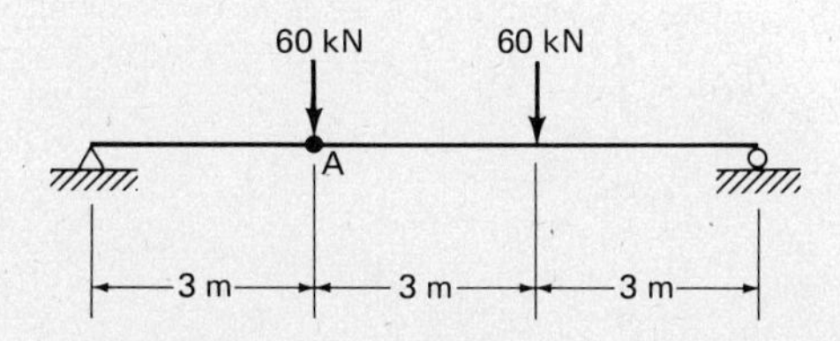

Problem 17.27 θ_A, δ_A: $E = 200\ 000$ MPa

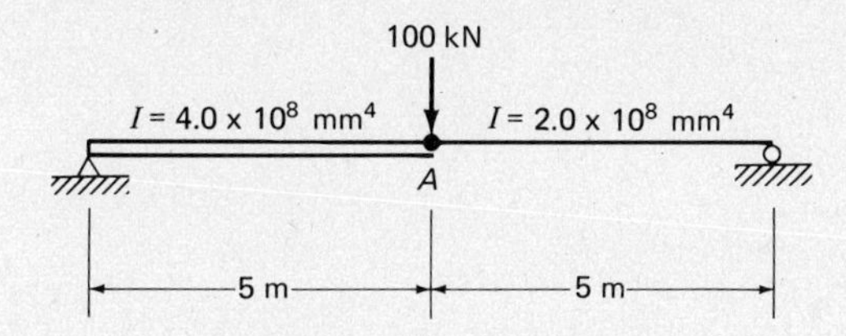

Problem 17.28 θ_A, δ_A: $E = 29 \times 10^6$ psi, $I = 1600$ in^4 (*Ans.* $\theta_A = 0.00535$ rad/, $\delta_A =$ 1.23 in↓)

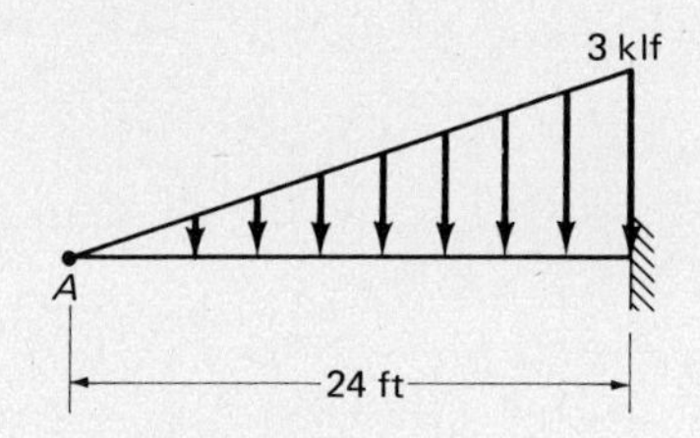

Problem 17.29 θ_A, δ_A: $E = 29 \times 10^6$ psi, $I = 513$ in^4

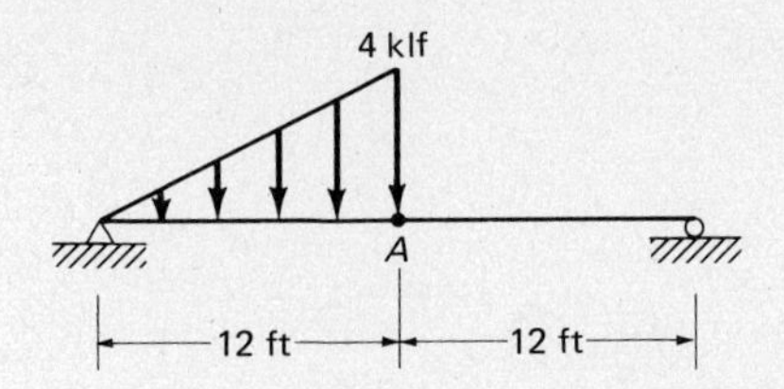

17.30. Repeat Prob. 17.13 if the beam consists of reinforced concrete with $I =$ 12,400 in^4. Assume that the uniform load is a long-term dead load and the 30-kip concentrated load is a short-term live load. Assume $E_{\text{short term}} = 3.12 \times 10^6$ psi and $E_{\text{long term}} = 1.56 \times 10^6$ psi. (*Ans.* $\delta_A = 2.98$ in↓)

Problem 17.31 $\theta_A, \theta_B, \delta_A, \delta_B$: $E = 29 \times 10^6$ psi

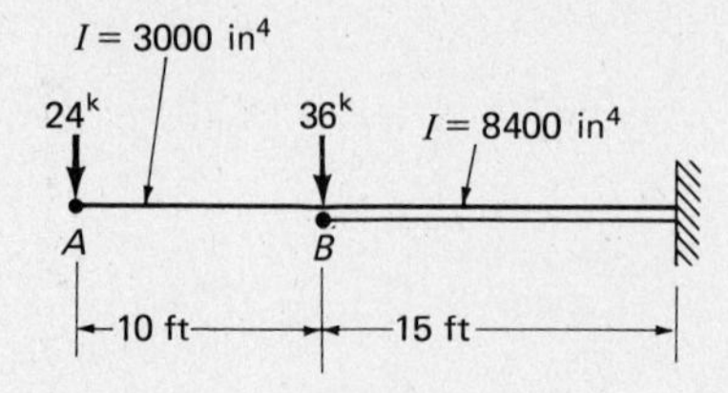

Problem 17.32 θ_A, δ_A: $E = 29 \times 10^6$ psi (*Ans.* $\theta_A = 0.00351$ rad\, $\delta_A = 0.588$ in↓)

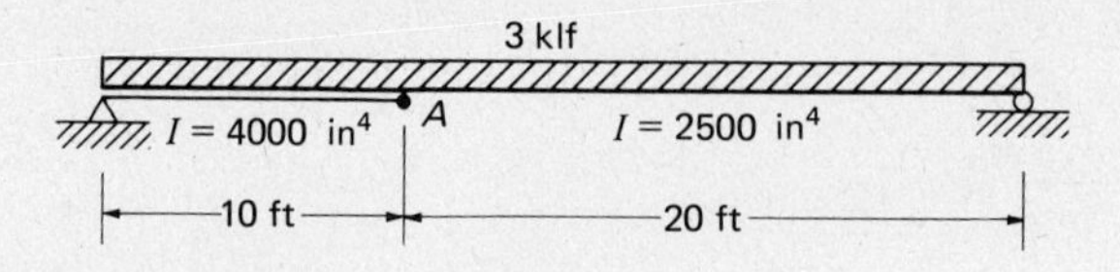

Problem 17.33 θ_A, δ_A: $E = 29 \times 10^6$ psi

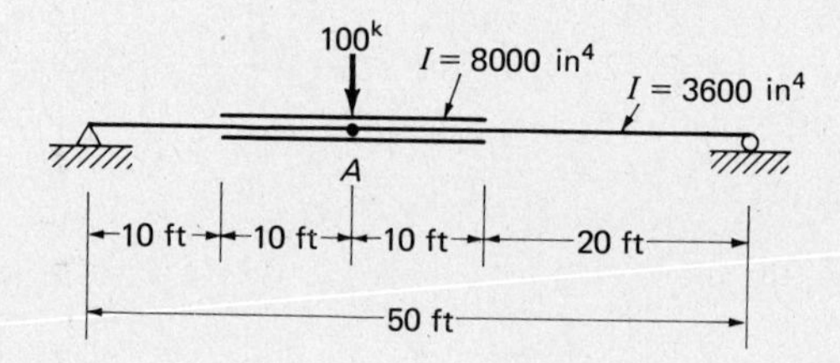

Problem 17.34 θ_A, δ_A: $E = 29 \times 10^6$ psi, $I = 1750$ in^4 (*Ans.* $\theta_A = 0.000532$ rad\, $\delta_A = -0.0493$ in↑)

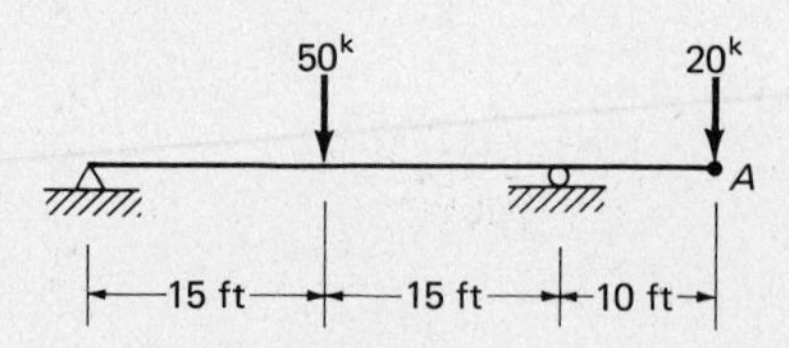

Problem 17.35 $\theta_A, \theta_B, \delta_A, \delta_B$: $E = 29 \times 10^6$ psi, $I = 1820$ in^4

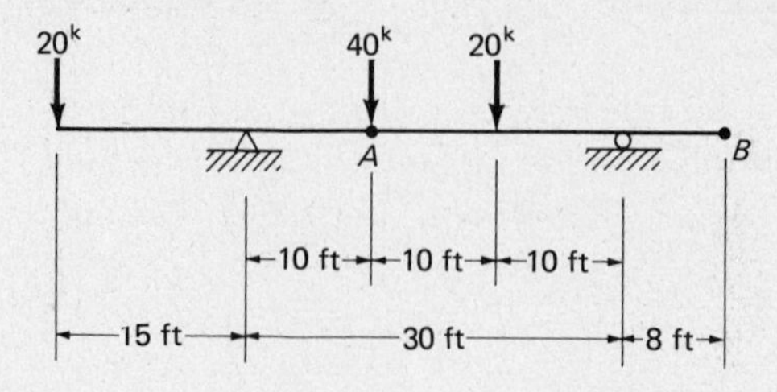

Problem 17.36 θ_A, δ_A: $E = 29 \times 10^6$ psi, $I = 1500$ in^4 (*Ans.* $\theta_A = 0.00234$ rad\, $\delta_A = 0.157$ in↓)

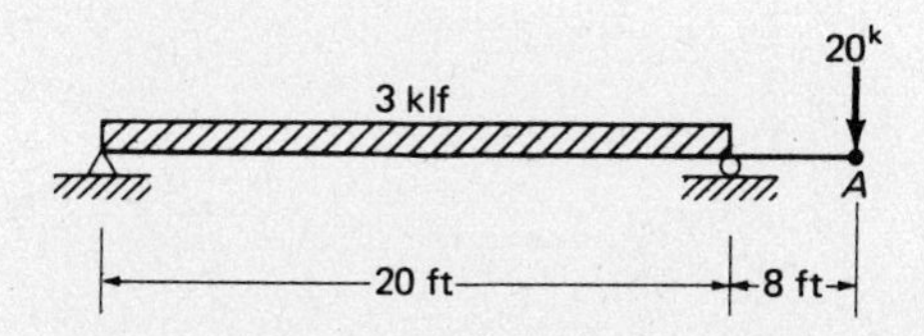

Problem 17.37 θ_A, θ_B, δ_A, δ_B: $E = 29 \times 10^6$ psi

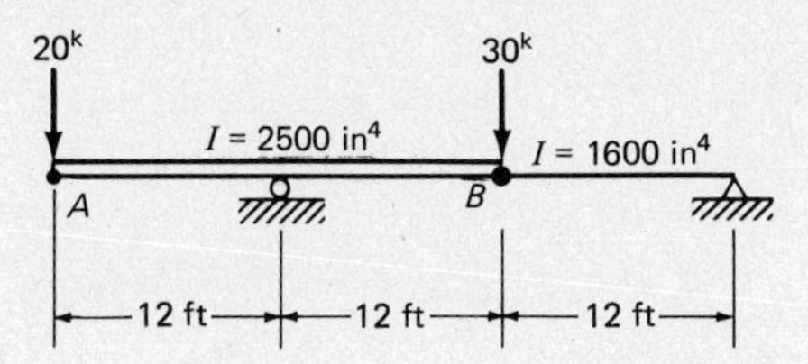

Problem 17.38 δ_A, θ_B: $E = 29 \times 10^6$ psi, $I = 1600$ in^4 (*Ans.* $\delta_A = 0.122$ in↓, $\theta_B = 0.00194$ rad\)

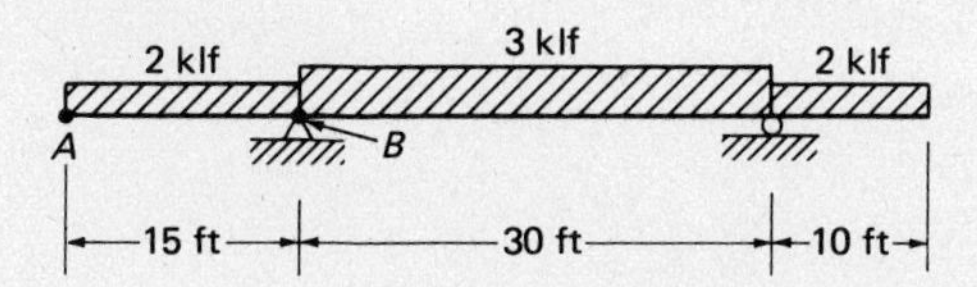

Problem 17.39 θ_A, θ_B, δ_A, δ_B: $E = 29 \times 10^6$ psi, $I = 1500$ in^4

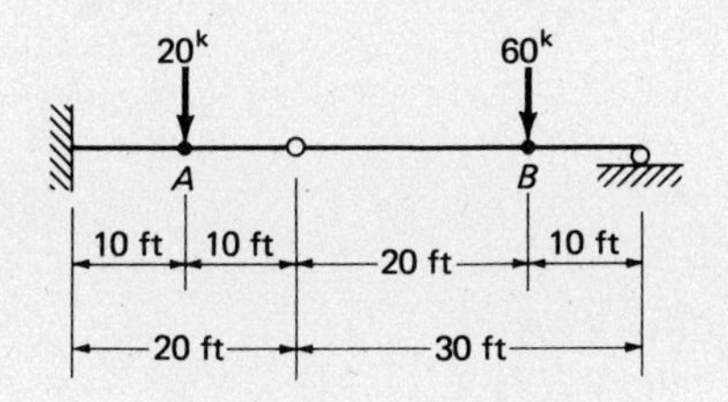

Problem 17.40 θ_A, δ_A: Reactions given, $E = 29 \times 10^6$ psi, $I = 2690$ in^4 (*Ans.* $\theta_A = 0.000383$ rad/, $\delta_A = 0.0262$ in↑)

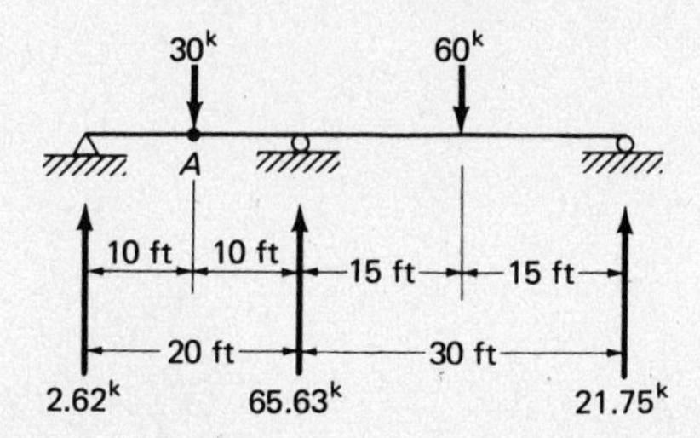

Problem 17.41 θ_A, δ_B: $E = 29 \times 10^6$ psi, $I = 2830$ in^4

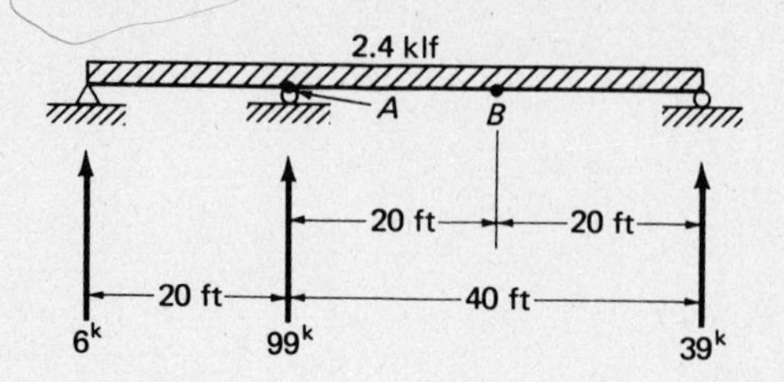

17.42. Determine slope and deflection at the 100-kip load in the beam shown in the accompanying illustration. The beam depth varies as shown. (Suggestion in drawing M/EI diagram: Take the part of the beam for which the depth varies and divide it into 2-ft sections and use the I at the center of each of the sections.) $E = 3.12 \times 10^6$ psi.

Problem 17.42 (*Ans.* $\theta_{100} = 0.0248$ rad\, $\delta_{100} = 2.57$ in$\downarrow$)

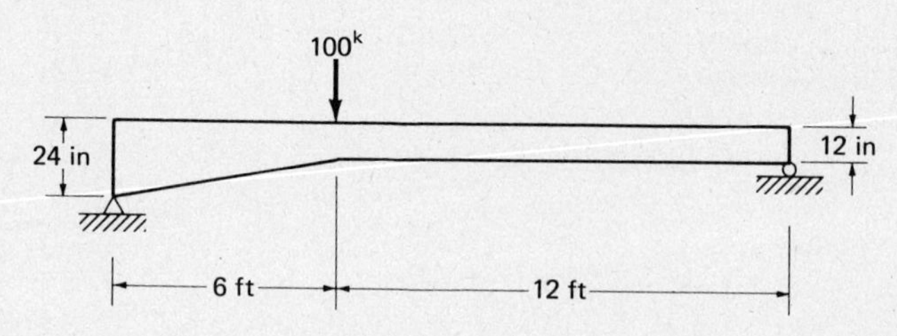

Problem 17.43 θ_A, δ_A: $E = 29 \times 10^6$ psi, $I = 1200$ in^4

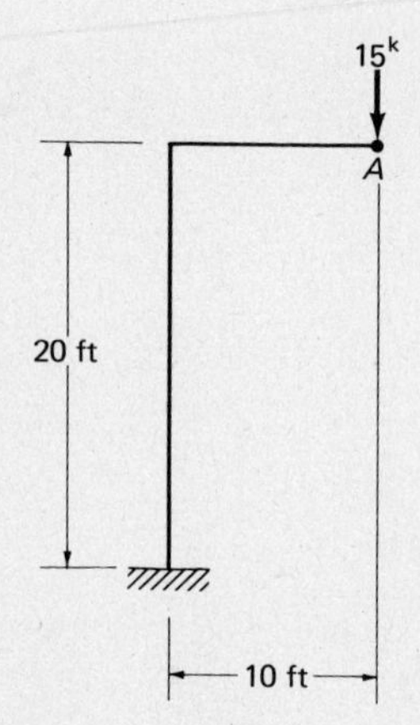

Problem 17.44 δ_{vert} at A, δ_{horiz} at B: $E = 29 \times 10^6$ psi, $I = 1400$ in^4 (*Ans.* $\delta_A = 1.21$ in$\downarrow$, $\delta_B = 0.382$ in$\leftarrow$)

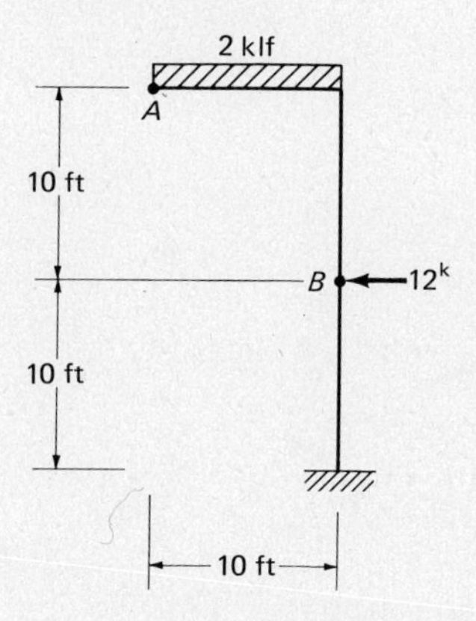

REFERENCES

[1] T. Y. Lin, *Design of Prestressed Concrete Structures* (New York: Wiley, 1963), p. 36.

[2] W. F. Scofield and W. H. O'Brien (revised by W. A. Oliver), *Modern Timber Engineering* (New Orleans, La.: Southern Pine Association, 1963), p. 97.

[3] J. S. Kinney, *Indeterminate Structural Analysis* (Reading, Mass.: Addison-Wesley), pp. 143–162.

18 Deflections, Continued

18.1 VIRTUAL WORK

The virtual-work method, often referred to as the method of work or dummy unit-load method, has the widest range of application of any of the deflection methods. It applies equally well to trusses, frames, and beams, being particularly applicable to trusses.

Virtual work is based on the law of conservation of energy, according to which the work done by a set of external loads gradually applied to a structure equals the internal elastic energy stored in the structure. To make use of this law in the derivations to follow, it is necessary that the following assumptions be made:

1. The external and internal forces are in equilibrium.
2. The elastic limit of the material is not exceeded.
3. There is no movement of the supports.

18.2 TRUSS DEFLECTIONS BY VIRTUAL WORK

The truss of Fig. 18.1 will be considered for this discussion. Loads P_1 to P_3 are applied to the truss, as shown, and cause forces in the truss members. Each member of the truss shortens or lengthens depending on the character of its force. These internal deformations cause external deflections, and each of the external loads moves through a short distance. The law of conservation of energy as it applies to the truss may now be stated in detail. The external work performed by the loads P_1 to P_3, as they move through their respective truss deflections, equals the internal work performed by the member forces as they move through their respective changes in length.

To write an expression for the internal work performed by a truss member, it is necessary to develop an expression for the deformation of the member. For this purpose the bar of Fig. 18.2 is considered.

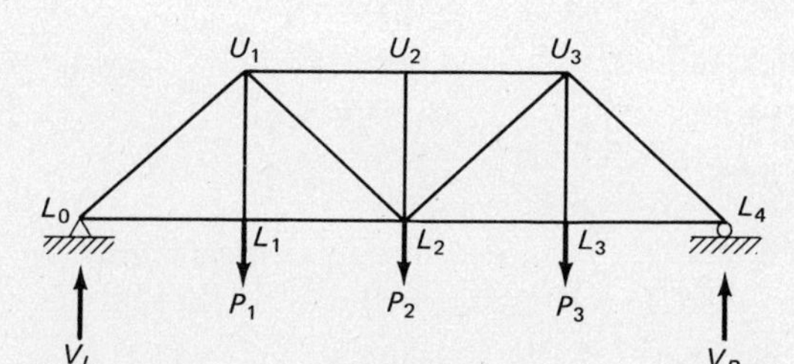

Figure 18.1

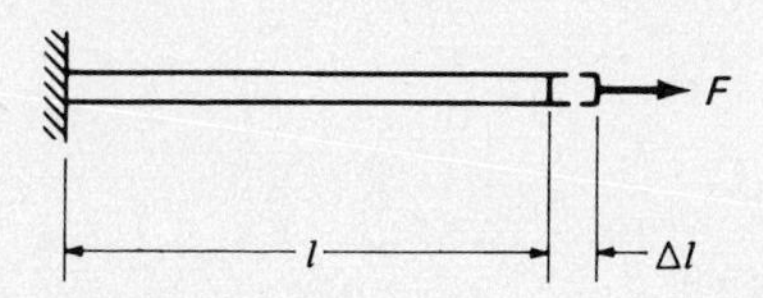

Figure 18.2

The force applied to the bar causes it to elongate by an amount Δl. The elongation may be computed from the properties of the bar. The unit elongation ϵ is equal to the total elongation divided by the length of the bar, and is also equal to the stress intensity divided by the modulus of elasticity. An expression for Δl may be developed as follows:

$$E = \frac{f}{\epsilon} = \frac{F/A}{\Delta l/l}$$

$$\Delta l = \frac{Fl}{AE}$$

In accordance with previous assumptions, the members of a truss have only axial forces. These forces are referred to as F forces, and each member will change in length by an amount equal to its Fl/AE value.

It is desired that an expression for the deflection at a joint in the truss of Fig. 18.1 be developed. A convenient means of developing such an expression is to remove the external loads from the truss, place a unit load at the joint where deflection is desired, replace the external loads, and write an expression for the internal and external work performed by the unit load and its forces when the external loads are replaced.

The forces caused in the truss members by the unit load are called μ forces. They cause small deformations of the members and small external deformations of the truss. When the external loads are returned to the truss, the force in each of the members changes by the appropriate F force, and the deformation of each member changes by its Fl/AE value. The truss deflects and the unit load is carried through a distance δ. The external work performed by the unit load when the external loads are returned to the structure may be expressed as follows:

$$W_e = 1 \times \delta$$

Internally the μ force in each member is carried through a distance $\Delta l = Fl/AE$. The internal work performed by all of the μ forces as they move through these distances is

$$W_i = \sum \mu \frac{Fl}{AE}$$

By equating the internal and external work, the deflection at a joint in the truss may be expressed as follows:

$$\delta = \sum \frac{F\mu l}{AE}$$

18.3 APPLICATION OF VIRTUAL WORK TO TRUSSES

Examples 18.1 and 18.2 illustrate the application of virtual work to trusses. In each case the forces due to the external loads are computed initially. Second, the external loads are removed, and a unit load is placed at the point and in the direction in which deflection is desired (not necessarily horizontal or vertical). The forces due to the unit load are determined, and, finally, the value of $F\mu l/AE$ for each of the members is found. To simplify the numerous multiplications, a table is used. The modulus of elasticity is carried through as a constant until the summation is made for all of the members, at which time its numerical value is used. Should there be members of different E's, it is necessary that their actual or their relative values be used for the individual multiplications. A positive value of $\Sigma\,(F\mu l/AE)$ indicates a deflection in the direction of the unit load.

Example 18.1

Determine the horizontal and vertical components of deflection at joint L_4 in the truss shown in Fig. 18.3. Circled figures are areas, in square inches. $E = 29 \times 10^6$ psi.

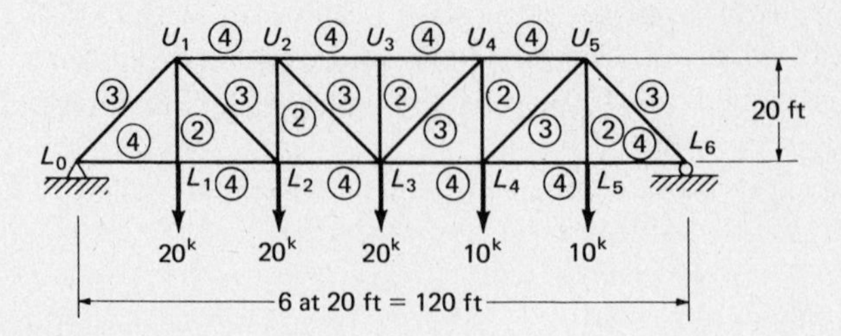

Figure 18.3

SOLUTION

Forces due to external loads:

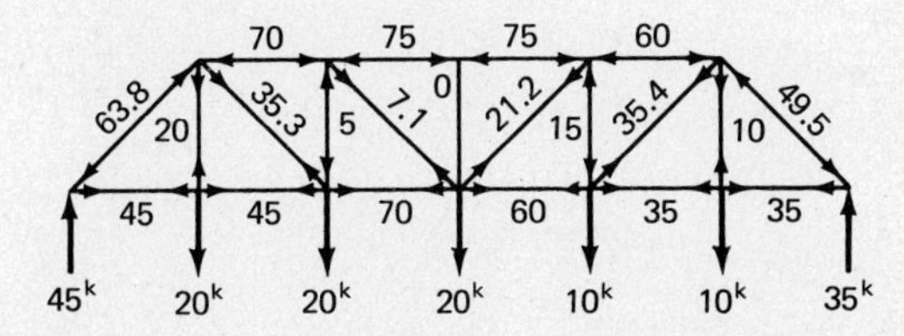

Forces due to a vertical unit load at L_4:

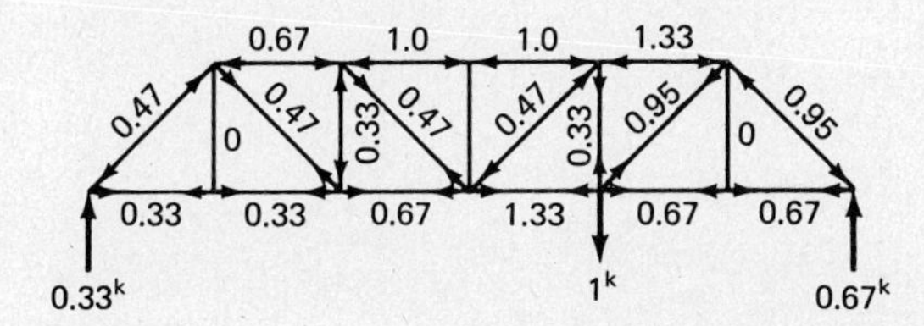

Forces due to a horizontal unit load at L_4:

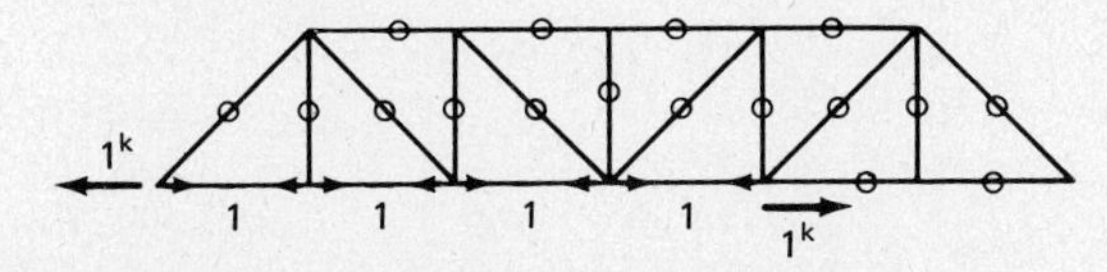

Vertical deflection:

$$E\delta_{L_4} = +42{,}078$$

$$\delta_{L_4} = \frac{42{,}078 \times 1000}{29{,}000{,}000} = +1.45 \text{ in}\downarrow$$

Horizontal deflection:

$$E\delta_{L_4} = +13{,}200$$

$$\delta_{L_4} = \frac{13{,}200 \times 1000}{29{,}000{,}000} = +0.455 \text{ in}\rightarrow$$

Table 18.1

MEMBER	l (in)	A (in²)	$\frac{l}{A}$	F (kip)	μ_V	$\frac{F\mu_V l}{A}$	μ_H	$\frac{F\mu_H l}{A}$
L_0L_1	240	4	60	+45	+0.33	+ 900	1.0	+2700
L_1L_2	240	4	60	+45	+0.33	+ 900	1.0	+2700
L_2L_3	240	4	60	+70	+0.67	+ 2800	1.0	+4200
L_3L_4	240	4	60	+60	+1.33	+ 4800	1.0	+3600
L_4L_5	240	4	60	+35	+0.67	+ 1400	0	0
L_5L_6	240	4	60	+35	+0.67	+ 1400	0	0
L_0U_1	340	3	113.3	−63.8	−0.47	+ 3400	0	0
U_1U_2	240	4	60	−70	−0.67	+ 2800	0	0
U_2U_3	240	4	60	−75	−1.0	+ 4500	0	0
U_3U_4	240	4	60	−75	−1.0	+ 4500	0	0
U_4U_5	240	4	60	−60	−1.33	+ 4800	0	0
U_5L_6	340	3	113.3	−49.5	−0.95	+ 5340	0	0
U_1L_1	240	2	120	+20	0	0	0	0
U_1L_2	340	3	113.3	+35.3	+0.47	+ 1880	0	0
U_2L_2	240	2	120	− 5	−0.33	+ 200	0	0
U_2L_3	340	3	113.3	+ 7.1	+0.47	+ 378	0	0
U_3L_3	240	2	120	0	0	0	0	0
L_3U_4	340	3	113.3	+21.2	−0.47	− 1130	0	0
U_4L_4	240	2	120	−15	+0.33	− 600	0	0
L_4U_5	340	3	113.3	+35.4	+0.95	+ 3810	0	0
U_5L_5	240	2	120	+10	0	0	0	0
Σ						+42,078		+13,200

TABLE 18.2

MEMBER	l(in)	A (in²)	$\frac{l}{A}$	F(kips)	μ	$\frac{F\mu l}{AE}$
L_0L_1	180	2	90	−40	−2.0	+7,200
L_1L_2	180	2	90	−40	−2.0	+7,200
L_2L_3	180	1	180	−40	−2.0	+14,400
L_3L_4	180	1	180	−40	−2.0	+14,400
L_0U_1	202	3	67.3	+44.7	+2.24	+6,730
U_1U_2	202	4	50.5	+67.05	+2.24	+7,575
U_2U_3	202	4	50.5	+67.05	+2.24	+7,575
U_3L_4	202	3	67.3	+44.7	+2.24	+6,730
U_1L_1	90	1	90	+20	0	0
U_1L_2	202	1.5	134.5	−22.35	0	0
U_2L_2	180	3	60	−60	−2.0	+7,200
L_2U_3	202	1.5	134.5	−22.35	0	0
U_3L_3	90	1	90	+20	0	0
Σ						$\frac{79{,}010}{E}$

Example 18.2

Determine the vertical component of deflection of joint L_4 in Fig. 18.4 by the virtual-work method. Circled figures are areas, in square inches. $E = 29 \times 10^6$ psi.

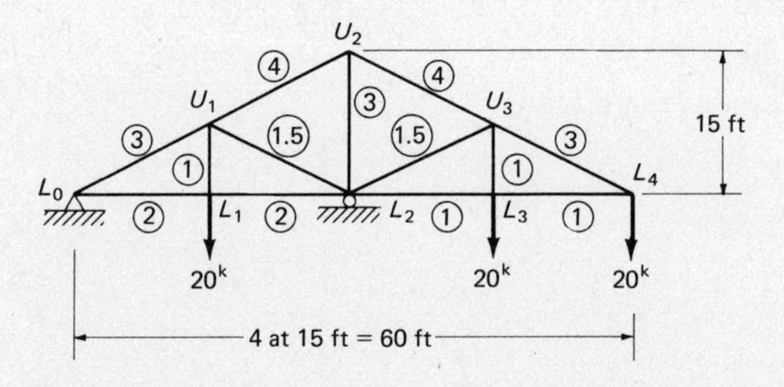

Figure 18.4

SOLUTION

Forces due to external loads:

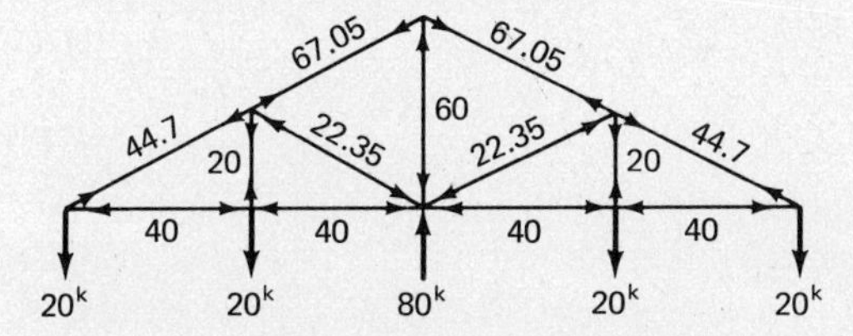

Forces due to a vertical unit load at L_4:

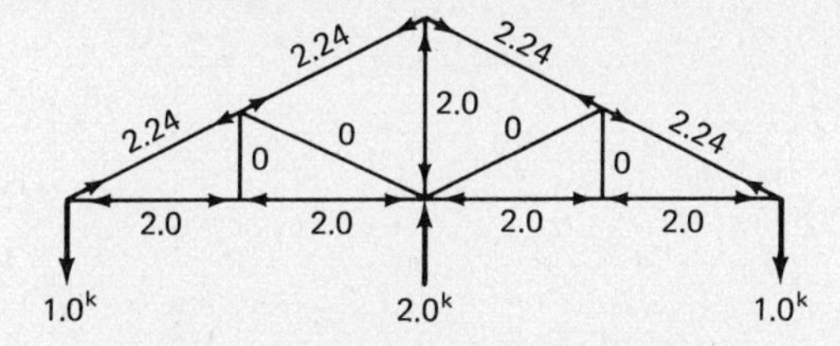

$$\text{vertical deflection at } L_4 = \frac{(79{,}010)(1000)}{29 \times 10^6} = 2.72 \text{ in}\downarrow$$

18.4 DEFLECTIONS OF BEAMS AND FRAMES BY VIRTUAL WORK

The law of conservation of energy may be used to develop an expression for the deflection at any point in a beam or frame. In the following derivation each fiber of the structure is considered to be a "bar" or member such as the members of the trusses considered in the preceding sections. The summation of the internal work performed by the force in each of the bars equals the external work performed by the loads.

For the following discussion the beam of Fig. 18.5(a) is considered. Part (b) of the figure shows the beam cross section. It is desired to know the vertical

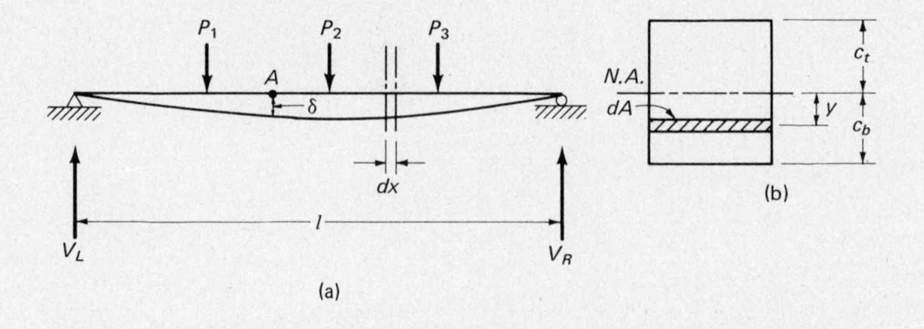

Figure 18.5

deflection δ at point A in the beam caused by the external loads P_1 to P_3. If the loads were removed from the beam and a vertical unit load placed at A, small forces and deformations would be developed in the bars, and a small deflection would occur at A. Replacing the external loads would cause increases in the bar forces and deformations, and the unit load at A would deflect an additional amount δ. The internal work performed by the unit load forces, as they are carried through the additional bar deformations, equals the external work performed by the unit load as it is carried through the additional deflection δ.

The following symbols are used in writing an expression for the internal work performed in a dx length of the beam: M is the moment at any section in the beam due to the external loads and m is the moment at any section due to the unit load. The stress in a differential area of the beam cross section due to the unit load can be found from the flexure formula as follows:

$$\text{unit stress in } dA = \frac{my}{I}$$

$$\text{total force in } dA = \frac{my}{I}\,dA$$

The dA area has a thickness or length of dx which deforms by $\epsilon\,dx$ when the external loads are returned to the structure. The deformation is as follows:

$$\text{unit stress due to external loads} = f = \frac{My}{I}$$

$$\text{deformation of } dx \text{ length} = \epsilon\,dx = \frac{f}{E}\,dx = \frac{My}{EI}\,dx$$

The total force in dA due to the unit load $(my/I)dA$ is carried through this deformation, and the work it performs is as follows:

$$\text{work in } dA = \left(\frac{my}{I}\,dA\right)\left(\frac{My}{EI}\,dx\right) = \frac{Mmy^2}{EI^2}\,dA\,dx$$

The total work performed on the cross section equals the summation of the work in each dA area in the cross section.

$$\text{work} = \int_{C_b}^{C_t} \frac{Mmy^2}{EI^2}\, dA\, dx = \frac{Mm}{EI^2} \int_{C_b}^{C_t} y^2\, dA\, dx$$

The expression $\int y^2\, dA$ is a familiar one, being the moment of inertia of the section, and the equation becomes

$$\text{work} = \frac{Mm}{EI}\, dx$$

It is now possible to determine the internal work performed in the entire beam, because it equals the integral from 0 to l of this expression:

$$W_i = \int_0^l m(d\theta) = \int_0^l m \left(\frac{M\, dx}{EI}\right)$$

The external work performed by the unit load as it is carried through the distance δ is $1 \times \delta$. By equating the external work and the internal work, an expression for the deflection at any point in the beam is obtained.

$$W_e = W_i$$

$$1 \times \delta = \int_0^l \frac{Mm}{EI}\, dx$$

$$\delta = \int_0^l \frac{Mm}{EI}\, dx$$

18.5 APPLICATION OF VIRTUAL WORK TO BEAMS AND FRAMES

Examples 18.3 to 18.7 illustrate the application of virtual work to beams and frames. To apply the method, a unit load is placed at the point and in the direction in which deflection is desired. Expressions are written for M and m throughout the structure, and the results are integrated from 0 to l. It is rarely possible to write one expression for M or one expression for m that is correct in all parts of the structure. As an illustration, consider the beam of Fig. 18.6 and let the deflection under P_2 be desired. A unit load is placed at this point in the figure. The reactions V_L and V_R are due to loads P_1 and P_2, whereas v_L and v_R are due to the unit load.

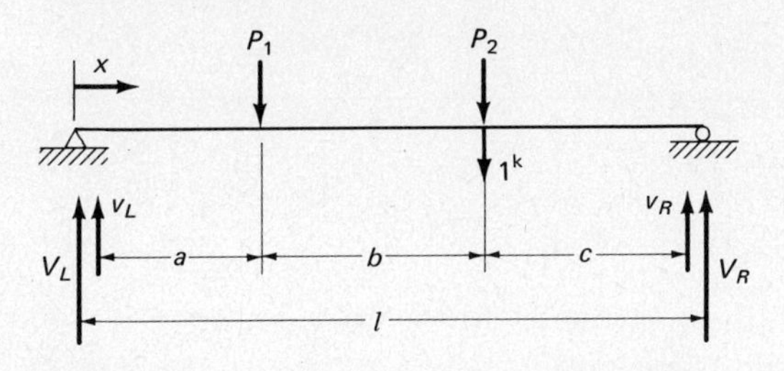

Figure 18.6

The values of M and m are written with respect to the distance x from the left support. From the left support to the unit load m may be represented by one expression, $v_L x$, but the expression for M is not constant for the full distance. Its value is $V_L x$ from the left support to P_1 and $V_L x - P_1(x - a)$ from P_1 to P_2. The integration will be made from 0 to a for $V_L x$ and $v_L x$, and from a to b for $V_L x - P_1(x - a)$ and $v_L x$. Moment expressions for all parts of the beam are shown as follows. The left support is used as the origin of x.

For $x = 0$ to a:

$$M = V_L x$$

$$m = v_L x$$

For $x = a$ to b:

$$M = V_L x - P_1(x - a)$$

$$m = v_L x$$

For $x = b$ to c:

$$M = V_L x - P_1(x - a) - P_2(x - a - b)$$

$$m = v_L x - 1(x - a - b)$$

$$\delta = \int_0^a \frac{Mm}{EI}\,dx + \int_a^{a+b} \frac{Mm}{EI}\,dx + \int_{a+b}^{l} \frac{Mm}{EI}\,dx$$

A positive sign is used for a moment that causes tension in the bottom fibers of a beam. If the result of integration is positive, the direction used for the unit load is the direction of the deflection.

Example 18.3

Determine the deflection at point A, Fig. 18.7, by virtual work.

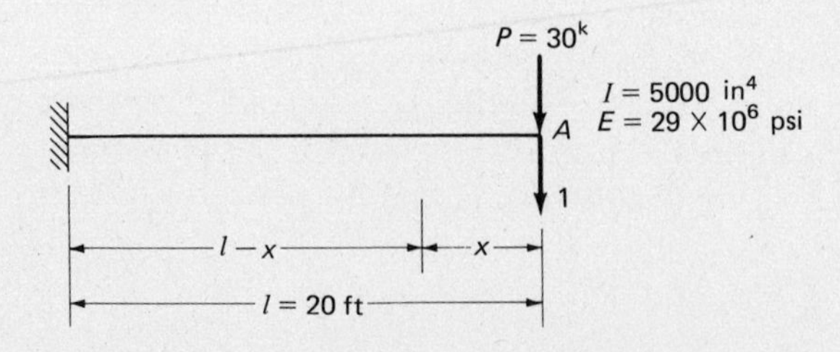

Figure 18.7

SOLUTION

An expression is written for M (the moment due to the 30-kip load) at any point a distance x from the free end. A unit load is placed at the free end, and an expression is written for the moment m it causes at any point. The origin of x

may be selected at any point as long as the same point is used for writing M and m for that portion of the beam.

$$M = -Px$$

$$m = -1x$$

$$\delta_A = \int_0^l \frac{Mm}{EI}\,dx = \int_0^l \frac{(-Px)(-1x)}{EI}\,dx = \int_0^l \frac{Px^2}{EI}\,dx$$

$$\delta_A = \frac{P}{EI}\left[\frac{x^3}{3}\right]_0^l = \frac{Pl^3}{3EI}$$

$$\delta_A = \frac{(30{,}000)(20 \times 12)^3}{(3)(29 \times 10^6)(5000)} = 0.955 \text{ in}\downarrow$$

Example 18.4

Determine the deflection at point A in Fig. 18.8.

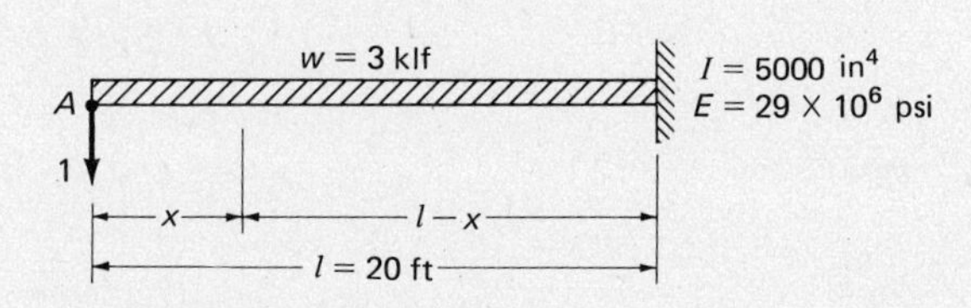

Figure 18.8

SOLUTION

$$M = -(w)(x)\left(\frac{x}{2}\right) = -\frac{wx^2}{2}$$

$$m = -1x$$

$$Mm = +\frac{wx^3}{2}$$

$$\delta_A = \int_0^l \frac{Mm}{EI}\,dx = \int_0^l \frac{wx^3}{2EI}\,dx$$

$$= \frac{w}{2EI}\left(\frac{x^4}{4}\right)_0^l = \frac{wl^4}{8EI}$$

$$= \frac{(3000/12)(20 \times 12)^4}{(8)(29 \times 10^6)(5000)} = 0.714 \text{ in}\downarrow$$

Example 18.5

Determine the deflection at point B in the beam shown in Fig. 18.9.

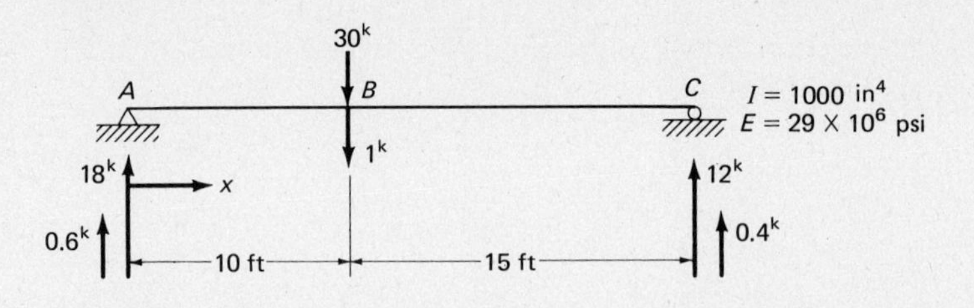

Figure 18.9

SOLUTION

It is necessary to write one expression for M from A to B and another from B to C. The same is true for m. Frequently it is possible to simplify the mathematics by using different origins of x for different sections of the beam. The same results would have been obtained if for the M and m expressions from B to C the origin had been taken at C. The small reactions at A and C are those due to the unit load.

For $x = 0$ to 10:

$$M = 18x$$

$$m = 0.6x$$

$$Mm = 10.8x^2$$

For $x = 10$ to 25:

$$M = 18x - (30)(x - 10) = -12x + 300$$

$$m = 0.6x - 1(x - 10) = -0.4x + 10$$

$$Mm = +4.8x^2 - 240x + 3000$$

$$\delta_B = \int_0^{10} \frac{10.8x^2}{EI} dx + \int_{10}^{25} \frac{(4.8x^2 - 240x + 3000)}{EI} dx$$

$$= \frac{1}{EI}[3.6x^3]_0^{10} + \frac{1}{EI}[1.6x^3 - 120x^2 + 3000x]_{10}^{25}$$

$$= \frac{3600}{EI} + \frac{25{,}000 - 19{,}600}{EI}$$

$$= \frac{9000 \text{ ft}^3\text{-k}}{EI} = \frac{(9000)(1728)(1000)}{(29 \times 10^6)(1000)} = 0.536 \text{ in}\downarrow$$

Example 18.6

Determine the deflection at point B in the beam of Example 17.12 which is reproduced in Fig. 18.10.

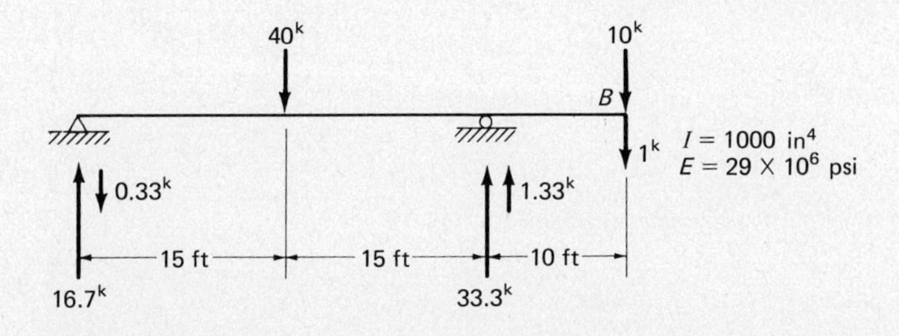

Figure 18.10

SOLUTION

In writing the moment expressions from the left support to the right support, the left support is used for the origin of x. For the overhanging portion of the beam, the right end is used as the origin.

For $x = 0$ to 15:

$$M = 16.7x$$
$$m = -0.33x$$
$$Mm = -5.56x^2$$

For $x = 15$ to 30:

$$M = 16.7x - (40)(x - 15)$$
$$M = -23.3x + 600$$
$$m = -0.33x$$
$$Mm = +7.77x^2 - 200x$$

For $x = 0$ to 10:

$$M = -10x$$
$$m = -x$$
$$Mm = +10x^2$$

$$\delta_B = \frac{1}{EI}\int_0^{15} (-5.56x^2)\,dx + \frac{1}{EI}\int_{15}^{30} (7.77x^2 - 200x)\,dx + \frac{1}{EI}\int_0^{10} (10x^2)\,dx$$

$$= \frac{1}{EI}[-1.85x^3]_0^{15} + \frac{1}{EI}[2.59x^3 - 100x^2]_{15}^{30} + \frac{1}{EI}[3.33x^3]_0^{10}$$

$$= \frac{-6250}{EI} + \frac{-20{,}000 + 13{,}750}{EI} + \frac{3330}{EI}$$

$$= -\frac{9170 \text{ ft}^3\text{-k}}{EI} = -0.546 \text{ in}\uparrow$$

Example 18.7

Find the horizontal deflection at D in the frame shown in Fig. 18.11.

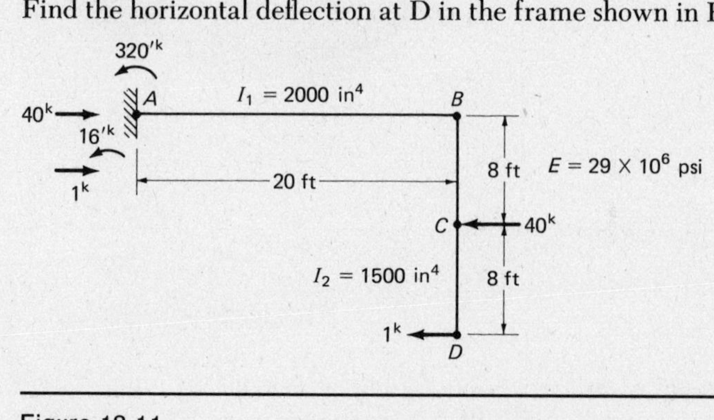

Figure 18.11

SOLUTION

A unit load acting horizontally to the left is placed at D. With A as the origin for x, moment expressions are written for member AB. For the vertical member, the origin is taken at D, one pair of M and m expressions is written for the portion of the member from D to C, and another pair is written for the portion of the beam from C to B. The members do not have the same moments of inertia, and it is necessary to carry them separately as shown in the calculations.

For $x = 0$ to 20:

$$M = -320$$
$$m = -16$$
$$Mm = +5120$$

For $x = 0$ to 8:

$$M = 0$$
$$m = x$$
$$Mm = 0$$

For $x = 8$ to 16:

$$M = -40(x - 8) = -40x + 320$$
$$m = -x$$
$$Mm = 40x^2 - 320x$$

$$\delta_D = \int_0^{20} \left(\frac{5120}{EI_1}\right) dx + \int_8^{16} \left(\frac{40x^2 - 320x}{EI_2}\right) dx$$
$$= \frac{1}{EI_1}[5120x]_0^{20} + \frac{1}{EI_2}\left(\frac{40}{3}x^3 - 160x^2\right)_8^{16}$$
$$= \frac{102{,}400 \text{ ft}^3\text{-k}}{EI_1} + \frac{17{,}066 \text{ ft}^3\text{-k}}{EI_2} = +3.73 \text{ in} \leftarrow$$

I-40, I-240 interchange, Oklahoma City, Oklahoma. (Courtesy of the State of Oklahoma Department of Transportation.)

18.6 ROTATIONS OR SLOPES BY VIRTUAL WORK

Virtual work may be used to determine the slope at various points in a structure. To find the slope at point A in the beam of Fig. 18.12, a unit couple is applied at A, the external loads being removed from the structure. The value of moment at any point in the beam caused by the couple is m. Replacing the external loads will cause an additional moment, at any point, of M.

If the application of the loads causes the beam to rotate through an angle θ at A, the external work performed by the couple equals $1 \times \theta$. The internal work or the internal elastic energy stored is $\int (Mm/EI)\,dx$.

$$\theta = \int \frac{Mm}{EI}\,dx$$

If a clockwise couple is assumed at the position where slope is desired and the result of integration is positive, the rotation is clockwise. Examples 18.8 and 18.9 illustrate the determination of slopes by virtual work. The slopes obtained are in radians.

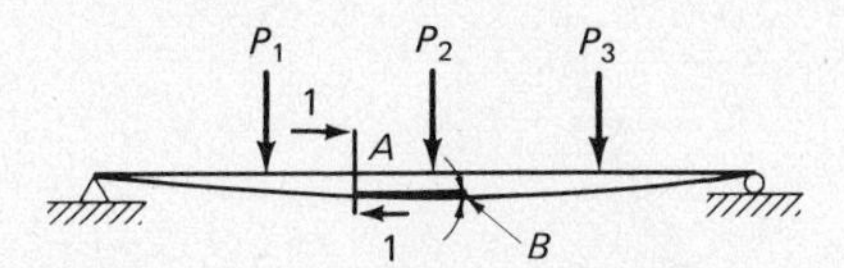

Figure 18.12

Example 18.8

Find the slope at the free end A, Fig. 18.13.

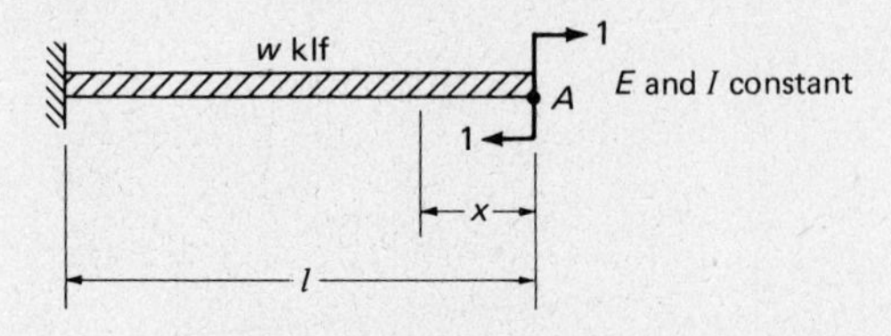

Figure 18.13

SOLUTION

$$M = -\frac{wx^2}{2}$$

$$m = -1$$

$$Mm = \frac{wx^2}{2}$$

$$\theta_A = \int_0^l \frac{wx^2}{2EI}\,dx$$

$$\theta_A = \left[\frac{wx^3}{6EI}\right]_0^l = +\frac{wl^3}{6EI} \text{ clockwise}$$

Example 18.9

Find the slope at the 30-kip load, Fig. 18.14.

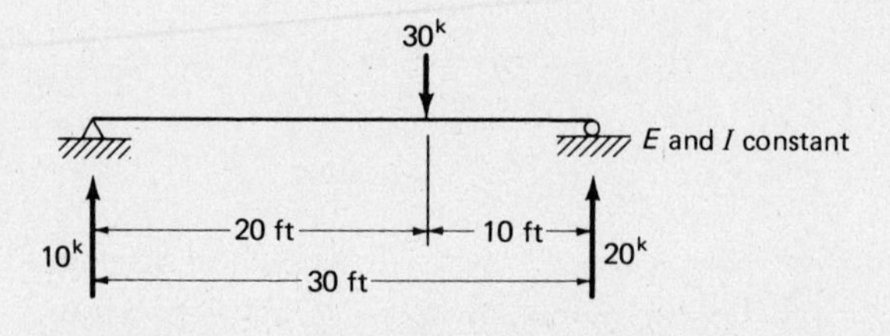

Figure 18.14

SOLUTION

A separate diagram is drawn for the couple and the reactions it causes. For the beam to the left of the load, the left support is used for the origin of x and the right support is used for the origin for the portion of the beam to the right of the 30-kip load.

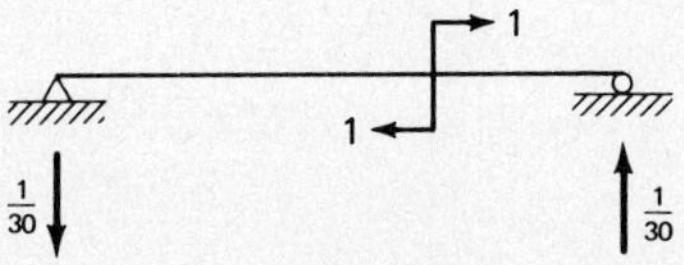

1
1
$\frac{1}{30}$
$\frac{1}{30}$

For $x = 0$ to 20:

$$M = 10x$$
$$m = -\tfrac{1}{30}x$$
$$Mm = -\tfrac{1}{3}x^2$$

For $x = 0$ to 10:

$$M = 20x$$
$$m = \tfrac{1}{30}x$$
$$Mm = \tfrac{2}{3}x^2$$

$$\theta = \int_0^{20}\left(-\frac{1x^2}{3EI}\right)dx + \int_0^{10}\left(\frac{2x^2}{3EI}\right)dx$$

$$= \left[-\frac{x^3}{9EI}\right]_0^{20} + \left[\frac{2x^3}{9EI}\right]_0^{10} = -\frac{888 \text{ ft}^2\text{-k}}{EI} + \frac{222 \text{ ft}^2\text{-k}}{EI}$$

$$= -\frac{666 \text{ ft}^2\text{-k}}{EI} \text{ counterclockwise}$$

18.7 MAXWELL'S LAW OF RECIPROCAL DEFLECTIONS

The deflections of two points in a beam have a surprising relationship to each other. The reader may have noticed this relationship, which was first published by James Clerk Maxwell in 1864. Maxwell's law may be stated as follows: **The deflection at one point *A* in a structure due to a load applied at another point *B* is exactly the same as the deflection at *B* if the same load is applied at *A*.** The rule is perfectly general and applies to any type of structure, whether it is truss, beam, or frame, which is made up of elastic materials following Hooke's law. The displacements may be caused by flexure, shear, or torsion. In preparing influence lines for continuous structures, in analyzing statically indeterminate structures, and in model-analysis problems this useful tool is frequently applied.

The law is not only applicable to the deflections in all of these types of structures but is also applicable to rotations. For instance, a unit couple at *A* will produce a rotation at *B* equal to the rotation caused at *A* if the same couple is applied at *B*.

Example 18.10 proves the law to be correct for a simple beam in which the deflections at two points are determined by the conjugate-beam method. Several applications of this law are made in Chaps. 19 and 20.

Example 18.10

Compute the deflections at points A and B for a load P applied first at B and then at A, Fig. 18.15.

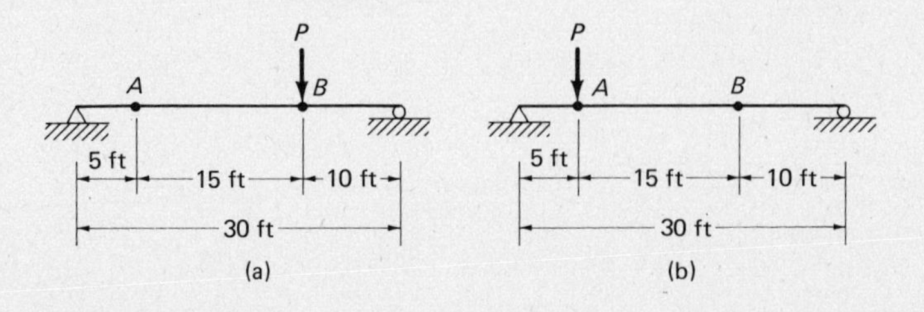

Figure 18.15

SOLUTION

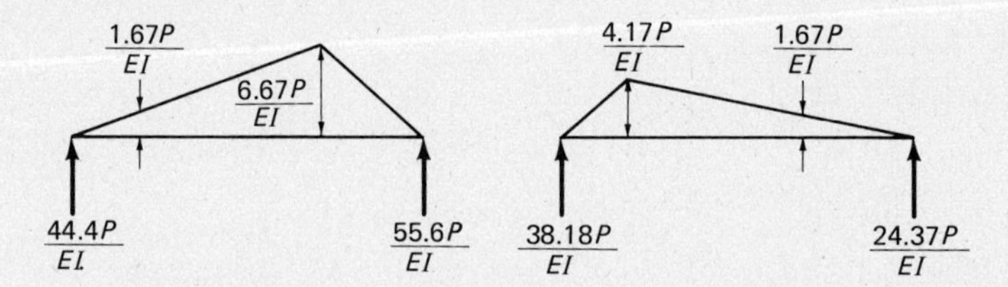

Deflection at A due to load at B:

$$\delta_A = \left(\frac{44.4P}{EI}\right)(5) - \left(\frac{1}{2}\right)\left(\frac{1.67P}{EI}\right)(5)(1.67)$$

$$= \frac{222P}{EI} - \frac{7P}{EI}$$

$$= \frac{215P}{EI}$$

Deflection at B due to load at A:

$$\delta_B = \left(\frac{24.37P}{EI}\right)(10) - \left(\frac{1}{2}\right)\left(\frac{1.67P}{EI}\right)(10)(3.33)$$

$$= \frac{243.7P}{EI} - \frac{28.7P}{EI}$$

$$= \frac{215P}{EI}$$

18.8 TRUSS DEFLECTIONS GRAPHICALLY

Truss deflections were determined by virtual work in Sec. 18.3. Virtual work is very satisfactory for calculating the deflections at one or two points in a truss, but should the deflection of all the joints in a large truss be desired, the process becomes so lengthy as to be impractical unless a computer is used. For this reason graphical methods, by which the deflection of all points in a truss can be determined simultaneously, are frequently used. Another problem conveniently solved graphically is that of finding the exact fabrication dimensions necessary to camber trusses.

The members of a truss become longer or shorter depending on the character of the stresses to which they are subjected. Since they are assumed to be free from bending, they must remain straight. In view of this it is seen that despite the changes in lengths of the members the triangles of a truss must remain perfect triangles. The change in length of a member due to a total axial force F has previously been found to equal

$$\Delta l = \frac{Fl}{AE}$$

The resulting deformations are exceedingly small for common structural materials, an increase or decrease in length of 1/1000 being an extreme value. Having calculated the deformations, it is theorectically possible to determine graphically the new position of each truss joint by plotting to scale the changed lengths of the members and drawing arcs with the new lengths as radii. The method, however, is completely impractical because it would require an enormous scale drawing for the deformations to be visible. Imagine a 30-ft member in a steel truss that elongates $\frac{3}{16}$ in. and the difficulty of showing the member and its deformation with the same scale on a normal-size drawing.

Nevertheless, this method is demonstrated in Figs. 18.16 to 18.18 as a background for the development of the Williot diagram. In Fig. 18.16(a) a simple structure and the deformations of its members are shown. It is noted that points A and C may not deflect horizontally or vertically; however, the changes in length of AB and BC will cause some movement of joint B. In Fig. 18.16(b) member AB is elongated by $\frac{1}{4}$ in to point a and member BC is shortened $\frac{1}{4}$ in. to point b. The deformations are drawn to a tremendously exaggerated scale as compared to the one used for the member lengths. Since points a and b must coincide because the members do not pull apart at B, an arc of radius Aa is drawn with A as the center, and an arc of radius Cb is drawn with C as a center. The intersection of the two arcs must be the new location of joint B, B', and the dotted lines represent the new positions of the members.

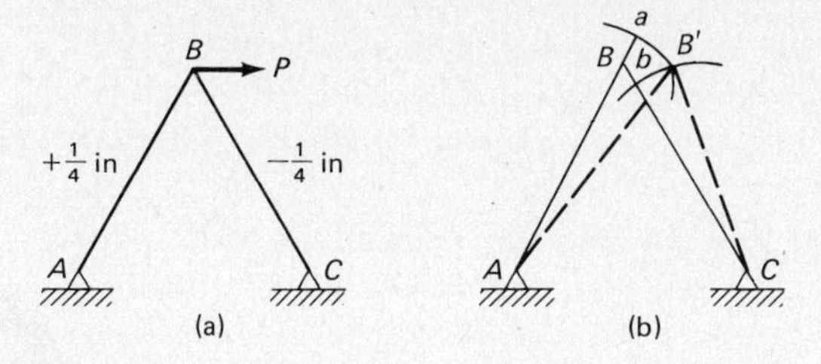

Figure 18.16

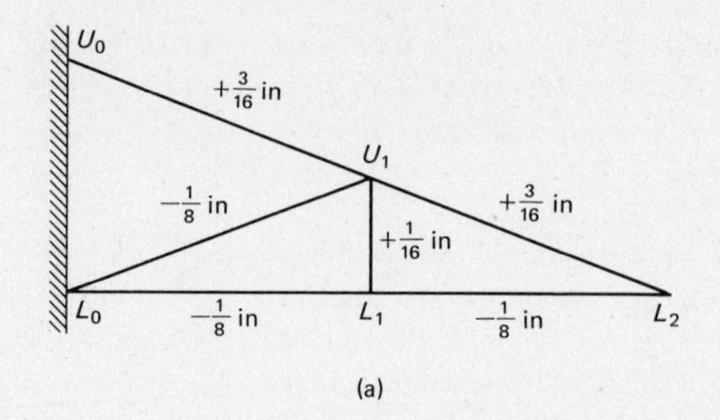

(a)

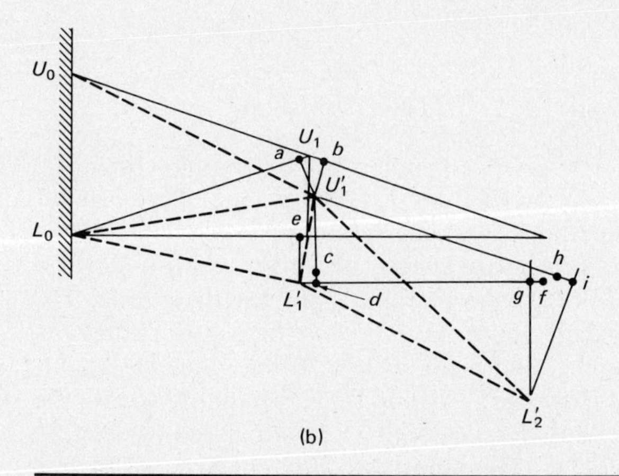

(b)

Figure 18.17

The distorted position of the cantilever truss of Fig. 18.17 is determined in a similar manner. To develop the drawing it is necessary that one point be assumed fixed in position and the direction of one line be assumed fixed. It is obvious that joints U_0 and L_0 cannot deflect and that the direction of a line through U_0 and L_0 remains vertical. In Fig. 18.17(b) the deflection of the other joints is shown with respect to the nondeflecting references.

To develop the distorted shape of the structure, the first member considered is L_0U_1, which shortens by $\frac{1}{8}$ in. Because L_0 is fixed in position, $\frac{1}{8}$ in is scaled off from U_1 toward L_0 to point a. Member U_0U_1 elongates by $\frac{3}{16}$ in, and this distance is measured to the right of U_1 to point b, because U_0 is unyielding. Points a and b must coincide at the new position of joint U_1; therefore with U_0 as a center an arc of radius, U_0b is drawn, and with L_0 as a center an arc of radius, L_0a is drawn. The intersection of the arcs gives U_1', the new position of U_1.

For locating the deflected positions of the joints in this truss and in subsequent trusses considered, perpendiculars are drawn to the radii, rather than using arcs. The distortions are so slight with respect to the radii that no appreciable error is caused by using perpendiculars, and results may be considered to be exact.

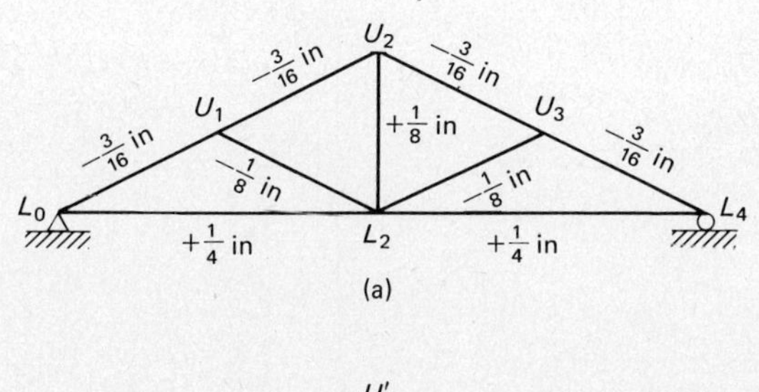

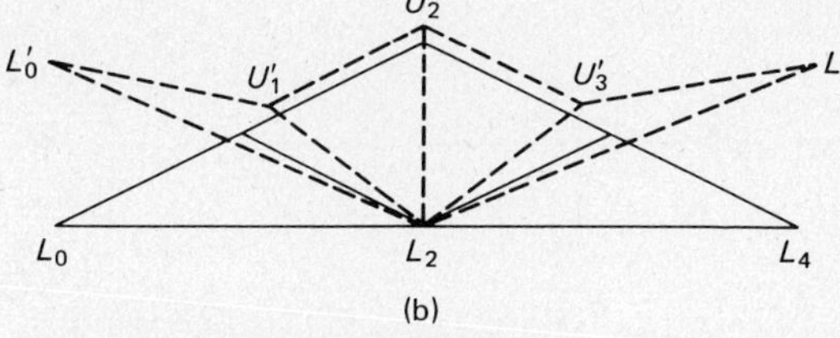

Figure 18.18

To locate the new position of joint L_1, member U_1L_1 is moved over so that U_1 coincides with U'_1, and the bottom end of the member is at c. The member U_1L_1 lengthens $\frac{1}{16}$ in down to point d. Similarly, L_0L_1 shortens $\frac{1}{8}$ in from L_1 to e. The new length of L_0L_1 is L_0e, and U'_1d is the new length of U_1L_1. The new position of L_1, L'_1 is found at the intersection of perpendiculars to the respective radii at e and d.

Finally, from L'_1 the original length of L_1L_2 is drawn to f. The member shortens $\frac{1}{8}$ in to g. Member U_1L_2 is laid off parallel to its original position from U'_1 to h, and it lengthens $\frac{3}{16}$ in to i. Perpendiculars are drawn at i and g, their intersection being L'_2. The new position of each of the truss members is shown by a dotted line.

The distorted shape of the truss of Fig. 18.18 was developed by the same process. The dimensions and member distortions were symmetrical about the centerline, and it was therefore assumed that U_2L_2 was fixed in direction and that joint L_2 was fixed in position. It is noted that joint L_2 does deflect vertically and that, contrary to the deflection diagram, joints L_0 and L_4 do not. To obtain the correct deflection of the joints, the deflected truss should be moved downward until L'_0 and L'_4 fall on a line drawn through L_0 and L_4.

18.9 THE WILLIOT DIAGRAM

The French engineer Williot discovered in 1877 that it is unnecessary to include member lengths in drawing truss-deflection diagrams. He found the same results are obtained for joint deflections if the original lengths of the members are assumed to be zero and if the changes in lengths of the members and perpendiculars to those changes are drawn exactly as before. His remarkable discovery enables the analyst to obtain excellent results, because he or she can draw large-scale diagrams with the actual distortions greatly exaggerated.

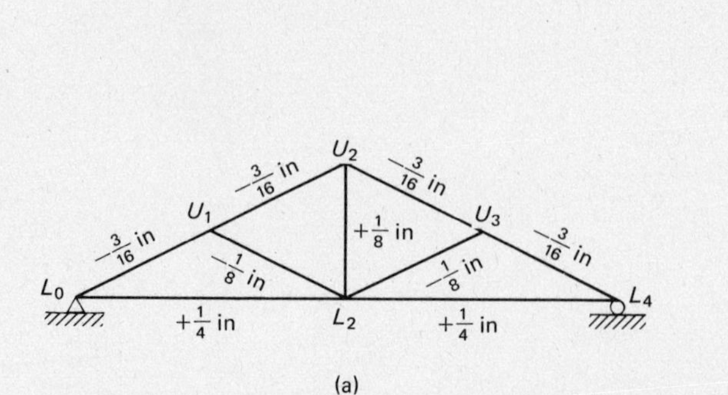

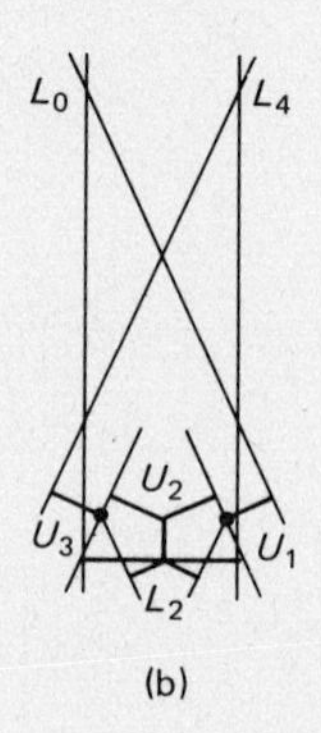

Figure 18.19

The distortion of the truss of Fig. 18.18 was developed by using the Williot diagram in Fig. 18.19. Joint L_2 is assumed fixed in position, and member U_2L_2 is assumed fixed in direction. The deflection of each of the joints is drawn with respect to L_2. If the distance from L_2 to a joint did not change in the distorted truss, that point would coincide with L_2 on the Williot diagram.

Starting with a point L_2 on the drawing, the position of U_2 is located. With respect to L_2, U_2 moves up $\frac{1}{8}$ in vertically, and that point is marked on the paper. By having the position of U_2 and L_2, the analyst can locate U_1. Member U_1U_2 shortens by $\frac{3}{16}$ in, meaning that U_1 moves $\frac{3}{16}$ in toward U_2. A line is drawn from U_2 up and to the right parallel to the member. Member U_1L_2 shortens by $\frac{1}{8}$ in, causing U_1 to move $\frac{1}{8}$ in down and to the right with respect to L_2. Perpendiculars are drawn at the end of each of the two distortions, their intersection being the deflected position of U_1.

Joint L_0 moves $\frac{3}{16}$ in to the right toward U_1 and parallel to L_0U_1, and with respect to L_2 it moves $\frac{1}{4}$ in to the left, parallel to L_0L_2. Perpendiculars to the two distortions will intersect at L_0. The locations of joints U_3 and L_4 are found by a similar process, and they are opposite U_1 and L_0, respectively, because of the symmetry of the truss.

It is customary, as illustrated in these figures, to draw the distortions with dark heavy lines and the perpendiculars with light construction lines. Better accuracy can probably be obtained in locating points by extending the perpendiculars until they pass the intersection points, as shown in the sample Williot diagrams herein.

Referring to the geometry of the truss, it is seen that a horizontal line from L_0 to L_4 must remain horizontal because the two supports do not deflect vertically. The vertical deflection of any joint in the truss may be found by measuring the vertical distance from a horizontal line through L_0 and L_4 in the Williot diagram. Similarly, it is seen that joint L_0 is a hinge and may not deflect horizontally. The horizontal deflection of any joint can be found by measuring the horizontal distance to the joint in the Williot diagram from a vertical line through L_0.

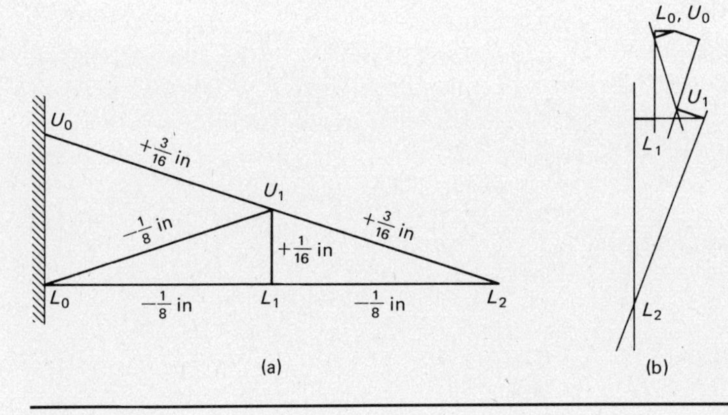

Figure 18.20

In Fig. 18.20 the Williot diagram is drawn for the truss of Fig. 18.17. In this truss, joints U_0 and L_0 are fixed in position and U_0L_0 is fixed in direction. With repect to U_0, U_1 moves down and to the right $\frac{3}{16}$ in, and with respect to L_0 it moves down and to the left $\frac{1}{8}$ in. Perpendiculars to these movements intersect at U_1. Joint L_1 moves to the left $\frac{1}{8}$ in with respect to L_0 and vertically down $\frac{1}{16}$ in with respect to U_1. Perpendiculars to these lines intersect at L_1. The deflected position of L_2 may be found in the same manner. The vertical deflection of a joint can be found by measuring the vertical distance on the diagram from a horizontal line through L_0, and the horizontal deflection is the horizontal distance from a vertical line through L_0. Examples 18.11 and 18.12 are two more illustrations of the application of the Williot diagram.

Example 18.11

Determine the vertical and horizontal components of deflection for each of the joints of the truss shown in Fig. 18.21.

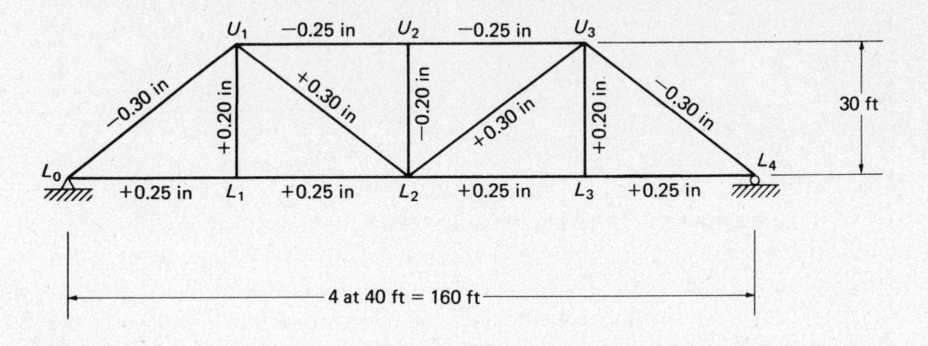

Figure 18.21

SOLUTION

The Williot diagram is drawn with respect to joint L_2, which is assumed fixed in position, and with respect to member U_2L_2, which is assumed fixed in direction. The vertical components of deflection are measured from a horizontal line through L_0 and L_4, and the horizontal components of deflection are measured from a vertical line through L_0.

Table 18.3 DEFLECTIONS (in)

JOINT	VERTICAL	HORIZONTAL
L_0	0	0
L_1	1.65↓	0.25→
L_2	2.34↓	0.50→
L_3	1.65↓	0.75→
L_4	0	0.96→
U_1	1.46↓	0.75→
U_2	2.54↓	0.50→
U_3	1.46↓	0.25→

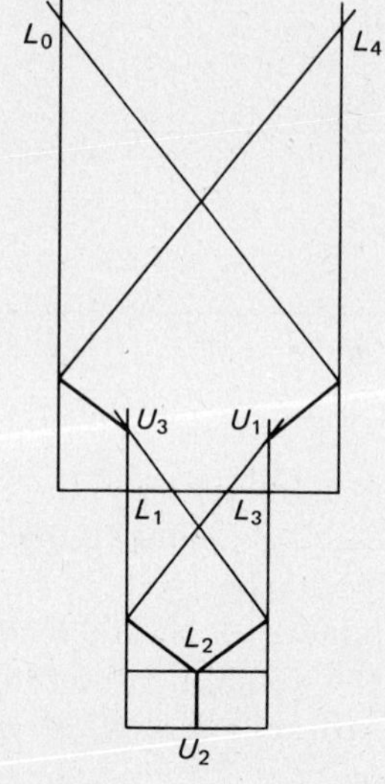

Example 18.12

Determine the increases in length of the top-chord and end-post members necessary to camber the truss of Fig. 18.22 by 4 in at joint L_3.

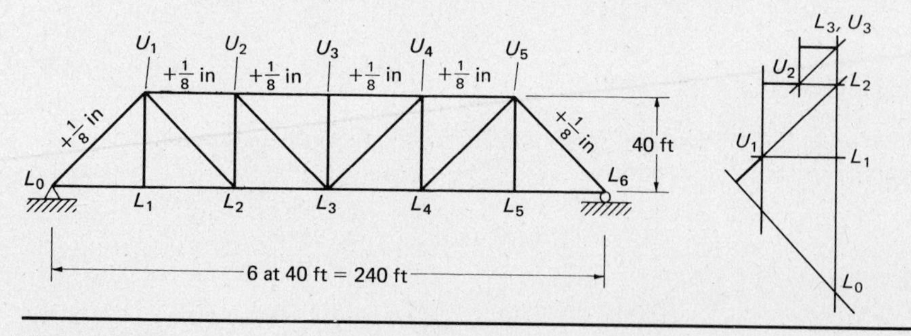

Figure 18.22

SOLUTION

Each of the members in question is assumed to be increased in length by $\frac{1}{8}$ in, and the resulting camber at L_3 is determined with the Williot diagram. A proportion may be written between the camber obtained and that desired as compared to the increases in length made and those necessary. In preparing the diagram, member U_3L_3 is assumed to remain vertical and joint L_3 is assumed fixed in position.

L_3 deflects upward 0.81 in ∴ top members should be lengthened

$$\frac{4}{0.81} \times \frac{1}{8} = 0.617 \text{ in}$$

say, $\frac{5}{8}$ in each.

NOTE: Trusses are often cambered for both dead and live loadings so that at no time will any of the lower-chord joints be below the horizontal. Another way to achieve camber is to change the length of each of the members from their normal length by an amount exactly equal and opposite to their shortenings or elongations under load. For the truss of Fig. 18.19, members U_1U_2 and U_2U_3 would each be made $\frac{3}{16}$ in longer, members L_0L_2 and L_2L_4 $\frac{1}{4}$ in shorter, and so on. This method has the disadvantage in fabrication of having to make so many different length changes.

18.10 THE WILLIOT-MOHR DIAGRAM

Graphical solutions for deflections have been made for trusses that had at least one member whose direction was constant. The center verticals in the trusses of Figs. 18.19, 18.21, and 18.22 remain vertical because of the symmetry of the trusses and loadings about centerlines. A line from U_0 to L_0 in the truss of Fig. 18.20 is fixed in direction and position. For cases such as these the Williot diagram will quickly and accurately give the joint deflections.

Should the diagram be drawn with respect to some other member that does rotate, the correct deflections may not be obtained directly. A correction must be made to take into account the rotation of that member. Trusses in which the directions of all the members change are often encountered, and the diagram must be drawn with respect to a rotating member.

The deflected shape of the truss of Fig. 18.19 is drawn in Fig. 18.23(b) with the assumption that member L_0L_2 (which obviously rotates) remains horizontal. A Williot diagram is drawn in Fig. 18.23(c) on the basis of the same assumption. Both diagrams seem to indicate that the deflected position of joint L_4 is well above joint L_0. Obviously this situation is not possible, because a horizontal line through the supports L_0 and L_4 must remain horizontal. The deflection diagram needs to be rotated clockwise until L_4 lies on a horizontal line through L_0.

In 1887 Otto Mohr presented a simple method of correcting the Williot diagram when the reference member, which was assumed to be fixed in direction, actually rotated. Referring to the truss of Fig. 18.23(a) and its deflected shape in part (b), it is obvious that the entire diagram needs to be rotated about L_0 until L_4' is at L_4'', a distance $L_4'L_4''$. A perpendicular to the radius L_0L_4' is used, because the difference between a perpendicular and an arc is negligible. The rotated positions of the other joints are indicated by L_2'', U_3'', and so on.

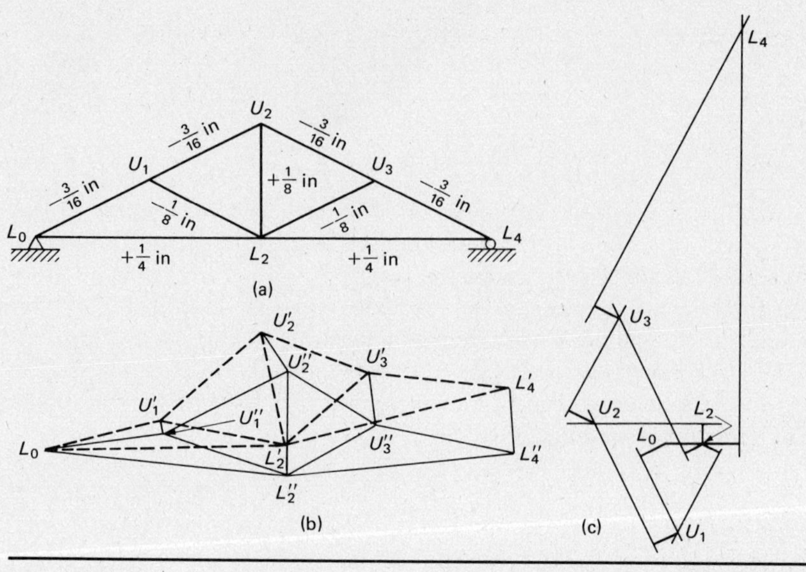

Figure 18.23

The amount of rotation of any joint about L_0 is in proportion to the rotation $L_4'L_4''$ as the distance of the joint from L_0 is to L_0L_4'. For example, the rotation of L_2' necessary to put it in its proper position is as follows:

$$\frac{L_2'L_2''}{L_4'L_4''} = \frac{L_0L_2'}{L_0L_4'}$$

$$L_2'L_2'' = L_4'L_4'' \times \frac{L_0L_2'}{L_0L_4'}$$

Should the truss be rotated 90° to the position shown in Fig. 18.24 and drawn to a scale such that AC equals the vertical distance from L_0 to L_4 on the Williot diagram, the rotation correction can be easily made.

The trusses are proportional, and the following relations may be written:

$$\frac{AE}{L_0U_2} = \frac{AC}{L_0L_4} \qquad AE = \frac{(L_0U_2)(AC)}{L_0L_4}$$

$$\frac{AD}{L_0U_1} = \frac{AC}{L_0L_4} \qquad AD = \frac{(L_0U_1)(AC)}{L_0L_4}$$

With reference to the distorted shape of the truss in Fig. 18.23(b), the distance from L_4' to L_4'' is the rotation of L_4', and the rotation of U_3' is $(L_0U_3'/L_0L_4') \times L_4'L_4''$. Let the truss of Fig. 18.24 be drawn to such a scale that AC equals the distance $L_4'L_4''$ in Fig. 18.23. Since AD is to L_0U_1 as AC is to L_0L_4

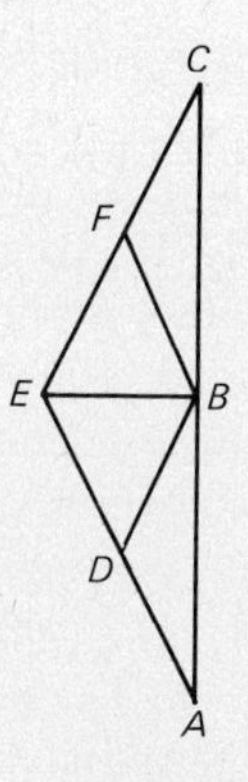

Figure 18.24

and since $U_1'U_1''$ is to $L_4'L_4''$ as L_0U_1' is to L_0L_4', AD equals the rotation correction that must be made to U_1', AF equals the correction of joint U_3', and so on.

Professor Mohr's ingenious idea was to draw a truss proportional to the real truss above the point in the Williot diagram corresponding to the assumed fixed support of the truss. A horizontal line is then drawn through the point on the diagram corresponding to the expansion support of the truss. The fictitious truss is at 90° with the actual truss, and its length between supports equals the distance from the fixed support to a horizontal line through the expansion support in the Williot diagram. The Williot diagram of Fig. 18.23 is repeated in Fig. 18.25, and the Mohr correction is made. The correct deflection of any

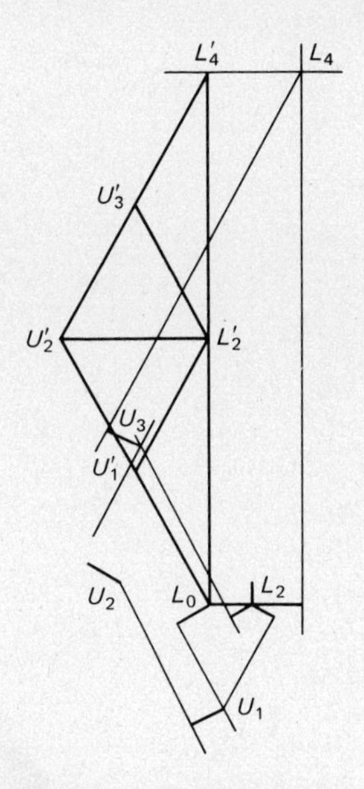

Figure 18.25

point in the truss is the distance from the joint number in the correction truss to the corresponding joint number in the regular Williot diagram.

The deflection of joint U_2, in magnitude and direction, in the actual truss is the distance from U_2' to U_2 in the figure. In Examples 18.13 to 18.16, joint deflections for four more trusses are determined with the Williot-Mohr diagram. An effort should be made to select a member whose direction rotates as little as possible to use as a reference in drawing the Williot diagram. The smaller the rotation, the smaller the Mohr correction diagram will be and the greater the accuracy of the deflection obtained, because the Williot diagram may be drawn to a larger scale on the same size paper.

There is one particular point that should be noted in Example 18.15. The roller joint L_3 falls below the pinned joint L_0 on the Williot diagram. A horizontal line is drawn through L_3, and the correction diagram is rotated clockwise so that L_3' falls on the horizontal line, which means that L_4' extends below as shown. The deflections are measured as usual from the joints on the correction diagram to the joints on the Williot diagram.

The supports of the truss of Example 18.16 are not on the same level, and the supporting surface beneath the roller is inclined. During the distortion of the truss under load, the roller will move along a plane parallel to the supporting surface; therefore a line is drawn parallel to the supporting surface through the position of the roller joint U_3 on the standard Williot diagram. The Mohr correction diagram is drawn by rotating the truss 90° about the hinge joint L_0 and fitting it to the proper size so that U_3' will fall on the line through the roller joint.

Example 18.13

Draw the Williot-Mohr diagram for the truss of Fig. 18.26; assume L_0 fixed in position and L_0L_1 fixed in direction.

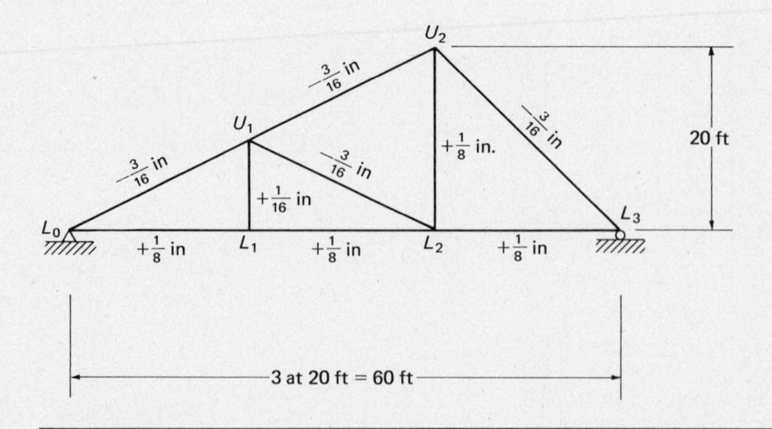

Figure 18.26

SOLUTION

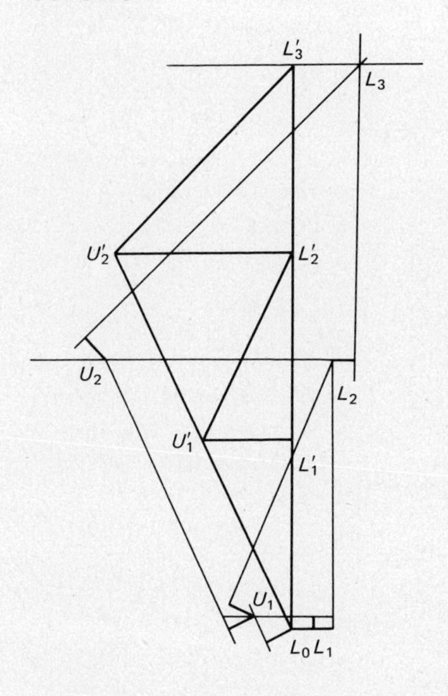

Example 18.14

By using the Williot-Mohr diagram, determine the vertical and horizontal components of deflection of joints U_2 and L_3, Fig. 18.27. Assume joint L_0 is fixed in position and member L_0L_1 is fixed in direction.

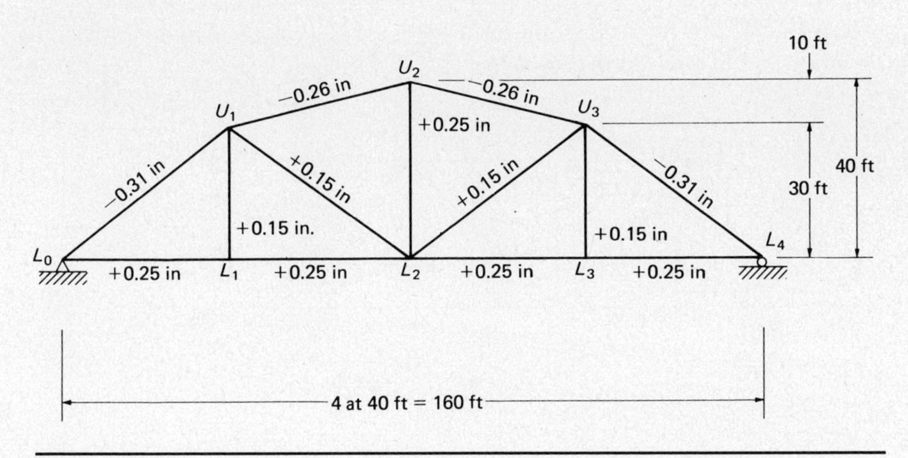

Figure 18.27

SOLUTION

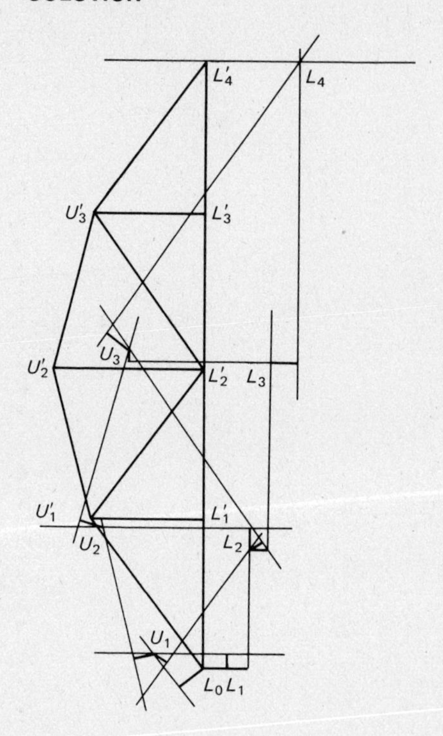

Deflections:

U_2 vertical = 1.70 in ↓
U_2 horizontal = 0.48 in →
L_3 vertical = 1.60 in ↓
L_3 horizontal = 0.77 in →

Example 18.15

Draw the Williot-Mohr diagram for the truss of Fig. 18.28. Assume L_0 fixed in position and L_0U_1 fixed in direction.

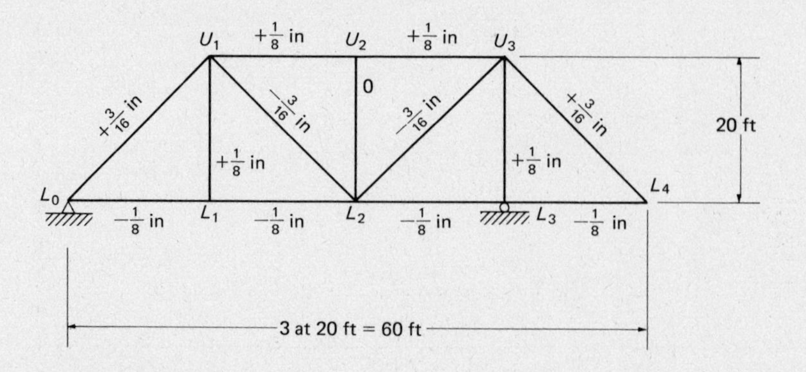

Figure 18.28

SOLUTION

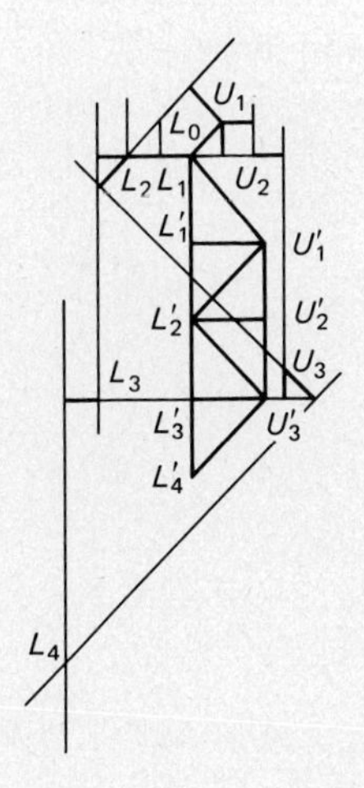

Example 18.16

Draw the Williot-Mohr diagram for the truss shown in Fig. 18.29; assume L_0 fixed in position and L_0U_1 fixed in direction.

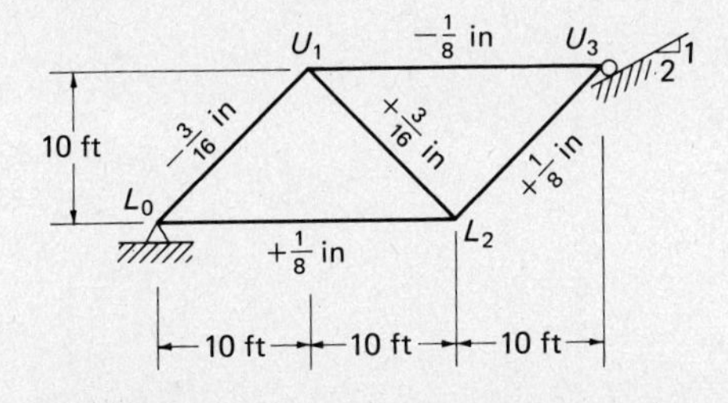

Figure 18.29

SOLUTION

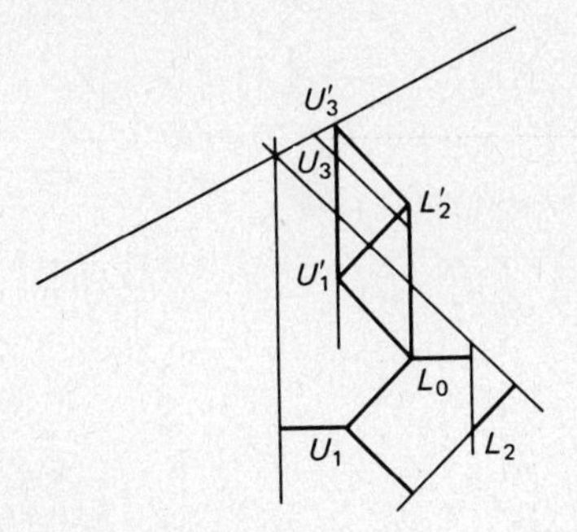

PROBLEMS

For Probs. 18.1 to 18.6, using the virtual-work method, determine the deflection of each of the joints marked on the trusses shown in the accompanying illustrations. The circled figures are areas in sq. in and $E = 29 \times 10^6$ psi unless otherwise indicated.

Problem 18.1 U_2 vertical, U_2 horizontal

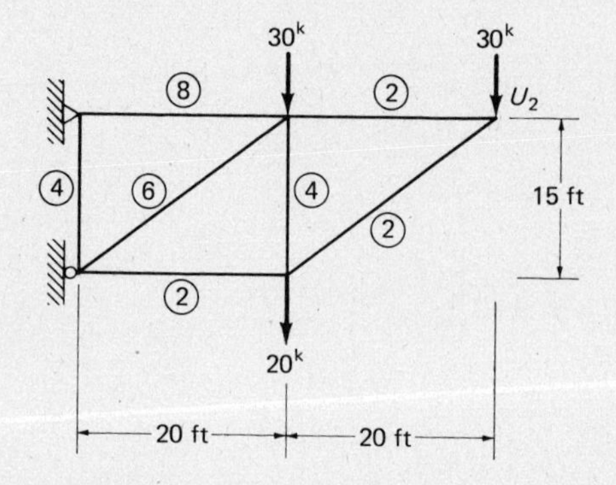

Problem 18.2 L_2 vertical, L_4 horizontal (*Ans.* $L_2 = 1.53$ in ↓, $L_4 = 0.55$ in →)

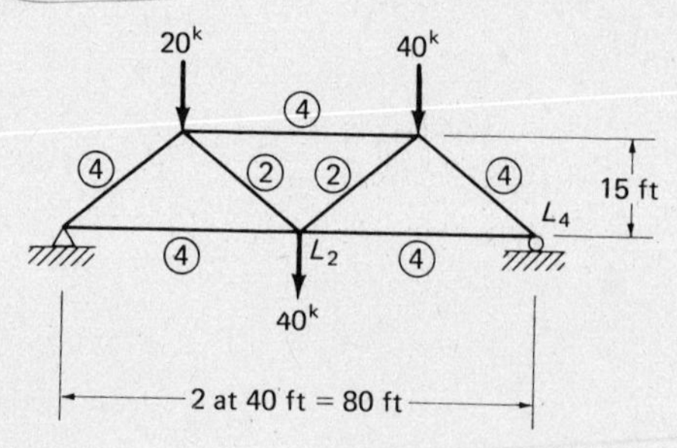

Problem 18.3 U_2 vertical, U_1 horizontal

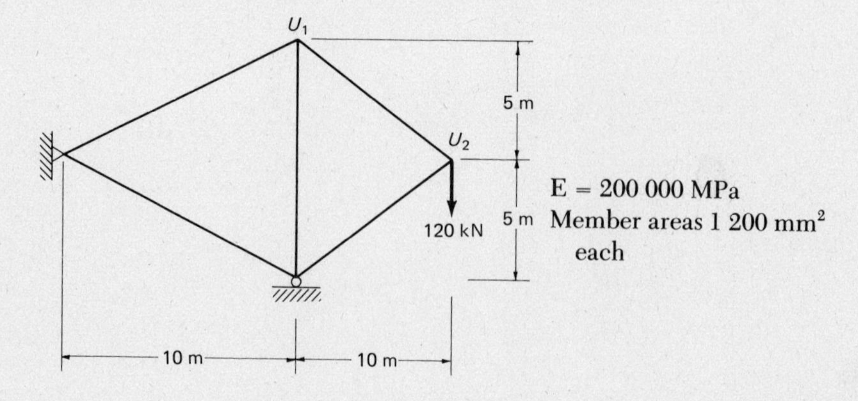

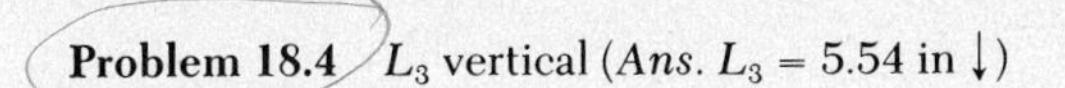

Problem 18.4 L_3 vertical (*Ans.* $L_3 = 5.54$ in $\downarrow$)

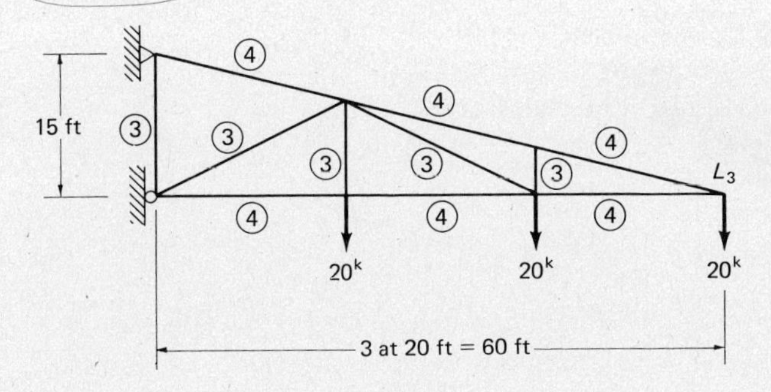

Problem 18.5 U_1 vertical, L_1 horizontal

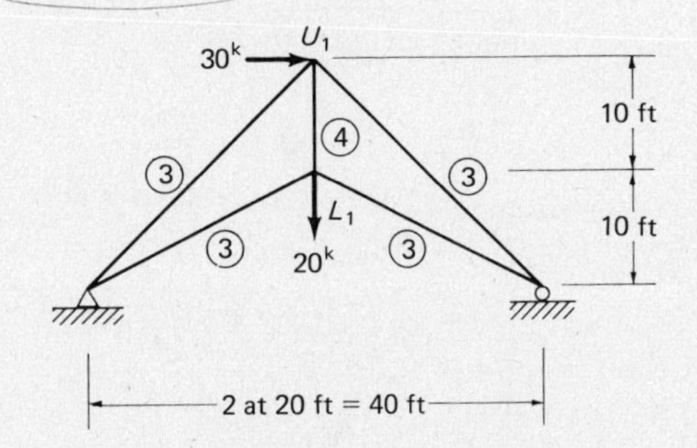

Problem 18.6 L_1 vertical, L_1 horizontal (*Ans.* $L_1 = 0.143$ in $\downarrow$, 0.130 in $\rightarrow$)

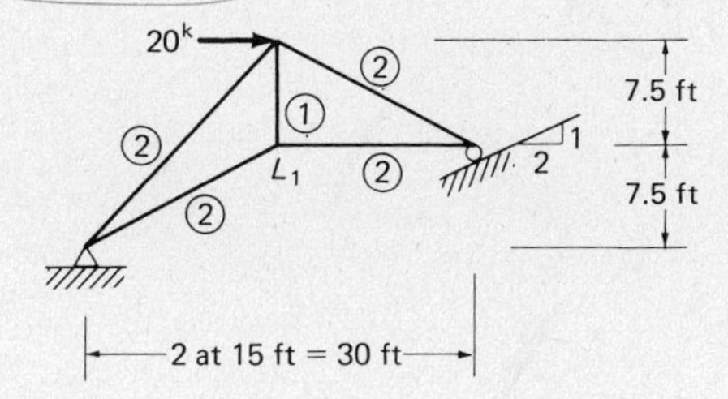

Use the virtual-work method for solving Probs. 18.7 to 18.31. $E = 29 \times 10^6$ psi for all problems unless otherwise indicated.

18.7. Determine the slope and deflection at points A and B of the structure shown in the accompanying illustration. $I = 2000$ in^4.

Problem 18.7

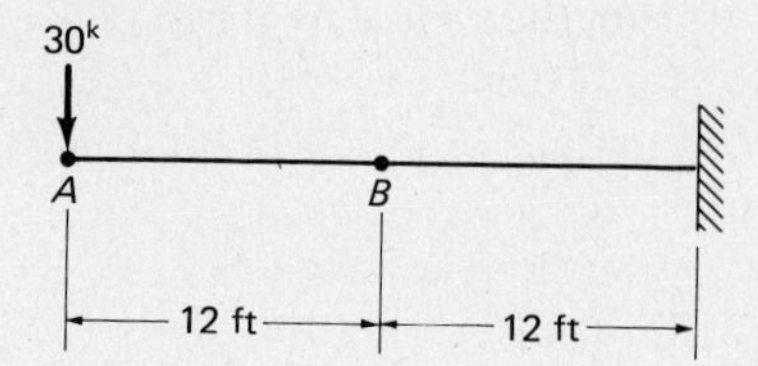

18.8. Determine the slope and deflection underneath each of the concentrated loads shown in the accompanying illustration. $I = 3500 \text{ in}^4$.

Problem 18.8 (*Ans.* $\theta_{20} = 0.00301$ rad\, $\theta_{30} = 0.00515$ rad\, $\delta_{20} = 0.0992$ in ↓, $\delta_{30} =$ 0.632 in ↓)

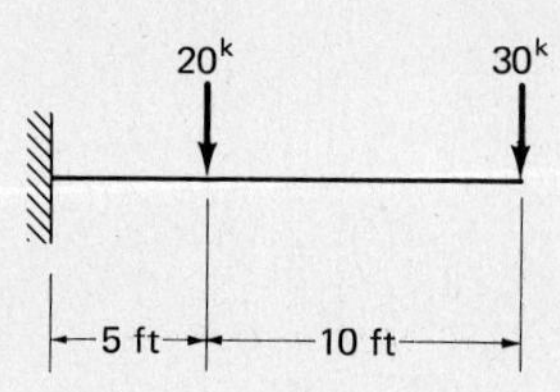

18.9. Find the slope and deflection at the free end of the beam shown in the accompanying illustration. $I = 2.35 \times 10^9 \text{ mm}^4$. $E = 200\,000$ MPa.

Problem 18.9

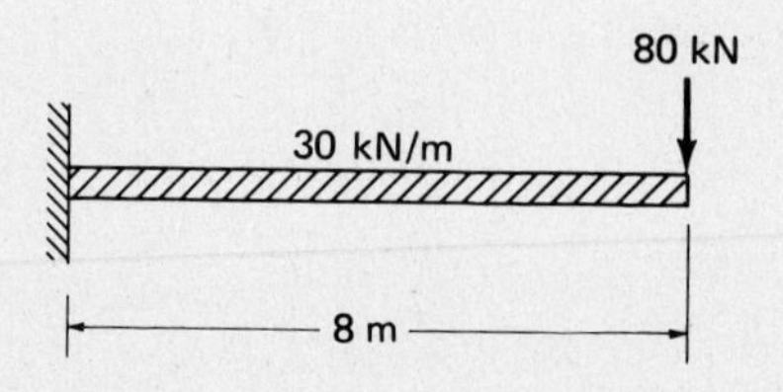

18.10. Find the deflections at points A and B on the beam shown in the accompanying illustration. $I = 3500 \text{ in}^4$.

Problem 18.10 (*Ans.* $\delta_A = 1.15$ in ↓, $\delta_B = 0.41$ in ↓)

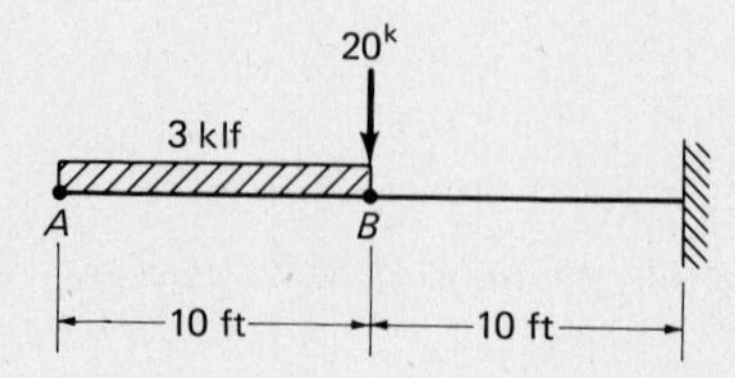

18.11. Rework Prob. 17.19.

18.12. Determine the slope and deflection 12 ft from the right end of the beam of Prob. 17.19. (*Ans.* $\theta_{12} = 0.00331$ rad/, $\delta_{12} = 0.667$ in ↓)

18.13. Find the slope and deflection at the 30-kip load in the beam of Prob. 17.21.

18.14. Calculate the slope and deflection at a point 3 m from the left support of the beam shown in the accompanying illustration. $I = 2.5 \times 10^8$ mm^4. $E = 200\,000$ MPa.

Problem 18.14 (Ans. $\theta = 0.018\,9$ rad ╲, $\delta = 84.7$ mm↓)

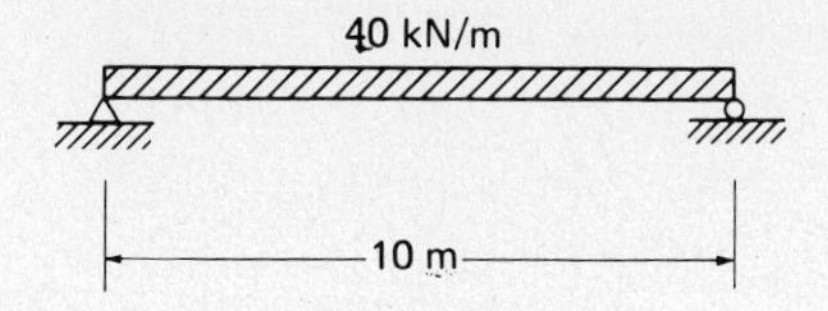

18.15. Rework Prob. 17.34.

18.16. Rework Prob. 17.36. $\theta = 0.00234$ rad╲, $\delta = 0.156$ in ↓.

18.17. Rework Prob. 17.39.

18.18. Determine the deflection at the centerline and the slope at the right end of the beam shown in the accompanying illustration. $I = 2100$ in^4.

Problem 18.18 (*Ans.* $\delta = 0.419$ in ↓, $\theta = 0.00244$ rad/)

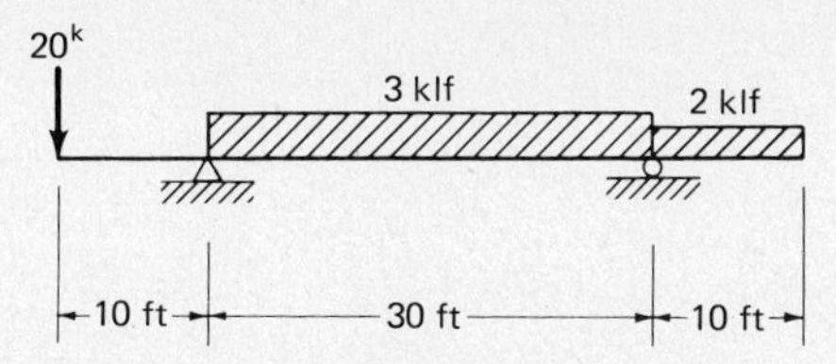

18.19. Calculate the deflection at the 60-kip load in the structure shown in the accompanying illustration.

Problem 18.19

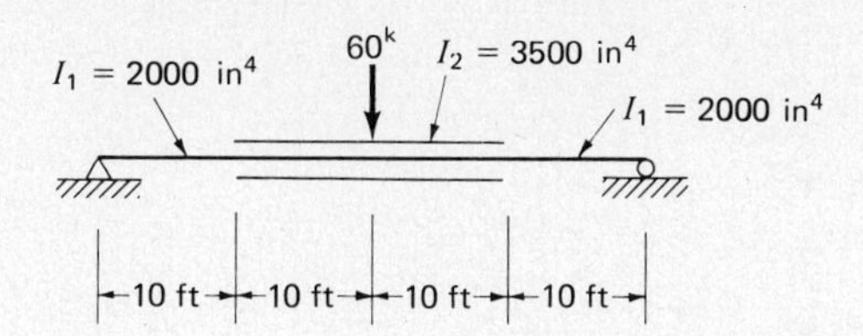

18.20. Calculate the slope and deflection at the 60-kip load on the beam shown in the accompanying illustration.

Problem 18.20 (*Ans.* $\theta_{60} = 0.00069$ rad/, $\delta_{60} = 0.0274$ in↓)

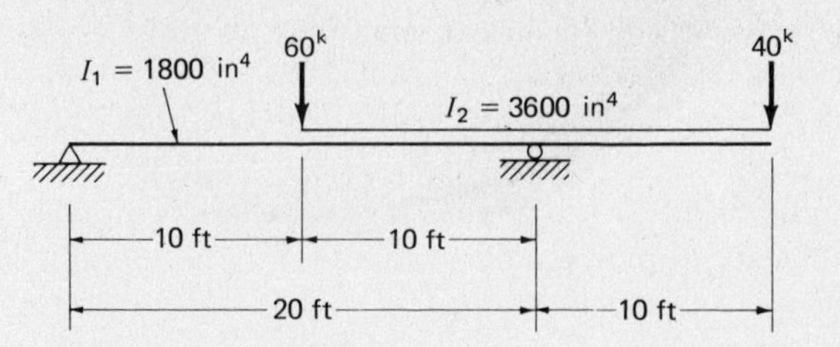

18.21. Calculate the slope and deflection at the 60-kN load on the structure shown in the accompanying illustration. $I = 1.46 \times 10^9$ mm⁴. $E =$ 200 000 MPa.

Problem 18.21

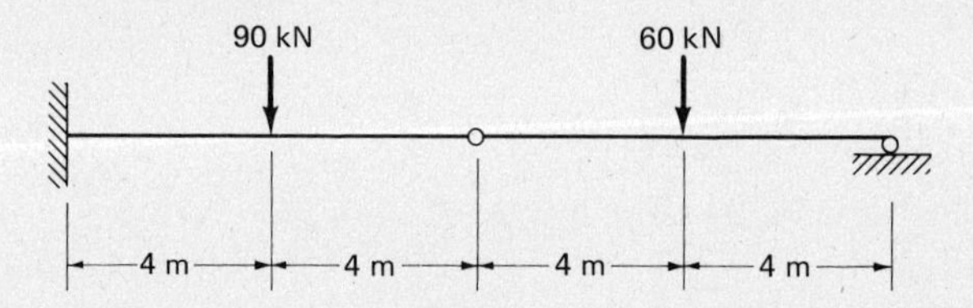

18.22. Find the slope and deflection 20 ft from the left end of the simple beam shown in the accompanying illustration. $I = 2250$ in⁴.

Problem 18.22 (*Ans.* $\theta_{20} = 0.00294$ rad/, $\delta_{20} = 0.612$ in↓)

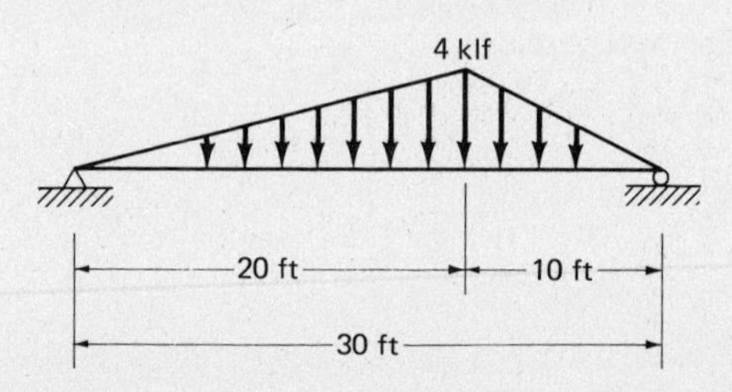

18.23. Calculate the slope and deflection at the free end of the beam shown in the accompanying illustration. $I = 3200$ in⁴.

Problem 18.23

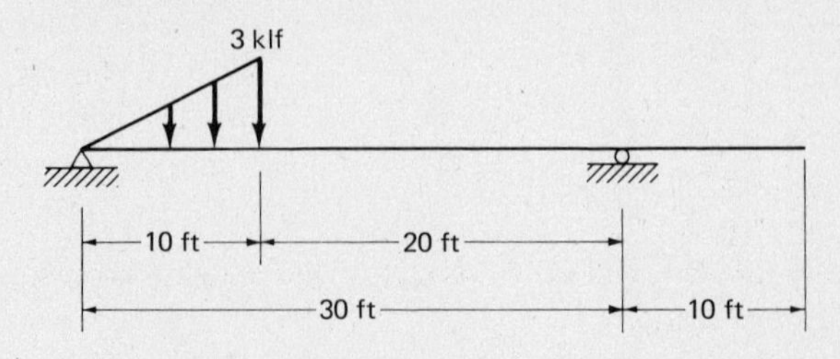

18.24. Find the vertical deflection at point A and the horizontal deflection at point B on the frame shown in the accompanying illustration. $I = 4000$ in^4.

Problem 18.24 (*Ans.* $\delta_A = 0.706$ in↓, $\delta_B = 0.218$ in←)

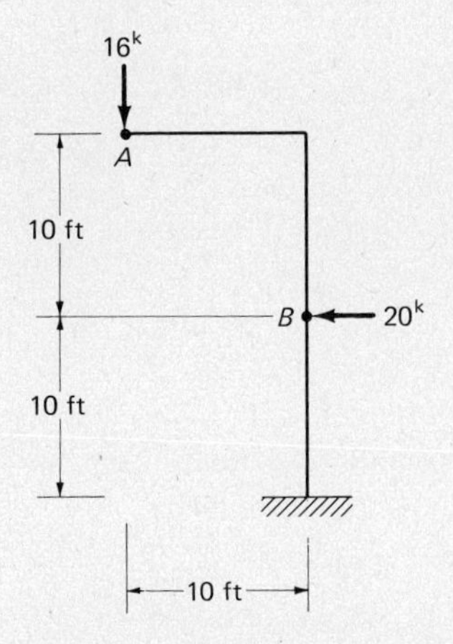

18.25. Determine the horizontal deflection at the 20-kip load and the slope at the 30-kip load on the frame shown in the accompanying illustration.

Problem 18.25

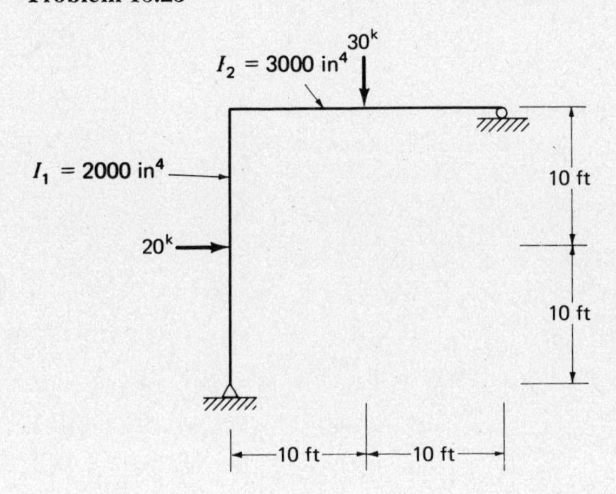

18.26. Find the vertical and horizontal components of deflection at the 20-kip load in the frame shown in the accompanying illustration. $I = 1500$ in^4.

Problem 18.26 (*Ans.* $\delta_{20} = 2.90$ in→, 1.59 in↓)

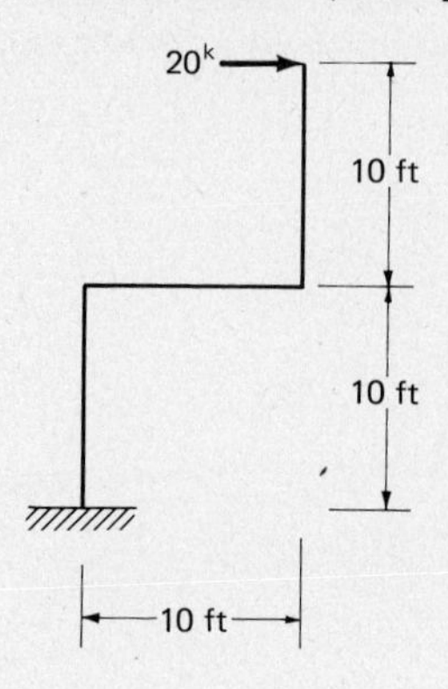

18.27. Determine the vertical and horizontal components of deflection at the free end of the frame shown in the accompanying illustration. $I = 2.60 \times 10^9$ mm^4. $E = 200\,000$ MPa.

Problem 18.27

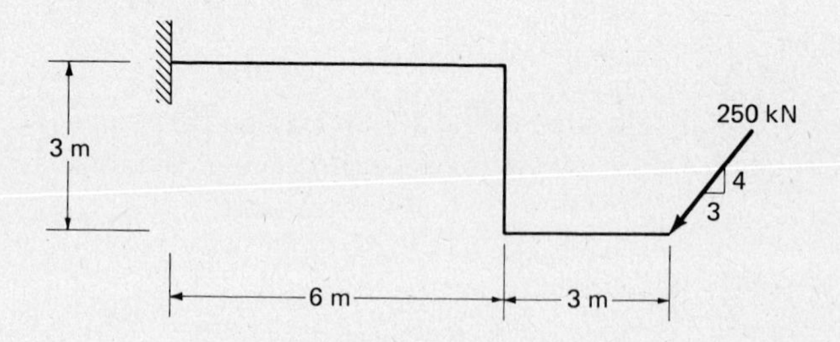

18.28. Find the horizontal deflection at point A and the vertical deflection at point B on the frame shown in the accompanying illustration. $I = 1200$ in^4.

Problem 18.28 (*Ans.* $\delta_A = 2.23$ in←, $\delta_B = 1.49$ in↑)

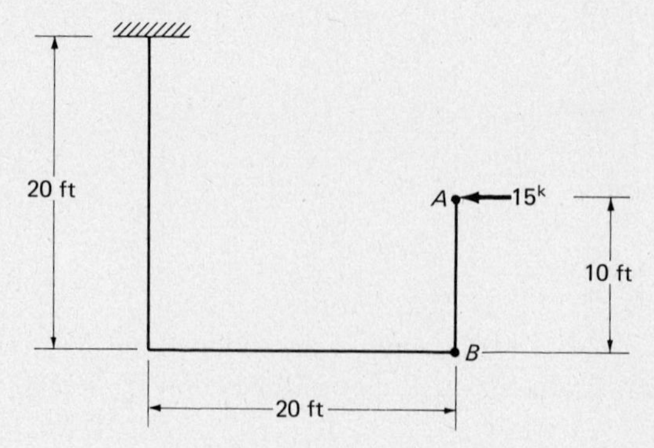

18.29. Calculate the horizontal deflection at the roller support and the vertical deflection at the 30-kip load for the frame shown in the accompanying illustration. $I = 1500\ \text{in}^4$.

Problem 18.29

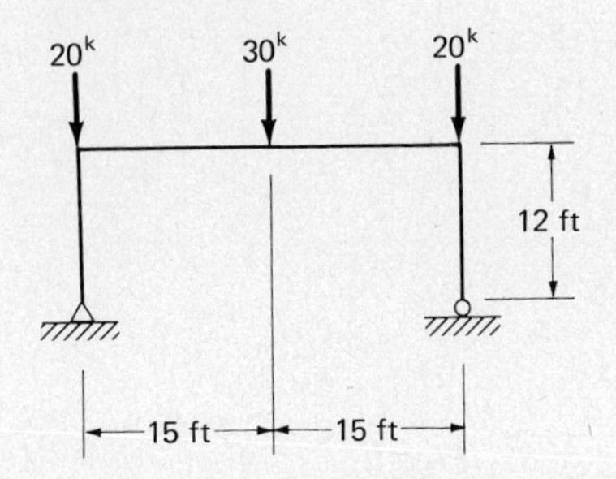

18.30. Find the horizontal deflections at points A and B on the frame shown in the accompanying illustration. $I = 3000\ \text{in}^4$.

Problem 18.30 (*Ans.* $\delta_A = 0.132$ in→, $\delta_B = 0.066$ in←)

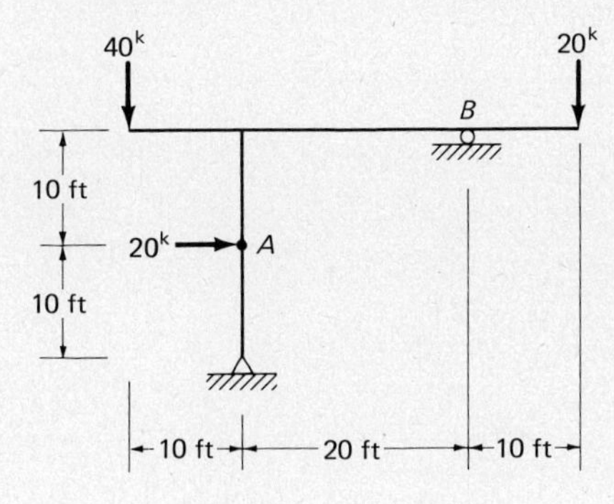

18.31. For the frame of Prob. 18.30 compute the slope and vertical deflection at the 40-kip load.

For Probs. 18.32 to 18.34 find the vertical and horizontal deflections at all of the joints of the trusses shown in the accompanying illustrations by drawing the Williot diagram.

Problem 18.32 (*Ans.* L_3 = 0.90 in→, 1.85 in↓; U_2 = 0.60 in→, 2.51 in↓)

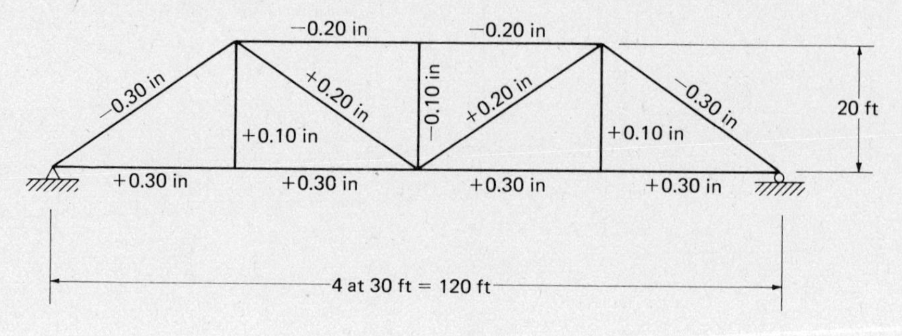

Problem 18.33

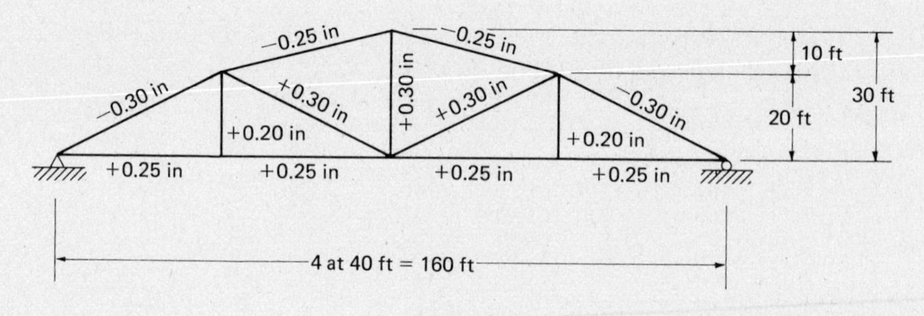

Problem 18.34 (*Ans.* L_2 = 3.45 in↓, 0.50 in←; U_1 = 1.20 in↓, 0.40 in→)

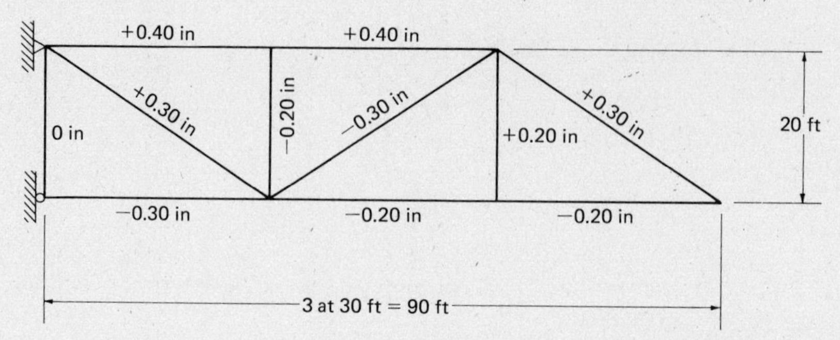

For Probs. 18.35 and 18.36, using the Williot diagram, determine the horizontal and vertical components of deflection for all of the joints of the trusses shown in the accompanying illustrations. Circled values are areas in square inches. $E = 29 \times 10^6$ psi.

Problem 18.35

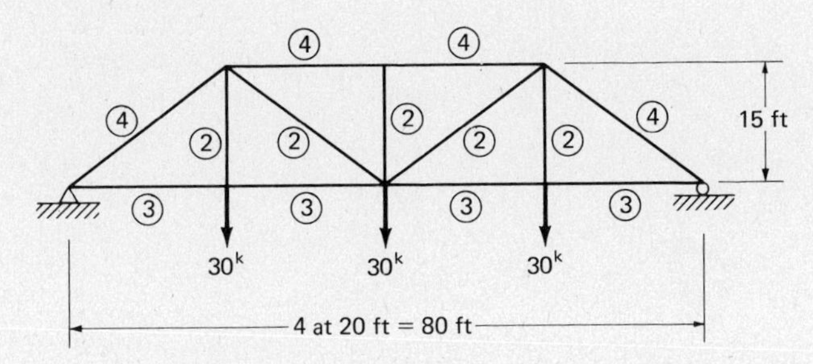

Problem 18.36 (*Ans.* U_1 = 1.70 in↓, 0.75 in←; L_3 = 0.248 in↓, 0.248 in←)

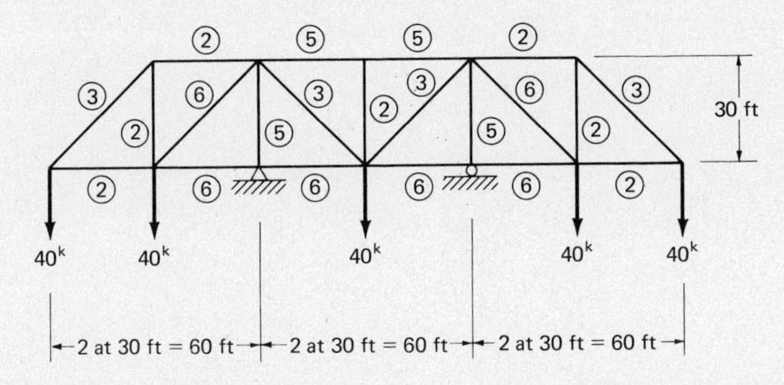

18.37. Determine equal increases in the lengths of the top chord and end diagonal members of the truss shown in the accompanying illustration which will camber the truss upwards 5 in at L_3. Use the Williot diagram.

Problem 18.37

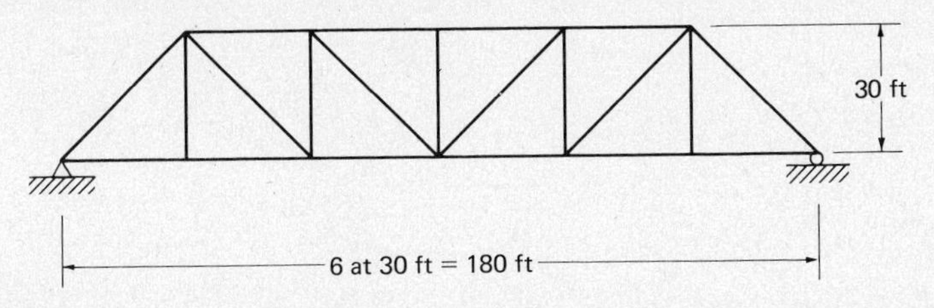

For Probs. 18.38 to 18.40 with the Williot-Mohr diagram determine the horizontal and vertical components of deflection for all of the joints of the trusses given.

18.38. For the truss of Prob. 18.2 assume L_0 fixed in position and L_0U_1 fixed in direction. (*Ans.* U_1 = 0.912 in↓, 0.442 in→; L_2 = 1.53 in↓, 0.248 in→)

18.39. For the truss of Prob. 18.3 assume L_0 fixed in position and L_0L_1 fixed in direction.

18.40. For the truss of Prob. 18.6 assume L_0 fixed in position and L_0L_1 fixed in direction. (*Ans.* L_1 = 0.143 in↓, 0.138 in→; U_1 = 0.120 in↓, 0.142 in→)

19 Deflection Methods of Statically Indeterminate Beams and Frames

19.1 INTRODUCTION

James Clerk Maxwell published in 1864 the first consistent method of analyzing statically indeterminate structures. His method was based on a consideration of deflections, but the presentation (which included the reciprocal deflection theorem) was rather brief and attracted little attention. Ten years later Otto Mohr independently extended the theory to almost its present stage of development. Analysis of redundant structures with the use of deflection computations is often referred to as the ***Maxwell-Mohr method*** or the ***method of consistent distortions*** [1,2].

Analysis of continuous beams and frames will be discussed in this chapter, and redundant trusses will be considered in Chap. 20. The process of analyzing a continuous beam consists in (1) the removal of enough supports to make the beam statically determinate, (2) the calculation of the deflections at the points where supports are removed, and (3) the determination of the forces necessary to push the support points back to their original nondeflected positions. These forces are equal to the redundant reactions.

19.2 BEAMS WITH ONE REDUNDANT

The two-span beam of Fig. 19.1(a) is assumed to consist of a material following Hooke's law. This statically indeterminate structure supports the loads P_1 and P_2 and is in turn supported by reaction components at points A, B, and C. Removal of support B would leave a statically determinate beam, proving the structure to be statically indeterminate to the first degree. It is a simple matter to find the deflection at B, δ_B in Fig. 19.1(b), caused by the external loads.

If the external loads are removed from the beam and a unit load is placed at B, a deflection at B equal to δ_{bb} will be developed, as indicated in Fig. 19.1(c). Deflections due to external loads are denoted with capital letters herein. The deflection at point C on a beam due to external loads would be δ_C. Deflections due to the imaginary unit load are denoted with two small letters. The first letter indicates the location of the deflection and the second letter indicates the location of the unit load. The deflection at E caused by a unit load at B would be δ_{eb}.

Support B is unyielding, and its removal is merely a convenient assumption. An upward force is present at B and is sufficient to prevent any deflection, or, continuing with the fictitious line of reasoning, there is a force at B that is large enough to push point B back to its original nondeflected position. The distance the support must be pushed is δ_B.

A unit load at B causes a deflection at B equal to δ_{bb}, and a 10-kip load at B will cause a deflection of $10\delta_{bb}$. Similarly, an upward reaction at B of V_B will push B up an amount $V_B\delta_{bb}$. The total deflection at B due to the external loads and the reaction V_B is zero and may be expressed as follows:

$$\delta_B + V_B\delta_{bb} = 0$$

$$V_B = -\frac{\delta_B}{\delta_{bb}}$$

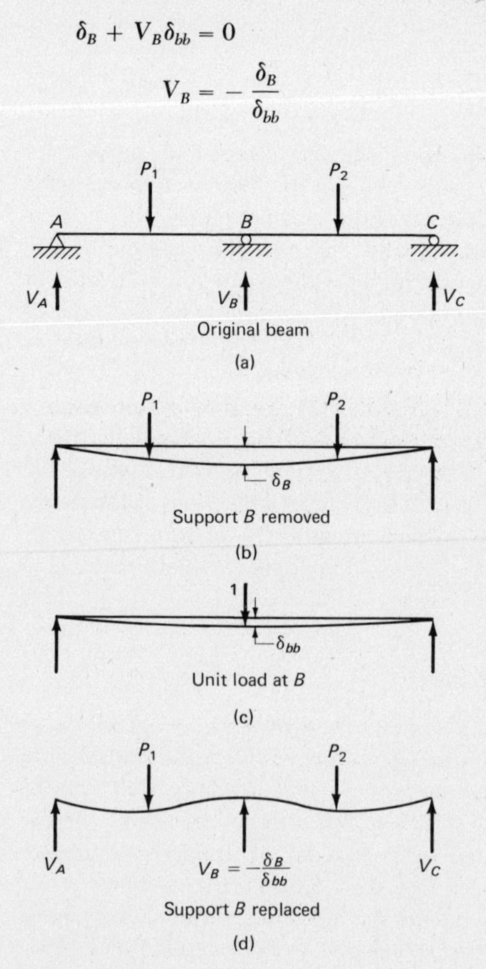

Figure 19.1

I-180 viaduct, Cheyenne, Wyoming. (Courtesy of the Wyoming State Highway Department.)

The minus sign in this expression indicates V_B is in the opposite direction from the downward unit load. If the solution of the expression yields a positive value, the reaction is in the same direction as the unit load. Examples 19.1 to 19.4 illustrate the deflection method of computing the reactions for statically indeterminate beams having one redundant reaction component. Example 19.5 shows that the method may be extended to include statically indeterminate frames as well. The necessary deflections for the first four examples are determined with the conjugate-beam procedure while those for Example 19.5 are determined by virtual work. After the value of the redundant reaction in each problem is found, the other reactions are determined by statics, and shear and moment diagrams are drawn.

Example 19.1

Determine the reactions and draw shear and moment diagrams for the two-span beam of Fig. 19.2; assume V_B to be the redundant; E and I are constant.

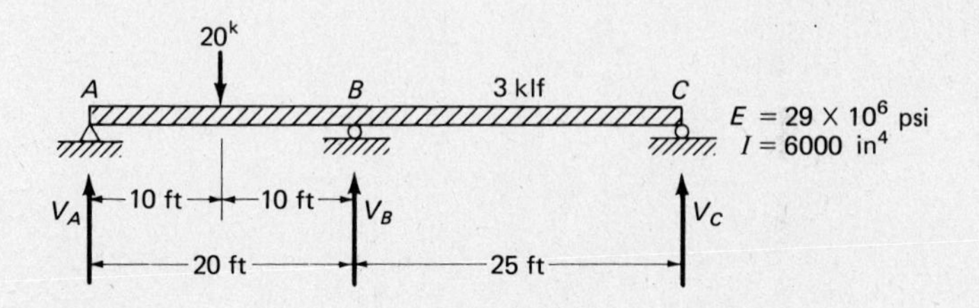

Figure 19.2

SOLUTION

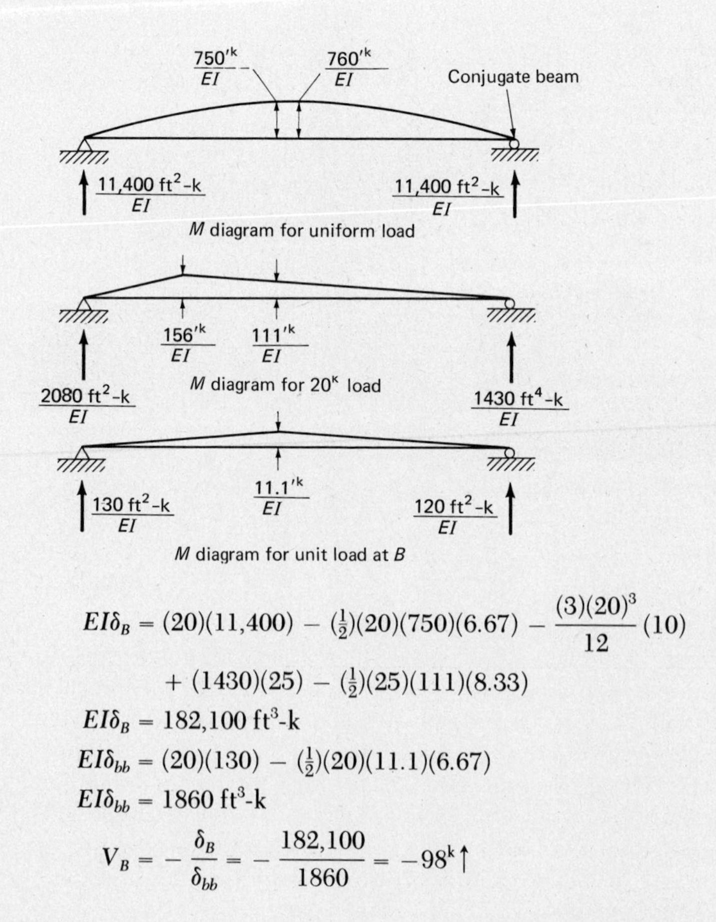

$$EI\delta_B = (20)(11{,}400) - (\tfrac{1}{2})(20)(750)(6.67) - \frac{(3)(20)^3}{12}(10)$$

$$+ (1430)(25) - (\tfrac{1}{2})(25)(111)(8.33)$$

$$EI\delta_B = 182{,}100 \text{ ft}^3\text{-k}$$

$$EI\delta_{bb} = (20)(130) - (\tfrac{1}{2})(20)(11.1)(6.67)$$

$$EI\delta_{bb} = 1860 \text{ ft}^3\text{-k}$$

$$V_B = -\frac{\delta_B}{\delta_{bb}} = -\frac{182{,}100}{1860} = -98^{\text{k}}\uparrow$$

Computing reactions at A and C by statics,

$$\Sigma M_A = 0$$

$$(20)(10) + (45)(3)(22.5) - (20)(98) - 45V_C = 0$$

$$V_C = 28.5^k \uparrow$$

$$\Sigma V = 0$$

$$20 + (3)(45) - 98 - 28.5 - V_A = 0$$

$$V_A = 28.5^k \uparrow$$

Shear and moment diagrams:

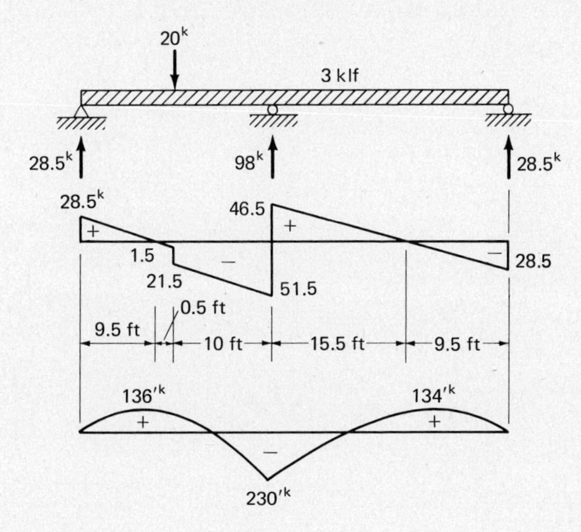

Example 19.2

Determine the reactions and draw shear and moment diagrams for the propped beam shown in Fig. 19.3. Consider V_B to be the redundant; E and I are constant.

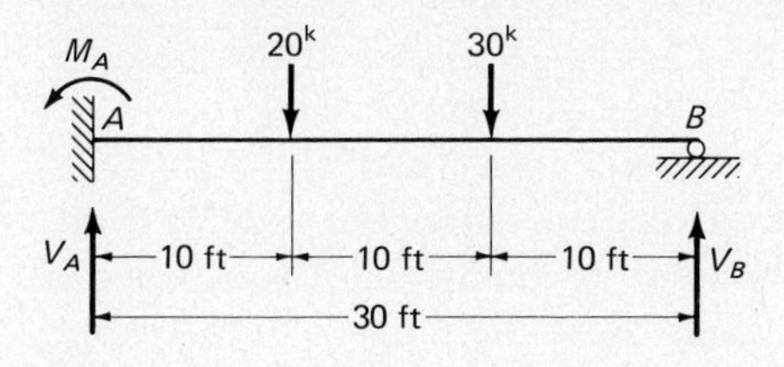

Figure 19.3

SOLUTION

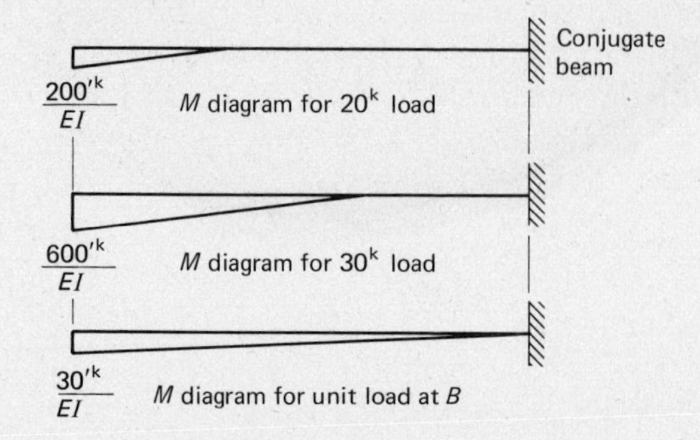

$$EI\delta_B = (\tfrac{1}{2})(200)(10)(26.67) + (\tfrac{1}{2})(600)(20)(23.33) = 166{,}670 \text{ ft}^3\text{-k}$$
$$EI\delta_{bb} = (\tfrac{1}{2})(30)(30)(20) = 9000 \text{ ft}^3\text{-k}$$
$$V_B = -\frac{166{,}670}{9000} = -18.5^k\uparrow$$

By statics

$$V_A = 31.5^k\uparrow \quad \text{and} \quad M_A = 245'^k$$

Shear and moment diagrams:

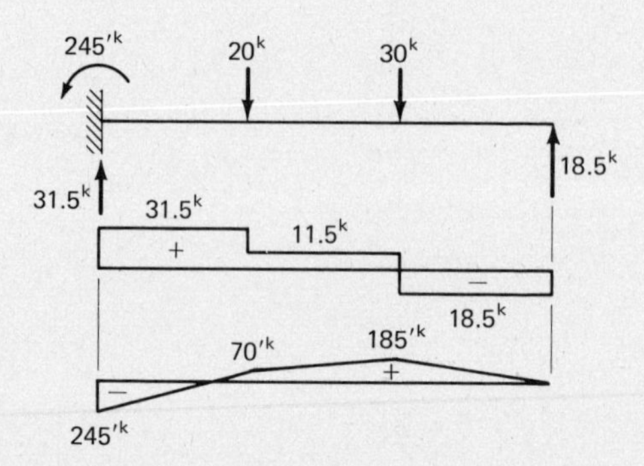

Example 19.3

Rework Example 19.2 by using the resisting moment at the fixed end as the redundant.

SOLUTION

Any one of the reactions may be considered to be the redundant and taken out, provided a stable structure remains. If the resisting moment at A is removed, a simple support remains, and the beam loads cause the tangent to the elastic curve to rotate an amount θ_A. A brief discussion of this condition will reveal a method of determining the moment.

The value of θ_A equals the shear at A in the conjugate beam loaded with the M/EI diagram. If a unit moment is applied at A, the tangent to the elastic curve will rotate an amount θ_{aa}, which can also be found from the conjugate beam. The tangent to the elastic curve at A actually does not rotate; therefore when M_A is replaced, it must be of sufficient magnitude to rotate the tangent back to its original horizontal position. The following expression equating θ_A to zero may be written and solved for the redundant M_A.

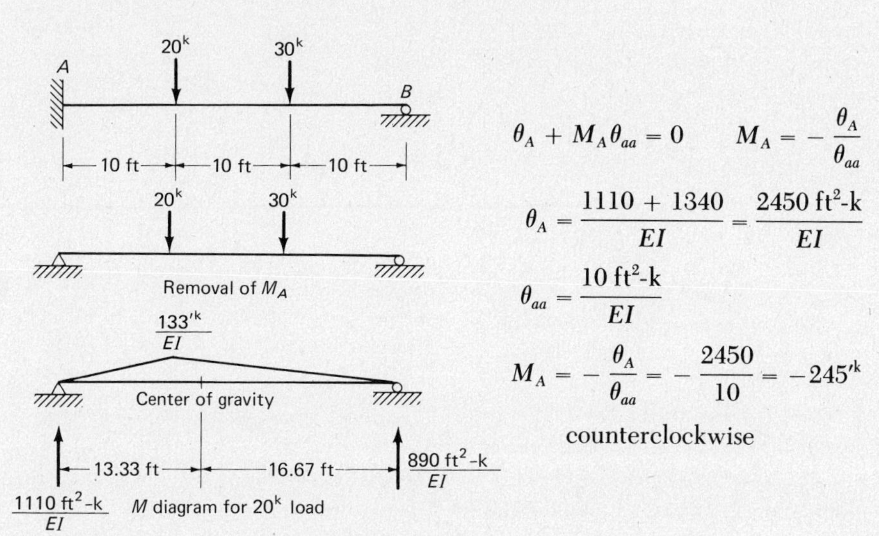

$$\theta_A + M_A\theta_{aa} = 0 \qquad M_A = -\frac{\theta_A}{\theta_{aa}}$$

$$\theta_A = \frac{1110 + 1340}{EI} = \frac{2450 \text{ ft}^2\text{-k}}{EI}$$

$$\theta_{aa} = \frac{10 \text{ ft}^2\text{-k}}{EI}$$

$$M_A = -\frac{\theta_A}{\theta_{aa}} = -\frac{2450}{10} = -245'^{k}$$

counterclockwise

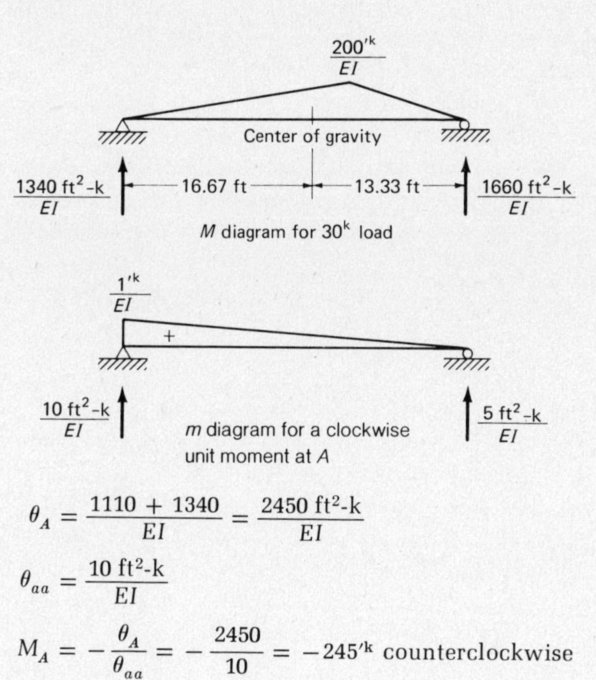

$$\theta_A = \frac{1110 + 1340}{EI} = \frac{2450 \text{ ft}^2\text{-k}}{EI}$$

$$\theta_{aa} = \frac{10 \text{ ft}^2\text{-k}}{EI}$$

$$M_A = -\frac{\theta_A}{\theta_{aa}} = -\frac{2450}{10} = -245'^{k} \text{ counterclockwise}$$

Example 19.4

Find the reactions and draw the shear and moment diagrams for the two-span beam shown in Fig. 19.4. Assume the moment at the interior support to be the redundant.

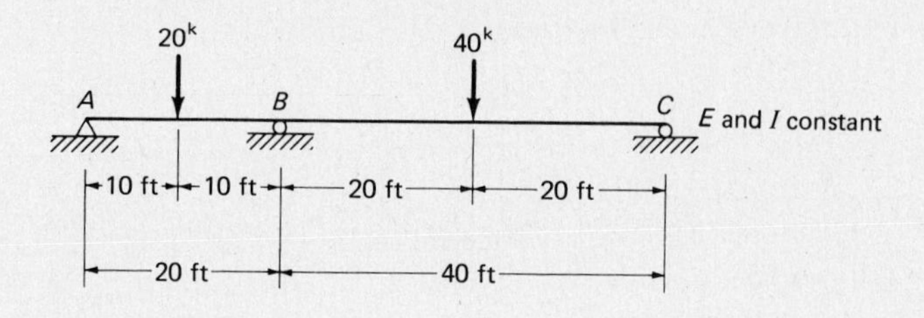

Figure 19.4

SOLUTION

Removal of moment from the interior support changes the support into a hinge, and the beam is free to slope independently on each side, as indicated by the angles θ_{b_1} and θ_{b_2} in the deflection curve shown. The numerical values of the angles can be found by placing the M/EI diagram on the conjugate structures and computing the shear on each side of the support. In the actual beam there is no change of slope of the tangent to the elastic curve from a small distance to the left of B to a small distance to the right of B.

The angle represented in the diagram by θ_B is the angle between the tangents to the elastic curve on each side of the support (that is, $\theta_{b_1} + \theta_{b_2}$). The actual moment M_B, when replaced, must be of sufficient magnitude to bring the tangents back together or reduce θ_B to zero. A unit moment applied on each side of the hinge produces a change of slope of θ_{bb}; therefore the following expression is applicable:

$$M_B = -\frac{\theta_B}{\theta_{bb}}$$

$$\theta_{b_1} = 500$$

$$\theta_{b_2} = 4000$$

$$\theta_B = \theta_{b_1} + \theta_{b_2} = 4500$$

$$\theta_{bb} = 6.67 + 13.33 = 20$$

$$M_B = -\frac{\theta_B}{\theta_{bb}} = -\frac{4500}{20} = -225'^{k}$$

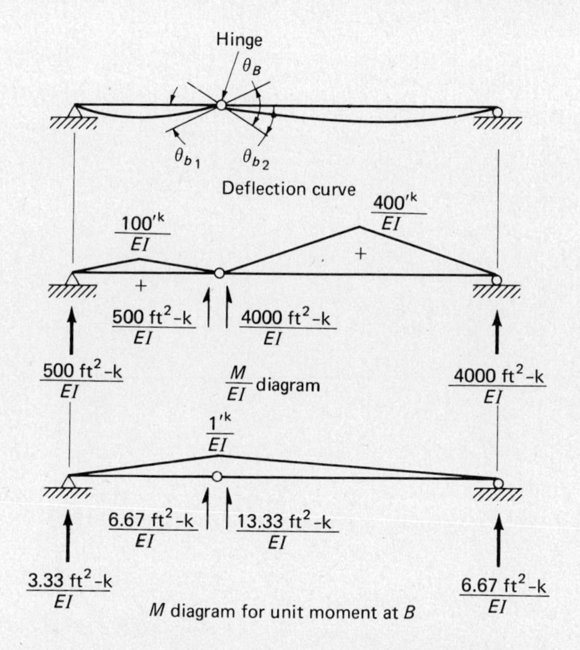

By statics the following reactions are found.

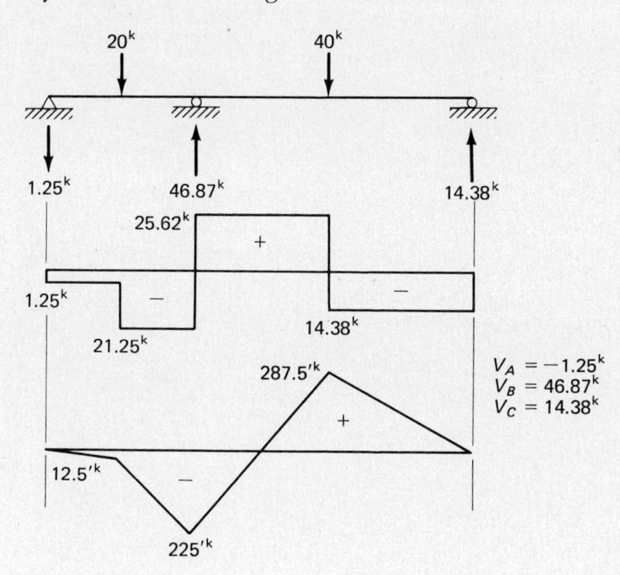

Example 19.5

Compute the reactions and draw the moment diagram for the structure shown in Fig. 19.5.

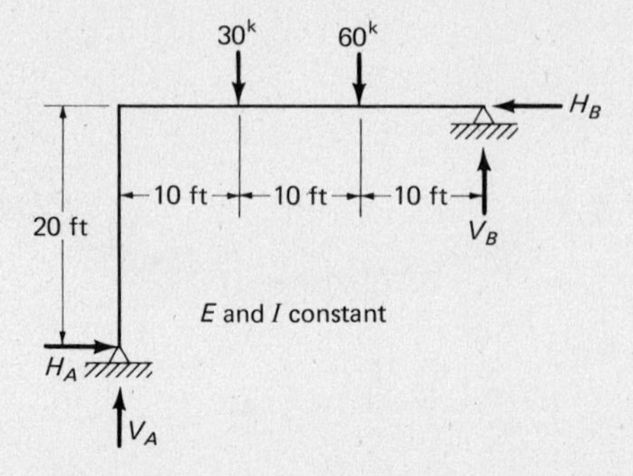

Figure 19.5

SOLUTION

Remove H_A as the redundant which changes A to a roller type of support.

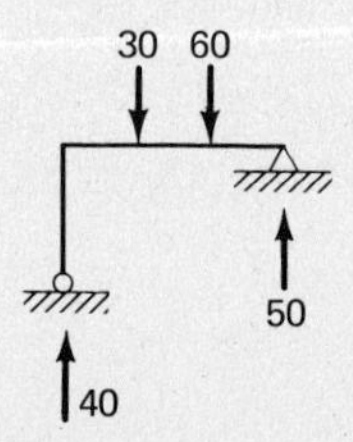

Compute the horizontal deflection at A by virtual work. The result is

$$\delta_A = \frac{86{,}600 \text{ ft}^{3k}}{EI} \leftarrow$$

Apply a unit horizontal load at A and determine the horizontal deflection δ_{aa}

The result is

$$\delta_{aa} = +\frac{6667 \text{ ft}^{3k}}{EI} \rightarrow$$

$$\delta_A + H_A\delta_{aa} = 0$$

$$H_A = -\frac{\delta_A}{\delta_{aa}} = -\frac{-86{,}660}{+6667} = +13^k \rightarrow$$

Raritan River Bridge in New Jersey. (Courtesy of Steinman, Boynton, Gronquist & Birdsall Consulting Engineers.)

Compute the remaining reactions by statics.

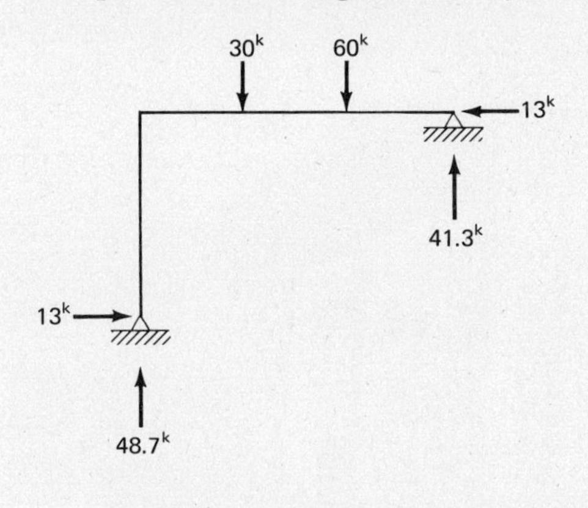

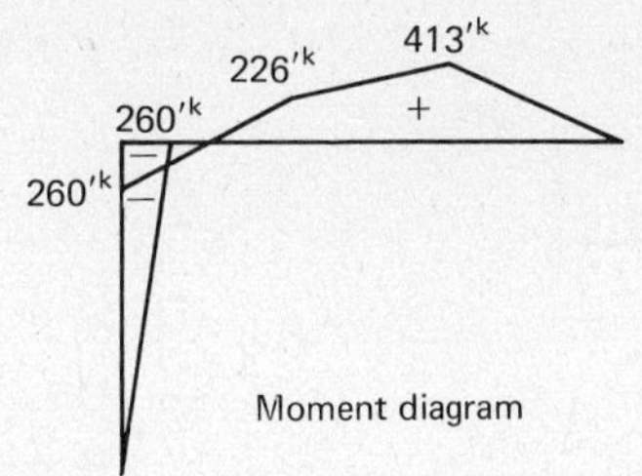

19.3 BEAMS WITH TWO OR MORE REDUNDANTS

The deflection method of analyzing beams with one redundant may be extended to beams having two or more redundants. The continuous beam of Fig. 19.6, which has two redundant reactions, is considered here.

To make the beam statically determinate, it is necessary to remove two supports. Supports B and C are assumed to be removed, and their deflections δ_B and δ_C due to the external loads are computed. The external loads are theoretically removed from the beam; a unit load is placed at B; and the deflections at B and C, δ_{bb}, and δ_{cb} are found. The unit load is moved to C, and the deflections at the two points δ_{bc} and δ_{cc} are again determined.

The reactions at supports B and C push these points up until they are in their original positions of zero deflection. The reaction V_B will raise B an amount $V_b\delta_{bb}$ and C an amount $V_B\delta_{cb}$. The reaction V_c raises C by $V_C\delta_{cc}$ and B by $V_C\delta_{bc}$.

An equation may be written for the deflection at each of the supports. Both equations contain the two unknowns, V_B and V_C, and their values may be obtained by solving the equations simultaneously.

$$\delta_B + V_B\delta_{bb} + V_C\delta_{bc} = 0$$
$$\delta_C + V_B\delta_{cb} + V_C\delta_{cc} = 0$$

The deflection method of computing redundant reactions may be extended indefinitely for beams with any number of redundants. The calcula-

I-81 river relief route interchange, Harrisburg, Pennsylvania. (Courtesy of Gannett Fleming.)

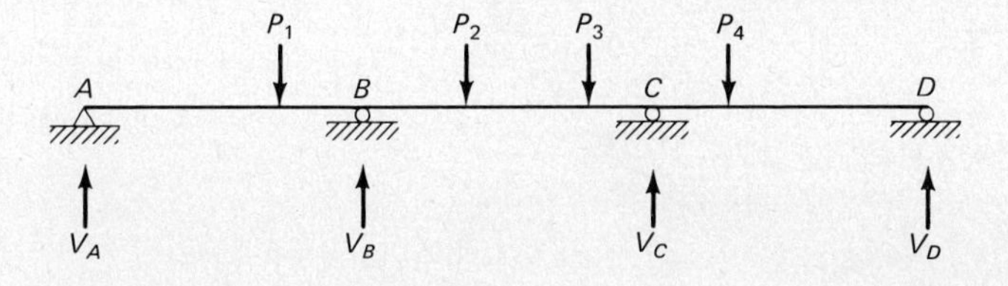

Figure 19.6

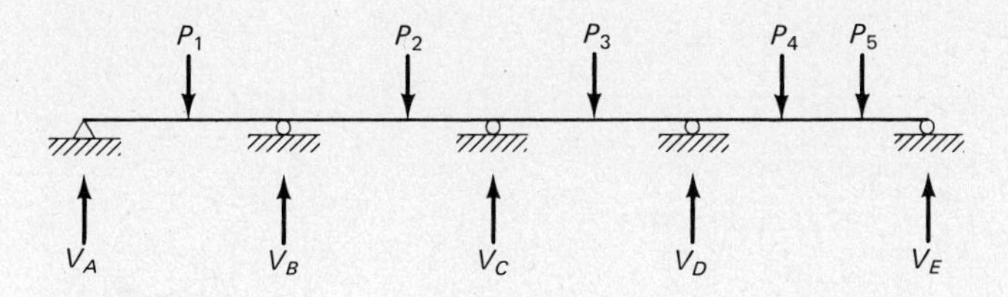

Figure 19.7

tions become quite lengthy, however, if there are more than two or three redundants. (Chapters 21 to 24 present more satisfactory methods for handling multiredundant structures.) Considering the beam of Fig. 19.7, the following expressions may be written:

$$\delta_B + V_B\delta_{bb} + V_C\delta_{bc} + V_D\delta_{bd} = 0$$
$$\delta_C + V_B\delta_{cb} + V_C\delta_{cc} + V_D\delta_{cd} = 0$$
$$\delta_D + V_B\delta_{db} + V_C\delta_{dc} + V_D\delta_{dd} = 0$$

Example 19.6 illustrates the analysis of a continuous beam with two redundants. The deflections necessary for the solution of the problem are determined with the conjugate-beam procedure. Since E and I are constant, they do not appear in the calculations.

Example 19.6

Find the reactions and draw shear and moment diagrams for the continuous beam shown in Fig. 19.8.

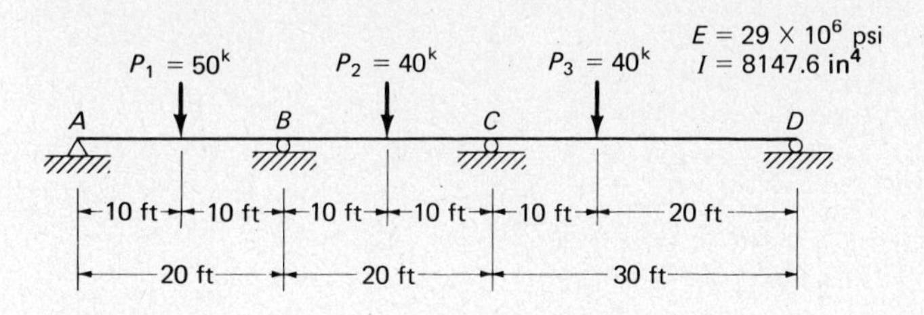

Figure 19.8

SOLUTION

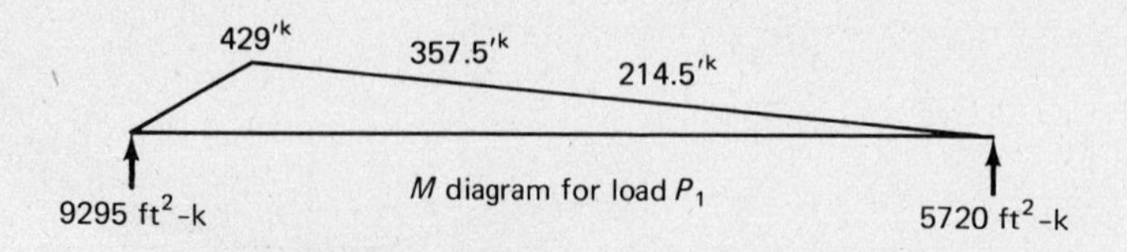

M diagram for load P_1

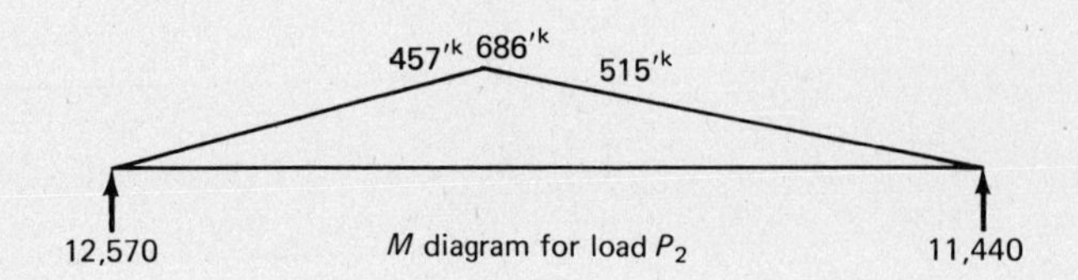

M diagram for load P_2

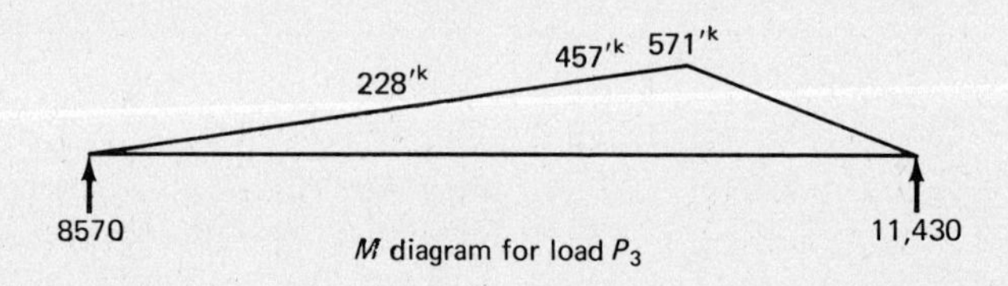

M diagram for load P_3

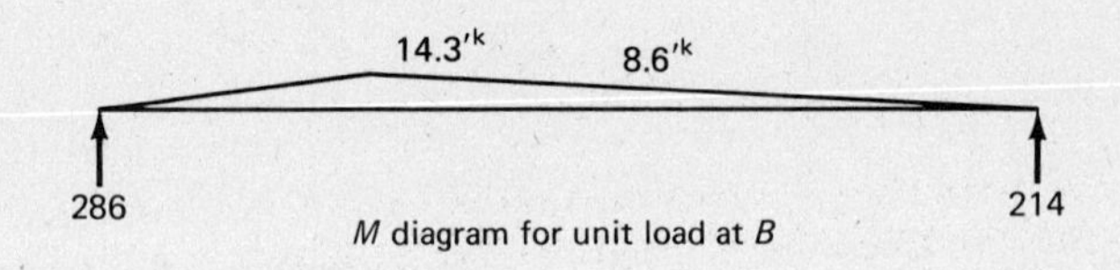

M diagram for unit load at B

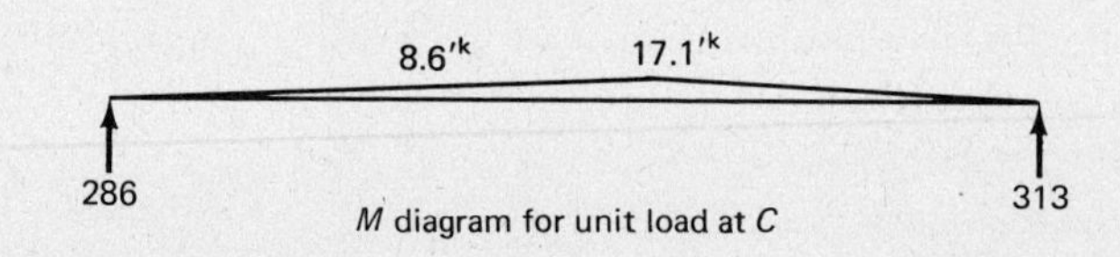

M diagram for unit load at C

$$\delta_B = (5720)(50) - (\tfrac{1}{2})(50)(357.5)(16.67) + (12{,}570 + 8570)(20) - (\tfrac{1}{2})(20)(457 + 228)(6.67)$$

$$\delta_B = 514{,}350$$

$$\delta_C = (5720 + 11{,}440)(30) - (\tfrac{1}{2})(30)(214.5 + 515)(10) + (8570)(40) - (\tfrac{1}{2})(40)(457)(13.33)$$

$$\delta_C = 625{,}300$$

$$\delta_{bb} = (20)(286) - (\tfrac{1}{2})(20)(14.3)(6.67) = 4765$$

$$\delta_{cc} = (30)(313) - (\tfrac{1}{2})(30)(17.1)(10) = 6820$$

$$\delta_{bc} = \delta_{cb} = (20)(286) - (\tfrac{1}{2})(20)(8.6)(6.67) = 5140$$

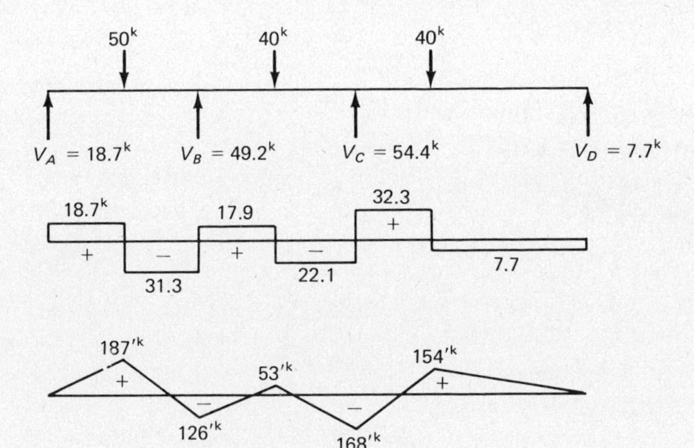

Writing the deflection equations,

$$\delta_B + V_B\delta_{bb} + V_c\delta_{bc} = 0$$

$$514{,}350 + 4765V_B + 5140V_C = 0 \tag{1}$$

$$\delta_C + V_B\delta_{cb} + V_C\delta_{cc} = 0$$

$$625{,}300 + 5140V_B + 6820V_C = 0 \tag{2}$$

Simultaneous solution of Eqs. (1) and (2) gives V_B and V_C; values of V_A and V_D are found by statics.

Continuous girder bridge over Wabash River, Lockport, Indiana. (Courtesy of the Portland Cement Association.)

19.4 SUPPORT SETTLEMENT

Continuous beams with unyielding supports have been considered in the preceding sections. Should the supports settle or deflect from their theoretical positions, major changes may occur in the reactions, shears, moments, and stresses. Whatever the factors causing displacement (weak foundations, temperature changes, poor erection or fabrication, and so on), analysis may be made with the deflection expressions previously developed for continuous beams.

An expression for deflection at point B in the two-span beam of Fig. 19.1 was written in Sec. 19.2. The expression was developed on the assumption that support B was temporarily removed from the structure, allowing point B to deflect, after which the support was replaced. The reaction at B, V_B, was assumed to be of sufficient magnitude to push B up to its original position of zero deflection. Should B actually settle 1.0 in, V_B will be smaller because it will only have to push B up an amount $\delta_B - 1.0$ in, and the deflection expression may be written as

$$\delta_B - 1.0 + V_B\delta_{bb} = 0$$

If three men are walking along with a log on their shoulders and one of them lowers his shoulder slightly, he will not have to support as much of the total weight as before. He has, in effect, backed out from under the log and thrown more of its weight to the other men. The settlement of a support in a continuous beam has the same effect.

The values of δ_B and δ_{bb} must be calculated in inches if the support movement is given in inches; they are calculated in feet if the support movement is given in feet; and so on. Example 19.7 illustrates the analysis of the two-span beam of Example 19.1 with the assumption of a $\frac{3}{4}$-in settlement of the interior support. The moment diagram is drawn after settlement occurs and is compared with the diagram before settlement. The seemingly small displacement has completely changed the moment picture.

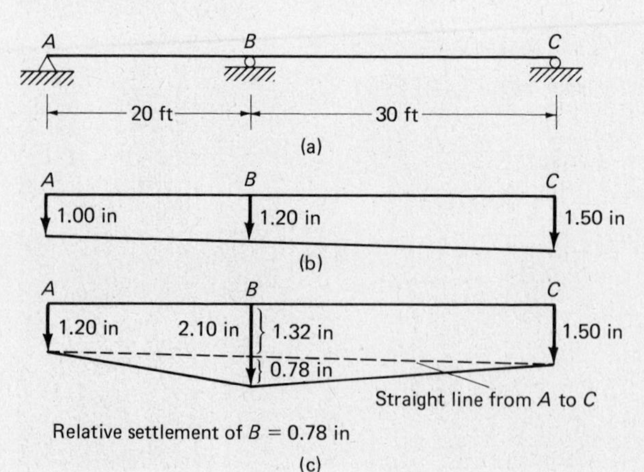

Figure 19.9

When several or all of the supports are displaced, the analysis may be conducted on the basis of relative settlement values. For example, if all of the supports of the beam of Fig. 19.9(a) were to settle 1.5 in, the stress conditions would be unchanged. If the supports settle different amounts but remain in a straight line, as illustrated in Fig. 19.9(b), the situation theoretically is the same as before settlement.

Where inconsistent settlements occur and the supports no longer lie on a straight line, the stress conditions change because the beam is distorted. The situation may be handled by drawing a line through the displaced positions of two of the supports, usually the end ones. The distances of the other supports from this line are determined and used in the calculations, as illustrated in Fig. 19.9(c) and Example 19.8.

Example 19.7

Determine the reactions and draw shear and moment diagrams for the beam of Example 19.1, which is reproduced in Fig. 19.10, if support B settles $\frac{3}{4}$ in.

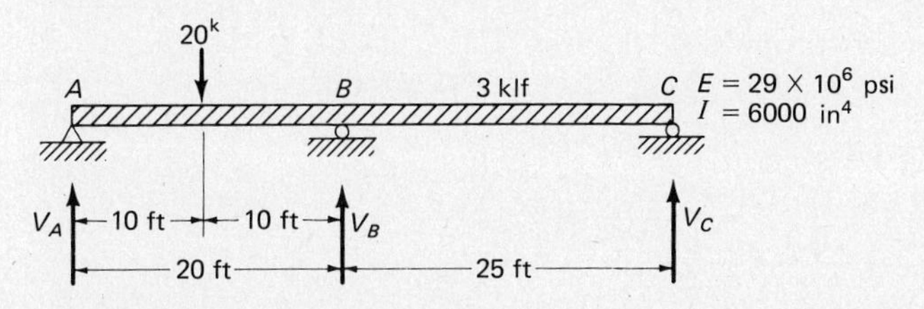

Figure 19.10

SOLUTION

The values of δ_B and δ_{bb} previously found are computed in inches, and the effect of the support settlement on V_B is determined. By statics the new values of V_A and V_B are found and the shear and moment diagrams are drawn. The moment diagram before settlement is repeated to illustrate the striking changes.

$$\delta_B \doteq \frac{182{,}100 \text{ ft}^3\text{-k}}{EI} = 1.81 \text{ in}$$

$$\delta_{bb} = \frac{1860 \text{ ft}^3\text{-k}}{EI} = 0.0185 \text{ in}$$

$$\delta_B - 0.750 + V_B\delta_{bb} = 0$$

$$V_B = -\frac{1.81 - 0.750}{0.0185} = -57.3^k \uparrow$$

$$V_A = 51.2^k \uparrow$$

$$V_C = 46.5^k \uparrow$$

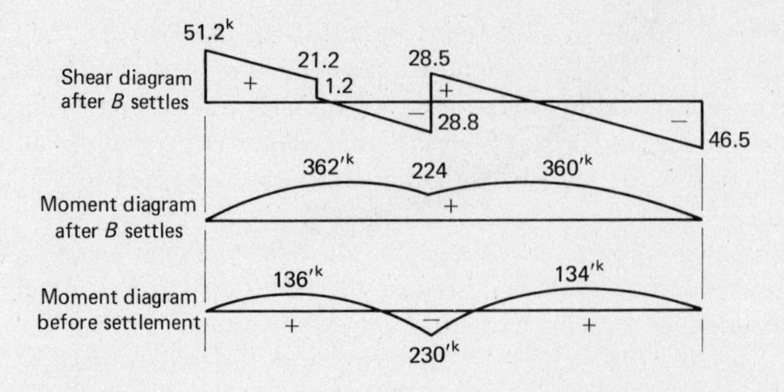

Example 19.8

Determine the reactions and draw shear and moment diagrams for the beam of Example 19.6, which is reproduced in Fig. 19.11, if the supports settle as follows: A is 1.25 in, B is 2.40 in, C is 2.75 in, and D is 1.10 in.

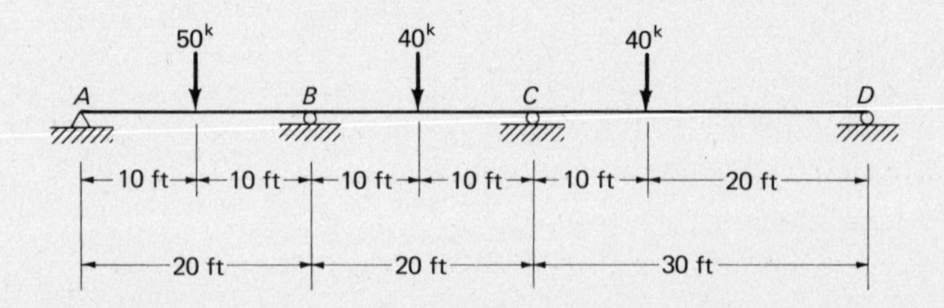

Figure 19.11

SOLUTION

The reactions at supports B and C are considered to be the redundants; therefore a diagram of the settlements is plotted to determine the relative settlements of B and C.

Settlement diagram:

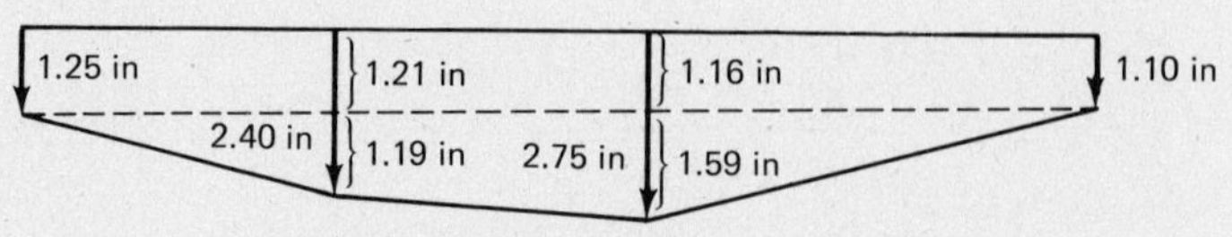

From Example 19.6:

$$\delta_B = \frac{514{,}350 \text{ ft}^3\text{-k}}{EI} = 3.76 \text{ in}$$

$$\delta_C = \frac{625{,}300 \text{ ft}^3\text{-k}}{EI} = 4.57 \text{ in}$$

$$\delta_{bb} = \frac{4765 \text{ ft}^3\text{-k}}{EI} = 0.0349 \text{ in}$$

$$\delta_{cc} = \frac{6820 \text{ ft}^3\text{-k}}{EI} = 0.0498 \text{ in}$$

$$\delta_{bc} = \delta_{cb} = \frac{5140 \text{ ft}^3\text{-k}}{EI} = 0.0374 \text{ in}$$

$$\delta_B + V_B\delta_{bb} + V_C\delta_{bc} = \Delta_B$$

$$3.76 + 0.0349V_B + 0.0374V_C = 1.19 \qquad (1)$$

$$\delta_C + V_B\delta_{cb} + V_C\delta_{cc} = \Delta_C$$

$$4.57 + 0.0374V_B + 0.0498V_C = 1.59 \qquad (2)$$

Solving Eqs. (1) and (2) simultaneously gives values of V_B and V_C, and by statics V_A and V_D are determined.

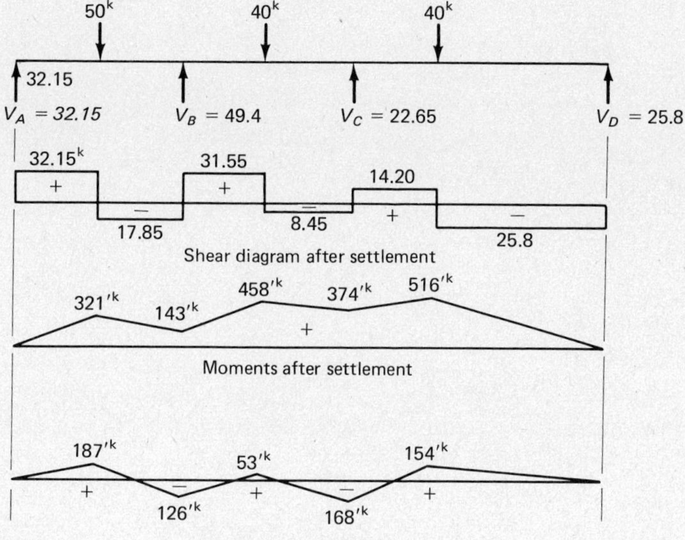

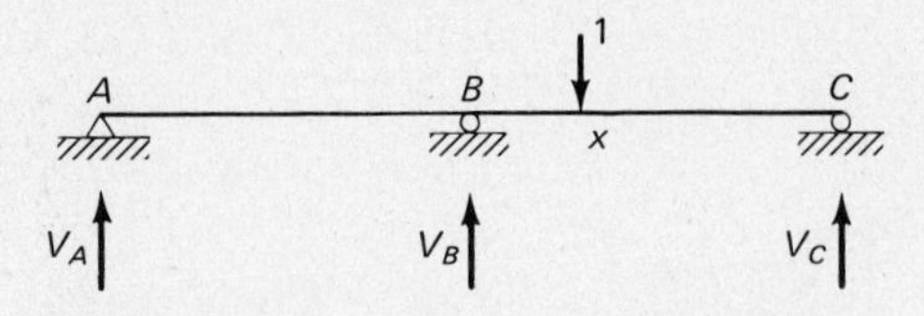

Figure 19.12

19.5 INFLUENCE LINES FOR STATICALLY INDETERMINATE BEAMS

The uses of influence lines for statically indeterminate structures are the same as those for statically determinate structures. They enable the designer to locate the critical positions for live loads and to compute forces for various positions of the loads. Influence lines for statically indeterminate structures are not as simple to draw as they are for statically determinate structures. For the latter case it is possible to compute the ordinates for a few controlling points and connect those values with a set of straight lines. Unfortunately, influence lines for continuous structures require the computation of ordinates at a large number of points because the diagrams are either curved or made up of a series of chords. The chord-shaped diagram occurs where loads can only be transferred to the structure at intervals, as at the panel points of a truss or at joists framing into a girder.

The problem of preparing the diagrams is not as difficult as the preceding paragraph seems to indicate because a large percentage of the work may be eliminated by applying Maxwell's law of reciprocal deflections. The preparation of an influence line for the interior reaction of the two-span beam of Fig. 19.12 is considered in the following paragraphs.

The procedure for calculating V_B has been to remove it from the beam and then compute δ_B and δ_{bb} and substitute their values in the usual formula. The same procedure may be used in drawing an influence line for V_B. A unit load is placed at some point x causing δ_B to equal δ_{bx}, from which the following expression is written:

$$V_B = -\frac{\delta_B}{\delta_{bb}} = -\frac{\delta_{bx}}{\delta_{bb}}$$

At first glance it appears that the unit load will have to be placed at numerous points on the beam and the value of δ_{bx} laboriously computed for each. A study of the deflections caused by a unit load at point x, however, proves these computations to be unnecessary. By Maxwell's law the deflection at B due to a unit load at $x(\delta_{bx})$ is identical with the deflection at x due to a unit load at $B(\delta_{xb})$. The expression for V_B becomes

$$V_B = -\frac{\delta_{xb}}{\delta_{bb}}$$

It is now evident that the unit load need only be placed at B, and the deflections at various points across the beam computed. Dividing each of these values by δ_{bb} gives the ordinates for the influence line. If a deflection curve is plotted for the beam for a unit load at B (support B being removed), an influence line for V_B may be obtained by dividing each of the deflection ordinates by δ_{bb}. Another way of expressing this principle is as follows: If a unit deflection is caused at a support for which the influence line is desired, the beam will draw its own influence line because the deflection at any point in the beam is the ordinate of the influence line at that point for the reaction in question.

Maxwell's presentation of his theorem in 1864 was so brief that its value was not fully appreciated until 1886 when Heinrich Müller-Breslau clearly showed its true worth as described in the preceding paragraph [3]. Müller-Breslau's principle may be stated in detail as follows: **The deflected shape of a structure represents to some scale the influence line for a function such as stress, shear, moment, or reaction component if the function is allowed to act through a unit displacement.** The principle is applicable to statically determinate and indeterminate beams, frames, and trusses.

The influence line for the reaction at the interior support of a two-span beam is presented in Example 19.9. Influence lines are also shown for the end reactions, the values for ordinates having been obtained by statics from those computed for the interior reaction. The conjugate-beam procedure is an excellent method of determining the beam deflections necessary for preparing the diagrams.

Example 19.9

Draw influence lines for reactions at each support of the structure shown in Fig. 19.13.

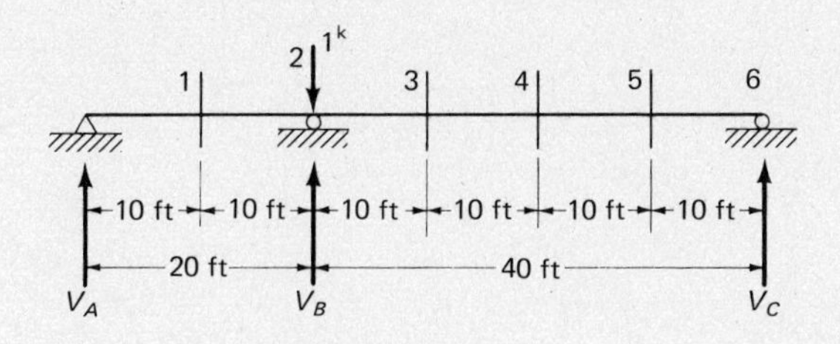

Figure 19.13

SOLUTION

Remove V_B, place a unit load at B, and compute the deflections caused at 10-ft intervals by the conjugate-beam method:

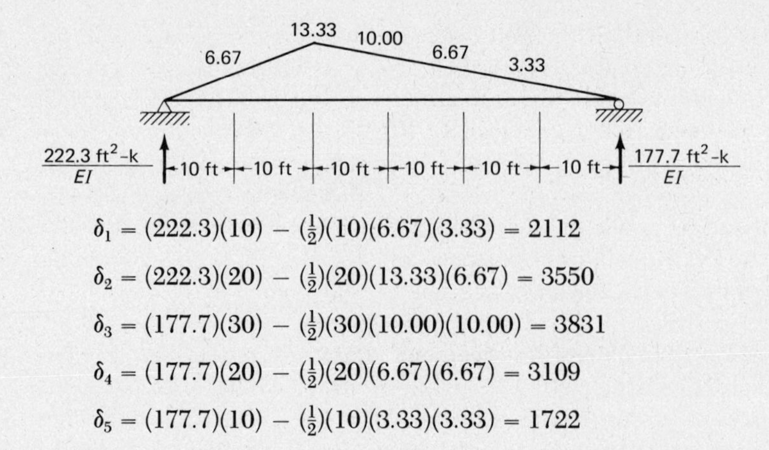

$$\delta_1 = (222.3)(10) - (\tfrac{1}{2})(10)(6.67)(3.33) = 2112$$

$$\delta_2 = (222.3)(20) - (\tfrac{1}{2})(20)(13.33)(6.67) = 3550$$

$$\delta_3 = (177.7)(30) - (\tfrac{1}{2})(30)(10.00)(10.00) = 3831$$

$$\delta_4 = (177.7)(20) - (\tfrac{1}{2})(20)(6.67)(6.67) = 3109$$

$$\delta_5 = (177.7)(10) - (\tfrac{1}{2})(10)(3.33)(3.33) = 1722$$

Noting $\delta_{bb} = \delta_2$, the values of the influence-line ordinates for V_B are found by dividing each deflection by δ_2.

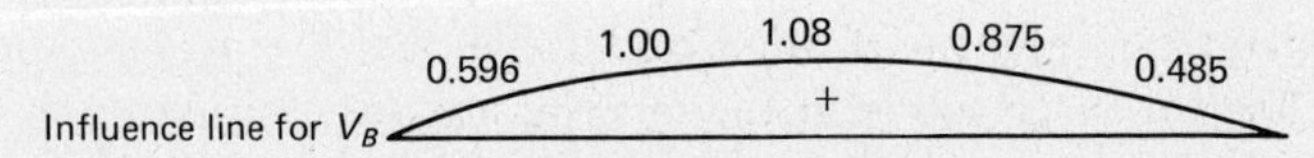

Having the values of V_B for various positions of the unit load the values of V_A and V_C for each load position can be determined by statics with the following results.

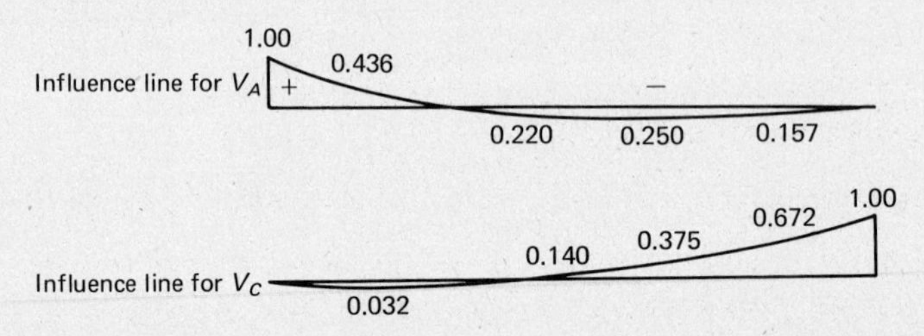

For checking results support C is removed, a unit load is placed there, and the resulting deflections are computed at 10-ft intervals.

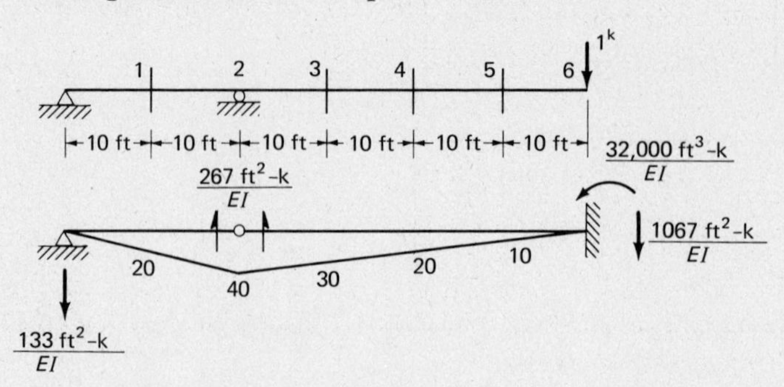

$$\delta_1 = -(133)(10) + (\tfrac{1}{2})(10)(20)(3.33) = -1000$$

$$\delta_2 = -(133)(20) + (\tfrac{1}{2})(20)(40)(6.67) = 0$$

$$\delta_3 = +32{,}000 - (1067)(30) + (\tfrac{1}{2})(30)(30)(10) = +4500$$

$$\delta_4 = +32{,}000 - (1067)(20) + (\tfrac{1}{2})(20)(20)(6.67) = +12{,}000$$

$$\delta_5 = +32{,}000 - (1067)(10) + (\tfrac{1}{2})(10)(10)(3.33) = +21{,}500$$

$$\delta_6 = +32{,}000$$

Since $\delta_{cc} = \delta_6$, the ordinates of the influence line for V_C are determined by dividing each deflection by δ_6.

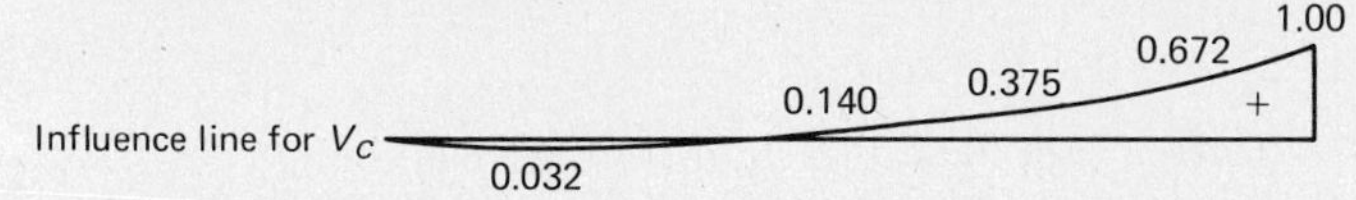

The next problem is to draw the influence lines for beams continuous over three spans, which have two redundants. For this discussion the beam of Fig. 19.14 is considered, and the reactions V_B and V_C are assumed to be the redundants.

It will be necessary to remove the redundants and compute the deflections at various sections in the beam for a unit load at B and also for a unit load at C. By Maxwell's law a unit load at any point x causes a deflection at $B(\delta_{bx})$ equal to the deflection at x due to a unit load at $B(\delta_{xb})$. Similarly, $\delta_{cx} = \delta_{xc}$. After computing δ_{xb} and δ_{xc} at the several sections, their values at each section may be substituted into the following simultaneous equations, whose solution will yield the values of V_B and V_C.

$$\delta_{xb} + V_B\delta_{bb} + V_C\delta_{bc} = 0$$

$$\delta_{xc} + V_B\delta_{cb} + V_C\delta_{cc} = 0$$

The simultaneous equations are solved quickly, even though a large number of ordinates are being computed, because the only variables in the equations are δ_{xb} and δ_{xc}. After the influence lines are prepared for the redundant reactions of a beam, the ordinates for any other function (moment, shear, and so on) can be determined by statics. Example 19.10 illustrates the calculations necessary for preparing influence lines for several functions of a three-span continuous beam.

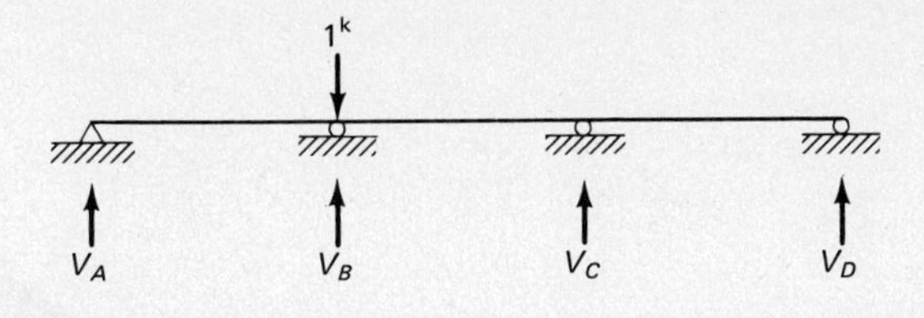

Figure 19.14

Example 19.10

Draw influence lines for V_B, V_C, V_D, M_7, and shear at section 6, Fig. 19.15.

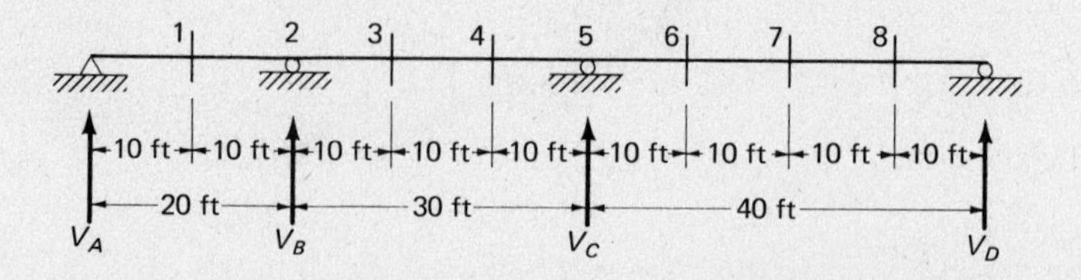

Figure 19.15

SOLUTION

Remove V_B and V_C, place a unit load at B, and load the conjugate beam with the M/EI diagram.

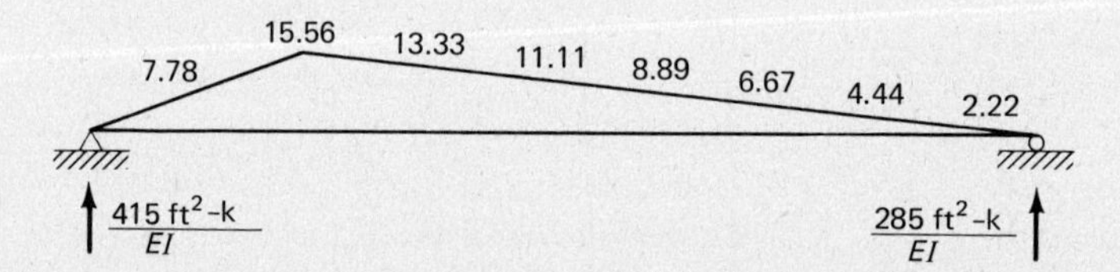

Place a unit load at C and load the conjugate beam with the M/EI diagram.

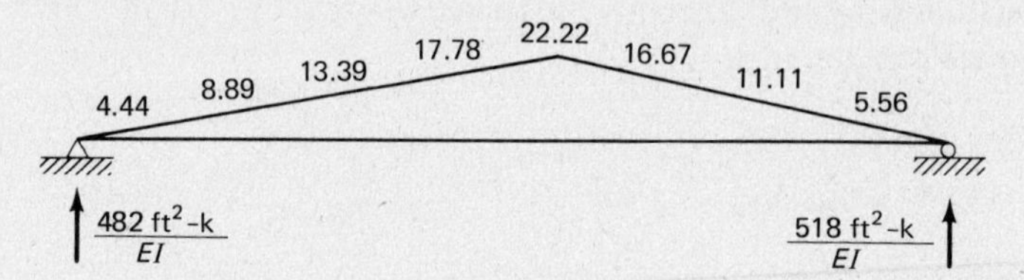

Compute the values of δ_{xb} and δ_{xc}, from which V_B and V_C are obtained by solving the simultaneous equations. Ordinates for M_7, V_D, and shear at section 6 are obtained by statics.

Table 19.1

SECTION	δ_{xb}	δ_{xc}	V_B	V_C	V_D	M_7	V_6
1	4020	4746	+0.646	−0.0735	+0.008	+0.160	−0.008
2	7255 = δ_{bb}	9040 = δ_{bc}	+1.00	0	0	0	0
3	9100	12,460	+0.855	+0.320	−0.0344	−0.688	+0.0344
4	9630	14,540	+0.432	+0.720	−0.0513	−1.026	+0.0513
5	9040 = δ_{cb}	14,825 = δ_{cc}	0	+1.00	0	0	0
6	7550	13,040	−0.227	+1.02	+0.1504	+3.008	−0.1504 +0.8496
7	5404	9618	−0.257	+0.805	+0.388	+7.76	+0.612
8	2813	5088	−0.159	+0.440	+0.688	+3.68	+0.316

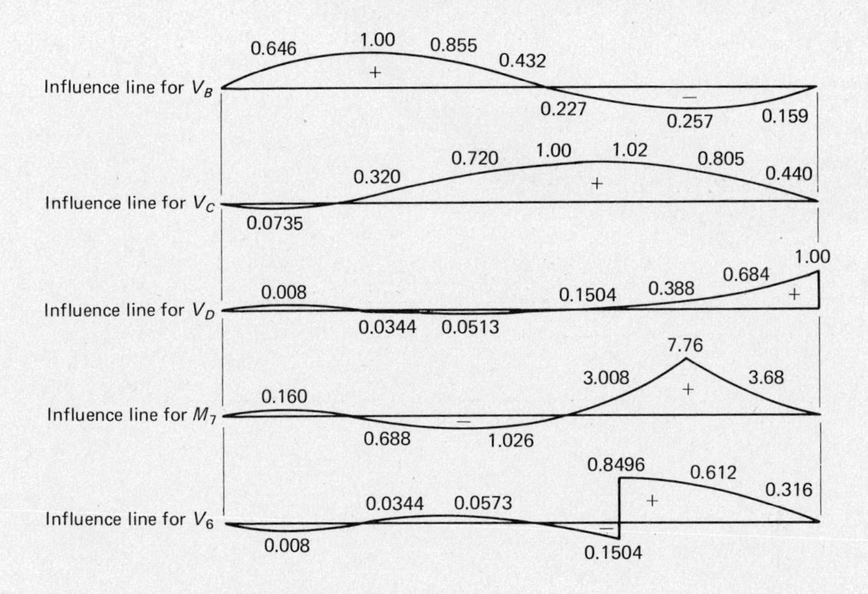

19.6 QUALITATIVE INFLUENCE LINES

Müller-Breslau's principle is of such importance that space is taken to emphasize its value. The shape of the usual influence line needed for continuous structures is so simple to obtain from his principle that in many situations it is unnecessary to perform the labor needed to compute the numerical values of the ordinates. It is possible to sketch roughly the diagram with sufficient accuracy to locate the critical positions for live load for various functions of the structure. This possibility is of particular importance for building frames, as will be illustrated in subsequent paragraphs.

If the influence line is desired for the left reaction of the continuous beam of Fig. 19.16(a), its general shape can be determined by letting the reaction act upward through a unit distance as shown in Fig. 19.16(b) of the figure. If the left end of the beam were pushed up, the beam would take the shape shown. This distorted shape can be easily sketched, remembering the other supports are unyielding. Influence lines obtained by sketching are said to be ***qualitative influence lines***, whereas the exact ones are said to be ***quantitative influence lines***. The influence line for V_C in Fig. 19.16(c) is another example of qualitative sketching for reaction components.

Figure 19.16(d) shows the influence line for positive moment at point x near the center of the left-hand span. The beam is assumed to have a pin or hinge inserted at x and a couple applied adjacent to each side of the pin that will cause compression on the top fibers (plus moment). Twisting the beam on each side of the pin causes the left span to take the shape indicated, and the deflected shape of the remainder of the beam may be roughly sketched. A similar procedure is used to draw the influence line for negative moment at point y in the third span, except that a moment couple is applied at the

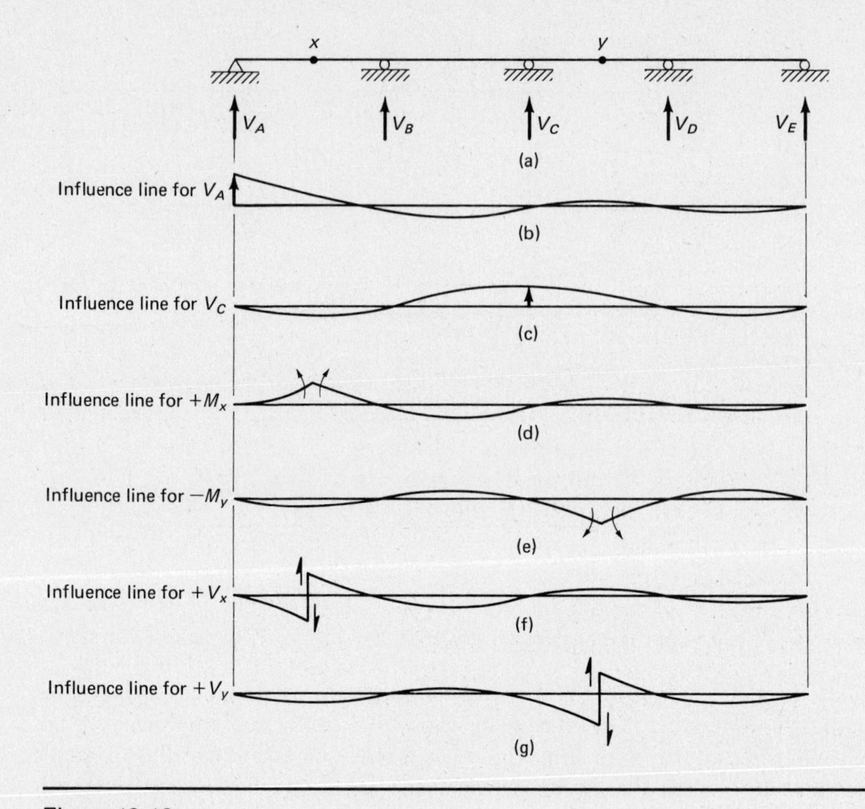

Figure 19.16

assumed pin which will tend to cause compression on the bottom beam fibers, corresponding with negative moment.

Finally, qualitative influence lines are drawn for positive shear at points x and y. At point x the beam is assumed to be cut, and two vertical forces of the nature required to give positive shear are applied to the beam on the sides of the cut section. The beam will take the shape shown in Fig. 19.16(f). The same procedure is used to draw the diagram for positive shear at point y.

From these diagrams considerable information is available concerning critical live-loading conditions. If a maximum positive value of V_A were desired for a uniform live load, the load would be placed in spans 1 and 3, where the diagram has positive ordinates; if maximum negative moment were required at point x, spans 2 and 4 would be loaded; and so on.

Qualitative influence lines are particularly valuable for determining critical load positions for buildings, as shown by the moment influence line for the building frame of Fig. 19.17. In drawing the diagrams for an entire frame, the joints are assumed to be free to rotate, but the members at each joint are assumed rigidly connected to each other so that the angles between them do not change during rotation. The diagram of this figure is sketched for positive moment at the center of beam AB.

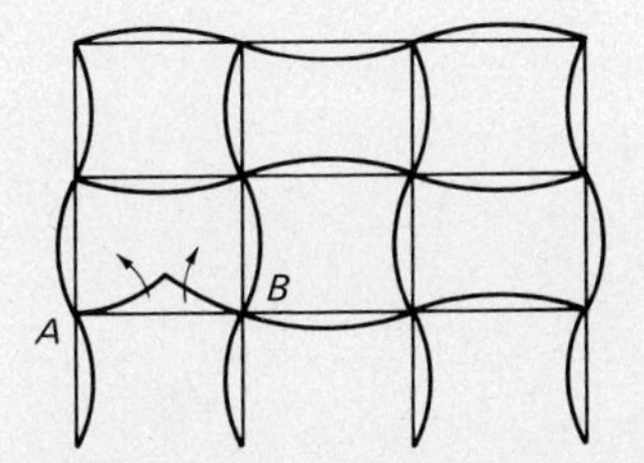

Figure 19.17

The spans that should be loaded to cause maximum positive moment are obvious from the diagram. It should be noted that loads on a beam more than approximately three spans away have little effect on the function under consideration. This fact can be seen in the influence lines of Example 19.10, where the ordinates even two spans away are quite small.

A warning should be given regarding qualitative influence lines. They should be drawn for functions near the center of spans or at the supports, but for sections near one-fourth points they should not be sketched without a good deal of study. Near the one-fourth point of a span is a point called the *fixed point* at which the influence line changes in type. The subject of fixed points is discussed at some length in the text *Continuous Frames of Reinforced Concrete* by H. Cross and N. D. Morgan [4].

PROBLEMS

For Probs. 19.1 to 19.24 compute the reactions and draw shear and moment diagrams for the continuous beams; E and I are constant unless noted otherwise. The method of consistent distortions is to be used.

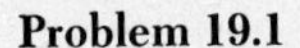

Problem 19.1

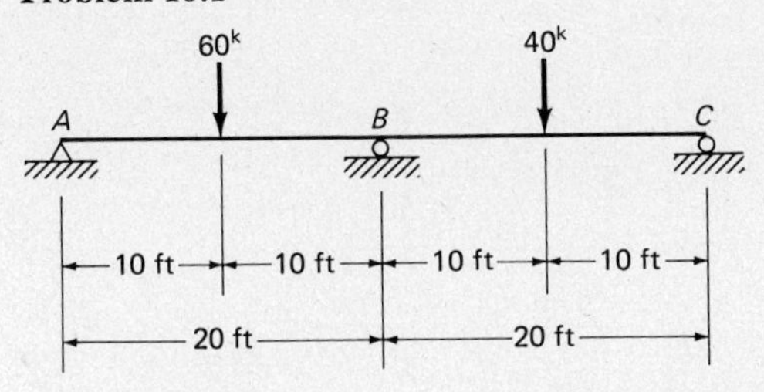

Problem 19.2 (*Ans.* $V_A = 53.75^k \uparrow$, $V_B = 26.25^k \uparrow$, $M_B = -125'^k$)

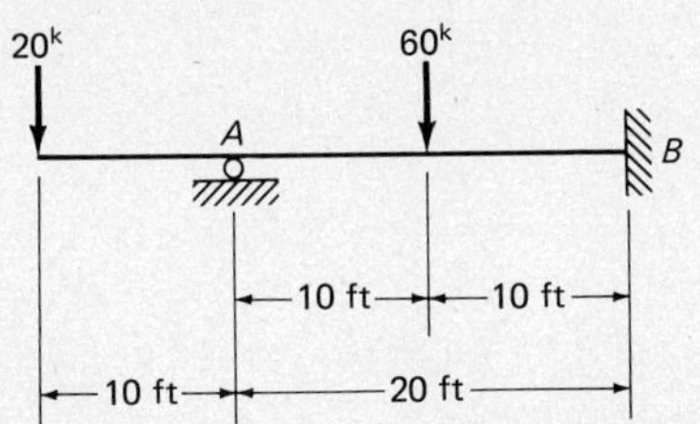

Problem 19.3

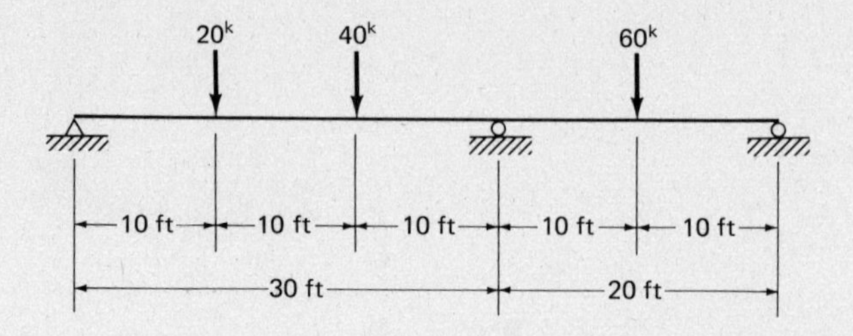

Problem 19.4 (*Ans.* $V_A = 15^k \uparrow$, $M_A = -60'^k$, $V_B = 9^k \uparrow$)

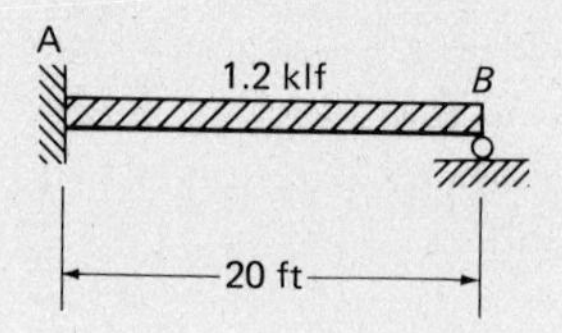

Problem 19.5

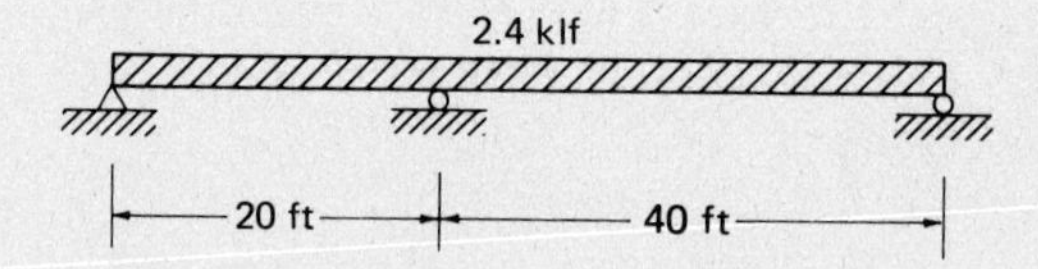

Problem 19.6 (*Ans.* $V_A = 7.53^k \uparrow$, $M_B = -111.7'^k$, $V_C = 14.27^k \uparrow$)

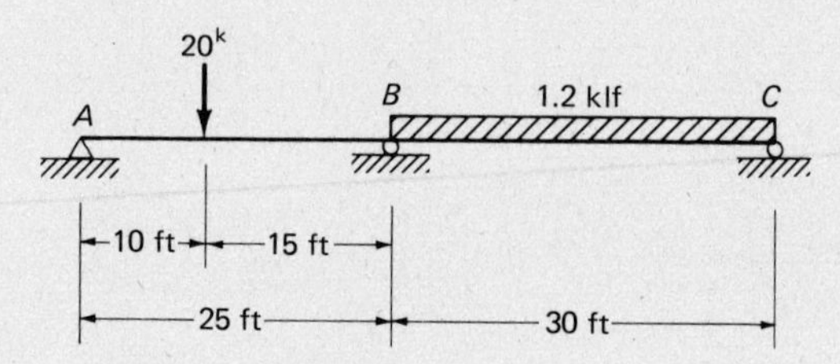

Problem 19.7

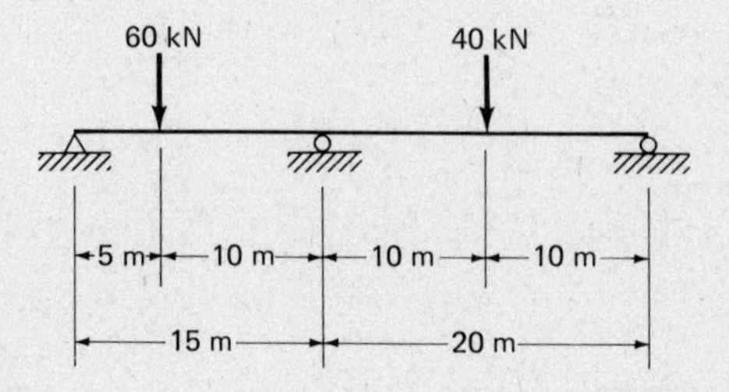

Problem 19.8 (*Ans.* $M_A = -125.0$ kN · m, $M_B = -250.0$ kN · m, $V_A = 87.5$ kN ↑, $V_B = 212.5$ kN ↑)

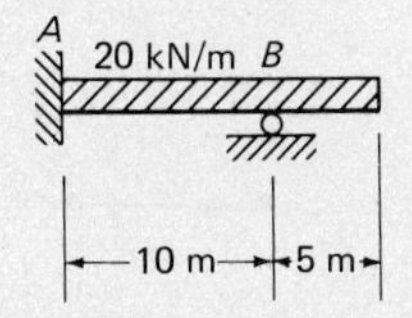

Problem 19.9

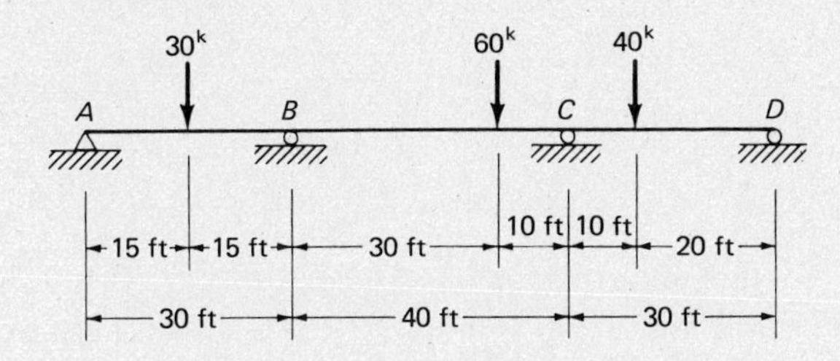

Problem 19.10 (*Ans.* $V_A = 30.19^k$ ↑, $V_B = 80.62^k$ ↑, $M_B = -176.2'^k$, $V_C = 15.19^k$ ↑)

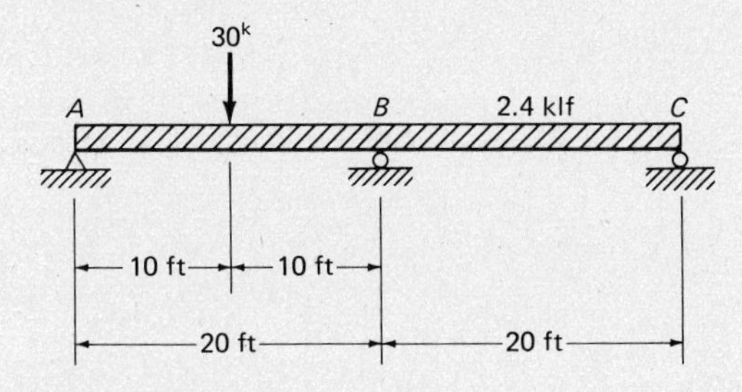

Problem 19.11

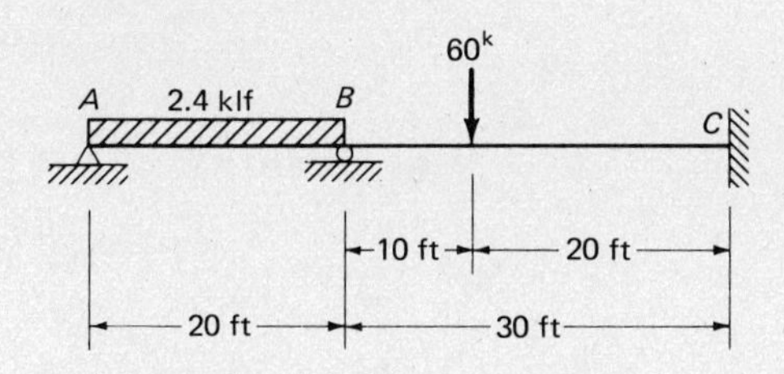

Problem 19.12 (*Ans.* $V_A = 3^k$ ↑, $M_A = +33.6'^k$, $V_B = 45.8^k$ ↑, $M_B = -240'^k$)

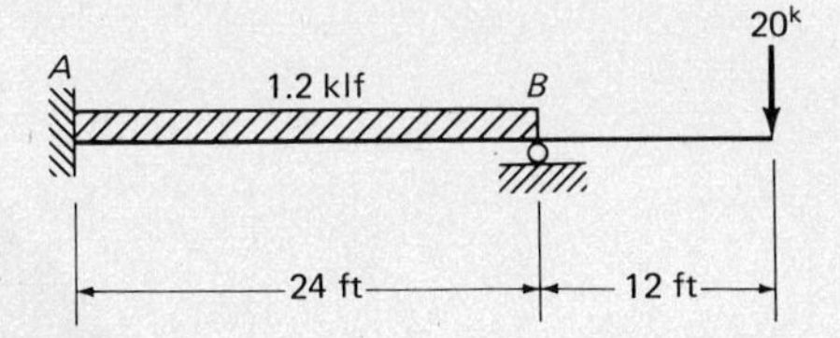

Problem 19.13

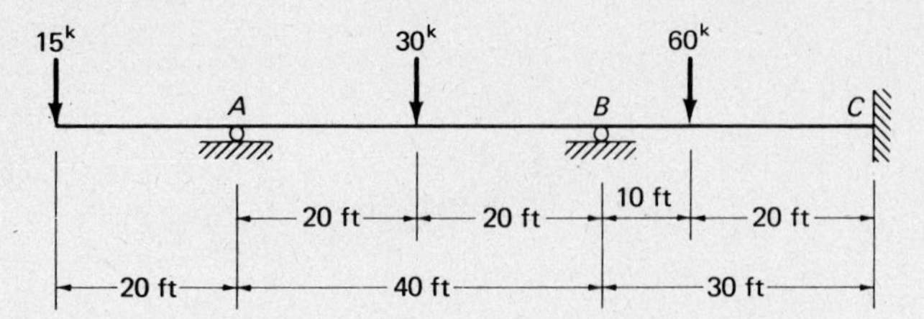

19.14. The beam of Problem 19.1 assuming support B settles 2.50 in. $E = 29 \times 10^6$ psi, $I = 1200$ in^4. (*Ans.* $V_A = 39.5^k \uparrow$, $V_B = 31^k \uparrow$, $M_B = +190'^k$, $V_C = 29.5^k \uparrow$)

19.15. The beam of Prob. 19.1 assuming the following support settlements: $A = 4.00$ in, $B = 2.00$ in, and $C = 3.50$ in. $E = 29 \times 10^6$ psi, $I = 1200$ in^4.

19.16. The beam of Prob. 19.9 assuming the following support settlements: $A = 1.00$ in, $B = 3.00$ in, $C = 1.50$ in, and $D = 2.00$ in. $E = 29 \times 10^6$ psi, $I = 3200$ in^4. (*Ans.* $V_A = 20.2^k \uparrow$, $V_C = 105^k \uparrow$, $M_B = +156'^k$)

19.17. The beam of Prob. 19.13 assuming support C settles 1.50 in, $E = 19 \times 10^6$ psi.

Problem 19.18 $E = 29 \times 10^6$ psi (*Ans.* $V_B = 63.3^k \uparrow$, $M_B = -313'^k$, $V_C = 19.5^k \uparrow$)

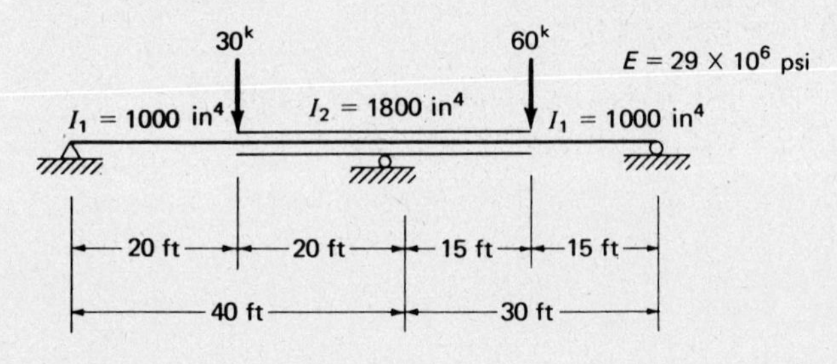

Problem 19.19

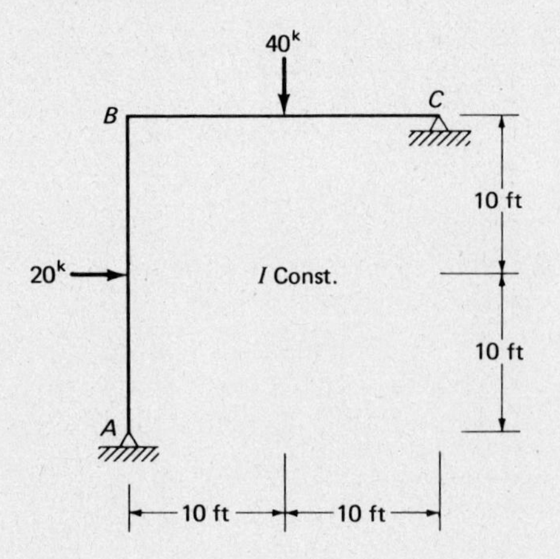

Problem 19.20 (*Ans.* $H_A = 13.70^k \leftarrow$, $M_B = -129'^k$, $V_C = 17.55^k \uparrow$)

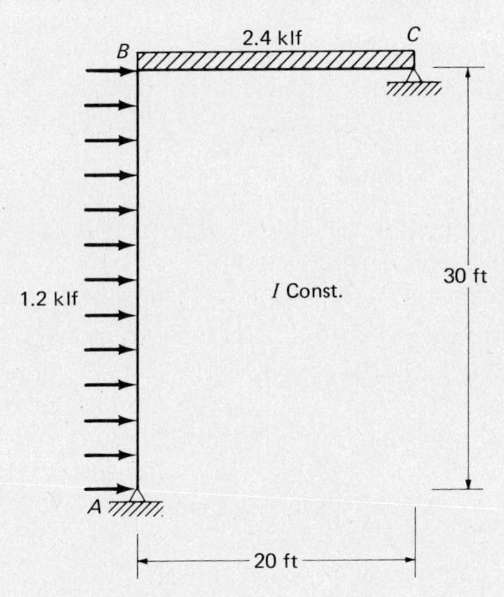

Problem 19.21

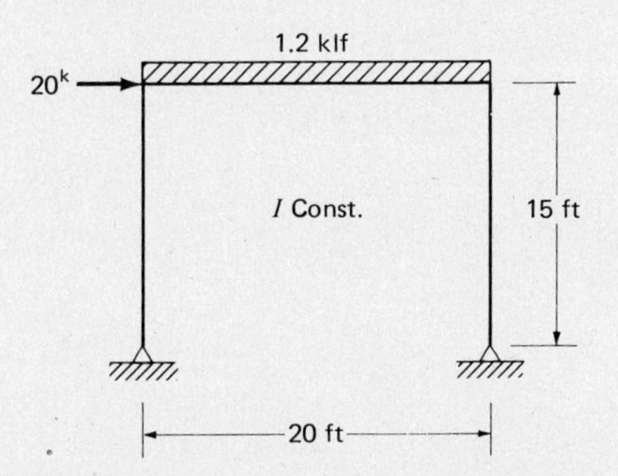

19.22. The frame of Prob. 19.19 assuming support C is fixed. (*Ans.* $V_A = 18.13^k \uparrow$, $M_B = 75.1'^k$, $M_C = -112.5'^k$)

19.23. The frame of Prob. 19.19 assuming supports A and C are fixed.

19.24. The frame of Prob. 19.19, assuming support A settles 1.00 in. $E = 29 \times 10^6$ psi, $I = 1200$ in^4 for beam and column. (*Ans* $H_A = 8.2^k \leftarrow$, $M_B = -36'^k$, $V_C = 18.2^k \uparrow$)

Draw quantitative influence lines for the situations listed in Probs. 19.25 to 19.28

19.25. Reactions for all supports of the beam of Prob. 19.1.

19.26. The left vertical reaction and the moment reaction at the fixed end for the beam of Prob. 19.4. (*Ans.* Load at ℄, $V_A = 0.687 \uparrow$, $M_A = -3.75$)

19.27. Shear immediately to the left of support *B* and moment at support *B* for the beam of Prob. 19.11.

19.28. Shear and moment at a point 20 ft to the left of the fixed-end support *C* of the beam of Prob. 19.13. (*Ans.* Load at left end: $V_{20} = -0.32$, $M_{20} = +3.2$; load halfway from *A* to *B*: $V_{20} = +0.24$, $M_{20} = -2.4$)

By using Müller-Breslau's principle, sketch influence lines qualitatively for the functions indicated in the structures of Probls. 19.29 to 19.32.

19.29. With reference to the accompanying illustration: (a) reactions at *A* and *C*, (b) positive moment at *x* and *y*, and (c) positive shear at *x*.

Problem 19.29

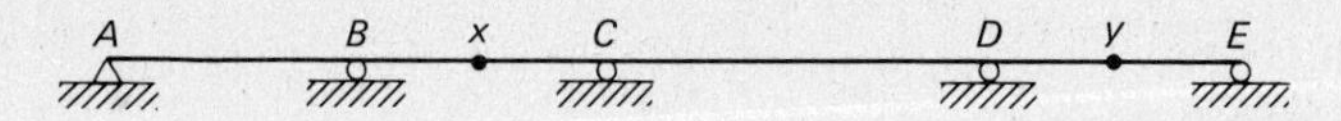

19.30. With reference to the accompanying illustration: (a) reaction at *A*, (b) positive and negative moments at *x*, and (c) negative moment at *B*.

Problem 19.30

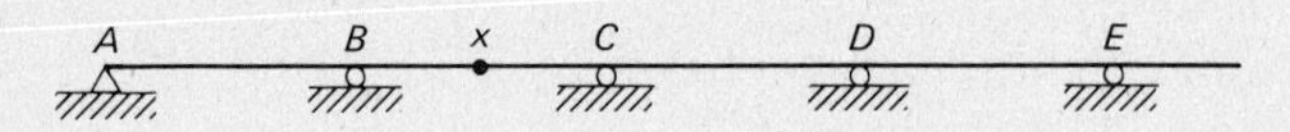

19.31. With reference to the accompanying illustration: (a) reaction at *A*, (b) moment at *x*, (c) positive shear at *y*, and (d) positive moment at *z*, assuming right side of column is bottom side.

Problem 19.31

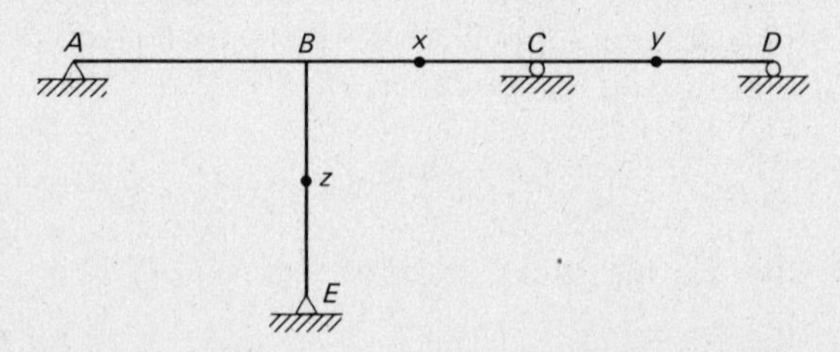

19.32. With reference to the accompanying illustration: (a) positive moment at *x*, (b) positive shear at *x*, and (c) negative moment just to the right of *y*.

Problem 19.32

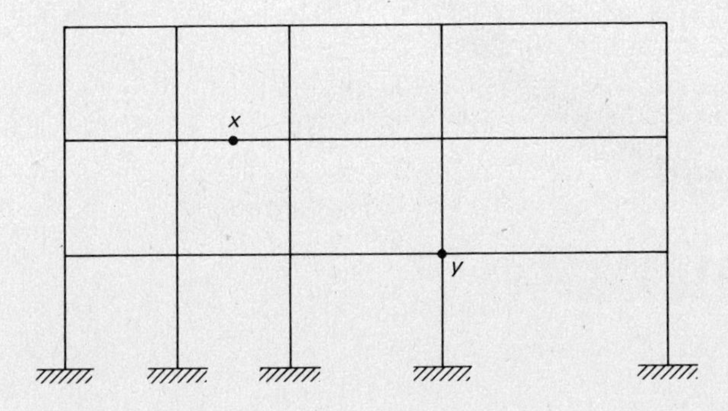

REFERENCES

[1] J. I. Parcel and R. B. B. Moorman, *Analysis of Statically Indeterminate Structures* (New York: Wiley, 1955), p. 48.
[2] J. S. Kinney, *Indeterminate Structural Analysis* (Reading, Mass.: Addison-Wesley, 1957), pp. 12–13.
[3] Ibid, p. 14.
[4] H. Cross and N. D. Morgan, *Continuous Frames of Reinforced Concrete* (New York: Wiley, 1932).

20 Deflection Methods of Analyzing Statically Indeterminate Trusses

20.1 ANALYSIS OF EXTERNALLY REDUNDANT TRUSSES

Trusses may be statically indeterminate because of redundant reactions, redundant members, or a combination of redundant reactions and members. Externally redundant trusses will be considered initially, and they will be analyzed on the basis of deflection computations in a manner closely related to the procedure used for statically indeterminate beams.

The two-span continuous truss of Example 20.1 is considered for the following discussion. One reaction component, for example, V_B, is removed, and the deflection at that point caused by the external loads is determined. Next the external loads are removed from the truss, and the deflection at the support point due to a unit load at that point is determined. The reaction is replaced, and it supplies the force necessary to push the support back to its original position. The familiar deflection expression is then written as follows:

$$\delta_B + V_B\delta_{bb} = 0$$

$$V_B = -\frac{\delta_B}{\delta_{bb}}$$

The forces in the truss members due to the external loads, when the redundant is removed, are not the correct final forces and are referred to as F' forces. The deflection at the removed support due to the external loads can be computed by $\Sigma(F'\mu l/AE)$. The deflection caused at the support by placing a

Rio Grande Gorge Bridge, Taos County, New Mexico. (Courtesy of the American Institute of Steel Construction, Inc.)

unit load there can be found by applying the same virtual work expression, except the unit load is now the external load and the forces caused are the same as the μ forces. The deflection at the support due to the unit load is $\Sigma(\mu^2 l/AE)$, and the redundant reaction may be expressed as follows:

$$V_B = -\frac{\Sigma(F'\mu_B l/AE)}{\Sigma(\mu_B^2 l/AE)}$$

Example 20.1 illustrates the complete analysis of a two-span truss by the method just described. After the redundant reaction is found, the other reactions and the final member forces may be determined by statics. Another method, however, is available for finding the final forces and should be used as a mathematics check. When the redundant reaction V_B is returned to the truss, it causes the force in each member to change by V_B times its μ force value. The final force in a member becomes

$$F = F' + V_B\mu$$

Example 20.1

Compute the reactions and member forces for the two-span continuous truss shown in Fig. 20.1. Circled figures are member areas, in square inches.

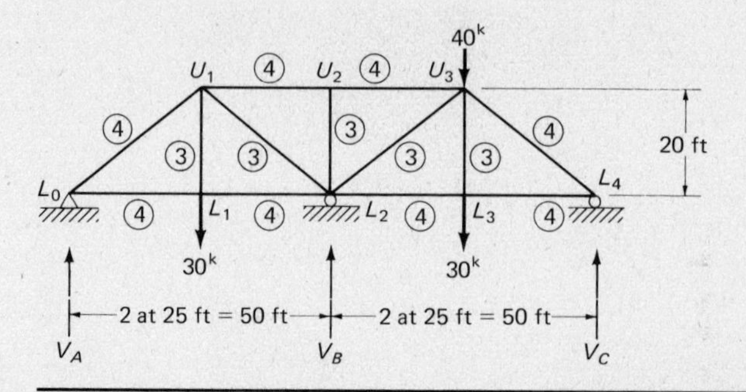

Figure 20.1

SOLUTION

Remove center support as the redundant and compute F' forces.

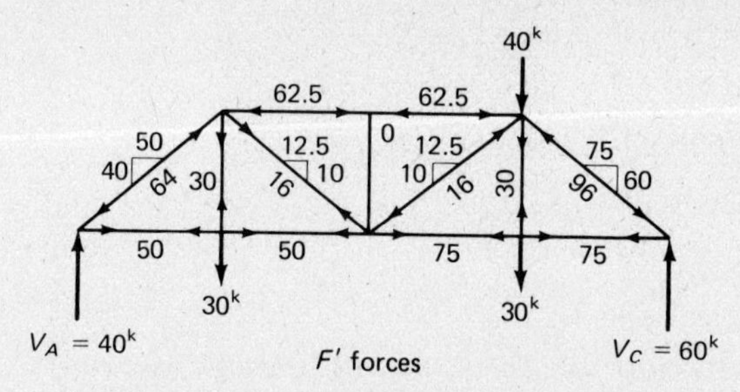

Remove the external loads and place a unit load at the center support; then compute the μ forces.

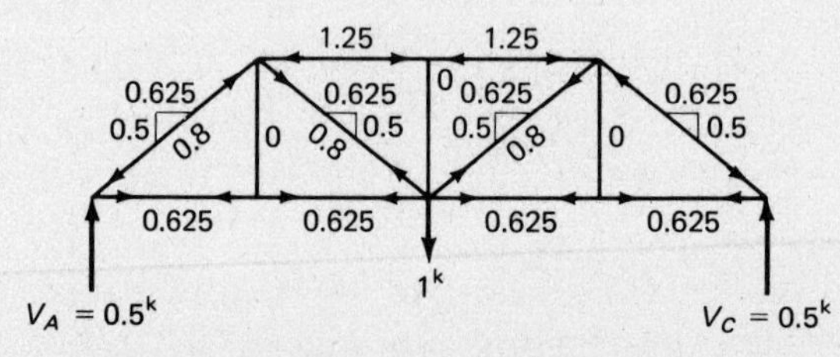

Computing the value of δ_B and δ_{ff} in Table 20.1.

$$V_B = -\frac{35{,}690}{637.6} = -56.0^k \uparrow$$

The reactions and member forces are found by statics in order to check the final values in Table 20.1.

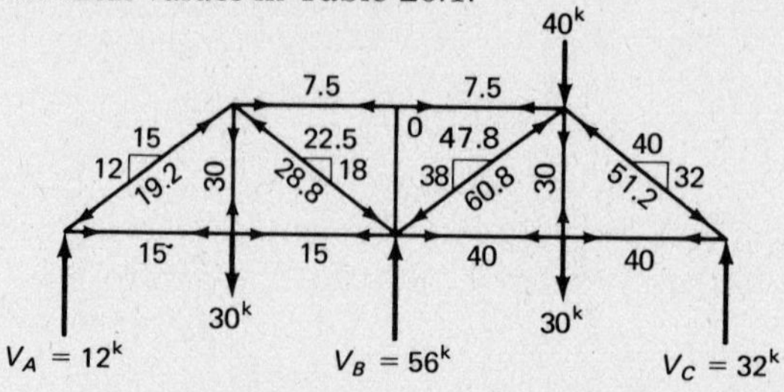

TABLE 20.1

MEMBER	L (in)	A in²	$\frac{l}{A}$	F' (kips)	μ	$\frac{F'\mu l}{AE}$	$\frac{\mu^2 l}{AE}$	$F = F' + V_B\mu$
L_0L_1	300	4	75	+50	+0.625	+2340	+ 29.2	+15.0
L_1L_2	300	4	75	+50	+0.625	+2340	+ 29.2	+15.0
L_2L_3	300	4	75	+75	+0.625	+3510	+ 29.2	+40.0
L_3L_4	300	4	75	+75	+0.625	+3510	+ 29.2	+40.0
L_0U_1	384	4	96	−64	−0.800	+4920	+ 61.4	−19.2
U_1U_2	300	4	75	−62.5	−1.25	+5850	+117.0	+ 7.5
U_2U_3	300	4	75	−62.5	−1.25	+5850	+117.0	+ 7.5
U_3L_4	384	4	96	−96	−0.800	+7370	+ 61.4	−51.2
U_1L_1	240	3	80	+30	0	0	0	+30.0
U_1L_2	384	3	128	+16	+0.800	+1640	+ 82.0	−28.8
U_2L_2	240	3	80	0	0	0	0	0
L_2U_3	384	3	128	−16	+0.800	−1640	+ 82.0	−60.8
U_3L_3	240	3	80	+30	0	0	0	+30.0
Σ						35,690	637.6	

It should be evident that the deflection procedure may be used to analyze trusses that have two or more redundant reactions. The truss of Example 20.2 is continuous over three spans, and the reactions at the interior supports V_B and V_C are considered to be the redundants. The following expressions, previously written for a three-span continuous beam, are applicable to the truss:

$$\delta_B + V_B\delta_{bb} + V_C\delta_{bc} = 0$$

$$\delta_C + V_B\delta_{cb} + V_C\delta_{cc} = 0$$

The forces due to a unit load at B are called the μ_B forces; the ones due to a unit load at C are called the μ_C forces. A unit load at B will cause a deflection at C equal to $\Sigma(\mu_B\mu_C l/AE)$, and a unit load at C causes the same deflection at B, $\Sigma(\mu_C\mu_B l/AE)$, which is thus another illustration of Maxwell's law. The deflection expressions become

$$\sum \frac{F'\mu_B l}{AE} + V_B \sum \frac{\mu_B^2 l}{AE} + V_C \sum \frac{\mu_B\mu_C l}{AE} = 0$$

$$\sum \frac{F'\mu_C l}{AE} + V_B \sum \frac{\mu_C\mu_B l}{AE} + V_C \sum \frac{\mu_C^2 l}{AE} = 0$$

A simultaneous solution of these equations will yield the values of the redundants. Should support settlement occur, the deflections would have to be worked out numerically in the same units given for the settlements.

Example 20.2

Analyze the three-span truss shown in Fig. 20.2. Assume the reactions at the interior supports to be the redundants. Circled figures are member areas, in square inches.

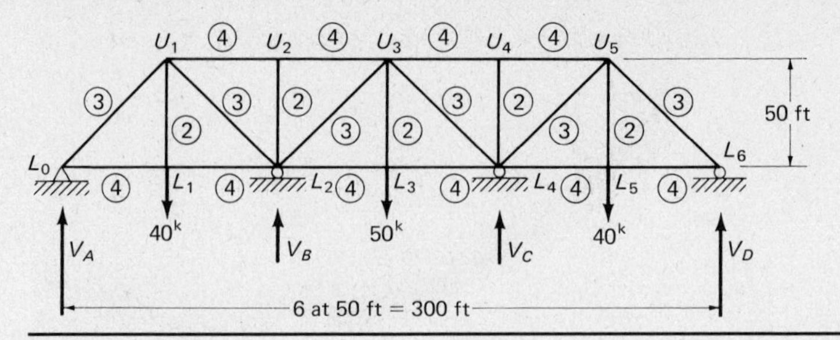

Figure 20.2

SOLUTION

Remove the redundants and compute the F' forces, the μ_B forces, and the μ_C forces.

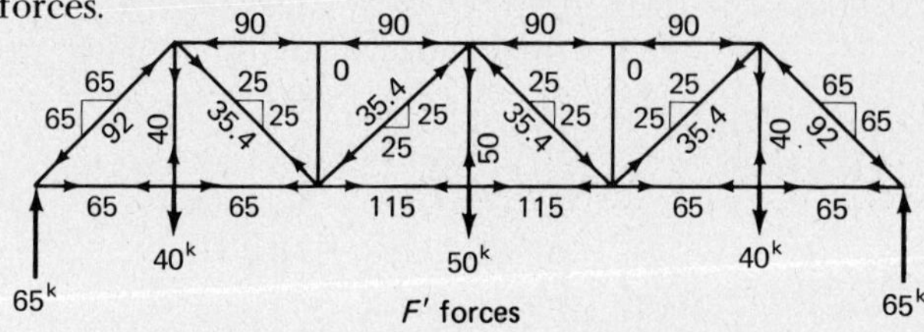

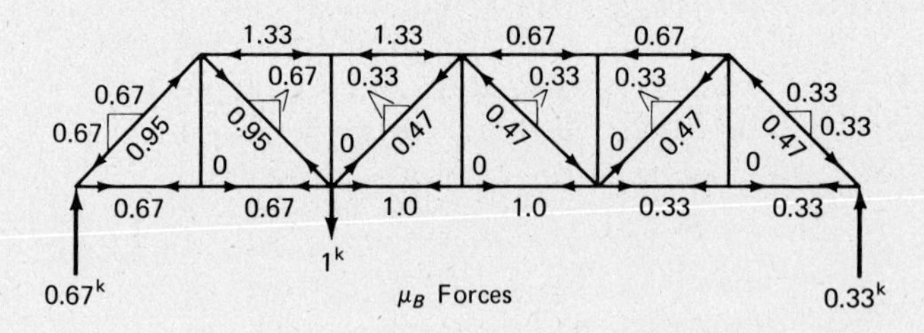

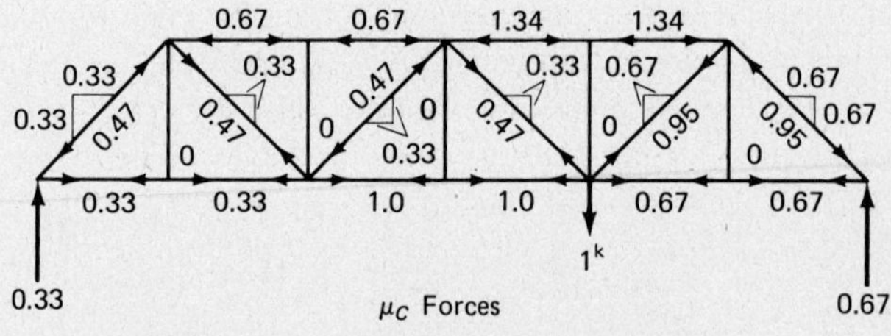

$$\sum \frac{F'\mu_B l}{AE} + V_B \sum \frac{\mu_B^2 l}{AE} + V_C \sum \frac{\mu_B \mu_C l}{AE} = 0 \tag{1}$$

$$159{,}480 + 1894.4V_B + 1347V_C = 0$$

$$\sum \frac{F'\mu_C l}{AE} + V_B \sum \frac{\mu_B \mu_C l}{AE} + V_C \sum \frac{\mu_C^2 l}{AE} = 0 \tag{2}$$

$$159{,}480 + 1347V_B + 1894.4V_C = 0$$

By solving Eqs. (1) and (2) simultaneously for V_B and V_C and finding V_A and V_D by statics,

$V_A = -15.7^k \uparrow \qquad V_C = -49.4^k \uparrow$

$V_B = -49.4^k \uparrow \qquad V_D = -15.7^k \uparrow$

Table 20.2

MEMBER	l (in)	A (in^2)	$\frac{l}{A}$	F'(kips)	μ_B	μ_C	$\frac{F'\mu_B l}{AE}$	$\frac{F'\mu_C l}{AE}$	$\frac{\mu_B^2 l}{AE}$	$\frac{\mu_C^2 l}{AE}$	$\frac{\mu_B \mu_C l}{AE}$	$F = V_B\mu_B + V_C\mu_C + F'$
L_0L_1	600	4	150	+ 65	+0.67	+0.33	+ 6500	+ 3250	+ 67.4	+ 16.4	+ 33.0	+15.6
L_1L_2	600	4	150	+ 65	+0.67	+0.33	+ 6500	+ 3250	+ 67.4	+ 16.4	+ 33.0	+15.6
L_2L_3	600	4	150	+115	+1.00	+1.00	+17,250	+17,250	+150.0	+150.0	+150.0	+16.4
L_3L_4	600	4	150	+115	+1.00	+1.00	+17,250	+17,250	+150.0	+150.0	+150.0	+16.4
L_4L_5	600	4	150	+ 65	+0.33	+0.67	+ 3250	+ 6500	+ 16.4	+ 67.4	+ 33.0	+15.6
L_5L_6	600	4	150	+ 65	+0.33	+0.67	+ 3250	+ 6500	+ 16.4	+ 67.4	+ 33.0	+15.6
L_0U_1	848	3	283	− 92	−0.95	−0.47	+24,800	+12,400	+256.0	+ 62.5	+126.5	−22.0
U_1U_2	600	4	150	− 90	−1.33	−0.67	+18,000	+ 9000	+265.0	+ 67.4	+133.5	+ 8.7
U_2U_3	600	4	150	− 90	−1.33	−0.67	+18,000	+ 9000	+265.0	+ 67.4	+133.5	+ 8.7
U_3U_4	600	4	150	− 90	−0.67	−1.33	+ 9000	+18,000	+ 67.4	+265.0	+133.5	+ 8.7
U_4U_5	600	4	150	− 90	−0.67	−1.33	+ 9000	+18,000	+ 67.4	+265.0	+133.5	+ 8.7
U_5L_6	848	3	283	− 92	−0.47	−0.95	+12,400	+24,800	+ 62.5	+255.0	+126.5	−22.0
U_1L_1	600	2	300	+ 40	0	0	0	0	0	0	0	+40.0
U_1L_2	848	3	283	+ 35.4	+0.95	+0.47	+ 9520	+ 4760	+256.0	+ 62.5	+126.5	−34.6
U_2L_2	600	2	300	0	0	0	0	0	0	0	0	0
L_2U_3	848	3	283	− 35.4	+0.47	−0.47	− 4760	+ 4760	+ 62.5	+ 62.5	− 62.5	−35.4
U_3L_3	600	2	300	+ 50	0	0	0	0	0	0	0	+50.0
U_3L_4	848	3	283	− 35.4	−0.47	+0.47	+ 4760	− 4760	+ 62.5	+ 62.5	− 62.5	−35.4
U_4L_4	600	2	300	0	0	0	0	0	0	0	0	0
L_4U_5	848	3	283	+ 35.4	+0.47	+0.95	+ 4760	+ 9520	+ 62.5	+255.0	+126.5	−34.6
U_5L_5	600	2	300	+ 40	0	0	0	0	0	0	0	+40.0
Σ							+159,480	+159,480	+1894.4	+1894.4	+1347	

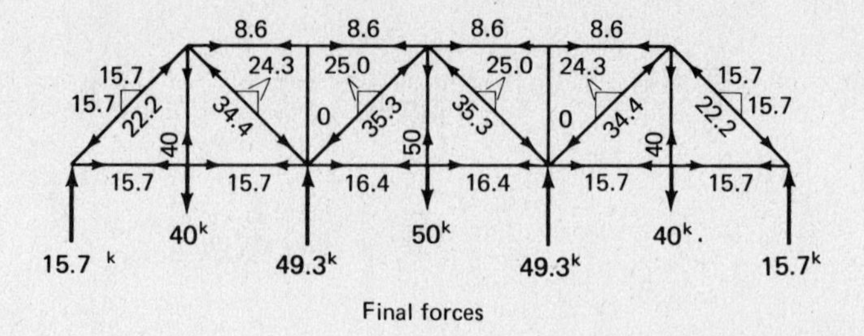

Final forces

20.2 ANALYSIS OF INTERNALLY REDUNDANT TRUSSES

The truss of Example 20.3 has one more member than necessary for stability and is therefore statically indeterminate to the first degree, as can be proved by applying the equation

$$m = 2j - 3$$

Internally redundant trusses may be analyzed in a manner closely related to the one used for externally redundant trusses. One member is assumed to be the redundant and is theoretically cut or removed from the structure. The remaining members must form a statically determinate and stable truss. The F' forces in these members are assumed to be of a nature causing the joints at the ends of the removed member to pull apart, the distance being $\Sigma(F'\mu l/AE)$.

The redundant member is replaced in the truss and is assumed to have a unit tensile force. The μ forces in each of the members caused by the redundant's force of $+1$ are computed and they will cause the joints to be pulled together an amount equal to $\Sigma(\mu^2 l/AE)$. If the redundant has an actual force of X, the joints will be pulled together an amount equal to $X\Sigma(\mu^2 l/AE)$.

If the member had been sawed in half, the F' forces would have opened a gap of $\Sigma(F'\mu l/AE)$; therefore X must be sufficient to close the gap, and the following expressions may be written:

$$X \sum \frac{\mu^2 l}{AE} + \sum \frac{F'\mu l}{AE} = 0$$

$$X = -\frac{\Sigma(F'\mu l/AE)}{\Sigma(\mu^2 l/AE)}$$

The application of this very common method of analyzing internally redundant trusses is illustrated by Example 20.3. After the truss force in the redundant member is found, the force in any other member equals its F' force plus X times its μ force. Final forces may also be calculated by statics as a check on the mathematics.

Example 20.3

Determine the forces in the members of the internally redundant truss shown in Fig. 20.3. Circled figures are member areas, in square inches.

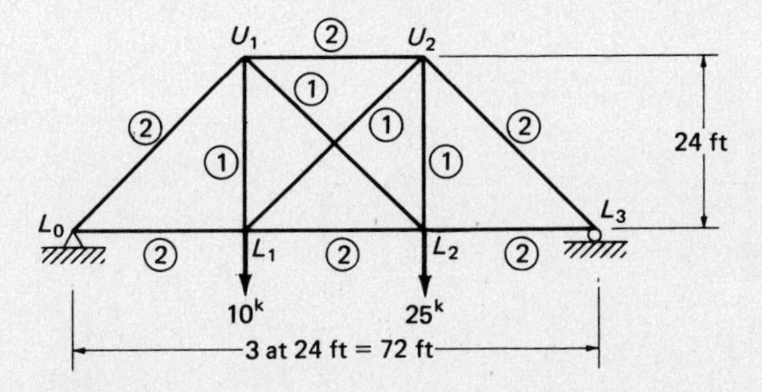

Figure 20.3

SOLUTION

Assume L_1U_2 to be the redundant, remove it, and compute the F' forces.

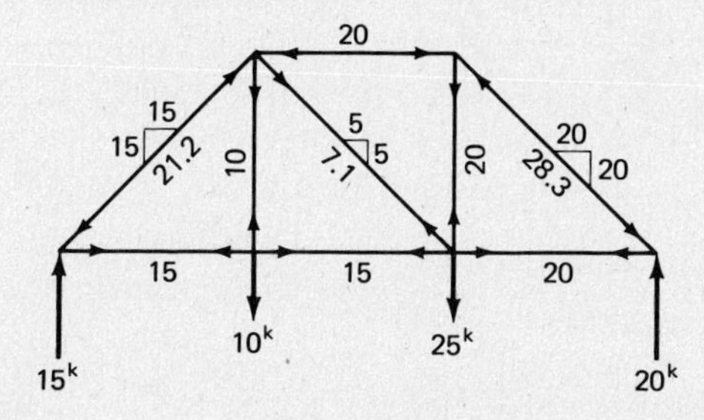

Replace L_1U_2 with a force of $+1$ and compute the μ forces.

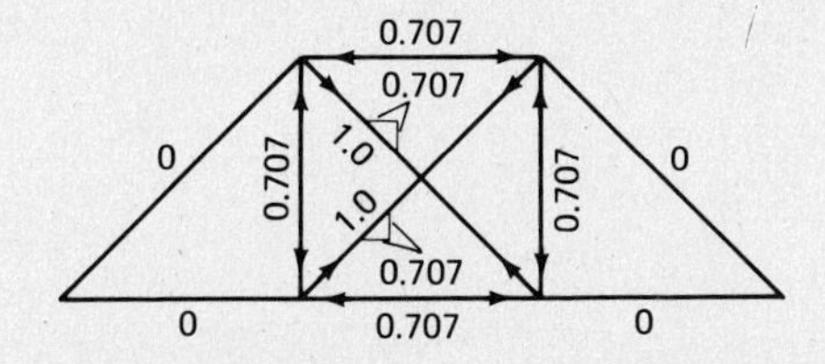

TABLE 20.3

MEMBER	l (in)	A (in^2)	$\frac{l}{A}$	F' (kips)	μ	$\frac{F'\mu l}{A}$	$\frac{\mu^2 l}{A}$	$F = F' + X\mu$ (kip)
L_0L_1	288	2	144	+15	0	0	0	+15.00
L_1L_2	288	2	144	+15	−0.707	−1530	+ 72	+13.47
L_2L_3	288	2	144	+20	0	0	0	+20.00
L_0U_1	408	2	204	−21.2	0	0	0	−21.20
U_1U_2	288	2	144	−20	−0.707	+2040	+ 72	−21.53
U_2L_3	408	2	204	−28.3	0	0	0	−28.30
U_1L_1	288	1	288	+10	−0.707	−2040	+ 144	+ 8.47
U_1L_2	408	1	408	+ 7.1	+1.0	+2900	+ 408	+ 9.26
L_1U_2	408	1	408	0	+1.0	0	+ 408	+ 2.16
U_2L_2	288	1	288	+20	−0.707	−4070	+ 144	+18.47
Σ						−2700	+1248	

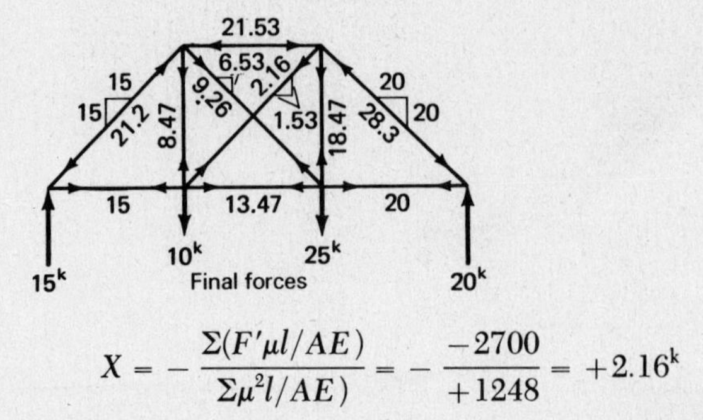

$$X = -\frac{\Sigma(F'\mu l/AE)}{\Sigma\mu^2 l/AE)} = -\frac{-2700}{+1248} = +2.16^k$$

For trusses that have more than one redundant internally, simultaneous equations are necessary in the solution. Example 20.4 illustrates the analysis of a truss that is statically indeterminate internally to the second degree. Two members, with forces of X_A and X_B, assumed to be the redundants are theoretically cut. The F' forces in the remaining truss members pull the cut places apart by $\Sigma(F'\mu_A l/AE)$ and $\Sigma(F'\mu_B l/AE)$, respectively. Replacing the first redundant member with a force of $+1$ causes μ_A forces in the truss members and causes the gaps to close by $\Sigma(\mu_A^2 l/AE)$ and $\Sigma\mu_A\mu_B l/AE$. Repeating the process with the other redundant causes μ_B forces and additional gap closings of $\Sigma(\mu_B\mu_A l/AE)$ and $\Sigma(\mu_B^2 l/AE)$. The redundant forces must be sufficient to close the gaps, permitting the writing of the following equations:

$$\sum \frac{F'\mu_A l}{AE} + X_A \sum \frac{\mu_A^2 l}{AE} + X_B \sum \frac{\mu_A \mu_B l}{AE} = 0$$

$$\sum \frac{F'\mu_B l}{AE} + X_A \sum \frac{\mu_B \mu_A l}{AE} + X_B \sum \frac{\mu_B^2 l}{AE} = 0$$

Example 20.4

Analyze the truss shown in Fig. 20.4; assume L_1U_2 and U_2L_3 to be the redundants. Circled figures are member areas, in square inches.

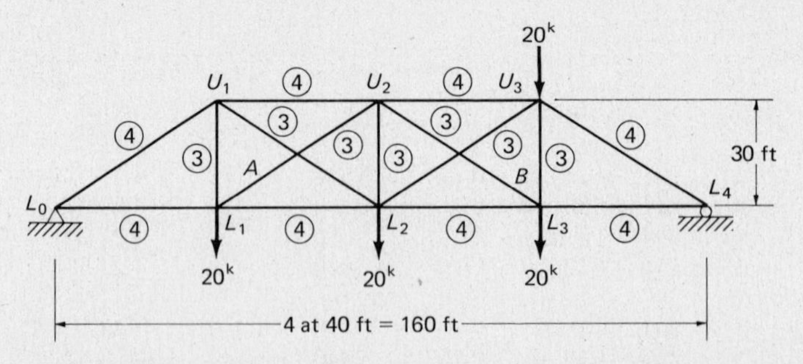

Figure 20.4

SOLUTION

Remove the redundants and compute the F' forces.

Delaware River Turnpike Bridge. (Courtesy of American Bridge.)

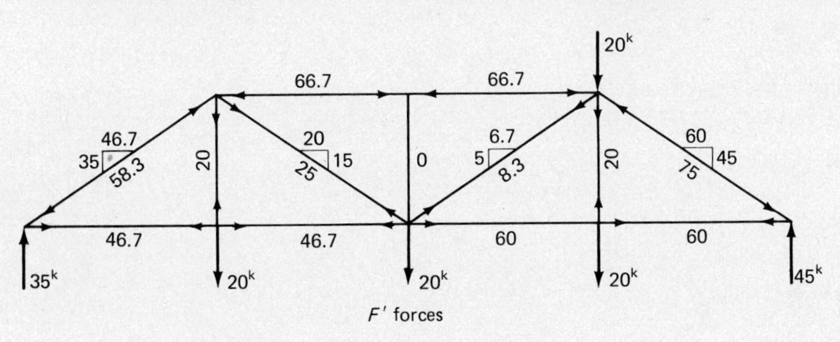

F′ forces

Replace L_1U_2 with a force of $+1$ and compute the μ_A forces.

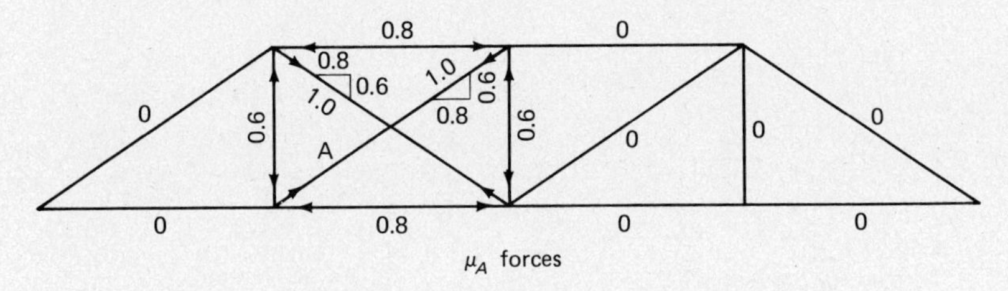

μ_A forces

Replace U_2L_3 with a force of $+1$ and compute the μ_B forces.

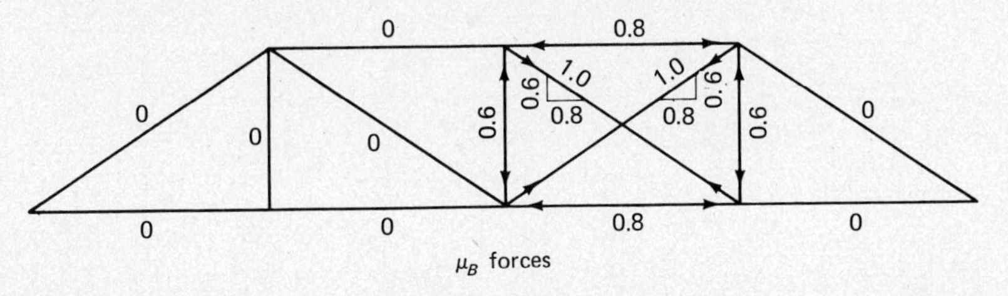

μ_B forces

Table 20.4

MEMBER	l (in)	A (in^2)	$\frac{l}{A}$	F'(kips)	μ_A	μ_B	$\frac{F'\mu_A l}{AE}$	$\frac{F'\mu_B l}{AE}$	$\frac{\mu_A^2 l}{AE}$	$\frac{\mu_B^2 l}{AE}$	$\frac{\mu_A\mu_B l}{AE}$	$F = F' + X_A\mu_A + X_B\mu_B$
L_0L_1	480	4	120	+46.7	0	0	0	0	0	0	0	+46.7
L_1L_2	480	4	120	+46.7	−0.8	0	−4490	0	+ 76.8	0	0	+53.5
L_2L_3	480	4	120	+60.0	0	−0.8	0	−5760	0	+ 76.8	0	+60.63
L_3L_4	480	4	120	+60.0	0	0	0	0	0	0	0	+60.0
L_0U_1	600	4	150	−58.3	0	0	0	0	0	0	0	−58.3
U_1U_2	480	4	120	−66.7	−0.8	0	+6410	0	+ 76.8	0	0	−59.9
U_2U_3	480	4	120	−66.7	0	−0.8	0	+6410	0	+ 76.8	0	−66.06
U_3L_4	600	4	150	−75.0	0	0	0	0	0	0	0	−75.0
U_1L_1	360	3	120	+20.0	−0.6	0	−1440	0	+ 43.2	0	0	+25.1
U_1L_2	600	3	200	+25.0	+1.0	0	+5000	0	+200.0	0	0	+16.5
L_1U_2	600	3	200	0	+1.0	0	0	0	+200.0	0	0	− 8.5
U_2L_2	360	3	120	0	−0.6	−0.6	0	0	+ 43.2	+ 43.2	+43.2	+ 5.57
U_2L_3	600	3	200	0	0	+1.0	0	0	0	+200.0	0	− 0.79
L_2U_3	600	3	200	+ 8.3	0	+1.0	0	+1660	0	+200.0	0	+ 7.51
U_3L_3	360	3	120	+20.0	0	−0.6	0	−1440	0	+ 43.2	0	+20.47
Σ							+5480	+ 870	+640	+640	+43.2	

$$\sum \frac{F'\mu_A l}{AE} + X_A \sum \frac{\mu_A^2 l}{AE} + X_B \sum \frac{\mu_A \mu_B l}{AE} = 0$$

$$+ 5480 + 640X_A + 43.2X_B = 0 \qquad (1)$$

$$\sum \frac{F'\mu_B l}{AE} + X_A \sum \frac{\mu_A \mu_B l}{AE} + X_B \sum \frac{\mu_B^2 l}{AE} = 0$$

$$+ 870 + 43.2X_A + 640X_B = 0 \qquad (2)$$

Simultaneous solutions of Eqs. (1) and (2) gives values of X_A and X_B of -8.5 and -0.79, respectively. These values are used in computing final forces by statics as a check on figures obtained in the last column of the table.

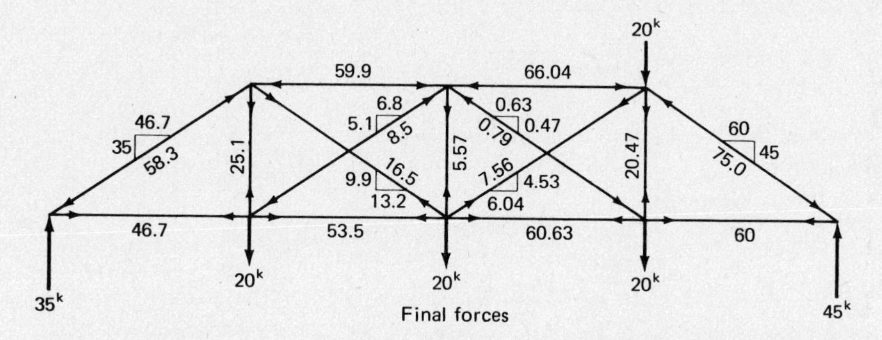

Final forces

20.3 ANALYSIS OF TRUSSES REDUNDANT INTERNALLY AND EXTERNALLY

Deflection equations have been written so frequently in the past few sections that the reader is probably able to set up his or her own equations for types of statically indeterminate beams and trusses not previously encountered. Nevertheless, one more group of equations is developed here, these being the ones necessary for the analysis of a truss that is statically indeterminate internally and externally. For the following discussion the truss of Fig. 20.5, which has two redundant members and one redundant reaction component, will be considered.

The diagonals lettered D and E and the interior reaction V_B are removed from the truss, which leaves a statically determinate structure. The openings of the gaps in the cut members and the deflections at the interior support due to the external loads may be computed from the following:

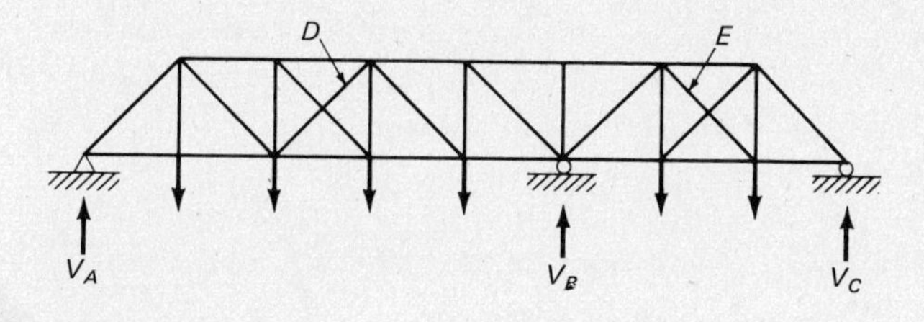

Figure 20.5

$$\delta_B = \sum \frac{F'\mu_B l}{AE} \qquad \delta_D = \sum \frac{F'\mu_D l}{AE} \qquad \delta_E = \frac{F'\mu_E l}{AE}$$

Placing a unit load at the interior support will cause deflections at the gaps in the cut members as well as at the point of application.

$$\delta_{bb} = \sum \frac{\mu_B^2 l}{AE} \qquad \delta_{db} = \sum \frac{\mu_B \mu_D l}{AE} \qquad \delta_{eb} = \sum \frac{\mu_B \mu_E l}{AE}$$

Replacing member D and assuming it to have a positive unit tensile force will cause the following deflections:

$$\delta_{bd} = \sum \frac{\mu_B \mu_D l}{AE} \qquad \delta_{dd} = \sum \frac{\mu_D^2 l}{AE} \qquad \delta_{ed} = \sum \frac{\mu_E \mu_D l}{AE}$$

Similarly, replacement of member E with a force of $+1$ will cause these deflections:

$$\delta_{be} = \sum \frac{\mu_B \mu_E l}{AE} \qquad \delta_{de} = \sum \frac{\mu_D \mu_E l}{AE} \qquad \delta_{ee} = \sum \frac{\mu_E^2 l}{AE}$$

Computation of these sets of deflections permits the calculation of the numerical values of the redundants, because the total deflection at each may be equated to zero.

$$\delta_B + V_B\delta_{bb} + X_D\delta_{bd} + X_E\delta_{be} = 0$$
$$\delta_D + V_B\delta_{db} + X_D\delta_{dd} + X_E\delta_{de} = 0$$
$$\delta_E + V_B\delta_{eb} + X_D\delta_{ed} + X_E\delta_{ee} = 0$$

Nothing new is involved in the solution of this type of problem, and space is not taken for the lengthy calculations necessary for an illustrative example.

20.4 INFLUENCE LINES FOR STATICALLY INDETERMINATE TRUSSES

For analyzing statically indeterminate trusses, influence lines are necessary to determine the critical positions for live loads as they were for statically determinate trusses.

The discussion of the details of construction of these diagrams for statically indeterminate trusses is quite similar to the one presented for statically indeterminate beams in the preceding chapter. To prepare the influence line for a reaction of a continuous truss, the support is removed and a unit load is placed at the support point. For this position of the unit load, the deflection at each of the truss joints is determined. For example, the preparation of an influence line for the interior reaction of the truss of Fig. 20.6 is considered. The value of the reaction when the unit load is at joint x may be expressed as follows:

$$V_B = -\frac{\delta_{xb}}{\delta_{bb}} = -\frac{\Sigma(\mu_x \mu_B l/AE)}{\Sigma(\mu_B^2 l/AE)}$$

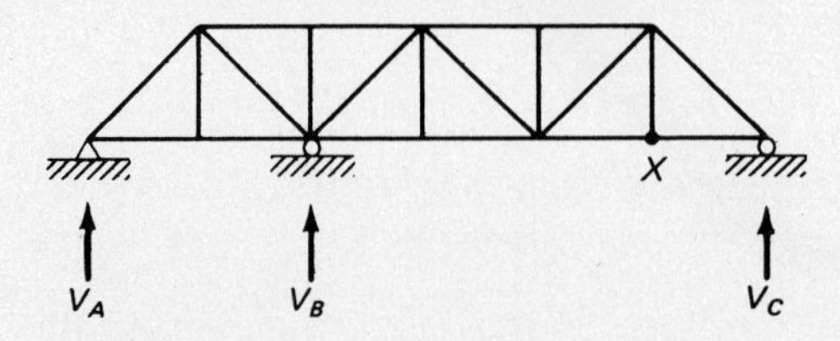

Figure 20.6

After the influence line for V_B has been plotted, the influence line for another reaction may be prepared by repeating the process of removing it as the redundant, introducing a unit load there, and computing the necessary deflections. A simpler procedure is to compute the other reactions, or any other functions for which influence lines are desired, by statics after the diagram for V_B is prepared. This method is employed in Example 20.5 for a two-span truss for which influence lines are desired for the reactions and several member forces.

Example 20.5

Draw influence lines for the three vertical reactions and for forces in members U_1U_2, L_0U_1, and U_1L_2 (Fig. 20.7). Circled figures are member areas, in square inches.

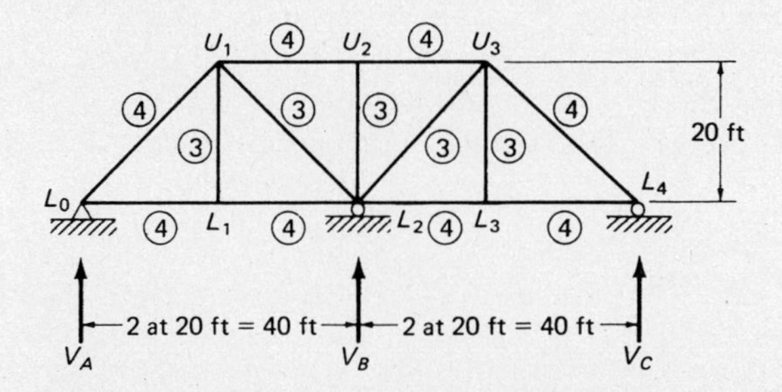

Figure 20.7

SOLUTION

Remove the interior support and compute the forces for a unit load at L_1 and at L_2. (Note that the deflection at L_1 caused by a unit load at L_2 is the same as the deflection caused at L_3 due to symmetry.)

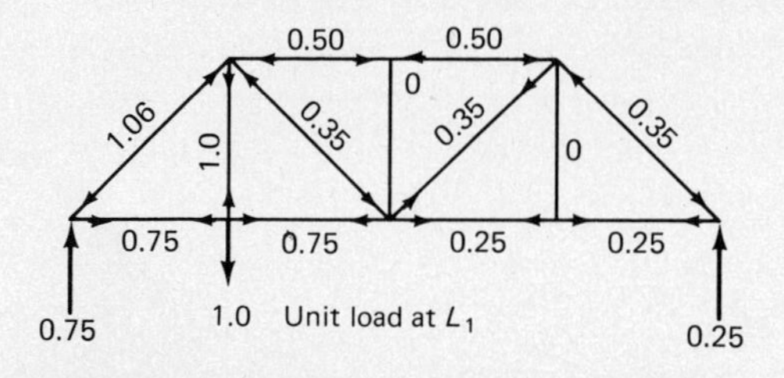

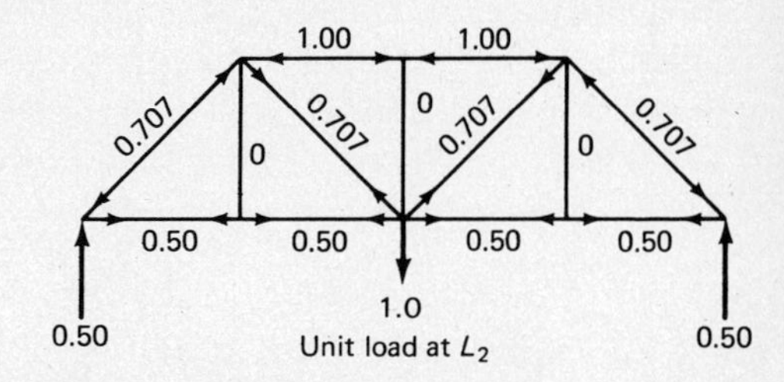

Table 20.5

MEMBER	l (in)	A (in^2)	$\frac{l}{A}$	μ_B	μ_A	$\frac{\mu_B^2 l}{AE}$	$\frac{\mu_B \mu_A l}{AE}$
L_0L_1	240	4	60	+0.50	+0.75	+15	+22.5
L_1L_2	240	4	60	+0.50	+0.75	+15	+22.5
L_2L_3	240	4	60	+0.50	+0.25	+15	+ 7.5
L_3L_4	240	4	60	+0.50	+0.25	+15	+ 7.5
L_0U_1	340	4	85	−0.707	−1.06	+42.5	+63.6
U_1U_2	240	4	60	−1.00	−0.50	+60	+30.0
U_2U_3	240	4	60	−1.00	−0.50	+60	+30.0
U_3L_4	340	4	85	−0.707	−0.35	+42.5	+21.0
U_1L_1	240	3	80	0	+1.00	0	0
U_1L_2	340	3	113	+0.707	−0.35	+56.5	−28.0
U_2L_2	240	3	80	0	0	0	0
L_2U_3	340	3	113	+0.707	+0.35	+56.5	+28.0
U_3L_3	240	3	80	0	0	0	0
Σ						$\frac{378}{E}$	$\frac{204.6}{E}$

Divide each of the values by δ_{bb} to obtain the influence-line ordinates for V_B and compute the ordinates for the other influence diagrams by statics.

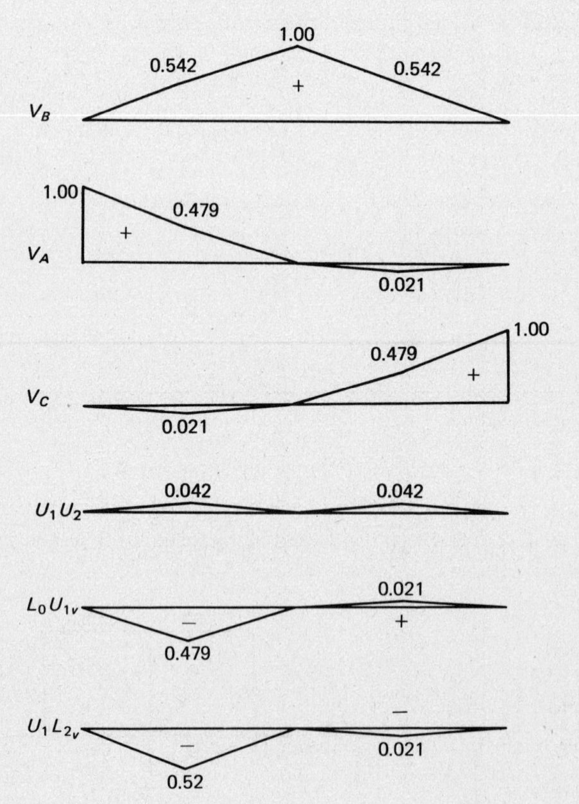

Example 20.6 shows that influence lines for members of an internally redundant truss may be prepared by an almost identical procedure. The member assumed to be the redundant is given a unit force, and deflections caused thereby at each of the joints are calculated. The ordinates of the diagram for the member are obtained by dividing each of these deflections by the deflection at the member. All other influence lines are prepared by statics.

Example 20.6

Prepare influence lines for force in members L_1U_2, U_1U_2, and U_2L_2 of the truss of Example 20.3, which is reproduced in Fig. 20.8. Circled figures are member areas, in square inches.

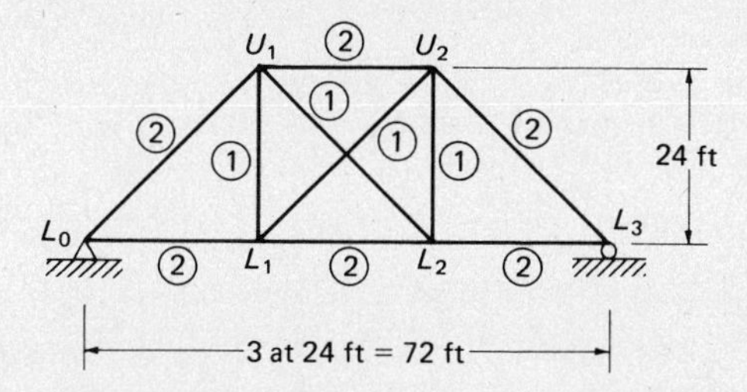

Figure 20.8

SOLUTION

Remove L_1U_2 as the redundant and compute the forces caused by unit loads at L_1 and L_2; replace L_1U_2 with a force of $+1$ and compute the forces in the remaining truss members.

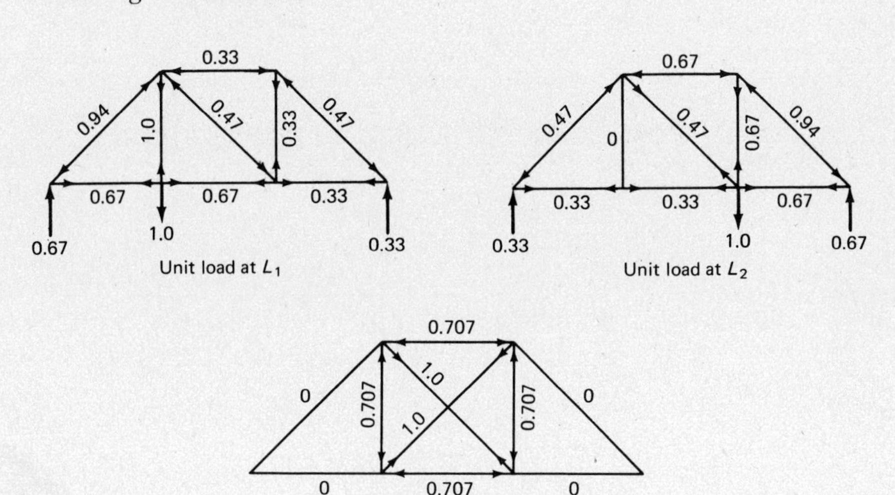

Table 20.6

MEMBER	l (in)	A (in²)	$\frac{l}{A}$	μ_{L_1}	μ_{L_2}	μ_A	$\delta_{AL_1} = \frac{\mu_{L_1}\mu_A l}{AE}$	$\delta_{AL_2} = \frac{\mu_{L_2}\mu_A l}{AE}$	$\delta_{aa} = \frac{\mu_A^2 l}{AE}$
L_0L_1	288	2	144	+0.67	+0.33	0	0	0	0
L_1L_2	288	2	144	+0.67	+0.33	−0.707	− 68	− 34	+ 72
L_2L_3	288	2	144	+0.33	+0.67	0	0	0	0
L_0U_1	408	2	204	−0.94	−0.47	0	0	0	0
U_1U_2	288	2	144	−0.33	−0.67	−0.707	+ 34	+ 68	+ 72
U_2L_3	408	2	204	−0.47	−0.94	0	0	0	0
U_1L_1	288	1	288	+1.00	0	−0.707	−204	0	+ 144
U_1L_2	408	1	408	−0.47	+0.47	+1.00	−192	+192	+ 408
L_1U_2	408	1	408	0	0	+1.00	0	0	+ 408
U_2L_2	288	1	288	+0.33	+0.67	−0.707	− 34	−136	+ 144
Σ							$\frac{-498}{E}$	$\frac{+90}{E}$	$\frac{+1248}{E}$

Draw the influence line for the redundant, L_1U_2, and note that the force in any other member equals $F' + X\mu$.

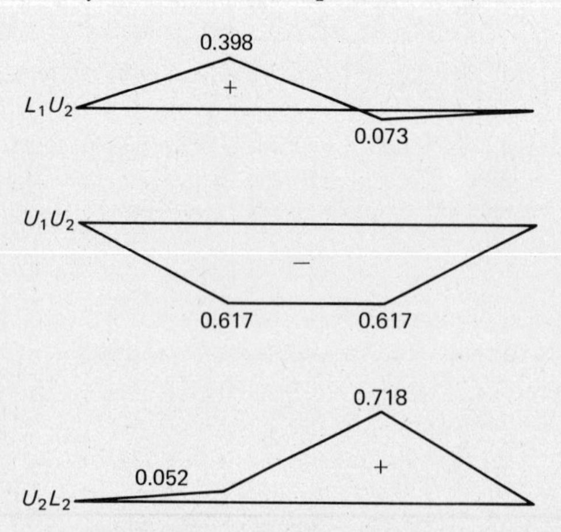

20.5 TEMPERATURE CHANGES, SHRINKAGE, FABRICATION ERRORS, AND SO ON

Structures are subject to deformations not only due to external loads but also due to temperature changes, support settlements, inaccuracies in fabrication dimensions, shrinkage in reinforced concrete members caused by drying and plastic flow, and so forth. Such deformations in statically indeterminate structures can cause the development of large additional forces in the members. For an illustration it is assumed that the top-chord members of the truss of Fig. 20.9 may be much more exposed to the sun than are the other members. As a result they may, on a hot sunny day, have much higher temperatures than those other members and the member forces may undergo some appreciable changes.

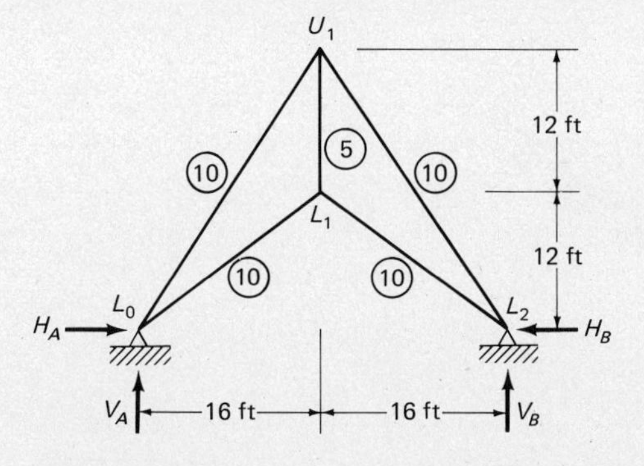

Figure 20.9

Problems such as these may be handled exactly as were the previous problems of this chapter. The changes in the length of each of the members due to temperature are computed. (These values which correspond to the $F'l/AE$ values each equal the temperature change times the coefficient of expansion of the material times the member length.) The redundant is removed from the structure and a unit load is placed at the support in the direction of the redundant and the μ forces are computed. Then the values $\Sigma(F'\mu l/AE)$ and $\Sigma(\mu^2 l/\text{AE})$ are determined in the same units and finally the usual deflection expression is written. Such a problem is illustrated in Example 20.7.

Example 20.7

The top-chord members of the statically indeterminate truss of Fig. 20.9 are assumed to increase in temperature by 60°F. If $E = 29 \times 10^6$ psi and the coefficient of linear temperature expansion is 0.0000065/°F, determine the forces induced in each of the truss members. The circled figures are areas in square inches.

SOLUTION

Assume H_B is the redundant and compute the μ forces.

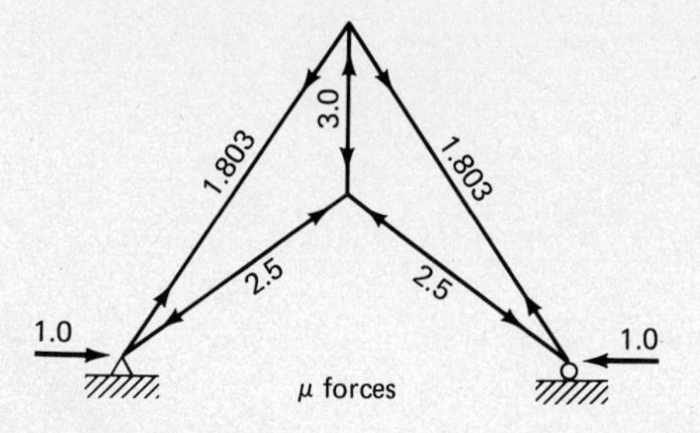

Table 20.7

MEMBER	l (in)	A (in^2)	$\frac{l}{A}$	μ	$\frac{\mu^2 l}{AE}$	$\Delta l = \Delta t \cdot \text{coeff} \cdot l$ $\left(\text{equivalent to } \frac{F'L}{AE}\right)$	$\mu(\Delta l) = \frac{F'\mu l}{AE}$
L_0L_1	240	10	24	−2.5	+150	—	—
L_1L_2	240	10	24	−2.5	+150	—	—
L_0U_1	346	10	34.6	+1.803	+ 62.38	(60)(0.0000065)(346) = 0.1349	0.2432
U_1L_2	346	10	34.6	+1.803	+ 62.38	(60)(0.0000065)(346) = 0.1349	0.2432
U_1L_1	144	5	28.8	−3.0	+259.2	—	—
					$\Sigma = +\frac{683.96}{E}$		$\Sigma = 0.4865$

$$\delta_{bb} = +\frac{(683.96)(1000)}{29 \times 10^6} = +0.02358 \text{ in}$$

$$\delta_B = -0.4864 \text{ in}$$

$$\delta_B + H_B\delta_{bb} = 0$$

$$H_B = -\frac{-0.4865}{0.02358} = 20.63^k$$

The final forces in the truss members due to the temperature change are as follows:

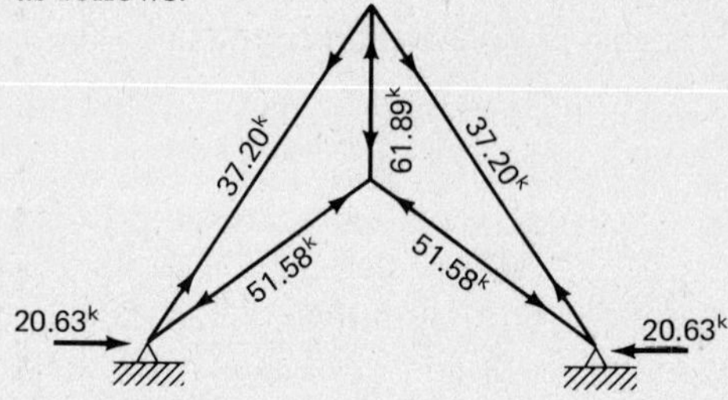

PROBLEMS

For Probs. 20.1 to 20.20 determine the reactions and member forces for the trusses. Circled figures are member areas, in square inches unless shown otherwise. E is constant.

Problem 20.1

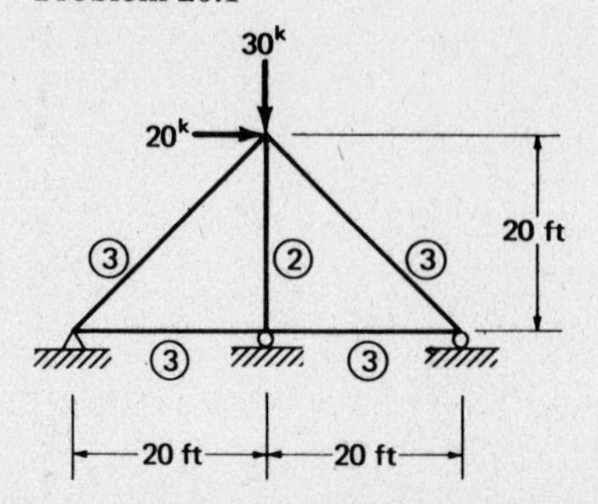

Problem 20.2 (*Ans.* $V_B = 104^k \uparrow$, $U_1L_2 = -78.2^k$, $U_2L_2 = -46^k$, $L_2L_3 = +6^k$)

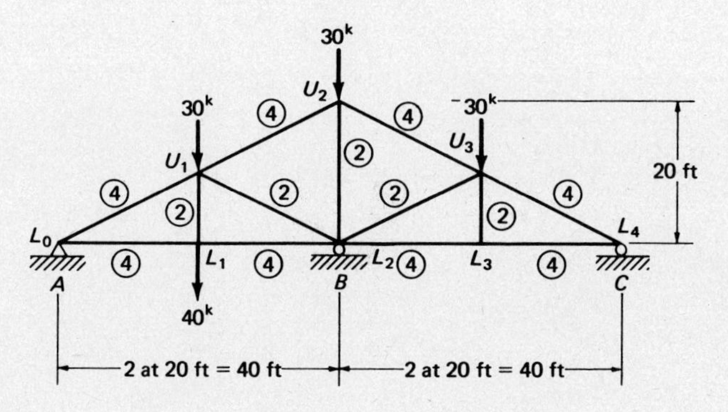

Problem 20.3 All areas are equal

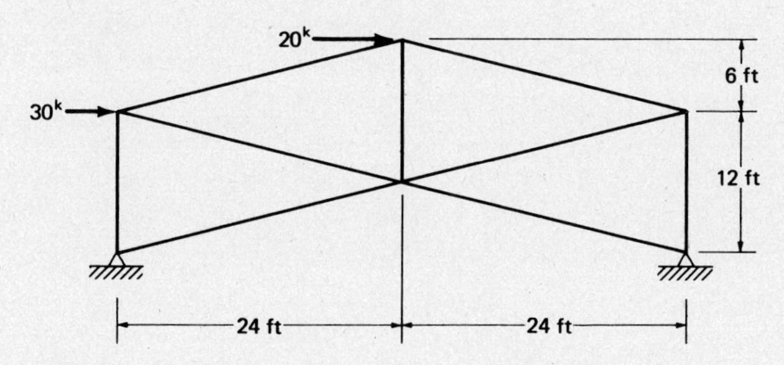

Problem 20.4 All areas are equal (*Ans.* $V_C = 76.9^k \uparrow$, $U_1U_2 = +64.7^k$, $L_2U_3 = +30.4^k$, $U_3L_3 = -16.9^k$)

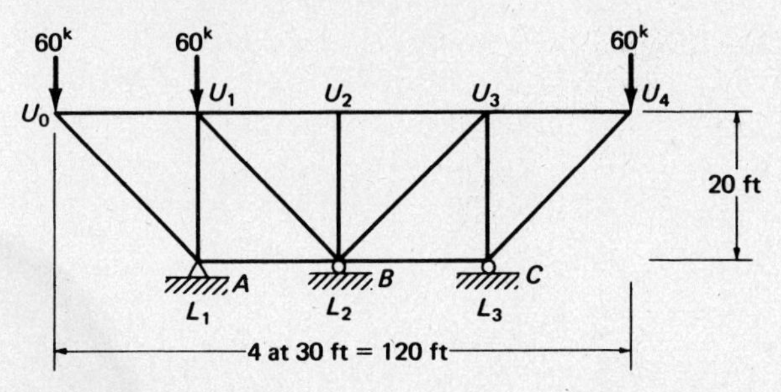

Problem 20.5

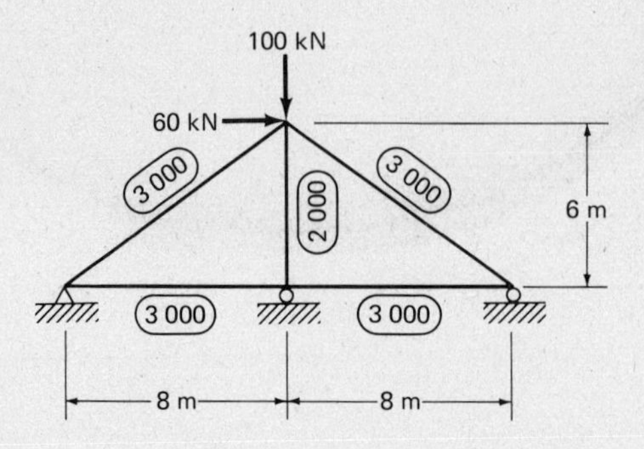

Problem 20.6 (*Ans.* $V_L = 47.88$ kN ↑, $L_0L_1 = +59.8$ kN, $U_1U_2 = -49.7$ kN)

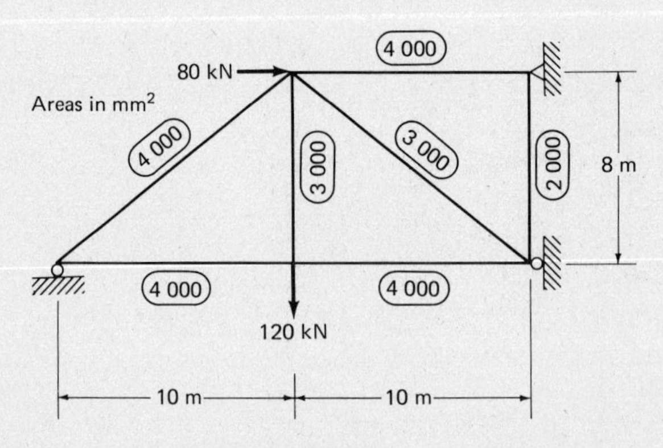

Problem 20.7

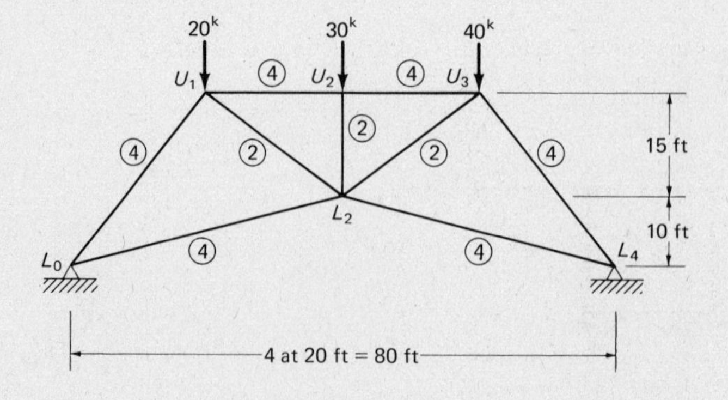

Problem 20.8 (*Ans.* $V_A = 48.4^k \uparrow$, $L_1L_2 = -80^k$, $U_1L_2 = -19.3^k$, $U_2U_3 = +95.5^k$)

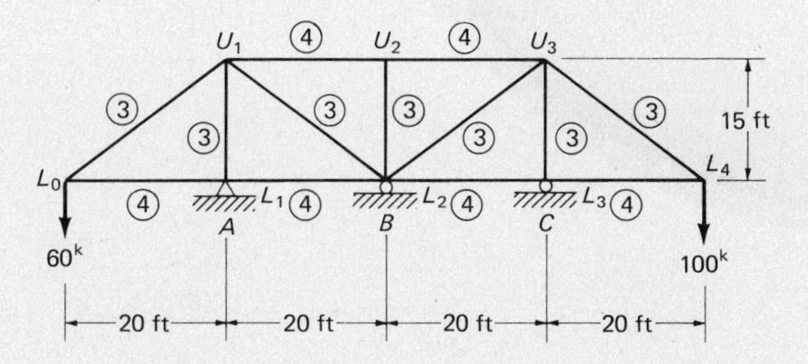

Problem 20.9

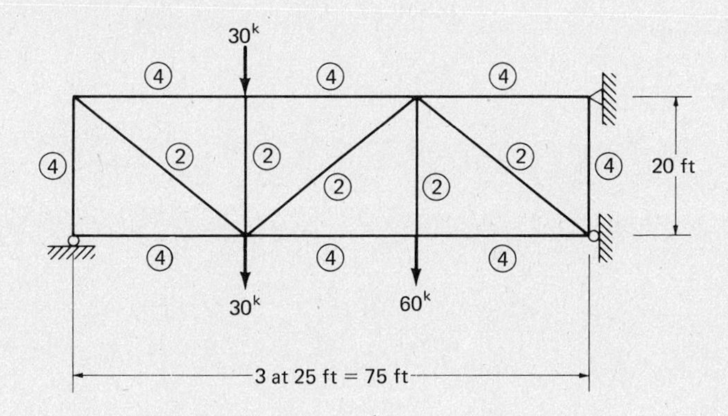

Problem 20.10 (*Ans.* $U_0U_1 = -14.94^k$, $U_0L_0 = +20.08^k$, $L_0U_1 = +24.9^k$)

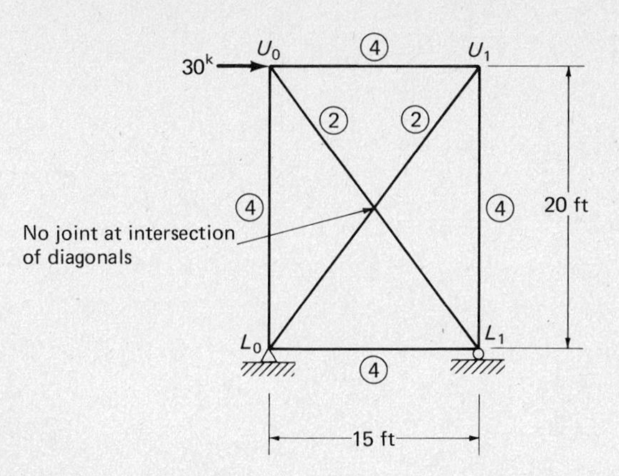

Problem 20.11

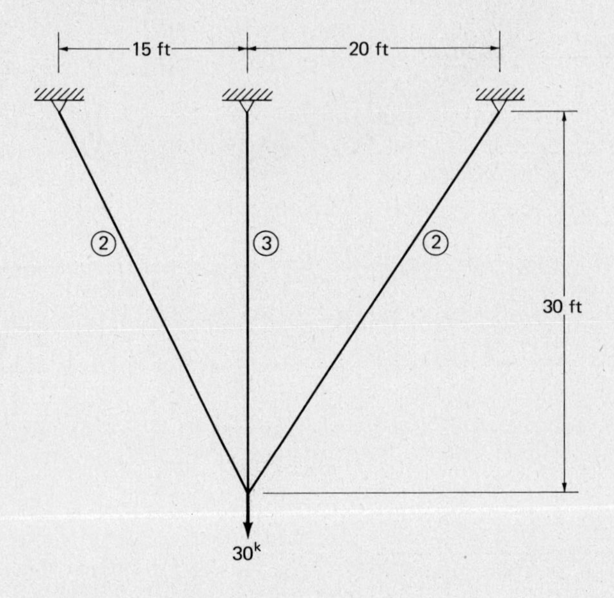

Problem 20.12 All areas are equal (*Ans.* $L_0U_1 = +19.6^k$, $U_1M_1 = +3.4^k$, $M_1L_2 = +4.5^k$)

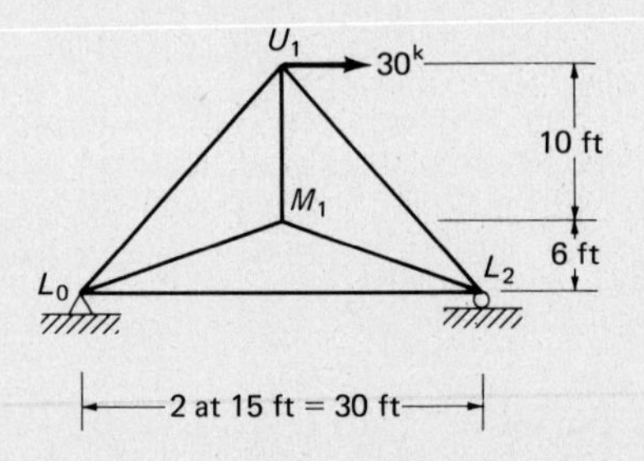

Problem 20.13 All areas are equal

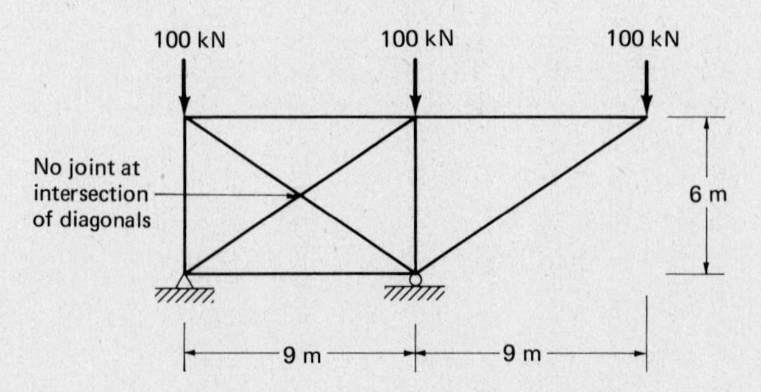

Problem 20.14 All areas are equal (*Ans.* $U_1L_2 = -16.8^k$, $L_1U_2 = -2.7^k$, $U_2L_2 = -38.1^k$)

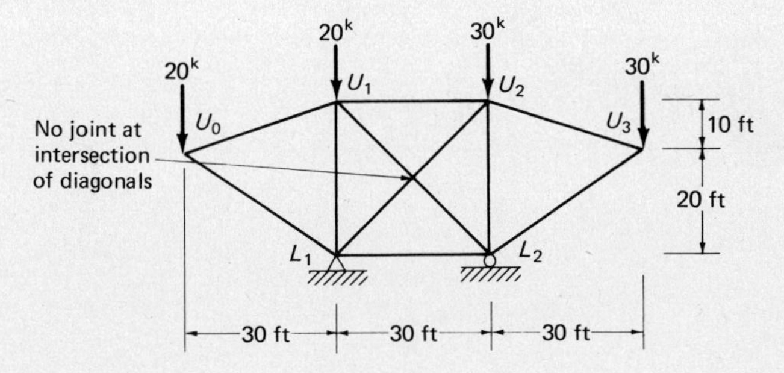

Problem 20.15

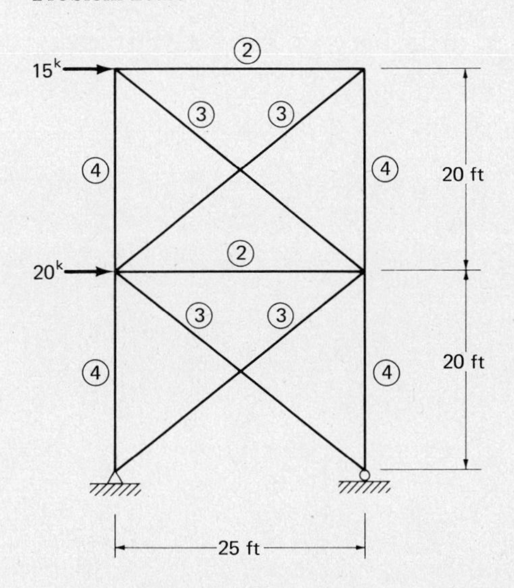

Problem 20.16 (*Ans.* $H_R = 79.6^k \leftarrow$, $U_1U_2 = -54.5^k$, $U_2L_3 = -6.1^k$, $L_4U_5 = +16.3^k$, $U_6L_6 = +8.2^k$)

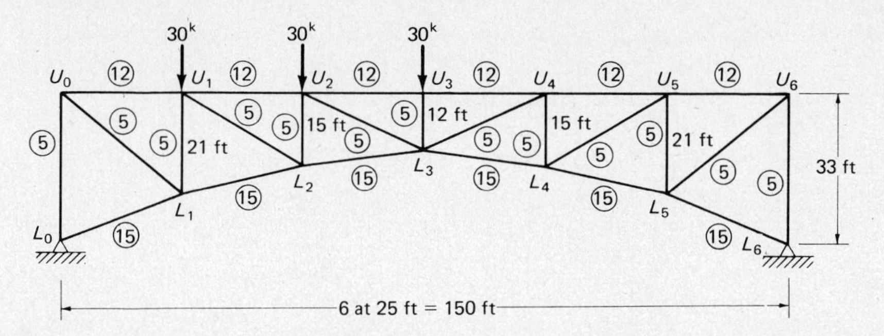

Problem 20.17

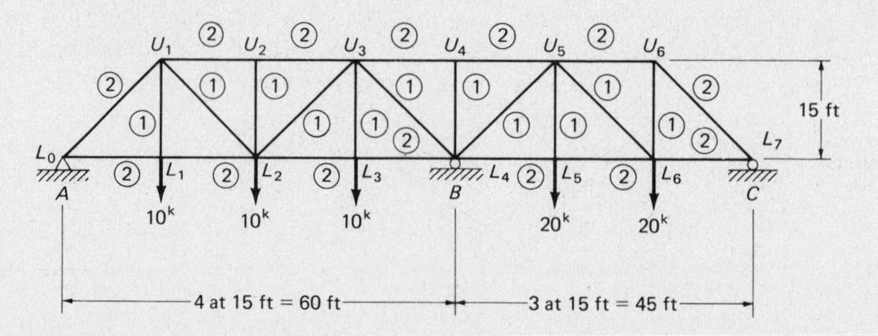

Problem 20.18 (*Ans.* $U_1U_2 = -53.3^k$, $L_2L_3 = +53.3^k$, $L_3U_4 = 0$, $L_3L_4 = +80^k$)

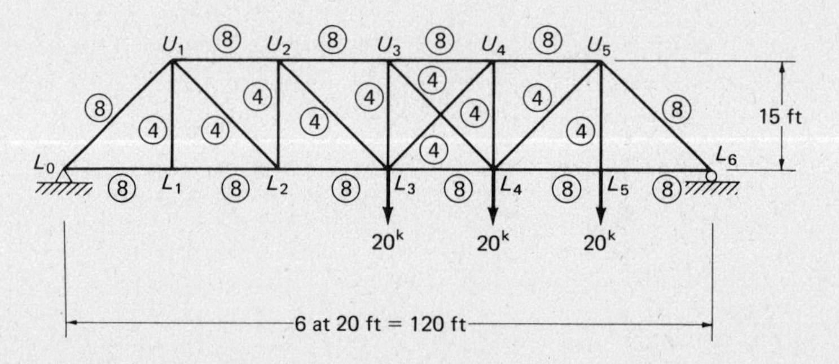

Problem 20.19

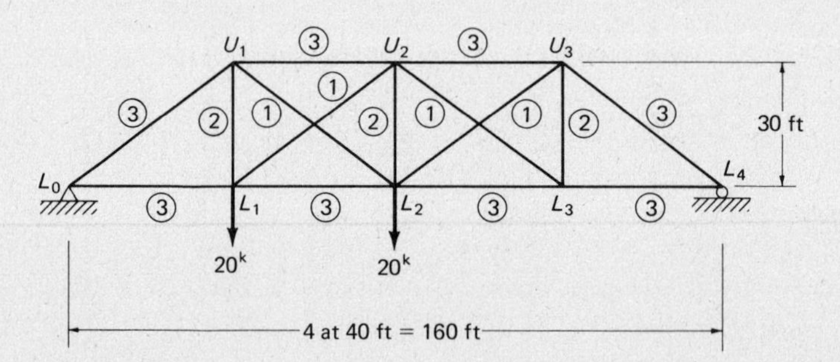

Problem 20.20 (*Ans.* $V_A = 1.1^k \downarrow$, $L_1L_2 = +18.9^k$, $U_2U_3 = -13.3^k$, $L_2U_3 = -15.0^k$)

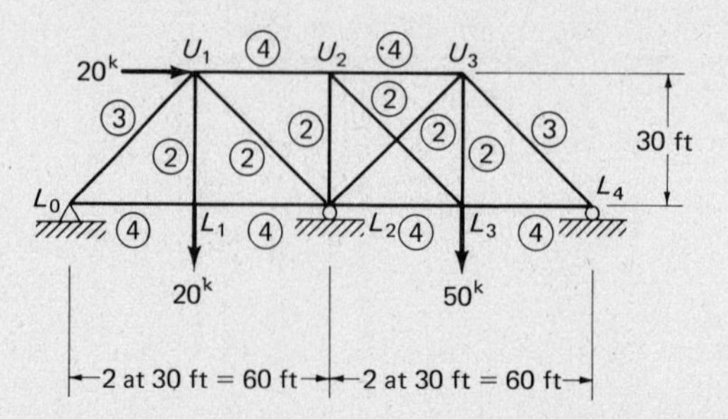

20.21. Determine the forces in all the members of the truss shown in the accompanying illustration if the top-chord members, U_0U_1 and U_1U_2, have an increase in temperature of 75°F. No change in temperature in other members. Coefficient of linear expansion $\epsilon = 0.0000065$.

Problem 20.21

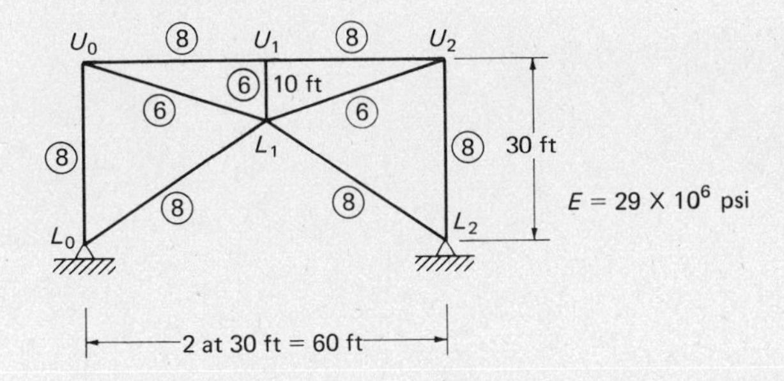

20.22. For Prob. 20.17 draw influence lines for each of the reactions. (*Ans.* Influence line for V_B: Load at $L_1 = +0.303$, at $L_3 = +0.867$, at $L_5 = +0.738$; Influence line for V_C: Load at $L_2 = -0.07$, at $L_5 = +0.293$)

20.23. Draw influence lines for force in members U_1L_2, U_3U_4, and L_4L_5 of the truss of Prob. 20.17.

20.24. Draw influence lines for the reactions of Prob. 20.4. Assume the loads are moving across the top of the structure. (*Ans.* Influence line for V_A: Load at $U_0 = +1.67$, at $U_1 = +0.951$, at $U_4 = -0.33$; Influence line for V_C: Load at $U_0 = -0.33$, at $U_1 = -0.049$, at $U_3 = +0.951$)

20.25. Draw influence lines for force in members L_3L_4, U_1U_2, and U_2L_2 of the truss of Prob. 20.18.

20.26. Draw influence lines for force in members L_1U_2 and U_2L_3 of the truss of Prob. 20.19. (*Ans.* Influence line for L_1U_2: Load at $L_1 = +0.270$, at $L_2 = -0.366$; Influence line for U_2L_3: Load at $L_1 = -0.202$, at $L_2 = -0.366$, at $L_3 = +0.270$).

20.27. Draw influence lines for force in member L_2U_3 and for the center reaction of the truss of Prob. 20.20.

20.28. Rework Prob. 20.4 for support settlement as follows: $A = 1.75$ in, $B = 1.25$ in, and $C = 1.50$ in, $E = 29 \times 10^6$ psi. (*Ans.* $V_B = 42.7^k \downarrow$, $U_1L_1 = -81.4^k$, $L_2U_3 = +38.5^k$)

20.29. Rework Prob. 20.17 for support settlement as follows; $A = 1.25$ in, $B = 2.25$ in, and $C = 0.75$ in, $E = 29 \times 10^6$ psi.

21 Castigliano's Theorems and the Three-Moment Theorem

21.1 INTRODUCTION TO CASTIGLIANO'S THEOREMS

An Italian railway engineer, Alberto Castigliano, published an original and elaborate book in 1879 on the study of statically indeterminate structures. In this book were included the two theorems that are known today as Castigliano's first and second theorems. Castigliano had previously presented them in a paper in 1876. As a matter of fact his theorem of least work (to be discussed) was first presented as a thesis for his engineering diploma at Turin in 1873 [1].

His first theorem provides an important method for computing deflections. Its application involves the equating of the deflection to the first partial derivative of the total internal work of the structure with respect to a load at the point where deflection is desired. Castigliano's second theorem, commonly known as the *method of least work*, provides a very important method for the analysis of statically indeterminate structures. In applying this later theorem the first partial derivative of the internal work with respect to each redundant is set equal to zero.

In presenting the virtual-work method for beams and frames in Sec. 18.4, the author derived the equation

$$W = \int \frac{Mm}{EI}\,dx$$

to express the total internal work accomplished by the stresses caused by a unit load placed at the point and in the direction of a desired deflection when the external loads have been replaced on the beam.

The internal work of the actual stresses caused by gradually applied external loads can be determined in a similar manner. The work in each fiber equals the average stress, $(My/2I)\,da$ (as the external loads vary from zero to full value), times the total strain in the fiber, $(My/EI)\,dx$. Integrating the

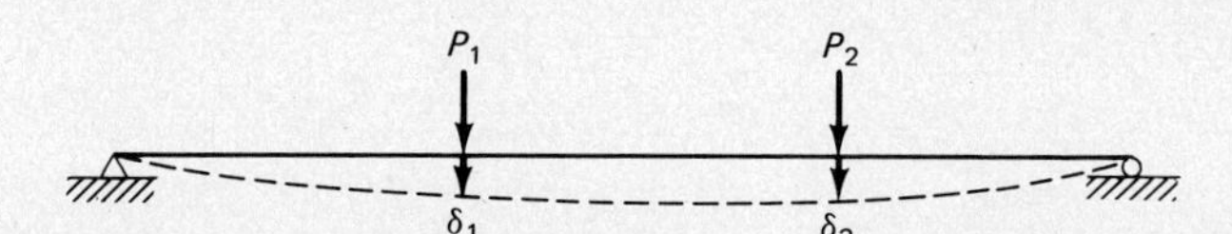

Figure 21.1

product of these two expressions for the cross section of the member and throughout the structure gives the total internal work or strain energy stored for the entire structure.

$$W = \int_0^l \frac{M^2 dx}{2EI}$$

In a similar manner the internal work in the members of a truss due to a set of gradually applied loads can be shown to equal

$$W = \sum \frac{F^2 l}{2AE}$$

These expressions will be used frequently in the following two sections in which a detailed discussion of Castigliano's theorems is presented.

21.2 CASTIGLIANO'S FIRST THEOREM

As a general rule the other methods of deflection computation (such as conjugate beam or virtual work) are a little easier to use and more popular than Castigliano's first theorem. For certain structures, however, this method is very useful and from the standpoint of the student its study serves as an excellent background for the extremely important second theorem.

The derivation of the first theorem (presented in this section) is very similar to that given by Kinney [2]. Figure 21.1 shows a beam that has been subjected to the gradually applied loads P_1 and P_2. These loads cause the deflections δ_1 and δ_2. It is desired to find the deflection δ_1 at load P_1.

The external work performed during the application of the loads must equal the average load times the deflection, and it also must equal the internal strain energy of the beam.

$$W = \frac{P_1 \delta_1}{2} + \frac{P_2 \delta_2}{2} \tag{21.1}$$

Should the load P_1 be increased by the small amount dP_1, the beam will deflect additionally. Figure 21.2 shows the additional deflections at each of the loads $d\delta_1$ and $d\delta_2$.

The additional work performed or strain energy stored during the application of dP_1 is as follows:

$$dW = \left(P_1 + \frac{dP_1}{2}\right) d\delta_1 + P_2\, d\delta_2$$

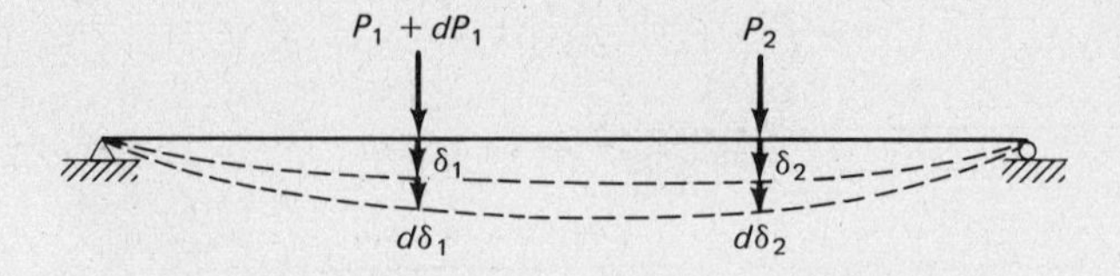

Figure 21.2

Performing the indicated multiplications and neglecting the product of the differentials,

$$dW = P_1\,d\delta_1 + P_2\,d\delta_2 \tag{21.2}$$

The same procedure is repeated except that the loads P_1, P_2, and dP_1 are all applied at the same time and the total strain energy is represented by W'.

$$W' = \left(\frac{P_1 + dP_1}{2}\right)(\delta_1 + d\delta_1) + \left(\frac{P_2}{2}\right)(\delta_2 + d\delta_2)$$

Performing the indicated multiplications and neglecting the product of the differentials,

$$W' = \frac{P_1\delta_1}{2} + \frac{P_1\,d\delta_1}{2} + \frac{dP_1\,\delta_1}{2} + \frac{P_2\delta_2}{2} + \frac{P_2\,d\delta_2}{2} \tag{21.3}$$

It is obvious that dW equals $W' - W$ and can be obtained by subtracting equation (21.1) from (21.3).

$$\begin{aligned} dW &= \frac{P_1\delta_1}{2} + \frac{P_1\,d\delta_1}{2} + \frac{dP_1\,\delta_1}{2} + \frac{P_2\delta_2}{2} + \frac{P_2\,d\delta_2}{2} - \frac{P_1\delta_1}{2} - \frac{P_2\delta_2}{2} \\ dW &= \frac{P_1\,d\delta_1}{2} + \frac{dP_1\delta_1}{2} + \frac{P_2\,d\delta_2}{2} \end{aligned} \tag{21.4}$$

From Eq. (21.2) the value of $P_2\,d\delta_2$ can be obtained as follows:

$$P_2\,d\delta_2 = dW - P_1\,d\delta_1$$

By substituting this value into Eq. (21.4) and solving for the desired deflection δ_1,

$$dW = \frac{P_1\,d\delta_1}{2} + \frac{dP_1\,\delta_1}{2} + \frac{dW}{2} - \frac{P_1\,d\delta_1}{2}$$

$$\delta_1 = \frac{dW}{dP_1}$$

Since more than one action is applied to a structure, this deflection is usually written as a partial derivative as follows:

$$\delta = \frac{\partial W}{\partial P}$$

Burro Creek Bridge, Wickieup, Mohave County, Arizona. (Courtesy of the American Institute of Steel Construction, Inc.)

To determine the deflection at a point in a beam or frame the first theorem would be written as

$$\delta = \frac{\partial}{\partial P} \int \frac{M^2}{2EI} dx$$

Should the procedure represented in the preceding expression be followed, M would be squared, integrated, and the first partial derivative would be taken. If M is rather complicated, however, as it frequently is, the process would be very tedious. For this reason it is usually simpler to differentiate under the integral sign with the following results:

$$\delta = \int M \left(\frac{\partial M}{\partial P}\right) \frac{dx}{EI}$$

For a truss a similar procedure is advisable with the following results:

$$\delta = \frac{\partial}{\partial P} \sum \frac{F^2 l}{2AE}$$

$$\delta = \sum F \left(\frac{\partial F}{\partial P}\right) \frac{l}{AE}$$

Examples 21.1 through 21.4 illustrate the application of Castigliano's first theorem. In applying the theorem, the load at the point where the deflection is desired is referred to as P. After the operations required in the equation are completed, the numerical value of P is replaced in the expression. Should there be no load at the point or in the direction in which deflection is desired, an imaginary force P will be placed there in the direction desired. After the operation is completed, the correct value of P (zero) will be substituted in the expression (see Example 21.2).

If slope or rotation is desired in a structure, the partial derivative is taken with respect to an assumed moment P acting at the point where rotation is desired (see Example 21.4). A plus sign on the answer indicates that the rotation is in the assumed direction of the moment P.

Example 21.1

Determine the vertical deflection at the 10-kip load in the beam shown in Fig. 21.3.

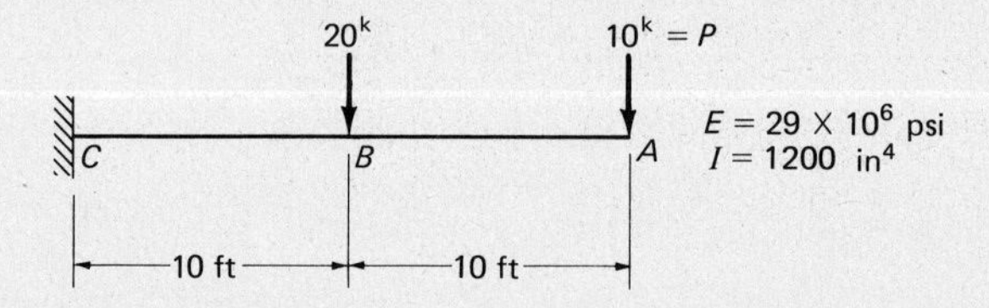

Figure 21.3

SOLUTION

By letting the 10-kip load be P,

Table 21.1

SECTION	M	$\frac{\partial M}{\partial P}$	$M\frac{\partial M}{\partial P}$	$\int M\left(\frac{\partial M}{\partial P}\right)\frac{dx}{EI}$
A to B	$-Px$	$-x$	$+Px^2$	$\int_0^{10}(Px^2)\frac{dx}{EI}=\frac{333P}{EI}$
B to C	$-P(x+10)$ $-20x$	$-x-10$	$+Px^2+20Px$ $+20x^2$ $+100P+200x$	$\int_0^{10}(Px^2+20Px+20x^2$ $+100P+200x)\frac{dx}{EI}$ $=\frac{2333P+16{,}667}{EI}$
Σ				$\frac{2666P+16{,}667}{EI}$

$$\delta=\frac{2666P+16{,}667}{EI}=\frac{43{,}333\ \text{ft}^{3\text{-k}}}{EI}=2.15\ \text{in}\downarrow$$

Example 21.2

Determine the vertical deflection at the free end of the cantilever beam shown in Fig. 21.4.

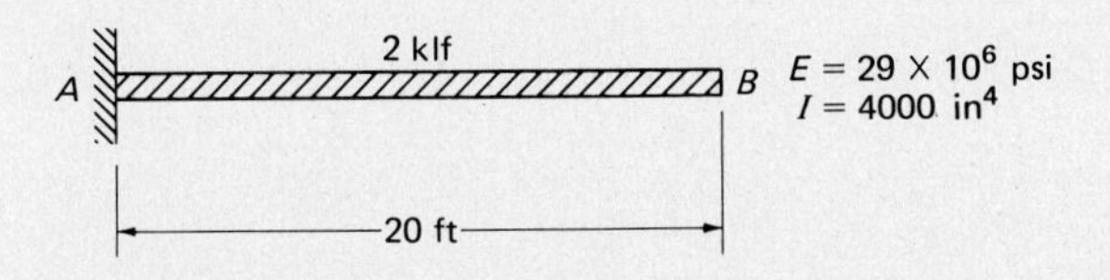

Figure 21.4

SOLUTION

Placing an imaginary load P at the free end,

Table 21.2

SECTION	M	$\frac{\partial M}{\partial P}$	$M\frac{\partial M}{\partial P}$	$\int M\left(\frac{\partial M}{\partial P}\right)\frac{dx}{EI}$
B to A	$-Px - x^2$	$-x$	$Px^2 + x^3$	$\int_0^{20}(Px^2 + x^3)\frac{dx}{EI} = \frac{40{,}000}{EI}$
Σ				$\frac{40{,}000}{EI}$

$$\delta = 0.596 \text{ in} \downarrow$$

Example 21.3

Determine the vertical deflection at the 30-kip load in the beam of Fig. 21.5.

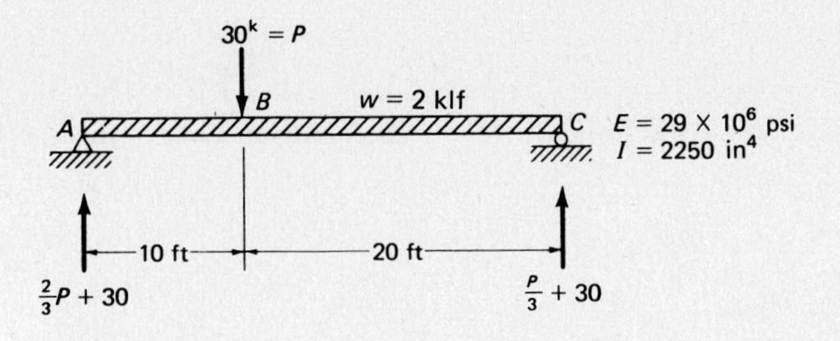

Figure 21.5

SOLUTION

The 30-kip load is replaced with P. Notice that the beam reactions are computed with a load of P at point B and not 30 kips.

Table 21.3

SECTION	M	$\frac{\partial M}{\partial P}$	$M \frac{\partial M}{\partial P}$	$\int M \left(\frac{\partial M}{\partial P}\right) \frac{dx}{EI}$
A to B	$\frac{2}{3}Px + 30x - x^2$	$\frac{2}{3}x$	$\frac{4}{9}Px^2 + 20x^2 - \frac{2}{3}x^3$	$\frac{1}{EI}\int_0^{10}\left(\frac{4}{9}Px^2 + 20x^2 - \frac{2}{3}x^3\right) dx$
C to B	$\frac{P}{3}x + 30x - x^2$	$\frac{x}{3}$	$\frac{Px^2}{9} + 10x^2 - \frac{x^3}{3}$	$\frac{1}{EI}\int_0^{20}\left(\frac{Px^2}{9} + 10x^2 - \frac{x^3}{3}\right) dx$

Performing the indicated operations $\delta = 0.838$ in. ↓

Example 21.4

Find the slope at the free end of the cantilever beam of Example 21.1. The beam is drawn again in Fig. 21.6.

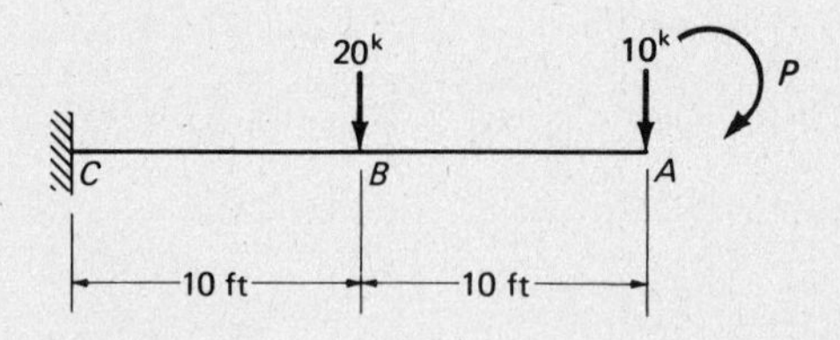

Figure 21.6

SOLUTION

A clockwise moment of P is assumed to be acting at the free end as shown in the figure.

Table 21.4

SECTION	M	$\frac{\partial M}{\partial P}$	$M \frac{\partial M}{\partial P}$	$\int M \left(\frac{\partial M}{\partial P}\right) \frac{dx}{EI}$
A to B	$-10x - P$	-1	$10x + P$	$\frac{1}{EI}\int_0^{10}(10x + P)\, dx$
B to C	$-30x - P - 100$	-1	$30x + P + 100$	$\frac{1}{EI}\int_0^{10}(30x + P + 100)\, dx$

Performing the indicated operations $\theta = +0.0124$ rad ↻

21.3 CASTIGLIANO'S SECOND THEOREM

Castigliano's second theorem, commonly known as the ***method of least work***, has played an important role in the development of structural analysis through the years and is occasionally used today. It is closely related to the method of consistent distortions discussed in the two preceding chapters and is very

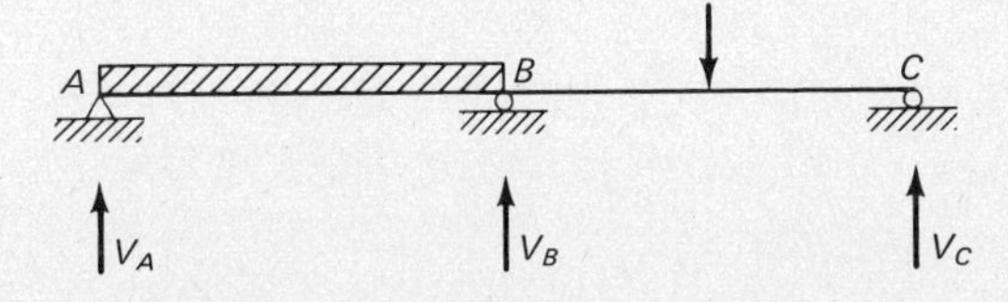

Figure 21.7

effective in the analysis of statically indeterminate structures, particularly trusses and composite structures. (Composite structures are defined here as structures that have some members with axial stress only and other members with axial stresses and bending moments.) Although applicable to beams and frames, other methods such as the moment-distribution method (Chaps. 23 and 24) are usually more satisfactory. The method of least work has the disadvantage that it is not applicable in its usual form to forces caused by displacements due to temperature changes, support settlements, and fabrication errors.

In the preceding section it was shown that the first partial derivative of the total internal work with respect to a load P (real or imaginary) applied at a point in a structure equaled the deflection in the direction of P. For this discussion the continuous beam of Fig. 21.7 and the vertical reaction at support B, V_B, are considered.

If the first partial derivative of the work in this beam is taken with respect to the reaction V_B the deflection at B will be obtained, but that deflection is zero.

$$\frac{\partial W}{\partial P} = 0$$

This is a statement of Castigliano's second theorem. Equations of this type can be written for each point of constraint of a statically indeterminate structure. A structure will deform in a manner consistent with its physical limitations or so that the internal work of deformation will be at a minimum.

The columns and girders meeting at a joint in a building will all deflect the same amount: the smallest possible value. Neglecting the effect of the other ends of these members, it can be seen that each member does no more work than necessary, and the total work performed by all of the members at the joint is the least possible.

From the foregoing discussion the theorem of least work may be stated: **The internal work accomplished by each member or each portion of a statically indeterminate structure subjected to a set of external loads is the least possible amount necessary to maintain equilibrium in supporting the loads.**

On some occasions (particularly for continuous beams and frames), the least-work method is very laborious to apply. Consequently readers often express rather strong opinions as to what they think of the term "least work."

To analyze a statically indeterminate structure with Castigliano's theorem, certain members are assumed to be the redundants and are considered removed from the structure. The removal of the members must be sufficient to leave a statically determinate and stable base structure. The F' forces in the structure are determined by means of the external loads; the redundants are replaced as loads X_1, X_2, and so on; and the forces the loads cause are determined.

The total internal work of deformation may be set up in terms of the F' forces and the forces caused by the redundant loads. The result is differentiated successively with respect to the redundants. The derivatives are made equal to zero in order to determine the values of the redundants.

Examples 21.5 to 21.10 illustrate the analysis of statically indeterminate structures by least work. Although the least-work and consistent-distortion methods are the most general methods for analyzing various types of statically indeterminate structures, they are not frequently used today because other methods to be discussed are more satisfactory.

Example 21.5

Determine the reaction at support C in the beam of Fig. 21.8 by least work; E and I are constant.

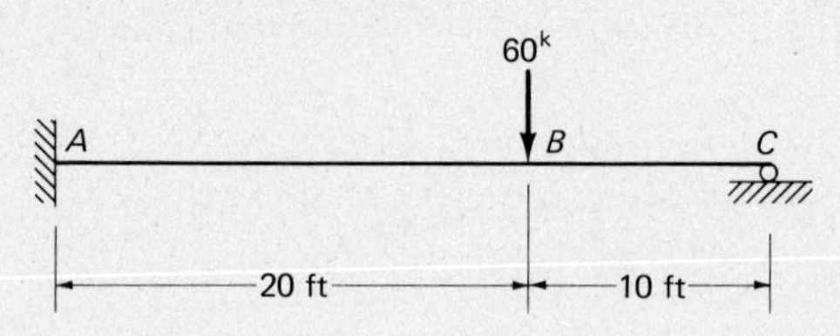

Figure 21.8

SOLUTION

The reaction at C is assumed to be V_c and the other reactions are determined as follows.

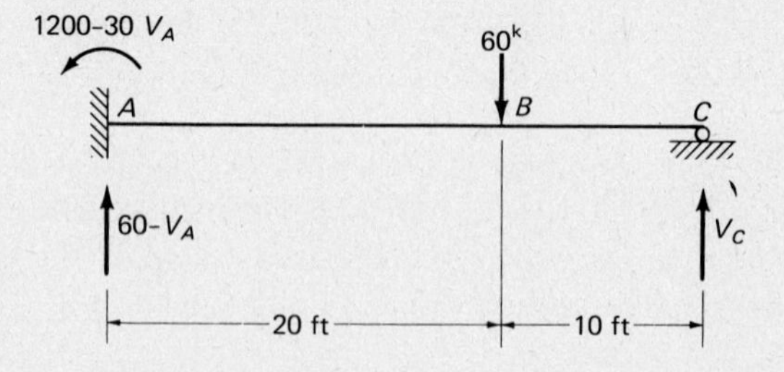

Table 21.5

SECTION	M	$\frac{\partial M}{\partial V_c}$	$M \frac{\partial M}{\partial V_c}$	$\int M \left(\frac{\partial M}{\partial V_c}\right) \frac{dx}{EI} = 0$
C to B	$V_c x$	x	$V_c x^2$	$\frac{1}{EI}\int_0^{10}(V_c x^2)dx$
B to A	$V_c x + 10V_c - 60x$	$x + 10$	$V_c x^2 + 20V_c x - 60x^2 + 100\ V_c - 600x$	$\frac{1}{EI}\int_0^{20}(V_c x^2 + 20V_c x - 60x^2 + 100V_c - 600x)dx$
Σ				$9000V_c - 280{,}000 = 0$

$$V_c = 31.1^{k}\uparrow$$

Example 21.6

Determine the value of the reaction at support C, Fig. 21.9, by the method of least work.

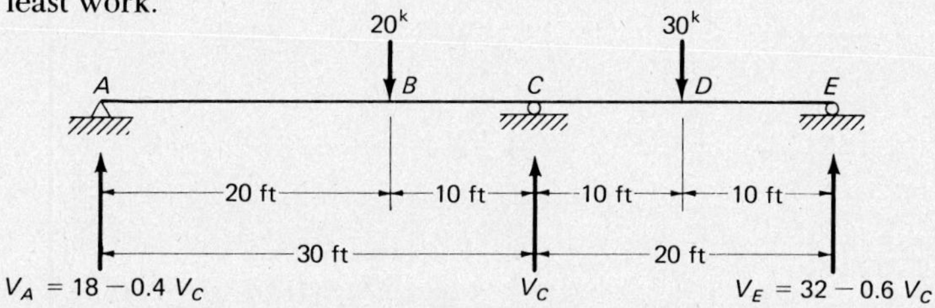

Figure 21.9

SOLUTION

Table 21.6

SECTION	M	$\frac{\partial M}{\partial V_c}$	$\int M \left(\frac{\partial M}{\partial V_B}\right) \frac{dx}{EI} = 0$
A to B	$18x - 0.4V_c x$	$-0.4x$	$\int_0^{20}(-7.2x^2 + 0.16V_c x^2)dx$
B to C	$-2x - 0.4V_c x - 8V_c + 360$	$-0.4x - 8$	$\int_0^{10}(+0.8x^2 + 0.16V_c x^2 + 6.4V_c x - 128x + 64V_c - 2880)\ dx$
E to D	$32x - 0.6V_c x$	$-0.6x$	$\int_0^{10}(-19.2x^2 + 0.36V_c x^2)dx$
D to C	$2x - 0.6V_c x - 6V_c + 320$	$-0.6x - 6$	$\int_0^{10}(-1.2x^2 + 0.36V_c x^2 + 7.2V_c x - 204x + 36V_c - 1920)dx$

By integrating the $\int M(\partial M/\partial V_B)(dx/EI)$ expressions and substituting the values of the proper limits, the result for the entire beam is:

$$-90{,}333 + 2400.3V_C = 0$$

$$V_C = +37.6^{k}\uparrow$$

Example 21.7

Determine the force in member CD of the truss shown in Fig. 21.10. Circled values are areas in square inches. E is constant.

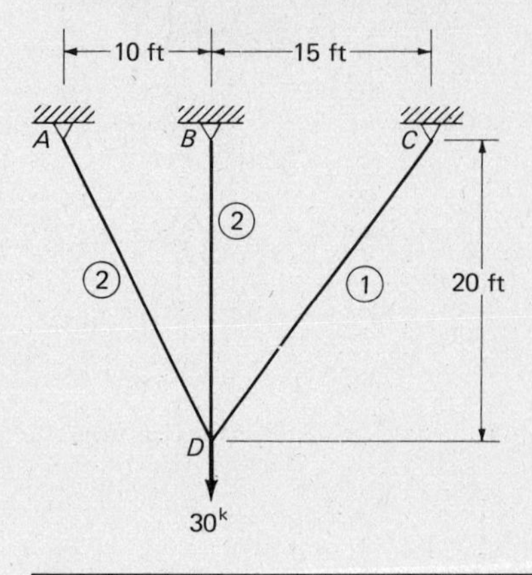

Figure 21.10

SOLUTION

Member CD is assumed to be the redundant and is assigned a force of T. The other forces are determined by joints with the following results.

$$\frac{\partial W}{\partial T} = \frac{\partial}{\partial T}\sum\frac{F^2 l}{2AE} = \sum F\frac{\partial F}{\partial T}\frac{l}{AE}$$

Table 21.7

MEMBER	l(in)	A(in^2)	$\frac{l}{A}$	F	$\frac{\partial F}{\partial T}$	$F\left(\frac{\partial F}{\partial T}\right)\frac{l}{AE}$
AD	268	2	134	$+1.34T$	$+1.34$	$+240T$
BD	240	2	120	$-2T+30$	-2	$+480T - 7200$
CD	300	1	300	$+T$	$+1$	$+300T$
Σ						$1020T - 7200 = 0$

$$T = +7.06^k$$

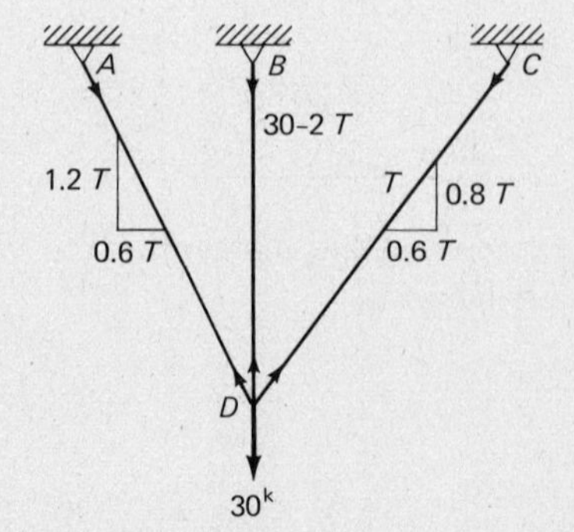

Example 21.8

Analyze the truss of Example 20.1, reproduced in Fig. 21.11, by the least-work method. Circled numbers are member areas, in square inches.

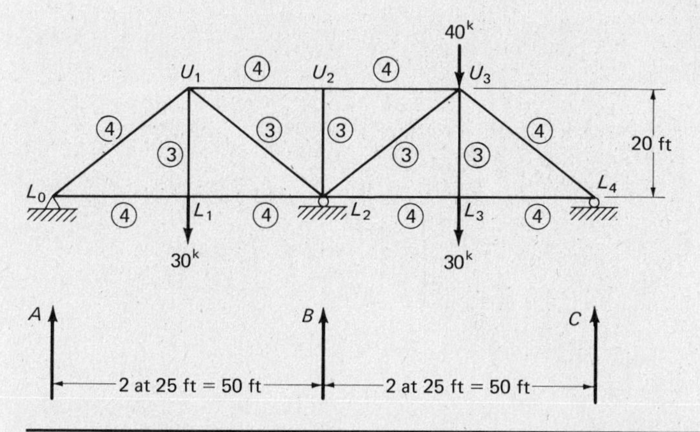

Figure 21.11

SOLUTION

Remove the center support and compute the F' forces.

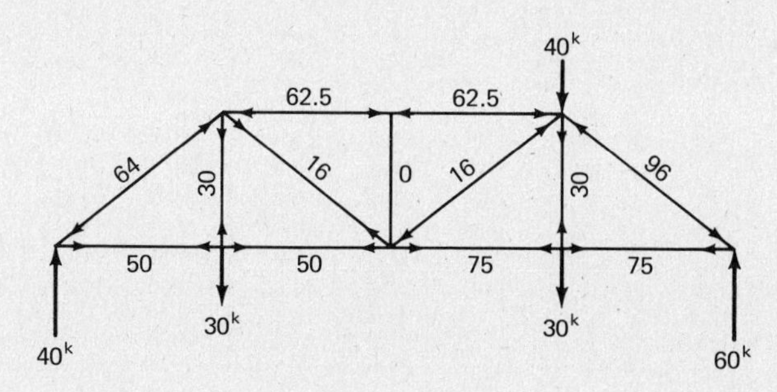

Replace the center support and determine its effect on member forces in terms of V_B.

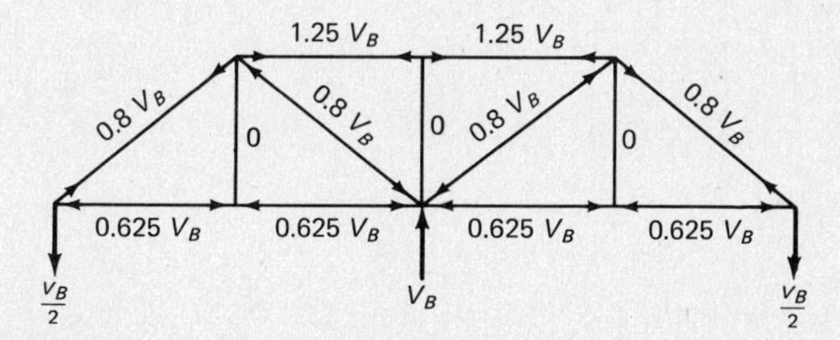

Tennessee River Bridge, Stevenson, Alabama. (Courtesy of the American Bridge Division, U.S. Steel Corporation.)

Table 21.8

MEMBER	l(in)	A(in²)	$\frac{l}{A}$	F(kips)	$\frac{Fl}{A}$	$\frac{\partial F}{\partial V_B}$	$\frac{Fl}{A}\frac{\partial F}{\partial V_B}$
L_0L_1	300	4	75	$+50 - 0.625V_B$	$+3750 - 46.9V_B$	-0.625	$-2345 + 29.3V_B$
L_1L_2	300	4	75	$+50 - 0.625V_B$	$+3750 - 46.9V_B$	-0.625	$-2345 + 29.3V_B$
L_2L_3	300	4	75	$+75 - 0.625V_B$	$+5625 - 46.9V_B$	-0.625	$-3520 + 29.3V_B$
L_3L_4	300	4	75	$+75 - 0.625V_B$	$+5625 - 46.9V_B$	-0.625	$-3520 + 29.3V_B$
L_0U_1	384	4	96	$-64 + 0.8V_B$	$-6144 + 76.8V_B$	$+0.8$	$-4915 + 61.4V_B$
U_1U_2	300	4	75	$-62.5 + 1.25V_B$	$-4687 + 93.8V_B$	$+1.25$	$-5860 + 117.2V_B$
U_2U_3	300	4	75	$-62.5 + 1.25V_B$	$-4687 + 93.8V_B$	$+1.25$	$-5860 + 117.2V_B$
U_3L_4	384	4	96	$-96 + 0.8V_B$	$-9216 + 76.8V_B$	$+0.8$	$-7373 + 61.4V_B$
U_1L_1	240	3	80	$+30 + 0$	$+2400 + 0$	0	$0 + 0$
U_1L_2	384	3	128	$+16 - 0.8V_B$	$+2048 - 102.4V_B$	-0.8	$-1638 + 81.9V_B$
U_2L_2	240	3	80	$0 + 0$	$0 + 0$	0	$0 + 0$
L_2U_3	384	3	128	$-16 - 0.8V_B$	$-2048 - 102.4V_B$	-0.8	$+1638 + 81.9V_B$
U_3L_3	240	3	80	$+30 + 0$	$+2400 + 0$	0	$0 + 0$
Σ							$-35{,}738$ $+638.2V_B$

$$E\frac{\partial W}{\partial V_B} = \frac{Fl}{A}\frac{\partial F}{\partial V_B}$$

$$-35{,}738 + 638.2V_B = 0$$

$$V_B = +56^{k}\uparrow$$

It should be obvious from the preceding example problems that the amount of work involved in the analysis of statically indeterminate trusses by the methods of least work and consistent distortions is about equal. For statically indeterminate beams and frames, the methods of analysis to be presented in Chaps. 22 through 24 will be seen to be much simpler to apply.

Least work, however, is particularly useful for analyzing composite structures, such as those considered in Examples 21.9 and 21.10. In these types of structures both bending and truss action take place. The reader will be convinced of the advantage of least work for the analysis of composite structures if he or she attempts to solve the following two problems by consistent distortions.

Example 21.9

Using the least-work procedure, calculate the force in the steel rod of the composite structure shown in Fig. 21.12.

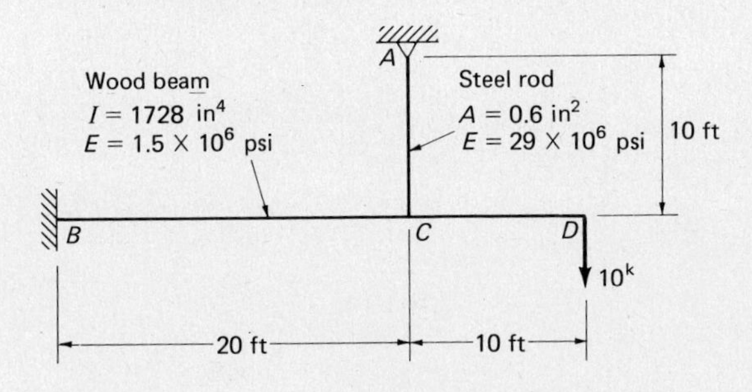

Figure 21.12

SOLUTION

The rod is assumed to be the redundant with a force of T and the values of $\Sigma F(\partial F/\partial T)(l/AE)$ and $\int M(\partial M/\partial T)(dx/EI)$ are computed. Relative values of E of 29 and 1.5 are used for the steel and wood, respectively.

Table 21.9

MEMBER	l(in)	A (in^2)	$\frac{l}{A}$	$\frac{l}{AE}$	F(kip)	$\frac{\partial F}{\partial T}$	$F\left(\frac{\partial F}{\partial T}\right)\frac{l}{AE}$
AC	120	0.6	200	6.90	$+T$	$+1$	$+6.90T$
Σ							$+6.90T$

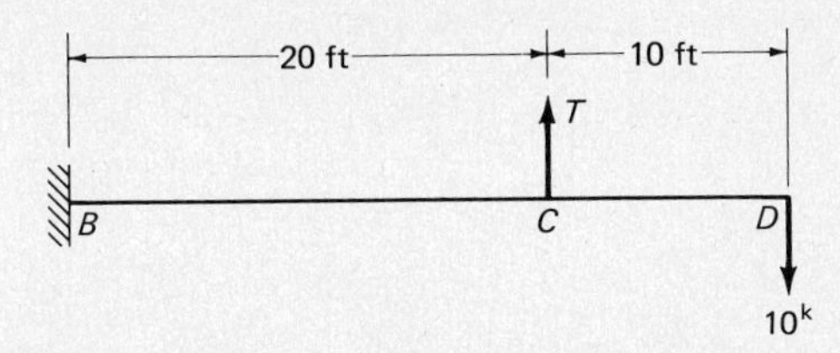

Table 21.10

SECTION	M	$\frac{\partial M}{\partial T}$	$M\frac{\partial M}{\partial T}$	$\int M\left(\frac{\partial M}{\partial T}\right)\frac{dx}{EI}$
D to C	$-10x$	0	0	0
C to B	$Tx - 10x - 100$	x	$Tx^2 - 10x^2 - 100x$	$\int_0^{20} \frac{(Tx^2 - 10x^2 - 100x)\, dx}{(1.5)(1728)}$
Σ				$1.03T - 18$

To change the value of $1.03T - 18$ to inches it is necessary to multiply it by 1728×1000 and divide by 1×10^6 as 29 was used for E instead of 29×10^6. To change the value $6.90T$ to inches it is necessary to multiply it by 1000 and divide by 1×10^6. The only difference in the two conversions is the 1728; therefore the expression can be written as follows to change them to the same units.

$$\int M\left(\frac{\partial M}{\partial T}\right)\frac{dx}{EI} + \Sigma F\left(\frac{\partial F}{\partial T}\right)\frac{l}{AE} = (1.03T - 18)(1728) + 6.90T = 0$$

$$T = +17.4^{k}$$

Example 21.10

Find the forces in all members of the king post truss shown in Fig. 21.13.

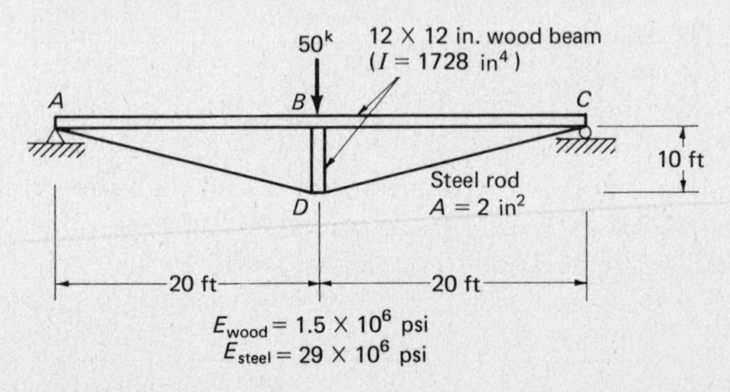

Figure 21.13

SOLUTION

By letting BD be the redundant with a force of F, the deflection of the beam at B is found in terms of F.

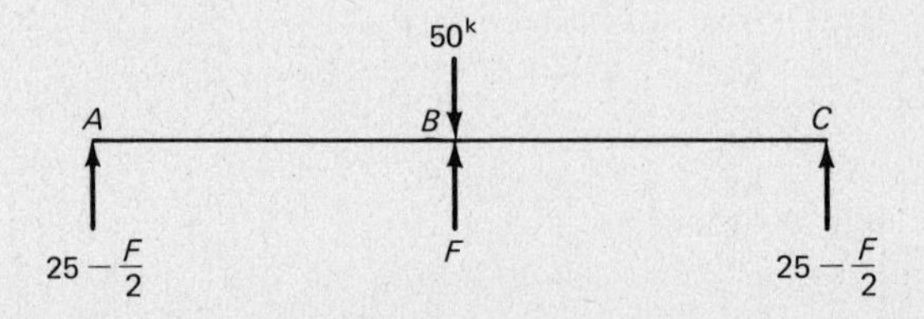

From A to B and from C to B:

$$M = 25x - \frac{F}{2}x \qquad \frac{\partial M}{\partial F} = -\frac{x}{2}$$

$$\int M \frac{\partial M}{\partial F}\frac{dx}{EI} = 2\int_0^{20} \frac{(-12.5x^2 + 0.25Fx^2)dx}{EI}$$

$$\frac{2(-12.5x^3/3 + 0.25Fx^3/3)_0^{20}}{EI} = \frac{-66{,}600 + 1333F}{EI}$$

$$\frac{-66{,}600 + 1333F}{1.5} = -44{,}400 + 889F$$

Determine the forces in the various members in terms of the unknown force F.

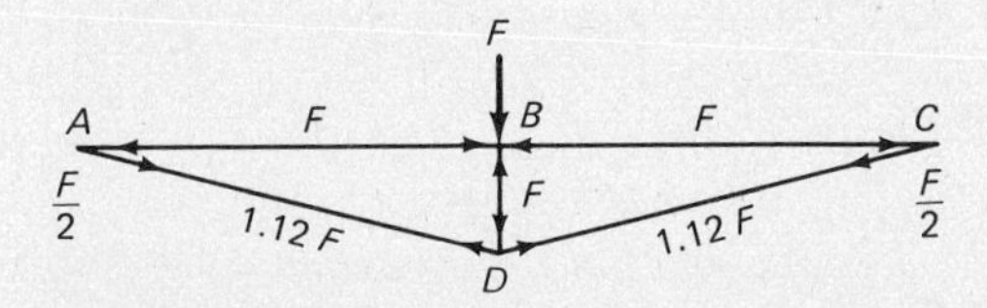

Table 21.11

MEMBER	l (in)	A (in^2)	E	$\frac{l}{AE}$	F'(kips)	$\frac{F'l}{AE}$	$\frac{\partial F'}{\partial F'}$	$\frac{\partial F'}{\partial F} \times \frac{F'l}{AE}$
AC	480	144	1.5×10^6	2.22	$-F$	$-2.22F$	-1	$+2.22F$
BD	120	144	1.5×10^6	0.555	$-F$	$-0.555F$	-1	$+0.555F$
AD	268	2	29×10^6	4.62	$+1.12F$	$+5.19F$	$+1.12$	$+5.80F$
DC	268	2	29×10^6	4.62	$+1.12F$	$+5.19F$	$+1.12$	$+5.80F$
Σ								$14.375F$

$$-44{,}400 + 889F + 14.375F = 0$$

By changing these values to equivalent units and solving for F,

$$F = 49.2^k$$

Final forces:

$$AC = -(1)(49.2) = -49.2^k \qquad AD = +(1.12)(49.2) = +55.1^k$$

$$BD = -(1)(49.2) = -49.2^k \qquad DC = +(1.12)(49.2) = +55.1^k$$

21.4 THE THREE-MOMENT THEOREM

Another subject in the study of the "classical" methods of analyzing statically indeterminate structures is the three-moment theorem, initiated in 1855 by the Frenchman, Bertot. Extensions of his original theorem were presented in 1857 by Clapeyron and in 1862 by Bresse, both Frenchmen [3].

The theorem, which presents a relationship between the moments at the supports in a continuous beam, is usually developed for beams of constant cross

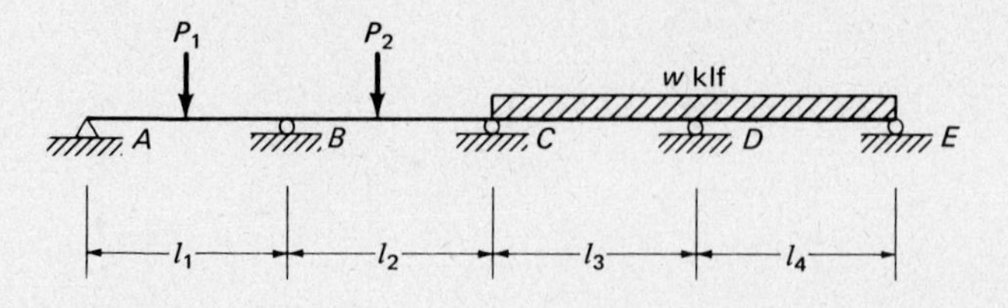

Figure 21.14

section between each pair of supports. Theoretically beams of varying sections may be considered, but the results are so complicated as to be of little practical value. The conjugate-beam method is convenient for developing the relationship between support moments. The resulting equation applies to the moments at any three consecutive supports, as long as the beam is continuous between those supports (that is, no internal hinges or other breaks in the continuity of the beam).

Considering the continuous beam of Fig. 21.14, the three-moment theorem can be written as follows: for the moments at supports *A*, *B*, and *C*; for the moments at supports *B*, *C*, and *D*; and, finally, for the moments at supports *C*, *D*, and *E*. Supports *A* and *E* are simple ends and must have zero moments. Three unknown moments (M_B, M_C, and M_D) remain, but three simultaneous equations are available from which their values may be determined. The theorem applies equally well to beams with fixed ends, as will be seen in the following sections.

21.5 DEVELOPMENT OF THE THEOREM

The elastic curve of a continuous beam has the same numerical value of slope (in radians), an infinitesimal distance on each side of an interior support, although the signs of slope are different. Considering three consecutive interior supports of a continuous beam and loading the beam with the M/EI diagrams, expressions may easily be written for the slope on each side of the center support. The two expressions, which are numerically equal, are equated to each other, the result being a statement of a relationship between the moments at the three supports.

Figure 21.15 shows a section of a uniformly loaded beam, the deflected shape of the elastic curve of the beam, and the M/EI diagrams for the section. The first M/EI diagram is for the negative support moments, and the second M/EI diagram is for the positive simple beam moments caused by the uniform loads.

Span 1

The slope θ_{BL} is equal to the shear or the upward force at the center support. By elastic weights the slope can be expressed as follows:

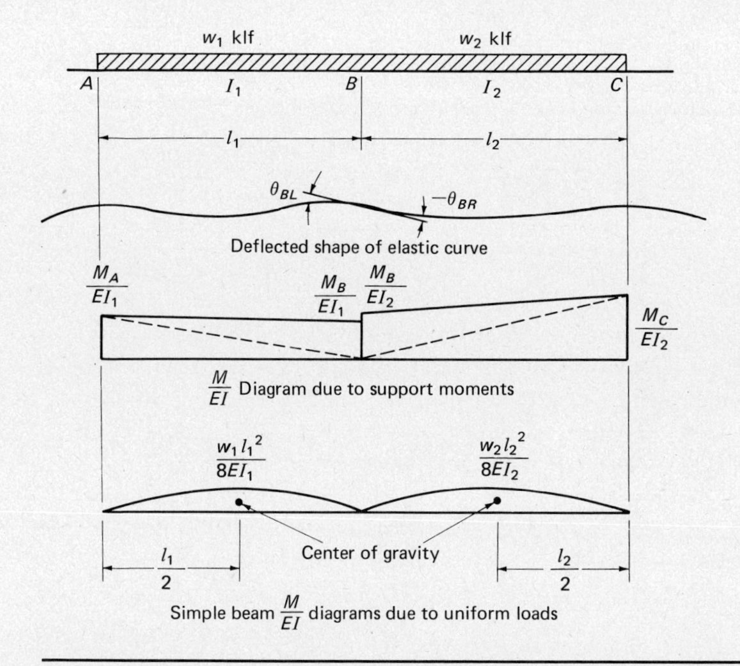

Figure 21.15

$$\theta_{BL} = \frac{\frac{1}{2}(M_A/EI_1)(l_1)(\frac{1}{3}l_1) + \frac{1}{2}(M_B/EI_1)(l_1)(\frac{2}{3}l_1) + \frac{2}{3}(w_1 l_1^2/8EI_1)(l_1)(l_1/2)}{l_1}$$

$$= \frac{M_A l_1}{6EI_1} + \frac{M_B l_1}{3EI_1} + \frac{w_1 l_1^3}{24EI_1}$$

Span 2

A similar expression is written for the slope θ_{BR} to the right of support B.

$$\theta_{BR} = \frac{\frac{1}{2}(M_C/EI_2)(l_2)(\frac{1}{3}l_2) + \frac{1}{2}(M_B/EI_2)(l_2)(\frac{2}{3}l_2) + \frac{2}{3}(w_2 l_2^2/8EI_2)(l_2)(l_2/2)}{l_2}$$

$$= \frac{M_C l_2}{6EI_2} + \frac{M_B l_2}{3EI_2} + \frac{w_2 l_2^3}{24EI_2}$$

The slope on the left of the center support is equal to minus the slope on the right, or $\theta_{BL} = -\theta_{BR}$. Equating the two expressions results in an equation for the three support moments.

$$\frac{M_A l_1}{6EI_1} + \frac{M_B l_1}{3EI_1} + \frac{w_1 l_1^3}{24EI_1} = -\frac{M_C l_2}{6EI_2} - \frac{M_B l_2}{3EI_2} - \frac{w_2 l_2^3}{24EI_2}$$

$$\frac{M_A l_1}{6EI_1} + \frac{M_B}{3E}\left(\frac{l_1}{I_1} + \frac{l_2}{I_2}\right) + \frac{M_C l_2}{6EI_2} = -\frac{w_1 l_1^3}{24EI_1} - \frac{w_2 l_2^3}{24EI_2}$$

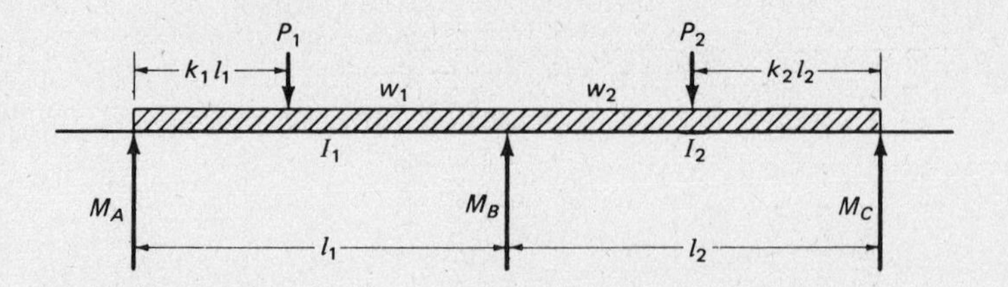

Figure 21.16

Assuming the beam to consist of the same material throughout results in the elimination of E. Multiplying through by six changes the expression into a more convenient form.

$$\frac{M_A l_1}{I_1} + 2M_B\left(\frac{l_1}{I_1} + \frac{l_2}{I_2}\right) + \frac{M_C l_2}{I_2} = -\frac{w_1 l_1^3}{4I_1} - \frac{w_2 l_2^3}{4I_2}$$

The derivation has considered the spans to be loaded with uniform loads only, but concentrated loads may be easily included by taking into account their triangular-shaped simple beam M/EI diagrams. The general equation for both concentrated and uniform loads (dimensions and symbols from Fig. 21.16) is as follows.

General Equation

$$\frac{M_A l_1}{I_1} + 2M_B\left(\frac{l_1}{I_1} + \frac{l_2}{I_2}\right) + \frac{M_C l_2}{I_2}$$
$$= -\sum \frac{P_1 l_1^2}{I_1}(k_1 - k_1^3) - \sum \frac{P_2 l_2^2}{I_2}(k_2 - k_2^3) - \frac{w_1 l_1^3}{4I_1} - \frac{w_2 l_2^3}{4I_2}$$

If the moments of inertia are the same for both spans, the expression becomes

Constant I

$$M_A l_1 + 2M_B(l_1 + l_2) + M_C l_2$$
$$= -\sum P_1 l_1^2(k_1 - k_1^3) - \sum P_2 l_2^2(k_2 - k_2^3) - \frac{w_1 l_1^3}{4} - \frac{w_2 l_2^3}{4}$$

If both span lengths and moments of inertia should be equal, the equation becomes

Constant I and equal spans

$$M_A + 4M_B + M_C = -\sum P_1 l(k_1 - k_1^3) - \sum P_2 l(k_2 - k_2^3) - \frac{w_1 l^2}{4} - \frac{w_2 l^2}{4}$$

21.6 APPLICATION OF THE THREE-MOMENT THEOREM

The determination of support moments by the three-moment theorem is illustrated by Examples 21.11 to 21.15. Little explanation is necessary for the analysis of the beams in the first three examples. These beams are simply supported on each end, indicating zero moments. It will be seen that application of the theorem to a simply end-supported continuous beam presents two less equations than supports, but the end-support moments are zero, and the equations may be solved simultaneously for the unknowns.

A slightly different approach is needed for the solution of Examples 21.14 and 21.15. The beam of Example 21.14 is fixed on the left end, and the beam of Example 21.15 is fixed on both ends. It would appear that there are not going to be enough equations for the unknowns, because the fixed-end supports have moments. The problem is solved, however, by assuming a fixed end to create another span beyond it having a length of zero. The three-moment theorem is written for the assumed span and the adjoining span, as shown in these two examples, giving the required number of equations.

Shear and moment diagrams are drawn for several of the examples to present a complete picture of the stress conditions throughout the beam. When the solution of an equation yields a negative support moment, it indicates the beam is bending over the support, which causes tension in the top fibers ($-M$).

Example 21.11

Determine the support moments of the structure in Fig. 21.17 by the three-moment theorem and draw shear and moment diagrams.

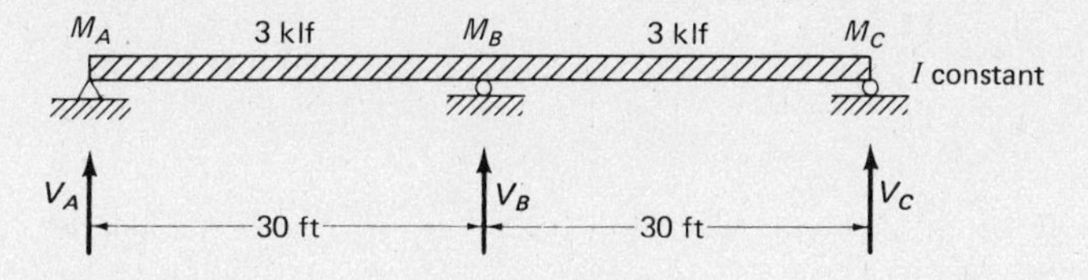

Figure 21.17

SOLUTION

For equal spans and constant I:

$$M_A + 4M_B + M_C = -\sum P_1 l(k_1 - k_1^3) - \sum P_2 l(k_2 - k_2^3) - \frac{w_1 l^2}{4} - \frac{w_2 l^2}{4}$$

$$M_A = M_C = 0$$

$$4M_B = -\frac{(3)(30)^2}{4} - \frac{(3)(30)^2}{4} = -675 - 675 = -1350$$

$$M_B = -337.5'^{k}$$

$$\Sigma M_B \text{ to left} = -337.5$$

$$-337.5 = (V_A)(30) - (3 \times 30)(15)$$

$$-337.5 = 30V_A - 1350$$

$$V_A = \frac{1012.5}{30} = +33.75^{k}$$

$$\Sigma M_B \text{ to right} = +337.5$$
$$+337.5 = -(V_C)(30) + (3)(30)(15)$$
$$+337.5 = -30V_C + 1350$$
$$V_C = \frac{1012.5}{30} = +33.75^k$$

$$\Sigma V = 0$$
$$33.75 + V_B + 33.75 = (3)(60) = 180$$
$$V_B = 112.5^k$$

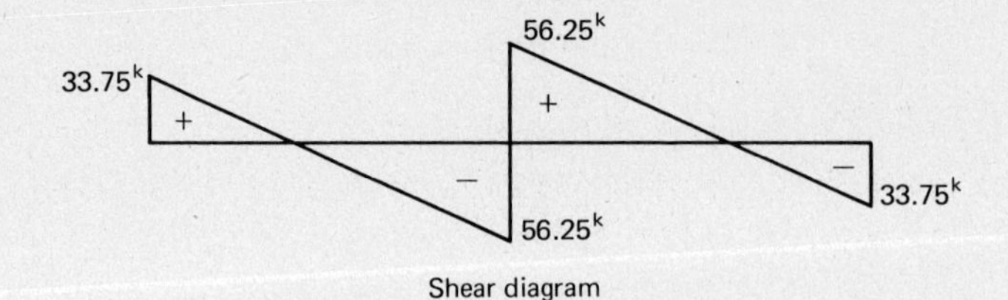

Shear diagram

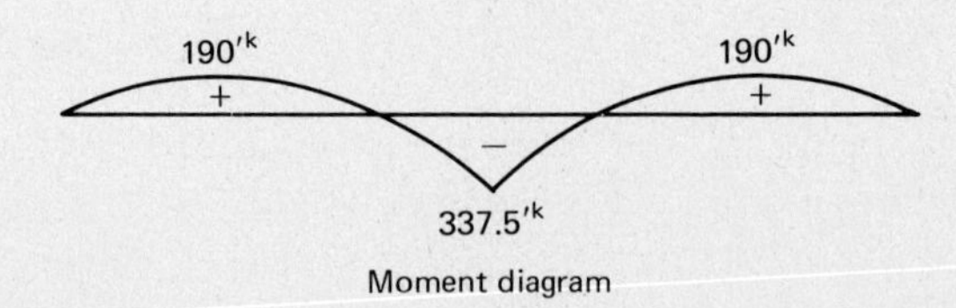

Moment diagram

Example 21.12

By using the three-moment theorem, determine all support moments for the structure in Fig. 21.18 and draw shear and moment diagrams.

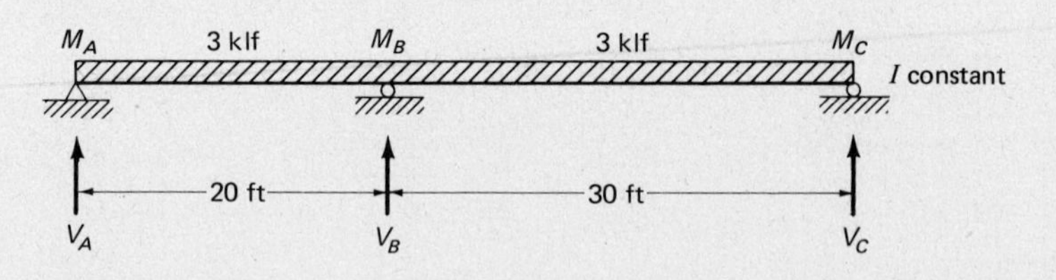

Figure 21.18

SOLUTION

For constant I:

$$M_Al_1 + 2M_B(l_1 + l_2) + M_Cl_2$$
$$= -\sum P_1l_1^2(k_1 - k_1^3) - \sum P_2l_2^2(k_2 - k_2^3) - \frac{w_1l_1^3}{4} - \frac{w_2l_2^3}{4}$$
$$M_A = M_C = 0$$

$$2M_B(20 + 30) = -\frac{(3)(20)^3}{4} - \frac{(3)(30)^3}{4} = -6000 - 20{,}250$$

$$100M_B = -26{,}250$$

$$M_B = -262.5'^k$$

ΣM_B to left $= -262.5$

$$-262.5 = 20V_A - (3)(20)(10)$$

$$-262.5 = 20V_A - 600$$

$$V_A = \frac{337.5}{20} = 16.9^k$$

ΣM_B to right $= +262.5$

$$+262.5 = -(30)(V_C) + (3 \times 30)(15)$$

$$+262.5 = -30V_C + 1350$$

$$V_C = \frac{1087.5}{30} = 36.2^k$$

$$\Sigma V = 0$$

$$16.9 + V_B + 36.2 - 150 = 0$$

$$V_B = 96.9^k$$

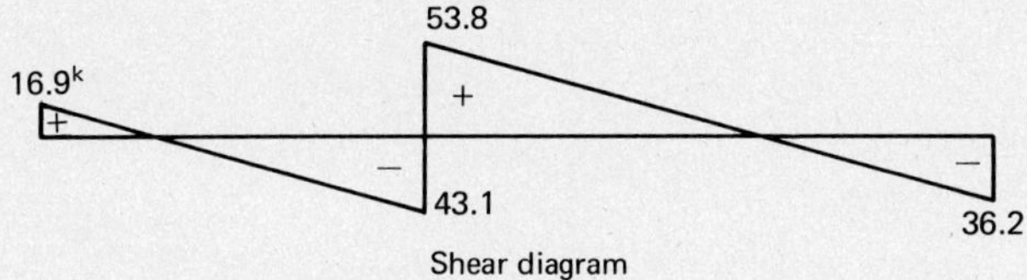

Shear diagram

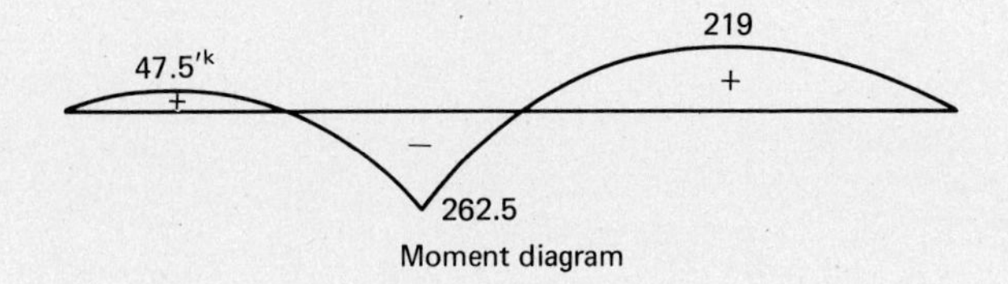

Moment diagram

Example 21.13

Draw shear and moment diagrams for the structure of Fig. 21.19.

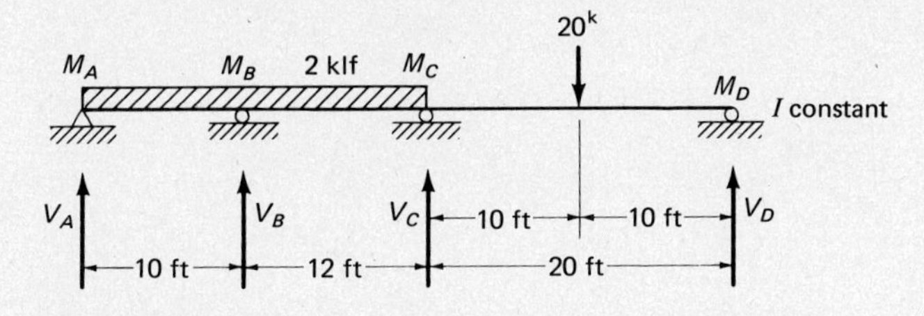

Figure 21.19

SOLUTION

For constant I:

$$M_A l_1 + 2M_B(l_1 + l_2) + M_C l_2$$

$$= -\sum P_1 l_1^2(k_1 - k_1^3) - \sum P_2 l_2^2(k_2 - k_2^3) - \frac{w_1 l_1^3}{4} - \frac{w_2 l_2^3}{4}$$

$$M_A = M_D = 0$$

$$2M_B(10 + 12) + M_C(12) = -\frac{(2)(10)^3}{4} - \frac{(2)(12)^3}{4}$$

$$= -500 - 864 = -1364$$

$$44M_B + 12M_C = -1364 \tag{1}$$

$$M_B l_2 + 2M_C(l_2 + l_3) + M_D l_3$$

$$= -\sum P_2 l_2^2(k_2 - k_2^3) - \sum P_3 l_3^2(k_3 - k_3^3) - \frac{w_2 l_2^3}{4} - \frac{w_3 l_3^3}{4}$$

$$M_B(12) + 2M_C(12 + 20) = -(20)(20)^2\left[\frac{1}{2} - \left(\frac{1}{2}\right)^3\right] - \frac{(2)(12)^3}{4}$$

$$12M_B + 64M_C = -3864 \tag{2}$$

By solving Eqs. (1) and (2) simultaneously for M_B and M_C,

$$M_B = -15.3'^k \qquad M_C = -57.5'^k$$

By determining reactions by statics,

$$V_A = 8.5^k \qquad V_C = 28.4^k$$
$$V_B = 20.0^k \qquad V_D = 7.1^k$$

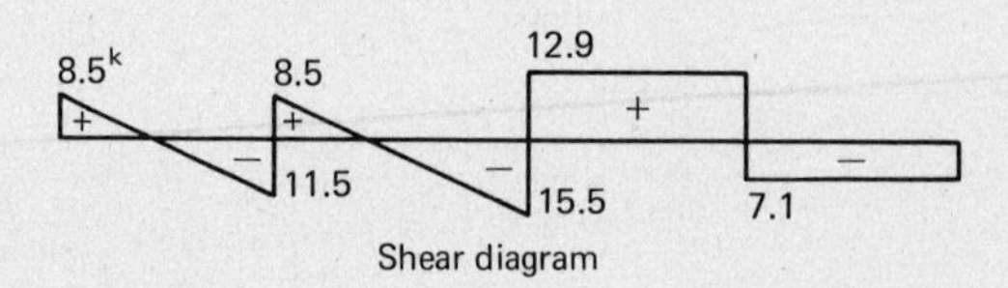

Shear diagram

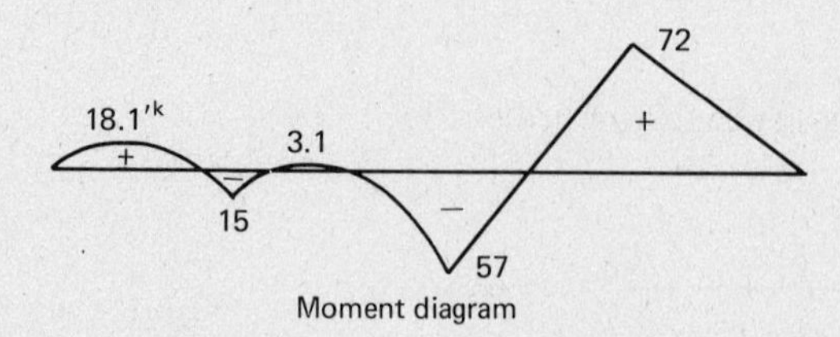

Moment diagram

Example 21.14

Draw shear and moment diagrams for the structure shown in Fig. 21.20.

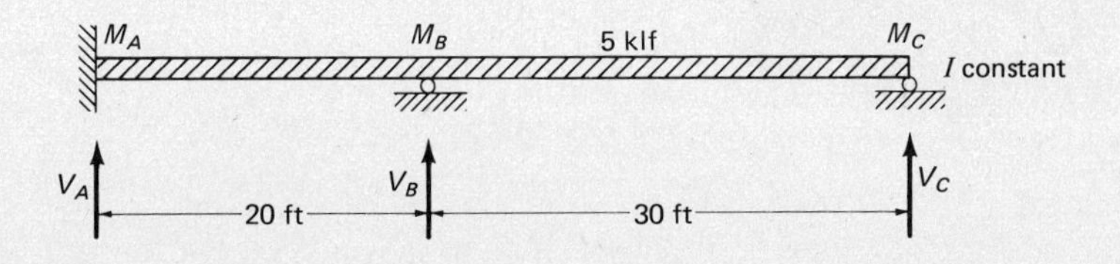

Figure 21.20

SOLUTION

For constant I:

$$M_A l_1 + 2M_B(l_1 + l_2) + M_C l_2$$

$$= -\sum P_1 l_1^2(k_1 - k_1^3) - \sum P_2 l_2^2(k_2 - k_2^3) - \frac{w_1 l_1^3}{4} - \frac{w_2 l_2^3}{4}$$

By assuming a span of zero length to left of fixed end,

$$M_0 l_0 + 2M_A(l_0 + l_1) + M_B l_1 = -\frac{w_0 l_0^3}{4} - \frac{w_1 l_1^3}{4}$$

$$M_0 = 0$$

$$2M_A(0 + 20) + M_B(20) = -\frac{(5)(20)^3}{4}$$

$$40M_A + 20M_B = -10{,}000 \tag{1}$$

$$M_A(20) + 2M_B(20 + 30) = -\frac{(5)(20)^3}{4} - \frac{(5)(30)^3}{4}$$

$$20M_A + 100M_B = -43{,}750 \tag{2}$$

By solving Eqs. (1) and (2) simultaneously,

$$M_A = -34.75'^{k}$$
$$M_B = -430.5'^{k}$$

By computing reactions by statics,

$$V_A = +\ 30.3^{k} \qquad V_C = +60.7^{k}$$
$$V_B = +159.0^{k}$$

89.3^{k} 30.3^{k} + + − − 69.7^{k} 60.7^{k}

Shear diagram

369 57 + − $34.75'^{k}$ 429

Moment diagram

Example 21.15

Determine the moments at the ends of the uniformly loaded fixed-ended beam shown in Fig. 21.21.

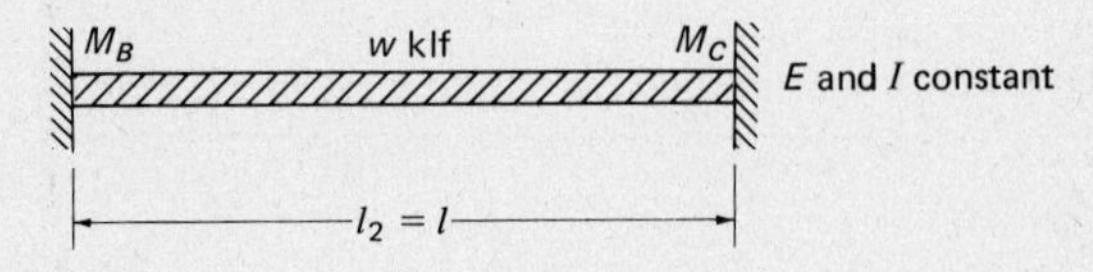

Figure 21.21

SOLUTION

By assuming spans of zero length (l_1 and l_3) outside each end with zero moments (M_A and M_D),

$$M_A l_1 + 2M_B(l_1 + l_2) + M_C l_2 = -\frac{w_1 l_1^3}{4} - \frac{w_2 l_2^3}{4}$$

$$M_A = 0 = M_D$$

$$2M_B(l) + M_C(l) = -\frac{wl^3}{4}$$

$$2M_B l + M_C l = -\frac{wl^3}{4} \tag{1}$$

$$M_B l_2 + 2M_C(l_2 + l_3) + M_D l_3 = -\frac{wl^3}{4}$$

$$M_B l + 2M_C l = -\frac{wl^3}{4} \tag{2}$$

By solving Eqs. (1) and (2) simultaneously

$$M_B = -\frac{wl^2}{12}$$

$$M_C = -\frac{wl^2}{12}$$

The moments at the ends of a fixed-ended beam loaded with a concentrated load can be determined in a similar manner, with the following results:

$$M_A = -\frac{Pab^2}{l^2} \qquad M_B = -\frac{Pa^2b}{l^2}$$

These expressions will be needed for solution of later problems.

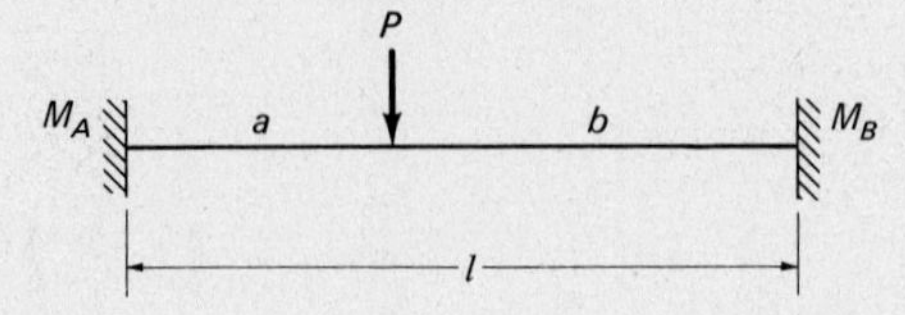

21.7 EFFECT OF SUPPORT SETTLEMENTS BY THE THREE-MOMENT THEOREM

The three-moment theorem as presented in Sec. 21.5 was developed on the basis of unyielding supports. Should the supports be displaced from their theoretical positions because of settlement, fabrication errors, and so on, the equation will have to be adjusted accordingly. The values of θ_{BL} and θ_{BR} were the slopes of the elastic curve of a beam on each side of the center support of any three consecutive supports being considered. Settlement of any of the three supports, provided they settle unequally and do not remain in a straight line, will change the values of the slopes.

Supports A, B, and C of the beam of Fig. 21.22 are assumed to settle δ_A, δ_B, and δ_C, respectively. If a horizontal line were drawn through the newly deflected position of support B, the angle θ_{BL} would be changed by $(\delta_A - \delta_B)/l_1$, and θ_{BR} would be changed by $(\delta_C - \delta_B)/l_2$. (It is to be remembered that the angle is small and the tangent of the angle change in question is the same as the angle in radians.)

By letting a downward value of the support settlements be positive, the values of θ_{BL} and θ_{BR} for a beam loaded with concentrated and uniform loads become

$$\theta_{BL} = \frac{M_A l_1}{6EI_1} + \frac{M_B l_1}{3EI_1} + \frac{w_1 l_1^3}{24EI_1} + \sum \frac{P_1 l_1^2}{EI_1}(k_1 - k_1^3) + \frac{\delta_A - \delta_B}{l_1}$$

$$\theta_{BR} = \frac{M_C l_2}{6EI_2} + \frac{M_B l_2}{3EI_2} + \frac{w_2 l_2^3}{24EI_2} + \sum \frac{P_2 l_2^2}{EI_2}(k_2 - k_2^3) + \frac{\delta_C - \delta_B}{l_2}$$

Equating these two expressions, because $\theta_{BL} = -\theta_{BR}$, gives the general equation when support displacement occurs.

$$\frac{M_A l_1}{I_1} + 2M_B\left(\frac{l_1}{I_1} + \frac{l_2}{I_2}\right) + \frac{M_C l_2}{I_2} = -\sum \frac{P_1 l_1^2}{I_1}(k_1 - k_1^3) - \sum \frac{P_2 l_2^2}{I_2}(k_2 - k_2^3) - \frac{w_1 l_1^3}{4I_1} - \frac{w_2 l_2^3}{4I_2} - 6E\left(\frac{\delta_A - \delta_B}{l_1} + \frac{\delta_C - \delta_B}{l_2}\right)$$

Examples 21.16 and 21.17 illustrate the application of the three-moment theorem to continuous beams having some support displacement. Keeping the units in terms of feet and kilopounds, as illustrated in these examples, makes the mathematics quite simple.

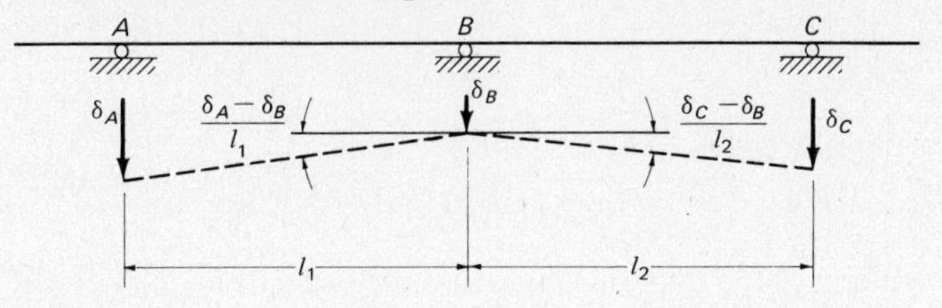

Figure 21.22

Example 21.16

Draw the shear and moment diagrams for the two-span beam shown in Fig. 21.23 if the interior support settles 0.25 in, or 0.0208 ft.

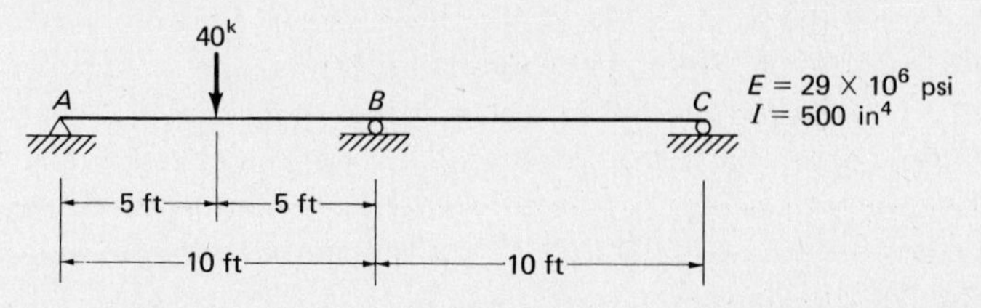

Figure 21.23

SOLUTION

$$2M_B\left(\frac{10}{I}+\frac{10}{I}\right)=-\frac{(40)(10)^2}{I}\left[\frac{1}{2}-\left(\frac{1}{2}\right)^3\right]-6E\left(\frac{0-0.0208}{10}+\frac{0-0.0208}{10}\right)$$

$$\frac{40M_B}{I}=-\frac{1500}{I}+0.02496E$$

$$40M_B=-1500+0.02496\text{EI}$$

$$M_B=-37.5+0.000624EI$$

$$M_B=-37.5+\frac{(0.000624)(29\times10^6)(500)}{(144)(1000)}$$

$$M_B=+25.3'^k$$

40k

22.53k 14.94 2.53

22.53k + 17.47 − 2.53

Shear diagram

112.65'k + 25.30'k

Moment diagram

Example 21.17

Determine the moments at the interior supports of the beam of Example 19.8, reproduced in Fig. 21.24, for which the following support settlements are assumed to occur:

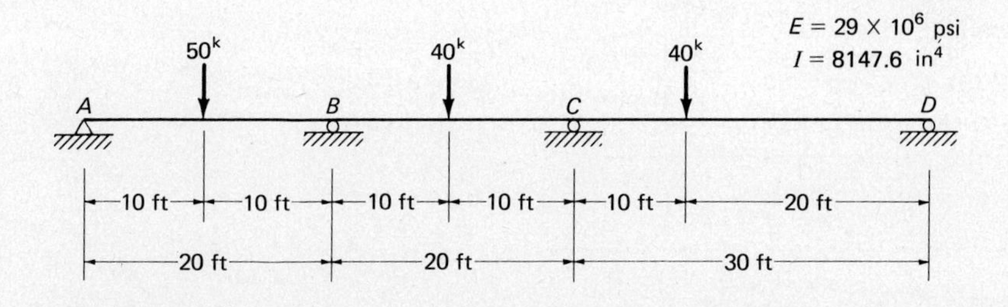

Figure 21.24

$A = 1.25 \text{ in} = 0.104 \text{ ft} \qquad C = 2.75 \text{ in} = 0.229 \text{ ft}$

$B = 2.40 \text{ in} = 0.200 \text{ ft} \qquad D = 1.10 \text{ in} = 0.0917 \text{ ft}$

SOLUTION

$$2M_B\left(\frac{20}{I}+\frac{20}{I}\right)+\frac{20M_C}{I}=-\frac{(50)(20)^2}{I}\left[\frac{1}{2}-\left(\frac{1}{2}\right)^3\right]-\frac{(40)(20)^2}{I}\left[\frac{1}{2}-\left(\frac{1}{2}\right)^3\right]$$
$$-6E\left(\frac{0.104-0.200}{20}+\frac{0.229-0.200}{20}\right)$$

$$4M_B + M_C = -375 - 300 + 0.001005EI$$
$$4M_B + M_C = +974 \qquad (1)$$

$$\frac{20M_B}{I}+2M_C\left(\frac{20}{I}+\frac{30}{I}\right)=-\frac{(40)(20)^2}{I}\left[\frac{1}{2}-\left(\frac{1}{2}\right)^3\right]$$
$$-\frac{(40)(30)^2}{I}\left[\frac{2}{3}-\left(\frac{2}{3}\right)^3\right]-6E\left(\frac{0.200-0.229}{20}+\frac{0.0917-0.229}{30}\right)$$

$$M_B + 5M_C = -997 + 0.001809EI$$
$$M_B + 5M_C = +2000 \qquad (2)$$

By solving Eqs. (1) and (2) simultaneously,

$$M_B = +151'^{k}$$
$$M_C = +370'^{k}$$

PROBLEMS

For Probs. 21.1 to 21.16 use Castigliano's first theorem to determine the slopes and deflections indicated. $E = 29 \times 10^6$ psi for the problems having customary units and 200 000 MPa for those with SI units. Similarly circled values are cross-sectional areas in square inches or in mm^2.

Problem 21.1 Deflection at each load; $I = 3250 \text{ in}^4$

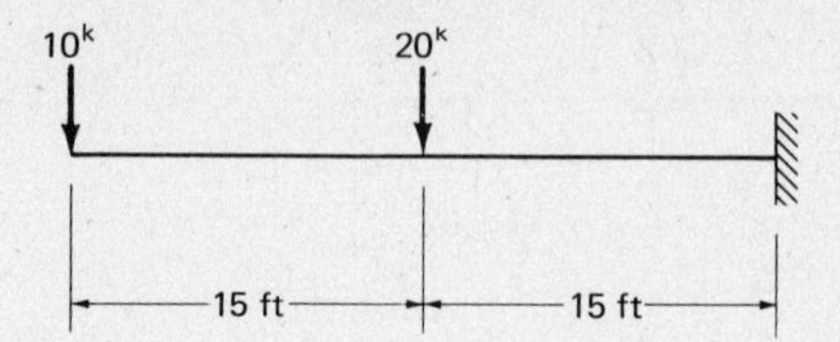

Problem 21.2 Slope and deflection at free end; $I = 1300 \text{ in}^4$ (*Ans.* $\theta = 0.0129$ rad\, $\delta = 3.09$ in↓)

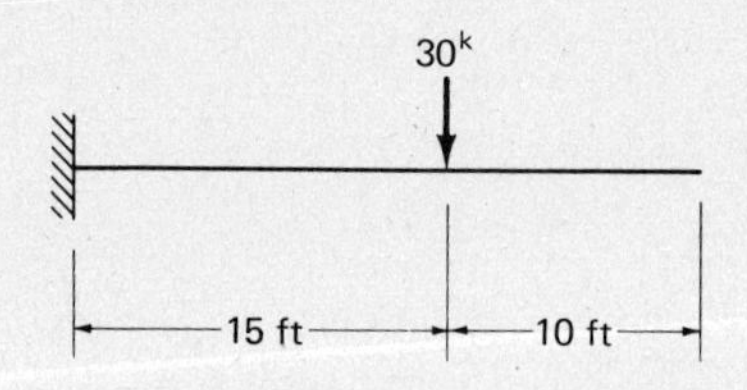

Problem 21.3 Slope and deflection at 20^k load; $I = 4750 \text{ in}^4$

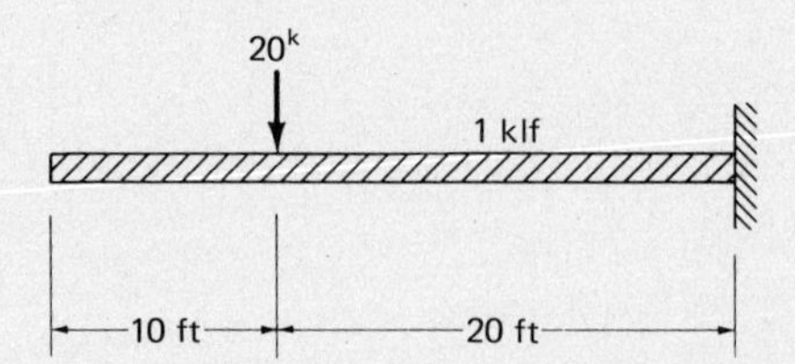

Problem 21.4 Deflection at each end of uniform load; $I = 1000 \text{ in}^4$ (*Ans.* $\delta_{20 \text{ ft}} =$ 2.035 in↓, $\delta_{10 \text{ ft}} = 0.692$ in↓)

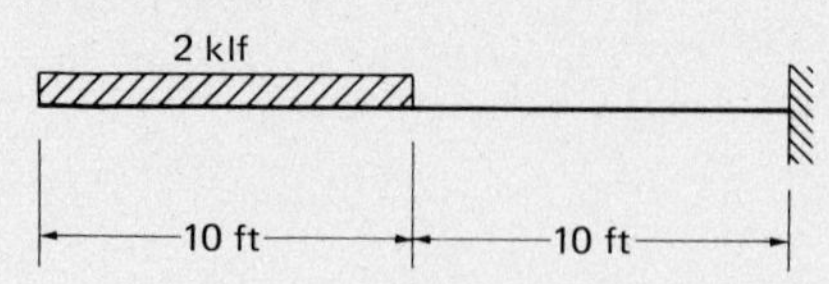

Problem 21.5 Deflection at each load. $I = 4 \times 10^8 \text{ mm}^4$

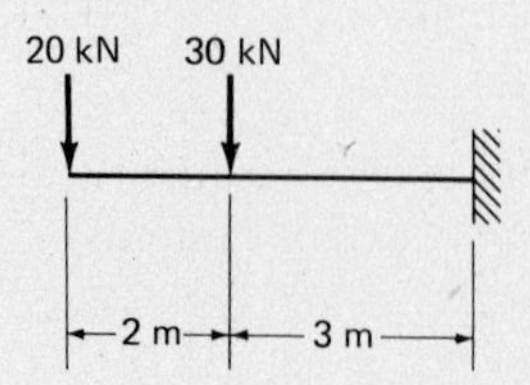

Problem 21.6 Slope and deflection at 100-kN load. $I = 1.798 \times 10^8$ mm^4 (*Ans.* $\theta = -0.003\,48$ rad/, $\delta = 26.07$ mm ↓)

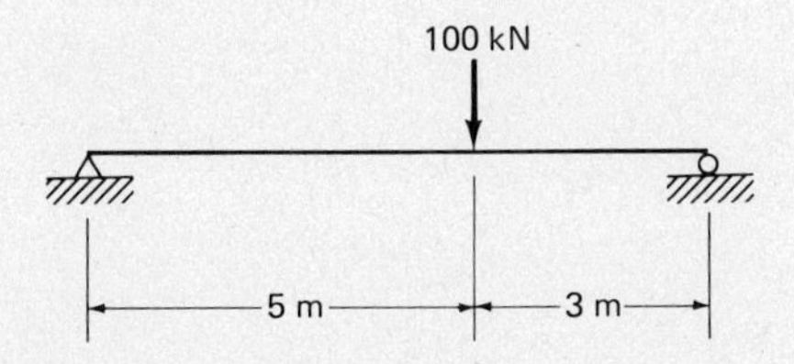

21.7. Problem 17.8

Problem 21.8 Deflection at each load; $I = 1750$ in^4 (*Ans.* $\delta_{30^k} = 1.98$ in ↓, $\delta_{20^k} = 1.45$ in ↓)

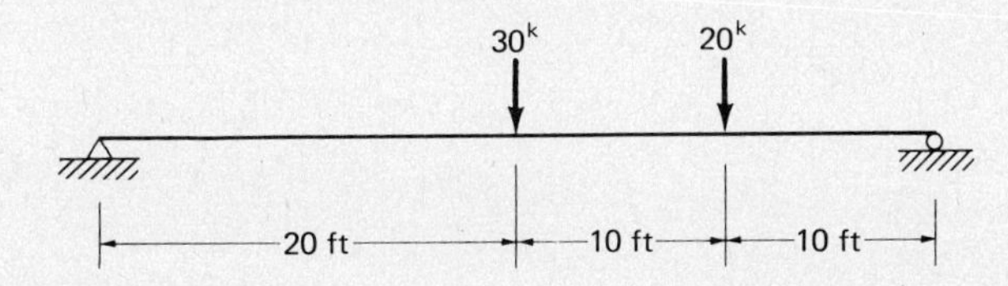

Problem 21.9 Slope and deflection at 60^k load; $I = 2700$ in^4

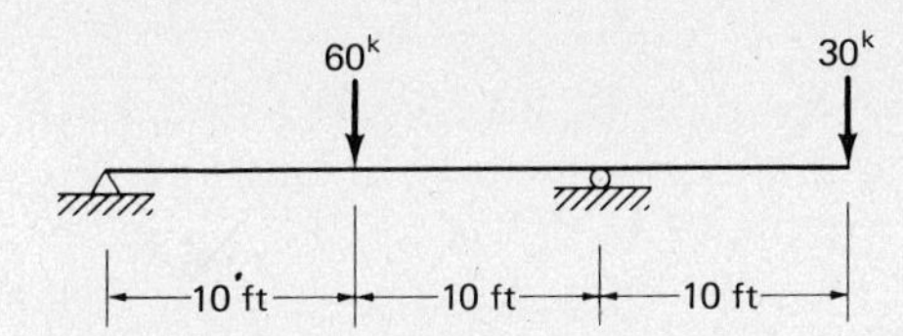

21.10. Problem 17.43. (*Ans.* $\theta = 0.0155$ rad ╲, $\delta = 1.74$ in ↓)
21.11. Problem 18.26.
21.12. Problem 18.23. (*Ans.* $\theta = -0.000772$ rad/, $\delta = 0.0868$ in ↑)
21.13. Problem 18.27.

Problem 21.14 Vertical deflection at joint L_1 (*Ans.* $\delta = 0.652$ in ↓)

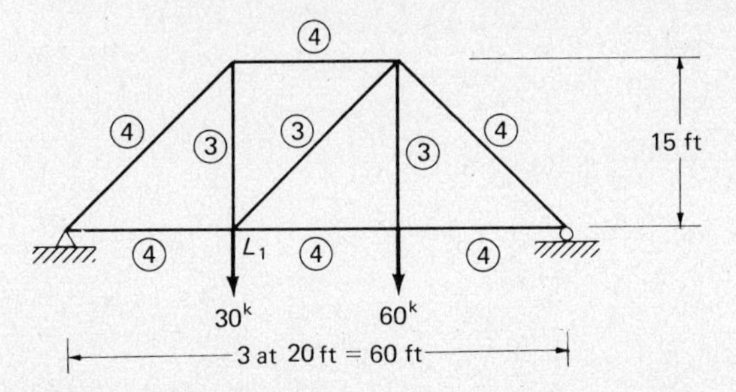

Problem 21.15 Vertical and horizontal deflection at L_0

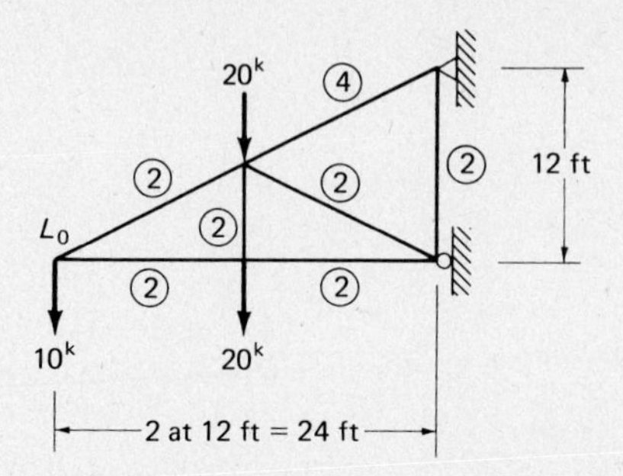

Problem 21.16 Vertical deflection at L_1 and horizontal deflection at L_2 (*Ans.* $\delta_V = 45.2$ mm ↓, $\delta_H = 21$ mm →)

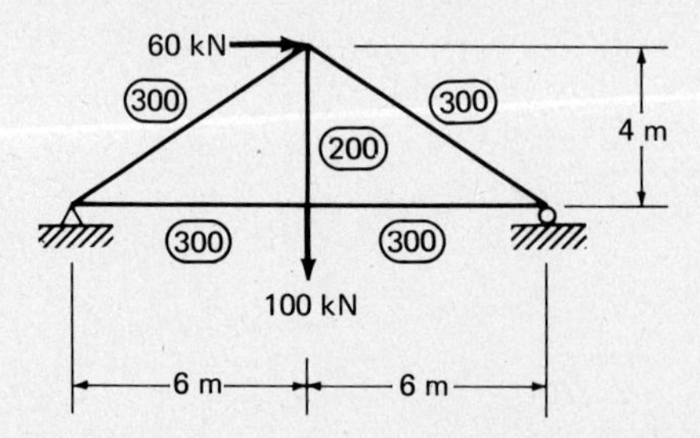

For Probs. 21.17 to 21.33 analyze the structures by least work; E and I are constant unless otherwise indicated. Circled values are areas, in square inches.

Problem 21.17

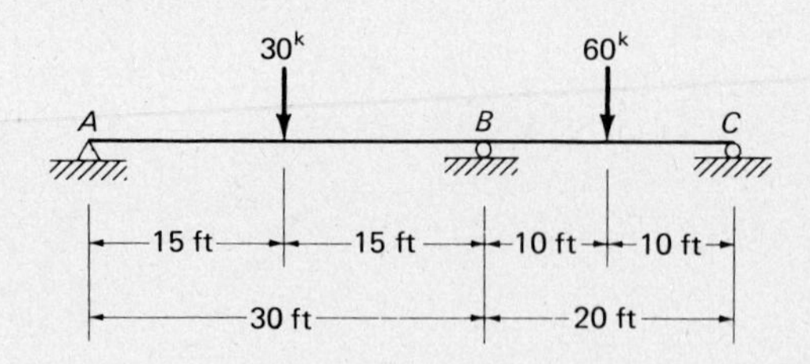

Problem 21.18 (*Ans.* $V_A = 18.25^k$↑, $V_C = 18.12^k$↑, $M_B = -235'^k$)

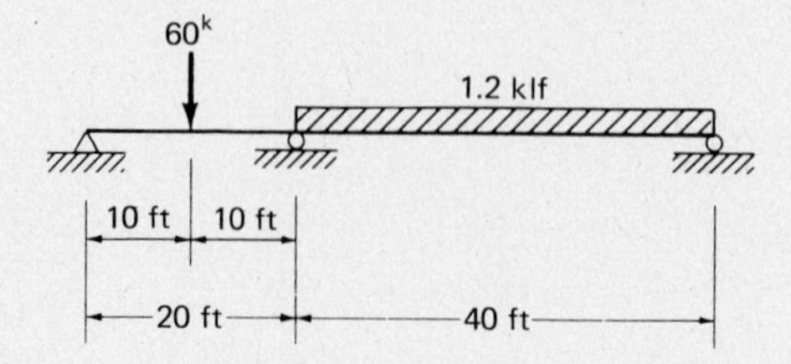

Problem 21.19

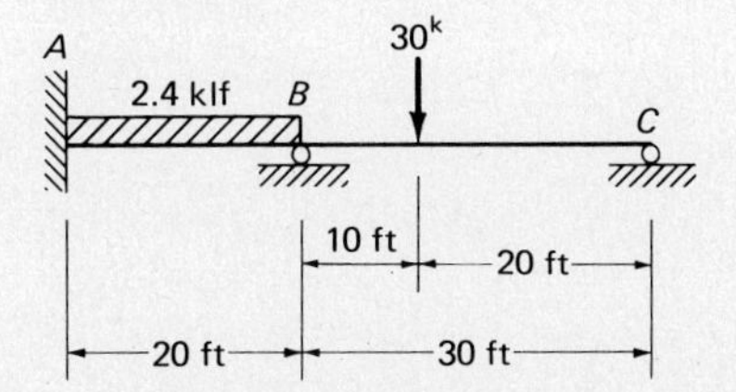

21.20. Problem 20.11. (*Ans.* Member forces left to right $+8.88^k$, $+16.10^k$, $+7.15^k$)

Problem 21.21

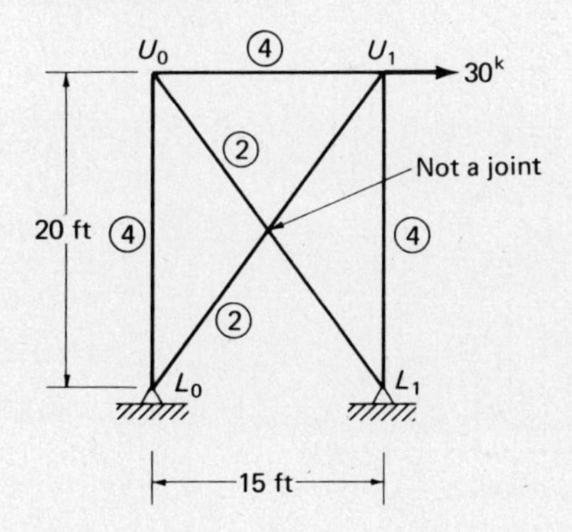

Problem 21.22 (*Ans.* $L_0L_1 = -209.0$ kN, $U_0L_1 = +161.3$ kN)

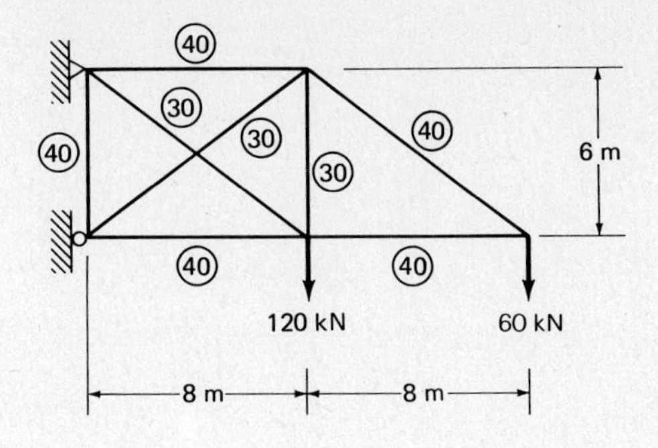

21.23. Problem 20.12.

Problem 21.24 Find force in tie (*Ans.* Cable tension = 50.7^k)

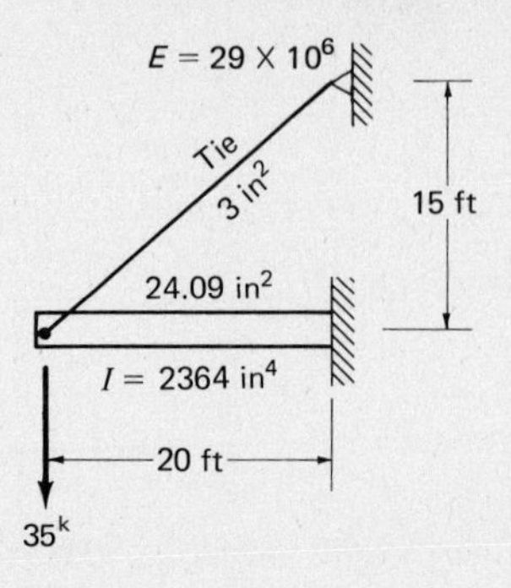

Problem 21.25

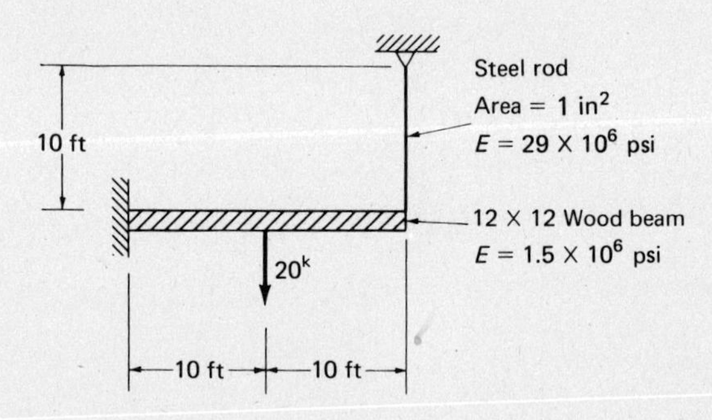

Problem 21.26 (*Ans.* $L_0U_1 = -39.2^k$, $L_1L_2 = -4.68^k$, cable tension $= +6.48^k$)

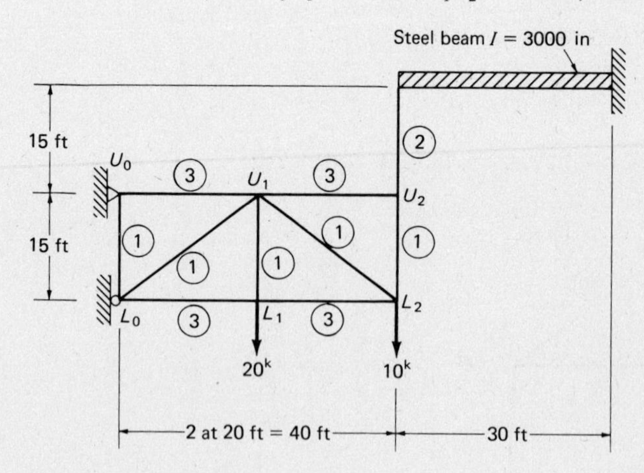

21.27. Problem 20.1
21.28. Problem 20.7. (*Ans.* $H_L = 58^k \rightarrow$, $U_1L_2 = +19.8^k$, $U_1U_2 = -41.3^k$, $L_2L_4 = -23.2^k$)
21.29. Problem 20.13.

21.30. Problem 20.14. (*Ans.* $L_1L_2 = -18.1^k$, $U_1U_2 = +31.9^k$, $U_1L_2 = -16.9^k$)

Problem 21.31

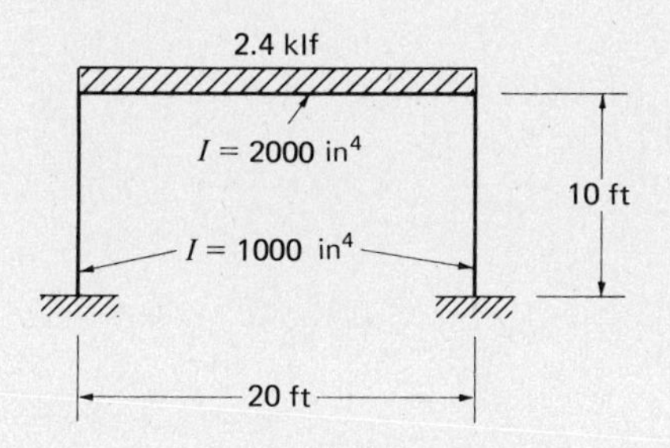

Problem 21.32 (*Ans.* $H_A = 20.50^k \rightarrow$, $M_B = -410'^k$, $M_C = +95.4'^k$)

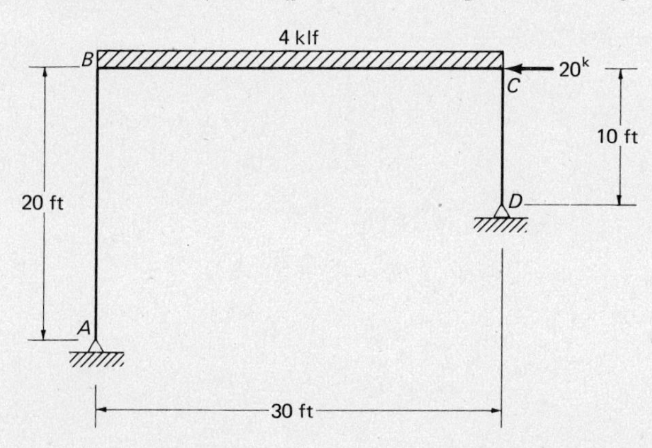

Problem 21.33

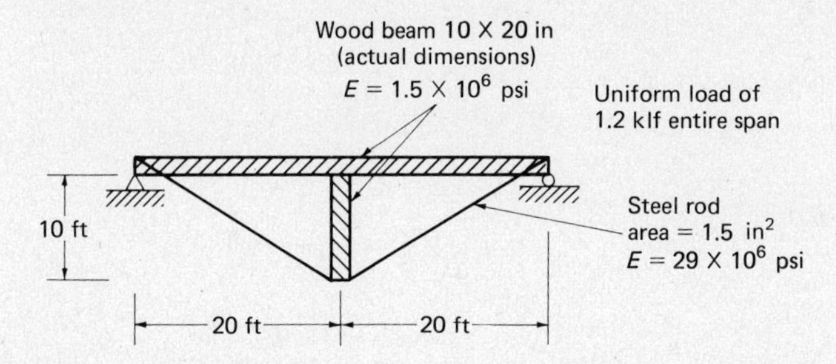

For Probs. 21.34 to 21.42 compute all moments for the continuous beams by using the three-moment theorem. Draw shear and moment diagrams.

Problem 21.34 (*Ans.* $V_A = 44.0^k \uparrow$, $V_B = 87.57^k \uparrow$, $M_B = -300'^k$, $M_C = -157.2'^k$)

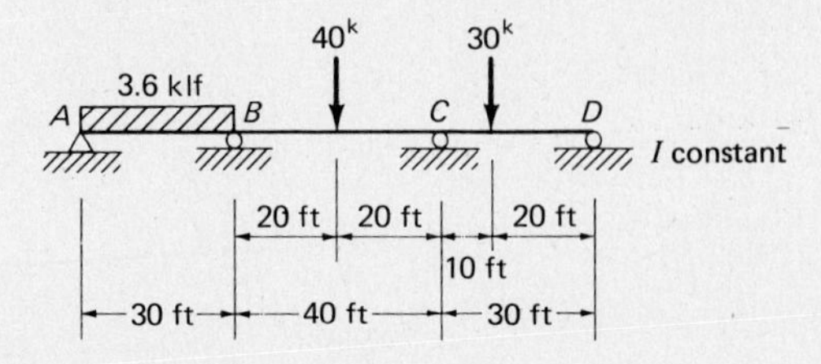

Problem 21.35

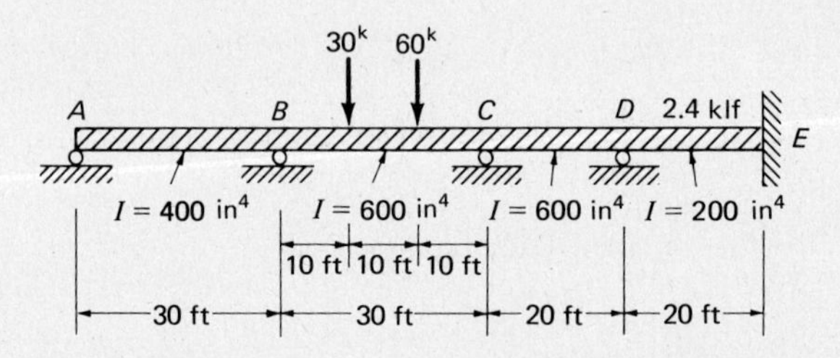

Problem 21.36 (*Ans.* $V_A = 148.5^k \uparrow$, $V_B = 31.5^k \uparrow$, $M_A = -720'^k$, $M_B = -45'^k$)

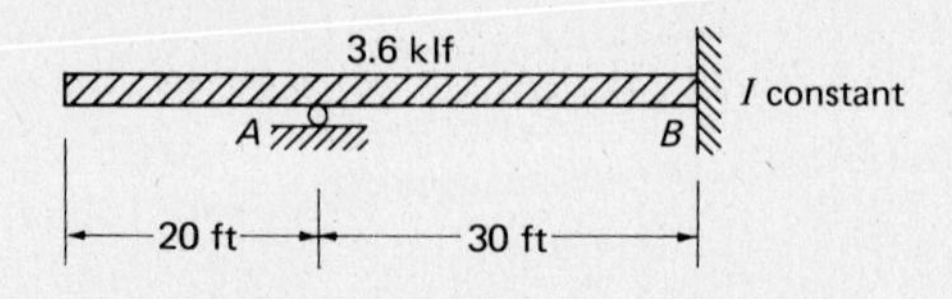

Problem 21.37

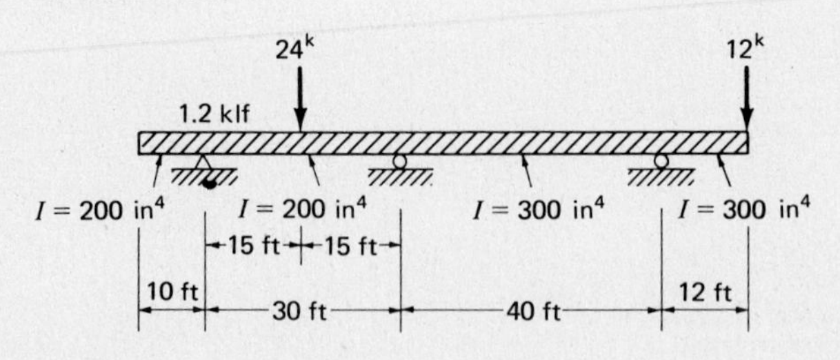

Problem 21.38 (*Ans.* $V_A = 131.25$ kN $\uparrow$, $V_C = 233.33$ kN, $M_B = -437.5$ kN · m)

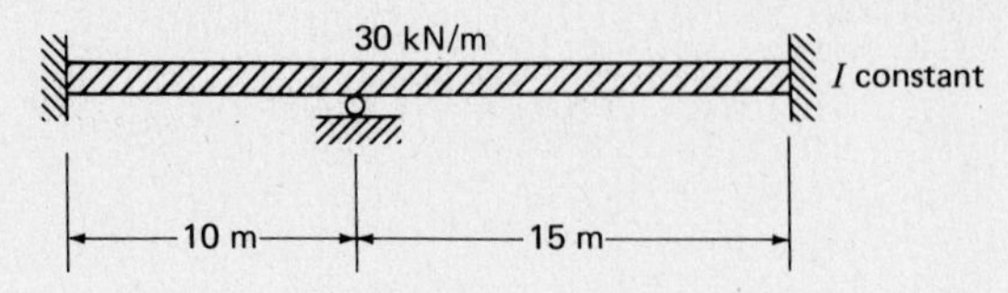

Problem 21.39

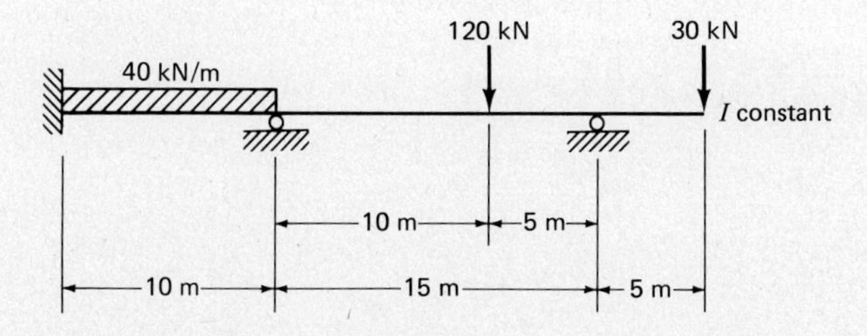

21.40. Problem 19.6. (*Ans.* $V_A = 7.53^k \uparrow$, $M_B = -111.7'^k$, $V_C = 14.27^k$)

21.41. Problem 19.9.

21.42. Problem 19.13. (*Ans.* $V_B = 49.41^k \uparrow$, $V_C = 21.69^k \uparrow$, $M_B = -144'^k$, $M_C = -194.7'^k$)

REFERENCES

[1] J. S. Kinney, *Indeterminate Structural Analysis* (Reading, Mass.: Addison-Wesley, 1957), p. 13.

[2] Ibid, pp. 84–86.

[3] H. Sutherland and H. L. Bowman, *Structural Theory* (New York: Wiley, 1954), Chap. 8.

22 Slope Deflection

22.1 INTRODUCTION

George A. Maney introduced slope deflection in a 1915 University of Minnesota engineering publication [1]. His work was an extension of earlier studies of secondary stresses by Manderla [2] and Mohr [3]. For nearly 15 years, until the introduction of moment distribution, slope deflection was the popular "exact" method used for the analysis of continuous beams and frames in the United States.

Slope deflection is a method which takes into account the flexural deformations of beams and frames (that is rotations, settlements, etc.), but which neglects shear and axial deformations. Although this classical method is generally considered to be obsolete its study can be useful for several reasons. These include:

1. Slope deflection is convenient for hand analysis of some small structures.
2. A knowledge of the method provides an excellent background for understanding the moment distribution method of the next two chapters.
3. It is a special case of the displacement method and provides a very effective introduction for the matrix formulation of structures described in Chap. 27.
4. The slopes and deflections determined by slope deflection enable the analyst to easily sketch the deformed shape of a particular structure. The result is that he or she has a better "feel" for the behavior of structures.

Hospital, Stamford, Connecticut. (Courtesy of the American Concrete Institute.)

22.2 DERIVATION OF SLOPE-DEFLECTION EQUATIONS

The name "slope deflection" comes from the fact that the moments at the ends of the members in statically indeterminate structures are expressed in terms of the rotations (or slopes) and deflections of the joints. For developing the equations, members are assumed to be of constant section between each pair of supports. Although it is possible to derive expressions for members of varying section, the results are so complex as to be of little practical value. It is further assumed that the joints in a structure may rotate or deflect, but the angles between the members meeting at a joint remain unchanged.

Span AB of the continuous beam of Fig. 22.1(a) is considered for the following discussion. If the span were completely fixed at each end, the slope of the elastic curve of the beam at the ends would be zero. External loads produce fixed-end moments, and these moments cause the span to take the shape shown in Fig. 22.1(b). Joints A and B are actually not fixed and will rotate slightly under load to some position such as the one shown in Fig. 22.1(c). In addition to the rotation of the joints there may possibly be some settlement of one or both of the supports which will cause a chord rotation of the member as shown in part (d), where support B is assumed to have settled an amount Δ.

From the study of Fig. 22.1 the values of the final end moments at A and B (M_{AB} and M_{BA}) are seen to be equal to the sum of the moments caused by the following:

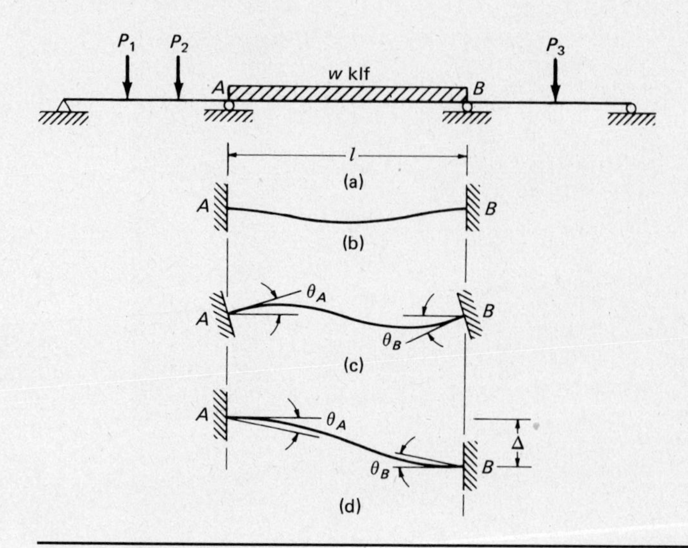

Figure 22.1

1. The fixed-end moments (FEM_{AB} and FEM_{BA}), which can be determined by the expressions developed in Example 21.15. A detailed list of fixed-end moment expressions are presented in Chap. 23 in Fig. 23.5.
2. The moments caused by the rotations of joints A and B (θ_A and θ_B).
3. The moments caused by chord rotation ($\psi = \Delta/l$) if one or both of the joints settles or deflects.

Rotations of the joints in a structure cause changes in the slopes of the tangents to the elastic curves at those points. For a particular beam the change in slope equals the end shear of the beam when it is loaded with the M/EI diagram. The beam is assumed to have the fixed-end moments FEM_{AB} and FEM_{BA}, shown in Fig. 22.2.

The end reactions or end slopes are as follows:

$$\theta_A = \frac{(\frac{1}{2})(M_{AB}/EI)(l)(\frac{2}{3}l) - (\frac{1}{2})(M_{BA}/EI)(l)(\frac{1}{3}l)}{l}$$

$$= \frac{l}{6EI}(2M_{AB} - M_{BA})$$

$$\theta_B = \frac{(\frac{1}{2})(M_{BA}/EI)(l)(\frac{2}{3}l) - (\frac{1}{2})(M_{AB}/EI)(l)(\frac{1}{3}l)}{l}$$

$$= \frac{l}{6EI}(2M_{BA} - M_{AB})$$

If one of the supports of the beam settled or deflected an amount Δ, the angles θ_A and θ_B caused by joint rotation would be changed by Δ/l (or ψ), as illustrated in Fig. 22.1(d). Adding chord rotation to the expressions results in

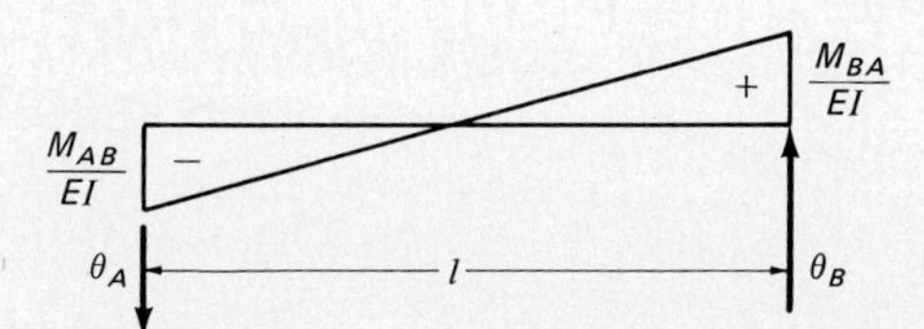

Figure 22.2

the following total values for the slopes of the tangents to the elastic curves at the ends of the beams.

$$\theta_A = \frac{l}{6EI}(2M_{AB} - M_{BA}) + \psi$$

$$\theta_B = \frac{l}{6EI}(2M_{BA} - M_{AB}) + \psi$$

Solving the equations simultaneously for M_{AB} and M_{BA} gives the values of the end moments due to slopes and deflections. In these expressions I/l has been replaced with K, the so-called ***stiffness factor*** (see Chap. 23).

$$M_{AB} = 2EK(2\theta_A + \theta_B - 3\psi)$$

$$M_{BA} = 2EK(\theta_A + 2\theta_B - 3\psi)$$

The final end moments are equal to the moments due to slopes and deflections plus the fixed-end moments. The slope-deflection equations are as follows:

$$M_{AB} = 2EK(2\theta_A + \theta_B - 3\psi) + FEM_{AB}$$

$$M_{BA} = 2EK(\theta_A + 2\theta_B - 3\psi) + FEM_{BA}$$

With these equations it is possible to express the end moments in a structure in terms of joint rotations and settlements. By previous methods it has been necessary to write one equation for each redundant in the structure. The number of unknowns in each equation totaled the number of redundants. The work of solving those equations was considerable for highly redundant structures. Slope deflection appreciably reduces the amount of work involved in analyzing multiredundant structures because the unknown moments are expressed in terms of only a few unknown joint rotations and settlements. Even for multistory frames, the number of unknown θ and ψ values appearing in any one equation is rarely more than five or six, whereas the degree of indeterminancy of the structure is several times that figure.

22.3 APPLICATION OF SLOPE-DEFLECTION EQUATIONS TO CONTINUOUS BEAMS

Examples 22.1 to 22.4 illustrate the analysis of statically indeterminate beams with the slope-deflection equations. Each member in the beams is considered individually; its fixed-end moments are computed; and one equation is written for the moment at each end of the member. For span AB of Example 22.1, equations for M_{AB} and M_{BA} are written; for span BC, equations for M_{BC} and M_{CB} are written; and so on.

The moment equations are written in terms of the unknown values of θ at the supports. The two moments at an interior support must total zero, as $M_{BA} + M_{BC} = 0$ at support B in Example 22.1. Expressions are therefore written for the total moment at each support, which gives a set of simultaneous equations from which the unknown θ values may be obtained. Two conditions that will simplify the solution of the equations may exist. These are fixed ends for which the θ values must be zero and simple ends for which the moment is zero. A special expression is derived at the end of Exmple 22.2 for end spans that are simply supported.

The beams of Examples 22.1 and 22.2 have unyielding supports, and ψ is zero for all of the equations. Some support settlement occurs for the beams of Examples 22.3 and 22.4, and ψ is included in the equations. Chord rotation is considered positive when the chord of a beam is rotated clockwise by the settlements, meaning that the sign is the same no matter which end is being considered as the entire beam rotates in that direction.

When span lengths, moduli of elasticity, and moments of inertia are constant for the spans of a continuous beam, the $2EK$ values are constant and may be canceled from the equations. Should the values of K vary from span to span, as they often do, it is convenient to express them in terms of relative values, as is done in Example 22.6.

The major difficulty experienced by readers in applying the slope-deflection equations lies in the use of correct signs. It is essential to understand these signs before attempting to apply the equations.

For previous work in this book a plus sign for the moment has indicated tension in the bottom fibers whereas a negative sign has indicated tension in the top fibers. This sign convention was necessary for drawing moment diagrams.

In applying the slope-deflection equations it is simpler to use a convention in which signs are given to clockwise and counterclockwise moments at the member ends. Once these moments are determined, their signs can easily be converted to the usual convention for drawing moment diagrams.

The following convention is used in this chapter for slope deflection and in Chap. 23 and 24 for moment distribution: Should a member cause a moment that tends to rotate a joint clockwise, the joint moment is considered negative; if counterclockwise, it is positive. In other words, a clockwise resisting moment on the member herein is considered positive, and a counterclockwise resisting moment is considered negative.

Figure 22.3 is presented to demonstrate this convention. The moment at the left end, M_{AB}, tends to rotate the joint clockwise and is considered negative. Notice that the resisting moment would be counterclockwise. At the right end of the beam the loads have caused a moment that tends to rotate the joint counterclockwise and M_{BA} is considered to be positive.

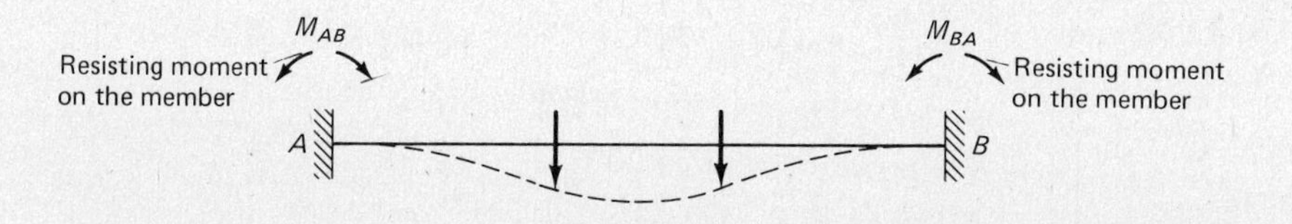

Figure 22.3

Example 22.1

Determine all support moments of the structure in Fig. 22.4 by the method of slope deflection.

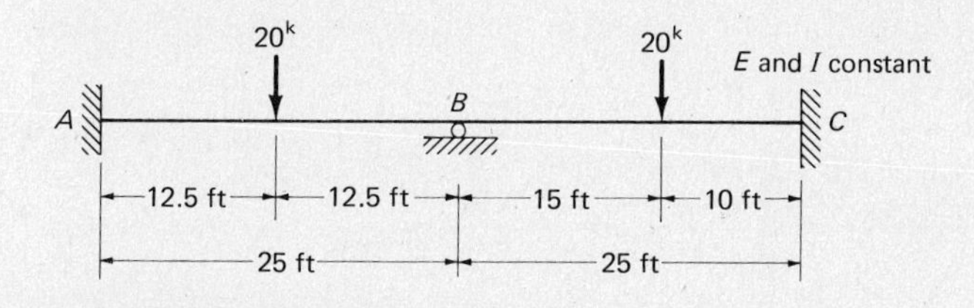

Figure 22.4

SOLUTION

By computing fixed-end moments,

$$FEM_{AB} = -\frac{(20)(12.5)(12.5)^2}{(25)^2} = -62.5'^{k}$$

$$FEM_{BA} = +\frac{(20)(12.5)(12.5)^2}{(25)^2} = +62.5'^{k}$$

$$FEM_{BC} = -\frac{(20)(15)(10)^2}{(25)^2} = -48'^{k}$$

$$FEM_{CB} = +\frac{(20)(10)(15)^2}{(25)^2} = +72'^{k}$$

By writing the equations, and noting $\theta_A = \theta_C = \psi = 0$,

$$M_{AB} = 2EK\theta_B - 62.5$$

$$M_{BA} = 4EK\theta_B + 62.5$$

$$M_{BC} = 4EK\theta_B - 48$$

$$M_{CB} = 2EK\theta_B + 72$$

$$\Sigma M_B = 0 = M_{BA} + M_{BC}$$

$$4EK\theta_B + 62.5 + 4EK\theta_B - 48 = 0$$

$$EK\theta_B = -1.8125$$

Final moments:

$$M_{AB} = (2)(-1.8125) - 62.5 = -66.125'^{k}$$
$$M_{BA} = (4)(-1.8125) + 62.5 = +55.25'^{k}$$
$$M_{BC} = (4)(-1.8125) - 48 = -55.25'^{k}$$
$$M_{CB} = (2)(-1.8125) + 72 = +68.375'^{k}$$

Example 22.2

Find all moments in the structure of Fig. 22.5 by slope deflection. Use the modified equation for simple ends developed in the discussion at the end of the problem.

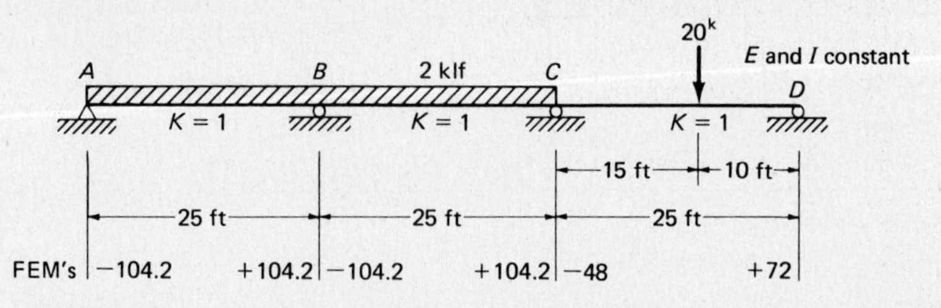

Figure 22.5

SOLUTION

$$M_{BA} = 3EK(\theta_B - \psi) + FEM_{BA} - \tfrac{1}{2}FEM_{AB}$$
$$M_{AB} = 2EK(2\theta_A + \theta_B - 3\psi) + FEM_{AB}$$

By writing the equations, and canceling EK because it is constant for all spans,

$$M_{AB} = M_{DC} = 0$$
$$M_{BA} = (3)(\theta_B) + 104.2 - (\tfrac{1}{2})(-104.2) = 3\theta_B + 156.3$$
$$M_{BC} = (2)(2\theta_B + \theta_C) - 104.2 = 4\theta_B + 2\theta_C - 104.2$$
$$M_{CB} = (2)(\theta_B + 2\theta_C) + 104.2 = 2\theta_B + 4\theta_C + 104.2$$
$$M_{CD} = (3)(\theta_C) - 48 - (\tfrac{1}{2})(+72) = 3\theta_C - 84$$
$$\Sigma M_B = 0 = M_{BA} + M_{BC}$$
$$3\theta_B + 156.3 + 4\theta_B + 2\theta_C - 104.2 = 0$$
$$7\theta_B + 2\theta_C = -52.1 \qquad (1)$$
$$\Sigma M_C = 0 = M_{CB} + M_{CD}$$
$$2\theta_B + 4\theta_C + 104.2 + 3\theta_C - 84 = 0$$
$$2\theta_B + 7\theta_C = -20.2 \qquad (2)$$

By solving Eqs. (1) and (2) simultaneously,

$$\theta_B = -7.2$$
$$\theta_C = -0.83$$

Final end moments:

$$M_{BA} = (3)(-7.2) + 156.3 = +134.7'^k$$
$$M_{BC} = (4)(-7.2) + (2)(-0.83) - 104.2 = -134.7'^k$$
$$M_{CB} = (2)(-7.2) + (4)(-0.83) + 104.2 = +\ 86.5'^k$$
$$M_{CD} = (3)(-0.83) - 84 = -86.5'^k$$

Discussion

For a beam with simply supported ends, it is obvious that the moments at those ends must be zero for equilibrium ($M_{AB} = M_{DC} = 0$ for the beam of Fig. 22.5). Application of the usual slope-deflection equations will yield zero moments, but it seems a waste of time to go through a process to determine the value of all of the moments in the beam when two of them by inspection are obviously zero. The usual slope-deflection equations are as follows:

$$M_{AB} = 2EK(2\theta_A + \theta_B - 3\psi) + FEM_{AB} \tag{1}$$

$$M_{BA} = 2EK(\theta_A + 2\theta_B - 3\psi) + FEM_{BA} \tag{2}$$

Assuming end A to be simply end supported, the value of M_{AB} is zero. Solving the two equations simultaneously by eliminating θ_A gives a simplified expression for M_{BA} which has only one unknown, θ_B. The resulting simplified equation will expedite considerably the solution of continuous beams with simple ends.

Twice Eq. (2)

$$2M_{BA} = 2EK(2\theta_A + 4\theta_B - 6\psi) + 2FEM_{BA}$$

minus Eq. (1)

$$0 = 2EK(2\theta_A + \theta_B - 3\psi) + FEM_{AB}$$

gives

$$2M_{BA} = 2EK(3\theta_B - 3\psi) + 2FEM_{BA} - FEM_{AB}$$
$$M_{BA} = 3EK(\theta_B - \psi) + FEM_{BA} - \tfrac{1}{2}FEM_{AB}$$

Example 22.3

Find the moment at support B in the beam of Example 21.16, reproduced in Fig. 22.6, assuming B settles 0.25 in, or 0.0208 ft.

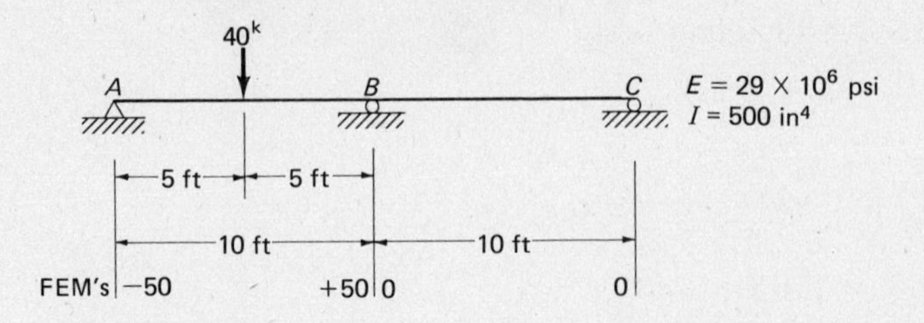

Figure 22.6

SOLUTION

By writing the equations, and noting that $M_{BA} = M_{CB} = 0$,

$$M_{BA} = 3EK\left(\theta_B - \frac{+0.0208}{10}\right) + 50 - (\tfrac{1}{2})(-50)$$

$$M_{BA} = 3EK\theta_B - 0.00624EK + 75$$

$$M_{BC} = 3EK\left(\theta_B - \frac{-0.0208}{10}\right)$$

$$M_{BC} = 3EK\theta_B + 0.00624EK$$

$$\Sigma M_B = 0 = M_{BA} + M_{BC}$$

$$3EK\theta_B - 0.0624EK + 75 + 3EK\theta_B + 0.0624EK = 0$$

$$6EK\theta_B + 75 = 0$$

$$EK\theta_B = -12.5$$

$$M_{BA} = (3)(-12.5) - 0.00624EK + 75$$

$$EK = \frac{(29 \times 10^6)(500)}{(12 \times 1000)(12 \times 10)} = 10{,}070$$

$$M_{BA} = -37.5 - 62.8 + 75 = -25.3'^k$$

$$M_{BC} = (3)(-12.5) + 62.8 = +25.3'^k$$

Example 22.4

Determine all moments for the beam of Example 21.17 (reproduced in Fig. 22.7), which has the following support settlements: $A = 1.25$ in $= 0.104$ ft, $B = 2.40$ in $= 0.200$ ft, $C = 2.75$ in $= 0.229$ ft, and $D = 1.10$ in $= 0.0917$ ft.

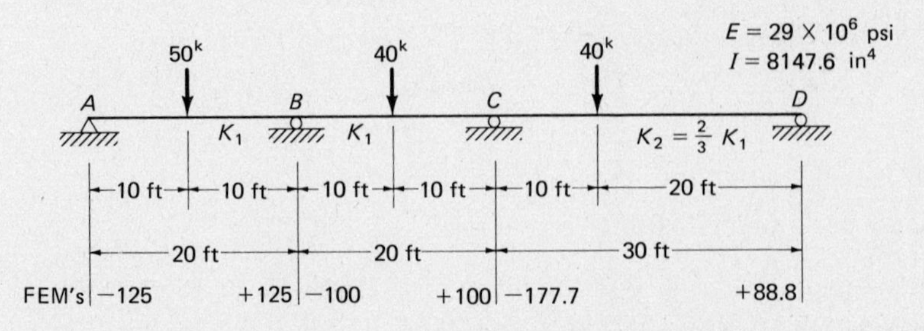

Figure 22.7

SOLUTION

For support B:

$$M_{BA} = 3EK_1\left(\theta_B - \frac{0.096}{20}\right) + 125 - (\tfrac{1}{2})(-125)$$

$$M_{BA} = 3EK_1\theta_B - 0.0144EK_1 + 187.5$$

$$M_{BC} = 2EK_1\left[2\theta_B + \theta_C - 3\,\frac{0.029}{20}\right] - 100$$

$$M_{BC} = 4EK_1\theta_B + 2EK_1\theta_C - 0.0087EK_1 - 100$$

For support C:

$$M_{CB} = 2EK_1\left[\theta_B + 2\theta_C - (3)\left(\frac{0.029}{20}\right)\right] + 100$$

$$M_{CB} = 2EK_1\theta_B + 4EK_1\theta_C - 0.0087EK_1 + 100$$

$$M_{CD} = 3EK_2\left(\theta_C - \frac{-0.1373}{30}\right) - 177.7 - (\tfrac{1}{2})(+88.8)$$

$$M_{CD} = 3EK_2\theta_C + 0.01373EK_2 - 222.1$$

$$M_{CD} = 2EK_1\theta_C + 0.00915EK_1 - 222.1$$

$$\Sigma M_B = 0 = M_{BA} + M_{BC}$$

$$3EK_1\theta_B - 0.0144EK_1 + 187.5 + 4EK_1\theta_B + 2EK_1\theta_C$$
$$-0.0087EK_1 - 100 = 0$$

$$7EK_1\theta_B + 2EK_1\theta_C - 0.0231EK_1 + 87.5 = 0 \qquad (1)$$

$$\Sigma M_C = 0 = M_{CB} + M_{CD}$$

$$2EK_1\theta_B + 4EK_1\theta_C - 0.0087EK_1 + 100 + 2EK_1\theta_C$$
$$+ 0.00915EK_1 - 222.1 = 0$$

$$2EK_1\theta_B + 6EK_1\theta_C + 0.00045EK_1 - 122.1 = 0 \qquad (2)$$

By solving Eqs. (1) and (2) simultaneously for $EK_1\theta_B$ and $EK_1\theta_C$,

$$EK_1\theta_B = 0.00367EK_1 - 20.2$$

$$EK_1\theta_C = -0.00129EK_1 + 27.1$$

$$EK_1 = \frac{(29 \times 10^6)(8147.6)}{(12 \times 20)(12 \times 1000)} = 82{,}040'^{k}$$

$$EK_1\theta_B = (0.00367)(82{,}040) - 20.2 = +280.9$$

$$EK_1\theta_C = -(0.00129)(81{,}970) + 27.1 = -78.7$$

Final moments:

$$M_{BA} = (3)(280.9) - (0.0144)(82{,}040) + 187.5 = -151'^{k}$$

$$M_{BC} = (4)(280.9) + (2)(-78.7) - (0.0087)(82{,}040) - 100 = +152'^{k}$$

$$M_{CB} = (2)(280.9) + (4)(-78.7) - (0.0087)(82{,}040) + 100 = -367'^{k}$$

$$M_{CD} = (2)(-78.7) + (0.00915)(82{,}040) - 222.1 = +371'^{k}$$

Generating station, South Carolina Public Service Authority, Cross, South Carolina. (Courtesy of the Owen Steel Company.)

22.4 ANALYSIS OF FRAMES—NO SIDESWAY

The slope-deflection equations may be applied to statically indeterminate frames in the same manner as they were to continuous beams if there is no possibility for the frames to lean or deflect asymmetrically. A frame theoretically will not lean to one side or sway if it is perfectly symmetrical about the centerline as to dimensions, loads, and moments of inertia, or if it is prevented from swaying by its supports. The frame of Fig. 22.8 cannot sway, because member AB is restrained against horizontal movement. Example 22.5 illustrates the analysis of a simple frame with no sidesway. Where sidesway occurs, the joints of the frame move, and that affects the θ values and the moments, as discussed in Sec. 22.5.

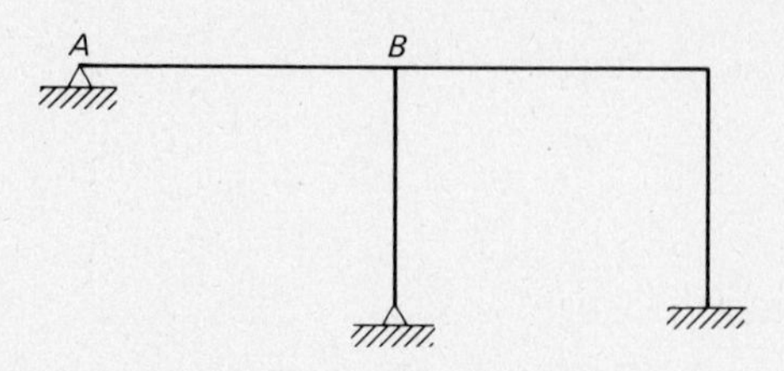

Figure 22.8

Example 22.5

Find all moments for the frame shown in Fig. 22.9, which has no sidesway because of symmetry.

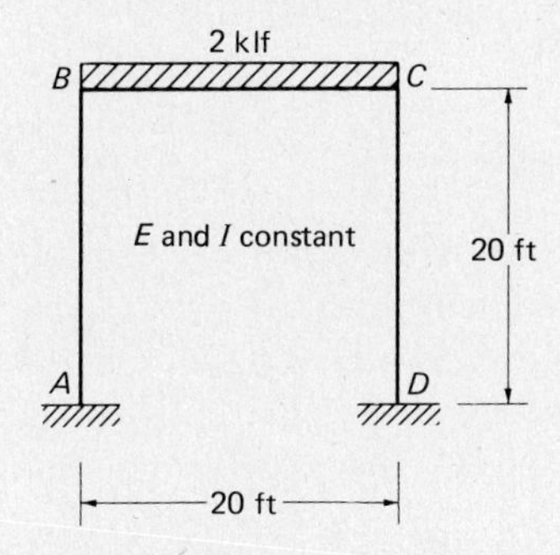

Figure 22.9

SOLUTION

By computing fixed-end moments.

$$FEM_{BC} = -\frac{(2)(20)^2}{12} = -66.7'^{k}$$

$$FEM_{CB} = +\frac{(2)(20)^2}{12} = +66.7'^{k}$$

By writing the equations, and noting $\theta_A = \theta_D = \psi = 0$ and $2EK$ is constant for all members, which permits it to be neglected,

$$M_{AB} = \theta_B$$

$$M_{BA} = 2\theta_B$$

$$M_{BC} = 2\theta_B + \theta_C - 66.7$$

$$M_{CB} = \theta_B + 2\theta_C + 66.7$$

$$M_{CD} = 2\theta_C$$

$$M_{DC} = \theta_C$$

$$\Sigma M_B = 0 = M_{BA} + M_{BC}$$

$$2\theta_B + 2\theta_B + \theta_C - 66.7 = 0$$

$$4\theta_B + \theta_C = 66.7 \qquad (1)$$

$$\Sigma M_C = 0 = M_{CB} + M_{CD}$$

$$\theta_B + 2\theta_C + 66.7 + 2\theta_C = 0$$

$$\theta_B + 4\theta_C = -66.7 \qquad (2)$$

By solving Eqs. (1) and (2) simultaneously,

$$\theta_B = +22.2$$

$$\theta_C = -22.2$$

Final end moments

$$M_{AB} = +22.2'^k \qquad M_{CB} = +44.4'^k$$
$$M_{BA} = +44.4'^k \qquad M_{CD} = -44.4'^k$$
$$M_{BC} = -44.4'^k \qquad M_{DC} = -22.2'^k$$

22.5 ANALYSIS OF FRAMES WITH SIDESWAY

The loads, moments of inertia, and dimensions of the frame of Fig. 22.10 are not symmetrical about the centerline, and the frame will obviously sway to one side. Joints B and C deflect to the right, which causes chord rotations in members AB and CD, there being theoretically no rotation of BC. Neglecting axial deformation of BC, each of the joints deflects the same horizontal distance Δ.

The chord rotations of members AB and CD, due to sidesway, can be seen to equal Δ/l_{AB} and Δ/l_{CD}, respectively. (Notice that the shorter a member for the same Δ, the larger its chord rotation and thus the larger the effect on its moment.) For this frame ψ_{AB} is three-halves as large as ψ_{CD} because l_{AB} is only two-thirds of l_{CD}. It is convenient to work with only one unknown chord rotation, and in setting up the slope-deflection equations, relative values are used. The value ψ is used for the two equations for member AB, whereas $\frac{2}{3}\psi$ is used for the equation for member CD.

Example 22.6 presents the analysis of the frame of Fig. 22.10. By noticing $\theta_A = \theta_D = 0$, it will be seen that the six end-moment equations for the entire structure contain a total of three unknowns: θ_B, θ_C, and ψ. There are present, however, three conditions that permit their determination. They are

1. The sum of the moments at B is zero ($\Sigma M_B = 0 = M_{BA} + M_{BC}$).
2. The sum of the moments at C is zero ($\Sigma M_C = 0 = M_{CB} + M_{CD}$).
3. The sum of the horizontal forces on the entire structure must be zero.

The only horizontal forces are the horizontal reactions at A and D, and they will be equal in magnitude and opposite in direction. Horizontal reactions may be computed for each of the columns by dividing the column moments by the

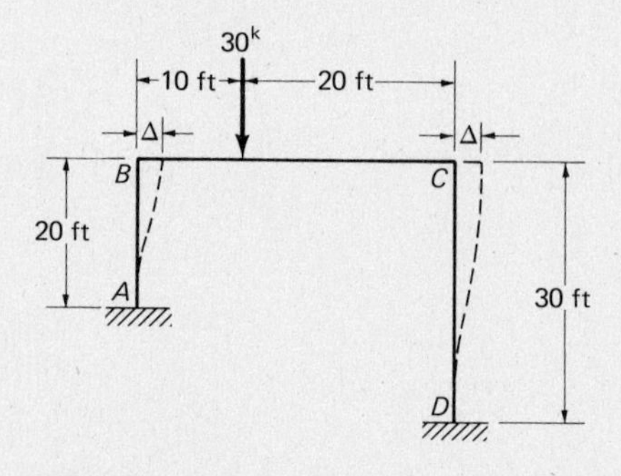

Figure 22.10

column heights or that is by taking moments at the top of each column. The sum of the two reactions must be zero.

$$H_A = \frac{M_{AB} + M_{BA}}{l_{AB}} \qquad H_D = \frac{M_{CD} + M_{DC}}{l_{DC}}$$

$$\sum H = 0 = \frac{M_{AB} + M_{BA}}{l_{AB}} + \frac{M_{CD} + M_{DC}}{l_{CD}}$$

Example 22.6

Determine all of the moments for the frame shown in Fig. 22.11, for which E and I are constant.

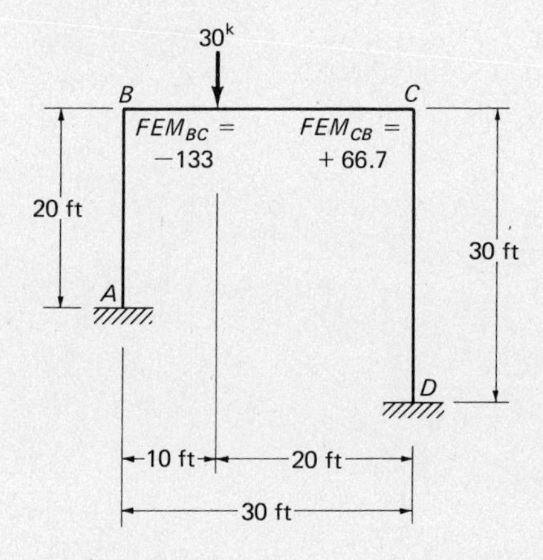

Figure 22.11

SOLUTION

By using relative ψ values.

$$\psi_{AB} = \frac{\Delta}{20} \qquad \psi_{CD} = \frac{\Delta}{30} \qquad \frac{\psi_{AB}}{\psi_{CD}} = \frac{3}{2}$$

By using relative $2EK$ values,

for AB $\quad 2EK = \dfrac{2E}{20}$ $\quad$ say a relative value of 3 with respect to $2EK$ values of BC and CD

for BC $\quad 2EK = \dfrac{2E}{30}$ $\quad$ use relative value = 2

for CD $\quad 2EK = \dfrac{2E}{30}$ $\quad$ use relative value = 2

By writing the equations, and noting $\theta_A = \theta_D = \psi_{BC} = 0$,

$$M_{AB} = 3(\theta_B - 3\psi) = 3\theta_B - 9\psi$$
$$M_{BA} = 3(2\theta_B - 3\psi) = 6\theta_B - 9\psi$$
$$M_{BC} = 2(2\theta_B + \theta_C) - 133 = 4\theta_B + 2\theta_C - 133$$
$$M_{CB} = 2(\theta_B + 2\theta_C) + 66.7 = 2\theta_B + 4\theta_C + 66.7$$
$$M_{CD} = 2[2\theta_C - (3)(\tfrac{2}{3}\psi)] = 4\theta_C - 4\psi$$
$$M_{DC} = 2[\theta_C - (3)(\tfrac{2}{3}\psi)] = 2\theta_C - 4\psi$$

$$\Sigma M_B = 0 = M_{BA} + M_{BC}$$
$$6\theta_B - 9\psi + 4\theta_B + 2\theta_C - 133 = 0$$
$$10\theta_B + 2\theta_C - 9\psi = 133 \qquad (1)$$

$$\Sigma M_C = 0 = M_{CB} + M_{CD}$$
$$2\theta_B + 4\theta_C + 66.7 + 4\theta_C - 4\psi = 0$$
$$2\theta_B + 8\theta_C - 4\psi = -66.7 \qquad (2)$$

$$\Sigma H = 0 = H_A + H_D$$
$$\frac{M_{AB} + M_{BA}}{l_{AB}} + \frac{M_{CD} + M_{DC}}{l_{CD}} = 0$$
$$\frac{3\theta_B - 9\psi + 6\theta_B - 9\psi}{20} + \frac{4\theta_C - 4\psi + 2\theta_C - 4\psi}{30} = 0$$
$$27\theta_B + 12\theta_C - 70\psi = 0 \qquad (3)$$

By solving Eqs. (1), (2), and (3) simultaneously,

$$\theta_B = +21.2$$
$$\theta_C = -10.5$$
$$\psi = +6.4$$

Final moments:

$$M_{AB} = +\ 6.0'^{k} \qquad M_{CB} = +67.4'^{k}$$
$$M_{BA} = +69.4'^{k} \qquad M_{CD} = -67.4'^{k}$$
$$M_{BC} = -69.4'^{k} \qquad M_{DC} = -46.6'^{k}$$

Slope deflection can be applied to frames with more than one condition of sidesway, such as the two-story frame of Fig. 22.12. Hand analysis of frames of this type is usually handled more conviently by the moment-distribution method, but a knowledge of the slope-deflection solution is valuable in understanding the moment-distribution solution.

The horizontal loads cause the structure to lean to the right; joints B and E deflect horizontally a distance Δ_1; and joints C and D deflect horizontally Δ_1 +

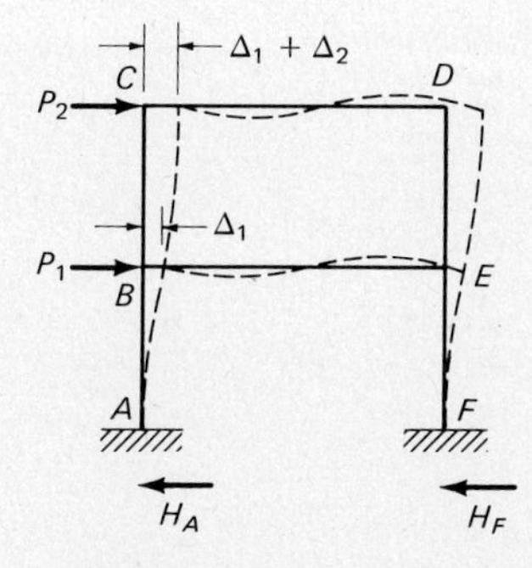

Figure 22.12

Δ_2, as shown in Fig. 22.12. The chord rotations for the columns ψ_1, and ψ_2 will therefore equal Δ_1/l_{AB} for the lower level and Δ_2/l_{BC} for the upper level.

The slope-deflection equations may be written for the moment at each end of the six members, and the usual joint-condition equations are available as follows:

$$\Sigma M_B = 0 = M_{BA} + M_{BC} + M_{BE} \tag{22.1}$$

$$\Sigma M_C = 0 = M_{CB} + M_{CD} \tag{22.2}$$

$$\Sigma M_D = 0 = M_{DC} + M_{DE} \tag{22.3}$$

$$\Sigma M_E = 0 = M_{ED} + M_{EB} + M_{EF} \tag{22.4}$$

These equations involve six unknowns (θ_B, θ_C, θ_D, θ_E, ψ_1, and ψ_2), noting that θ_A and θ_F are equal to zero. Two more equations are necessary for determining the unknowns, and they are found by considering the horizontal forces or shears on the frame. It is obvious that the sum of the horizontal resisting forces on any level must be equal and opposite to the external horizontal shear on that level. For the bottom level the horizontal shear equals $P_1 + P_2$, and the reactions at the base of each column equal the end moments divided by the column heights. Therefore

$$P_1 + P_2 - H_A - H_F = 0$$

$$H_A = \frac{M_{AB} + M_{BA}}{l_{AB}}$$

$$H_F = \frac{M_{EF} + M_{FE}}{l_{EF}} \tag{22.5}$$

$$\frac{M_{AB} + M_{BA}}{l_{AB}} + \frac{M_{EF} + M_{FE}}{l_{EF}} + P_1 + P_2 = 0$$

A similar condition equation may be written for the top level, which has an external shear of P_2. The moments in the columns produce shears equal and opposite to P_2, which permits the writing of the following equation for the level:

$$\frac{M_{BC} + M_{CB}}{l_{BC}} + \frac{M_{DE} + M_{ED}}{l_{DE}} + P_2 = 0 \tag{22.6}$$

Six condition equations are available for determining the six unknowns in the end-moment equations, and the problem may be solved as illustrated in Example 22.7. It is desirable to assume all θ and ψ values to be positive in setting up the equations. In this way the signs of the answers will take care of themselves.

No matter how many floors the building has, one shear-condition equation is available for each floor. The slope-deflection procedure is not very practical for multistory buildings. For a six-story building four bays wide there will be 6 unknown ψ values and 30 unknown θ values, or a total of 36 simultaneous equations to solve. (Their solution is not quite as bad as it may seem to the reader because each equation would contain only a few unknowns and not 36.)

Example 22.7

Find the moments for the frame of Fig. 22.13 by the method of slope deflection.

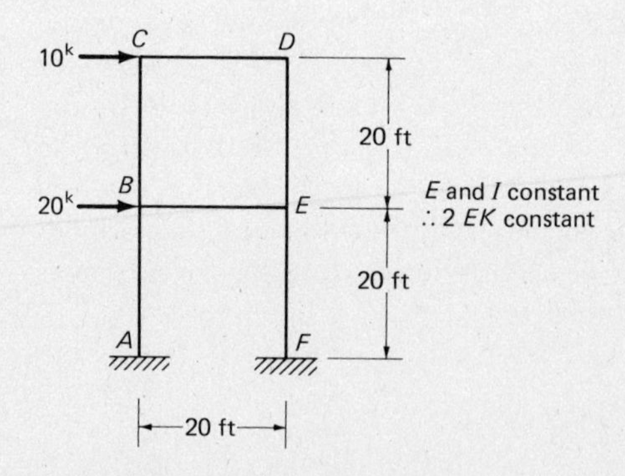

Figure 22.13

SOLUTION

By noting that chords BE and CD do not rotate,

$$\psi_{AB} = \psi_{EF} = \psi_1 \qquad \psi_{BC} = \psi_{DE} = \psi_2$$

By writing the equations, and noting that $\theta_A = \theta_F = 0$,

$$M_{AB} = \theta_B - 3\psi_1 \qquad M_{DC} = \theta_C + 2\theta_D$$
$$M_{BA} = 2\theta_B - 3\psi_1 \qquad M_{DE} = 2\theta_D + \theta_E - 3\psi_2$$
$$M_{BC} = 2\theta_B + \theta_C - 3\psi_2 \qquad M_{ED} = \theta_D + 2\theta_E - 3\psi_2$$
$$M_{BE} = 2\theta_B + \theta_E \qquad M_{EB} = \theta_B + 2\theta_E$$
$$M_{CB} = \theta_B + 2\theta_C - 3\psi_2 \qquad M_{EF} = 2\theta_E - 3\psi_1$$
$$M_{CD} = 2\theta_C + \theta_D \qquad M_{FE} = \theta_E - 3\psi_1$$

$$\Sigma M_B = 0 = M_{BA} + M_{BC} + M_{BE}$$
$$2\theta_B - 3\psi_1 + 2\theta_B + \theta_C - 3\psi_2 + 2\theta_B + \theta_E = 0$$
$$6\theta_B + \theta_C + \theta_E - 3\psi_1 - 3\psi_2 = 0 \qquad (1)$$

$$\Sigma M_C = 0 = M_{CB} + M_{CD}$$
$$\theta_B + 2\theta_C - 3\psi_2 + 2\theta_2 + \theta_D = 0$$
$$\theta_B + 4\theta_C + \theta_D - 3\psi_2 = 0 \qquad (2)$$

$$\Sigma M_D = 0 = M_{DC} + M_{DE}$$
$$\theta_C + 2\theta_D + 2\theta_D + \theta_E - 3\psi_2 = 0$$
$$\theta_C + 4\theta_D + \theta_E - 3\psi_2 = 0 \qquad (3)$$

$$\Sigma M_E = 0 = M_{ED} + M_{EB} + M_{EF}$$
$$\theta_D + 2\theta_E - 3\psi_2 + \theta_B + 2\theta_E + 2\theta_E - 3\psi_1 = 0$$
$$\theta_B + \theta_D + 6\theta_E - 3\psi_1 - 3\psi_2 = 0 \qquad (4)$$

$\Sigma H = -10$, top level:

$$\frac{M_{BC} + M_{CB}}{20} + \frac{M_{DE} + M_{ED}}{20} = -10$$

$$\frac{2\theta_B + \theta_C - 3\psi_2 + \theta_B + 2\theta_C - 3\psi_2}{20} + \frac{2\theta_D + \theta_E - 3\psi_2 + \theta_D + 2\theta_E - 3\psi_2}{20} = -10$$

$$3\theta_B + 3\theta_C + 3\theta_D + 3\theta_E - 12\psi_2 = -200 \qquad (5)$$

$\Sigma H = 30$, bottom level:

$$\frac{M_{AB} + M_{BA}}{20} + \frac{M_{EF} + M_{FE}}{20} = -30$$

$$\frac{\theta_B - 3\psi_1 + 2\theta_B - 3\psi_1}{20} + \frac{2\theta_E - 3\psi_1 + \theta_E - 3\psi_1}{20} = -30$$

$$3\theta_B + 3\theta_E - 12\psi_1 = -600 \qquad (6)$$

By solving equations simultaneously,

$$\theta_B = \theta_E = +52.75 \qquad \psi_1 = +76.37$$
$$\theta_C = \theta_D = +21.82 \qquad \psi_2 = +53.95$$

Final moments

$$M_{AB} = M_{FE} = -176.4'^{k} \qquad M_{BE} = M_{EB} = +158.2'^{k}$$
$$M_{BA} = M_{EF} = -123.6'^{k} \qquad M_{CB} = M_{DE} = -\ 65.5'^{k}$$
$$M_{BC} = M_{ED} = -\ 34.6'^{k} \qquad M_{CD} = M_{DC} = +\ 65.5'^{k}$$

22.6 ANALYSIS OF FRAMES WITH SLOPING MEMBERS

The frames with possible sidesway discussed in Sec. 22.5 were assumed to have horizontal members that had no chord rotations even though the columns had them. When a frame with sloping legs is encountered, such as the trapezoidal frame of Fig. 22.14(a), the chord of each horizontal member may rotate. Similarly, the chords of all of the members of the gabled structure of Fig. 22.14(b) may rotate.

Another item of importance is the fact that the vertical reactions need to be considered in writing the expressions for shear for sloping members. The deflected shape of a trapezoidal shaped structure, acted on by a horizontal load, is shown in Fig. 22.15.

Neglecting axial deformation, joints B and C move through the same distance z. To be absolutely correct, the deflected distances of the joints should be shown as arcs, but the distances are small, and perpendiculars are perfectly satisfactory. Each of the members has a chord rotation that needs to be included in applying the slope-deflection equations. A study of the trigonome-

Transit shed, Port of Long Beach, California. (Courtesy of the American Institute of Steel Construction.)

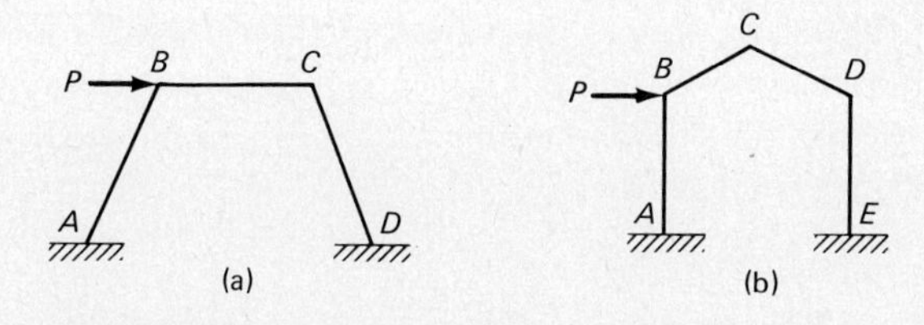

Figure 22.14

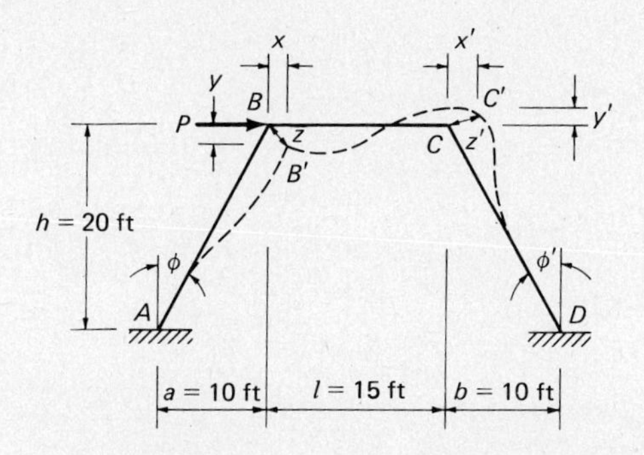

Figure 22.15

try involved shows that the chord rotations may be expressed quite simply in relation to each other.

Suppose joint B moves a distance horizontally x and a vertical distance y and let the angle between member AB and the vertical be ϕ. The values of z and y will then be

$$z = \frac{x}{\cos \phi} \qquad y = x \tan \phi$$

Therefore the chord rotation of AB is as follows, noting $l_{AB} = h/\cos \phi$:

$$\psi_{AB} = \frac{z}{l_{AB}} = \frac{x/\cos \phi}{h/\cos \phi} = \frac{x}{h}$$

Similarly, the rotation of CD is x'/h. The value of y' is $x' \tan \phi'$ and the chord rotation of BC is developed as follows:

$$\psi_{BC} = \frac{y + y'}{l_{BC}} = \frac{x(\tan \phi + \tan \phi')}{l_{BC}}$$

But $\tan \phi = a/h$ and $\tan \phi' = b/h$, and ψ_{BC} becomes

$$\psi_{BC} = \frac{x(a/h + b/h)}{l_{BC}} = \frac{x(a + b)}{hl_{BC}}$$

For the dimensions shown on the frame, the values of the three chord rotations may all be expressed in terms of one unknown, x. Similar relations may be derived trigonometrically for chord rotation for more complicated structures, but space is not taken to include them.

$$\psi_{AB} = \frac{x}{h} = \frac{x}{20}$$

$$\psi_{BC} = \frac{x(a+b)}{hl_{BC}} = \frac{x}{15}$$

$$\psi_{CD} = \frac{x'}{h} = \frac{x}{20}$$

A much simpler method of obtaining relative rotations for sloping leg frames is presented in Sec. 24.4 where a complete analysis of such a structure is illustrated by moment distribution. Analysis of these types of frames by slope deflection is usually a very tedious job unless a computer is available. For analysis with a hand-held calculator the moment-distribution procedure is much more practical.

PROBLEMS

For Probs. 22.1 to 22.16 compute the end moments of the beams with the slope-deflection equations. Draw shear and moment diagrams. E is constant for all problems.

Problem 22.1

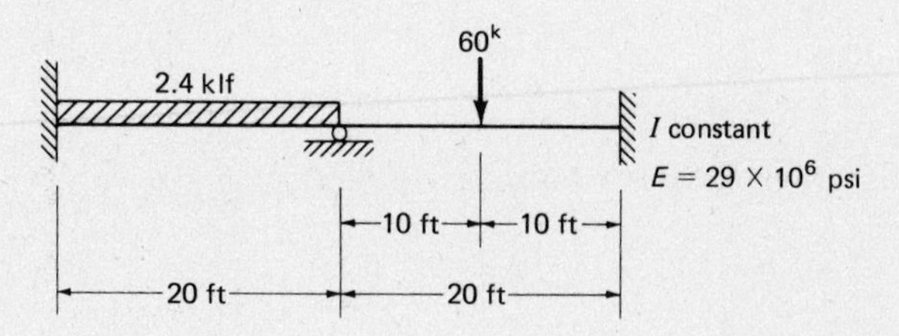

Problem 22.2 (*Ans.* $M_A = -131.1'^k$, $M_B = -71.1'^k$, $M_C = -121.0'^k$, $V_D = 21.45^k \uparrow$)

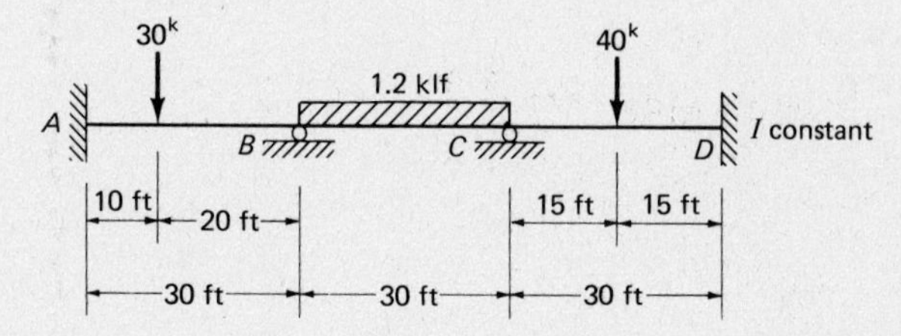

Problem 22.3

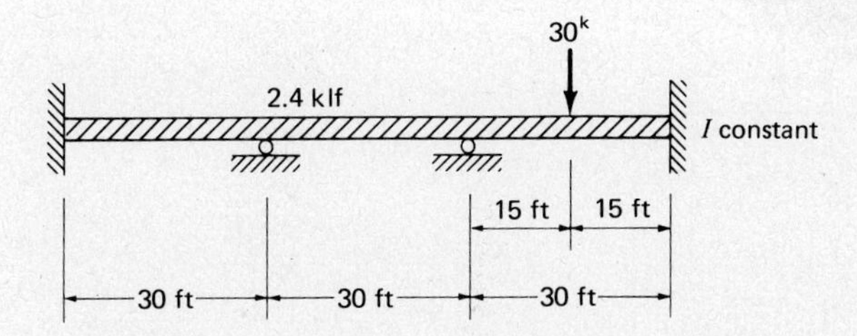

Problem 22.4 (*Ans.* $V_A = 18.09^k \uparrow$, $M_B = -375.4'^k$, $M_C = -305.4'^k$, $V_D = 29.74^k \uparrow$)

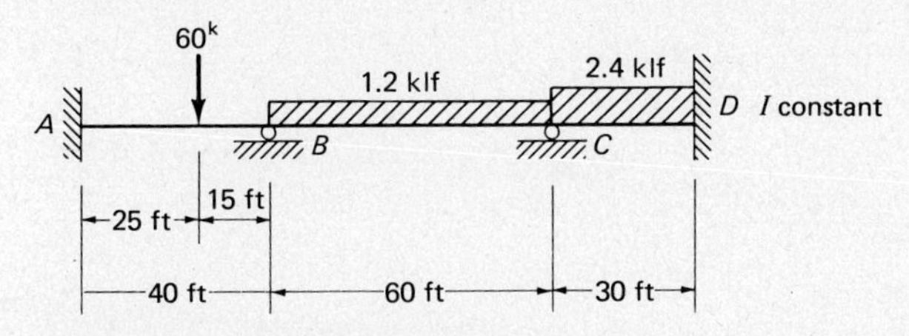

Problem 22.5

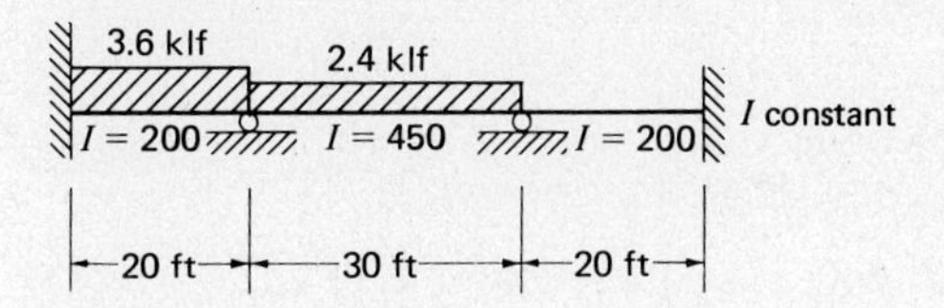

Problem 22.6 (*Ans.* $V_A = 30.0^k \uparrow$, $V_B = 18.0^k \uparrow$, $M_A = -120'^k$)

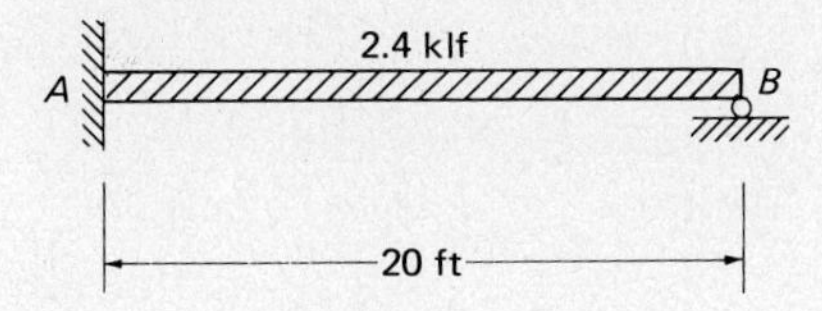

Problem 22.7

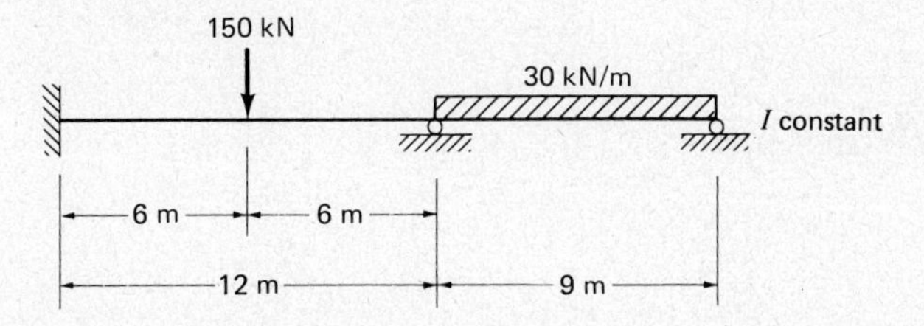

Problem 22.8 (*Ans.* $V_A = 12.08^k \uparrow$, $V_B = 74.17^k \uparrow$, $M_B = -237.6'^k$, $M_C = -112.6'^k$)

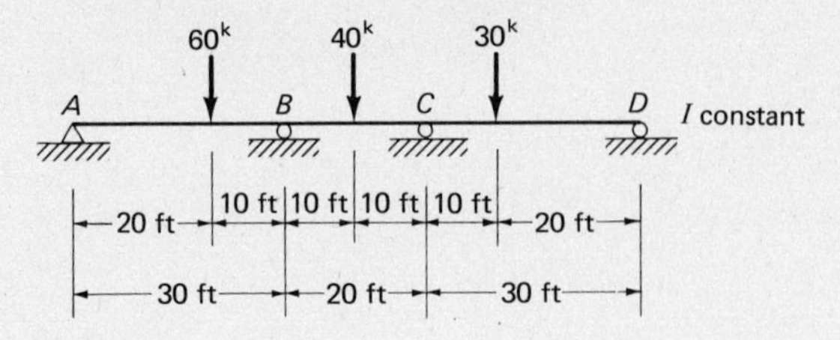

Problem 22.9

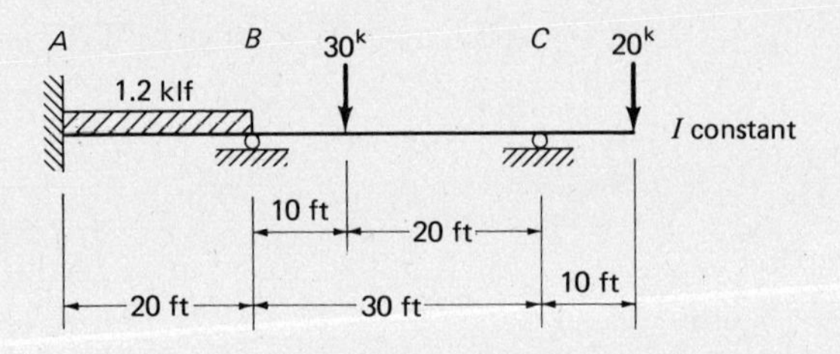

22.10. The beam of Prob. 22.2 if both ends are simply supported. (*Ans.* $V_A = 16.83^k \uparrow$, $M_B = -95.1'^k$, $V_C = 45.25^k \uparrow$, $M_C = -156.4'^k$)
22.11. The beam of Prob. 22.4 if the right end is simply supported.
22.12. Problem 21.34. (*Ans.* $V_B = 87.57^k \uparrow$, $V_D = 4.76^k \uparrow$, $M_B = -300.0'^k$, $M_C = -157.2'^k$)
22.13. Problem 21.36.
22.14. Problem 21.35. (*Ans.* $V_A = 23.71^k \uparrow$, $V_D = 27.79^k \uparrow$, $M_B = -368.8'^k$, $M_C = -372.3'^k$)
22.15. The beam of Prob. 22.1 assuming support B settles 2.50 in, $I = 1250$ in^4.
22.16. The beam of Prob. 19.9 assuming the following support settlements: $A = 1.00$ in, $B = 2.00$ in, $C = 1.50$ in, and $D = 2.00$ in. $E = 29 \times 10^6$ psi; $I = 3200$ in^4. (*Ans.* $M_B = -14.7'^k$, $M_C = -390.14'^k$, $V_C = 94.07^k \uparrow$)

For Probs. 22.17 to 22.25 determine the end moments for the members of all of the structures by using the slope deflection equations.

Problem 22.17

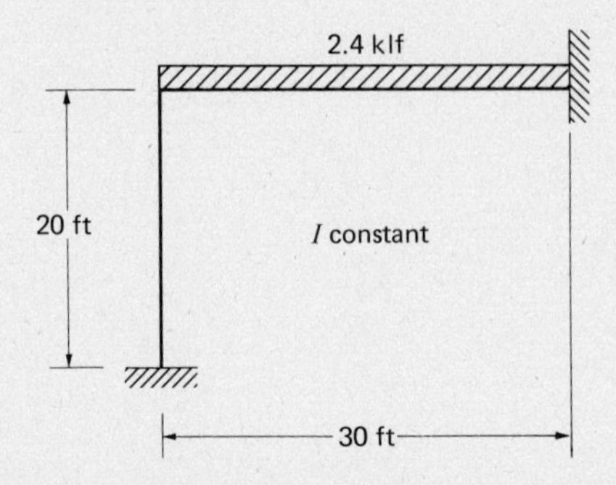

Problem 22.18 (*Ans.* $M_{BA} = 114.7'^{k}$, $M_C = 21.3'^{k}$, $M_D = 18.7'^{k}$)

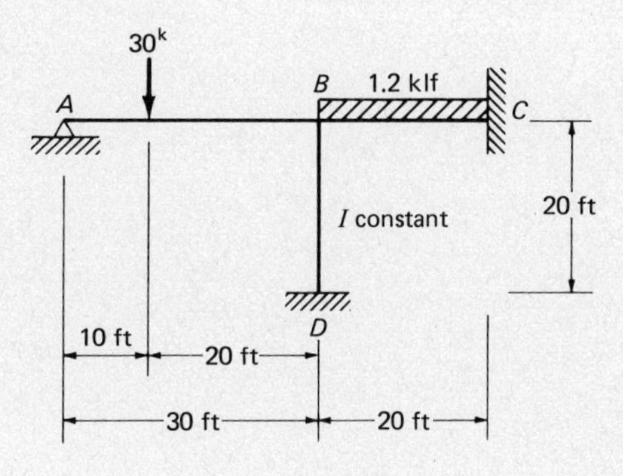

Problem 22.19

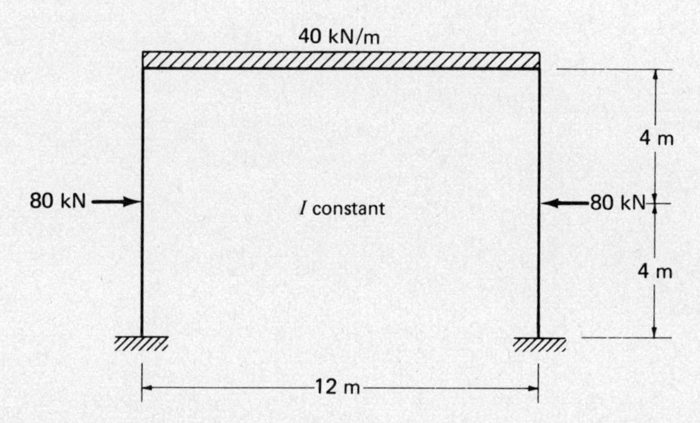

Problem 22.20 (*Ans.* $M_A = 141.4'^{k}$, $M_B = 33.6'^{k}$, $M_C = 94.6'^{k}$, $M_D = 98.0'^{k}$)

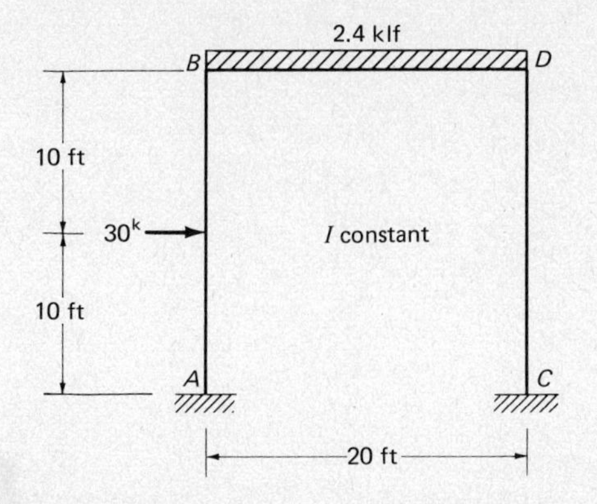

22.21. The frame of Prob. 22.20 if the column bases are pinned.

Problem 22.22 (*Ans.* $M_A = 4.3'^k$, $M_C = 70.7'^k$, $M_{DB} = 43.9'^k$, $M_F = 130.7'^k$)

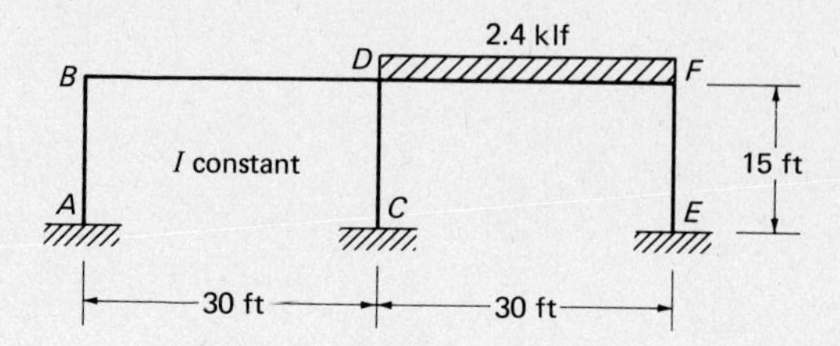

Problem 22.23

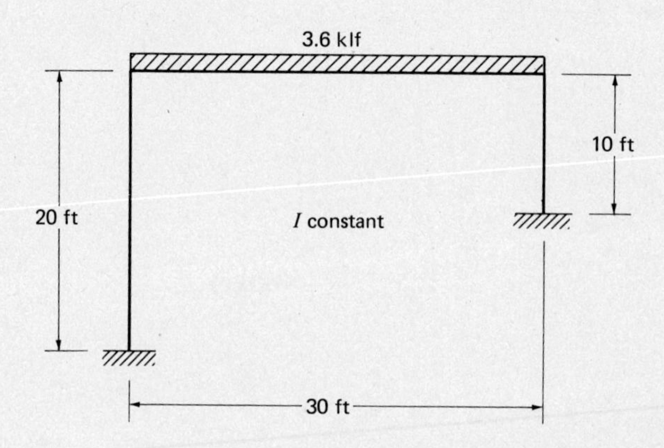

Problem 22.24 (*Ans.* $M_A = 83.2'^k$, $M_B = 106.8'^k$, $M_C = 43.2'^k$, $M_D = 27.0'^k$)

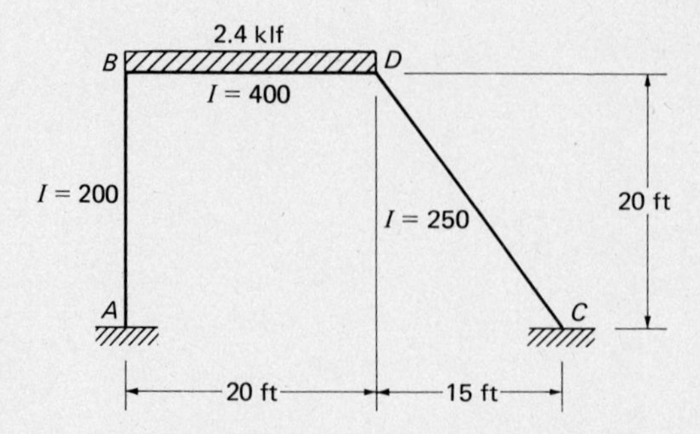

Problem 22.25

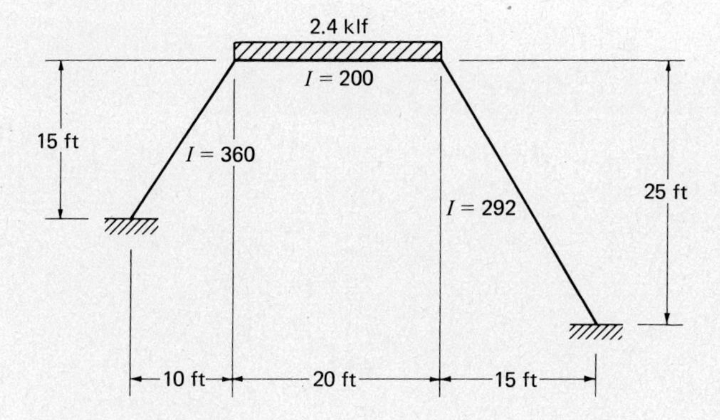

REFERENCES

[1] G. A. Maney, *Studies in Engineering,* No. 1 (Minneapolis: University of Minnesota, 1915).

[2] H. Manderla, "Die Berechnung der Sekundarspannungen," *Allg. Bautz* 45 (1880): 34.

[3] O. Mohr, "Die Berechnung der Fachwerke mit starren knotenverbingungen," *Zivilinginieur* (1892).

23 Moment Distribution for Beams

23.1 GENERAL

The late Hardy Cross published the moment-distribution method in the *Proceedings of the American Society of Civil Engineers* in May, 1930 [1], after having taught the subject to his students at the University of Illinois since 1924. His paper began a new era in the analysis of statically indeterminate frames and gave added impetus to their use. The moment-distribution method of analyzing continuous beams and frames involves little more labor than the approximate methods but yields accuracy equivalent to that obtained from the infinitely more laborious "exact" methods previously studied.

The analysis of statically indeterminate structures in the preceding chapters frequently involved the solution of inconvenient simultaneous equations. These equations are not necessary in solutions by moment distribution except in a few rare situations for complicated frames. The Cross method involves successive cycles of computation, each cycle drawing closer to the "exact" answers. The calculations may be stopped after two or three cycles, giving a very good approximate analysis, or they may be carried on to whatever degree of accuracy is desired. When these advantages are considered in the light of the fact that the accuracy obtained by the lengthy "classical" methods is often of questionable value, the true worth of this quick and practical method is understood.

23.2 INTRODUCTION

The beauty of moment distribution lies in its simplicity of theory and application. Readers will be able to grasp quickly the principles involved, and will clearly understand what they are doing and why they are doing it.

Largest curved "horizontal skyscraper" in the United States, Boston, Massachusetts. (Courtesy of the Bethlehem Steel Corporation.)

The following discussion pertains to structures having members of contant cross section throughout their respective lengths (that is, prismatic members). It is assumed that there is no joint translation where two or more members frame together, but that there can be some joint rotation (that is, the members may rotate as a group but may not move with respect to each other). Finally, axial deformation of members is neglected.

Considering the frame of Fig. 23.1(a), joints A to D are seen to be fixed. Joint E, however, is not fixed, and the loads on the structure will cause it to rotate slightly, as represented by the angle θ_E in Fig. 23.1(b).

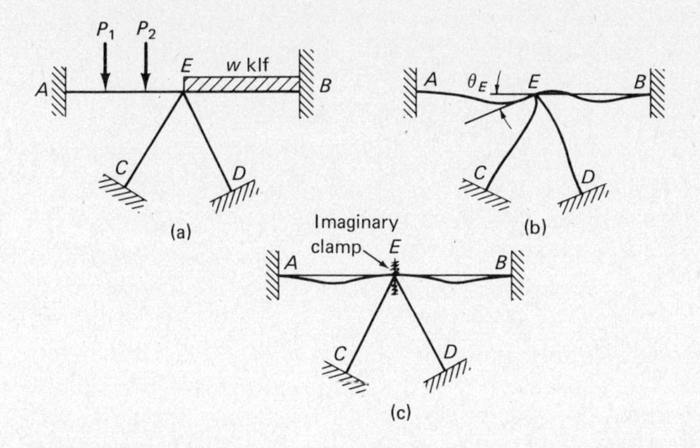

Figure 23.1

If an imaginary clamp is placed at E, fixing it so that it cannot be displaced, the structure will under load take the shape of Fig. 23.1(c). For this situation, with all ends fixed, the end moments can be calculated with little difficulty by the usual expressions ($wl^2/12$ for uniform loads and Pab^2/l^2 or Pa^2b/l^2 for concentrated loads).

If the clamp at E is removed, the joint will rotate slightly, twisting the ends of the members meeting there and causing a redistribution of the moments in the member ends. The changes in the moments or twists at the E ends of members AE, BE, CE, and DE cause some effect at their other ends. When a moment is applied to one end of a member, the other end being fixed, there is some effect or ***carry-over*** to the fixed end.

After the fixed-end moments are computed, the problem to be handled may be briefly stated as consisting of the calculation of (1) the moments caused in the E ends of the members by the rotation of joint E, (2) the magnitude of the moments carried over to the other ends of the members, and (3) the addition or subtraction of these latter moments to the original fixed-end moments.

These steps can be simply written as being the fixed-end moments plus the moments due to the rotation of joint E.

$$M = M_{\text{fixed}} + M_{\theta_E}$$

23.3 BASIC RELATIONS

There are two questions that must be answered in order to apply the moment-distribution method to actual structures. They are:

1. What is the moment developed or carried over to a fixed end of a member when the other end is subjected to a certain moment?
2. When a joint is unclamped and rotates, what is the distribution of the unbalanced moment to the members meeting at the joint, or how much resisting moment is supplied by each member?

Carry-Over Moment

To determine the carry-over moment, the unloaded beam of constant cross section in Fig. 23.2(a) is considered. If a moment M_1 is applied to the left end of the beam, a moment M_2 will be developed at the right end. The left end is at a joint that has been released and the moment M_1 causes it to rotate an amount θ_1. There will, however, be no deflection or translation of the left end with respect to the right end.

The second moment-area theorem may be used to determine the magnitude of M_2. The deflection of the tangent to the elastic curve of the beam at the left end with respect to the tangent at the right end (which remains horizontal) is equal to the moment of the area of the M/EI diagram taken about the left end and is equal to zero. By drawing the M/EI diagram in Fig. 23.2(b) and

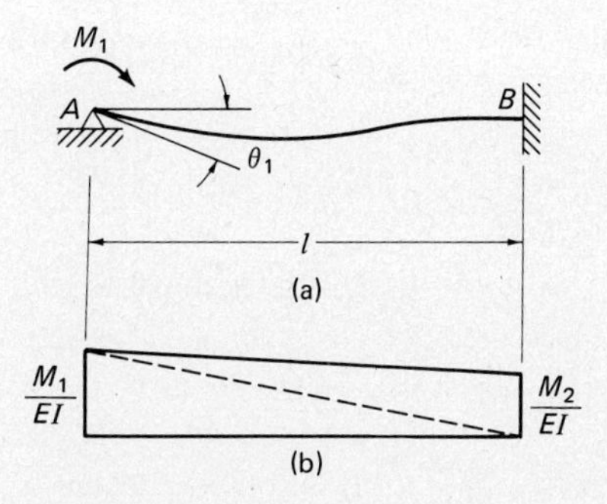

Figure 23.2

dividing it into two triangles to facilitate the area computations, the following expression may be written and solved for M_2:

$$\delta_A = \frac{(\frac{1}{2} \times M_1 \times l)(\frac{1}{3}l) + (\frac{1}{2} \times M_2 \times l)(\frac{2}{3}l)}{EI} = 0$$

$$\frac{M_1 l^2}{6EI} + \frac{M_2 l^2}{3EI} = 0$$

$$M_2 = -\tfrac{1}{2}M_1$$

A moment applied at one end of a prismatic beam, the other end being fixed, will cause a moment half as large and of opposite sign at the fixed end. The carry-over factor is $-\frac{1}{2}$. The minus sign refers to strength-of-materials sign convention: A distributed moment on one end causing tension in bottom fibers must be carried over so that it will cause tension in the top fibers of the other end. A study of Figs. 23.2 and 23.3 shows that carrying over with a $+\frac{1}{2}$ value with the moment-distribution sign convention automatically takes care of the situation, and it is unnecessary to change signs with each carryover.

Distribution Factors

Usually a group of members framed together at a joint have different stiffnesses. When a joint is unclamped and begins to rotate under the unbalanced moment, the resistance to rotation varies from member to member. The problem is to determine how much of the unbalanced moment will be taken up by each of the members. It seems reasonable to assume the unbalance will be resisted in direct relation to the respective resistance to end rotation of each member.

The beam and M/EI diagram of Fig. 23.2 are redrawn in Fig. 23.3, with the proper relationship between M_1 and M_2, and an expression is written for the amount of rotation caused by a moment M_1.

Using the first moment-area theorem, the angle θ_1 may be represented by the area of the M/EI diagram between A and B, the tangent at B remaining horizontal.

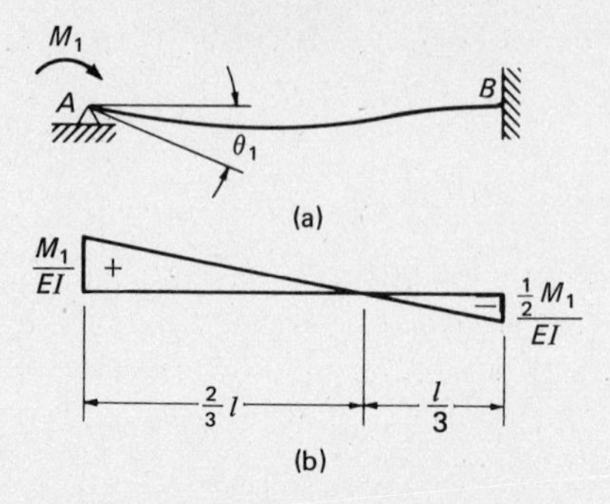

Figure 23.3

$$\theta_1 = \frac{(\frac{1}{2})(M_1)(\frac{2}{3}l) - (\frac{1}{2})(\frac{1}{2}M_1)(\frac{1}{3}l)}{EI}$$

$$= \frac{M_1 l}{4EI}$$

Assuming that all of the members consist of the same material, having the same E values, the only variables in the foregoing equation affecting the amount of end rotation are the l and I values. The amount of rotation occurring at the end of a member obviously varies directly as the l/I value for the member. The larger the rotation of the member, the less moment it will carry. The moment resisted varies inversely as the amount of rotation or directly as the I/l value. This latter value is referred to as the *stiffness factor K*.

$$K = \frac{I}{l}$$

To determine the unbalanced moment taken by each of the members at a joint, the stiffness factors at the joint are totaled, and each member is assumed to carry a proportion of the unbalanced moment equal to its K value divided by the sum of all the K values at the joint. These proportions of the total unbalanced moment carried by each of the members are the ***distribution factors.***

$$DF_1 = \frac{K_1}{\Sigma K} \qquad DF_2 = \frac{K_2}{\Sigma K}$$

23.4 DEFINITIONS

The following terms are constantly used in discussing moment distribution.

Fixed-End Moments

When all of the joints of a structure are clamped to prevent any joint rotation, the external loads produce certain moments at the ends of the members to which they are applied. These moments are referred to as fixed-end moments.

Unbalanced Moments

Initially the joints in a structure are considered to be clamped. When a joint is released, it rotates because the sum of the fixed-end moments at the joint is not zero. The difference between zero and the actual sum of the end moments is the unbalanced moment.

Distributed Moments

After the clamp at a joint is released, the unbalanced moment causes the joint to rotate. The rotation twists the ends of the members at the joint and changes their moments. In other words, rotation of the joint is resisted by the members and resisting moments are built up in the members as they are twisted. Rotation continues until equilibrium is reached—when the resisting moments equal the unbalanced moment—at which time the sum of the moments at the joint is equal to zero. The moments developed in the members resisting rotation are the distributed moments.

Carry-Over Moments

The distributed moments in the ends of the members cause moments in the other ends, which are assumed fixed, and these are the carry-over moments.

23.5 SIGN CONVENTION

The moments at the end of a member are assumed to be negative when they tend to rotate the member clockwise about the joint (the resisting moment of the joint would be counterclockwise). The continuous beam of Fig. 23.4, with all joints assumed to be clamped, has clockwise (or −) moments on the left end of each span and counterclockwise (or +) moments on the right end of each span. (The usual sign convention used in strength of materials shows fixed-ended beams to have negative moments on both ends for downward loads, because tension is caused in the top fibers of the beams at those points.) It should be noted that this sign convention, to be used for moment distribution, is the same as the one used in Chap. 22 for slope deflection.

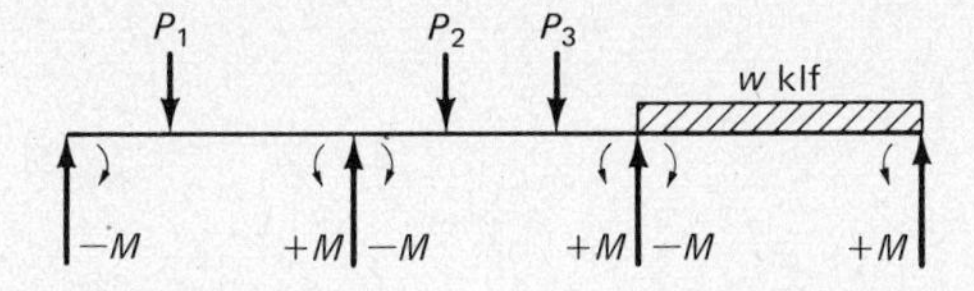

Figure 23.4

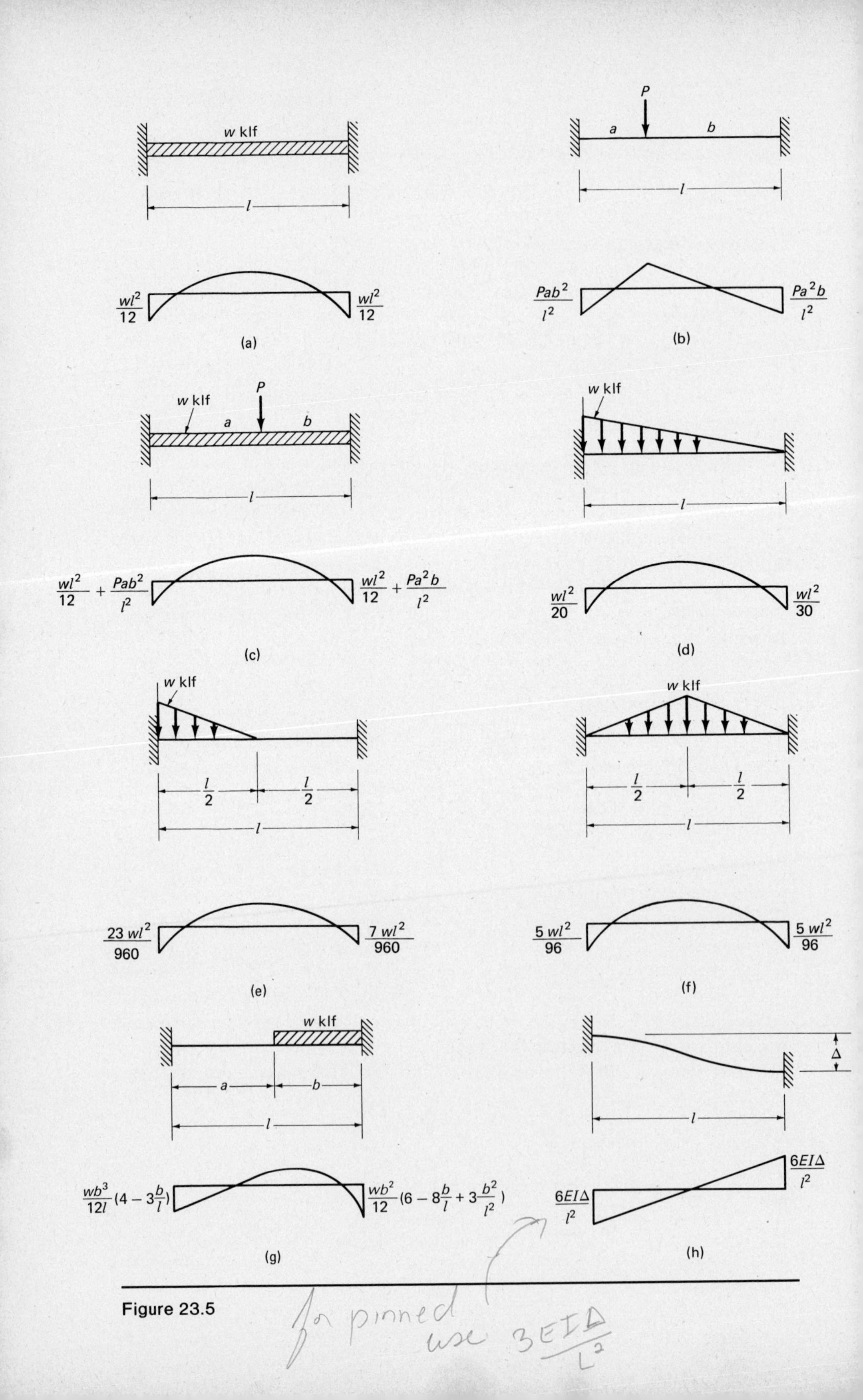
w klf
l
wl²/12
wl²/12
(a)
P
a
b
l
Pab²/l²
Pa²b/l²
(b)
w klf
P
a
b
l
wl²/12 + Pab²/l²
wl²/12 + Pa²b/l²
(c)
w klf
l
wl²/20
wl²/30
(d)
w klf
l/2
l/2
l
23 wl²/960
7 wl²/960
(e)
w klf
l/2
l/2
l
5 wl²/96
5 wl²/96
(f)
w klf
a
b
l
wb³/12l (4 − 3 b/l)
wb²/12 (6 − 8 b/l + 3 b²/l²)
(g)
Δ
l
6EIΔ/l²
6EIΔ/l²
(h)
for pinned
use 3EIΔ/L²

Figure 23.5

23.6 FIXED-END MOMENTS FOR VARIOUS LOADS

Figure 23.5 presents expressions for computing the fixed-end moments for several types of loading conditions. The first three parts of the figure cover the large percentage of practical cases whereas the other parts are for a little more unusual loading conditions. Particular attention should be given to part (c) of the figure in which the superposition idea is presented. If fixed-end moments are needed in the same span for different loading conditions, they are each computed separately and added together as shown in the figure.

For cases not shown in Fig. 23.5 fixed-end moments can be obtained from various tables or they may be calculated with the moment area method as illustrated with Example 17.6.

23.7 APPLICATION OF MOMENT DISTRIBUTION

The very few tools needed for applying moment distribution are now available, and the method of applying them is described, reference being made to Fig. 23.6

Figure 23.6(a) shows a beam and the several loads applied to it. In (b) the interior joints *B* and *C* are assumed to be clamped, and the fixed-end moments are computed. At joint *B* the unbalanced moment is computed and the clamp is removed, as seen in (c). The joint rotates, thus distributing the unbalanced moment to the *B* ends of *BA* and *BC* in proportion to their distribution factors. The values of these distributed moments are carried over at the one-half rate to the other ends of the members. When equilibrium is reached, joint *B* is clamped in its new rotated position and joint *C* is released, as shown in (d). Joint *C* rotates under its unbalanced moment until it reaches equilibrium, the rotation causing distributed moments in the *C* ends of members *CB* and *CD*

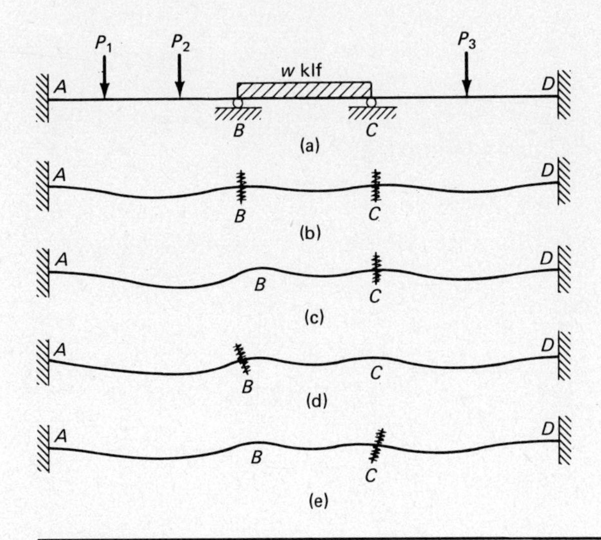

Figure 23.6

and their resulting carry-over moments. Joint C is now clamped and joint B is released, Fig. 23.6(e).

The same procedure is repeated again and again for joints B and C, the amount of unbalanced moment quickly diminishing, until the release of a joint causes only negligible rotation. This process, in brief, is moment distribution.

Examples 23.1 to 23.3 illustrate the procedure used for analyzing relatively simple continuous beams. The stiffness factors and distribution factors are computed as follows for Example 23.1.

$$DF_{BA} = \frac{K_{BA}}{\Sigma K} = \frac{\frac{1}{20}}{\frac{1}{20} + \frac{1}{15}} = 0.43$$

$$DF_{BC} = \frac{K_{BC}}{\Sigma K} = \frac{\frac{1}{15}}{\frac{1}{20} + \frac{1}{15}} = 0.57$$

A simple tabular form is used for Examples 23.1 and 23.2 to introduce the reader to moment distribution. (For subsequent examples a slightly varying but quicker solution much preferred by the author is used.) The tabular procedure may be summarized as follows:

1. The fixed-end moments are computed and recorded on one line (line FEM in Examples 23.1 and 23.2).
2. The unbalanced moments at each joint are balanced in the next line (Dist 1).
3. The carry-overs are made from each of the joints on the next line (CO 1).
4. The new unbalanced moments at each joint are balanced (Dist 2), and so on. (As the beam of Example 23.1 has only one joint to be balanced, only one balancing cycle is necessary.)

When the distribution has reached the accuracy desired, a double line is drawn under each column of figures. The final moment in the end of a member equals the sum of the moments opposite its position in the table. Unless a joint is fixed, the sum of the final end moments in the ends of the members meeting at the joint must total zero.

Example 23.1

Determine the end moments of the structure shown in Fig. 23.7 by moment distribution.

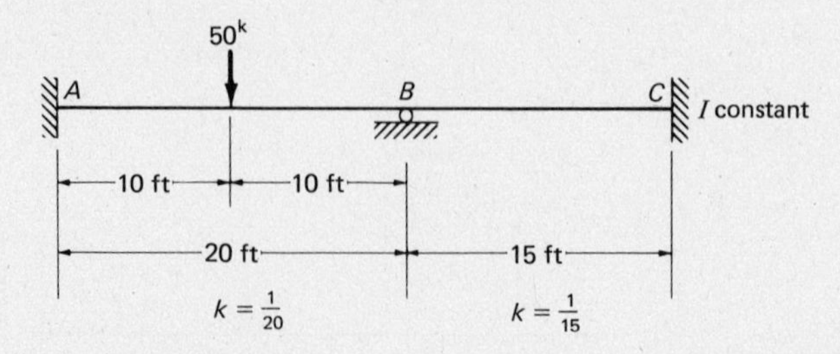

Figure 23.7

SOLUTION

	0.43	0.57		
−125	+125			FEM
	−53.8	−71.2		Dist 1
−26.9			−35.6	CO 1
−151.9	+71.2	−71.2	−35.6	Final moments

Example 23.2

Determine the end moments of the structure shown in Fig. 23.8.

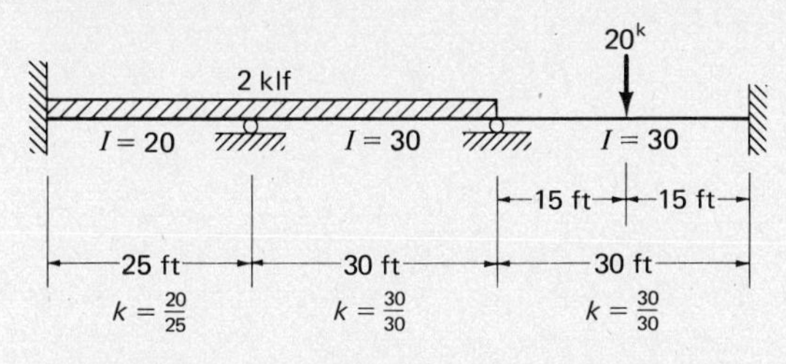

Figure 23.8

SOLUTION

	0.44	0.56	0.5	0.5		
−104.2	+104.2	−150.0	+150.0	−75.0	+75.0	FEM
	+20.2	+25.6	−37.5	−37.5		Dist 1
+10.1		−18.8	+12.8		−18.8	CO 1
	+8.3	+10.5	−6.4	−6.4		Dist 2
+4.2		−3.2	+5.3		−3.2	CO 2
	+1.4	+1.8	−2.7	−2.7		Dist 3
+0.7		−1.3	+0.9		−1.3	CO 3
	+0.6	+0.7	−0.4	−0.4		Dist 4
+0.3		−0.2	+0.4		−0.2	CO 4
	+0.1	+0.1	−0.2	−0.2		Dist 5
−88.9	+134.8	−134.8	+122.2	−122.2	+51.5	Final moments

Beginning with Example 23.3, a slightly different procedure is used for distributing the moments. Only one joint at a time is balanced and the required carry-overs are made from that joint. Generally speaking it is desirable (but not necessary) to balance the joint that has the largest unbalance, make the carry-overs, balance the joint with the largest unbalance, and so on, because such a process will result in the quickest convergence. This procedure is quicker than the tabular method used for Examples 23.1 and 23.2, and it follows along exactly with the description of the behavior of a continuous beam (with imaginary clamps) as pictured by the author in Fig. 23.6.

Example 23.3

Compute the end moments in the beam shown in Fig. 23.9.

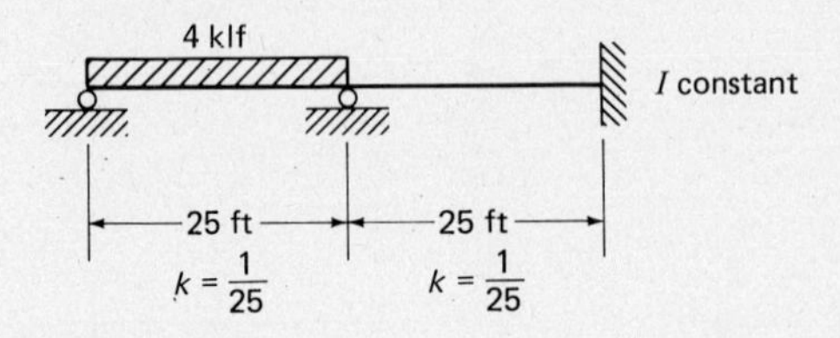

Figure 23.9

SOLUTION

	0.5	0.5	
−208.3	+208.3		
+208.3	+104.2		
− 78.1	−156.2	−156.2	−78.1
+ 78.1	+ 39.0		
− 9.8	− 19.5	− 19.5	− 9.8
+ 9.8	+ 4.9		
− 1.2	− 2.5	− 2.5	− 1.2
+ 1.2	+ 0.6		
− 0.2	− 0.3	− 0.3	− 0.2
+ 0.2	+178.5	−178.5	−89.3
0			

23.8 MODIFICATION OF STIFFNESS FOR SIMPLE ENDS

The carry-over factor was developed for carrying over to fixed ends, but it is applicable to simply supported ends, which must have final moments of zero. The simple end of Example 23.3 was considered to be clamped; the carry-over was made to the end; and the joint was freed and balanced back to zero. This procedure repeated over and over is absolutely correct, but it involves a little unnecessary work which may be eliminated by studying the stiffness of members with simply supported ends.

Figure 23.10(a) and (b) compares the relative stiffness of a member subjected to a moment M_1 when the far end is fixed and when it is simply supported. In part (a) the conjugate beam for the beam with the far end fixed is loaded with the M/EI diagram, and the reactions are determined. The slope at the left end is represented by θ_1 and equals the shear when the conjugate beam is loaded with the M/EI diagram. Its value is $M_1l/4EI$.

The conjugate beam for the simple end-supported beam is loaded with the M/EI diagram and its reactions are determined in Fig. 23.10(b). The moment M_1 is found to cause a slope of $\theta_1 = M_1l/3EI$; therefore the slope caused by the moment M_1 when the far end is fixed is only three-fouths as large ($M_1l/4EI \div M_1l/3EI = \frac{3}{4}$) when the far end is simply supported. The beam simply supported at the far end is only three-fourths as stiff as the one that is fixed. If the stiffness factors for end spans that are simply supported are modified by

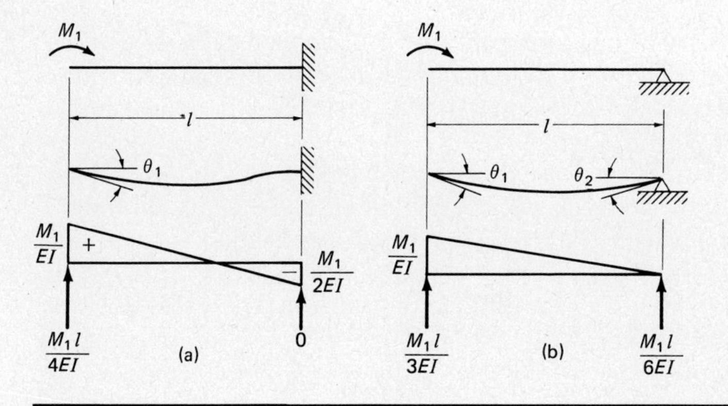

Figure 23.10

three-fourths the simple end is initially balanced to zero, no carry-overs are made to the end afterward, and the same results will be obtained. The stiffness modification is used for the beam of Example 23.3 in Example 23.4.

Example 23.4

Determine the end moments of the structure shown in Fig. 23.11 by using the simple end-stiffness modification for the left end.

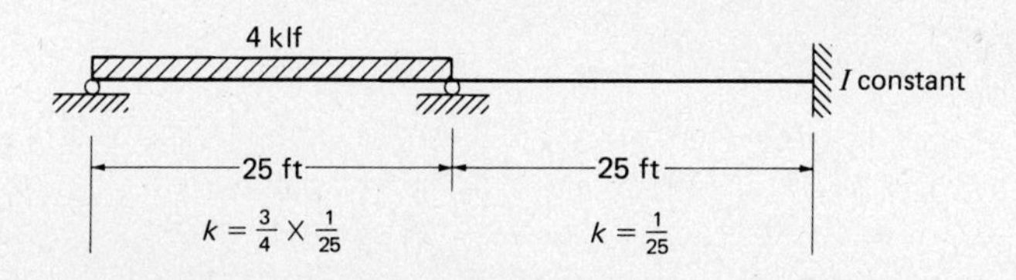

Figure 23.11

SOLUTION

	0.43	0.57	
−208.3	+208.3		
+208.3	+104.2		
0	−134.4	−178.1	−89.1
	+178.1	−178.1	−89.1

23.9 SHEAR AND MOMENT DIAGRAMS

The drawing of shear and moment diagrams is an excellent way to check the final moments computed by moment distribution and to obtain an overall picture of the stress condition in the structure.

Before preparing the diagrams, it is necessary to consider a few points relating the shear and moment diagram sign convention to the one used for moment distribution. The usual conventions for drawing the diagrams will be used (tension in bottom fibers of beam is positive moment and upward shear to left is positive shear).

The relationship between the signs of the moments for the two conventions is shown with the beams of Fig. 23.12. Part (a) of the figure illustrates a fixed-end beam for which the result of moment distribution is a negative moment. The clockwise moment bends the beam as shown, causing tension in the top fibers or a negative moment for the shear and moment diagram convention. In Fig. 23.12(b) the result of moment distribution is a positive moment, but again the top beam fibers are in tension, indicating a negative moment for the moment diagram.

An interior simple support is represented by Fig. 23.12(c) and (d). In part (c) moment distribution gives a negative moment to the right and a positive moment to the left, which causes tension in the top fibers. Part (d) shows the effect of moments of opposite character at the same support considered in part (c).

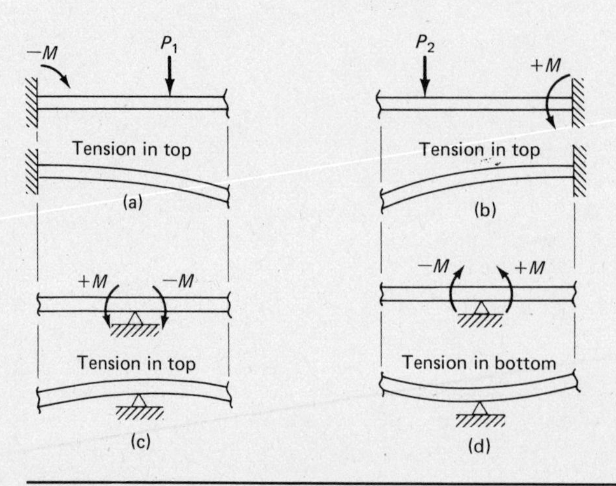

Figure 23.12

To draw the diagrams for a vertical member, the right side is often considered to be the bottom side. Moments are distributed for a continuous beam in Example 23.5, and shear and moment diagrams are drawn.

The reactions shown in the solution of this problem were obtained by computing the reactions as though each span were simply supported and by adding to them the reactions due to the moments at the beam supports.

Example 23.5

Distribute moments and draw shear and moment diagrams for the structure shown in Fig. 23.13.

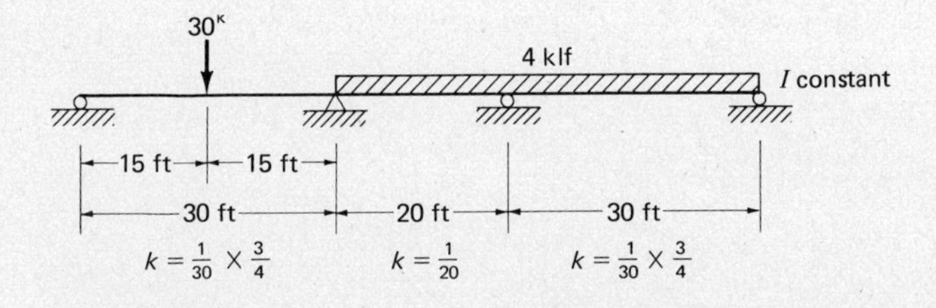

Figure 23.13

SOLUTION

	A	B	B	C	C	D
DF		0.33	0.67	0.67	0.33	
	−112.5	+112.5	−133.3	+133.3	−300.0	+300.0
	+112.5	+ 56.2			−150.0	−300.0
	0		+105.6	+211.1	+105.6	0
		− 47.0	− 94.0	− 47.0		
			+ 15.7	+ 31.3	+ 15.7	
		− 5.2	− 10.5	− 5.2		
			+ 1.7	+ 3.5	+ 1.7	
		− 0.6	− 1.1	− 0.6		
			+ 0.2	+ 0.4	+ 0.2	
		− 0.1	− 0.1			
		+115.8	−115.8	+326.8	−326.8	
Reactions	↑ 15.00	15.00 ↑	↑ 40.00	40.00 ↑	↑ 60.00	60.00 ↑
	↓ 3.86	3.86 ↑	↓ 10.55	10.55 ↑	↑ 10.89	10.89 ↓
	↑ 11.14	18.86 ↑	↑ 29.45	50.55 ↑	↑ 70.89	49.11 ↑

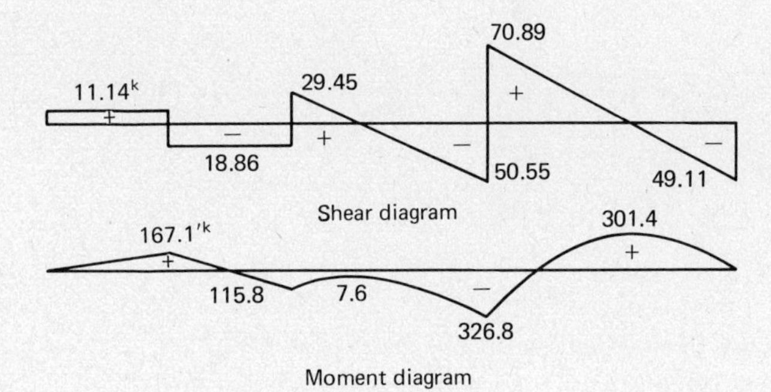

PROBLEMS

For Probs. 23.1 to 23.20 analyze the structures by the moment-distribution method and draw shear and moment diagrams.

Problem 23.1

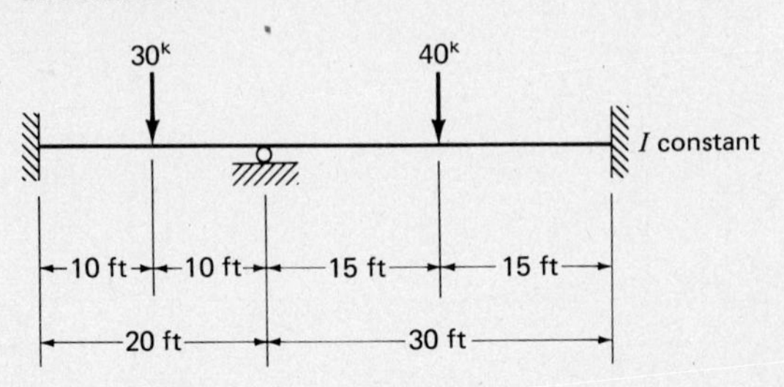

Problem 23.2 (*Ans.* $M_A = M_D = -190'^k$, $M_B = M_C = -100'^k$)

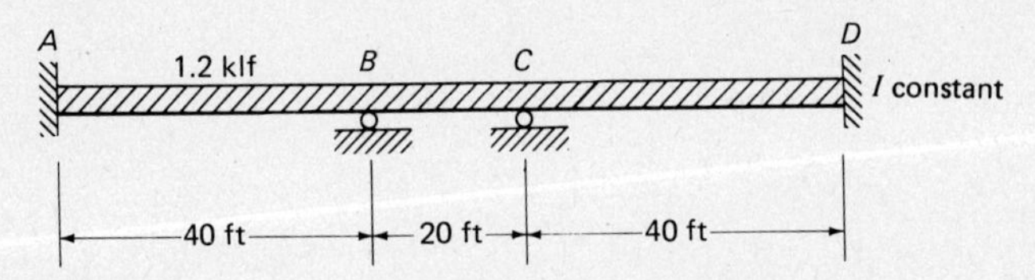

Problem 23.3

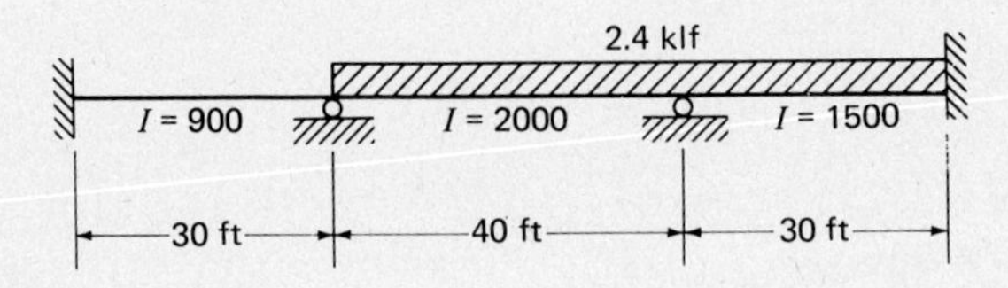

Problem 23.4 (*Ans.* $M_A = -64.3'^k$, $M_B = -141.5'^k$, $M_C = -247.5'^k$, $M_D = -142.9'^k$)

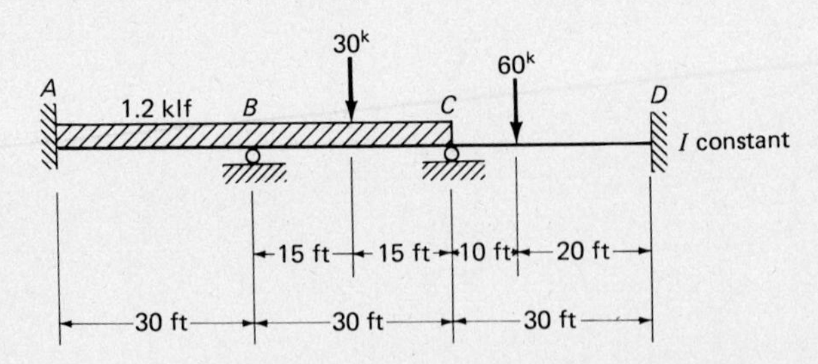

23.5. Rework Prob. 23.3 if a 40-kip load is added at the centerline of each of the right two spans.

Problem 23.6 (*Ans.* $M_A = M_D - 234.6$ kN · m, $M_B = M_C = -138.2$ kN · m)

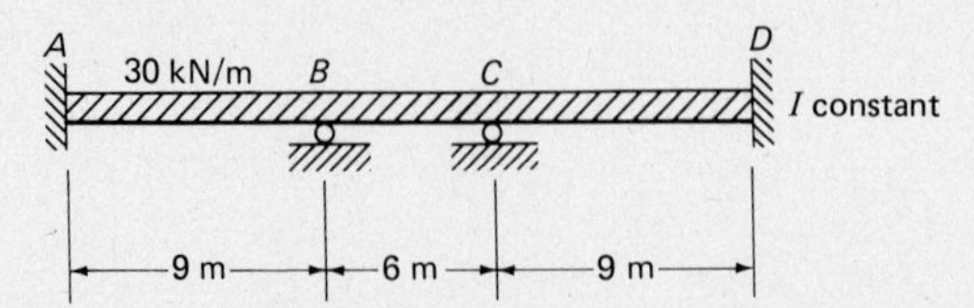

Problem 23.7

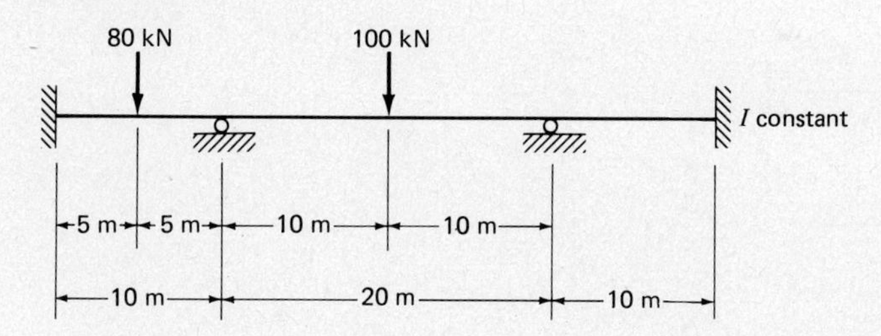

Problem 23.8 (*Ans.* $M_A = -131.0'^k$, $M_B = -98.0'^k$, $M_C = -125.3'^k$, $M_D = -207.3'^k$)

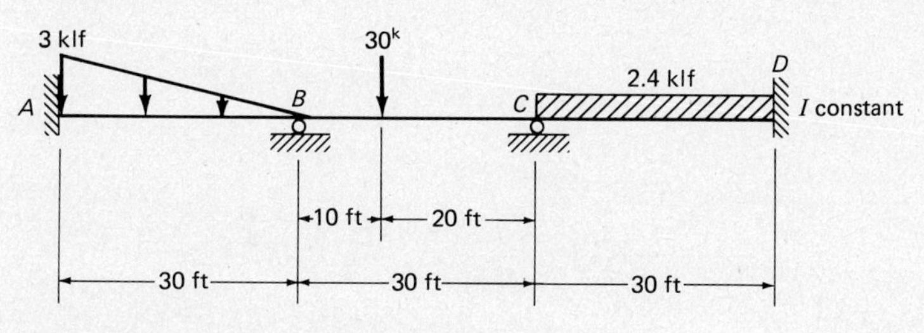

Problem 23.9

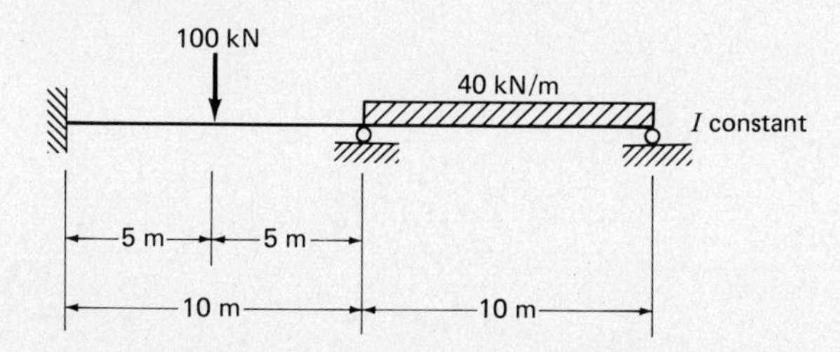

Problem 23.10 (*Ans.* $M_A = -158.3'^k$, $M_B = -163.3'^k$, $M_C = -148.3'^k$, $M_D = 0$)

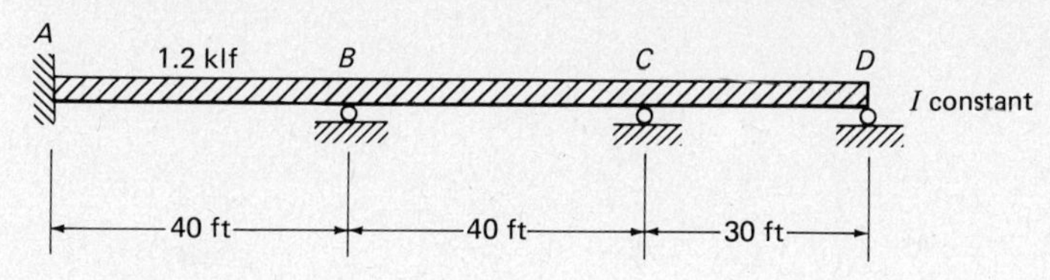

Problem. 23.11

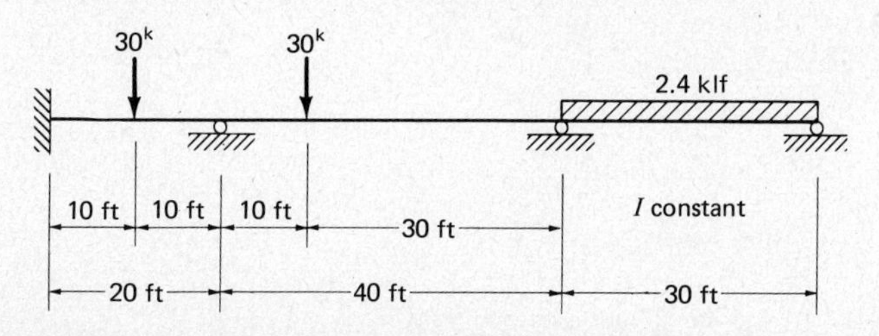

Problem 23.12 (*Ans.* $M_B = -158.7'^k$, $M_C = -177.4'^k$, $V_A = 9.71^k \uparrow$)

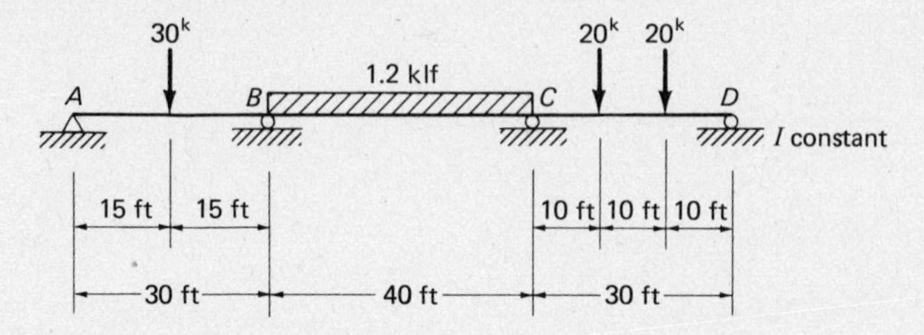

23.13 Rework Prob. 22.8

Problem 23.14 ($M_B = -500.9'^k$, $M_C = -428.2'^k$, $M_C = -265.9'^k$)

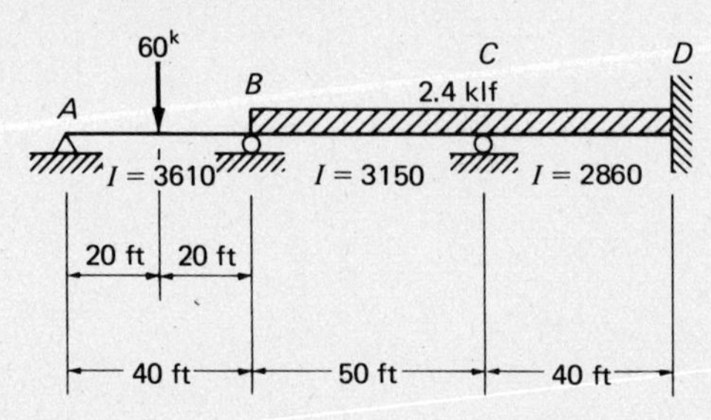

Problem 23.15

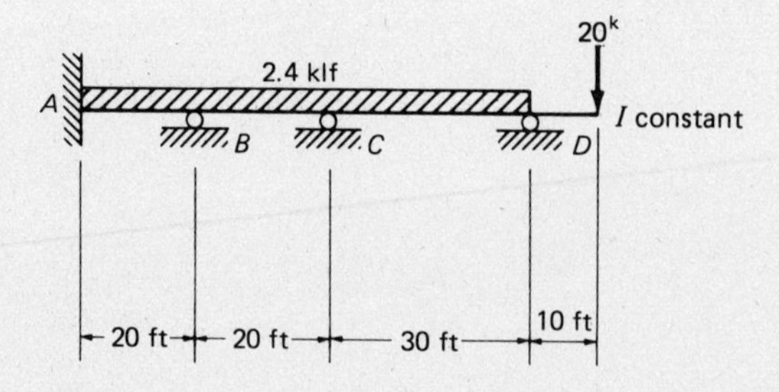

23.16. Rework Prob. 22.9. (*Ans.* $M_A = -31.1'^k$, $M_B = -57.7'^k$, $M_C = -200'^k$)

Problem 23.17

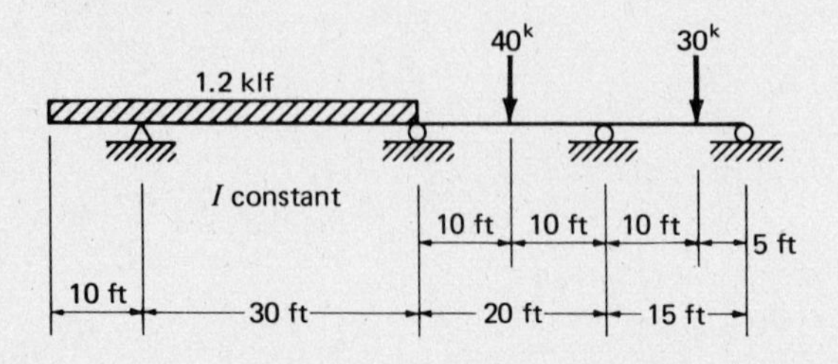

Problem 23.18 (*Ans.* $M_B = -215.3'^k$, $M_C = -156.3'^k$)

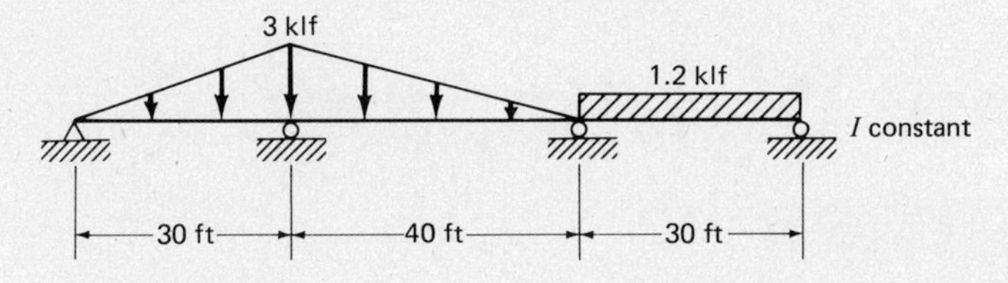

Problem 23.19

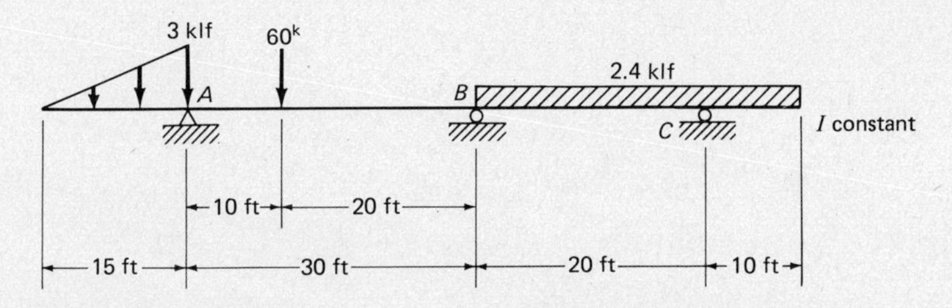

Problem 23.20 (*Ans.* $M_A = -114'^k$, $M_B = -192.0'^k$, $M_C = -212.0'^k$)

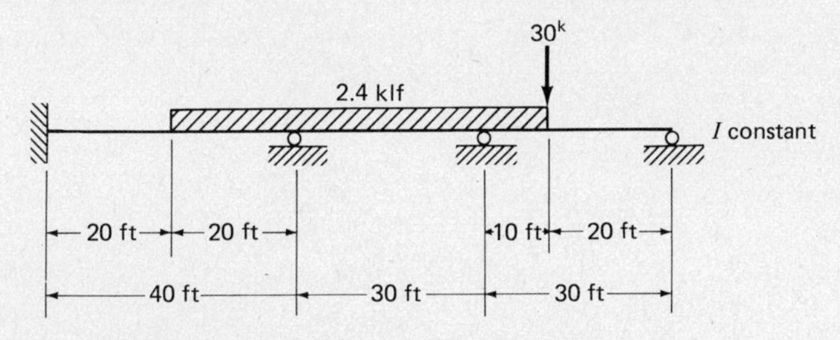

REFERENCES

[1] Hardy Cross. "Analysis of Continuous Frames by Distributing Fixed-End Moments," *Proceedings of the American Society of Civil Engineers* (May 1930): 919–928.

24 Moment Distribution for Frames

24.1 FRAMES WITH SIDESWAY PREVENTED

Moment distribution is handled in the usual manner for frames if sidesway or movement laterally is prevented. Analysis of frames without sidesway is illustrated by Examples 24.1 and 24.2. Where sidesway is possible, however, it must be taken into account, because the movements or deflections cause twisting and affect the moments in the rotated members.

As the structures being analyzed become more complex, it is necessary to use some method of recording the figures so they will not interfere or run into each other. In this chapter a system is used for frames whereby the moments are recorded below beams on their left ends and above them on their right ends. For columns the same system is used, the right sides being considered the bottom sides.

Example 24.1

Determine the end moments for the frame shown in Fig. 24.1.

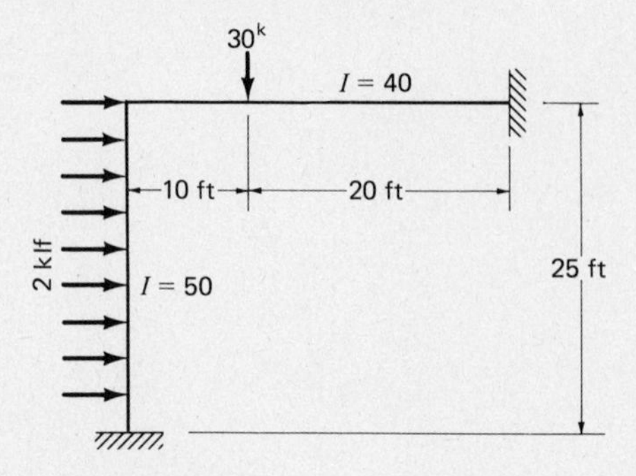

Figure 24.1

SOLUTION

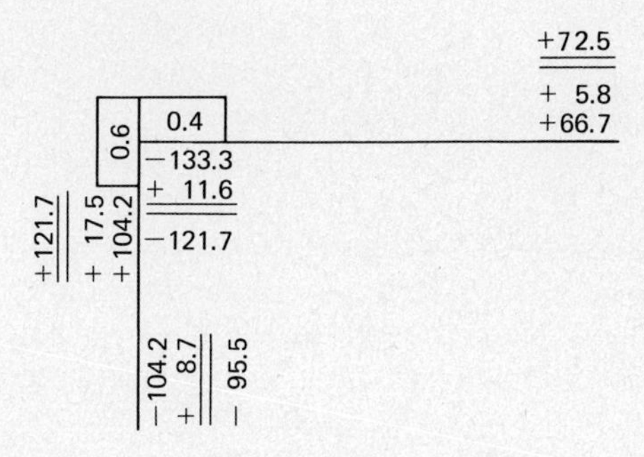

Example 24.2

Compute the end moments of the structure shown in Fig. 24.2.

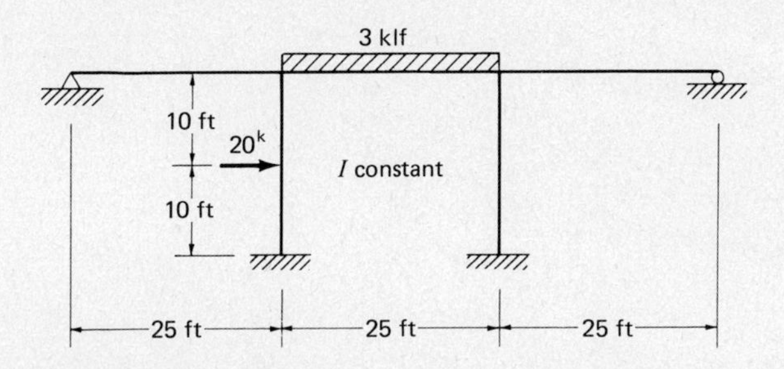

Figure 24.2

SOLUTION

+33.7
+ 0.7
+ 6.5
+26.5
0.25
0.33
0.42
−156.2
+ 35.4
− 26.0
+ 8.7
− 3.0
+ 1.0
−140.1
+106.4
+ 1.3
+ 10.8
+ 44.3
+ 50.0
−50.0
+22.2
+ 5.4
−22.4

+118.8
− 1.5
+ 4.4
− 5.9
+ 17.7
− 52.1
+156.2
0.33
0.25
0.42
−39.0
− 4.4
− 1.1
−44.5
−74.3
− 1.8
− 7.4
−65.1
−32.6
− 3.7
− 0.9
−37.2

24.2 FRAMES WITH SIDESWAY

Structural frames, similar to the one shown in Fig. 24.3, are usually so constructed that they may possibly sway to one side or the other under load. The frame in this figure is symmetrical, but it will tend to sway because the load P is not centered. Analysis of the frame by the usual procedure, illustrated in the figure, gives inconsistent results.

The values of the horizontal components at the supports are computed, and from the results obtained it can be seen that the sum of the horizontal forces on the structure is not equal to zero. The sum of the forces to the right is 0.89 kip more than the sum of the forces to the left. If the structure were

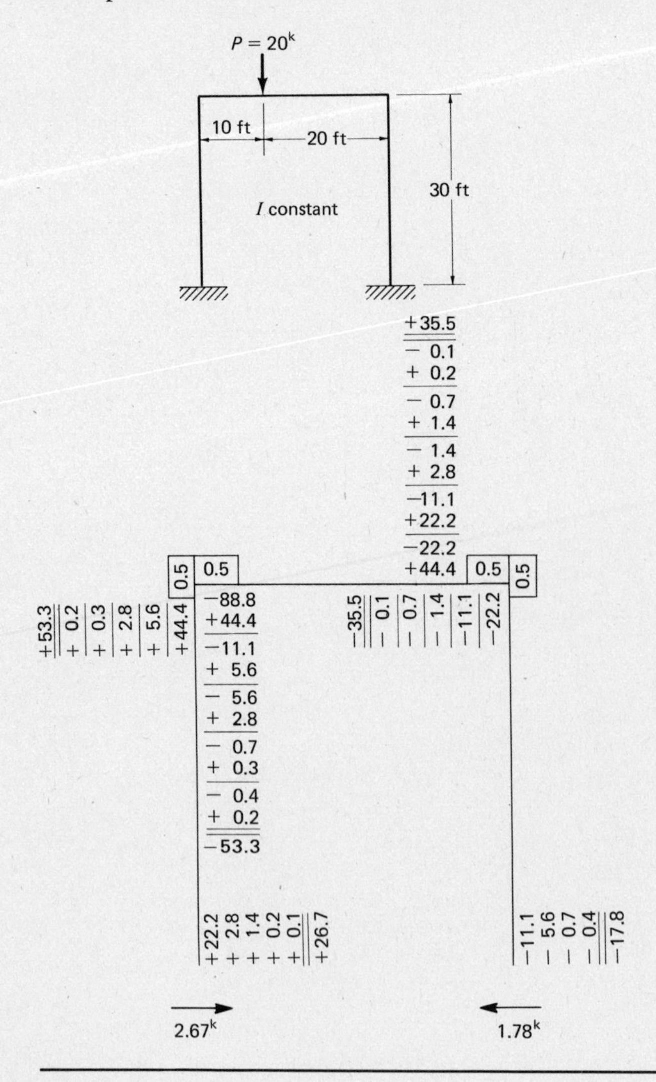

Figure 24.3

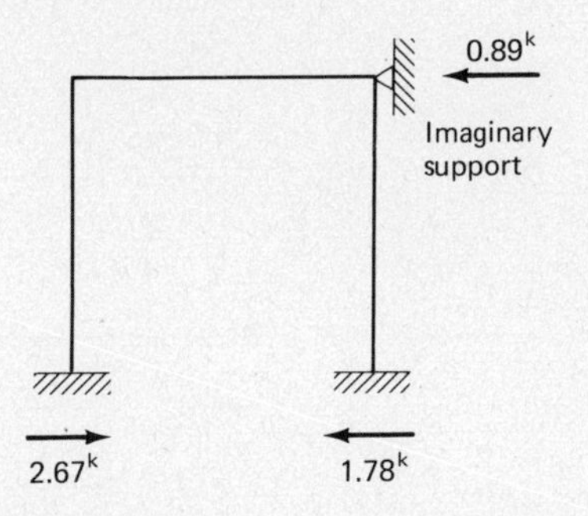

Figure 24.4

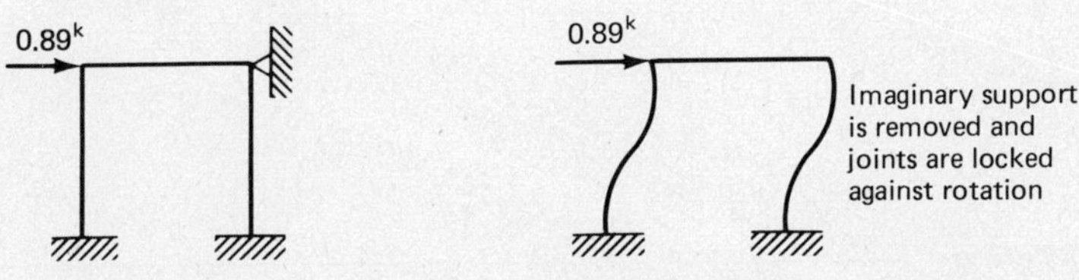

Figure 24.5

subjected to an unbalanced force system such as this one, it would not be in equilibrium.

The usual analysis does not yield consistent results, because the structure actually sways or deflects to one side, and the resulting deflections affect the moments. One possible solution is to compute the deflections caused by applying a force of 0.89 kip acting to the right at the top of the bent. The moments caused could be obtained for the computed deflections and added to the originally distributed fixed-end moments, but the method is so difficult as to be impractical.

A much more convenient method is to assume the existence of an imaginary support that prevents the structure from swaying, as shown in Fig. 24.4. The fixed-end moments are distributed, and the force the imaginary support must supply to hold the frame in place is computed. For the frame of Fig. 24.3 the fictitious support must supply 0.89 kip pushing to the left.

The support is imaginary and if removed will allow the frame to sway to the right. As the structure sways to the right the joints are assumed to be locked against rotation. The ends of the columns rotate in a clockwise direction and produce clockwise or negative moments at the joints (Fig. 24.5). Assumed values of these sidesway moments can be placed in the columns. The necessary relations between these moments are discussed in the section to follow.

24.3 SIDESWAY MOMENTS

Should a frame have columns all of the same length and the same moments of inertia, the assumed sidesway moments will be the same for each column. Should, however, the columns have differing lengths and/or moments of

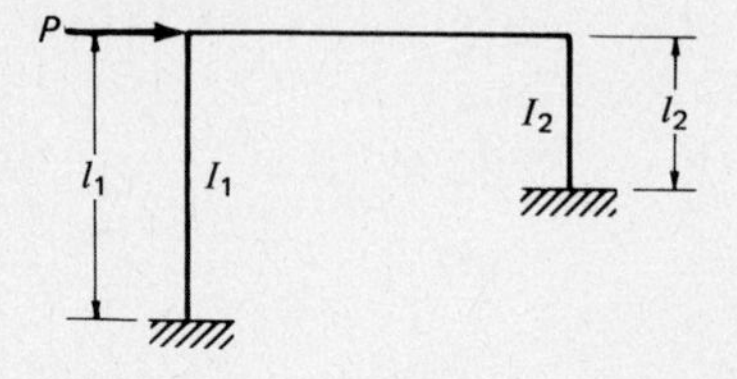

Figure 24.6

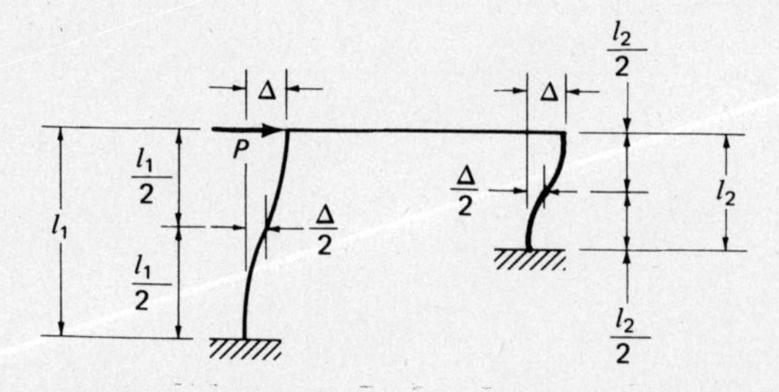

Figure 24.7

inertia, this will not be the case. It will be proved in the following paragraphs that the assumed sidesway moments should vary from column to column in proportion to their I/l^2 values.

If the frame of Fig. 24.6 is pushed laterally an amount Δ by the load P, it will take the deflected shape of Fig. 24.7.

Theoretically both columns will become perfect S curves, the beam being considered rigid and unbending. At their middepths the deflection for both columns will equal $\Delta/2$. Middepths of the columns may be considered points of contraflexure; and the bottom halves may be handled as though they were cantilever beams. The expression for deflection of a cantilever beam with a concentrated load at its end is $Pl^3/3EI$. Since the deflections are the same for both columns, the following expressions may be written:

$$\frac{\Delta}{2} = \frac{(P_1)(l_1/2)^3}{3EI_1} = \frac{P_1 l_1^3}{24EI_1}$$

$$\frac{\Delta}{2} = \frac{(P_2)(l_2/2)^3}{3EI_2} = \frac{P_2 l_2^3}{24EI_2}$$

By solving these deflection expressions for P_1 and P_2, the forces pushing on the cantilevers, we have

$$P_1 = \frac{12EI_1\Delta}{l_1^3}$$

$$P_2 = \frac{12EI_2\Delta}{l_2^3}$$

The moments caused by the two forces at the ends of their respective cantilevers are equal to the force times the cantilever length. These moments are written and the values of P_1 and P_2 are substituted in them.

$$M_1 = P_1 \frac{l_1}{2}$$

$$= \left(\frac{12EI_1\Delta}{l_1^3}\right)\frac{l_1}{2} = \frac{6EI_1\Delta}{l_1^2}$$

$$M_2 = P_2 \frac{l_2}{2}$$

$$= \left(\frac{12EI_2\Delta}{l_2^3}\right)\frac{l_2}{2} = \frac{6EI_2\Delta}{l_2^2}$$

From these expressions a proportion may be written between the moments as follows:

$$\frac{M_1}{M_2} = \frac{6EI_1\Delta/l_1^2}{6EI_2\Delta/l_2^2}$$

$$= \frac{I_1/l_1^2}{I_2/l_2^2}$$

This relationship must be used for assuming sidesway moments for the columns of a frame. Any convenient moments may be assumed, but they must be in proportion to each other, as their I/l^2 values. Should their I and l values be equal, the assumed moments will be equal.

The analysis of the frame of Fig. 24.3 is completed in Fig. 24.8 with the sidesway method. First, I/l^2 values are computed for each of the columns (equal in this case) in part (a) of the figure. Then, in part (b) sidesway moments in proportion to the I/l^2 values are assumed and distributed throughout the frame; after which the horizontal reactions at the column bases are calculated. The assumed sidesway moments of $-10'^{k}$ each are found to produce horizontal reactions to the left totaling 0.92 kip. Only 0.89 kip was needed, and if 0.89/0.92 times the values of the distributed moments are added to the originally distributed fixed-end moments, the final moments will be obtained. The results are shown in Fig. 24.8(c). The horizontal reactions at the column bases are also calculated and shown.

Examples 24.3 to 24.5 present the solutions for additional sidesway problems. It will be noted in Example 24.4 that the column bases are pinned. Such a situation does not alter the method of solution. For the balancing of fixed-end moments or for balancing assumed sidesway moments, the bases are balanced to their correct zero values as was done in continuous beams. The author used the three-quarter factor for computing the column stiffnesses for this example.

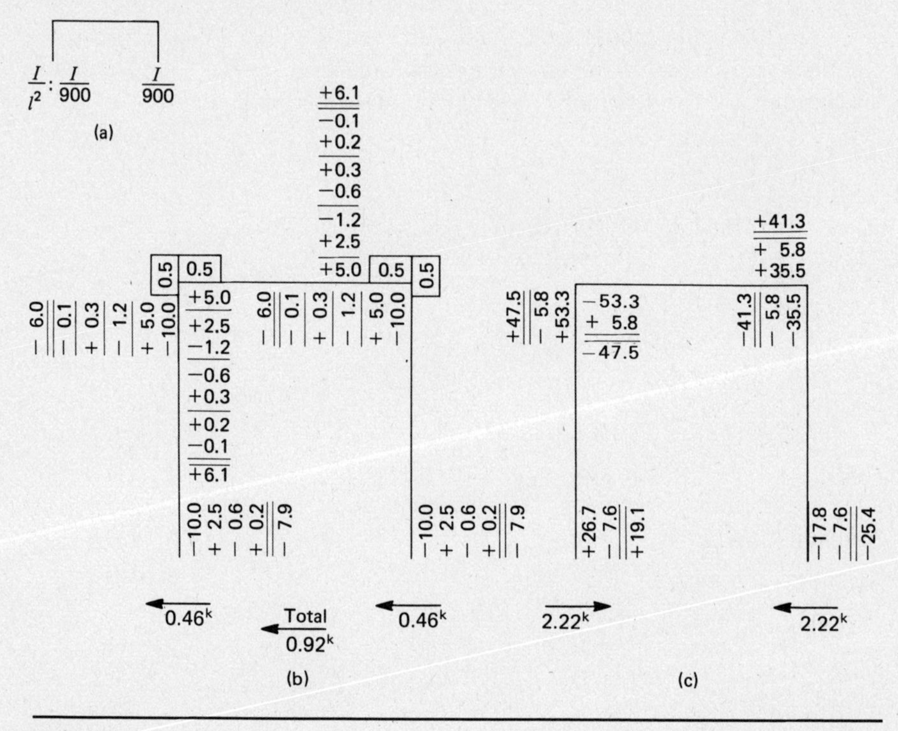

Figure 24.8

Example 24.3

Find all moments in the structure shown in Fig. 24.9. Use the sidesway method.

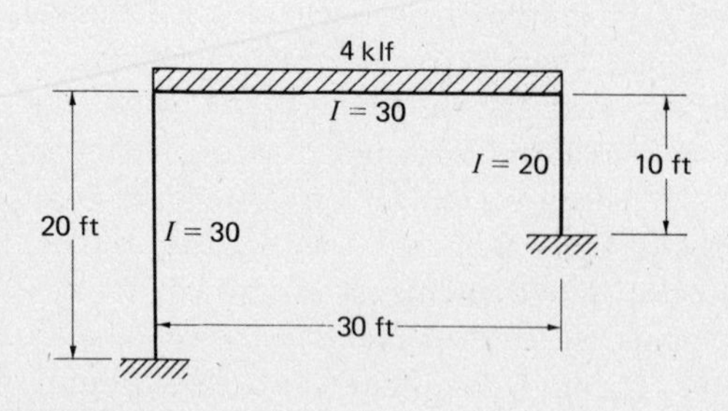

Figure 24.9

SOLUTION

Distribute fixed-end moments and compute horizontal reactions at column bases.

The imaginary support must supply a force of 20.93 kip to the right. Removal of the support will allow the frame to sway to the left; therefore

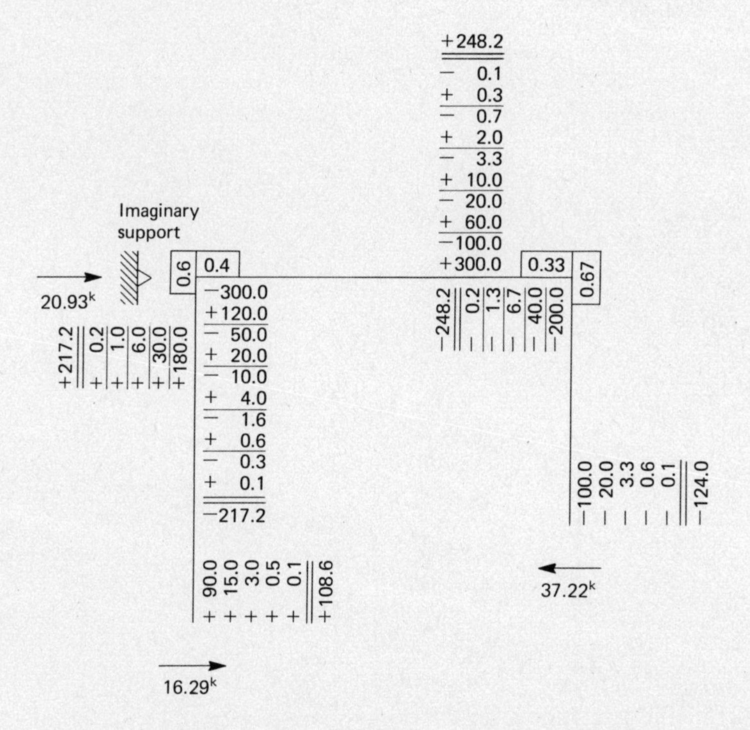

counterclockwise or positive moments are assumed in the columns in proportion to their I/l^2 values.

$$\frac{M_1}{M_2} = \frac{I_1/l_1^2}{I_2/l_2^2}$$

$$= \frac{30/20^2}{20/10^2} = \frac{3}{8}$$

−29.0
+ 0.1
− 0.2
− 0.9
+ 2.7
+ 2.0
− 6.0
−26.7
0.6
0.4
0.33
0.67
−12.0
−13.3
+ 5.3
+ 1.0
− 0.4
− 0.5
+ 0.2
−19.7
+19.7
+ 0.3
− 0.6
+ 8.0
−18.0
+30.0
+29.0
+ 0.1
− 1.8
+ 4.0
−53.3
+80.0
+80.0
−26.6
+ 2.0
− 0.9
+54.5
+30.0
− 9.0
+ 4.0
− 0.3
+ 0.1
+24.8
8.35ᵏ
2.23ᵏ
Total
10.58ᵏ

The assumed moments develop total reactions to the right of 10.58 kip; however, 20.93 kip was needed, and 20.93/10.58 times these moments is added to the results obtained initially by distributing the fixed-end moments.

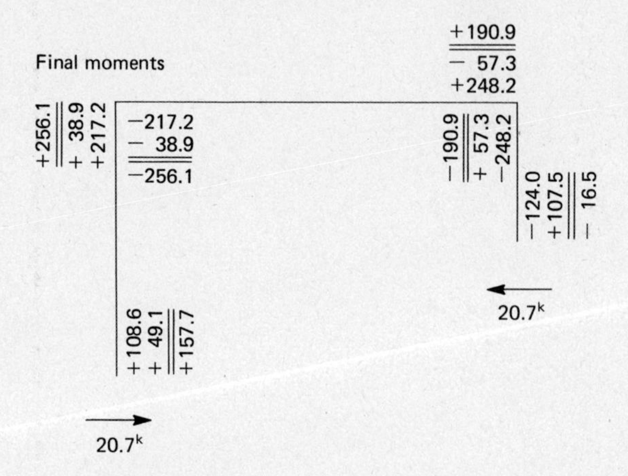

Example 24.4

Determine the final end moments for the frame shown in Fig. 24.10. Circled values are moments of inertia.

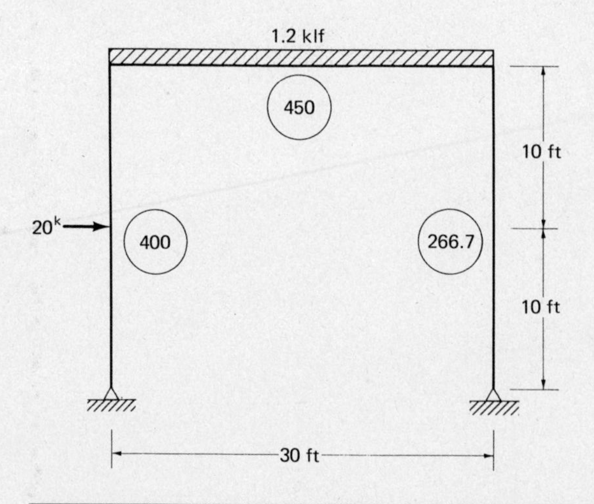

Figure 24.10

SOLUTION

Distribute fixed-end moments and compute horizontal reactions at column bases.

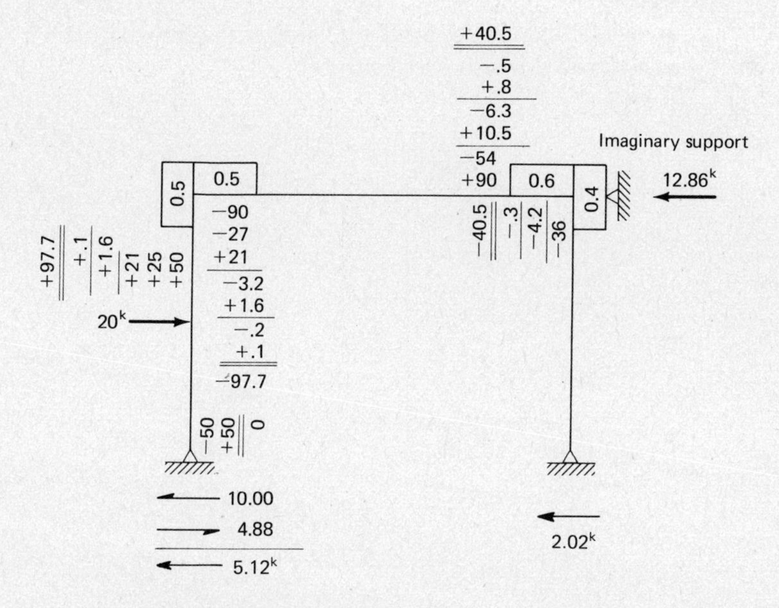

The structure sways to the right and assumed sidesway moments are negative and in the following proportion:

$$M_1 : M_2$$

$$\frac{400}{(20)^2} : \frac{266.7}{(20)^2}$$

$$1.00 : 0.667.$$

Distribute assumed sidesway moments.

+24.4
+1.0
−1.6
+12.5
+12.5
0.5
0.5
0.6
0.4
+25
+6.2
−3.1
+.5
−.2
+28.4
−28.3
−.2
−3.1
+25
+50
−100
−24.4
+.6
+8.4
+33.3
−66.7
−100
+100
0
−66.7
+66.7
0
1.42k
1.22k
Total
2.64k

Final moments equal distributed fixed-end moments plus 12.86/2.64 times the distributed assumed sidesway moments.

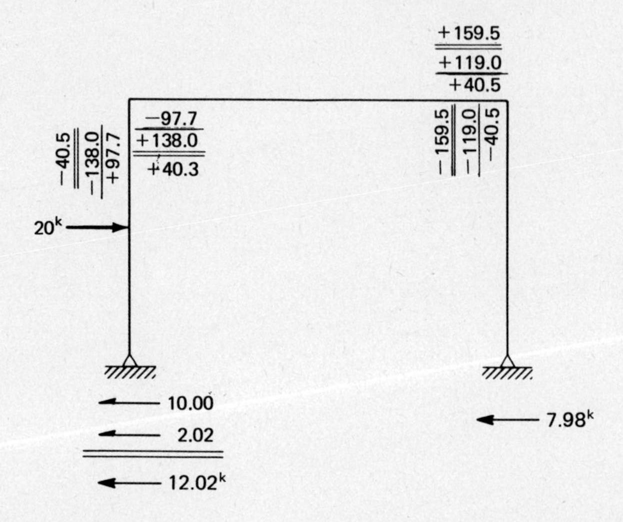

Example 24.5

Compute the final end moments for the frame shown in Fig. 24.11.

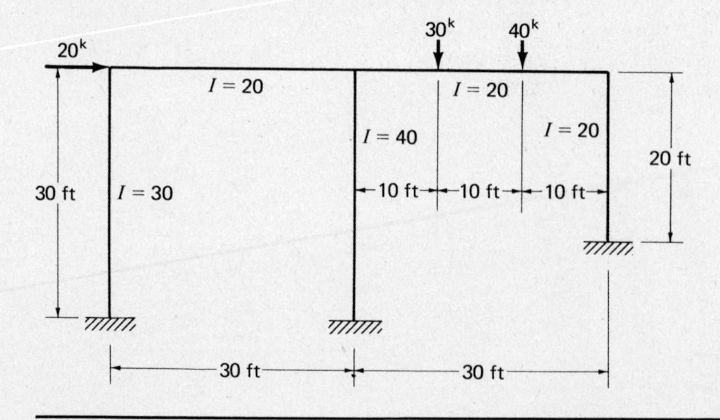

Figure 24.11

SOLUTION

Distribute fixed-end moments and compute horizontal reactions at column bases.

The structure sways to the right; therefore negative moments are assumed in the columns in proportion to the I/l^2 values.

$$M_1 : M_2 : M_3$$

$$\frac{30}{30^2} : \frac{40}{30^2} : \frac{20}{20^2}$$

$$30 : 40 : 45$$

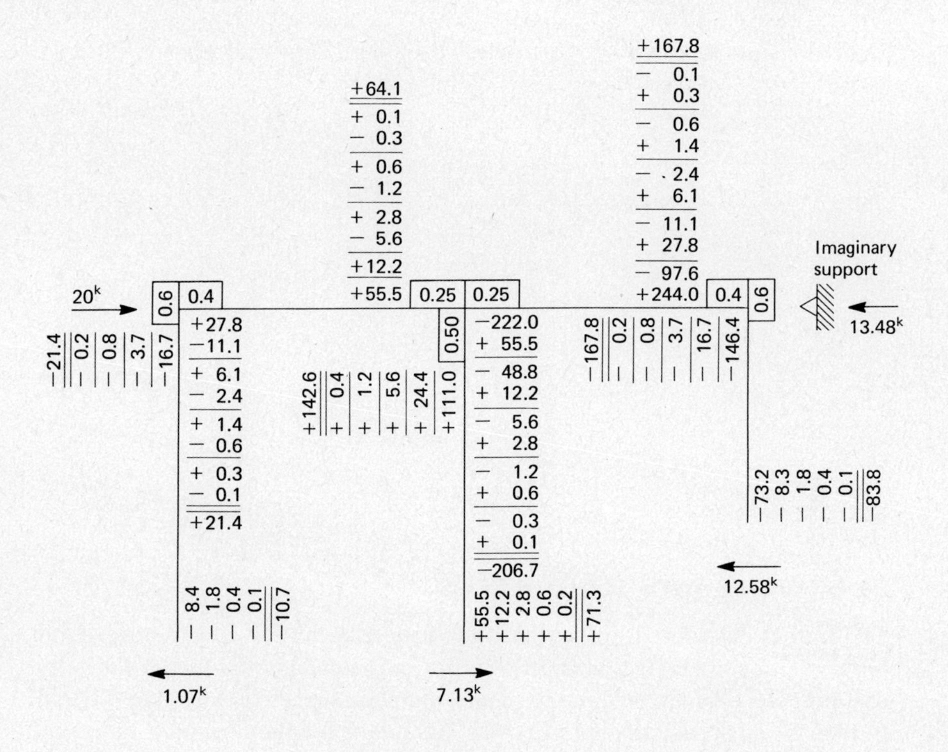

Distribute assumed sidesway moments.

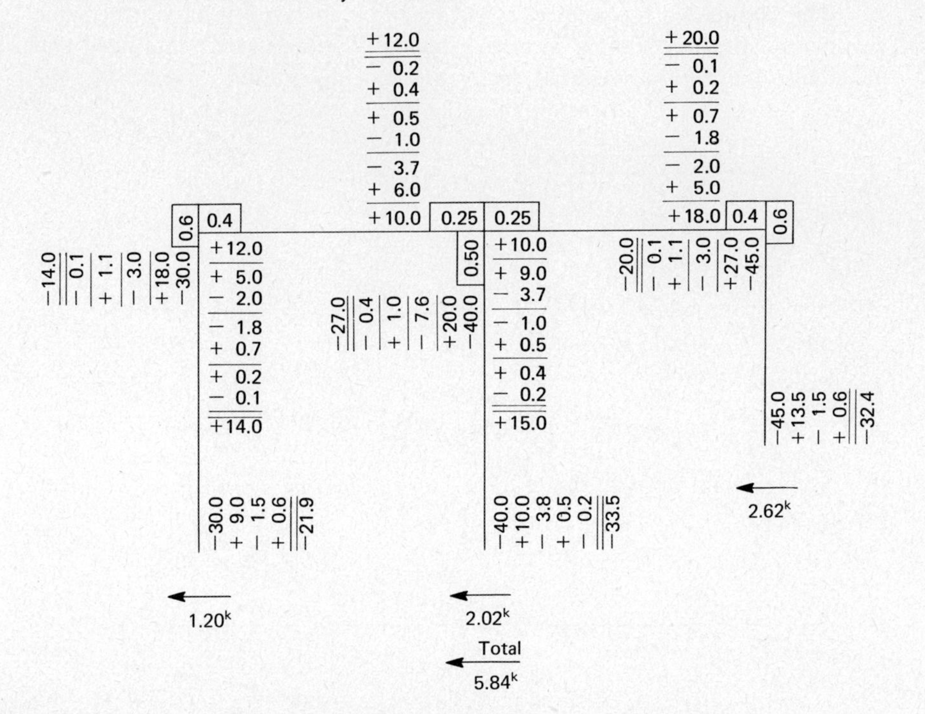

Final moments:

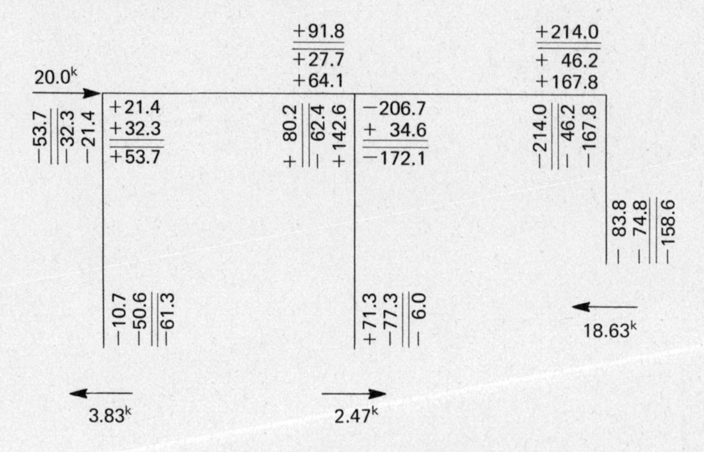

24.4 FRAMES WITH SLOPING LEGS

The frames considered up to this point have been made up of vertical and horizontal members. It was proved earlier in this section that when sidesway occurs in such frames, it causes fixed-end moments in the columns proportional to their $6EI\Delta/l^2$ values. (As $6E\Delta$ was a constant for these frames, moments were assumed proportional to their I/l^2 values.) Furthermore, lateral swaying did not produce fixed-end moments in the beams.

The sloping-leg frame of Fig. 24.12 can be analyzed in much the same manner as the vertical-leg frames previously considered. The fixed-end moments due to the external loads are calculated and distributed; the

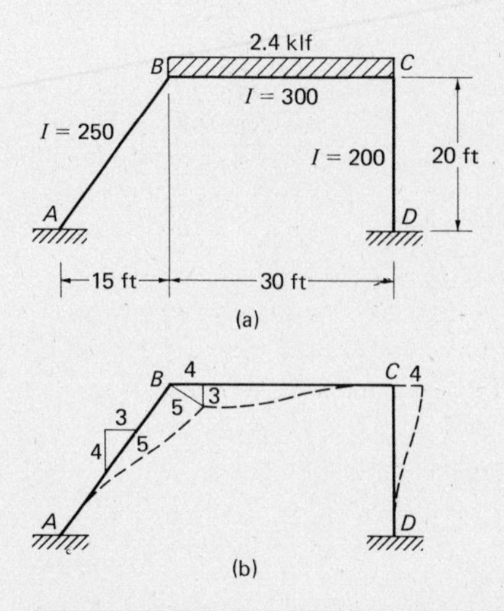

Figure 24.12

I-91 bridge, Lyndon, Vermont. (Courtesy of the Vermont Agency of Transportation.)

horizontal reactions are computed (a little more difficult than before); and the horizontal force needed at the imaginary support is determined.

As the Δ values will usually be unequal in the various members, sidesway moments are assumed in proportion to the $I\Delta/l^2$ values. These assumed moments are distributed, the horizontal reactions computed, and the necessary moments needed for balancing are calculated and superimposed on the distributed fixed-end moments.

For this discussion the frame of Fig. 24.12(a) is considered and it is assumed to sway to the right as shown in part (b) of the figure. It can be seen that lateral movement of the frame will cause Δ's and thus moments in the beam as well as in the columns. To determine the $I\Delta/l^2$ values for each member it is necessary to determine the relative Δ values in each member.

As the frame sways to the right, joint B moves in an arc about joint A and joint C moves in an arc about joint D. As these arcs are very short, they are considered to be straight lines perpendicular to the respective columns. Rather than developing complex trigonometric formulas for relative Δ's, as was done in Chap. 22, deformation triangles are drawn as shown in the figure.

Column AB has a slope of four vertically, three horizontally, or five inclined. If the relative movement of joint B perpendicular to AB is assumed to be five then its vertical movement will be three and its horizontal movement will be four.

Joint C moves perpendicular to CD, which is in a horizontal direction in this frame. If the change in length of member BC is neglected, joint C must

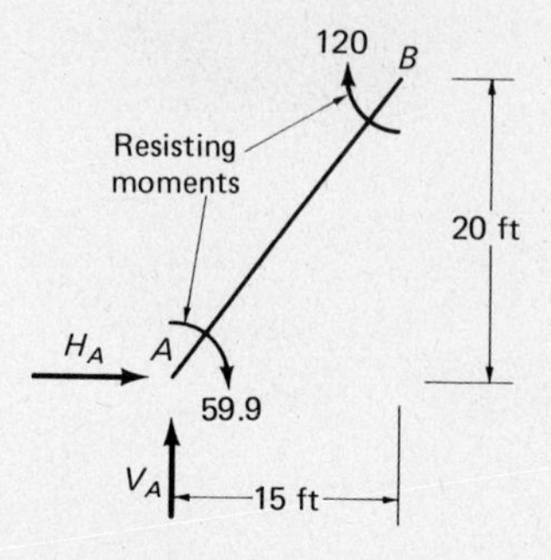

Figure 24.13

move horizontally the same distance as joint B or a distance of four. The relative Δ values are now available as follows: $\Delta_{AB} = 5$, clockwise; $\Delta_{BC} = 3$, counterclockwise; and $\Delta_{CD} = 4$, clockwise. A clockwise rotation produces a counterclockwise or negative resisting moment. These values are given at the end of this paragraph together with the $I\Delta/l^2$ values.

Relative Δ values	*Relative sidesway moments $I\Delta/l^2$*
$\Delta_{AB} = -5$	$M_{AB} = \dfrac{(250)(-5)}{(25)^2} = -2$
$\Delta_{BC} = +3$	$M_{BC} = \dfrac{(300)(+3)}{(30)^2} = +1$
$\Delta_{CD} = -4$	$M_{CD} = \dfrac{(200)(-4)}{(20)^2} = -2$

Example 24.6 illustrates the analysis of the frame of Fig. 24.12. The reader will need to study the method used carefully to determine the horizontal reactions at the column bases. Each column in previous sidesway examples has been considered to be a free body, and moments taken at its top were used to determine the horizontal reaction at its base.

The left column AB of the frame of Example 24.6 is considered as a free body after the fixed-end moments are distributed and shown in Fig. 24.13. Taking moments at B to determine H_A the following equation results:

$$+59.9 + 120 + 15V_A - 20H_A = 0$$

It is noted that it is necessary to compute V_A before the equation can be solved for H_A. The author finds it convenient to remove the beam as a free body and to compute the vertical reaction applied at each end by the columns. Once the value at B is determined, the sum of the vertical forces on column AB can be equated to zero and V_A can be determined. The same procedure is followed after the assumed sidesway moments are distributed.

Example 24.6

Determine the final end moments for the frame shown in Fig. 24.12.

SOLUTION

Distribute fixed-end moments.

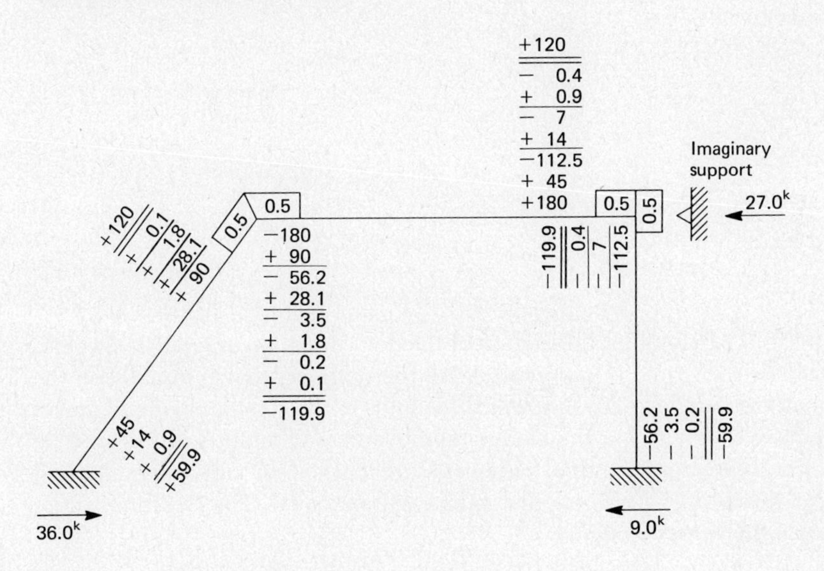

Distribute assumed sidesway moments.

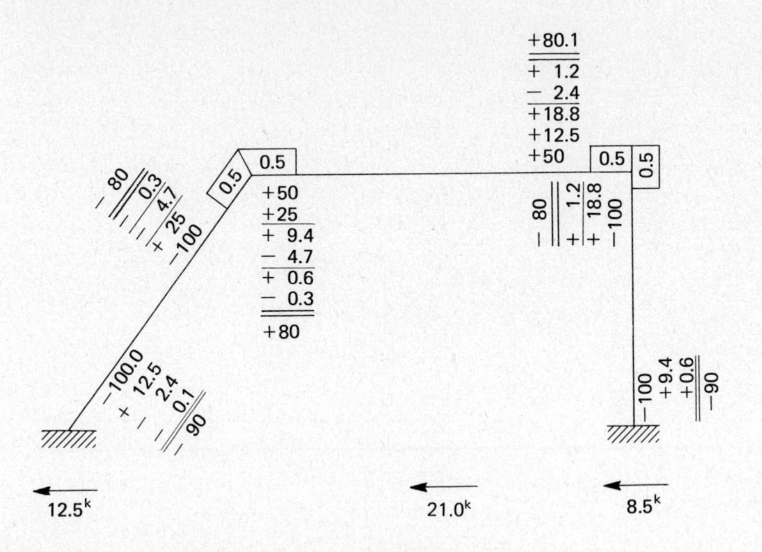

Final moments:

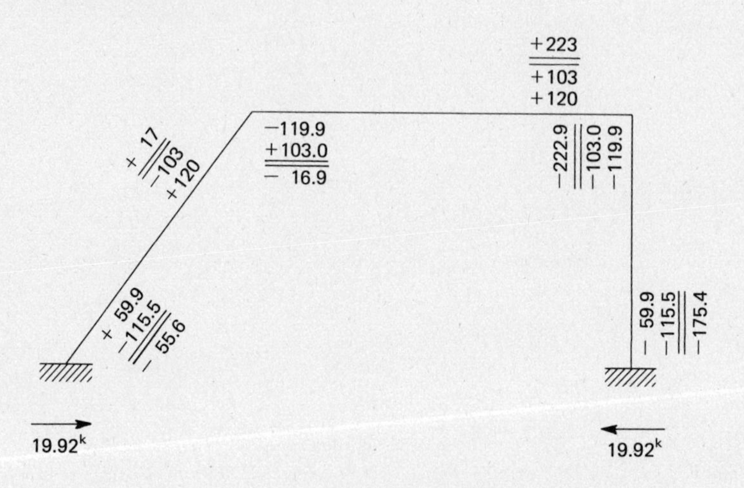

In Fig. 24.14 another illustration of the determination of relative Δ values for a sloping leg frame is presented. With the numbers shown initially on the deformation triangles, joint *B* moves three units to the right and joint *C* moves four units to the right, but the horizontal movement of joints *B* and *C* must be equal. For this reason the initial values at *C* are marked through and multiplied by three-quarters so the horizontal values will be equal. The resulting values are shown following the figure.

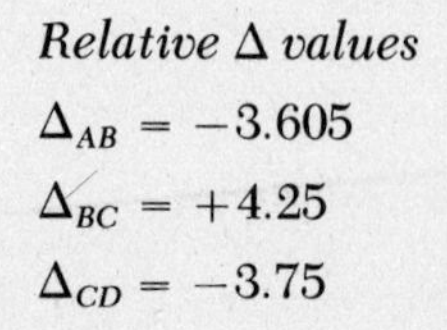

Relative Δ *values*

$\Delta_{AB} = -3.605$

$\Delta_{BC} = +4.25$

$\Delta_{CD} = -3.75$

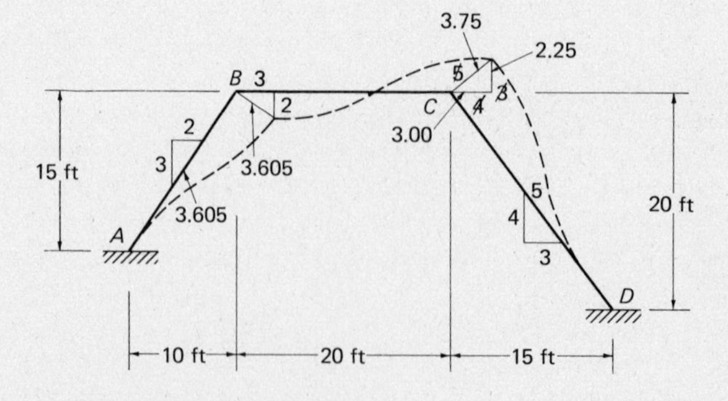

Figure 24.14

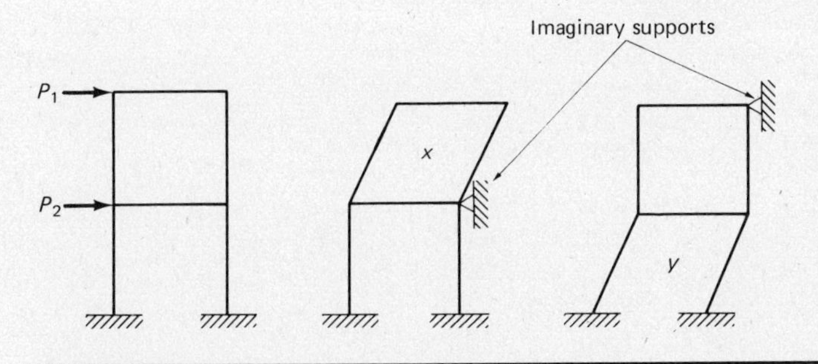

Figure 24.15

24.5 MULTISTORY FRAMES

There are two possible ways in which the frame of Fig. 24.15 may sway. The loads P_1 and P_2 will obviously cause both floors of the structure to sway to the right, but it is not known how much of the swaying is going to occur in the top floor (x condition) or how much will occur in the bottom floor (y condition). There are two sidesway conditions that need to be considered.

To analyze the frame by the usual sidesway procedure would involve (1) an assumption of moments in the top floor for the x condition and the distribution of the moments throughout the frame and (2) an assumption of moments in the lower floor for the y condition and the distribution of those moments throughout the frame. One equation could be written for the top floor by equating x times the horizontal forces caused by the x moments plus y times the horizontal forces caused by the y moments to the actual total shear on the floor, P_1. A similar equation could be written for the lower floor by equating the horizontal forces caused by the assumed moments to the shear on that floor, $P_1 + P_2$. Simultaneous solution of the two equations would yield the values of x and y. The final moments in the frame equal x times the x distributed moments plus y times the y distributed moments.

The sidesway method is not difficult to apply for a two-story frame, but for multistory frames it becomes unwieldy because each additional floor introduces another sidesway condition and another simultaneous equation.

Professor C. T. Morris of Ohio State University introduced a much simpler method for handling multistory frames which involves a series of successive corrections [1]. His method is also based on the total horizontal shear along each level of a building. In considering the frame of Fig. 24.16(a), which is being deflected laterally by the loads P_1 and P_2, each column is assumed to take the approximate S shape shown in (b).

At the middepth of the columns there is assumed to be a point of contraflexure. The column may be considered to consist of a pair of cantilever beams, one above the point and one below the point as shown in (c). The moment in each cantilever will equal the shear times $h/2 = Vh/2$, and the total

Office building, 99 Park Avenue, New York City. (Courtesy of the American Institute of Steel Construction, Inc.)

moment top and bottom is equal to $Vh/2 + Vh/2 = Vh$. The total moment in a column must be equal to the shear carried by the column times the column height.

Similarly, on any one level the total moments top and bottom of all the columns will equal the total shear on that level multiplied by the column height. The method consists in initially assuming this total for the column moments on a floor and distributing it between the columns in proportion to their I/l^2 values. The moment taken by each column is divided half to top end

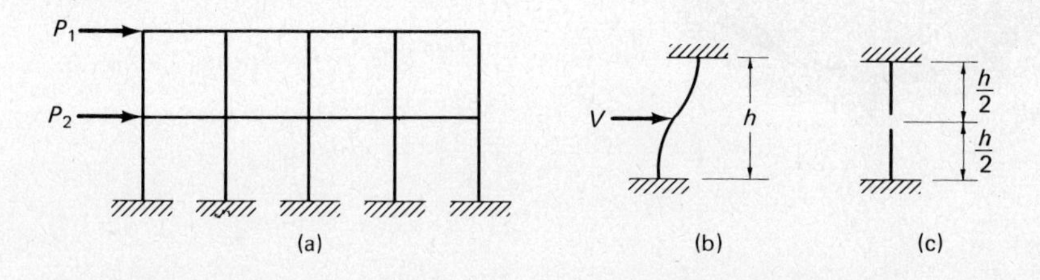

Figure 24.16

and half to bottom end. The joints are balanced, including the fixed-end moments, making no carry-overs until all joints are balanced. The column moments are corrected to their initial and final total, and the joints are balanced again. Successive corrections work exceptionally well for multistory frames, as illustrated by Examples 24.7 and 24.8. The procedure used is as follows:

1. Compute fixed-end moments.
2. Compute total moments in columns (equal to shear on the level times column height) for each level and distribute between the columns in proportion to their I/l^2 values and divide each by half, one-half to top of column and one-half to bottom.
3. Balance all joints throughout the structure, making no carry-overs.
4. Make carry-overs for entire frame.
5. The total of the column moments on each level has been changed and will not equal the shear times the column height. Determine the difference and add or subtract the amount back to the columns in proportion to their I/l^2 values.
6. Steps 3 to 5 are repeated over and over until the amount of the corrections to be made is negligible.

Example 24.7

Determine final moments for the structure in Fig. 24.17. Use the successive-correction method developed by Professor Morris.

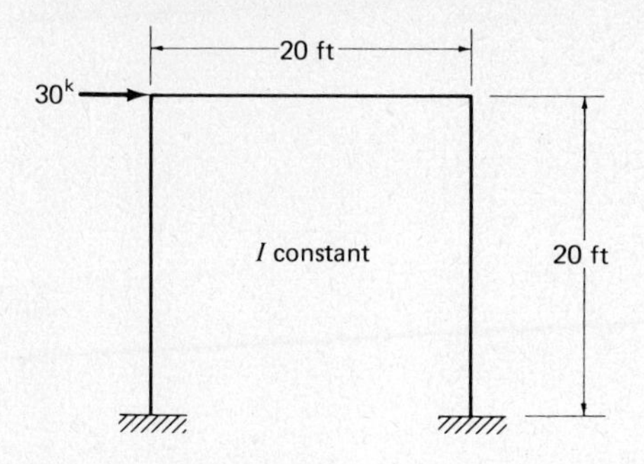

Figure 24.17

SOLUTION

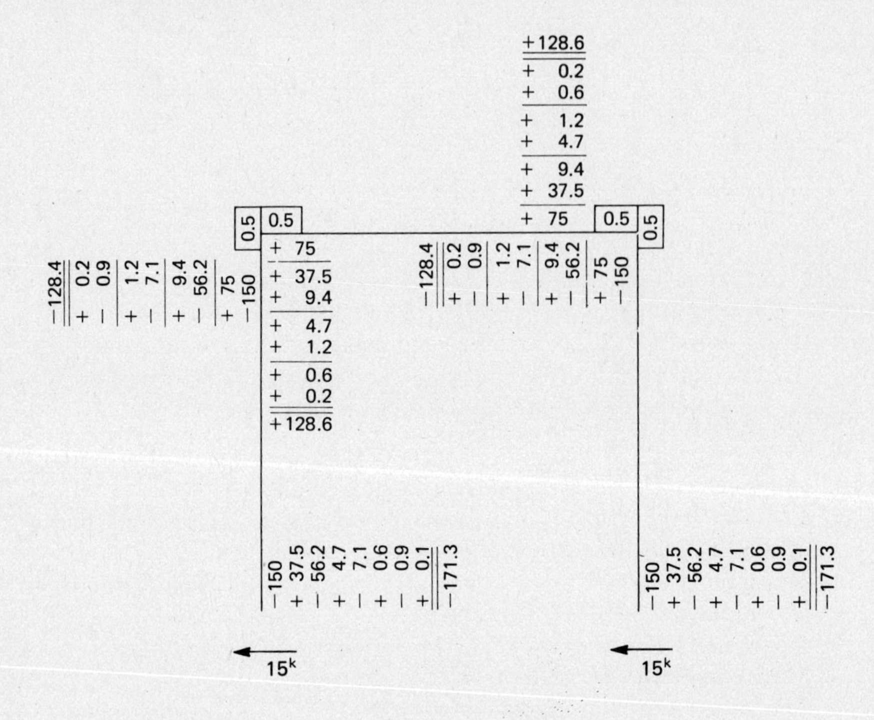

Example 24.8

Determine the final moments for the frame of Example 22.7, reproduced in Fig. 24.18, by successive corrections.

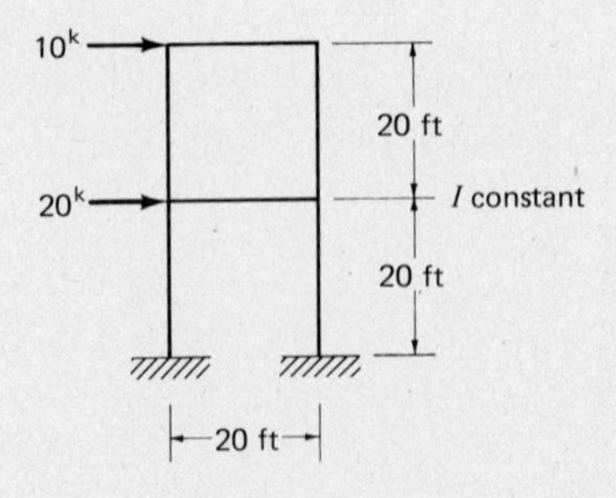

Figure 24.18

SOLUTION

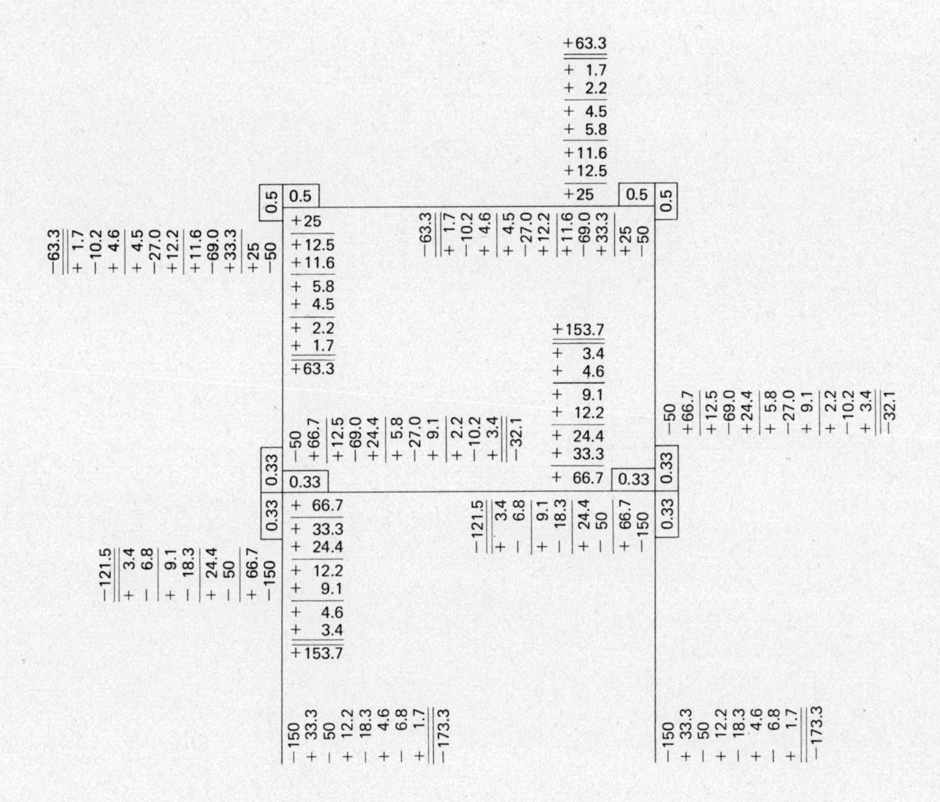

PROBLEMS

Balance moments and calculate horizontal reactions at bases of columns for Probs. 24.1 to 24.6.

Problem 24.1

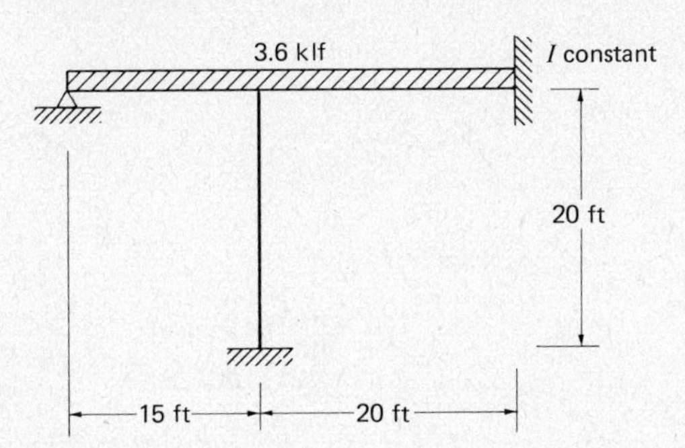

Problem 24.2 (*Ans.* $M_B = 7.2'^k$, $M_{CA} = 142.2'^k$, $M_D = 7.5'^k$, $H_B = 1.08^k\rightarrow$)

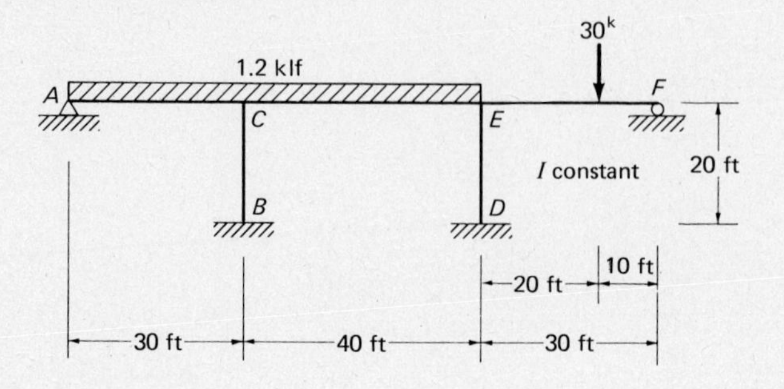

Problem 24.3

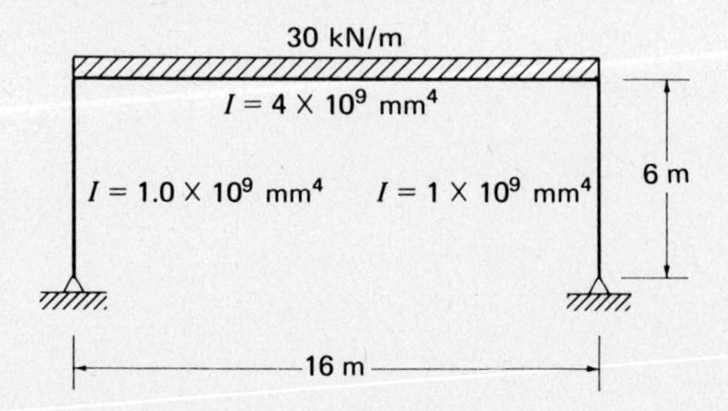

Problem 24.4 (*Ans.* $M_A = M_C = 35.6'^k$, $M_B = M_D = 153.7'^k$, $H_A = 9.10^k\leftarrow$)

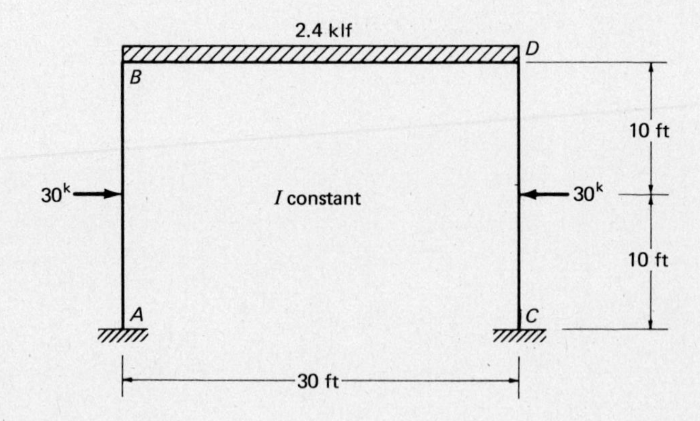

Problem 24.5

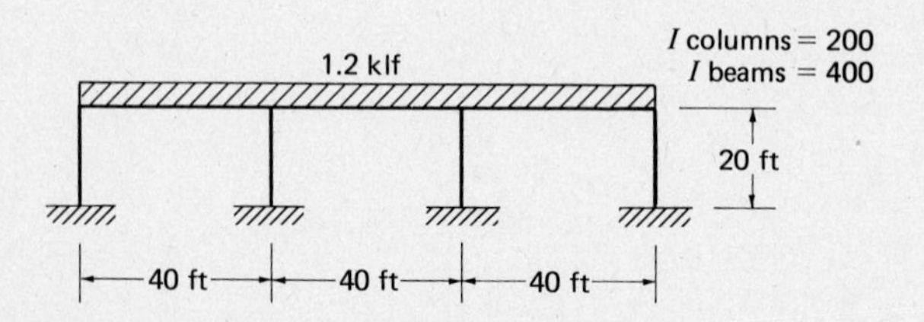

Problem 24.6 (*Ans.* $M_B = 389.7$ kN · m, $M_{DB} = 537.6$ kN · m, $M_E = 36.9$ kN · m)

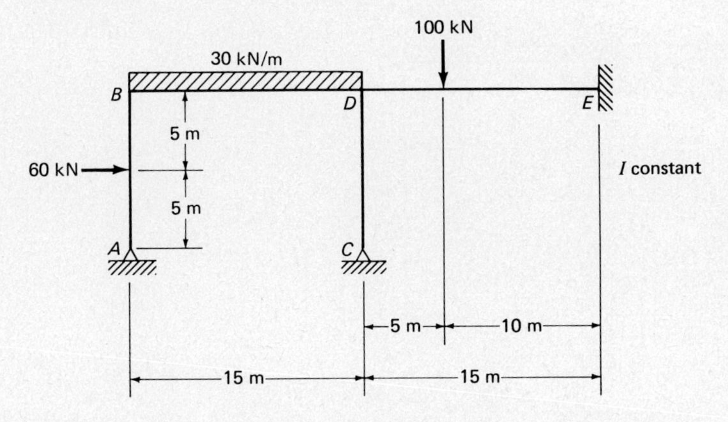

For Probs. 24.7 to 24.20 determine final moments with the sidesway method.

Problem 24.7

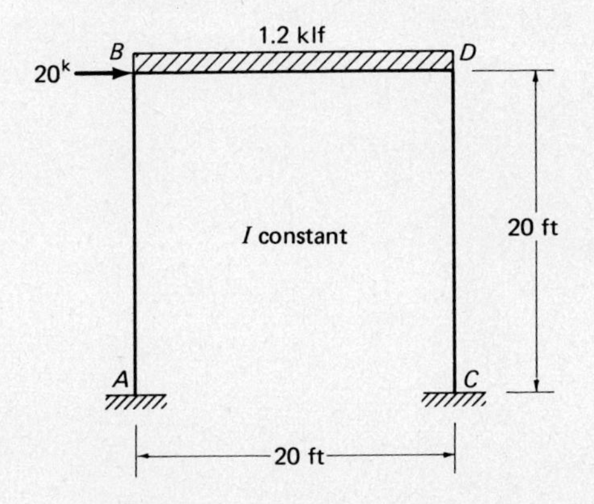

Problem 24.8 (*Ans.* $M_A = 72.7'^k$, $M_B = 70.9'^k$, $M_D = 110.2'^k$)

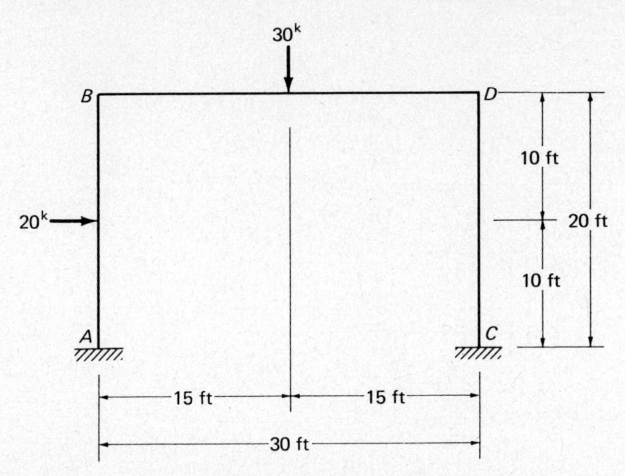

Problem 24.9

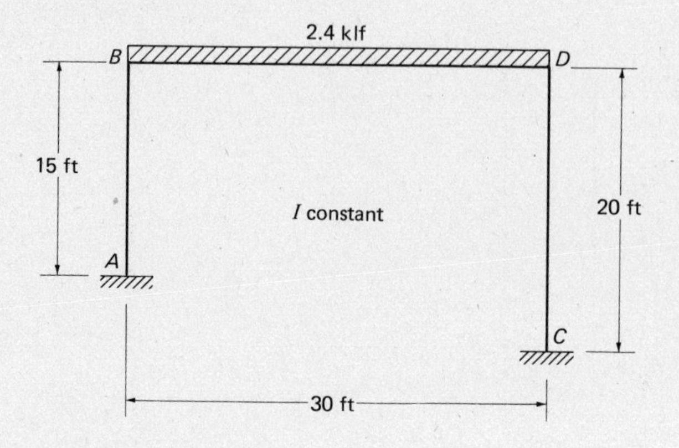

24.10. Rework Prob. 24.8 if column bases are pinned. (*Ans.* $M_B = 10.5'^k$, $M_C = 189.4'^k$)

24.11. Rework Prob. 24.7 if column CD is pinned at its base.

Problem 24.12 (*Ans.* $M_A = 22.9'^k$, $M_B = 305.0'^k$, $M_D = 315.6'^k$)

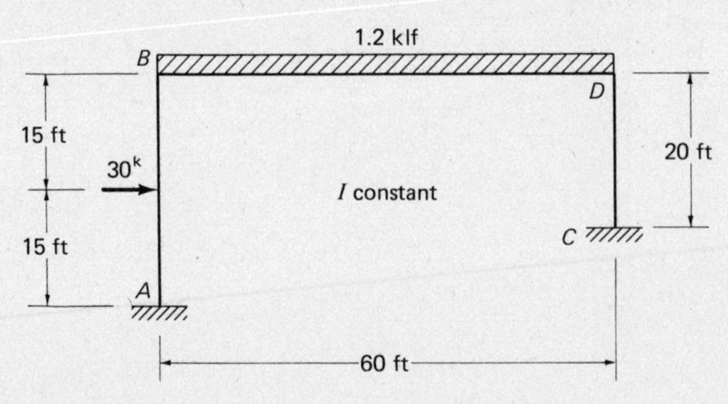

Problem 24.13

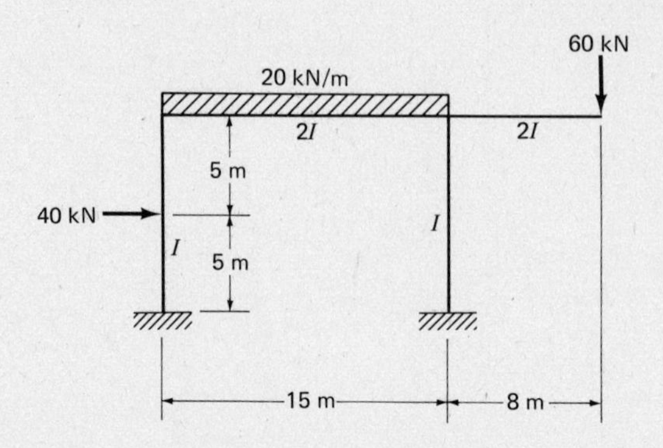

Problem 24.14 (*Ans.* $M_A = 129.5'^k$, $M_B = 149.5'^k$, $M_{DB} = 150.8'^k$, $M_E = 22.9'^k$)

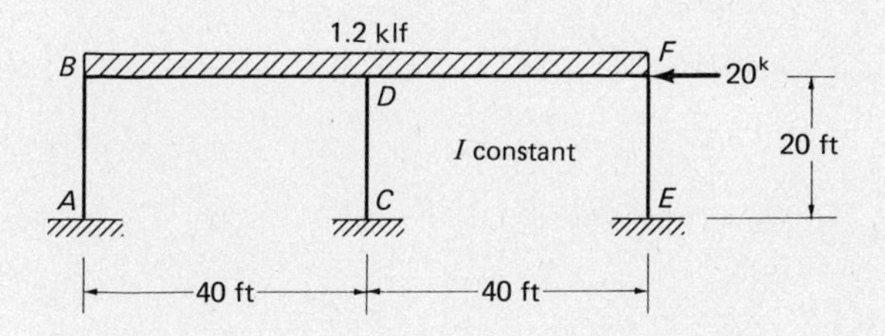

Problem 24.15

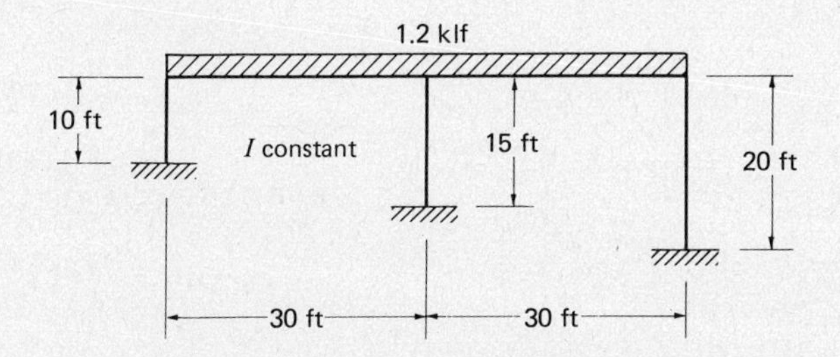

Problem 24.16 (*Ans.* $M_A = 147.8'^k$, $M_{CD} = 83.5'^k$, $M_D = 133.3'^k$, $M_E = 102.8'^k$)

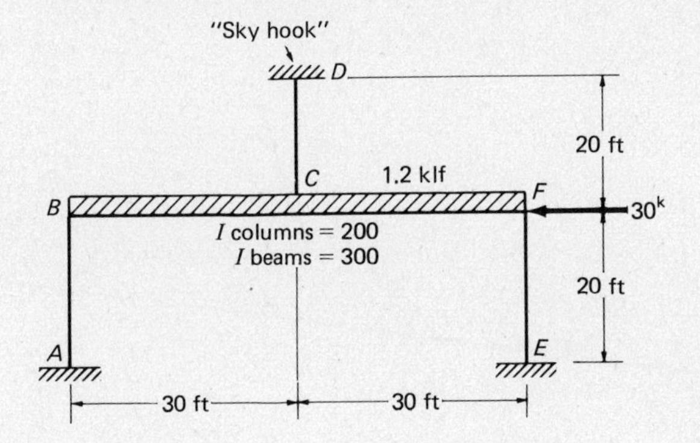

Problem 24.17

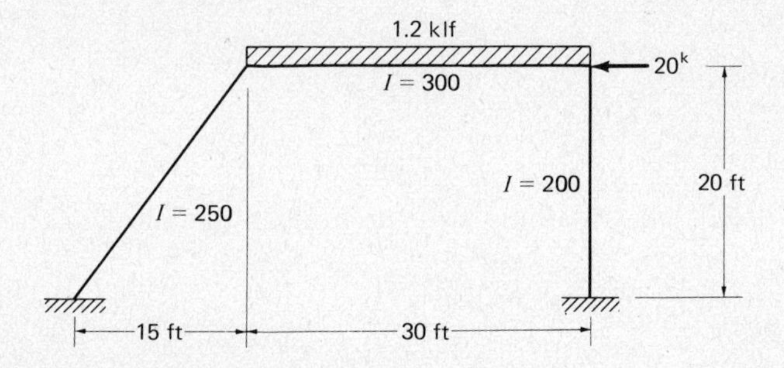

Problem 24.18 (*Ans.* $M_A = 21.1'^k$, $M_B = 14.7'^k$, $M_C = 41.0'^k$, $M_D = 54.4'^k$)

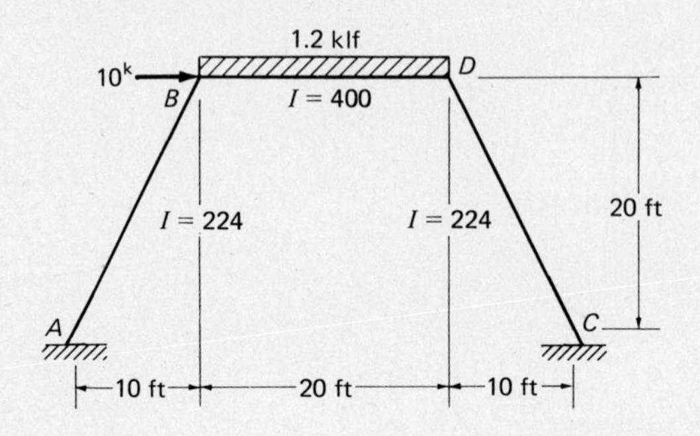

Problem 24.19

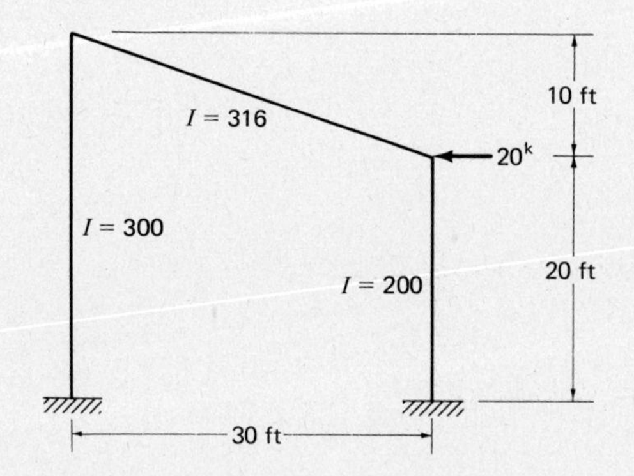

Problem 24.20 (*Ans.* $M_A = 18.9'^k$, $M_B = 21.1'^k$, $M_C = 24.4'^k$, $M_D = 23.9'^k$)

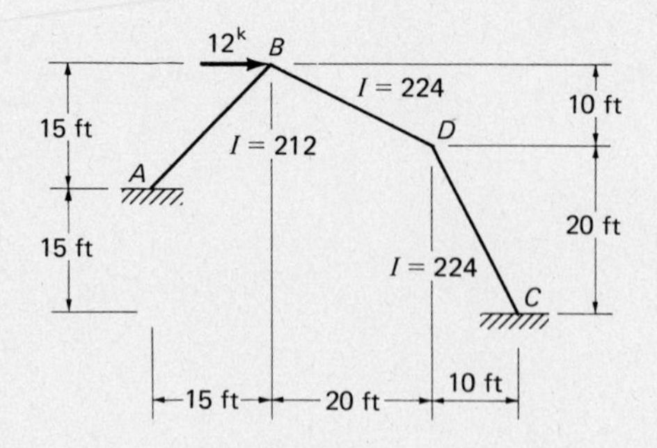

For Probs. 24.21 to 24.27 analyze these structures using the successive-correction method.

24.21. Rework Prob. 24.7.

Problem 24.22 (*Ans.* $M_A = 1.5'^k$, $M_B = 7.6'^k$, $M_{DF} = 106.3'^k$, $M_F = 67.5'^k$)

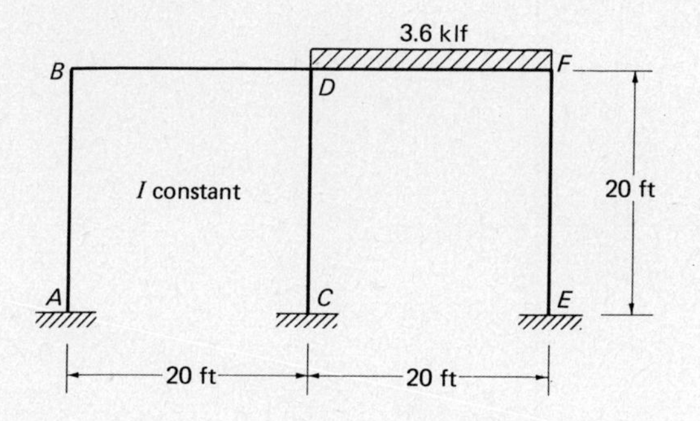

Problem 24.23

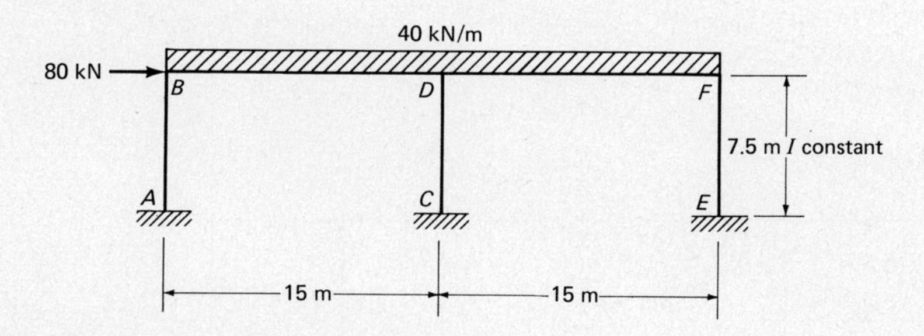

Problem 24.24 (*Ans.* $M_{BA} = 151'^k$, $M_{BE} = 222'^k$, $M_{CF} = 96'^k$, $M_{ED} = 94'^k$)

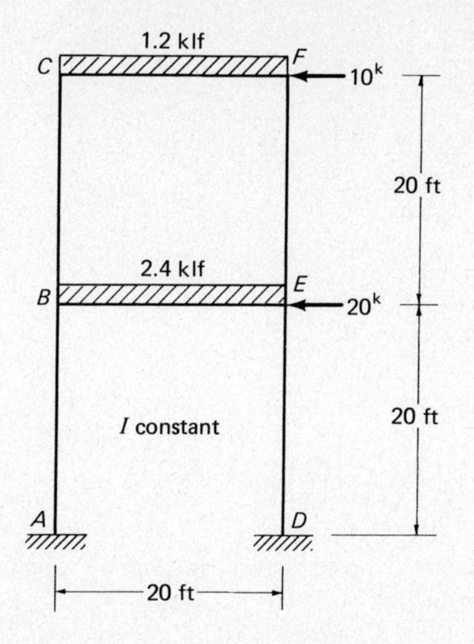

Problem 24.25

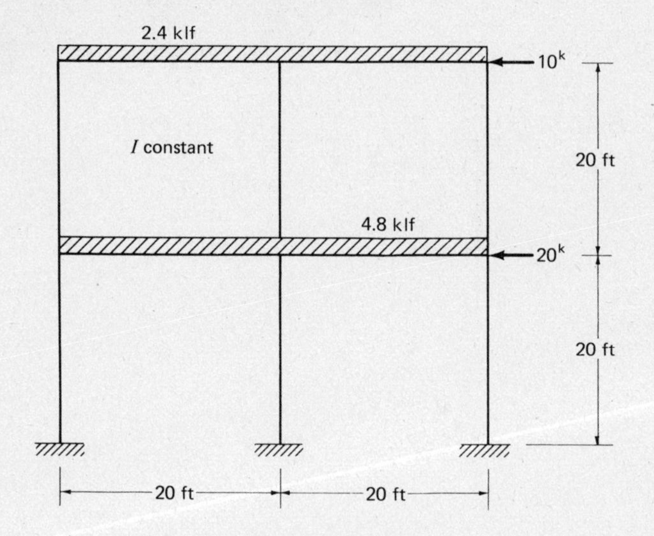

Problem 24.26 (*Ans.* $M_A = 131'^k$, $M_B = 86'^k$, $M_{DC} = 94'^k$, $M_{JG} = 86'^k$)

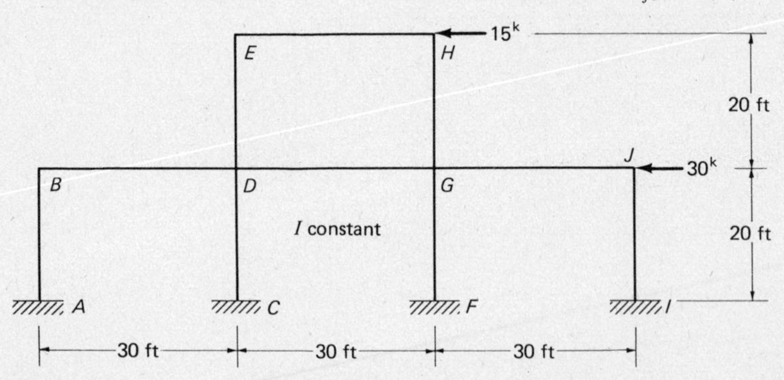

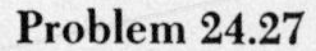

Problem 24.27

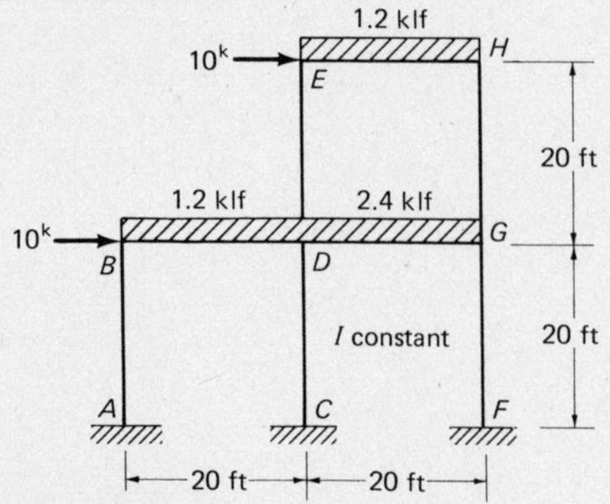

REFERENCE

[1] C. T. Morris, "Morris on Analysis of Continuous Frames," *Transactions of the American Society of Civil Engineers* 96 (1932): 66–69.

25 Column Analogy

25.1 GENERAL

The conjugate-beam procedure is an analogy in that a convenient viewpoint was taken of the mechanical operations involved in figuring the deflections in members subject to bending moment. The M/EI diagram was placed on a fictitious structure in which the resulting shear and moments coincided exactly with the slopes and deflections in the original structure.

In a similar manner the column-analogy procedure takes a convenient view of the operations involved in figuring the moments in a statically indeterminate structure. The analogy pertains to the identities existing between the moments produced in a statically indeterminate structure and the stresses produced in an eccentrically loaded short column. The computations are reduced to a nearly mechanical procedure, and the comment, "I just put the figures in a table, push the button of the calculator, and out comes the answer," is sometimes heard.

Column analogy was the second outstanding contribution to the structural field made by the late Hardy Cross. The method is applicable to the analysis of structures statically indeterminate to not more than the third degree, including fixed-ended beams, single-span arches and frames, and closed boxes. Another application of column analogy, and perhaps the most useful today, is the calculation of the carry-over factors, stiffness factors, and fixed-end moments necessary for analyzing structures with members of varying moments of inertia by moment distribution.

25.2 DEVELOPMENT OF THE METHOD

For the following discussion a short prismatic column, for which the deflections caused by bending are negligible, is considered. The column is shown in Fig. 25.1 loaded with an eccentric load P, and a stress diagram showing the

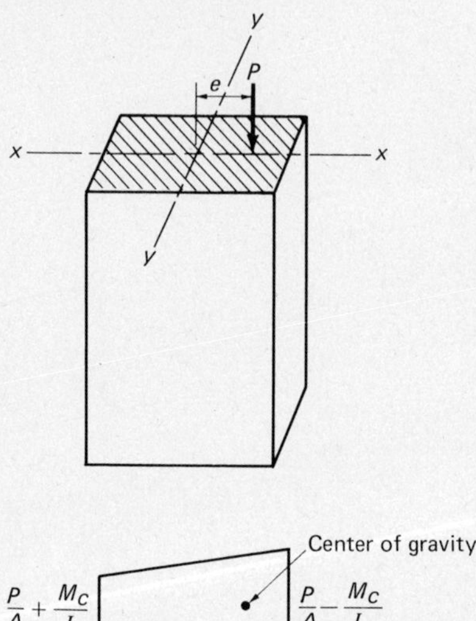

Figure 25.1

variation of stress across the cross section of the column is drawn. The load is assumed to be applied at a point along the X axis of the cross section a distance e from the Y axis. The following expression may be written for the stress at any point a perpendicular distance y from the Y axis:

$$f = \frac{P}{A} \pm \frac{Mc}{I_y} = \frac{P}{A} \pm \frac{Pey}{I_y}$$

An important fact to notice is that the load P coincides with the centroid of the forces produced in the column cross section or with the centroid of the stress diagram.

The fixed-ended beam of Fig. 25.2, which supports a concentrated load P, is now considered, and the magnitude of the end moments is desired. The beam is assumed to be replaced with a short column having a cross section with a centerline the same in shape and length as the beam. Its width at any section equals the $1/EI$ value of the beam at the corresponding section.

The load P produces a bending-moment diagram which can be broken down into two parts as previously demonstrated in Example 17.6. These are the simple beam moment and the end moment. A review of the moment-area theorems will reveal the following two important points about these diagrams:

1. There is no change in slope of the tangent at A from the one at B; therefore the total area of the M/EI diagram from A to B is zero. If the simple beam-moment diagram is considered as a downward load (plus moment), the end-moment diagram must act as an upward load (minus moment) and be equal in area.

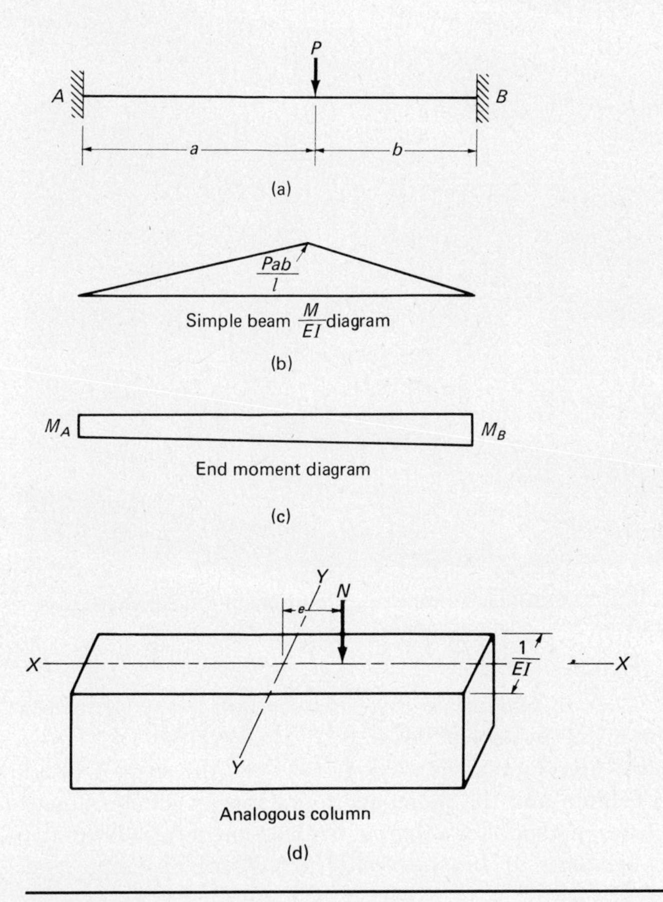

Figure 25.2

2. The deflection of the tangent at one end of the beam from the tangent at the other is zero, as must be the moment of the total area of the M/EI diagram about either end; therefore the centroid of the end-moment diagram must coincide with the centroid of the simple beam-moment diagram.*

If the total area of the simple beam M/EI diagram, represented by N, is applied to the analogous column at a point corresponding to the centroid of the diagram, the "stress" at any point on the cross section will be

$$f = \frac{N}{A} \pm \frac{Mc}{I_y} = \frac{N}{A} \pm \frac{Ney}{I_y}$$

The relation of the moments in the fixed-ended beam to the "stresses" in the analogous column should now be evident. The load N is opposed by a stress

*This discussion is quite similar to the one presented in C. D. Williams, *Analysis of Statically Indeterminate Structures* [1].

East Belt Freeway (I-440). Little Rock, Arkansas. (Courtesy of the Arkansas State Highway and Transportation Department.)

diagram that coincides as to magnitude and centroid and has as ordinates stresses equal to the moments in the real beam.

Another method of developing the analogy between the stresses in an eccentrically loaded column and the moments in a statically indeterminate frame is the elastic-center method. For a thorough discussion of this subject the reader may refer to *Analysis of Indeterminate Structures* by Parcel and Moorman [2].

25.3 COMPUTATION OF FIXED-END MOMENTS

Examples 25.1 and 25.2 illustrate the application of column analogy to the calculation of moments for fixed-ended beams. The calculations involved include obtaining certain properties of the analogous column and applying the bending and direct stress formula. The first of the two examples pertains to a beam that is prismatic, but the second one considers a beam with increased moments of inertia near the supports. It is necessary in this latter case to show the width of the cross section of the analogous cross section corresponding to the different $1/EI$ values present. If the size of the section should be constantly varying (see Example 25.5) over some little distance, it is necessary to divide the beam into short segments and figure the $1/EI$ widths for each of the segments.

Example 25.1

Find the fixed-end moments of the structure in Fig. 25.3. Use column analogy.

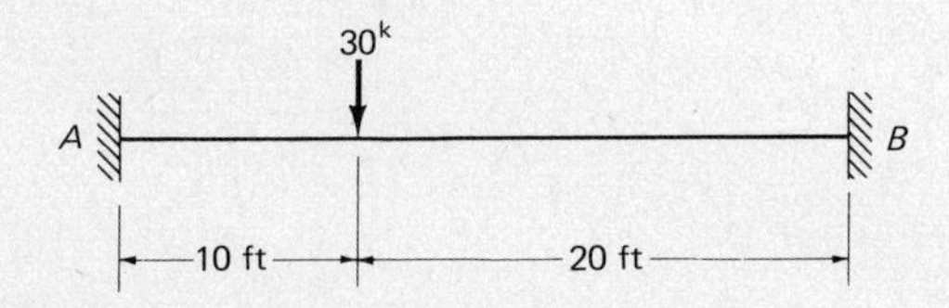

Figure 25.3

SOLUTION

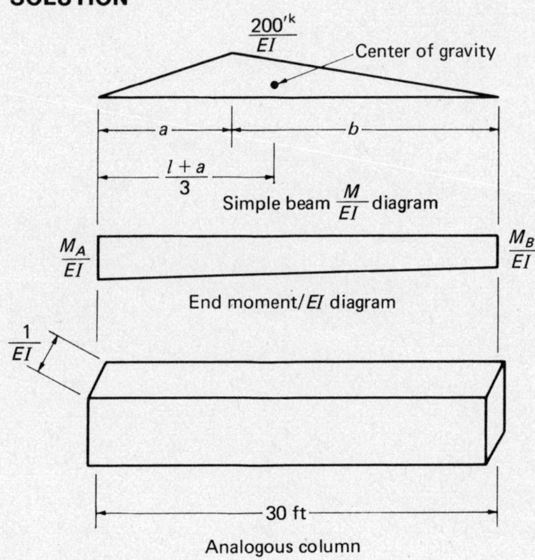

$$M_A = \frac{N}{A} + \frac{Ney}{I}$$

$$M_B = \frac{N}{A} - \frac{Ney}{I}$$

$$N = (\tfrac{1}{2})(30)\left(\frac{200}{EI}\right) = \frac{3000}{EI}$$

$$A = \frac{l}{EI} = \frac{30}{EI}$$

$$e = \frac{l}{2} - \frac{l + a}{3} = 15 - \frac{30 + 10}{3} = 1.67$$

$$I = \frac{l^3}{12EI} = \frac{30^3}{12EI} = \frac{2250}{EI}$$

$$y = \frac{l}{2} = 15$$

$$M_A = \frac{3000/EI}{30/EI} + \frac{(3000/EI)(1.67)(15)}{2250/EI} = 133.3'^{k}$$

$$M_B = \frac{3000/EI}{30/EI} - \frac{(3000/EI)(1.67)(15)}{2250/EI} = 66.7'^{k}$$

Example 25.2

Compute the fixed-end moments for the beam shown in Fig. 25.4.

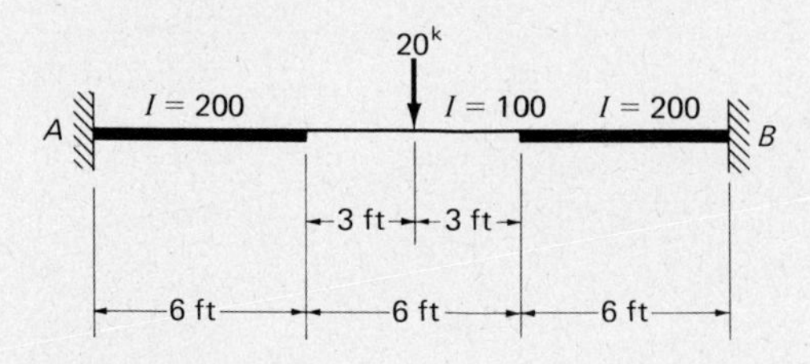

Figure 25.4

SOLUTION

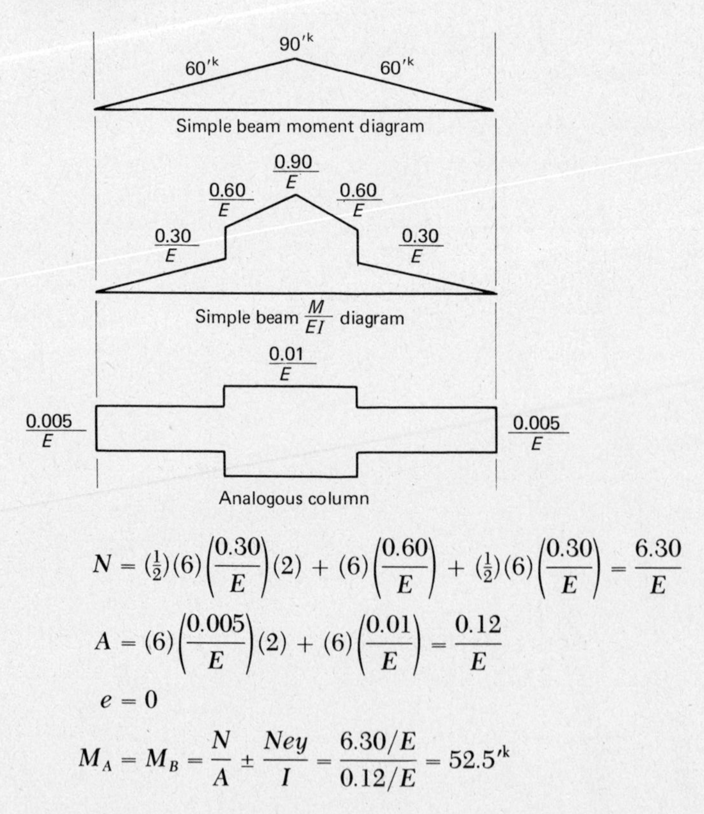

$$N = (\tfrac{1}{2})(6)\left(\frac{0.30}{E}\right)(2) + (6)\left(\frac{0.60}{E}\right) + (\tfrac{1}{2})(6)\left(\frac{0.30}{E}\right) = \frac{6.30}{E}$$

$$A = (6)\left(\frac{0.005}{E}\right)(2) + (6)\left(\frac{0.01}{E}\right) = \frac{0.12}{E}$$

$$e = 0$$

$$M_A = M_B = \frac{N}{A} \pm \frac{Ney}{I} = \frac{6.30/E}{0.12/E} = 52.5'^k$$

25.4 COMPUTATION OF CARRY-OVER AND STIFFNESS FACTORS

The structures analyzed in Chaps. 23 and 24 by moment distribution consisted of prismatic members for which the stiffnesses, fixed-end moments, and carry-over factors were easily obtainable (the carry-over factors were always

Chetco River Bridge, Brookings, Oregon. (Courtesy of the Oregon Department of Transportation.)

$-\frac{1}{2}$). Although the application of moment distribution to structures containing nonprismatic members is exactly the same for those consisting entirely of prismatic members, the properties necessary for analysis are not so easily obtained. One of the better methods of computing these properties is the column-analogy procedure.

The calculations necessary for computing the fixed-end moments have been described in Sec. 25.3 and will be further illustrated in Example 25.5. The first problem considered in this section is the calculation of carry-over factors. With reference to Fig. 25.5 it is remembered that a moment M_A applied at the A end of the member shown will induce a moment in the B end equal to the carry-over factor times M_A or $C_{AB}M_A$. In this figure the A end is considered to be hinged so that it may rotate. A hinge is theoretically a point of no stiffness and the moment of inertia is zero; therefore the width of the analogous column $(1/EI)$ is infinite and the Y axis of the cross section will pass through the hinge.

The A end of the member is assumed to be rotated through an angle of 1 rad; therefore the total area of the M/EI diagram from A to B must be 1.0. Since point A does not deflect vertically with respect to B, the moment of the M/EI diagram about A must be zero. The effect of the facts stated in the preceding two sentences is that an equivalent load of 1.0 can be placed on the analogous column at point A because it represents the net effect of the diagram. This load could not be placed anywhere else because the effect would be a moment about A, falsely indicating deflection. Whatever the shape of the beam, 1.0 is placed at the end which is rotated. The moments or "stresses" at

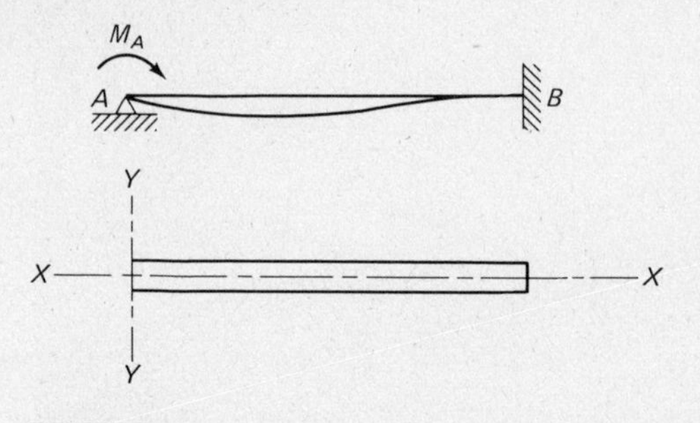

Figure 25.5

each end of the analogous column are determined; and from a ratio of the two, the carry-over factor is found.

The stiffness of one end of a member may be defined as the moment required to rotate that end of the member through a unit angle when the other end is fixed. This value is computed in making the carry-over calculations.

Examples 25.3 to 25.5 illustrate the calculations necessary for computing these beam properties. Fortunately, there are available in various publications tables and curves that give the properties needed for some groups of beams. The *Handbook of Frame Constants,* published by the Portland Cement Association, is probably the most widely known. Example 25.6 illustrates the analysis of a continuous beam of constantly varying moments of inertia by moment distribution.

Example 25.3

Compute the carry-over factor from A to B for the beam of Fig. 25.6, which has a constant moment of inertia.

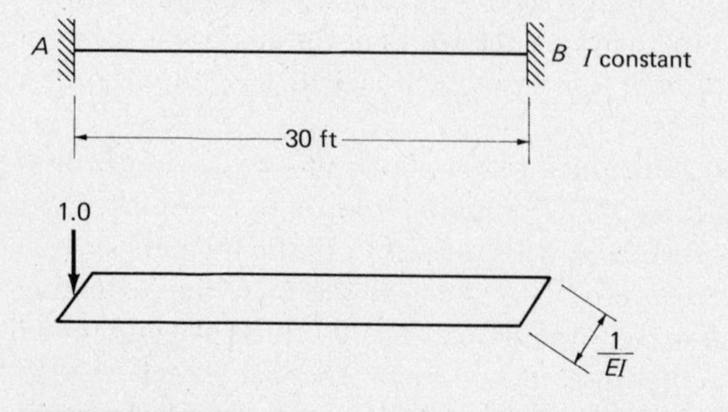

Figure 25.6

SOLUTION

$$N = 1.0$$

$$A = \frac{30}{EI}$$

$$I = \frac{l^3}{12EI} = \frac{2250}{EI}$$

$$e = \frac{l}{2} = 15$$

$$y = \frac{l}{2} = 15$$

$$M_A = \frac{N}{A} + \frac{Ney}{\mathrm{I}} = \frac{1.0}{30/EI} + \frac{(1.0)(15)(15)}{2250/EI} = +0.1333EI$$

$$M_B = \frac{N}{A} - \frac{Ney}{\mathrm{I}} = \frac{1.0}{30/EI} - \frac{(1.0)(15)(15)}{2250/EI} = -0.0667EI$$

$$C_{AB} = \frac{M_B}{M_A} = \frac{-0.0667EI}{0.1333EI} = -\tfrac{1}{2}$$

Example 25.4

Compute the stiffness at A and the carry-over from A to B for the beam of Example 25.2, the analogous column of which is reproduced in Fig. 25.7.

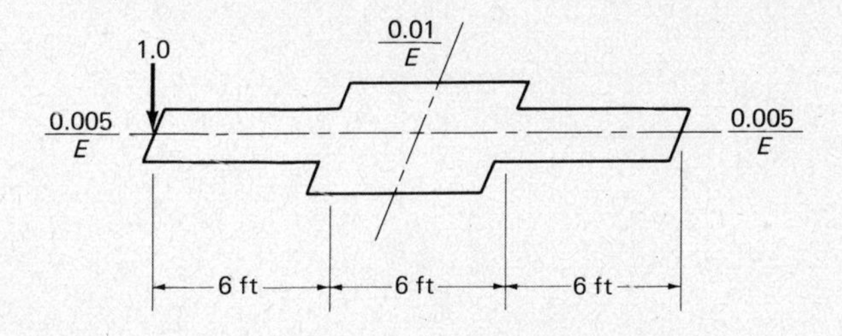

Figure 25.7

SOLUTION

$$N = 1.0$$

$$A = \frac{0.12}{E}$$

$$I = (\tfrac{1}{12})\left(\frac{0.005}{E}\right)(18)^3 + (\tfrac{1}{12})\left(\frac{0.005}{E}\right)(6)^3 = \frac{2.52}{E}$$

$$e = 9.0$$

$$M_A = \frac{1.0}{0.12/E} + \frac{(1.0)(9.0)(9.0)}{2.52/E} = +40.48E$$

$$M_B = \frac{1.0}{0.12/E} - \frac{(1.0)(9.0)(9.0)}{2.52/E} = -23.81E$$

$$C_{AB} = \frac{M_B}{M_A} = \frac{-23.81E}{+40.48E} = -0.588$$

$$K_A = 40.48E$$

NOTE: For members that are not symmetrical, Maxwell's law presents an interesting method of checking the carry-over and stiffness values obtained at the two ends. The twisting of the A end of a member through an angle of 1 rad produces a moment at the B end equal to the moment produced at the A end if the B end is twisted through an angle of 1 rad. This relationship may be expressed as $C_{AB}K_A = C_{BA}K_B$.

Example 25.5

Compute fixed-end moments, carry-over factors, and stiffnesses for the beam shown in Fig. 25.8. This varying depth beam has a rectangular cross section with a constant width of 6 in. Divide the analogous column into 2-ft sections for computing values of I and A, although more precise values could be obtained by using smaller divisions.

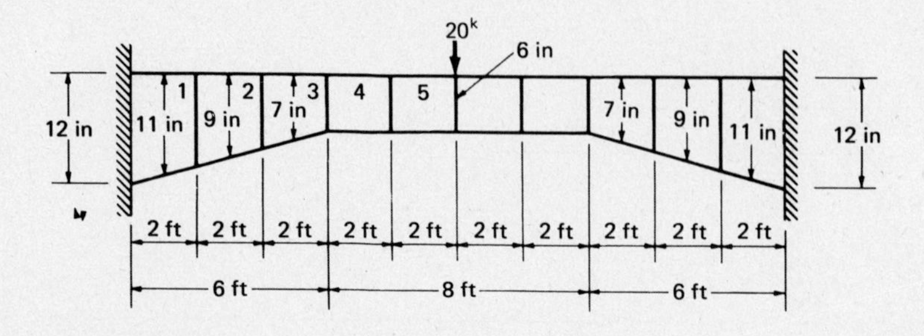

Figure 25.8

SOLUTION

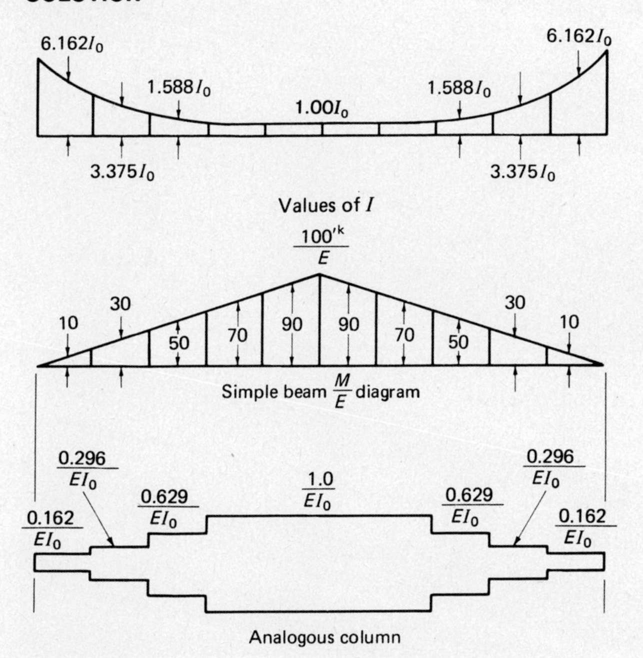

Table 25.1

SECTION	N	A
1	$\frac{(10)(2)}{6.162EI_0} = \frac{3.24}{EI_0}$	$\frac{(2)(0.162)}{EI_0} = \frac{0.324}{EI_0}$
2	$\frac{(30)(2)}{3.375EI_0} = \frac{17.80}{EI_0}$	$\frac{(2)(0.296)}{EI_0} = \frac{0.592}{EI_0}$
3	$\frac{(50)(2)}{1.588EI_0} = \frac{63.00}{EI_0}$	$\frac{(2)(0.629)}{EI_0} = \frac{1.258}{EI_0}$
4	$\frac{(70)(2)}{1.0EI_0} = \frac{140.00}{EI_0}$	$\frac{(2)(1.0)}{EI_0} = \frac{2.000}{EI_0}$
5	$\frac{(90)(2)}{1.0EI_0} = \frac{180.00}{EI_0}$	$\frac{(2)(1.0)}{EI_0} = \frac{2.000}{EI_0}$
Σ	$\frac{404.04}{EI_0}$	$\frac{6.174}{EI_0}$

$$N = 2 \times \frac{404.04}{EI_0} = \frac{808.08}{EI_0}$$

$$A = 2 \times \frac{6.174}{EI_0} = \frac{12.348}{EI_0}$$

$$I = (\tfrac{1}{12})\left(\frac{0.162}{EI_0}\right)(20)^3 + (\tfrac{1}{12})\left(\frac{0.134}{EI_0}\right)(16)^3 + (\tfrac{1}{12})\left(\frac{0.333}{EI_0}\right)(12)^3$$

$$+ (\tfrac{1}{12})\left(\frac{0.371}{EI_0}\right)(8)^3 = \frac{217.3}{EI_0}$$

Rigid-frame bridge on the Henry G. Shirley Memorial Highway in Fairfax County, Virginia. (Courtesy of the Portland Cement Association.)

Fixed-end moments:

$$e = 0$$

$$M_A = M_B = \frac{808.08/EI_0}{12.348/EI_0} = 65.4'^{k}$$

Carry-overs and stiffness factors:

$$N = 1.0$$

$$e = 10.0$$

$$M_A = \frac{1.0}{12.348/EI_0} + \frac{(1.0)(10.0)(10.0)}{217.3/EI_0}$$

$$= +0.081EI_0 + 0.46EI_0 = +0.541EI_0$$

$$M_B = +0.081EI_0 - 0.46EI_0 = -0.379EI_0$$

$$C_{AB} = C_{BA} = \frac{-0.379EI_0}{+0.541EI_0} = -0.701$$

$$K_A = K_B = 0.541EI_0$$

Example 25.6

Distribute the fixed-end moments for the beam shown in Fig. 25.9, for which stiffnesses, carry-over factors, and fixed-end moments have been calculated.

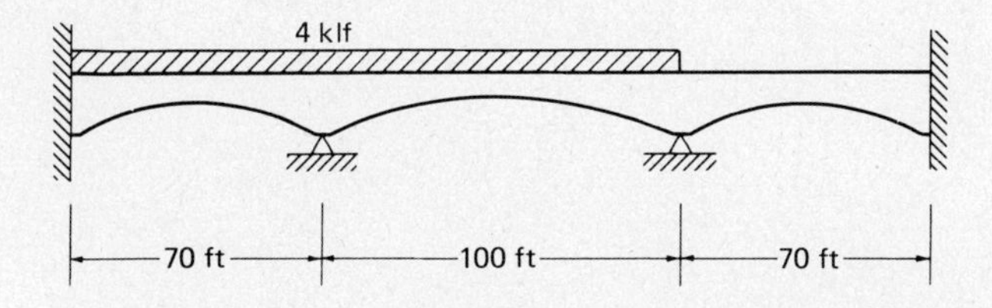

Figure 25.9

SOLUTION

	←0.75	0.78→	←0.78	0.75→		CO
	17.1	22.8	22.8	17.1		K
−2100	+2100	−4400	+4400			FEM
	0.43	0.57	0.57	0.43		
−2100	+2100	−4400	+4400			
		−1956	−2508	−1892	−1419	
+1372	+1830	+2426	+1890			
		− 840	−1078	− 812	− 609	
+ 270	+ 360	+ 480	+ 374			
		− 166	− 213	− 161	− 121	
+ 53	+ 71	+ 95	+ 74			
		− 33	− 42	− 32	− 24	
+ 10	+ 14	+ 19	+ 15			
		− 7	− 9	− 6	− 4	
+ 2	+ 3	+ 4				
− 393'k	+4378	−4378	+2903	−2903	−2177	

NOTE: A booklet entitled *Continuous Concrete Bridges*, published by the Portland Cement Association, presents an excellent method of analyzing and designing concrete bridges whose floors or girders have parabolic soffits such as the ones of this problem.

25.5 ANALYSIS OF FRAMES

The analogous columns that would be used for several types of frames are shown in Fig. 25.10. Each member of the frame is drawn to a width equal to its respective $1/EI$ value, and the principal axes are placed according to the principles previously discussed.

The center of gravity of the loads (that is, the center of gravity of the simple beam moment diagrams) will rarely fall along one of the principal axes, and the "stress" developed is caused by bending about both axes. The "stress" may be calculated with the following formula, in which x is the perpendicular distance from the point where "stress" is desired to the Y axis and y is the perpendicular distance from the point to the X axis.

$$M = \frac{N}{A} \pm \frac{Nex}{I_y} \pm \frac{Ney}{I_x}$$

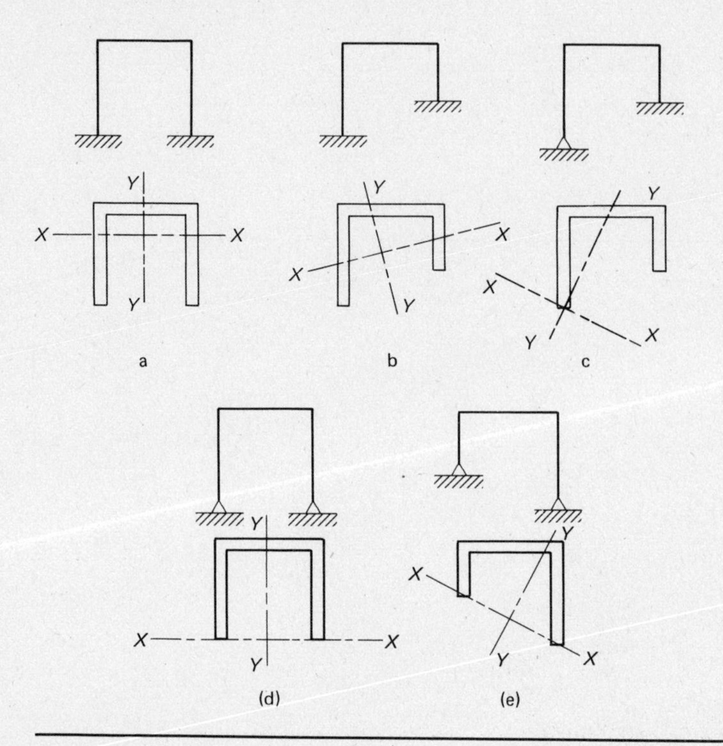

Figure 25.10

The analogous column is drawn with the proper dimensions, and its properties (A, I_x, I_y, and so on) are computed. The real frame is statically indeterminate and the number of reaction components required to leave a statically determinate and stable structure are removed. A moment diagram is drawn for the structure remaining, and its parts are placed in the appropriate position along the centerlines of the analogous column. It is again obvious that for equilibrium the resultant of the simple beam moment diagrams and the end-moment diagrams must be equal and opposite and their center of gravities must coincide. The final moment at any point equals the sum of the moment at the point in the simple beam moment diagram and the "stresses" it causes at the point in the analogous column. Only one example (25.7) is given, but it should be sufficient to illustrate the procedure. The procedure is rather lengthy and tedious for structures consisting of prismatic members as compared to a solution by the moment-distribution method, but it has some advantage for frames of varying moments of inertia. A simple beam moment diagram causing tension on the outside of the frame is considered to be a positive load.

The principal axes for the analogous columns for some frames are not vertical and horizontal, and symmetry is not present about either axis. The frames of Fig. 25.10(b), (c), and (e) fall into this class. For this situation it is necessary to use the principal moments of inertia and the products of inertia in the solution.

Example 25.7

Determine the moments at C and D in the frame shown in Fig. 25.11. Remove the fixed end at A to leave a statically determinate structure.

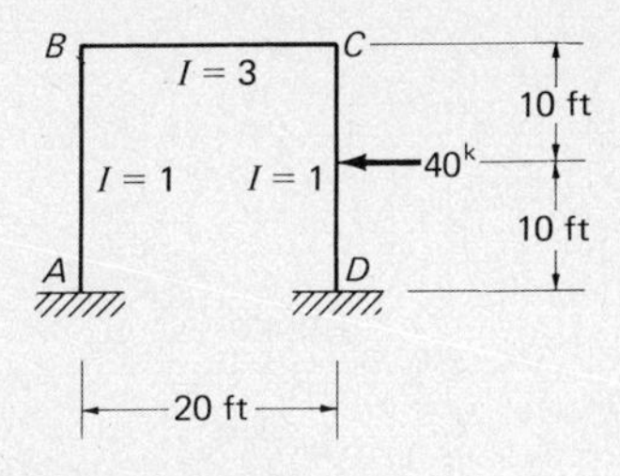

Figure 25.11

SOLUTION

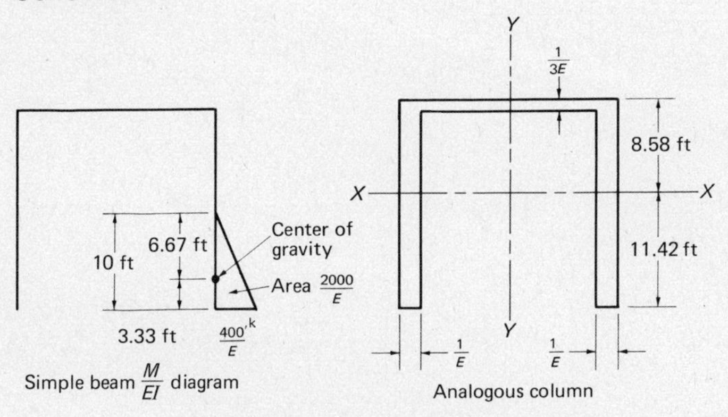

Properties of analogous column:

$$A = (20)\left(\frac{1}{E}\right) + (20)\left(\frac{1}{E}\right) + (20)\left(\frac{3}{E}\right) = \frac{46.67}{E}$$

$$\bar{y} = \frac{(20/E)(10) + (20/E)(10)}{46.67/E} = 8.58 \text{ ft}$$

$$I_x = \left(\frac{1}{3}\right)\left(\frac{1}{E}\right)(8.58^3 \times 2 + 11.42^3 \times 2) + \left(\frac{20}{3E}\right)(8.58)^2 = \frac{1904}{E}$$

$$I_y = \left(\frac{1}{12}\right)\left(\frac{1}{3E}\right)(20)^3 + (2)\left(\frac{20}{E}\right)(10)^2 = \frac{4222}{E}$$

Moment at C:

$$M_C = \frac{2000/E}{46.67/E} + \frac{(2000/E)(10)(10)}{4222/E} - \frac{(2000/E)(8.09)(8.58)}{1904/E} + M_{\text{simple beam}}$$

$$M_C = +42.8 + 47.4 - 72.8 + 0 = 17.4'^{k}$$

Moment at D:

$$M_D = \frac{2000/E}{46.67/E} + \frac{(2000/E)(10)(10)}{4222/E} + \frac{(2000/E)(8.09)(11.42)}{1904/E} - 400$$

$$M_D = +42.8 + 47.4 + 96.9 - 400 = -212.9'^{k}$$

PROBLEMS

25.1. Determine the fixed-end moments, carry-over factors, and stiffness for the beam shown in the accompanying illustration.

Problem 25.1

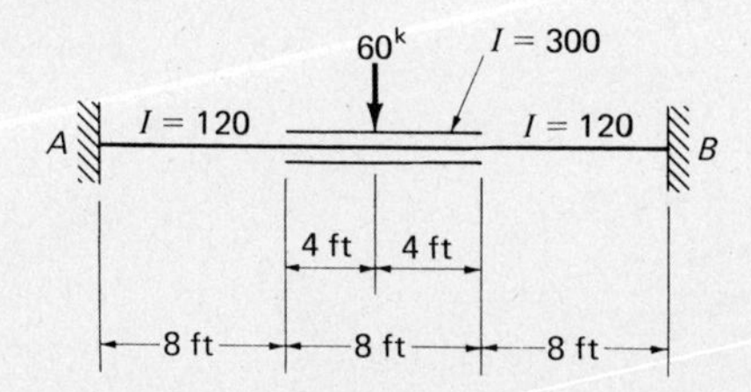

25.2. Repeat Prob. 25.1 for the 60-kip load moved to a point 8 ft from the left end of the span. (*Ans.* $M_A = -188.4'^{k}$, $M_B = -91.6'^{k}$, $C_{AB} = C_{BA} = -0.419$, $K_A = K_B = 21.5E$)

25.3. Determine the fixed-end moments, carry-over factors, and stiffnesses for both ends of the beam shown in the accompanying illustration.

Problem 25.3

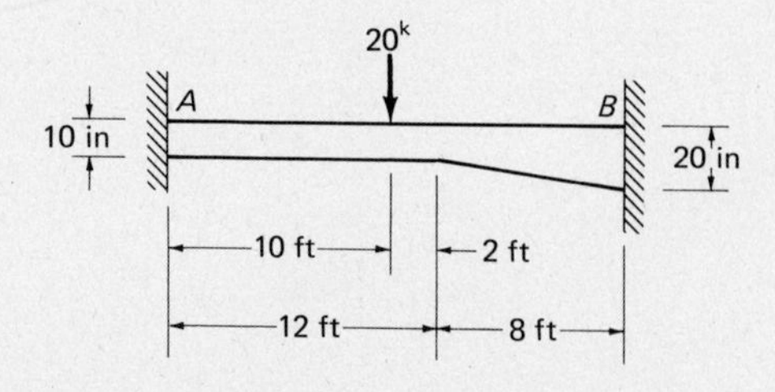

25.4. Compute the fixed-end moments in Prob. 25.3 for the 20-kip load removed and the beam loaded with a 4-klf uniform load for the entire span. (*Ans.* $M_A = -98.2'^{k}$, $M_B = -217.2'^{k}$)

25.5. Compute carry-over and stiffness factors for the beam shown in the accompanying illustration. Divide the beam into 3-ft sections for the calculations. The beam soffit is laid out on a curve in accordance with the formula $y = ax^2$.

Problem 25.5

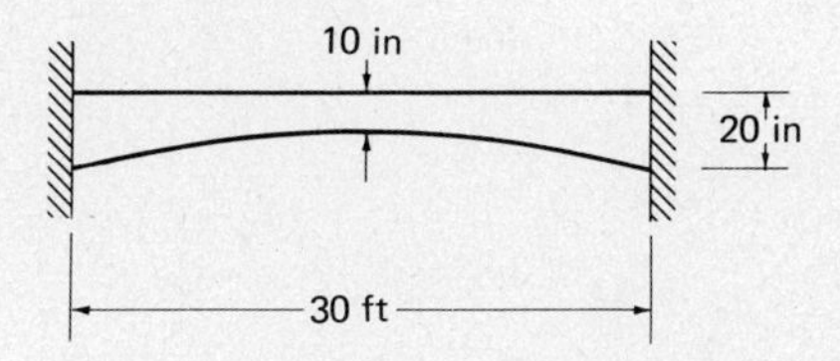

Calculate the moments at the ends of the members for the frames of Probs. 25.6 to 25.7 by the column-analogy method.

25.6. Problem 24.8. (*Ans.* $M_A = 72.3'^k$, $M_B = 70.4'^k$, $M_C = 87.9'^k$, $M_D = 110.7'^k$)

25.7. Problem 24.7.

REFERENCES

[1] C. D. Williams, *Analysis of Statically Indeterminate Structures* (New York and London: Intext Educational Publishers, 1959), pp. 292–294.

[2] J. I. Parcel and B. B. Moorman, *Analysis of Indeterminate Structures* (New York: Wiley, 1955).

26 Plastic Analysis

26.1 INTRODUCTION

The preceding 25 chapters have presented material concerning the behavior of structures which is valid only as long as the materials follow Hooke's law. In recent years structural engineers have become increasingly interested in the ultimate strengths of structures. As a good many structural materials are ductile, it is necessary to consider the behavior of those materials beyond the elastic range.

Mild structural steels exhibit the most desirable of the postelastic or plastic properties from the standpoint of plastic principles. For this reason the discussion of analysis and structural behavior presented here is confined to these steels. A few remarks are made in Sec. 26.2 concerning the properties of such steels.

26.2 PROPERTIES OF STRUCTURAL STEEL

If a piece of mild structural steel is subjected to a tensile force, it will begin to elongate. If the tensile force is increased at a constant rate the amount of elongation will increase constantly within certain limits. In other words, elongation will double when the stress goes from 6000 to 12,000 psi. When the tensile stress reaches a value roughly equal to one-half of the ultimate strength of the steel, the elongation will begin to increase at a greater rate without a corresponding increase in the stress.

The largest stress for which Hooke's law applies or the highest point on the straight-line portion of the stress-strain diagram is the ***proportional limit.*** The largest stress that a material can withstand without being permanently deformed is called the ***elastic limit.*** This value is seldom actually measured and for most engineering materials including structural steel is synonomous with the proportional limit. For this reason the term ***proportional elastic limit*** is sometimes used.

The stress at which there is a decided increase in the elongation or strain without a corresponding increase in stress is said to be the ***yield point.*** It is the first point on the stress-strain diagram where a tangent to the curve is horizontal. The yield point is probably the most important property of steel to the designer as the elastic design procedures are based on this value (with the exception of compression members where buckling may be a factor). The allowable stresses used in these methods are usually taken as some percentage of the yield point. Beyond the yield point there is a range in which a considerable increase in strain occurs without increase in stress. The strain that occurs before the yield point is referred to as the ***elastic strain;*** the strain that occurs after the yield point, with no increase in stress, is referred to as the ***plastic strain.*** These latter strains usually vary from 10 to 15 times the elastic strains.

Yielding of steel without stress increase may be thought to be a severe disadvantage when in actuality it is a very useful characteristic. It has often performed the wonderful service of preventing failure due to omissions or mistakes on the designer's part. Should the stress at one point in a ductile steel structure reach the yield point, that part of the structure will yield locally without stress increase, thus preventing premature failure. This ductility allows the stresses in a steel structure to be readjusted. Another way of describing this phenomenon is to say that very high stresses caused by fabrication, erection, or loading will tend to equalize themselves. It might also be said that a steel structure has a reserve of plastic strain that enables it to resist overloads and sudden shocks. If it did not have this ability, it might suddenly fracture, like glass or other vitreous substances.

Following the plastic strain there is a range where additional stress is necessary to produce additional strain and this is called ***strain hardening.*** This portion of the diagram is not too important to today's designer. A familiar stress-strain diagram for mild structural steel is shown in Fig. 26.1. Only the initial part of the curve is shown here because of the great deformation that occurs before failure. At failure in the mild steels the total strains are from 150 to 200 times the elastic strains. The curve will actually continue up to its maximum stress value and then "tail off" before failure. A sharp reduction in the cross section of the member takes place (called "necking") followed by failure.

The stress-strain curve of Fig. 26.1 is typical of the usual ductile structural steel and is assumed to be the same for members in tension or compression. (The compression members must be short because long compression members subjected to compression loads tend to bend laterally and their properties are greatly affected by the bending moments so produced.) The shape of the diagram varies with the speed of loading, the type of steel, and the temperature. One such variation is shown in the figure by the dotted line marked

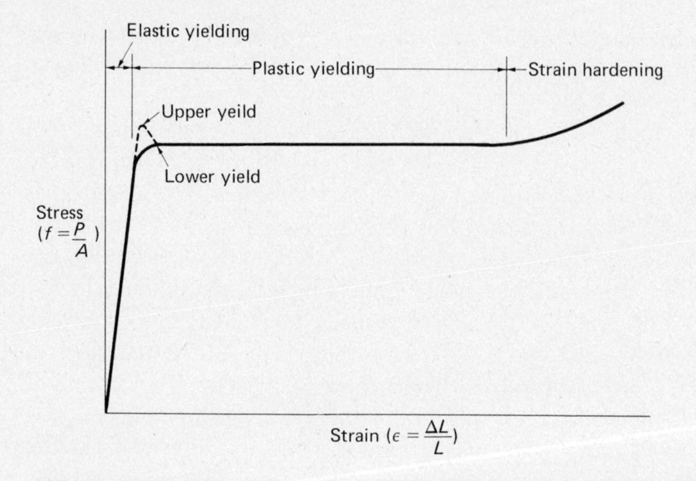

Figure 26.1

upper yield. This shape stress-strain curve is the result when a mild steel has the load applied rapidly whereas the lower yield is the case for slow loading.

As early as 1914 a Hungarian, Gabor Kazinczy, recognized that the ductility of steel permitted a redistribution of stresses in an overloaded statically indeterminate structure [1]. In the United States J. A. Van den Broek introduced his plastic theory which he called *limit design*. This theory was published in a paper entitled "Theory of Limit Design" in February 1939 in the *Proceedings of the ASCE*.

In the plastic theory, rather than basing designs on the allowable-stress method the problem is handled by considering the greatest load that can be carried by the structure acting as a unit. The resulting designs are quite interesting to the structural designer as they offer several advantages. These include the following:

1. When the plastic-design procedure is used there can be considerable savings in steel (perhaps as high as 10 or 15% for some structures) as compared to a similar design made by the elastic procedure. The necessary continuity connections required in plastic design may reduce the actual money savings somewhat.
2. Plastic design permits the designer to make a more accurate estimate of the maximum load that a structure can support, enabling him or her to have a better idea of the actual safety factor of the structure. The elastic method works very well for computing the stresses and strains for loads in the elastic range but gives a very poor estimate of the actual collapse strength of a structure.
3. For many complicated structures plastic analysis is easier to apply than is elastic analysis.
4. Structures are often subjected to large stresses that are difficult to predict such as those caused by settlement, erection, and so forth.

Plastic design provides for such situations by permitting plastic deformation.

Despite these several important advantages the acceptance of plastic design has been rather slow. Until recent years there has not been a great deal of information available concerning the ductility of steel. If a particular steel is brittle, the plastic theory does not apply. A few decades ago when the design profession was just becoming interested in this theory there were several disastrous brittle failures of welded tanks and ships. (That also was the time when welded structures were beginning to gain popularity.) The acceptance of plastic design was slowed down to a walk for a long time, but in recent years much research has been performed in these areas, and plastic design is today gaining considerable acceptance.

Despite the great progress made in the field of plastic design, the method still has some disadvantages with which the designer should be completely familiar. These include the following:

1. Plastic design is of little value for the high-strength brittle steels. (The method is just as applicable to high-strength steels as it is to lower strength structural grade steels as long as the steels have the required ductility.)
2. Plastic design today is not satisfactory for situations in which fatigue stresses are a problem.
3. Columns designed by the plastic theory provide little savings.
4. Although for many statically indeterminate structures plastic analysis is simpler than elastic analysis, it should be realized that unstable plastic structures are more difficult to detect than are unstable elastic structures.

26.3 THEORY OF PLASTIC ANALYSIS

The basic theory of plastic analysis has been shown to be a major change in the distribution of stresses after the stresses at certain points in a structure reach the yield point. The theory is that those parts of the structure that have been stressed to the yield point cannot resist additional stresses. They instead will yield the amount required to permit the load or stresses to be transferred to other parts of the structure where the stresses are below the yield stress and thus in the elastic range and able to resist increased stress.

For this discussion the stress-strain diagram is assumed to have the idealized shape shown in Fig. 26.2. The yield point and the proportional limit are assumed to occur at the same point for this steel, and the stress-strain diagram is assumed to be a perfectly straight line in the plastic range. Beyond the plastic range there is the range of strain hardening. This latter range could theoretically permit steel members to withstand additional stress, but from a practical standpoint the strains and deflections occurring are so large that they cannot be considered.

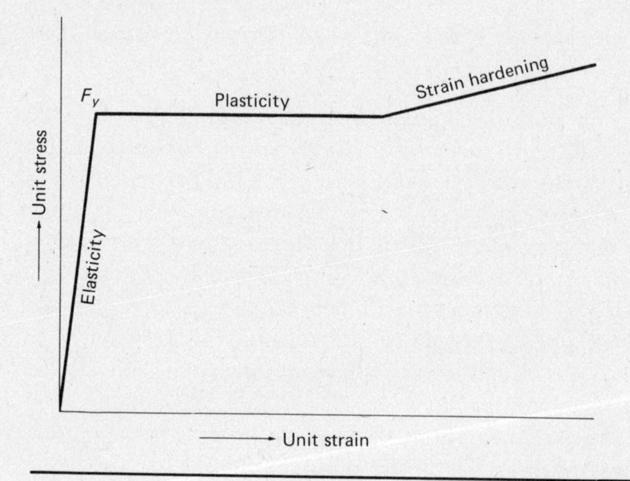

Figure 26.2

26.4 THE PLASTIC HINGE

As the bending moment is increased at a particular section of a beam there will be a linear variation of stress until the yield stress is reached in the outermost fibers. An illustration of stress variation for a rectangular beam in the elastic range is shown in part (b) of Fig. 26.3. The *yield moment* of a cross section is defined as the moment that will just produce the yield stress in the outermost fiber of the section.

If the moment is increased beyond the yield moment, the outermost fibers that had previously been stressed to their yield point will continue to have the same stress but will yield, and the duty of providing the necessary additional resisting moment will fall on the fibers nearer to the neutral axis. This process will continue with more and more parts of the beam cross section stressed to the yield point as shown by the stress diagrams of parts (c) and (d) of the figure, until, finally, a fully plastic distribution is approached as shown in part (e). When the stress distribution has reached this stage, a *plastic hinge* is said to

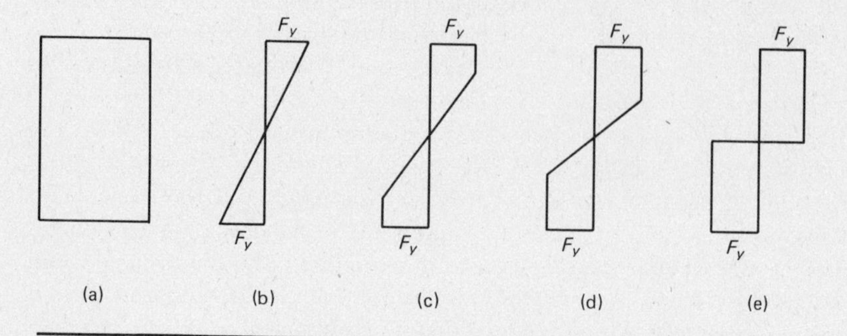

Figure 26.3

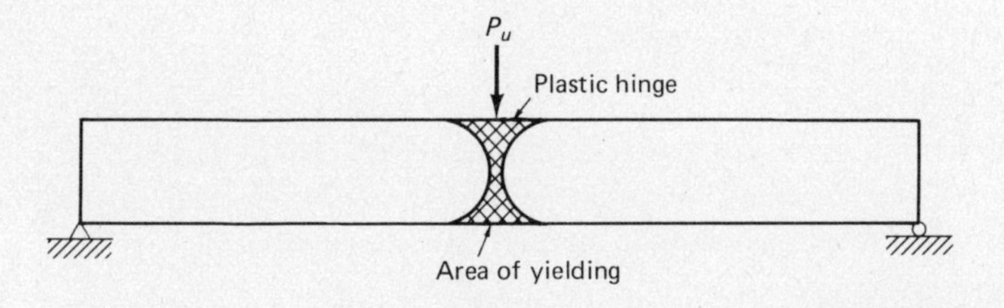

Figure 26.4

have formed because no additional moment can be resisted at the section. Any additional moment applied at the section will cause the beam to rotate with little increase in stress.

The *plastic moment* is the moment that will produce full plasticity in a member cross section and create a plastic hinge. The ratio of the plastic moment M_p to the yield moment M_y is called the *shape factor*. The shape factor equals 1.50 for rectangular sections and varies from about 1.10 to 1.20 for standard rolled-beam sections.

This paragraph is devoted to a description of the development of a plastic hinge in the simple beam shown in Fig. 26.4. The load shown is applied to the beam and increased in magnitude until the yield moment is reached and the outermost fiber is stressed to the yield point. The magnitude of the load is further increased with the result that the outer fibers begin to yield. The yielding spreads out to the other fibers away from the section of maximum moment as indicated in the figure. The length in which this yielding occurs away from the section in question is dependent on the loading conditions and the member cross section. For a concentrated load applied at the centerline of a simple beam with a rectangular cross section, yielding in the extreme fibers at the time the plastic hinge is formed will extend for one third of the span. For a W shape in similar circumstances, yielding will extend for approximately one-eighth of the span. During this same period the interior fibers at the section of maximum moment yield gradually until nearly all of them have yielded and a plastic hinge is formed as shown in Fig. 26.4.

When steel frames are loaded to failure, the points where rotation is concentrated (plastic hinges) become quite visible to the observer before collapse occurs.

26.5 THE PLASTIC MODULUS

The yield moment M_y can be defined as equaling the yield stress F_y times the elastic modulus S. The elastic modulus equals I/c or $bd^2/6$ for a rectangular section and the yield moment equals $F_y bd^2/6$. The same value can be obtained by considering the resisting internal couple shown in Fig. 26.5.

The resisting moment equals T or C times the lever arm between them, as follows:

$$M_y = \left(\frac{F_y bd}{4}\right)\left(\frac{2}{3}d\right) = \frac{F_y bd^2}{6}$$

The elastic section modulus can again be seen to equal $bd^2/6$ for a rectangular beam. The resisting moment of a rectangular section at full plasticity can be determined in a similar manner (see Fig. 26.6).

$$M_p = \left(F_y \frac{d}{2} b\right)\frac{d}{2} = \frac{F_y bd^2}{4}$$

The plastic moment is said to equal the yield stress times the plastic modulus. From the foregoing expression for a rectangular section, the plastic modulus Z can be seen to equal $bd^2/4$. The shape factor, which equals M_p/M_y, F_yZ/F_yS, or Z/S, is $(bd^2/4)/(bd^2/6) = 1.50$ for a rectangular section.

A study of the plastic modulus determined here shows that it equals the statical moment of the tension and compression areas about the neutral axis. Unless the section is symmetrical, the neutral axis for the plastic condition will not be in the same location as for the elastic condition. The total internal compression must equal the total internal tension. As all fibers are considered to have the same stress (F_y) in the plastic condition, the areas above and below the neutral axis must be equal. This situation does not hold for unsymmetrical sections in the elastic condition. Example 26.1 illustrates the calculations

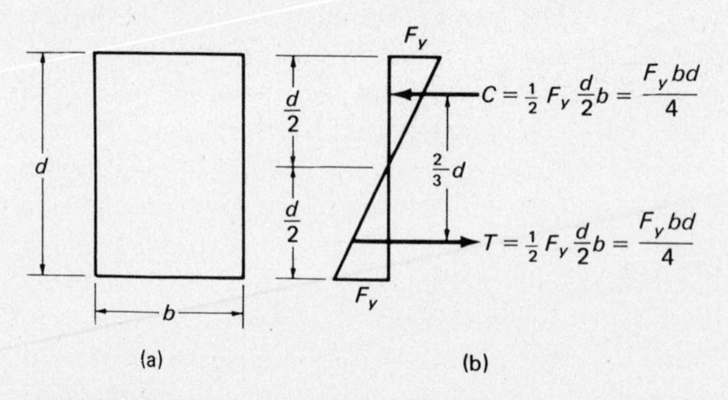

Figure 26.5

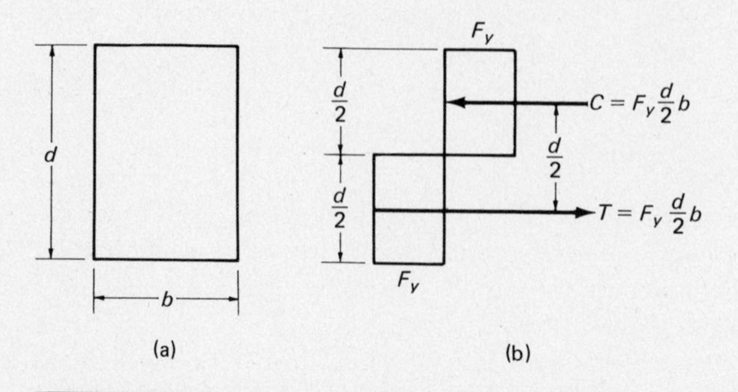

Figure 26.6

necessary to determine the shape factor for a tee beam and the ultimate uniform load w_u that the beam can support.

Example 26.1

Determine M_y, M_p, and Z for the steel tee beam shown in Fig. 26.7. Also calculate the shape factor and the ultimate uniform load w_u that can be placed on the beam for a 12-ft simple span. $F_y = 36$ ksi.

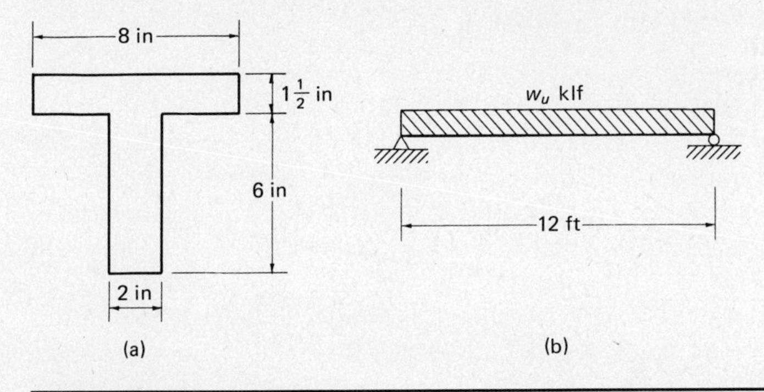

Figure 26.7

SOLUTION

Elastic calculations:

$$A = (8)(1\tfrac{1}{2}) + (6)(2) = 24 \text{ in}^2$$

$$\bar{y} = \frac{(12)(0.75) + (12)(4.5)}{24} = 2.625 \text{ in from top flange}$$

$$I = (\tfrac{1}{3})(2)(1.125^3 + 4.875^3) + (\tfrac{1}{12})(8)(1\tfrac{1}{2})^3 + (12)(1.875)^2$$
$$= 122.4 \text{ in}^4$$

$$S = \frac{I}{C} = \frac{122.4}{4.875} = 25.1 \text{ in}^3$$

$$M_y = F_y S = \frac{(36)(25.1)}{12} = 75.3'^{k}$$

Plastic calculations:

Neutral axis is at base of flange,

$$Z = (12)(0.75) + (12)(3) = 45 \text{ in}^3$$

$$M_p = F_y Z = \frac{(36)(45)}{12} = 135'^{k}$$

$$\text{shape factor} = \frac{M_p}{M_y} \quad \text{or} \quad \frac{Z}{S} = \frac{45}{25.1} = 1.79$$

$$M_p = \frac{w_u L^2}{8}$$

$$w_u = \frac{(8)(135)}{(12)^2} = 7.5 \text{ klf}$$

The values of the plastic moduli for the standard steel beam sections are tabulated in the *Manual of Steel Construction*, 8th ed. [2] in the Plastic Design Selection Table. These values will be frequently used in the pages to follow. All of the beams considered in this chapter are assumed to consist of A36 steel for which $F_y = 36$ ksi.

26.6 THE COLLAPSE MECHANISM

A statically determinate beam will fail if one plastic hinge develops. To illustrate this fact, the simple beam of constant cross section loaded with a concentrated load at midspan shown in Fig. 26.8(a) is considered. Should the load be increased until a plastic hinge is developed at the point of maximum moment (underneath the load in this case), an unstable structure will have been created as shown in part (b) of the figure. Any further increase in load will cause collapse.

The plastic theory is of little advantage for statically determinate beams and frames but it may be of decided advantage for statically indeterminate beams and frames. For a statically indeterminate structure to fail it is necessary for more than one plastic hinge to form. The number of plastic hinges required for failure of statically indeterminate structures will be shown to vary from structure to structure, but may never be less than two. The fixed-end beam of Fig. 26.9 cannot fail unless the three plastic hinges shown in the figure are developed.

Although a plastic hinge may have formed in a statically indeterminate structure, the load can still be increased without causing failure if the geometry of the structure permits. The plastic hinge will act like a real hinge insofar as increased loading is concerned. As the load is increased, there is a redistribution of moment because the plastic hinge can resist no more moment. As more plastic hinges are formed in the structure, there will eventually be a sufficient number of them to cause collapse. Actually some additional load can be carried after this time before collapse occurs as the stresses go into the strain hardening range, but the deflections that would occur are too large to be permissible.

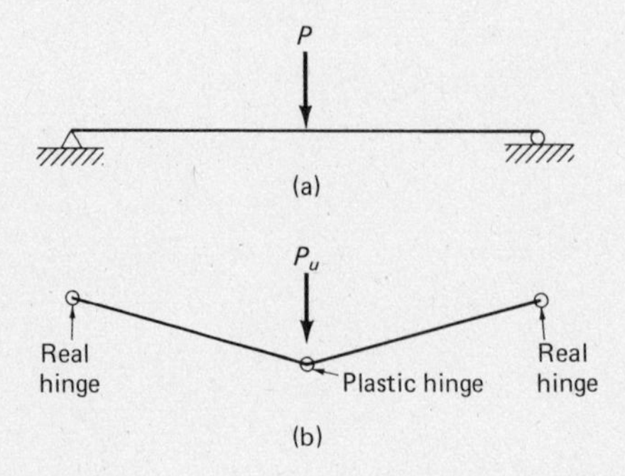

Figure 26.8

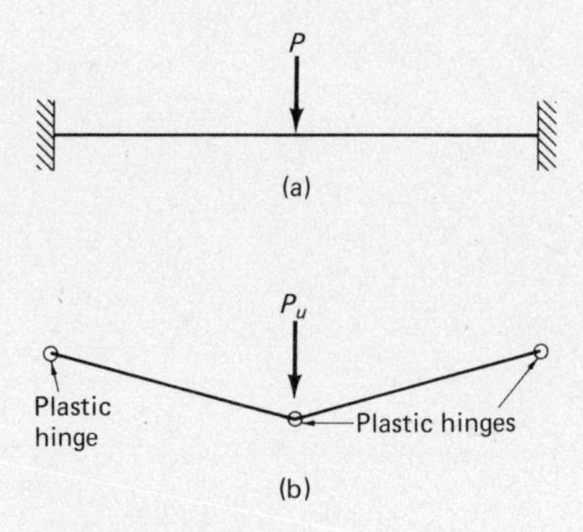

Figure 26.9

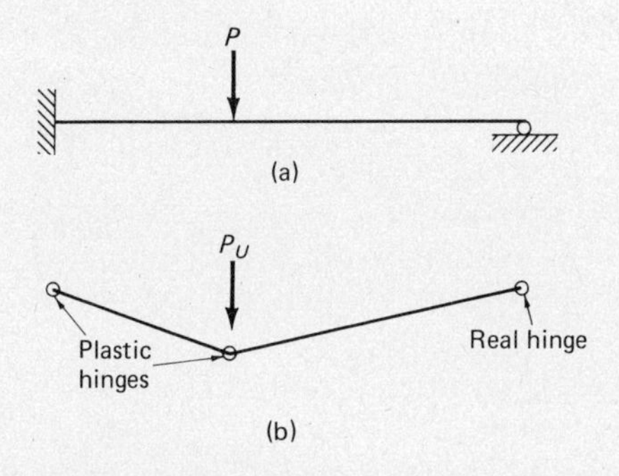

Figure 26.10

The propped beam of Fig. 26.10 is an example of a structure that will fail after two plastic hinges develop. Three hinges are required for collapse, but there is a real hinge on the right end. In this beam the largest elastic moment caused by the design concentrated load is at the fixed end. As the magnitude of the load is increased, a plastic hinge will form at that point.

The load may be further increased until the moment at some other point (here it will be at the concentrated load) reaches the plastic moment. Additional load will cause the beam to collapse. The arrangement of plastic hinges and perhaps real hinges that permit collapse in a structure is called the *mechanism*. Parts (b) of Figs. 26.8, 26.9, and 26.10 show mechanisms for various beams.

After observing the large number of fixed-end and propped beams used for illustration in this text, the reader may form the mistaken idea that such beams will be frequently encountered in engineering practice. These types of beams, truthfully, are difficult to find in actual structures but are very convenient to use in illustrative examples. They are particularly convenient for introducing plastic analysis before continuous beams and frames are considered.

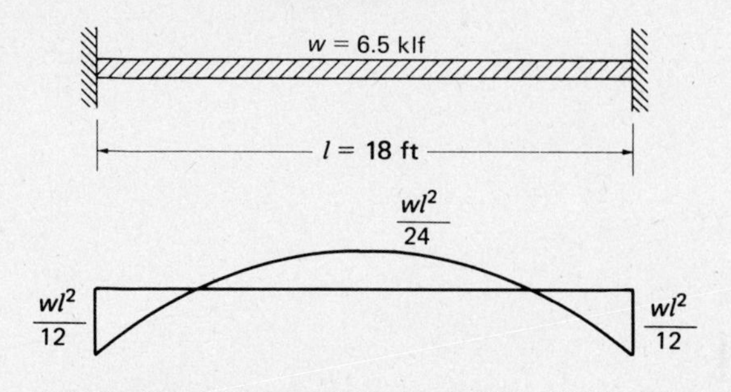

Figure 26.11

26.7 PLASTIC ANALYSIS BY THE EQUILIBRIUM METHOD

The method of plastic analysis known as the *equilibrium method* will be illustrated in this section for several beams. The analysis includes the computations of the plastic moments, the consideration of load redistribution after plastic hinges have formed, and the calculation of the ultimate loads that exist when the collapse mechanism is created.

As the first illustration the fixed-end beam of Fig. 26.11 is considered. This beam is assumed to support a load of 6.5 klf, including its own estimated weight. A W 18 × 50 (S = 88.9 in^3) has been selected by the elastic procedure.

It is desired to determine the value of w_u, the ultimate uniform load which this W section can support before collapse.

The maximum moments in a uniformly loaded fixed-end beam in the elastic range occur at the fixed ends as shown in Fig. 26.11. If the magnitude of the uniform load is increased, plastic moments will eventually be developed at the beam ends as shown in Fig. 26.12(b). Although the plastic moment has been reached at the ends and plastic hinges formed, the beam cannot fail as it has, in effect, become a simple end-supported beam as shown in part (c) of the figure.

The load can now be increased on this "simple" beam; and the moments

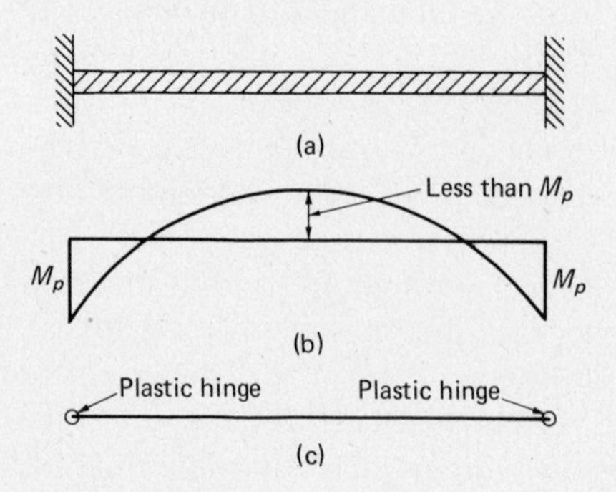

Figure 26.12

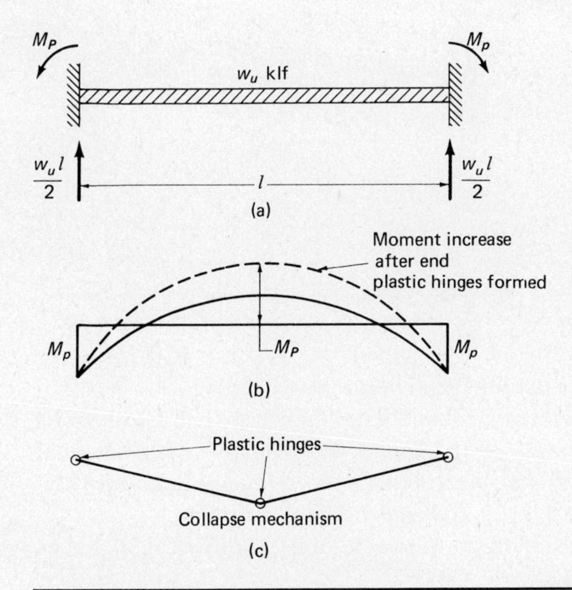

Figure 26.13

at the ends will remain constant; but the moment out in the span will increase as it would in a uniformly loaded simple beam. This increase is shown by the dotted line in Fig. 26.13(b). The load may be increased until the moment at some other point (here the beam centerline) reaches the plastic moment. When this happens, a third plastic hinge will have developed and a mechanism will have been created permitting collapse.

One method of determining the value of w_u is to take moments at the centerline of the beam (knowing the moment there is M_p at collapse). Reference is made here to Fig. 26.13(a) for the beam reactions. The value of Z (101 in^3) was obtained from the *Manual of Steel Construction*.

$$M_p = -M_p + \left(w_u \frac{l}{2}\right)\left(\frac{l}{2} - \frac{l}{4}\right)$$

$$= \frac{w_u l^2}{16}$$

$$w_u = \frac{16M_p}{l^2}$$

$$M_p = F_y Z = \frac{(36)(101)}{(12)} = 303'^{k}$$

$$w_u = \frac{(16)(303)}{(18)^2} = 14.96 \text{ klf}$$

The plastic safety factor equals 14.96/6.50 = 2.30. This value appears to be a more realistic value for a ductile steel structure than one computed on the basis of the yield stress (approximately 1.50).

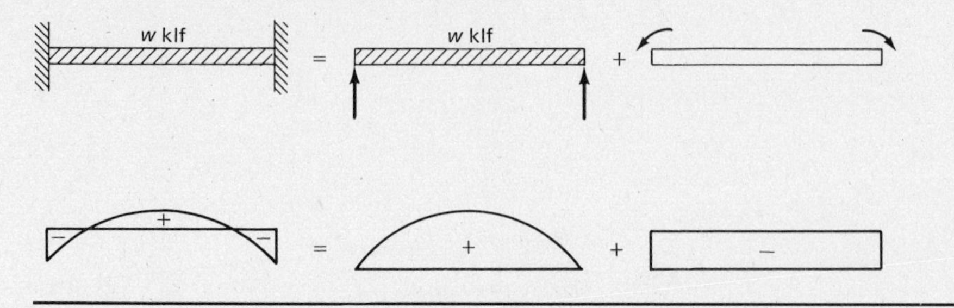

Figure 26.14

The same values could be obtained by considering the diagrams shown in Fig. 26.14. The reader will remember that a fixed-end beam can be replaced with a simply supported beam plus a beam with end moments. Thus the final moment diagram for the fixed-end beam equals the moment diagram if the beam had been simply supported plus the end-moment diagram.

For the beam under consideration the value of M_p can be calculated as follows after studying Fig. 26.15:

$$2M_p = \frac{w_u l^2}{8}$$

$$M_p = \frac{w_u l^2}{16}$$

The propped beam of Fig. 26.16(a) has been designed by the elastic method to support a 50 ft-kip concentrated load at midspan. The W 18 × 55 ($S = 98.3$ in^3, $Z = 112$ in^3) beam of A36 steel selected is now to be considered as a second illustration of plastic analysis. The elastic-moment diagram for a propped beam loaded with a concentrated load P at midspan is shown in part (b) of the figure. From this diagram the maximum moment can be seen to occur at the fixed end. The concentrated load is assumed to be increased until the plastic moment is reached at the fixed end and a plastic hinge formed.

After this plastic hinge is formed the beam will act as though it is simply supported insofar as load increases are concerned, because it will have a plastic hinge at the left end and a real hinge at the right end. An increase in the magnitude of the load P will not increase the moment at the left end but will increase the moment out in the beam as it would in a simple beam. The increasing simple beam moment is indicated by the dotted line in part (c) of Fig. 26.16. Eventually the moment at the concentrated load will reach M_p and

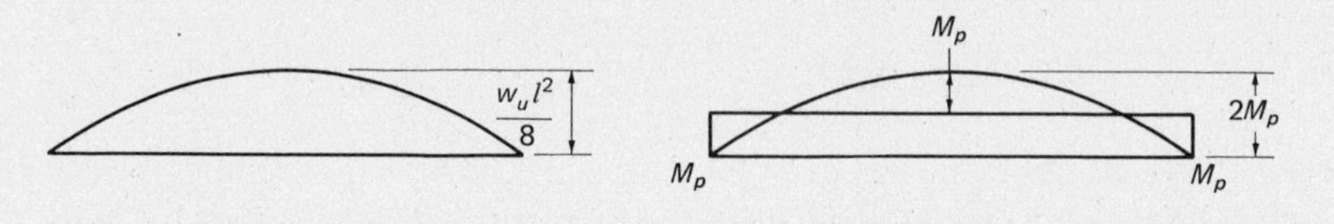

Figure 26.15

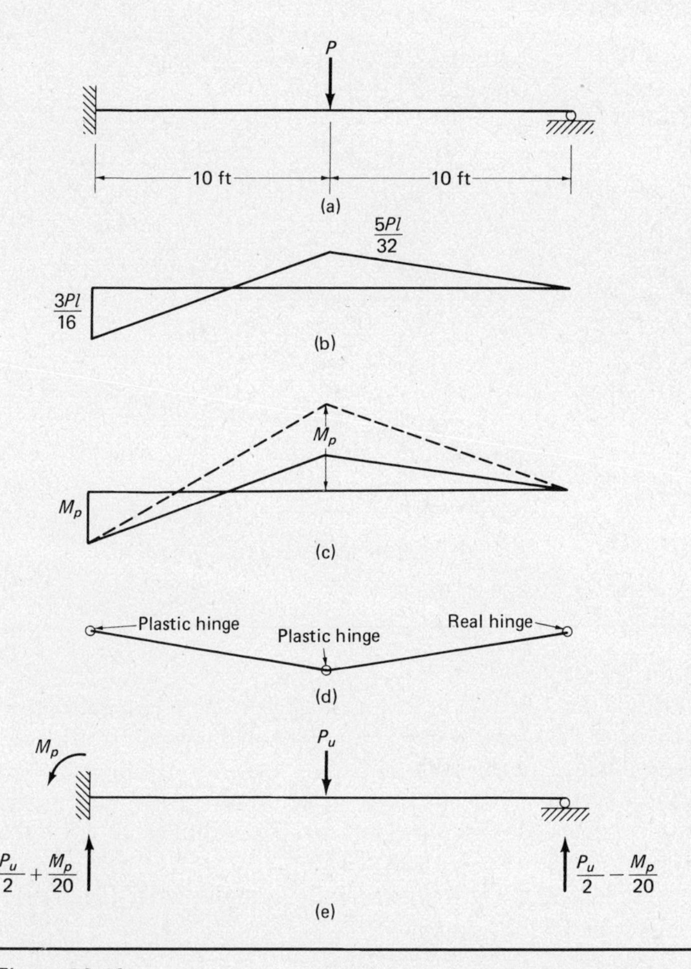

Figure 26.16

a mechanism will form, consisting of two plastic hinges and one real hinge as shown in part (d).

The value of the maximum concentrated load P_u that the beam can support can be determined by taking moments to the right or left of the load. Part (e) of Fig. 26.16 shows the beam reactions for the conditions existing just before collapse. Moments are taken to the right of the load as follows:

$$M_p = \left(\frac{P_u}{2} - \frac{M_p}{20}\right) 10$$

$$= 3.33\ P_u$$

$$P_u = 0.3\ M_p$$

$$M_p = F_y Z = \frac{(36)(112)}{12} = 336'^{k}$$

$$P_u = 100.8^{k}$$

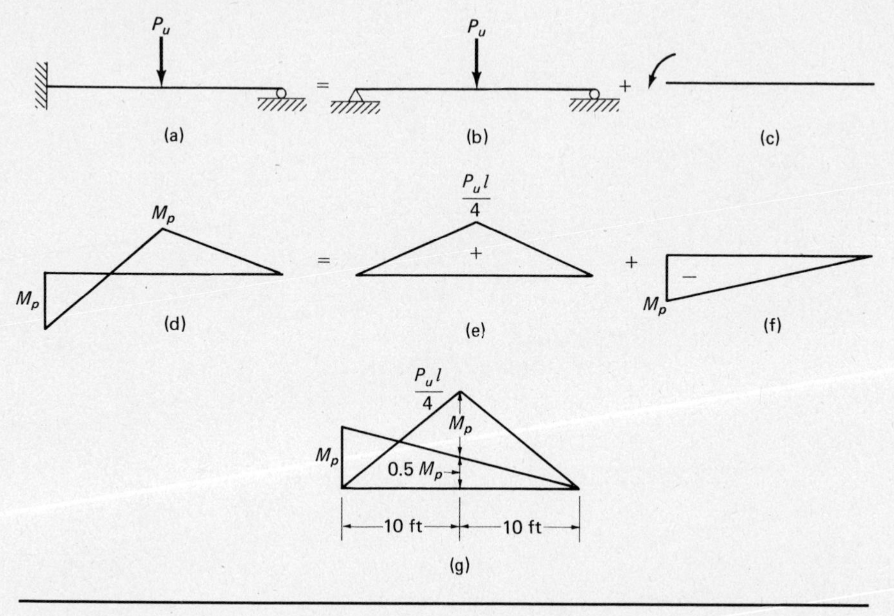

Figure 26.17

The other method described for handling the plastic analysis of a structure involved the drawing of the plastic-moment diagram on the structure after the structure had been made statically determinate by the creation of a sufficient number of plastic hinges and the equating of the resulting diagram to the simple-beam moment diagram. This procedure is repeated in Fig. 26.17 for the beam previously considered in Fig. 26.16.

From part (g) of Fig. 26.17 the following expression can be written for the moment at the centerline of the beam:

$$M_p + 0.5M_p = \frac{P_u l}{4}$$

$$P_u = 0.3M_p$$

Another example is handled by a similar procedure in Fig. 26.18.

$$2Mp = \frac{P_u ab}{l} = \frac{(P_u)(10)(20)}{30} = 6.67P_u$$

$$P_u = 0.3Mp$$

A propped beam with two concentrated loads is shown in Fig. 26.19(a). Design was performed by the elastic procedure for the 30 and 50-kip loads shown and a W 24 × 84 (S = 196 in^3, Z = 224.0 in^3) was selected. It is now desired to determine the ultimate values of the two concentrated loads if they are increased at such a rate that they remain in the same proportion to each other. In part (b) of the figure the larger load at collapse is said to equal P_u and the smaller one $0.6P_u$.

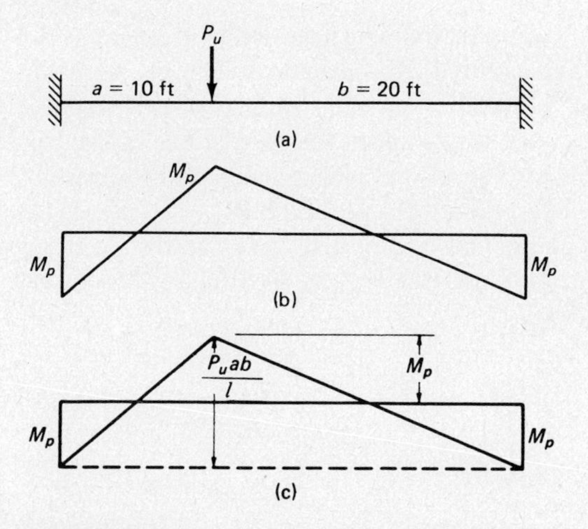

Figure 26.18

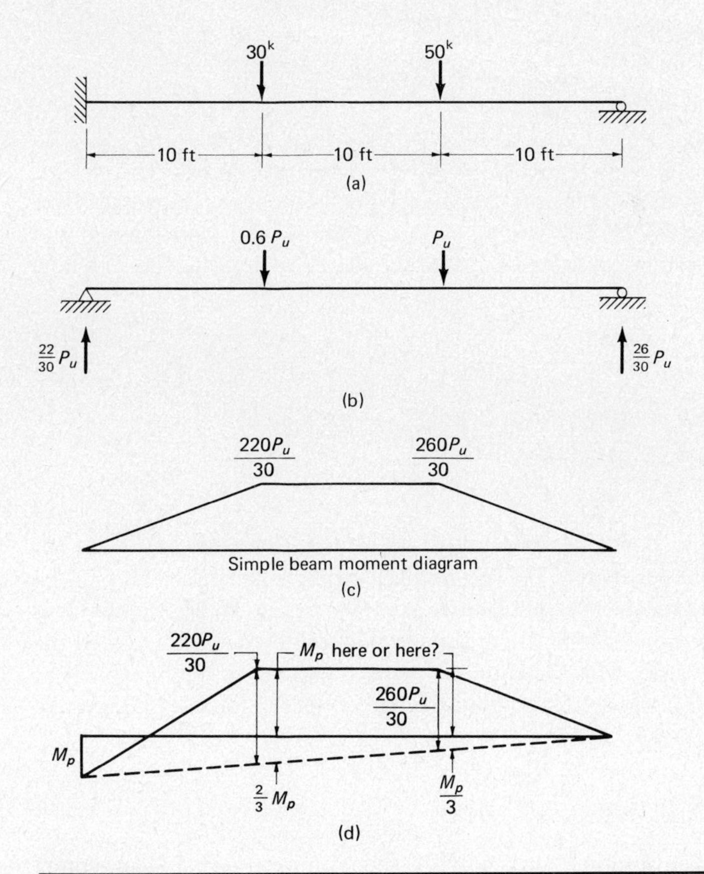

Figure 26.19

As the loads are increased, the plastic moment will first be reached at the left end. After this plastic hinge forms, the structure will be statically determinate since the right end is a real hinge. The loads may be increased until eventually another plastic hinge forms out in the span at one of the two concentrated loads. The point where the second plastic hinge will form is not obvious and it will be necessary to consider both possibilities.

Should the second plastic hinge be assumed to form at the point of application of the $0.6P_u$ concentrated load [see Fig. 26.19(d)] the values of M_p and P_u can be determined as follows:

$$M_p + \tfrac{2}{3}M_p = \frac{220P_u}{30}$$

$$M_p = 4.4P_u$$

$$P_u = 0.227M_p$$

If the second plastic hinge is assumed to form at the point of application of P_u, the values of M_p and P_u would be as follows:

$$M_p + \frac{M_p}{3} = \frac{260P_u}{3}$$

$$M_p = 6.5P_u$$

$$P_u = 0.154M_p$$

The moment at the concentrated load P_u will obviously be greater than the moment at the concentrated load $0.6P_u$ and the second plastic hinge will occur at P_u. The numerical value of P_u and the plastic factor of safety can now be calculated.

$$M_p = F_yZ = \frac{(36)(224)}{12} = 672'^{k}$$

$$P_u = (0.154)(672) = 103.5^{k}$$

$$\text{factor of safety} = \frac{103.5}{50} = 2.07$$

If there is any doubt in the reader's mind as to whether he or she has selected the correct location of the second plastic hinge, he or she can plot the moment diagrams for the two possibilities as shown in Fig. 26.20. Should he or she have selected the wrong location, the moment somewhere will exceed the calculated value of M_p. This situation is shown in part (c) of the figure where the assumed location of the second plastic hinge under the left load is shown to be impossible.

26.8 THE VIRTUAL-WORK METHOD

As beams or frames become more complex, the equilibrium method becomes more and more tedious to apply. A method is introduced in this section, called the ***virtual-work method***, which will usually prove to be simpler and quicker

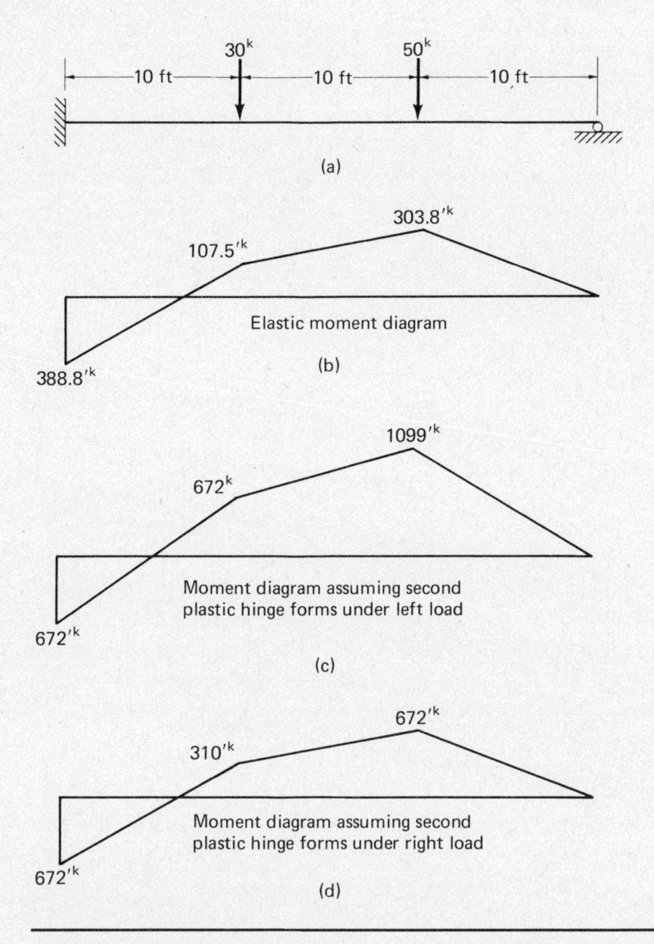

Figure 26.20

than the equilibrium method. This procedure will be used to rework the problems previously considered in Sec. 26.7.

Each structure is assumed to deflect through a small additional displacement after the ultimate load is reached. The work performed by the external loads during the displacement is equated to the internal work absorbed by the hinges. As a first illustration the uniformly loaded fixed-ended beam of Fig. 26.11 is considered. This beam and its collapse mechanism are redrawn in Fig. 26.21. Owing to symmetry, the rotations at the end plastic hinges are equal and they are represented by θ in the figure; thus the rotation at the middle plastic hinge will be 2θ.

The work performed by the total external load ($w_u l$) is equal to $w_u l$ times the average deflection of the mechanism. The average deflection equals one-half the deflection at the center plastic hinge ($\frac{1}{2} \times \theta \times l/2$). The external work is equated to the internal work absorbed by the hinges or to the sum of M_p at each plastic hinge times the angle through which it works. The resulting expression can be solved for M_p and w_u as follows:

$$M_p(\theta + 2\theta + \theta) = w_u l \left(\theta \times \frac{l}{2} \times \frac{1}{2}\right)$$

$$w_u = \frac{16M_p}{l^2}$$

or

$$M_p = \frac{w_u l^2}{16}$$

For the 18-ft span used in Fig. 26.11 these values become

$$M_p = \frac{(w_u)(18)^2}{16} = 20.25w_u$$

and

$$w_u = \frac{M_p}{20.25}$$

Plastic analysis can be handled in a similar manner for the propped beam of Fig. 26.16. This beam is redrawn in Fig. 26.22 together with its collapse mechanism. Again the end rotations are equal and are assumed to equal θ.

The work performed by the external load P_u as it moves through the distance $\theta l/2$ is equated to the internal work performed by the plastic moments at the hinges, noting that there is no moment at the real hinge on the right end of the beam.

$$M_p(\theta + 2\theta) = P_u\left(\theta \frac{l}{2}\right)$$

$$M_p = \frac{P_u l}{6} \quad \text{or } 3.33\, P_u \text{ for the 20-ft beam shown}$$

$$P_u = \frac{6M_p}{l} \quad \text{or } 0.3M_p \text{ for the 20-ft beam shown}$$

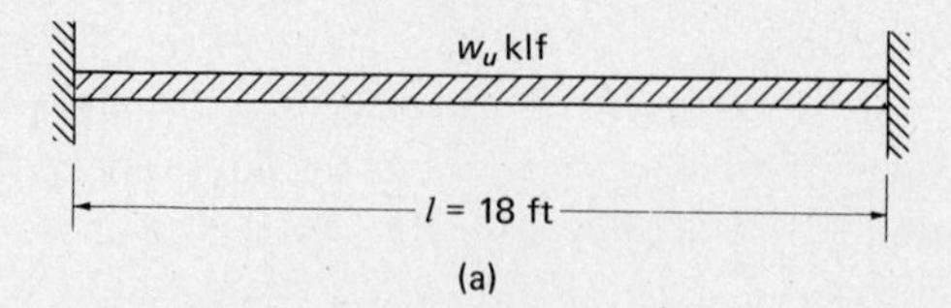

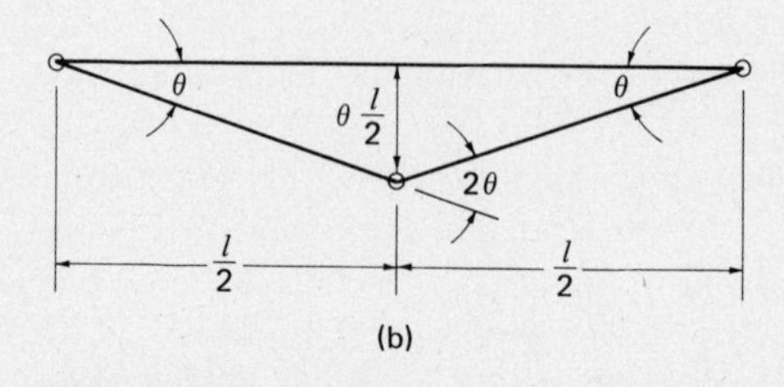

Figure 26.21

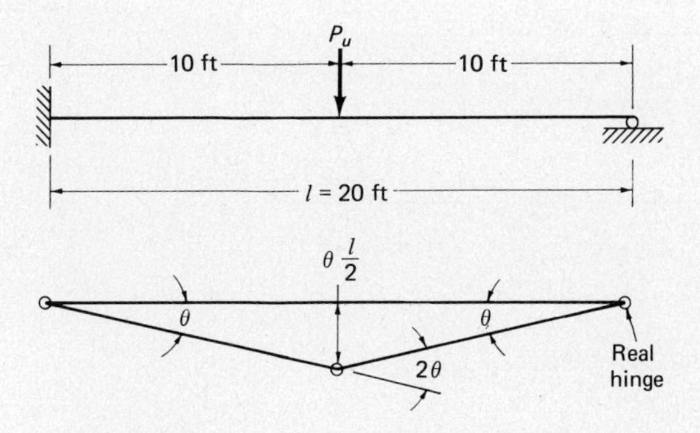

Figure 26.22

The fixed-end beam of Fig. 26.18 is redrawn in Fig. 26.23 together with its collapse mechanism and the assumed angle rotations. From this figure the values of M_p and P_u can be determined by virtual work as follows:

$$M_p(2\theta + 3\theta + \theta) = P_u\left(2\theta \times \frac{l}{3}\right)$$

$$M_p = \frac{P_u l}{9} \qquad \text{or } 3.33\ P_u \text{ for this beam}$$

$$P_u = \frac{9M_p}{l} \qquad \text{or } 0.3\ M_p \text{ for this beam}$$

The plastic analysis of the propped beam of Fig. 26.19 by the virtual-work method is now considered. The beam with its two concentrated loads is redrawn in Fig. 26.24(a). The location of the second plastic hinge is not obvious and a trial-and-error procedure is necessary. In part (b) of Fig. 26.24 the second hinge is assumed to form at the left concentrated load, and the collapse mechanism with its assumed angles in terms of θ is drawn. In part (c) of the figure the second hinge is assumed to form at the right concentrated load, and the collapse mechanism is drawn.

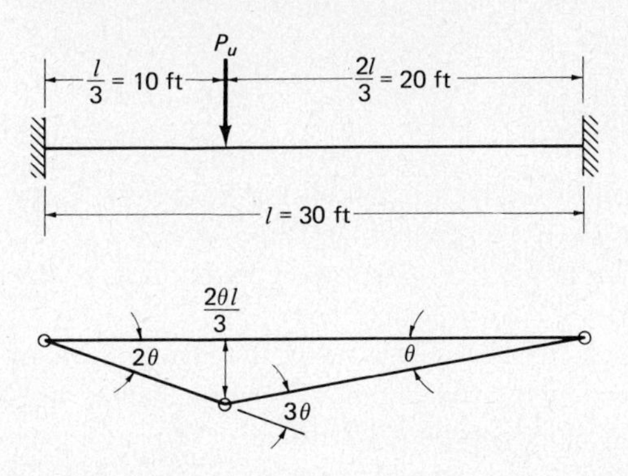

Figure 26.23

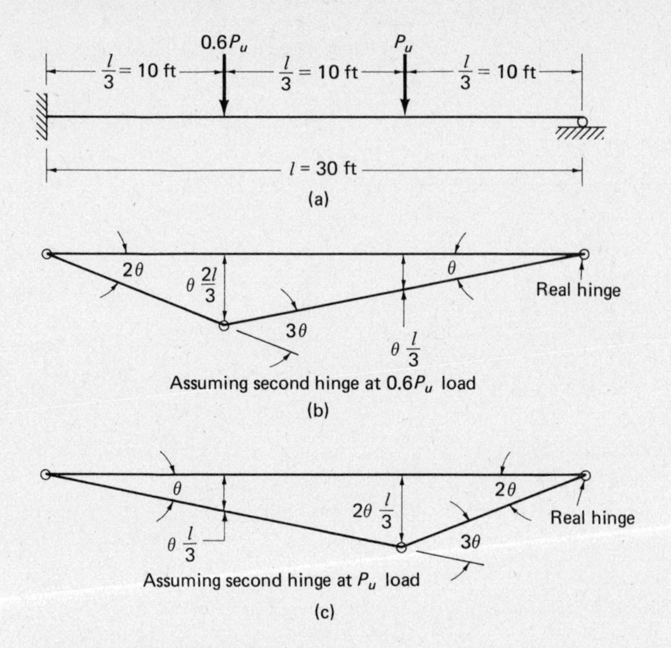

Figure 26.24

If the second plastic hinge is assumed to form at the $0.6P_u$ load, the value of M_p can be found as follows:

$$M_p(2\theta + 3\theta) = 0.6P_u\left(\theta\frac{2l}{3}\right) + P_u\left(\theta\frac{l}{3}\right)$$

$$M_p = 0.1466P_u l$$

$$P_u = \frac{6.82M_p}{l} = \frac{6.82M_p}{30} = 0.227M_p$$

If the second plastic hinge is assumed to form at the P_u load, the value of M_p can be determined as follows:

$$M_p(\theta + 3\theta) = 0.6P_u\left(\theta\frac{l}{3}\right) + P_u\left(\frac{2\theta l}{3}\right)$$

$$M_p = 0.2167P_u l$$

$$P_u = \frac{4.61M_p}{l} = \frac{4.61M_p}{30} = 0.154M_p$$

The value for which the collapse load is the smallest in terms of M_p is the correct value (or the value where M_p is the greatest in terms of P_u). For this beam the second plastic hinge forms at the P_u concentrated load and P_u equals $4.61M_p/l$.

A person beginning the study of plastic analysis needs to learn to think of all the possible ways in which a particular structure may collapse. Such a habit is of the greatest importance when more complex structures are encountered. In this chapter the author has attempted to explain how to handle different mechanisms and has not tried to consider every possible situation. For instance upon examining Fig. 26.24 the reader might decide that plastic hinges could form at the concentrated loads before one formed at the fixed end. The result would be a mechanism for the part of the beam to the right of the $0.6P_u$ load. The virtual work expression could easily be written for this case. (It doesn't control.)

26.9 LOCATION OF PLASTIC HINGE FOR UNIFORM LOADINGS

There was no difficulty in locating the plastic hinge for the uniformly loaded fixed-end beam, but for other beams with uniform loads, such as propped or continuous beams, the problem is rather difficult. For this discussion the uniformly loaded propped beam of Fig. 26.25(a) is considered.

The elastic-moment diagram for this beam is shown in part (b) of the figure. As the uniform load is increased in magnitude, a plastic hinge will first form at the fixed end. At this time the beam will, in effect, be a "simple" beam with a plastic hinge on one end and a real hinge on the other. Subsequent increases in the load will cause the moment to change as represented by the dotted line in part (b) of the figure. This process will continue until the moment at some other point (a distance x from the right support in the figure) reaches M_p and creates another plastic hinge.

The virtual work expression for the collapse mechanism of this beam shown in part (c) of Fig. 26.25 is written as follows:

$$M_p\left(\theta + \theta + \frac{l - x}{x}\theta\right) = (w_u l)(\theta)(l - x)\left(\frac{1}{2}\right)$$

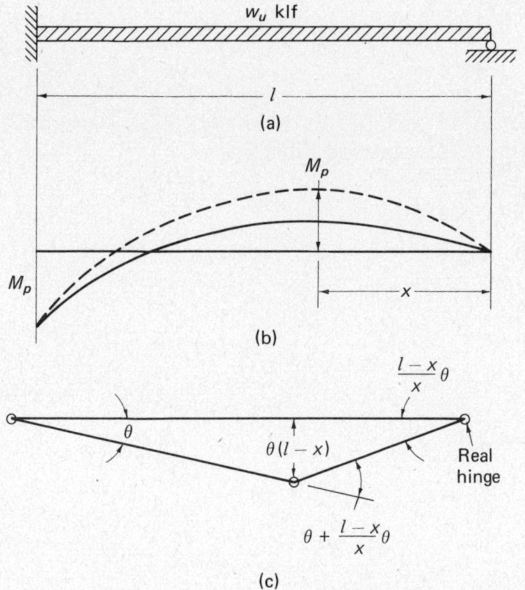

Figure 26.25

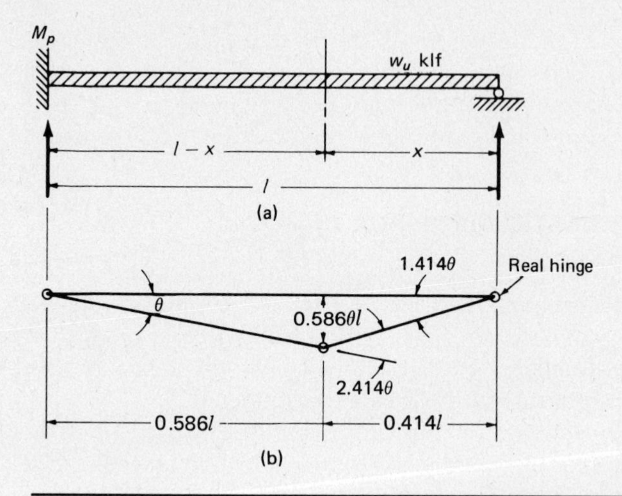

Figure 26.26

Solving this equation for M_p, taking $dM_p/dx = 0$, the value of x can be calculated to equal $0.414l$. This value is also applicable to uniformly loaded end spans of continuous beams with simple end supports.

The beam and its collapse mechanism are redrawn in Fig. 26.26 and the following expression for the plastic moment is written using the virtual-work procedure:

$$M_p(\theta + 2.414\theta) = (w_u l)(0.586\theta l)(\tfrac{1}{2})$$

$$M_p = 0.0858 w_u l^2$$

PROBLEMS

For Probs. 26.1 to 26.9, find the values of S, Z, and the shape factor about the x axes for the sections shown in the accompanying illustrations.

Problem 26.1

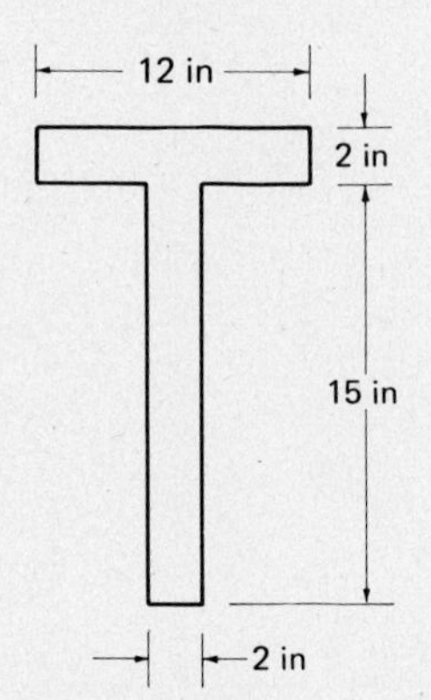

Problem 26.2 (*Ans.* $S = 229.1\ \text{in}^3$, $Z = 271\ \text{in}^3$, shape factor = 1.18)

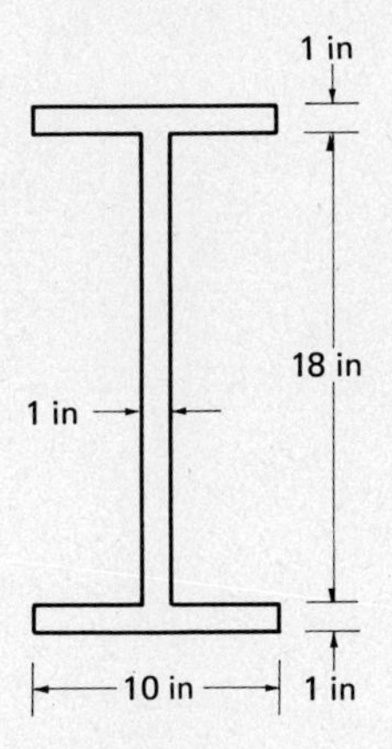

Problem 26.3

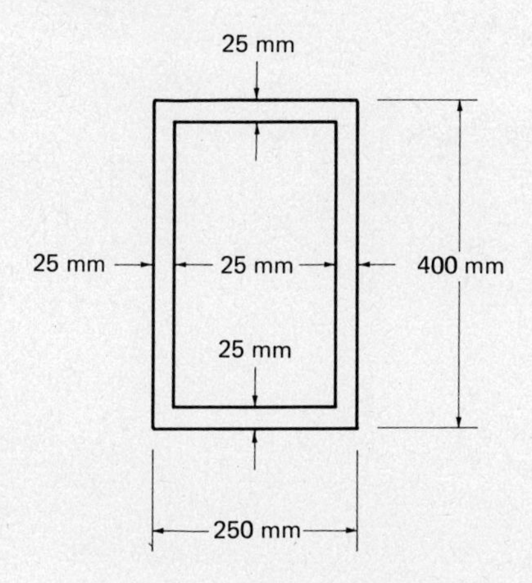

Problem 26.4 (*Ans.* $S = 209\ \text{in}^3$, $Z = 281\ \text{in}^3$, shape factor = 1.345)

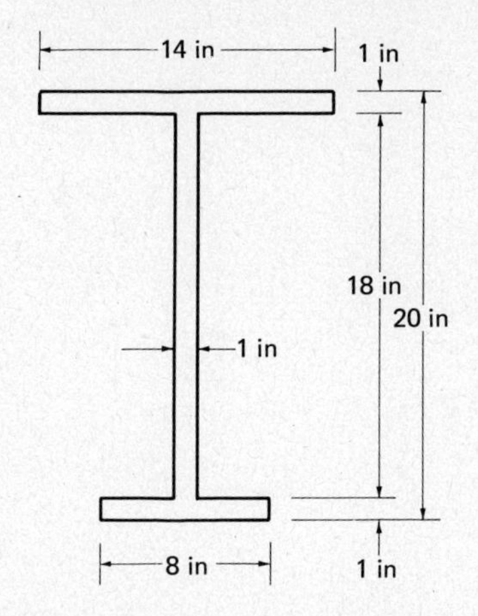

Problem 26.5

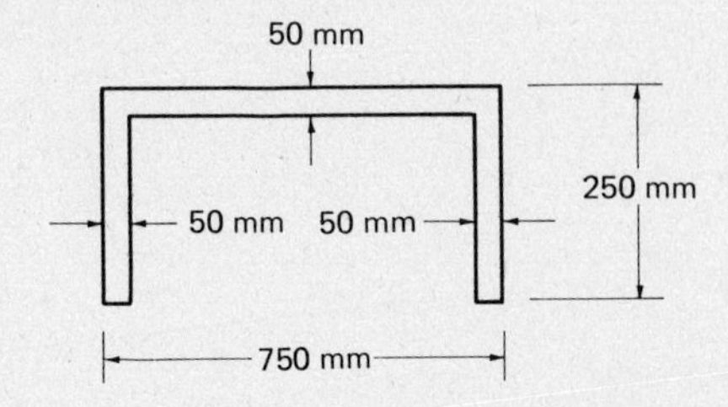

Problem 26.6 (*Ans.* $S = 359 \text{ in}^3$, $Z = 654 \text{ in}^3$, shape factor = 1.83)

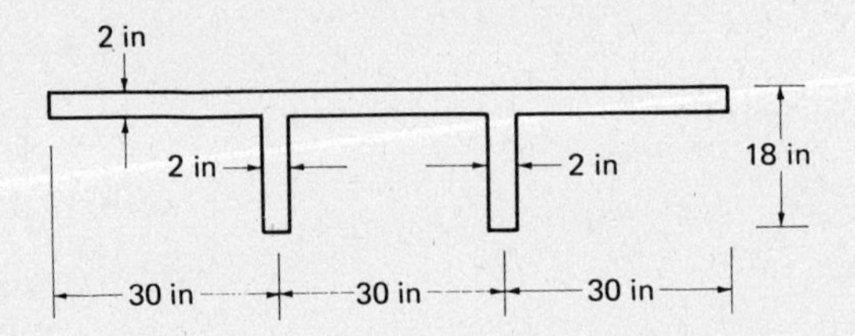

Problem 26.7

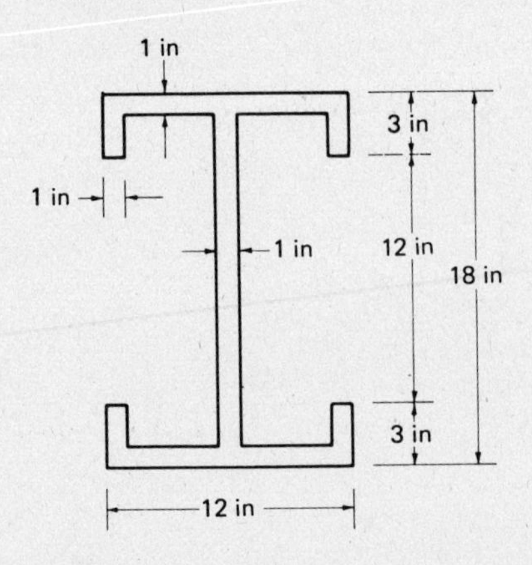

Problem 26.8 (*Ans.* $S = 98.2 \text{ in}^3$, $Z = 166 \text{ in}^3$, shape factor = 1.69)

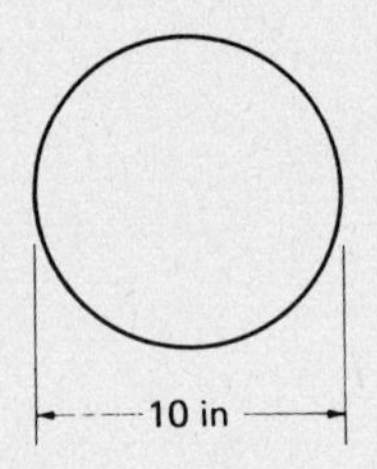

Problem 26.9

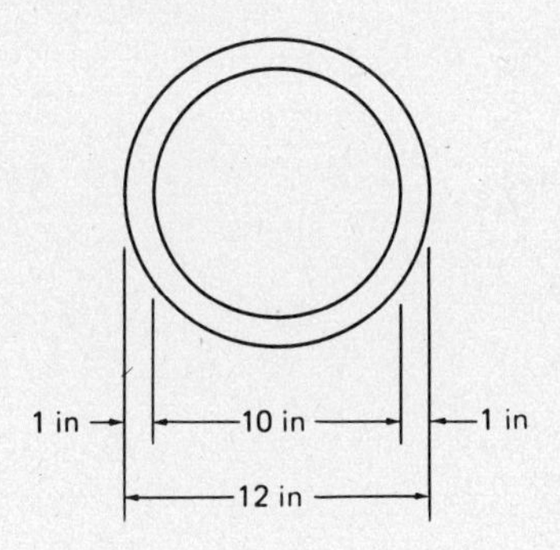

For Probs. 26.10 to 26.15 determine the values of S, Z, and the shape factor about the x axes for the situation described. Average thicknesses of flanges and webs are to be used.

26.10. A *W* 18 × 40. (*Ans.* $S = 67.3$ in^3, $Z = 77.2$ in^3, shape fator = 1.145)
26.11. A *W* 36 × 135.
26.12. An *S* 10 × 25.4 (*Ans.* $S = 24.5$ in^3, $Z = 28.1$ in^3, shape factor = 1.145)
26.13. A *W* 24 × 62 with one *PL* $\frac{1}{2}$ × 12 welded to each flange.
26.14. Two *C* 12 × 30 sections back to back. (*Ans.* $S = 53.8$ in^3, $Z = 67.3$ in^3, shape factor = 1.25)
26.15. Two *Ls* 8 × 6 × $\frac{3}{4}$ long legs vertical and back to back.
26.16. Rework Prob. 26.1 considering the y axis. (*Ans.* $S = 49.7$ in^3, $Z = 87.0$ in^3, shape factor = 1.75)
26.17. Rework Prob. 26.13 considering the y axis.
26.18. Rework Prob. 26.14 considering the y axis. (*Ans.* $S = 7.00$ in^3, $Z = 12.9$ in^3, shape factor = 1.84.)

For Probs. 26.19 to 26.28 calculate the factors of safety for the beam sections according to the elastic and plastic theories. The loads shown in each figure are assumed to include the estimated beam weights $F_y = 36$ ksi in all problems. Use the AISC Manual of Steel Construction for determining the values of S and Z.

Problem 26.19

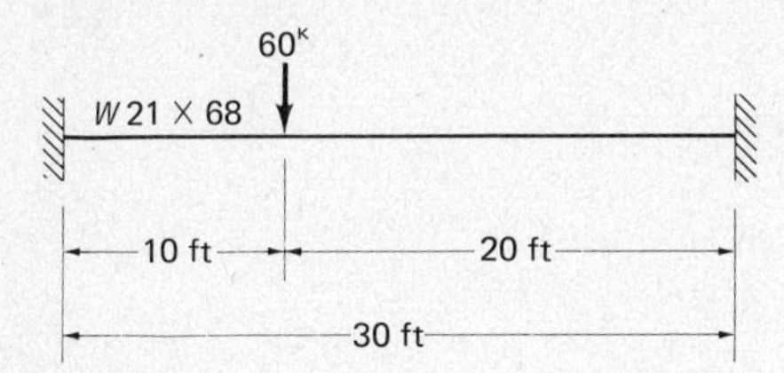

Problem 26.20 (*Ans.* Elastic = 1.54, plastic = 2.34)

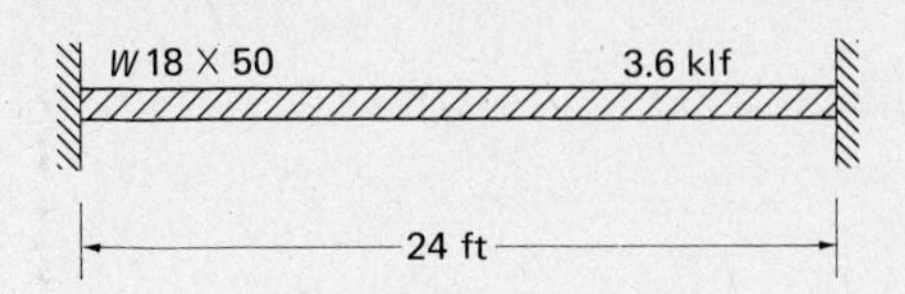

Problem 26.21

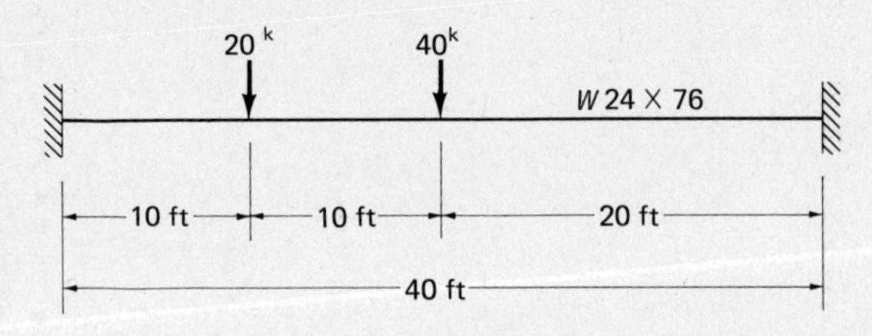

Problem 26.22 (*Ans.* Elastic = 1.43, plastic = 2.24)

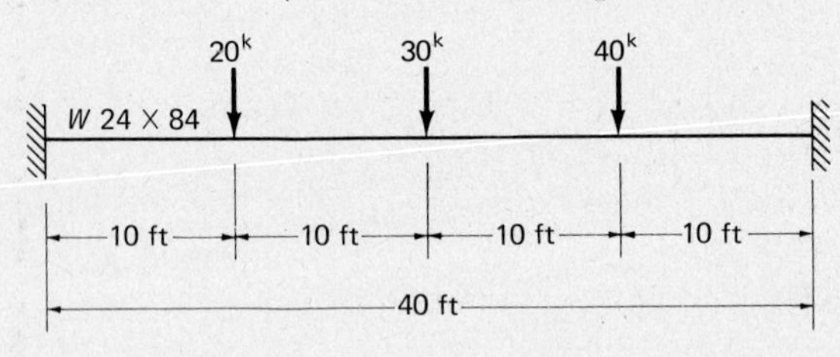

Problem 26.23

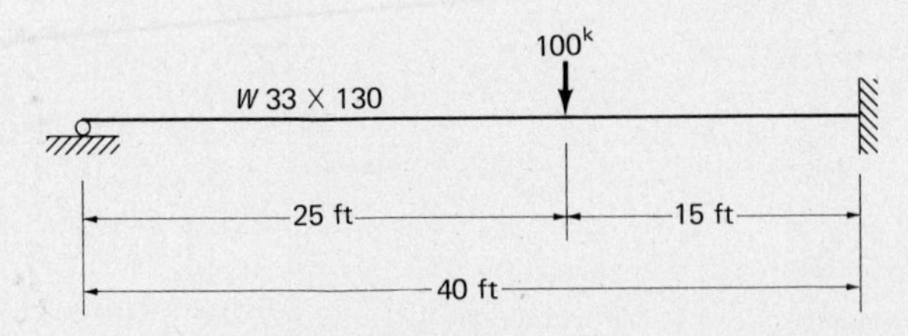

Problem 26.24 (*Ans.* Elastic = 1.55, plastic = 2.74)

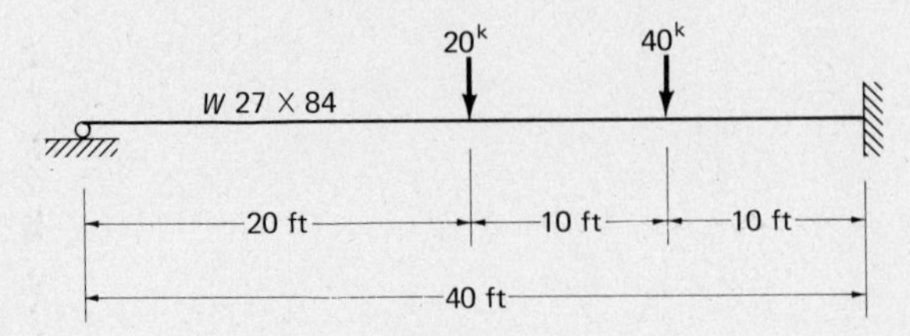

Problem 26.25

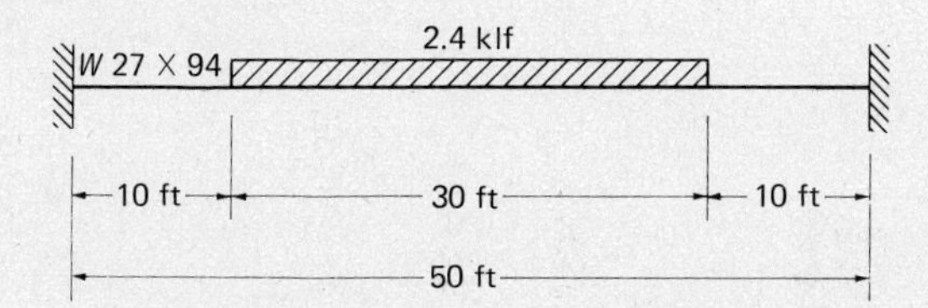

Problem 26.26 (*Ans.* Elastic = 1.53, plastic = 2.16)

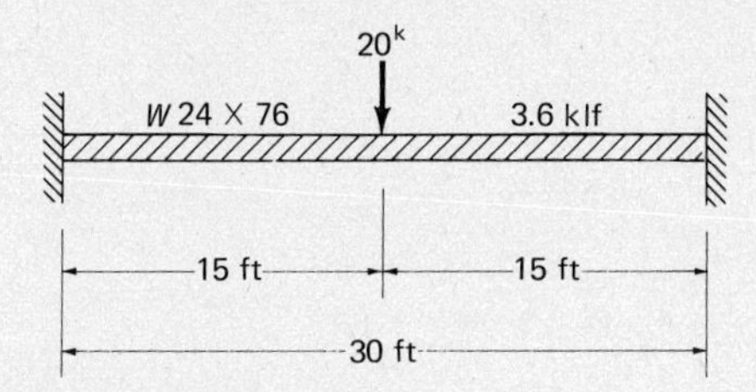

Problem 26.27

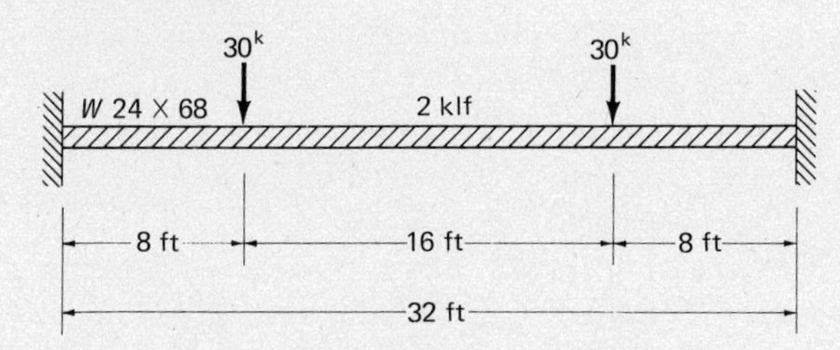

Problem 26.28 (*Ans.* Elastic = 1.42, plastic = 2.41)

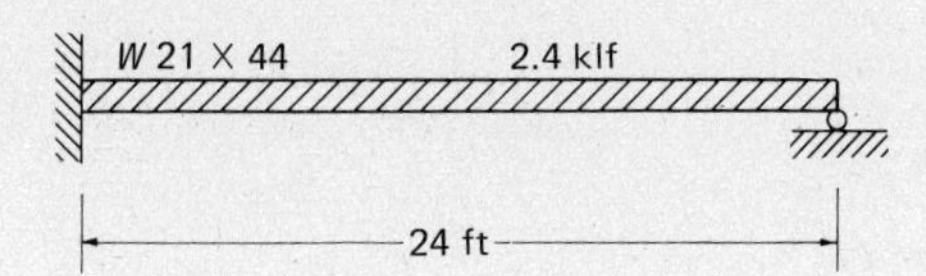

REFERENCES

[1] Lynn S. Beedle, *Plastic Design of Steel Frames* (New York: Wiley, 1958), p. 3.
[2] Plastic Design Selection Table, *Manual of Steel Construction* (New York: American Institute of Steel Construction, 1980).

27 Matrix Methods of Structural Analysis*

27.1 GENERAL

During the past several decades there have been tremendous changes in the structural analysis methods used in engineering practice. These changes have primarily occurred due to the great developments made with high-speed digital computers and due to the increasing use of very complex structures. Matrix methods of analysis provide a convenient mathematical language for describing complex structures and the matrix manipulations can easily be handled with computers.

Structural engineers have been attempting to handle analysis problems for a good many years by applying the methods used by mathematicians in linear algebra. Although many structures could be analyzed with the resulting equations the work was tedious, at least until large-scale computers became available.

Today matrix analysis is increasingly replacing the classical methods of analysis in engineering offices. As a result engineering educators and structural analysis textbook writers are faced with a difficult decision. Should they require a thorough study of the classical methods followed by a study of modern methods; should they require students to study both at the same time in an integrated approach; or should they just present a study of the modern methods?

The reader can see from the preceding 26 chapters that the author feels an initial study of the classical methods followed by a study of the matrix methods will result in an engineer who has a better understanding of structural behavior. (The author, however, often has second thoughts on this matter.)

Matrix methods usually require the use of digital computers to complete their solution. In fact, the usual matrix equations are not manageable by hand calculations unless the most elementary structures are involved. As a result due

*The author is particularly grateful to Dr. J. C. Smith of N.C. State University for his considerable assistance in preparing this chapter.

to the seeming complexity of matrix methods, members of the structural engineering profession may be frightened away from using them. These engineers need to realize that the application of digital computers to everyday structural problems is today a reality and that matrix algebra provides the simplest approach presently available for preparing these problems for computer solution.

This chapter presents only a brief introduction to matrix methods of structural analysis. Its purpose is to give the reader a "feel" for the subject and prepare him or her for further study elsewhere. The simple examples solved herein could be solved more quickly by classical methods using a pocket calculator rather than a matrix approach. However, as structures become more complex and as more loading patterns are considered matrix methods become increasingly useful.

Any method of analysis involving linear algebraic equations can be put into matrix notation and matrix operations may be used to obtain their solution. The possibility of application of matrix methods by the structural engineer is very important because all statically indeterminate structures are governed by systems of linear equations.

Standard digital computer programs are readily available for handling matrices and should be considered for structural problems that involve the solution of large numbers of simultaneous equations. Furthermore, many computers can be programmed directly for these operations and if matrix formulation is used, the programming can be appreciably shortened.

27.2 MATRIX ALGEBRA

For the purposes of this chapter it is assumed that the student has had some previous exposure to matrix algebra. Thus the author has not included an overall discussion of the subject. He has, however, attempted to review in this section a very few terms and procedures of matrix algebra which are used later in this chapter.

A *matrix* is defined as a rectangular block or array of elements. The block is enclosed within square brackets to show that it is a matrix. It can be represented with a single symbol which is bracketed, although the brackets can be omitted if no ambiguity results. Each element of the matrix is represented by a letter with two subscripts, the first denoting the row, the second the column in which the element appears. The element of a matrix in the ith row and jth column is represented by the notation a_{ij}. A typical matrix follows. Commas are used to separate the subscript values because they are necessary when the numbers exceed one digit.

$$[A] = \begin{bmatrix} a_{1,1} & a_{1,2} & a_{1,3} & a_{1,4} \\ a_{2,1} & a_{2,2} & a_{2,3} & a_{2,4} \\ a_{3,1} & a_{3,2} & a_{3,3} & a_{3,4} \\ a_{4,1} & a_{4,2} & a_{4,3} & a_{4,4} \end{bmatrix}$$

For the purposes of this discussion the following three simultaneous equations are considered.

$$\begin{aligned} 2X_1 + 3X_2 + 4X_3 &= 20 \\ X_1 - 4X_2 + 5X_3 &= 8 \\ 4X_1 + 3X_2 - 2X_3 &= 4 \end{aligned}$$

The coefficients in these equations can be represented by matrix $[A]$. This is a square matrix which has the same number of rows as columns.

$$[A] = \begin{bmatrix} 2 & 3 & 4 \\ 1 & -4 & 5 \\ 4 & 3 & -2 \end{bmatrix}$$

A *row matrix* is one having a single row. Its order is $1 \times n$. The following is a 1×3 row matrix,

$$[B] = [1 \quad 3 \quad -16]$$

Or as sometimes written

$$\lfloor B \rfloor = \lfloor 1 \quad 3 \quad -16 \rfloor$$

A *column matrix* has a single column and thus the order $m \times 1$. Column and row matrices are often referred to as *vectors*. It is common to enclose rectangular and row matrices in brackets and column matrices in braces as illustrated for the 4×1 column matrix to follow:

$$\{C\} = \begin{Bmatrix} 3 \\ 4 \\ 2 \\ 7 \end{Bmatrix}$$

A *unit* or *identity matrix* is a diagonal matrix all of whose elements on the main diagonal equal 1 and whose other elements equal zero. This type of matrix, usually represented by the symbol I, serves the same function in matrix algebra as does unity in ordinary algebra.

$$[I] = \begin{bmatrix} 1 & 0 & 0 & 0 \\ 0 & 1 & 0 & 0 \\ 0 & 0 & 1 & 0 \\ 0 & 0 & 0 & 1 \end{bmatrix}$$

To conclude this abbreviated review a few comments are made about the multiplication of matrices and the inverse matrix.

Matrix Multiplication

There are many ways in which two matrices might be multiplied, such as by simple multiplication of corresponding elements, but the results so obtained would serve no practical purposes. As a large part of matrix application involves the solution of linear simultaneous equations, the definition of matrix multiplication is one that facilitates such operations. Two matrices can be multiplied only if the number of columns in the premultiplier matrix equals the number of rows in the postmultiplier matrix. The row elements of the premultiplier are multiplied by the column elements of the postmultiplier. The matrix that results has the same number of rows as the premultiplier and the same number of columns as the postmultiplier. Some examples follow.

$$AB = D$$

$$\begin{bmatrix} a_{11} & a_{12} & a_{13} \\ a_{21} & a_{22} & a_{23} \end{bmatrix} \begin{bmatrix} b_{11} & b_{12} \\ b_{21} & b_{22} \\ b_{31} & b_{32} \end{bmatrix} = \begin{bmatrix} d_{11} & d_{12} \\ d_{21} & d_{22} \end{bmatrix}$$

The elements of matrix D can be obtained as shown:

$$d_{11} = [a_{11} \quad a_{12} \quad a_{13}] \begin{bmatrix} b_{11} \\ b_{21} \\ b_{31} \end{bmatrix}$$

$$= a_{11}b_{11} + a_{12}b_{21} + a_{13}b_{31}$$

$$d_{21} = [a_{21} \quad a_{22} \quad a_{23}] \begin{bmatrix} b_{11} \\ b_{21} \\ b_{31} \end{bmatrix}$$

$$= a_{21}b_{11} + a_{22}b_{21} + a_{23}b_{31}$$

The following two numerical examples illustrate the multiplication of square matrices. These examples demonstrate a peculiar but important property of matrices. They are noncommutative in multiplication. In other words, in general, AB does not equal BA.

$$AB = D$$

$$\begin{bmatrix} 2 & 6 \\ 3 & 1 \end{bmatrix} \begin{bmatrix} 3 & 4 \\ 5 & 2 \end{bmatrix} = \begin{bmatrix} 36 & 20 \\ 14 & 14 \end{bmatrix}$$

$$BA = E$$

$$\begin{bmatrix} 3 & 4 \\ 5 & 2 \end{bmatrix} \begin{bmatrix} 2 & 6 \\ 3 & 1 \end{bmatrix} = \begin{bmatrix} 18 & 22 \\ 16 & 32 \end{bmatrix}$$

The following example illustrates the multiplication of rectangular matrices.

$$\begin{bmatrix} 4 & 6 & 1 \\ 3 & 1 & 5 \end{bmatrix} \begin{bmatrix} 2 & 1 \\ 3 & 0 \\ 1 & 2 \end{bmatrix} = \begin{bmatrix} 27 & 6 \\ 14 & 13 \end{bmatrix}$$

The *inverse* (or reciprocal) of a matrix is a matrix which when multiplied by the original matrix produces the unit or identity matrix. Thus

$$[A][A]^{-1} = [I]$$

Matrix inversion (only square matrices can be inverted) performs a function in matrix algebra which is similar to that performed by division in ordinary algebra. For instance if a set of matrix equations are given as

$$[F]\{R\} + \{\Delta\} = \{0\}$$

They can be written as

$$\{R\} = -[F]^{-1}\{\Delta\}$$

where

$[F]^{-1}$ is the inverse of $[F]$

27.3 FORCE AND DISPLACEMENT METHODS OF ANALYSIS

The methods of analyzing statically indeterminate structures presented in Chaps. 19 through 25 are often placed into two general classifications. These are the force and displacement methods. Both of these methods have been developed to a stage where they can be applied to almost any structures such as trusses, beams, frames, plates, shells, and so on. The displacement procedure, however, is by far the more commonly used since it can be routinely programmed for a general computer solution.

Force Method of Analysis

In this method of analysis, also called the *flexibility* or *compatibility method*, redundants are selected and removed from the structure so that a stable and statically determinate structure remains. An equation of deformation compatibility is written at each location where a redundant has been removed. These equations are written in terms of the redundants and the resulting equations are solved for the numerical values of the redundants. After the redundants are determined statics can be used to compute all other desired internal forces, moments, and so on. The method of consistent distortions, Castigliano's second theorem, and the three-moment theorem are force methods.

Displacement Method of Analysis

In the displacement method of analysis, also called the *stiffness* or *equilibrium method*, the displacements of the joints necessary to describe fully the deformed shape of the structure are used in the simultaneous equations. When the simultaneous equations are solved for these displacements, they are substituted into the force-deformation relations of each member to determine the various internal forces. Slope deflection is a displacement method whereas moment distribution is a successive approximation procedure based on the same general theory as the displacement methods.

The sections to follow present a brief introduction to the force and displacement methods.

27.4 THE FORCE OR FLEXIBILITY METHOD

This method is actually the method of consistent distortions (previously described in Chaps. 19 and 20) cast in matrix form. The steps involved in applying the method are as follows:

1. A sufficient number of redundants is chosen to make the structure statically determinate. The remaining structure often called the *primary structure* or the *released structure* must be stable.
2. The primary structure is analyzed to find the deformations at and in the direction of the redundants that were removed.
3. A unit value for a redundant is applied to the primary structure at the point and in the direction of one of the redundants and the deformations at all of the redundants are determined. This same procedure is followed with a unit value of a redundant applied at each of the other redundant locations.
4. Finally, simultaneous equations of deformation compatibility are written at each of the redundant locations. The unknowns in these equations are the redundant forces. The equations are expressed in matrix form and solved for the redundants. Example 27.1 shows the application of the force method to a continuous beam. The detailed calculations of the deflections and rotations are not included. These values can easily be calculated by the methods described in earlier chapters.

Example 27.1 illustrates the analysis of a three-span beam using the force method.

Example 27.1

Determine the magnitude of each of the reactions for the beam of Fig. 27.1 for which E and I are constant for both spans.

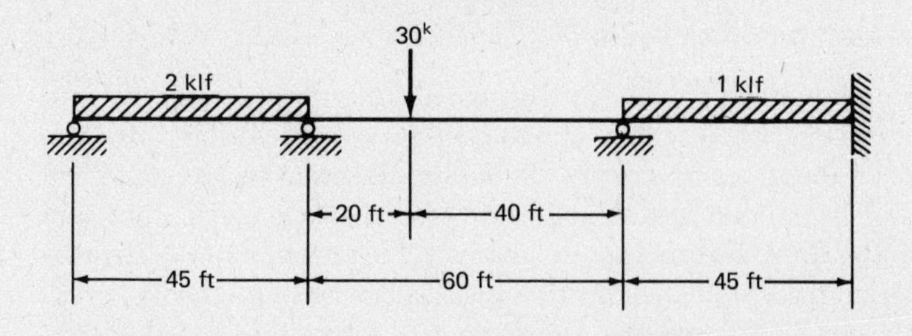

Figure 27.1

SOLUTION

The vertical reaction at B (R_1), the moment at C (R_2) and the internal moment at D (R_3) are removed as the redundants leaving the primary structure to follow.

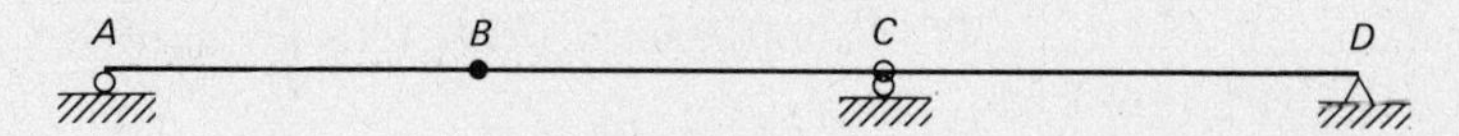

The vertical deflection at B, and the rotations at C and D due to the external loads are computed with the results shown. It will be noted that the removal of the moment C in effect cuts the beam so that it is free to rotate on each side (θ_{CA} and θ_{CD}, respectively, in the figure to follow). In other words it is as though the C end of AC and the C end of CD are separately simply supported.

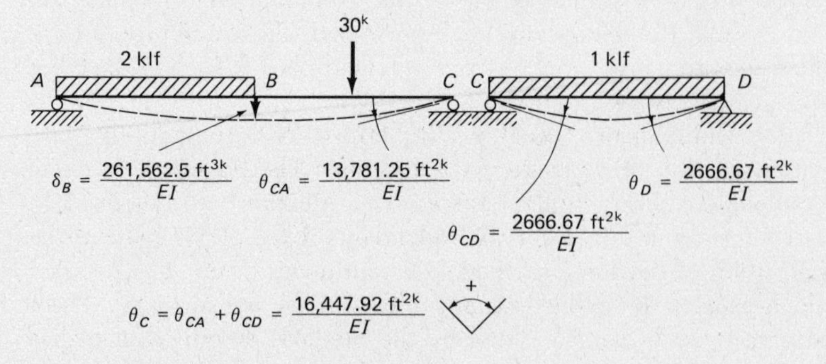

A unit load is applied downward at B and the deflection there and the rotation at C are determined. If the load had been R_1 the values would have been R_1 times as large.

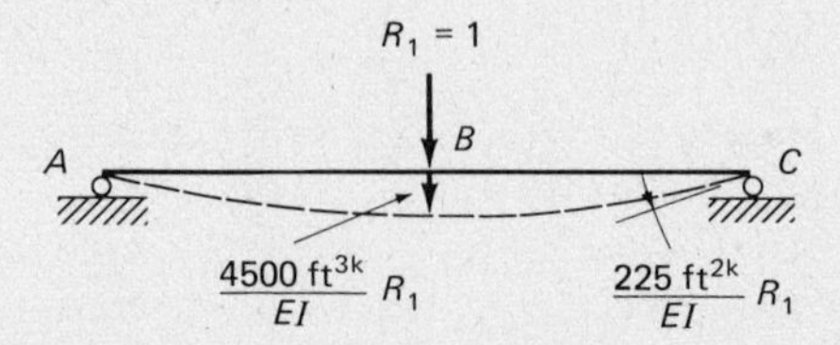

In a similar manner a unit value of the internal moment is applied at C and the resulting deformations at redundant locations are determined. If the applied moment was equal to R_2 the deformations would be R_2 times as large.

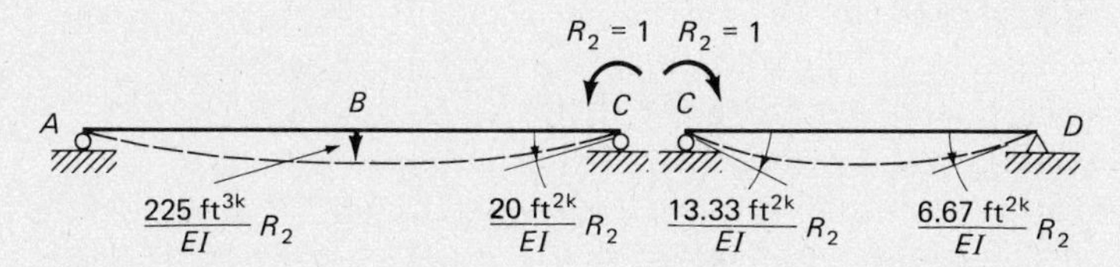

Finally a unit value of moment is applied at D and the resulting deformations determined. It can be seen that the use of Maxwell's law of reciprocal deflections will simplify these calculations.

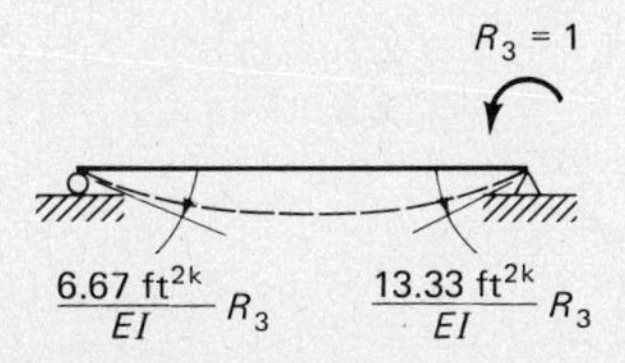

In the real structure the deflection at B is zero as is the slope at D and there is a common nonzero slope at C. Thus the equations of consistent distortion or the compatibility equations can be written as follows:

$$\overset{+}{\downarrow} \Delta_B = 261{,}562.5 + 4500R_1 + 225R_2 = 0$$

$$+\theta_C = 16{,}447.92 + 225R_1 + 33.33R_2 + 6.67R_3 = 0$$

$$+\theta_D = 2{,}667.67 + 6.67R_2 + 13.33R_3 = 0$$

Writing these in matrix form

$$\frac{1}{EI}\begin{bmatrix} 4500 & 225 & 0 \\ 225 & 33.33 & 6.67 \\ 0 & 6.67 & 13.33 \end{bmatrix} \begin{Bmatrix} R_1 \\ R_2 \\ R_3 \end{Bmatrix} = -\frac{1}{EI}\begin{Bmatrix} 261{,}562.5 \\ 16{,}447.92 \\ 2{,}666.67 \end{Bmatrix}$$

Solving these equations

$$R_1 = -52.69^k \qquad R_1 = 52.69^k \uparrow$$

$$R_2 = -108.61'^k \quad \text{or} \quad R_2 = 108.61'^k$$

$$R_3 = -145.70'^k \qquad R_3 = 145.70'^k$$

The remaining reactions can be determined by statics.

27.5 REVIEW OF THE STEPS INVOLVED IN THE FORCE METHOD

Now that a numerical example has been presented the steps in its solution are summarized with the objective of better understanding the matrix application. First a sufficient number of redundants were selected and removed from the structure to render it statically determinate but yet stable. For the remaining or primary structure the redundants each were zero or that is $\{R\} = \{0\}$.

A vector direction was chosen for each of the redundants. Then the displacements $\{D\}$ at and in the directions of the redundants $\{R\}$ were determined. For the primary structure unit redundants were applied one at a time and the displacements were found at all the redundant locations.

Next a *flexibility matrix* was prepared for the displacements caused by the unit redundants. The units of this matrix are the displacements per unit of redundant force. The value f_{ij} in this matrix is a *flexibility coefficient* which corresponds to the displacement at point i caused by a unit force at point j. Column 1 gives the displacements at R_1, R_2, . . . , R_n due to $R_1 = 1$. Similar values are given in column 2 due to $R_2 = 1$.

$$[F] = \begin{bmatrix} f_{1,1} & f_{1,2} & \cdots & f_{1,n} \\ f_{2,1} & f_{2,2} & \cdots & f_{2,n} \\ \cdot & \cdot & \cdots & \cdot \\ \cdot & \cdot & & \cdot \\ \cdot & \cdot & \cdots & \cdot \\ f_{n,1} & f_{n,2} & \cdots & f_{n,n} \end{bmatrix}$$

The compatibility equations were written as follows:

$$\underset{\substack{\text{a square flexibility} \\ \text{matrix} \\ [F]}}{\begin{bmatrix} f_{1,1} & f_{1,2} & \cdots & f_{1,n} \\ f_{2,1} & f_{2,2} & \cdots & f_{2,n} \\ \cdot & \cdot & \cdots & \cdot \\ \cdot & \cdot & \cdots & \cdot \\ \cdot & \cdot & \cdots & \cdot \\ f_{n,1} & f_{n,2} & \cdots & f_{n,n} \end{bmatrix}} \underset{\substack{\text{a vector of} \\ \text{redundant forces} \\ \{R\}}}{\begin{Bmatrix} R_1 \\ R_2 \\ \cdot \\ \cdot \\ \cdot \\ R_n \end{Bmatrix}} + \underset{\substack{\text{a vector of displacements} \\ \text{caused by the real} \\ \text{loads on the primary} \\ \text{structure} \\ \{D\}}}{\begin{Bmatrix} \delta_1 \\ \delta_2 \\ \cdot \\ \cdot \\ \cdot \\ \delta_n \end{Bmatrix}} = \begin{Bmatrix} 0 \\ 0 \\ \cdot \\ \cdot \\ \cdot \\ 0 \end{Bmatrix}$$

or

$$[F]\{R\} + \{D\} = \{0\}$$

Solving for the redundants,

$$\{R\} = -[F]^{-1}\{D\}$$

For Example 27.1,

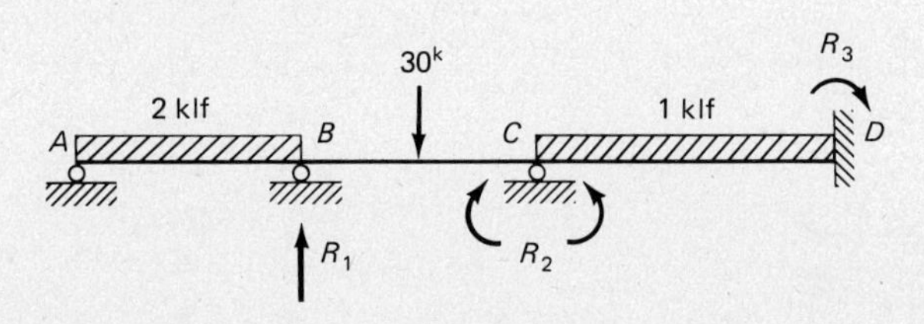

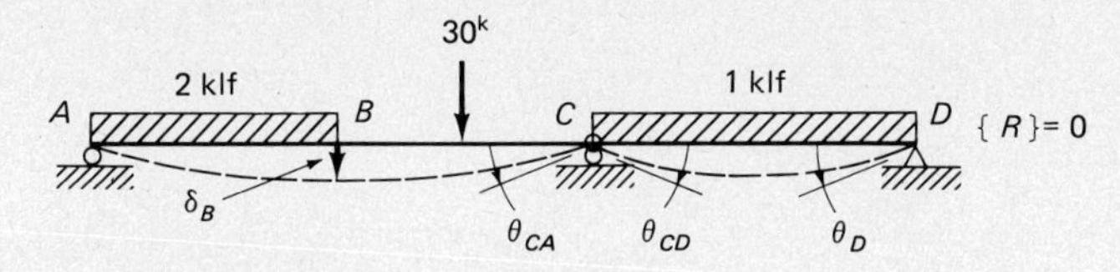

$$D_1 = -\delta_B$$
$$D_2 = -\theta_C$$
$$D_3 = -\theta_D$$

$$\{D\} = -\frac{1}{EI}\begin{Bmatrix} 261{,}562.5 \\ 16{,}447.92 \\ 2{,}667.67 \end{Bmatrix}$$

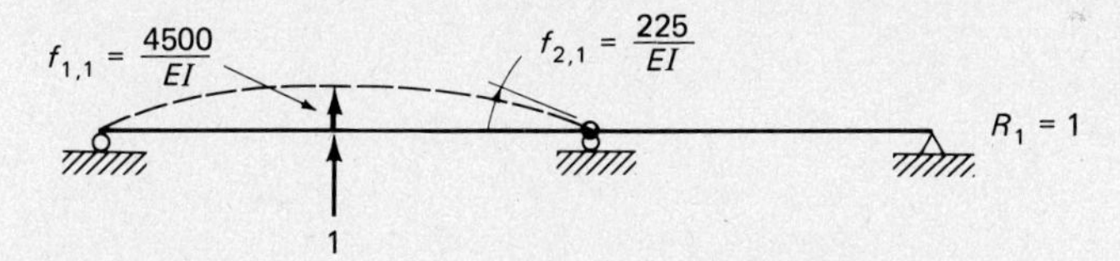

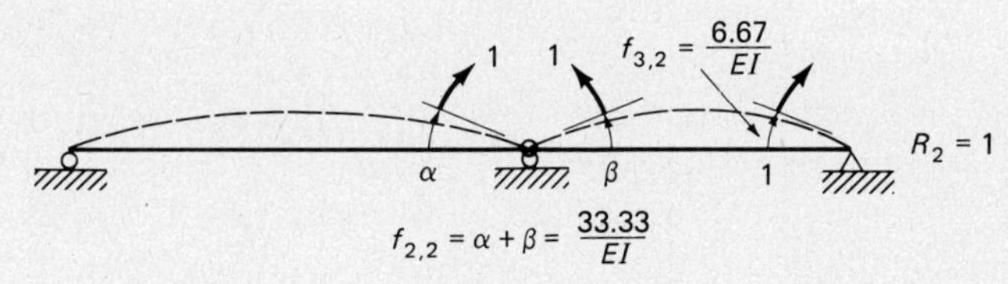

$f_{2,3} = \dfrac{6.67}{EI}$ 1 $f_{3,3} = \dfrac{13.33}{EI}$

$R_3 = 1$

Then

$$[F]\{R\} = -\{D\}$$

$$\frac{1}{EI}\begin{bmatrix} 4500 & 225 & 0 \\ 225 & 33.33 & 6.67 \\ 0 & 6.67 & 13.33 \end{bmatrix}\begin{Bmatrix} R_1 \\ R_2 \\ R_3 \end{Bmatrix} = -\frac{1}{EI}\begin{Bmatrix} -261{,}562.5 \\ -16{,}447.92 \\ -2667.67 \end{Bmatrix}$$

Therefore

$$\{R\} = -[F]^{-1}\{D\}$$

and solving

$$\{R\} = \begin{Bmatrix} 52.69^{k} \\ 108.61'^{k} \\ 145.70'^{k} \end{Bmatrix}$$

27.6 GENERAL FORCE METHOD

Up to this point the author has described the method of consistent distortions with the resulting equations cast in matrix form. A more general matrix approach is necessary for computerizing the force method.

The general force method is not included in this chapter because it is not as useful as the displacement method discussed in the remaining sections of this chapter and because it would require quite a few pages to describe. The interested reader can find a description of this method in several texts such as the one by H. C. Martin [1].

27.7 THE DISPLACEMENT METHOD

The basic theory used with the displacement method is the development of *equilibrium* equations expressed as a function of the unknown joint displacements for a structural system. The known coefficients in these simultaneous equations of system equilibrium are stiffness values. Thus the reader can understand why the method is referred to as the *displacement method* or the *equilibrium method* or the *stiffness method.*

This method is of the greatest importance because it is the matrix analysis method that can be computerized most readily for general usage. To provide a complete theoretical background for its application to beams, trusses, frames and other types of structures would require the addition of several chapters to this book. To keep the material down to a few pages the author has greatly limited the type of structures considered. Though the understanding of this discussion will not enable the reader to immediately handle more complicated structures it is felt that he or she will be able to quickly pick up such material with a little more study.

Stiffness

The stiffness of a member is usually defined as the force that has to be applied at some point to produce a unit displacement of that point when all other displacements are constrained to be zero. It is assumed for the purposes of this discussion that elastic materials are being considered for which Hooke's law applies. Consider a member subjected only to an axial tension load. If a plot is made of load versus stretching displacement for varying loads applied to the member the result will be a straight line. The slope of this line is called the *stiffness* k (not to be confused with the stiffness factor K used in moment distribution). It can be determined as follows where P is the load and δ is the displacement.

$$k = \frac{P}{\delta}$$

Another way to define the factor k is to say that it is the force required at a particular point to cause a unit displacement at that point. If a particular member has a stiffness k, and changes in length by an amount δ_1 the force P_1 that produces δ_1 can be determined as follows:

$$P_1 = k_1\delta_1$$

The preceding discussion has been presented for the load-displacement situation at a single point. For most practical structures, however, stiffness has to be defined in a little more detailed manner because it is necessary to describe the force-displacement relationships for more than one point on the structure.

A structure is usually thought to consist of a network of members and *node points*. These node points or nodes may be introduced at member ends, supports, intersections with other members, and changes in cross sections. Furthermore, they may be introduced at other places where displacements are needed. It is usually assumed that members are straight and prismatic between node points. The components of the displacements at the nodes are referred to as the *degrees of freedom*. To continue the discussion of stiffnesses the beam of Fig. 27.2 and the three node points (1, 2 and 3) are considered.

If a unit displacement is induced at point 1 while points 2 and 3 are restrained from deflecting up or down the forces at the three points will be as shown in Fig. 27.3. It will be noted in this figure that $\delta_1 = 1$ and thus $P_1 = k_{1,1}\,\delta_1 = k_{1,1}$. In a related fashion $P_2 = k_{2,1}\,\delta_1 = k_{2,1}$. (For the situations described herein the nodal force P_i is represented by the quantity k_{ij}. The value of k_{ij} is the force at point i caused by a unit displacement at point j.)

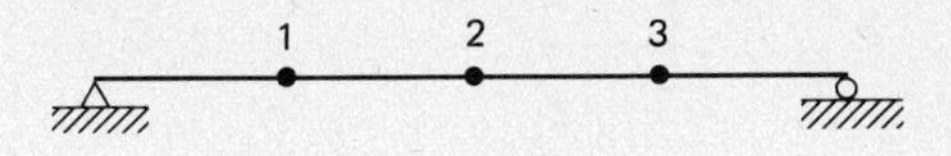

Figure 27.2

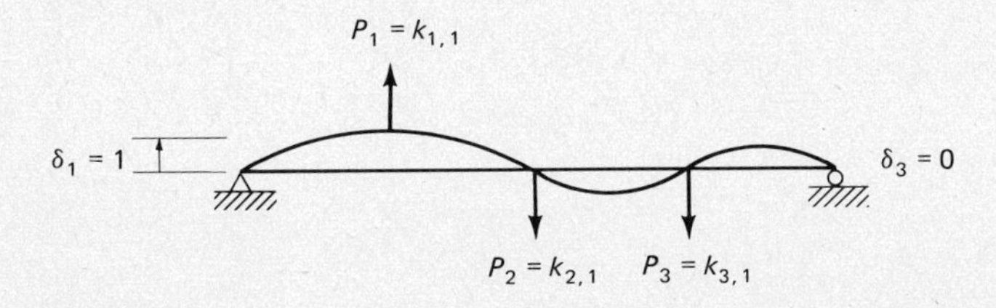

Figure 27.3

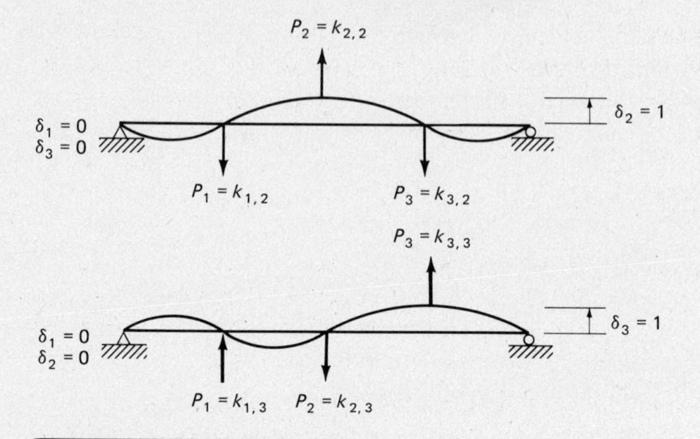

Figure 27.4

In a similar fashion unit displacements are applied at points 2 with points 1 and 3 restrained, and at point 3 with points 1 and 2 restrained as shown in Fig. 27.4.

Stiffness Matrix and Stiffness Coefficients

The stiffnesses described in the last few paragraphs can be gathered together in a *stiffness matrix.*

$$[K] = \begin{bmatrix} k_{1,1} & k_{1,2} & k_{1,3} \\ k_{2,1} & k_{2,2} & k_{2,3} \\ k_{3,1} & k_{3,2} & k_{3,3} \end{bmatrix}$$

Each of the elements of the stiffness matrix is called a *stiffness coefficient*. The units of such a coefficient are force per unit of displacement. In other words the stiffness coefficients are the forces at the nodes caused by inducing one at a time the various unit displacements. Thus for linearly elastic structures, the forces P_1, P_2, and P_3 are proportional to the displacements and the following equations can be written:

$$P_1 = k_{1,1}\delta_1 + k_{1,2}\delta_2 + k_{1,3}\delta_3$$

$$P_2 = k_{2,1}\delta_1 + k_{2,2}\delta_2 + k_{2,3}\delta_3$$

$$P_3 = k_{3,1}\delta_1 + k_{3,2}\delta_2 + k_{3,3}\delta_3$$

or in matrix form

$$\begin{Bmatrix} P_1 \\ P_2 \\ P_3 \end{Bmatrix} = \begin{vmatrix} k_{1,1} & k_{1,2} & k_{1,3} \\ k_{2,1} & k_{2,2} & k_{2,3} \\ k_{3,1} & k_{3,2} & k_{3,3} \end{vmatrix} \begin{Bmatrix} \delta_1 \\ \delta_2 \\ \delta_3 \end{Bmatrix}$$

This expression, which expresses the equilibrium at each of the node points in terms of the stiffness coefficients and the unknown nodal displacements, can be written as

$$\{P\} = [K]\{\Delta\}$$

where K is the stiffness matrix and $\{P\}$ and $\{\Delta\}$ are load and displacement vectors, respectively.

Some Relations Previously Developed

There are three relations which have previously been developed in this book which are needed to apply the displacement method. These are:

1. When an axial force is applied to a prismatic member the member will change in length by an amount $\delta = Pl/AE$. If the change in length is a unit one the applied force P can be determined as follows:

$$1 = \frac{Pl}{AE}$$

$$P = \frac{EA}{l}$$

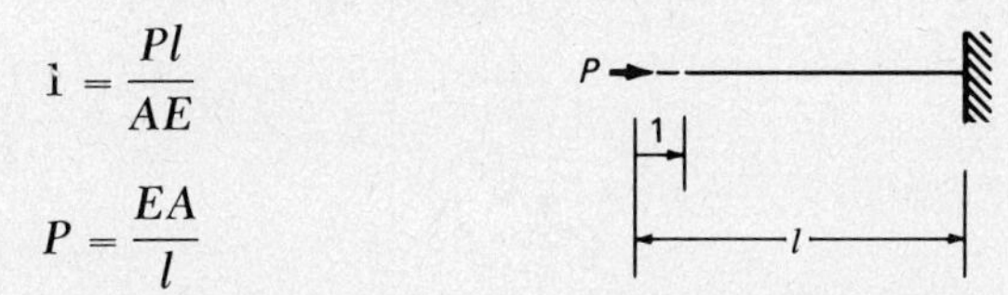

2. Should one end of a fixed-ended beam of constant EI be moved up a unit distance the end shears or reactions would each equal $12EI/l^3$.

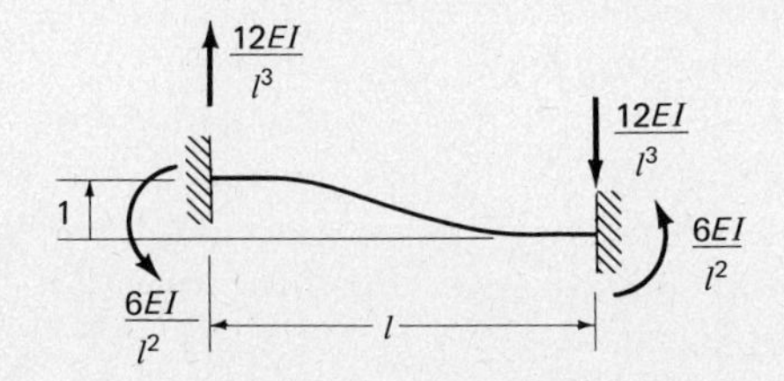

3. When one end of a prismatic member is subjected to a unit rotation, the far end being fixed, the applied moment is $4EI/l$ and the restraining moment is $2EI/l$.

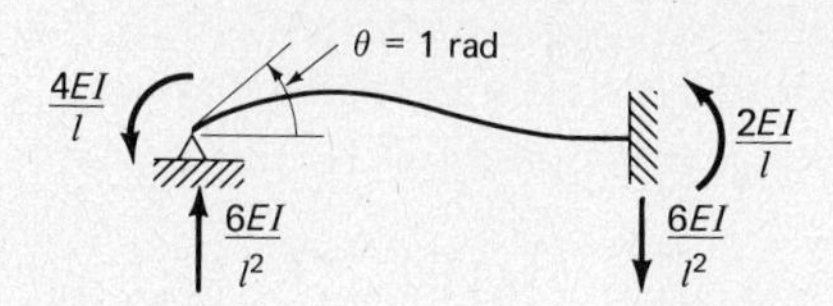

Coordinate Systems

When the equations of equilibrium for the analysis of a structure are being set up it is logical to select a set of coordinate axes (to define positive directions for the forces and displacements) that will simplify the work as much as possible. For instance, if an initially straight member is being considered it seems sensible to make the x axis coincide with the long direction of the member and

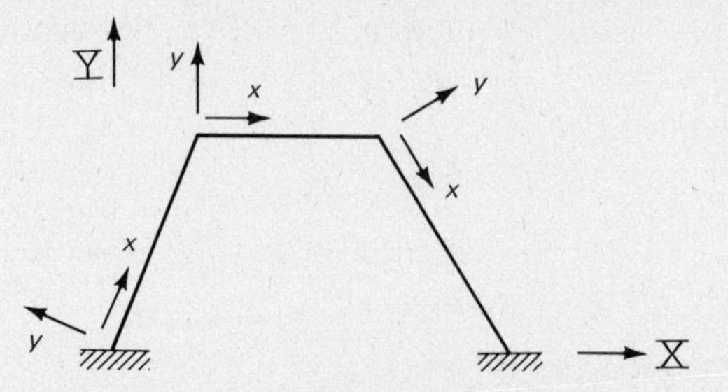

Figure 27.5 Coordinate systems for a plane frame.

to let the y and z axes coincide with the principal axes of the member cross section. Such a set of coordinates is usually called a *member coordinate system*. For instance, for the frame of Fig. 27.5 lower case x, y, and z are used to define the member coordinates.

When members are connected together to form a structure it is necessary to use a common coordinate system to describe the equilibrium and displacement situations at various points in the structure. Such a system is normally referred to as a *global coordinate system*. It consists of a single coordinate system for the entire structure and is represented in Fig. 27.5 by the capital letters X, Y, and Z.

If the member stiffnesses are expressed in a member coordinate system they cannot be used in an overall structure matrix until they are transformed into the global system. Forces and displacements are vector quantities and their components cannot be added unless they are in the same direction.

For some structures such as continuous beams, the member and global coordinates will coincide. In this chapter, as only beams are considered, only the member coordinate system is presented. For other types of structures it will be necessary for the reader to learn to change from member to global coordinates.

Member Stiffness Matrix in Member Coordinates

A prismatic member with length l, area A, moment of inertia I, and modulus of elasticity E is shown in Fig. 27.6. It is desired to develop a stiffness matrix for the member. It is assumed that each end of the member has three degrees of freedom. For instance, at the left end these are represented by the positive

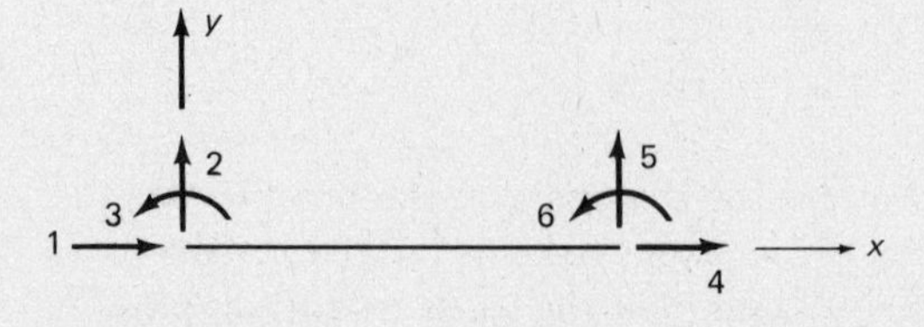

Figure 27.6

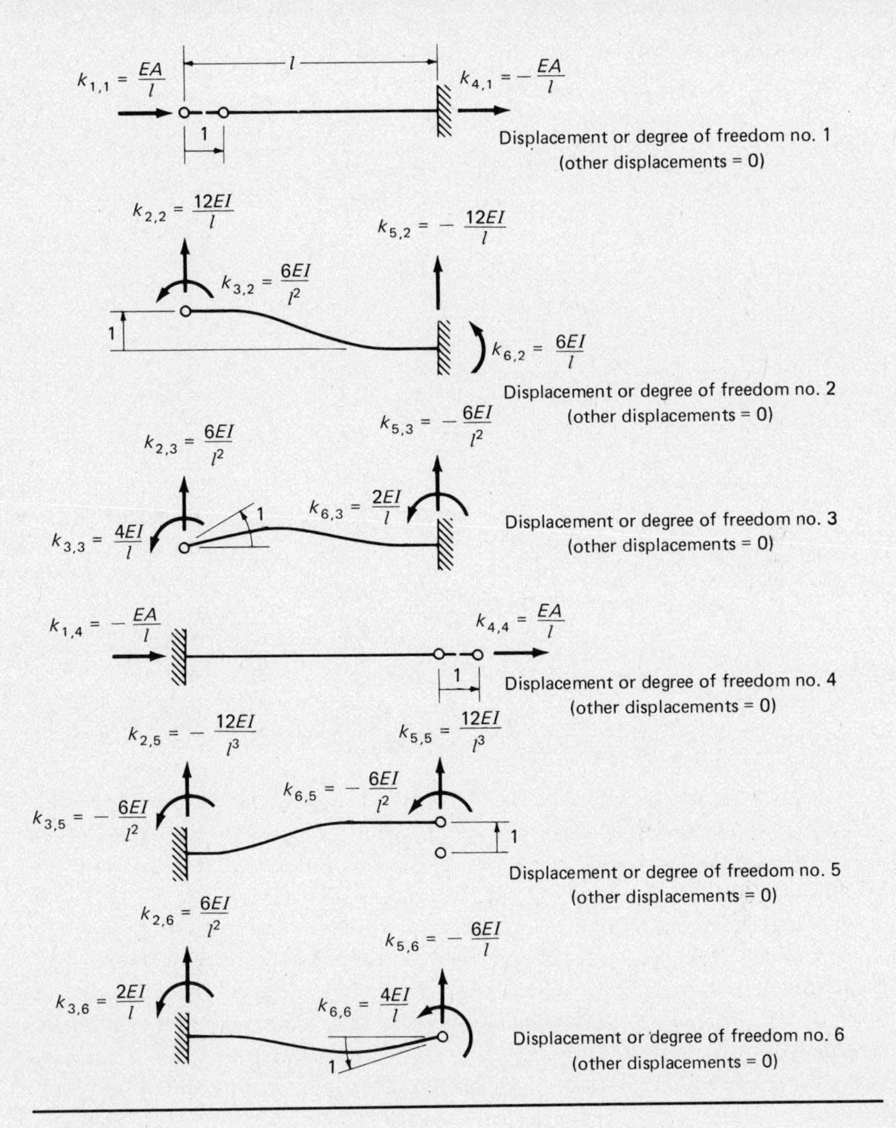

Figure 27.7

directions 1, 2, and 3, and at the right end by the positive directions 4, 5, and 6. In other words there are six displacement directions corresponding to the six degrees of freedom.

A unit displacement corresponding to each of these six degrees of freedom is induced one at a time. From these applications the member stiffness matrix is developed. Figure 27.7 shows the member of Fig. 27.6 subjected in turn to each of the six displacements. In addition the figure shows the axial forces, shears, and moments which are produced when those displacements are applied. Each force or couple is shown in the assumed positive direction in the figure. Should it actually act in the opposite direction the computed value is negative.

From the information given in Fig. 27.7 the 6 × 6 member stiffness matrix to follow can be developed.

$$
\begin{array}{l}
\text{Column} \\
\text{numbers}
\end{array}
\qquad
[K] = \left[
\begin{array}{cccccc}
1 & 2 & 3 & 4 & 5 & 6 \\
\frac{EA}{l} & 0 & 0 & -\frac{EA}{l} & 0 & 0 \\
0 & \frac{12EI}{l^3} & \frac{6EI}{l^2} & 0 & -\frac{12EI}{l^3} & \frac{6EI}{l^2} \\
0 & \frac{6EI}{l^2} & \frac{4EI}{l} & 0 & -\frac{6EI}{l^2} & -\frac{2EI}{l} \\
-\frac{EA}{l} & 0 & 0 & \frac{EA}{l} & 0 & 0 \\
0 & -\frac{12EI}{l^3} & -\frac{6EI}{l^2} & 0 & \frac{12EI}{l^3} & -\frac{6EI}{l^2} \\
0 & \frac{6EI}{l^2} & \frac{2EI}{l} & 0 & -\frac{6EI}{l^2} & \frac{4EI}{l}
\end{array}
\right]
\begin{array}{c}
\text{Row numbers} \\
1 \\ 2 \\ 3 \\ 4 \\ 5 \\ 6
\end{array}
$$

If a two-dimensional or planar framed structure consists of straight prismatic members each of those members will have the same stiffness matrix definition. Of course the E, I, A, and l terms may change from member to member.

Consider the continuous beam of Fig. 27.8 for the following discussion.

The only degrees of freedom considered are the rotations at the joints. As a result the member stiffness matrix can be greatly reduced by deleting the columns and rows that do not apply. In other words, vertical and horizontal displacements of the joints are ignored. (It is further assumed that axial and shear deformations are negligible.) Striking out the columns and rows numbered 1, 2, 4, and 5 from the 6 × 6 member stiffness matrix leaves the stiffness matrix to follow which applies to each of the members of the continuous beam of Fig. 27.8.

$$
[K] = \begin{bmatrix}
\frac{4EI}{l} & \frac{2EI}{l} \\
\frac{2EI}{l} & \frac{4EI}{l}
\end{bmatrix}
$$

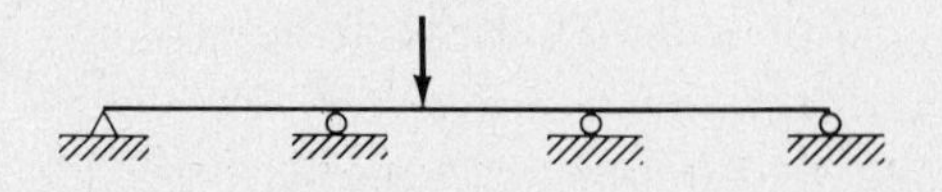

Figure 27.8

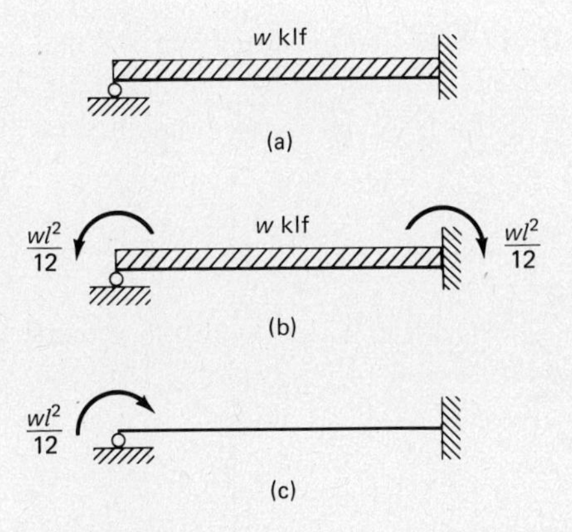

Figure 27.9

27.8 EXAMPLE ANALYSIS OF CONTINUOUS BEAM BY DISPLACEMENT METHOD

To illustrate the procedure that will be followed in the example problem to follow the beam of Fig. 27.9(a) is considered. In part (b) of the figure are shown the resisting moments that are necessary at the supports to prevent any rotation of those joints.

The moment at the right end is a fixed-end moment which can be supplied by the wall—but at the left end the fixed-end moment cannot be supplied by the support and it needs to be removed as shown in part (c) of the figure. The superposition of the values shown in parts (b) and (c) will provide the full solution.

It is desired to analyze the three-span continuous beam of Fig. 27.10 by the displacement method. Part (a) of Fig. 27.11 shows the degrees of freedom of the structure. In part (b) are shown the resisting moments at supports 1, 2, and 3 which would be necessary to prevent any rotation of these joints. These "fixed-end moments" were hand calculated. Part (c) of the solution shows the fixed-end moments applied to the structure. The displacement method is to be used to analyze the structure for this case and the final moments will be determined by superimposing the moments produced in part (b) on those produced in part (c).

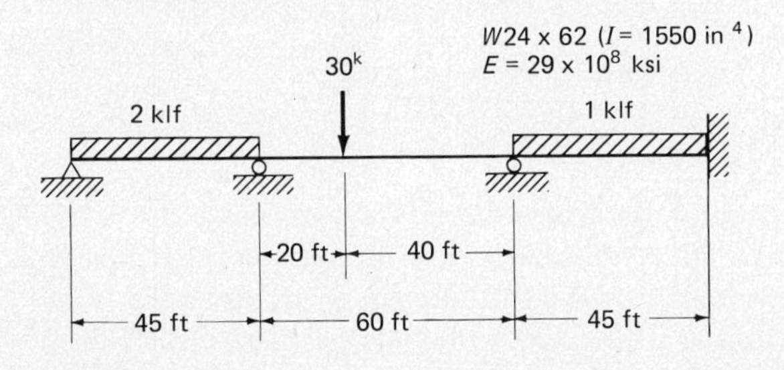

Figure 27.10

Further study of matrix analysis will show the reader that he or she can choose additional node points that are not located at the supports. For example, it may be desired to choose a node point at the position of the concentrated load.

Example 27.2

Analyze the three-span continuous beam of Fig. 27.10 using the displacement method. This is the same beam previously analyzed in Example 27.1 by the force method.

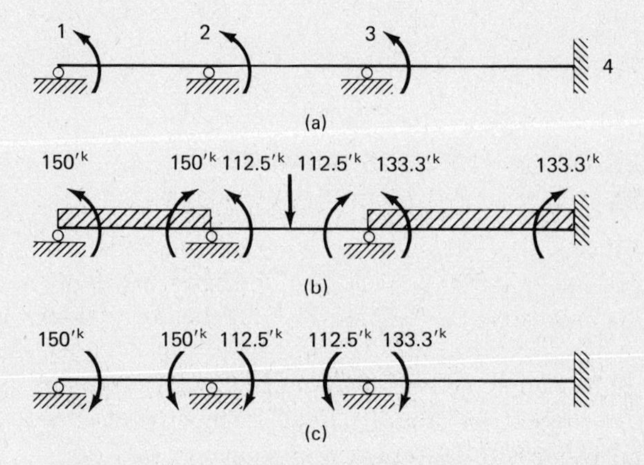

Figure 27.11 (a) Degrees of freedom. (b) Resisting member-end moments needed to prevent rotations at nodes 1 to 4 (that is, $\theta_1 = \theta_2 = \theta_3 = \theta_4 = 0$). (c) But moments in figure (b) are not present (nodes are not fixed against rotations) and it is necessary to reverse them and solve this problem for the nodal rotations.

SOLUTION

As a first step the joints are numbered, the degrees of freedom shown, and the fixed-end moments computed as shown in Fig. 27.11.

The individual member stiffness matrixes are written for the three members from left to right as follows:

$$\begin{bmatrix} \dfrac{4EI}{l_{1,2}} & \dfrac{2EI}{l_{1,2}} \\[2ex] \dfrac{2EI}{l_{1,2}} & \dfrac{4EI}{l_{1,2}} \end{bmatrix} \quad \begin{bmatrix} \dfrac{4EI}{l_{2,3}} & \dfrac{2EI}{l_{2,3}} \\[2ex] \dfrac{2EI}{l_{2,3}} & \dfrac{4EI}{l_{2,3}} \end{bmatrix} \quad \begin{bmatrix} \dfrac{4EI}{l_{3,4}} & \dfrac{2EI}{l_{3,4}} \\[2ex] \dfrac{2EI}{l_{3,4}} & \dfrac{4EI}{l_{3,4}} \end{bmatrix}$$

Next the member stiffness matrix arrays are added together to form the total structure stiffness array. They may be added directly as described in several references. In other words, the nodal stiffness at a particular point is the sum of the corresponding member stiffnesses. As $\theta_4 = 0$ the values corresponding to that end of the member are deleted.

$$\begin{bmatrix} \dfrac{4EI}{l_{1,2}} & \dfrac{2EI}{l_{1,2}} & 0 \\ \dfrac{2EI}{l_{1,2}} & \dfrac{4EI}{l_{1,2}} + \dfrac{4EI}{l_{2,3}} & \dfrac{2EI}{l_{2,3}} \\ 0 & \dfrac{2EI}{l_{2,3}} & \dfrac{4EI}{l_{2,3}} + \dfrac{4EI}{l_{3,4}} \end{bmatrix}$$

The EI/l values are computed as follows:

$$\frac{EI}{l_{1,2}} = \frac{EI}{l_{2,3}} = \frac{(29 \times 10^3)(843)}{(12)(30)} = 6.7908 \times 10^4 \ ''^{k}$$

$$\frac{EI}{l_{3,4}} = \frac{(29 \times 10^3)(843)}{(12)(40)} = 5.0931 \times 10^4 \ ''^{k}$$

and substituted into the total structure stiffness matrix

$$[K] = \begin{bmatrix} 27.163 & 13.582 & 0 \\ 13.582 & 54.326 & 13.582 \\ 0 & 13.582 & 47.536 \end{bmatrix} \times 10^4 \ ''^{k}$$

The moments at the node points have been previously calculated as follows in

$$M_1 = -150'^{k} = -1800''^{k}$$

$$M_2 = -37.5'^{k} = +450''^{k}$$

$$M_3 = -20.8'^{k} = -249.6''^{k}$$

or

$$M = \begin{Bmatrix} -1.8000 \\ -0.4500 \\ -0.2496 \end{Bmatrix} \times 10^3 \ ''^{k}$$

The displacements can be determined from the overall expression

$$\begin{Bmatrix} M_1 \\ M_2 \\ M_3 \end{Bmatrix} = [K] \begin{Bmatrix} \theta_1 \\ \theta_2 \\ \theta_3 \end{Bmatrix}$$

$$\{\theta\} = \begin{bmatrix} 27.163 & 13.582 & 0 \\ 13.582 & 54.326 & 13.582 \\ 0 & 13.582 & 47.536 \end{bmatrix}^{-1} \times 10^4 \ ''^{k} \begin{Bmatrix} -1.8 \\ +0.45 \\ -0.2496 \end{Bmatrix} \times 10^3 \ ''^{k}$$

Solving the matrix equations yields the following displacements:

$$\theta_1 = -0.00825464 \text{ rad}$$

$$\theta_2 = +0.00325584 \text{ rad}$$

$$\theta_3 = -0.00145534 \text{ rad}$$

Back substitution into the matrix stiffness equation for each member produces the end moments.

For member 1–2

$$\begin{Bmatrix} m_1 \\ m_2 \end{Bmatrix} = \begin{bmatrix} 27.163 & 13.582 \\ 13.582 & 27.163 \end{bmatrix} \times 10^4 \text{ ''k} \begin{Bmatrix} -0.00825464 \\ +0.00325584 \end{Bmatrix}$$

$$= \begin{Bmatrix} -1800''\text{k} \\ -236''\text{k} \end{Bmatrix} = \begin{Bmatrix} -150'\text{k} \\ -19.7'\text{k} \end{Bmatrix}$$

For member 2–3

$$\begin{Bmatrix} m_3 \\ m_4 \end{Bmatrix} = \begin{bmatrix} 27.163 & 13.582 \\ 13.582 & 27.163 \end{bmatrix} \times 10^4 \text{ ''k} \begin{Bmatrix} 0.00325584 \\ -0.00825464 \end{Bmatrix}$$

$$= \begin{Bmatrix} +687''\text{k} \\ +47''\text{k} \end{Bmatrix} = \begin{Bmatrix} +57.3'\text{k} \\ +3.9'\text{k} \end{Bmatrix}$$

For member 3–4

$$\begin{Bmatrix} m_3 \\ m_4 \end{Bmatrix} = \begin{bmatrix} 20.3724 & 10.1862 \\ 10.1862 & 20.3724 \end{bmatrix} \times 10^4 \text{ ''k} \begin{Bmatrix} -0.00145534 \\ 0 \end{Bmatrix}$$

$$= \begin{Bmatrix} -296''\text{k} \\ -148''\text{k} \end{Bmatrix} = \begin{Bmatrix} -24.7'\text{k} \\ -12.4'\text{k} \end{Bmatrix}$$

The final moments in the beam can be determined by superimposing the moments of parts (a) and (b) of Fig. 27.12.

Obviously this problem could have been solved much more quickly with one of the methods of analysis previously described such as moment distribution. The purpose of this example was to give the reader an idea of the items that are involved in matrix analysis with the displacement method. The displacement method of analysis is very practical for structures with large numbers of members when computers are available to solve the resulting equations. The solution of large matrices is usually handled with one of the packaged computer programs readily available around the country.

If the beam of Example 27.2 was modified to have both ends fixed as shown in Fig. 27.13 ($\theta_1 = \theta_4 = 0$) the calculations would be abbreviated

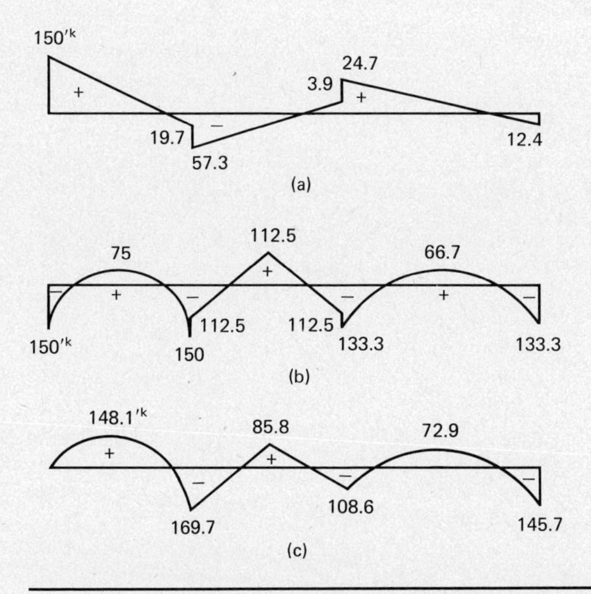

Figure 27.12 (a) Moments obtained due to displacements. (b) Fixed-end moments. (c) Final moments.

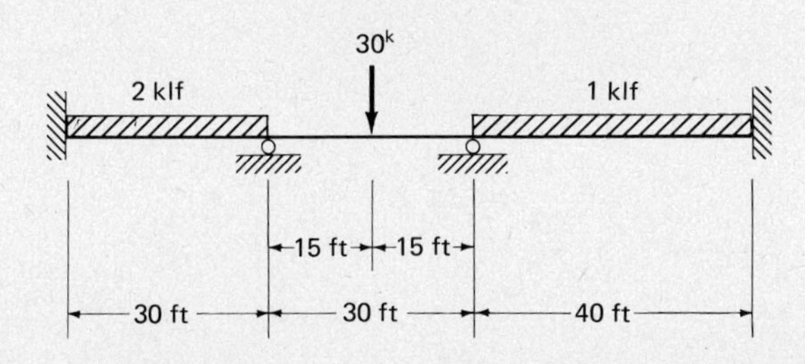

Figure 27.13

somewhat because the rows and columns of both θ_1 and θ_4 in the previous stiffness matrix would be marked out.

Thus the following matrix remains for the three-span member.

$$\{\theta\} = \begin{bmatrix} 54.326 & 13.582 \\ 13.582 & 54.326 \end{bmatrix}^{-1} \times 10^4\ {}^{\prime\prime k} \begin{Bmatrix} +0.45 \\ -0.2946 \end{Bmatrix} \times 10^3\ {}^{\prime\prime k}$$

Solving these equations yields the following values:

$= +0.00103343$ rad

$= -0.00082035$ rad

Back substitutions into the stiffness equations for each member produces the end moments.

$$\begin{Bmatrix} m_1 \\ m_2 \end{Bmatrix} = \begin{Bmatrix} +11.7'^{k} \\ +23.4'^{k} \end{Bmatrix}$$

$$\begin{Bmatrix} m_2 \\ m_3 \end{Bmatrix} = \begin{Bmatrix} +14.1'^{k} \\ -6.9'^{k} \end{Bmatrix}$$

$$\begin{Bmatrix} m_3 \\ m_4 \end{Bmatrix} = \begin{Bmatrix} -13.9'^{k} \\ -7.0'^{k} \end{Bmatrix}$$

These moments when superimposed onto the diagram of "resisting moments necessary to prevent rotation" of Example 27.2 will yield the correct final moments.

The displacement method can easily be expanded to include frames where there may be as many as six degrees of freedom at each joint. Space, however, is not taken for such a presentation.

Mackinac Bridge, St. Ignace, Michigan. (Courtesy of H.D. Ellis, St. Ignace, Michigan.)

27.9 CONCLUSION

The heading of this section is entitled "Conclusion," but a more appropriate title might be "End of the Beginning," because there are so many topics of further possible study in the field of structural analysis. To mention a very few, there are secondary stresses, model analysis, statically indeterminate space frames, suspension bridges, and more advanced study of several of the topics introduced in this text, particularly matrix methods.

The use of continuous structures becomes more common each year, but the emphasis of the analysis in undergraduate schools does not increase correspondingly. Some civil engineering schools are transferring all or part of indeterminate study from the required to the elective lists to permit the inclusion of more of the so-called broadening courses.

This trend makes it more difficult for the student to have a thorough background in the subject on graduation. Only a small percentage of those persons entering the structural field can continue to graduate school to remove the deficiency. The solution to the problem lies in a program of continuing study by the individual. It is difficult to believe that the knowledge displayed by the outstanding engineers of our time was achieved only in the classroom and during the 40 or so hours spent on the job each week. Their accomplishments are surely based to no small extent upon many hours of painstaking *self-instruction*.

REFERENCES

[1] H. C. Martin, *Introduction to Matrix Methods of Structural Analysis* (New York: McGraw-Hill, 1966), pp. 233–291.

Index

84 85 86 87 9 8 7 6 5 4 3